Designer's Handbook of Instrumentation and Control Circuits

Designer's Handbook of Instrumentation and Control Circuits

JOSEPH J. CARR

ACADEMIC PRESS, INC.
Harcourt Brace Jovanovich, Publishers
San Diego   New York   Boston   London
Sydney   Tokyo   Toronto

Academic Press, Inc.
San Diego, California 92101

United Kingdom Edition published by
Academic Press Limited
24–28 Oval Road, London NW1 7DX

Library of Congress Cataloging-in-Publication Data

Carr, Joseph J.
    Designer's handbook of instrumentation and control
circuits /
    Joseph J. Carr.
        p.   cm.
    Includes bibliographical references.
    ISBN 0-12-160640-6 (alk. paper)
    1. Electronic istruments--Design and construction. 2.
Integrated
    circuits.   I. Title.
    TK7878.4.C358   1990
    621.381--dc20   90-510
                                                            CIP

Printed in the United States of America
91 92 93 94   9 8 7 6 5 4 3 2 1

Contents

CHAPTER 3

CHAPTER 4

CHAPTER 5

CHAPTER 6

Preface

Most of modern science and engineering uses electronic instrumentation for purposes of understanding some process, measuring values of physical parameters, monitoring events, and controlling processes. The electronic instrument is so ubiquitous as to seemingly need no introduction to any but the most inexperienced technical workers. There are, however, certain aspects of the design and construction of electronic instruments that must be learned in a more formal manner. That is the underlying purpose of this book.

The field of electronic instruments is very broad and therefore impossible to cover in its entirety here. In this book, we take a look at certain fundamental issues such as the process of design, the nature of electronic signals and noise, and the sensors used to create electrical signals that represent physical parameters. We then turn the majority of our attention to the circuits that make up both the stand-alone analog instrument and the analog subsystem of digital or computer-based instruments. The details of these circuits and their design criteria will be discussed in depth.

Many of the chapters of this book that cover various types of circuits are based on the integrated circuit operational amplifier device. While there are other choices for basic circuit elements (and, indeed, others are also offered in the text), the "op-amp" is so utterly useful, and so widespread in actual instruments, that it is given a large share of the space in this book. One main feature of the op-amp is that the circuit design reduces to (often) simple manipulation of the negative feedback network. Thus, according to one early commentator on the subject, op-amps make " ... the contriving of contrivances a game for all," not just engineers.

Several groups of readers were kept in mind when this book was written (although due care was exercised to avoid overuniversalization of the book—books written for everyone rarely serve anyone well).

Electronic engineers and engineering students will want to keep this book handy as a reference guide. Practicing engineers know the need for a ready reference to "keep their heads above water." Engineering students often need a practical guidebook to complement their theoretical texts, especially in the laboratory when working on the dreaded "Senior Project." I have noticed many a senior engineering student hiding a practical volume in his or her briefcase or knapsack so that a disdainful professor does not find out that they are leaning on the practical experiences of those who have gone ahead of them.

Another class of readers who will find this book useful are senior electronics technicians or technologists. These people will find the book especially useful if they are studying engineering part time or if they are in an independent working environment where there are no engineers to guide their efforts. Examples of the latter include electronic technicians working in medical schools, research laboratories, or certain industrial plants where they must constantly be concerned with electronic instrumentation problems. These technicians, if properly motivated and educated, can often work at the more professional levels associated with engineers.

In addition, repair and maintenance activities are often easier for technicians who understand in depth how the circuits and sensors of the instrument work. Perhaps it is then possible to return to a components level of troubleshooting—a seemingly lost skill among service technicians.

The last group of readers anticipated by the author are scientists or engineers whose high level of sophistication is in a field other than electronics, generally, or electronic instrumentation, specifically. While such people must guard against the tendency to believe themselves able to "walk on water," the simple truth is that some minimal electronics ability is often found among such professionals. I have known physicists, physiologists, physicians, chemists, and biologists who knew enough about electronics, either from formal training or avocational interests, to get a great deal of benefit from this book. They can, under the right circumstances, learn to design and build their own instruments. In any event, the material in this book will help them understand their instruments better—and that advantage will lead to their being able to specify and use instruments in a wiser manner.

The reader will find that many of the examples used in this book are derived from the general field of medical, biological, and other life sciences instrumentation. For that I make no apologies, because that is the field in which a large portion of my professional life is spent. The underlying technology, however, is applicable over a broad range of scientific and engineering fields.

Joseph J. Carr, MSEE
Certified Clinical Engineer (AAMI)

1

Analog Instrumentation

The term *analog instrumentation* is a rubric that is almost too broad for practical use. It covers so many things that it is often necessary to define the term operationally when writing a book such as this one. For our present purposes, analog instrumentation includes whatever sensors or transducers are needed to detect some physical parameter, the input circuitry needed to acquire the signal, the signal processing or signal conditioning circuits needed to manipulate the signal, and whatever display or data storage devices are needed to present the result. This book is about designing the electronic circuits needed to form either stand-alone analog instruments or the analog subsystems of digital computer instruments. Toward serving that end, we will examine the nature of signals and noise, some typical examples of sensors (but not an exhaustive treatment), and in this chapter, a conceptual framework for either selecting or designing both typical analog instruments and the analog subsystems of digital systems.

Figure 1-1 shows a block diagram for a generic analog instrument. While this figure is merely a "mind example," it could easily serve for a large class of actual analog and digital instruments presently seen on the market. The principal components are the input parameter (stimulus); the sensor or transducer; the input functions; the amplification and signal processing functions; the output functions; and the display, recording, or other medium used to present the data to the external world.

Physical Stimulus

The physical stimulus that is sensed in an instrumentation system may be the temperature, displacement, flow, electrical resistance, electrical potential, or any of a host of other physical parameters. That which is intended is not important in a discussion of generic instruments, but it becomes important when specifics are defined. (Chapters

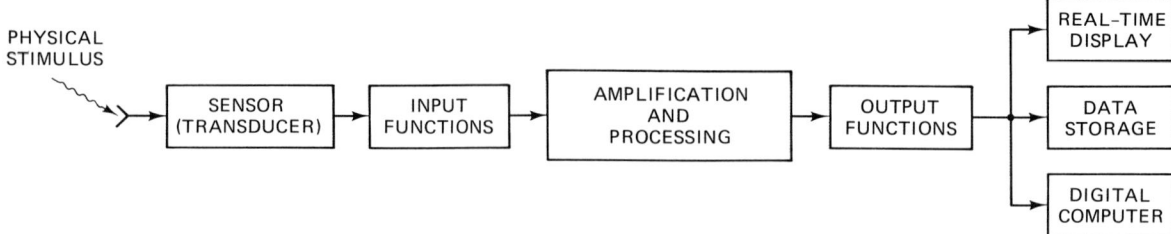

Fig. 1-1   Basic instrumentation system.

3 through 9 will deal with specific sensors for specific forms of physical parameters or stimuli.)

Sensors

The sensor or transducer is a device that is capable of responding to the applied stimulus and producing an electrical output signal that corresponds to the value of the applied stimulus. There is often some ambiguity in the use of the words *sensor* and *transducer*, and in some cases, they are properly used interchangeably. A transducer is a device that converts energy from one form to another (e.g., pressure to an electrical potential), while a sensor may or may not make some sort of conversion. Thus, an electrode used in a medical electrocardiograph (ECG) recording is a sensor but not a transducer, while a pressure transducer is both a sensor and a transducer. A loudspeaker transduces an electrical audio frequency signal to a mechanical acoustical vibration. It is an output transducer and not a sensor at all.[1]

Some physical parameter or stimulus (e.g., temperature, flow, pressure, or displacement) affects the output of a sensor; the sensor is a device that produces an output signal that is proportional to the applied parameter. Thus, the output of the sensor will be either a voltage or current that represents the parameter being measured (e.g., a temperature sensor that outputs a voltage V of 10 mV/K). More often than not, the magnitude of the voltage or current from the sensor represents the magnitude of the parameter at the instant of measurement. Over time, this voltage or current represents the time history of the parameter.

The most desirable sensors have an output signal characteristic that is linear with respect to the stimulus parameter. However, there are also many useful transducers that are either quasilinear (i.e., linear

---

[1]For those who know more than is necessary for this discussion, it is recognized that certain loudspeakers, notably those of the dynamic permanent magnet moving coil (PMMC) design, can also serve as microphones, so they are both sensors and output devices.

only over a portion of their total range) or nonlinear. Such transducers are often used over a limited range, or they must be artificially linearized.

Input Functions

The purpose of the input circuit is to receive the signal from the transducer and convert it into whatever form (usually a voltage) that is required by the circuits to follow. In this section of the instrument, interfacing becomes terribly important. The input functions usually include amplification, and they can also include an AC or DC excitation voltage (especially in the case of Wheatstone bridge sensors), DC level shifting, and isolation of the input circuit from the remainder of the instrument (common in medical instruments because of patient safety considerations).

Signal Processing and Amplification

The output signal from most sensors is not usually suitable for immediate display. Rather, some form of signal conditioning is usually needed. This conditioning may be only amplification, or it might also include frequency-selective filtering; mathematical operations, such as differentiation, integration, "logging" or "antilogging"; or something simple, such as DC level translation. In other cases, signal processing uses the analog circuit, in effect, as a fixed program, dedicated analog computer to solve for a mathematical expression. Some of these functions are most reasonably apportioned to analog circuits, while others are most reasonably apportioned to either digital circuits or computer software. In each case, the designer must decide which is the proper choice for the problem at hand.

Output Functions

The output of the instrument must often be processed in some manner before it can be displayed. The output functions may include power amplification (as in the case of a control system motor driver), digitization for input to a computer, and voltage scaling, so that the display is easily read by a human operator.

Display

Finally, for an instrument to be useful, there must be a display, data storage, or control function to perform. The display device might be a DC meter movement, an oscilloscope, a strip-chart recorder, a video terminal, or a simple "GO–NO GO" lamp. In some systems, the signal will have to be digitized before it can be stored as data in a digital computer system. In still other systems, the output signal is simply stored in a mass storage device for later use.

DIRECT, INDIRECT, AND INFERENTIAL MEASUREMENTS

One way to categorize instrumentation systems is according to how a measurement is derived. It is reasonable to speak of three different categories: direct measurements, indirect measurements, and inferential measurements (although some authorities generally lump together the latter two categories).

The direct measurement is, as the term implies, a measurement that is made of the parameter itself without the need for interpretation, calculation, or any form of interpolation. An example of the direct type

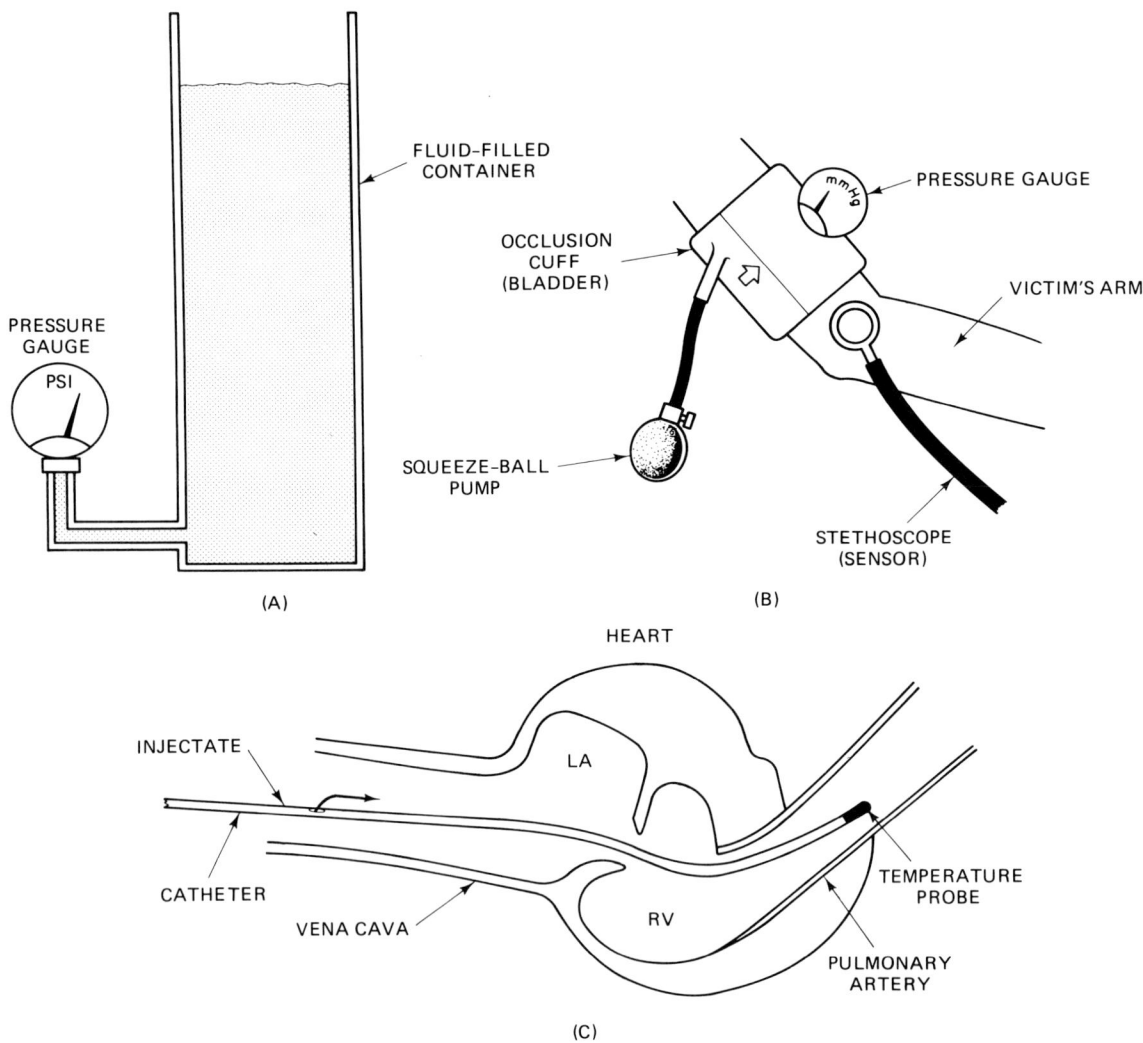

Fig. 1-2 (A) The pressure gauge is used to directly measure the fluid tank pressure; (B) the blood pressure cuff is an indirect measurement; (C) in cardiac output measurements, the measured parameter is derived from other data.

of measurement is a pressure gauge in a hydraulic system (Fig. 1-2A). The pressure at the bottom of the tank is measured directly by the gauge connected to the tank (and by Pascal's principle, if this were a closed system, everywhere else in the tank as well).

An indirect measure of pressure is seen in Fig. 1-2B. In this case, it is human blood pressure that is measured. The measurement is made by occluding the underlying arteries against the bone of the upper arm by inflating a bladder called a blood pressure cuff. It is known that the onset of certain key Korotkoff sounds (detected by a stethoscope or ultrasonic detector) occur when the cuff pressure drops below the peak arterial pressure (systolic) value, and these sounds cease when the cuff pressure equals the diastolic (minima) value. The indirect measurement of blood pressure, called sphygmomanometry, was discovered by Nicolas Korotkoff in 1905, but it did not become widespread until the 1930s, after a long validation period.

An example of an inferential measurement is cardiac output (Fig. 1-2C). The output of the heart, measured in liters of blood per minute (l/min), is not easily measured directly or indirectly, except in the case of an open-heart patient. However, it is possible to thread a thermistor or other temperature sensor through the circulatory system (from an outside site) to a location in the pulmonary artery, just outside the output chamber of the right side of the heart. By injecting a cold saline solution into the vena cava (at the input side of the heart) and then measuring the temperature profile at the output side, it is possible to calculate the cardiac output from the temperature data. Thus, we can state that an inferential measurement is one in which there is either a correlation (but not necessarily causal) or statistical relationship between an easily (or safely) measured parameter and a parameter that is not easily measured, cannot be safely measured, or is impossible to measure. In this type of measurement, the validation process is of critical importance, and it is bound to be the most controversial issue about the design.

PURPOSE OF THE INSTRUMENT

One of the principal problems faced by engineers and other designers of electronic instruments (and other products, for that matter) is to keep in mind the purpose of the device. What is the end use of the device? Why does the user (or customer) want or need this instrument? To what purpose will the end user put this instrument? What are the reasonable (or unreasonable, as is often the case) expectations of the user? It is a common error to lose sight of these factors when designing the instrument. It is all too easy for the designer to become engrossed in the process, the complexity, the technology, or any of a number of factors that are only peripherally relevant to the process of designing and building a useful tool for another person to use.

Mindfully considering the purpose of the instrument is basic to making decisions regarding the design requirements of the instrument. There are several different categories of purpose found in the instrument design field: (1) basic understanding, (2) empirical understanding, (3) data logging, (4) monitoring, and (5) controlling.[2]

The need for basic understanding is seen most acutely in research applications. In most (but not all) research situations, the accuracy and precision of instruments is of primary importance because of the inherent need to reduce uncontrolled variables to a minimum. Thus, instrument errors or biases can seriously affect the understanding sought by the scientist and can easily undermine the entire research project. A less rigorous level of understanding is what might be called empirical understanding. This category may require fewer instruments than the research case, but nonetheless may require good quality instruments. Diagnostic recordings of the human ECG waveform fall into the empirical category.

Data logging is used in many industrial and scientific applications. The accuracy and precision of the instrument will depend mostly on the application. The important criteria here are the repeatability and stability of the measurement process. The instrument, and its related measurement process, must be in a state of long-term statistical control[3] for it to be valid.

Monitoring involves keeping track of the input parameter, usually with the purpose of being alerted when it goes outside established control limits. For example, the hydraulics system might be monitored to make sure that the water pressure does not either drop below a minimum or exceed a maximum (both of these conditions could indicate a problem). Similarly, the physician in an intensive care unit will order monitoring of the patient's heart rate (a function of the bedside ECG monitor) in such a manner that low-rate (bradycardia) and high-rate (tachycardia) alarm limits are set. Again, either condition could be pathological and require immediate intervention. In such a case, the frequency response of the system is less important (ECG waveform fidelity is not important), but the accuracy of the heart rate counter is acutely important.

Finally, control systems often use analog electronic instruments as part of the signal acquisition system. The nature of a control system is that it is active and will take some action (or not), depending on the state of the input signal. Thus, a thermistor used in a heating or cooling system will provide data that will, in turn, be used to determine whether the system should turn on the air conditioner, ignite the furnace, or remain dormant.

---

[2] This list is expanded from a list given by Richard S. C. Cobbold, *Transducers for Biomedical Applications*, Wiley, New York, 1974.

[3] *Out of the Crisis*, W. Edwards Deming, MIT—CAES, Cambridge, Massachusetts, 1982, 1986.

DESIGNING THE INSTRUMENT

The design of analog instruments (or analog subsystems to digital instruments) is not an arcane art, open only to a few initiates. Rather, it is a logical, step-by-step process that can be learned. Like any skill, design is improved with practice, so one is cautioned against both excessive expectations and discouragement if the process does not work out exactly as planned the first time.

Some of the material in this chapter may fairly be called philosophic (by which the author usually means not true philosophy, but rather some other person's unappreciated belief). Although many technical people claim to disdain philosophy, all of us philosophize. It is just that some people think about it a lot, some think about it either a little or at only a few times during their lives, while others do not think about it at all—they take actions and form opinions based not on a considered viewpoint, but by default. So, if some of this material seems as though I am philosophizing, then I hope that it is at least mindfully considered.

A DESIGN PROCEDURE

The procedure that one adopts to designing instruments may well be different from what is presented here, and that is all right. The purpose of offering a procedure is to systematize the process, not necessarily to force a particular set of guidelines on everyone. While it is conceivable that one can design an instrument by a process similar to Brownian motion (as, theoretically, a Tom Clancy novel can be written by an infinite collection of monkeys randomly pounding on an infinite number of personal computer keyboards), it is the systematic approach that most often yields success in day-to-day engineering.

The procedure below assumes a product that is a one-of-a-kind instrument, as in a scientific laboratory or plant. Designing a product for production and sale follows a similar procedure, but it involves consideration of marketing information and production problems, as well as purely design issues. The steps in the procedure, some of which are iterative with respect to each other, are offered here.

1. Define and tentatively solve the problem.
2. Determine the critical attributes required of the final product and incorporate these into a specification and test plan that determines objective criteria for acceptance or rejection.
3. Determine the critical parameters and requirements.
4. Attempt a block diagram solution.
5. Apportion requirements (e.g., gain, frequency response) to the various blocks.

6. Perform analysis and do simulations on the block diagram to test the validity of the approach.

7. Design specific circuits to fill in the blocks.

8. Build and test the circuits.

9. Combine the circuits with each other on a breadboard.

10. Test the breadboarded circuit according to a fixed test plan.

11. Build a brassboard that incorporates all changes made in the previous steps.

12. Test the brassboard and correct problems.

13. Design and construct final configuration.

14. Test final configuration.

15. Ship the product.

Solve the Right Problem

The purpose of the designer is to solve some problem or another using analog circuits, digital circuits, a personal computer, or whatever else is available in the armamentarium. There are two related problems often seen in the efforts of novices.

First, it is often the case that the designer will have a tentative favorite approach in mind before the problem is properly understood. Decisions are made based on what the designer is most comfortable doing. For example, many younger designers are likely to select the digital solution in a knee-jerk manner that excludes any consideration of the analog solution. All available approaches should be evaluated, and the one that best fits the need selected. Keep in mind that your job is to solve the problem in an economical and scientifically valid manner. It is not necessarily to produce a dazzling display of pyrotechnics designed to impress others with your technical acumen.

Second, be sure you understand the problem being solved. While this advice seems trivial, it is also true that failure to understand the problem at hand often sinks designs before they have a chance to mature. There are several facets to this problem. For example, a natural tendency of many engineers is to think that an elegant solution is always complex and large scale (as it sometimes is). If this mistake is made, then it is likely that the product will be overdesigned, or it will have too many "whistles and bells." It was, after all, designed to solve a much harder problem than was actually presented. Therefore, the most elegant solution is the one that is scaled to the actual problem at hand, plus a reasonable growth capability for future expansion.

Another aspect to understanding the problem is to understand the final end customer's actual use for the product. It is all too easy to get caught up in the specification, or our own ideas about the job, and overlook altogether what task the user needs to accomplish with the product. An example is derived (like many of the examples in this book) from medical instrumentation. A physiologist requested a pressure amplifier that would measure blood pressures over a range of 0 to

300 mm Hg (Torr). What the salesperson never told the plant was (1) it would be used on humans (safety and regulatory issues), (2) blood would come in contact with the diaphragm (cleaning, microbiological isolation, and liquid–material compatibility issues), and (3) it would occasionally be used for measuring 1 to 5 mm Hg central venous pressures (which means low-end linearity issues).

Part of the problem in determining the level of complexity, or the specific design's function, is miscommunication between the end user and the designer. Although miscommunications occur frequently between in-house designers and their clients, it is probably most common between distant customers and engineers in the plant. Of course, marketing people may never let the engineer and customer get together (either from ignorance, or from a fear that little fibs will surface: "The reason I hate engineers is that, under duress, they tend to blurt out the truth").

The proper role of the designer is to scope out the problem at hand and understand what the circuit or instrument is supposed to do, how and where the user is going to use it, and exactly what the user wants and expects from the product.

Determine Critical Attributes

This step is basically harvesting the fruit of understanding and correctly solving the right problem. From the solution of the problem, one can determine and write a set of attributes, characteristics, parameters, and other indices of the product's final nature.

It is at this point that one must write a specification that accurately documents what the final product is supposed to do. The specification must be clearly written so that others can understand it. A concept or idea does not really exist, except in the mind of the originator. One must create operational definitions for the attributes of the product.[4] One cannot simply say "it must measure pressure to a linearity of 1% over a range of 0 to 100 psi." Rather, it might be necessary instead to specify a rational and reasonable test method under which this requirement can be met. There might be, after all, more than one standard for pressure and measurement, and there is certainly more than one definition of linearity. The operational definition serves the powerful function of providing everyone with the same set of rules; basically, it levels the playing field.

Part of this step, and of making an operational definition, is to write a test plan for the final product. It is here that one determines (and often contractually agrees) exactly what the final product will do and defines the objective criteria of goodness or badness that will be used to judge the product.

---

[4]*Out of the Crisis*, W. Edwards Deming, MIT—CAES, Cambridge, Massachusetts, 1982, 1986.

Determine Critical Parameters and Requirements

Once the product is properly scoped, it is time to determine the critical technical parameters that are needed to meet the test requirements (and hopefully the user's needs—if the test requirements are properly written). Parameters, such as frequency response and gain, tend to vary in multimode instruments, so one must determine the worst case for each specification item and design for it.

Attempt a Block Diagram Solution

The block diagram is a signal flow or function diagram that represents stages, or collections of stages, in the final instrument. In large instruments, there might be several indexed levels of a block diagram, each one becoming finer in detail.

Apportion Requirements to the Blocks

Once the block diagram solution is on paper, tentatively apportion system requirements to each block. Distribute gain, frequency response, and the other attributes to each block. Keep in mind that factors such as gain distribution, for example, can have a profound effect on other factors (such as dynamic range). Also, the noise factor and drift of any one amplifier can have a tremendous effect on the final performance, and it is in these types of parameters (where critical placement of one high-quality stage may be sufficient) that added cost and complexity often arises.

Analyze and Simulate

Once the block diagram is set and the requirements apportioned to the various stages, it is time to analyze the circuit and run simulations to see if it will actually work over the entire range of input values. A little "desk checking" goes a long way toward eliminating problems later, when the design is first prototyped in the workshop. Plug in typical input values, and see what happens on a stage by stage basis. Check for the reasonableness of outputs at each stage. For example, if the input signal should drive an output signal to 17 V and the operational amplifiers are only operated from ± 15-V power supplies, then something is seriously wrong with the design, and it will have to be corrected before proceeding.

Design Specific Circuits for Each Block

It is at this point that the remainder of this book is of the most use to you, because specific circuits are the main subject of the text. In this step, fill in those blocks with real circuit diagrams.

Build and Test the Circuits

At this point, one must actually construct the individual circuits and test them to make sure they work as designed (unless, of course, the circuit is so familiar that no testing is needed). Keep in mind that some of your best ideas for simplified circuits may not actually work—and this is the place to find out. Use a benchtop breadboard that allows circuit construction using plug-in stripped end wires.

Combine the Circuits in a Formal Breadboard

Once the validity of the individual circuits is determined, combine them together in a formal breadboard. Whether built on a benchtop breadboard or on a prototyping board, make sure that the layout is similar to that expected in the final form.

Test the Breadboard

Test the overall circuit according to formally established objective criteria. This test plan should be developed early in step two. As problems arise and are solved, make changes and corrections and document the results. It is, perhaps, the main failing of inexperienced designers that they do not properly document their work, even in an engineer's or scientist's notebook.

Build and Test a Brassboard Version

A brassboard is a version made as close as possible to the final configuration. While breadboarding techniques can be a little sloppy, the brassboard should be a properly printed circuit board. The test criteria should be the same as before, updated only for changes that occurred. If problems turn up, they should be corrected prior to proceeding further. Keep in mind that the most common problems that occur in leaping from breadboard to brassboard are layout (e.g., coupling between stages), power distribution, and ground plane noise (these are the principal areas of difference between the two configurations).

Design, Build, and Test the Final Version

Once all of the problems are known and solved and the resultant changes (if any) are incorporated, it is time to build the final product as it will be given to the end user. It is at this point that the reputation of the designer is made or broken, because it is here that the product is finally evaluated by the client.

HUMAN FACTORS

The designer of an instrument must consider the issues of who will use this instrument, how it will be used, and how its output data will be interpreted by the user. These issues are generally considered under the rubric *human factors*. While it is impossible to discuss everything about human factors design, one must at least consider this aspect. Issues such as knob or control placement, which functions need to be externally controlled and which can be internalized or automated, the display format, and a score of related matters are components of human factors design. If the displayed data are not in the correct format or if the display is difficult to either read or understand, then the instrument design, no matter how elegant technically, is basically a failure.

Let us consider an example that shows how human factors were included in a design versus one where the issue was not addressed. There are two clinical blood pressure monitors in a hospital. Both instruments are digital. The physician orders an intensive care nurse to medicate the patient if the diastolic blood pressure drops below 90 mm Hg. Unfortunately, in digital instruments, there is a phenomenon called last digit bobble. Because of the digitization process, the last digit is always ambiguous. When the real value is between the two allowed steps, then the display may switch back and forth (bobble) between them. In the case of one blood pressure monitor, the nurse was confronted with the situation where the display bobbled back and forth between 89 and 90. Does he medicate or not medicate? Two monitors are available. One of them measures to only two significant figures and displays both digits—this instrument throws the nurse into a quandry because of the bobble. The other instrument is better designed. It measures the pressure to three or more significant figures, but only displays the two most significant digits. Thus, the bobble takes place (most of the time) in the undisplayed digits—the band of ambiguity is reduced one order of magnitude.

INTERFACING

One of the key reasons why certain circuits and sensors do not play well together in an attempted instrument design is due to improper interfacing between the elements of the circuit. While the issues involved in interfacing are much greater than can be presented here, the main issue is to make the destination for a signal compatible with the source of the signal. Perhaps the most common problems, which will be used here for illustration, involve the nature of the signal source and the input of the destination circuit or device. Figure 1-3 shows three scenarios: current sources (Fig. 1-3A), voltage sources (Fig. 1-3B), and power sources (Fig. 1-3C).

The current source (Fig. 1-3A) is a circuit or sensor that produces a current output signal (I_{S1}). The internal resistance (R_{S1}) of the current source might be quite high. The input resistance of the following stage (R_{i1}) must be very low compared with the source resistance ($R_{i1} \ll R_{S1}$), or it will cause the signal current to diminish from its correct value.

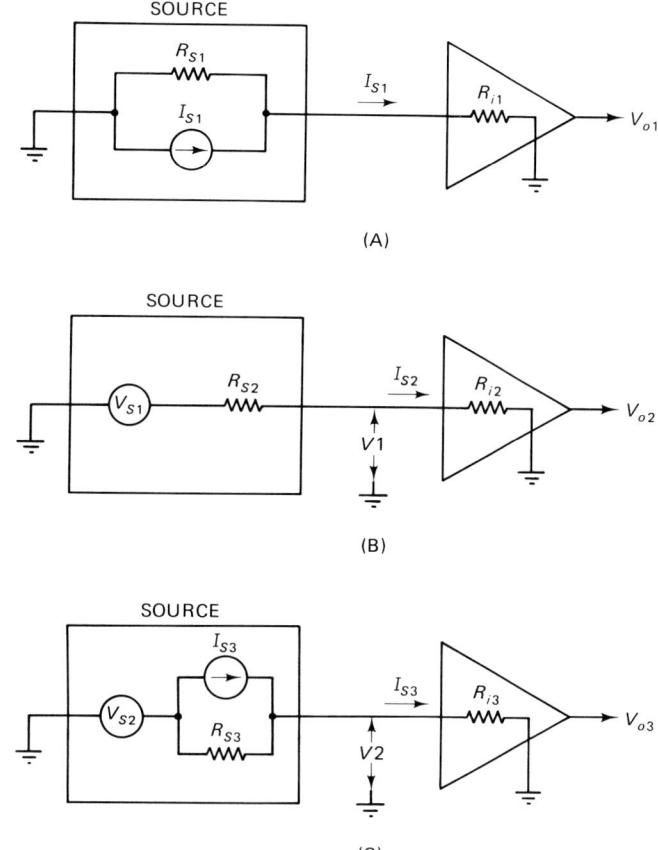

Fig. 1-3 Circuit models for types of amplifiers: (A) current amplifier, (B) voltage amplifier, and (C) power amplifier.

The voltage source (Fig. 1-3B) represents the most common form of sensor and signal processing circuit. In this case, the source resistance (R_{S2}) tends to be very low. The designer's goal is to make the impedance of the input circuitry of the following stage very high ($R_{i2} \gg R_{S2}$). Otherwise, the action is that of a voltage divider—the source output voltage ($V1$) will be reduced according to the ratio $R_{i2}/(R_{i2} + R_{S2})$.

The power source (Fig. 1-3C) differs from both the current and voltage sources. In this case, the maximum efficiency of the signal transfer occurs when the source resistance (R_{S3}) is matched to the destination input resistance (R_{i3}).

We can sum up the relationship of the signal source impedance to the destination input impedance:

1. current sources: $R_{in} \ll R_{s}$;
2. voltage sources: $R_{in} \gg R_{s}$; and
3. power sources: $R_{in} = R_{s}$.

Other issues to consider when dealing with interface issues are the frequency responses of the source and destination circuits, the relative voltage levels, dynamic range, DC offset levels, and gain accuracies.

CONCLUSION

In this chapter, we discussed the basics of designing either an analog instrument or the analog subsection of the digital computer based instrument. Before proceeding with the discussions of typical sensors and actual circuits, however, we need to take a look at the nature of signals and noise.

2

Signals and Noise

Signals are the grist for the instrumentation designer's mill, and it is to the processing of signals that we will soon direct our attention. But, the nature of signals, and their relationship to noise and interfering signals, determines the appropriate design, from the system level to the component selection level. In this chapter, we will take a look at signals and noise and how each affects the design of instrumentation circuits.

Signals can be categorized several ways, but one of the most fundamental is according to time domain behavior (the other major category is the frequency domain). We will therefore consider signals of the form $v = f(t)$ or $i = f(t)$. The time domain classes of signals include static, quasistatic, periodic, repetitive, transient, random, and chaotic categories. Each of these categories has certain properties that can profoundly influence appropriate design decisions.

Static and Quasistatic Signals

A static signal (Fig. 2-1A) is, by definition, unchanging over a very long period of time (T_{long} in Figs. 2-1A and 2-1B). Such a signal is essentially a DC level, so it must be processed in low-drift DC amplifier circuits. The term quasistatic means nearly unchanging, so a quasistatic signal (Fig. 2-1B) refers to a signal that changes so slowly over long times that it possesses characteristics more like static signals than dynamic (i.e., rapidly changing) signals.

Periodic Signals

A periodic signal (Fig. 2-1C) is one that repeats itself on a regular —that is, periodic—basis. Examples of periodic signals include sine waves, square waves, sawtooth waves, and triangle waves. The nature of the periodic waveform is such that each waveform is identical at like points along the time line. In other words, if you advance along the

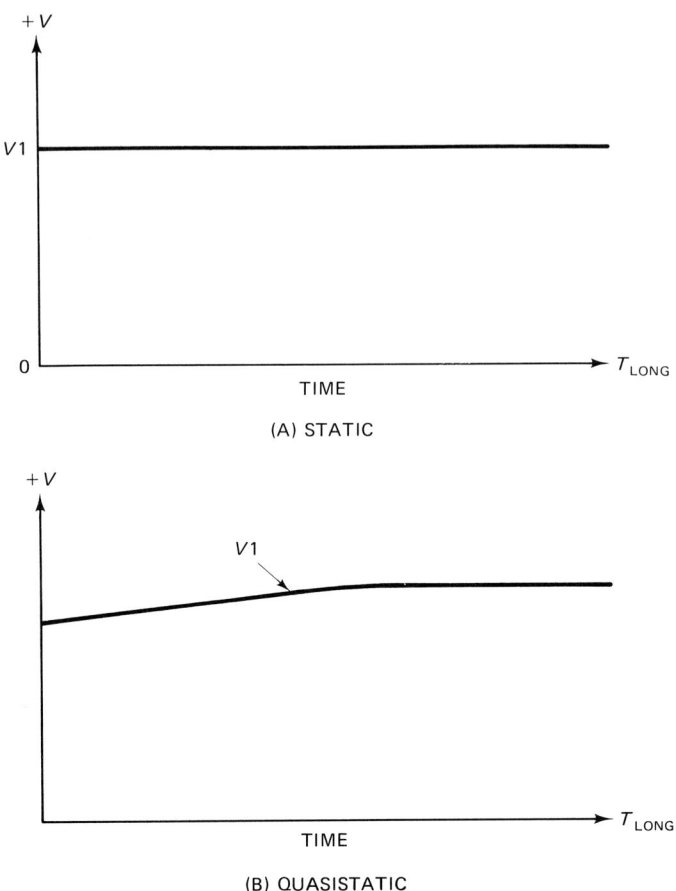

Fig. 2-1 Types of signals: (A) static signal, (B) quasistatic signal, (C) periodic signal, (D) quasiperiodic signal, (E) transient signal, (F) quasitransient signal, and (G) chaotic or random signal. (*Figure continues on the following two pages.*)

time line by exactly one period (T), then the voltage, polarity, and direction of change of the waveform will be repeated. That is, for a voltage waveform, $V(t) = V(t + T)$.

Repetitive Signals

A repetitive signal (Fig. 2-1D) is quasiperiodic in nature, so it bears some similarity to the periodic waveform. The principal difference between repetitive and periodic signals is seen by comparing the signal at $f(t)$ and $f(t + T)$, where T is the period of the signal. These points might not be identical in repetitive signals, but they are identical in periodic signals. The repetitive signal might contain either transient or stable features that vary from period to period. An example is the human arterial blood pressure waveform (Fig. 2-1D). While the waveform tends to vary from a minima (diastolic) to a maxima (systolic) in a quasiperiodic manner, there are both normal and pathological anomalies from one cycle to another. For example, the ampli-

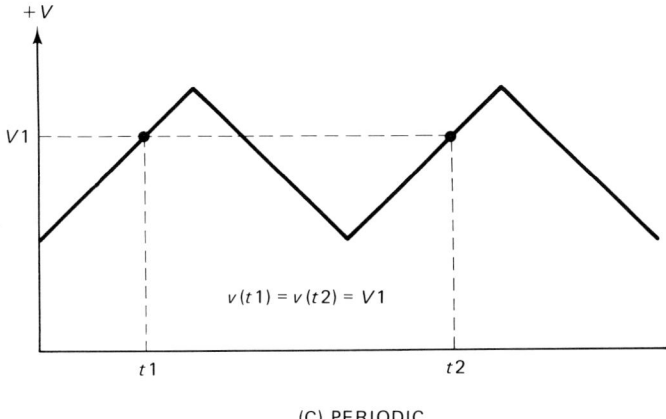

(C) PERIODIC

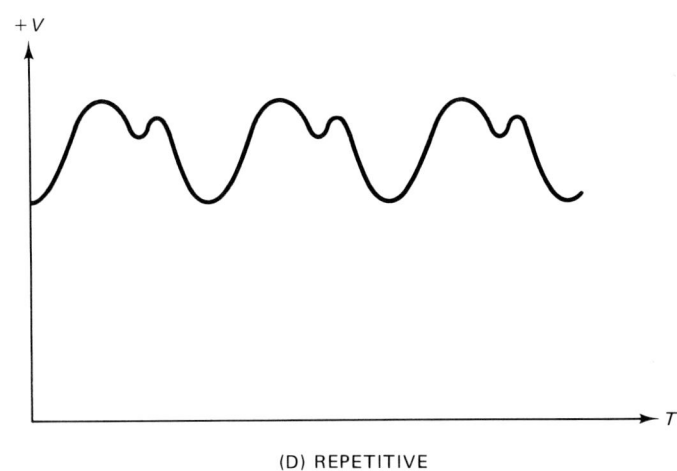

(D) REPETITIVE

Fig. 2-1 (*continued*)

tudes of the maxima and minima, and the repetition rate (i.e., heart rate) tend to vary quite normally in healthy humans. In addition, events such as premature ventricular contractions (PVC) are anomalies that may be pathological (PVCs tend to be purely transient events superimposed on the repetitive signal). Thus, the repetitive signal may bear characteristics of both transient and periodic signals.

Transient Signals

A transient signal (Fig. 2-1E) is either a one-time event or a periodic event in which the event duration is very short compared with the period of the waveform. In terms of Fig. 2-1F, the latter definition means that $t_1 \ll t_2$. These signals can be treated as if they are transients.

Random and Chaotic Signals

A random signal (Fig. 2-1G) is one that is unpredictable and has either or both of the following properties: (1) one or more inputs to the

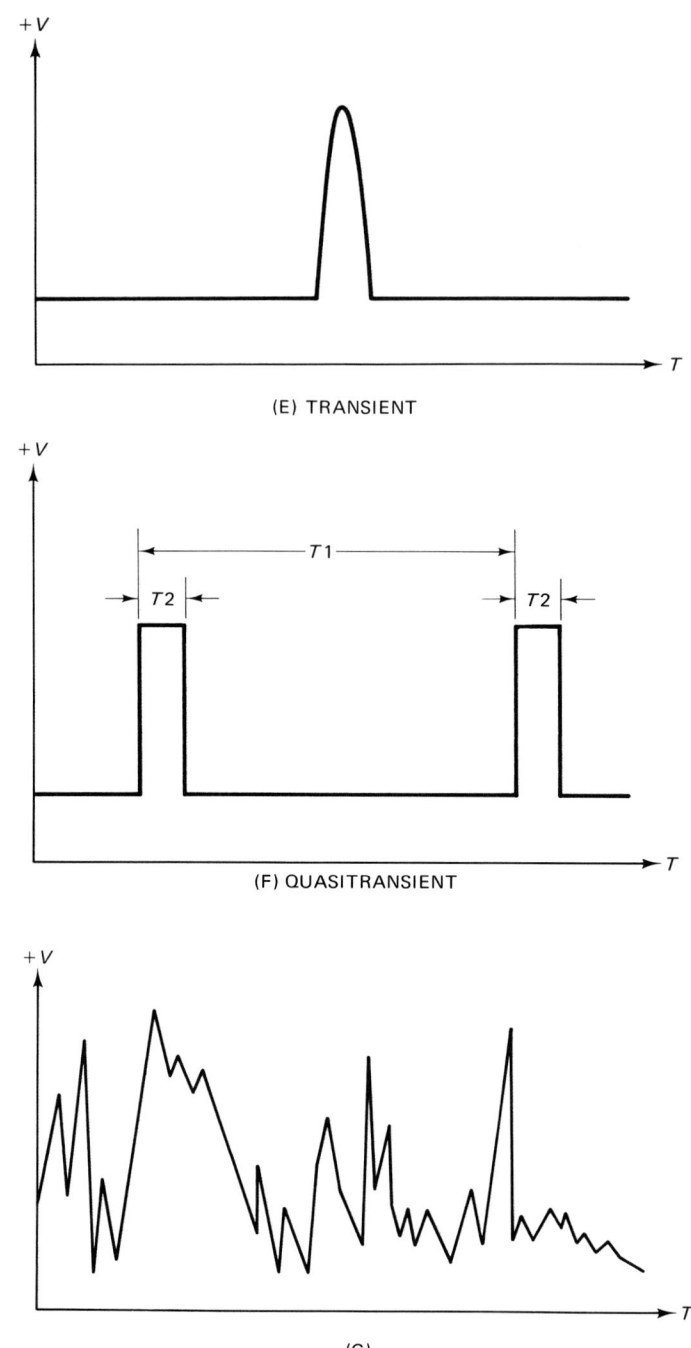

(E) TRANSIENT

(F) QUASITRANSIENT

(G)

Fig. 2-1 (*continued*)

system are unknown, and (2) the rules by which the system affects the output in response to the inputs are unknown.

A new form of signal recently recognized is the chaotic signal. Many of these signals were previously labeled random, but there is a critical difference between chaotic and random signals. The difference is that even though all inputs and transfer function rules are known, the output signal nonetheless remains unpredictable even after only a short time.

The characteristics of these various types of signal often drive the design of the instrumentation system. An important consideration of all dynamic signals (whether periodic or otherwise) is the frequency response required to faithfully reproduce the signal. For that reason, we will turn our attention to the frequency domain characteristics of signals.

FOURIER SERIES

All continuous periodic signals can be represented by a fundamental frequency sine wave and a collection of harmonics of that fundamental sine wave that are summed together linearly. These frequencies comprise the Fourier series of the waveform. The elementary sine wave (Fig. 2-2) is described by

$$v = V_m \sin(2\omega t) \tag{2-1}$$

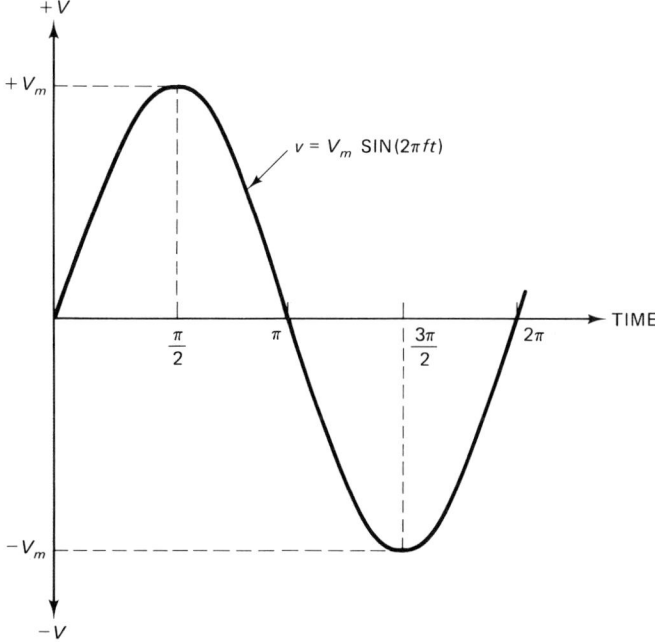

Fig. 2-2 Sine wave signal showing time and amplitude relationships.

where v is the instantaneous amplitude of the sine wave, V_m the peak amplitude of the sine wave, ω the angular frequency $(2\pi F)$ of the sine wave, and t the time in seconds.

The period of the sine wave is the time between recurrence of identical events, or $T = 2\pi/\omega = 1/F$ (where F is the frequency in cycles per second).

The Fourier series that makes up a waveform can be found if a given waveform is decomposed into its constituent frequencies, either by a bank of frequency selective filters or a digital signal processing algorithm called the fast Fourier transform (FFT). The Fourier series can also be used to construct a waveform from the ground up. Figure 2-3 shows square wave (Fig. 2-3A), sawtooth wave (Fig. 2-3B), and peaked wave (Fig. 2-3C) signals constructed from fundamental sine waves and their harmonic sine and cosine functions.

The Fourier series for any waveform can be expressed in the form

$$f(t) = \frac{a_0}{2} + \sum_{n=1}^{\infty} [a_n \cos(n\omega t) + b_n \sin(n\omega t)] \qquad (2\text{-}2)$$

where a_n and b_n represent the amplitudes of the harmonics, n is an integer, and other terms are as previously defined.

The amplitude coefficients (a_n and b_n) are expressed by

$$a_n = \frac{2}{T} \int_0^T f(t) \cos(n\omega t)\, dt \qquad (2\text{-}3)$$

and

$$b_n = \frac{2}{T} \int_0^T f(t) \sin(n\omega t)\, dt \qquad (2\text{-}4)$$

The amplitude terms are nonzero at the specific frequencies determined by the Fourier series. Because only certain frequencies, determined by integer n, are allowable, the spectrum of the periodic signal is said to be discrete.

The term $a_0/2$ in the Fourier series expression [Eq. (2-2)] is the average value of $f(t)$ over one complete cycle (one period) of the wave form. In practical terms, it is also the DC component of the waveform. When the waveform possesses halfwave symmetry (i.e., the peak amplitude above zero is equal to the peak amplitude below zero at every point in t, or $+V_m = |-V_m|$), there is no DC component, so $a_0 = 0$.

An alternative Fourier series expression replaces the $a_n \cos(n\omega t) + b_n \sin(n\omega t)$ with an equivalent expression of another form:

$$f(t) = \frac{2}{T} \sum_{n=1}^{\infty} c_n [n\omega t - \phi_n] \qquad (2\text{-}5)$$

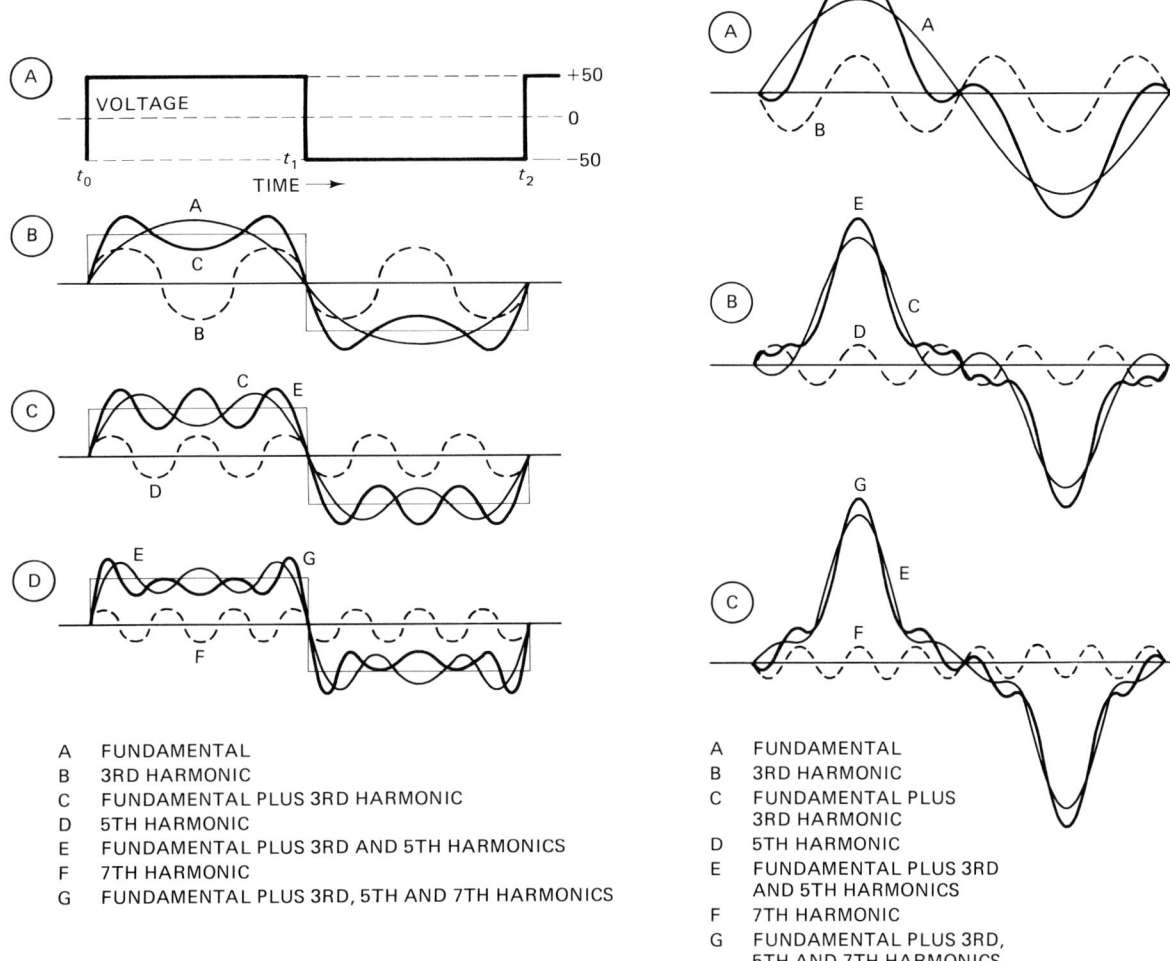

A   FUNDAMENTAL
B   3RD HARMONIC
C   FUNDAMENTAL PLUS 3RD HARMONIC
D   5TH HARMONIC
E   FUNDAMENTAL PLUS 3RD AND 5TH HARMONICS
F   7TH HARMONIC
G   FUNDAMENTAL PLUS 3RD, 5TH AND 7TH HARMONICS

(A) SQUARE WAVE

A   FUNDAMENTAL
B   3RD HARMONIC
C   FUNDAMENTAL PLUS
     3RD HARMONIC
D   5TH HARMONIC
E   FUNDAMENTAL PLUS 3RD
     AND 5TH HARMONICS
F   7TH HARMONIC
G   FUNDAMENTAL PLUS 3RD,
     5TH AND 7TH HARMONICS

(C) PEAKED WAVE

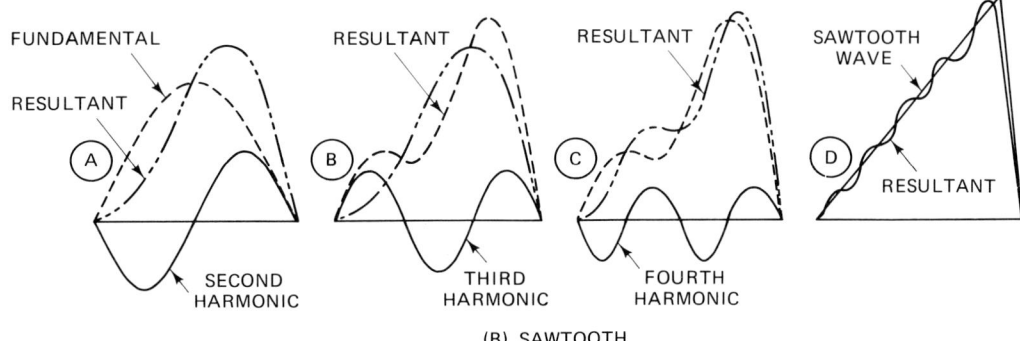

(B) SAWTOOTH

Fig. 2-3 Waveforms constructed from sine and cosine waves: (A) square wave, (B) sawtooth wave, and (C) peaked wave.

where $c_n = [(a_n)^2 + (b_n)^2]^{1/2}$, $\phi_n = \arctan(a_n/b_n)$, and all other terms are as previously defined.

One can infer certain things about the harmonic content of a waveform by examination of its symmetries. One would conclude from the above equations that the harmonics extend to infinity on all waveforms. Clearly, in practical systems, a much less than infinite bandwidth is found, so some of those harmonics will be removed by the normal action of the electronic circuits. Also, it is sometimes found that higher harmonics might not be truly significant, so they can be ignored. As n becomes larger, the amplitude coefficients, a_n and b_n, tend to become smaller. At some point, the amplitude coefficients are reduced sufficiently that their contribution to the shape of the wave is either negligible for the practical purpose at hand or totally unobservable in practical terms. The value of n at which this occurs depends partially on the rise time of the waveform. Rise time is usually defined

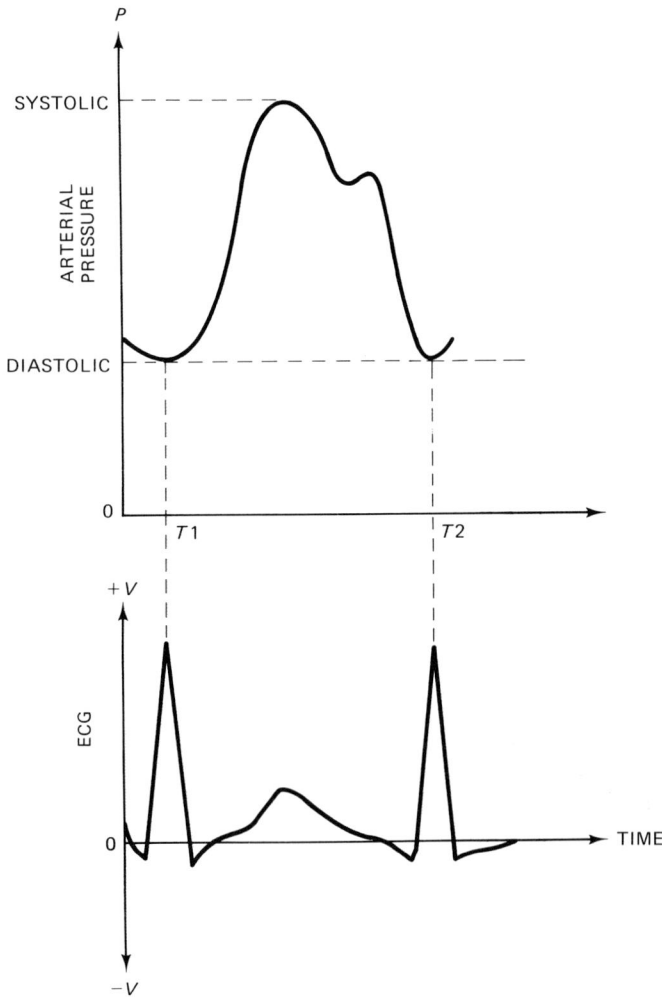

Fig. 2-4 Human arterial blood pressure waveform and ECG on same time line.

as the time required for the waveform to rise from 10 to 90% of its final amplitude. Let us consider a practical example from biomedical instrumentation.

Figure 2-4 shows the human arterial pressure waveform superimposed on the same time line as the ECG waveform. These waveforms are time correlated because they both represent different views of the same physical event, for example, the beating of the human heart. Suppose that the heart rate is 72 beats per minute (BPM), or 1.2 Hz. The pressure waveform has a slower rise time than the ECG waveform, so it contains a smaller number of harmonics. The pressure waveform can be accurately reproduced with about 25 harmonics (e.g., approximately a 30-Hz bandwidth), while the ECG waveform requires 70 to 80 harmonics for faithful reproduction (e.g., a bandwidth of about 100 Hz). In order to adequately process these two waveforms, the instrument must have upper 3-dB frequency responses of 30 and 100 Hz for the pressure and ECG channels, respectively. Because both pressure and ECG waveforms have significantly rounded features, the lower 3-dB frequency response (a function of subharmonic content) must be 0.05 Hz.

The square wave represents another case altogether because it has a very fast rise time. Theoretically, the square wave contains an

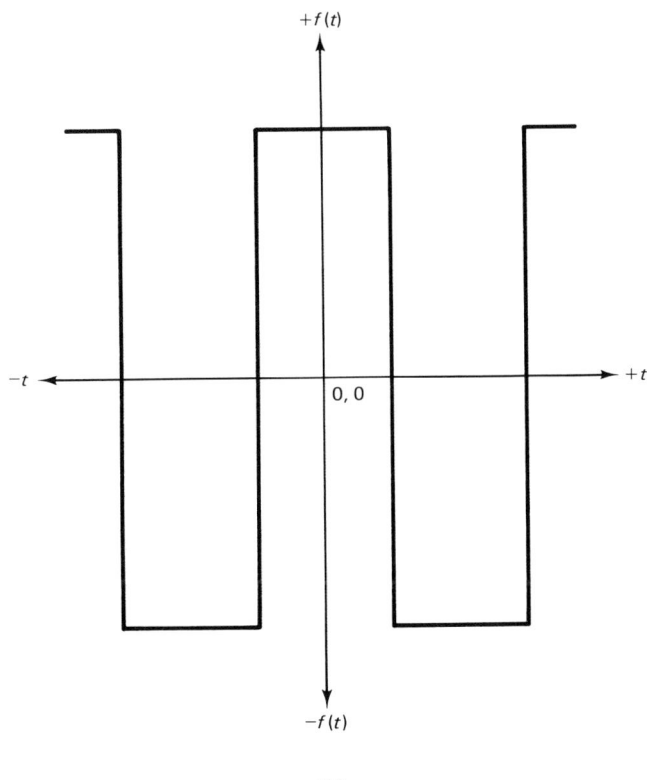

(A)

Fig. 2-5 (A) Odd waveform, and (B) even waveform. (*Figure continues.*)

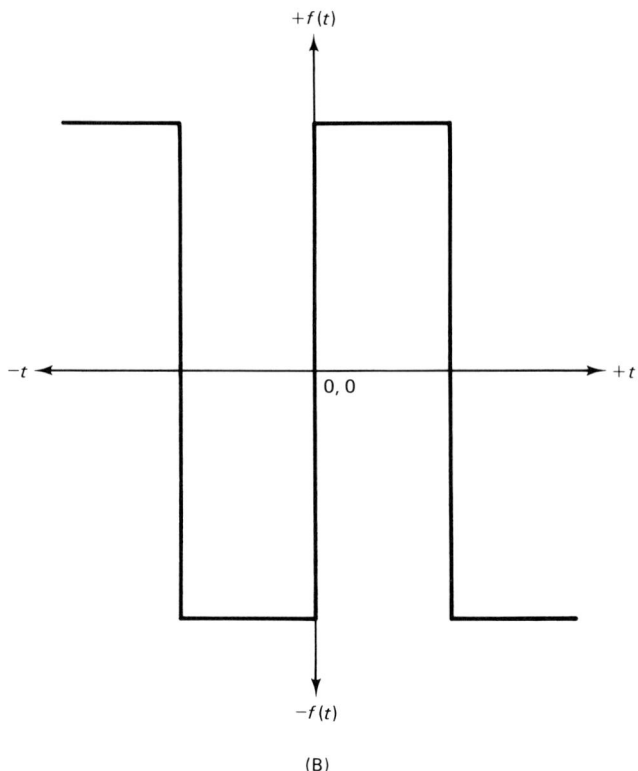

Fig. 2-5 (*continued*)

infinite number of harmonics, but not all of the possible harmonics are present. For example, in the case of the square wave, only the odd harmonics are typically found (e.g., $3, 5, 7$). According to some standards, accurately reproducing the square wave requires 100 harmonics, while others claim that 1000 harmonics are needed. Which standard to use may depend on the specifics of the application.

Another factor that determines the profile of the Fourier series of a specific waveform is whether the function is odd or even. Figure 2-5A shows an odd-function square wave, and Fig. 2-5B shows an even-function square wave. The even function is one in which $f(t) = f(-t)$, while for the odd function, $-f(t) = f(-t)$. In the even function, only cosine harmonics are present, so the sine amplitude coefficient b_n is zero. Similarly, in the odd function, only sine harmonics are present, so the cosine amplitude coefficient a_n is zero.

Waveform Symmetry

Both symmetry and asymmetry can occur in several ways in a waveform (Fig. 2-6), and those factors can affect the nature of the Fourier series of the waveform. In Fig. 2-6A, we see the case of a waveform with a DC component, or in terms of the Fourier series equation, the term a_0 is nonzero. The DC component represents a case of asymmetry in a signal. This offset can seriously affect electronic instrumentation circuits that are DC coupled, and thereby result in a serious artifact.

Two different forms of symmetry are shown in Fig. 2-6B. Zero-axis symmetry occurs when, on a point-for-point basis, the wave shape and amplitude above the zero baseline are equal to the amplitude below the baseline (or $|+V_m| = |-V_m|$). When a waveform possesses zero-axis symmetry, it will usually not contain even harmonics, only odd harmonics are present. This situation is found in square waves, see, for

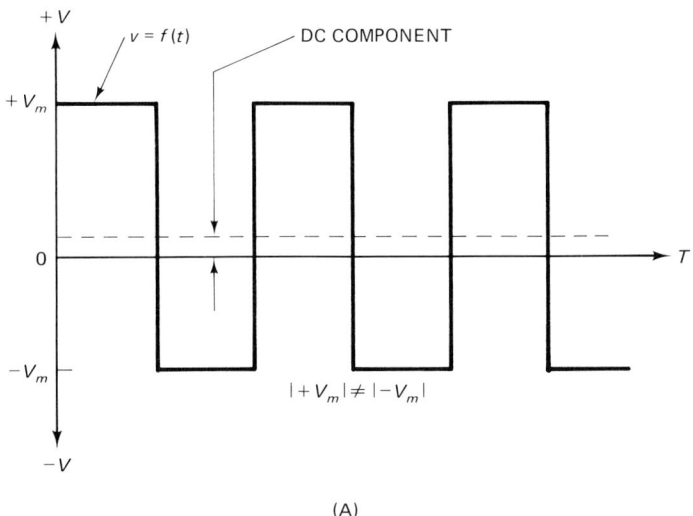

(A)

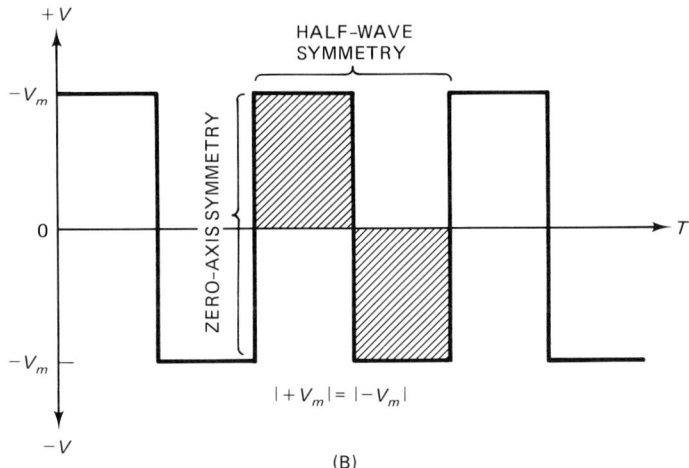

(B)

Fig. 2-6 (A) Square wave with a DC component that produces asymmetry, (B) a symmetrical square wave, (C) a sawtooth wave forms a mirror image across the baseline, and (D) quarterwave symmetry. (*Figure continues.*)

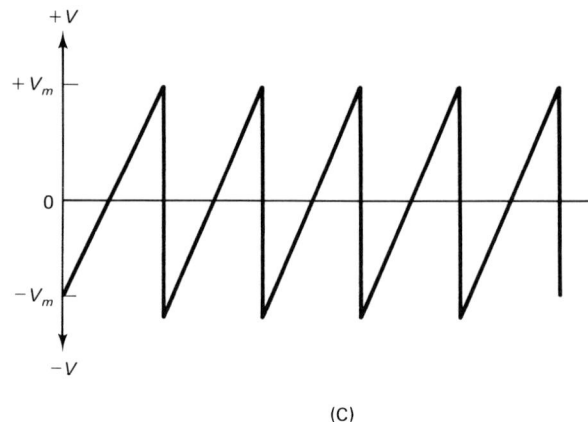

(C)

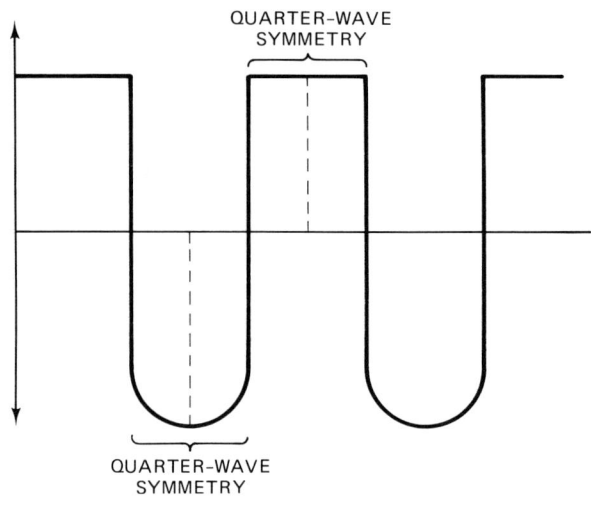

(D)

Fig. 2-6 (*continued*)

example, Fig. 2-7A. Zero-axis symmetry is not found only in sine and square waves, however, as the sawtooth waveform in Fig. 2-6C demonstrates.

An exception to the "no even harmonics" general rule is that there will be even harmonics present in the zero-axis symmetrical waveform (Fig. 2-7B) if the even harmonics are in-phase with the fundamental sine wave. This condition will neither produce a DC component nor disturb the zero-axis symmetry.

Also shown in Fig. 2-6B is halfwave symmetry. In this type of symmetry, the shape of the wave above the zero baseline is a mirror image of the shape of the wave below the baseline (see shaded region in Fig. 2-6B). Halfwave symmetry also implies a lack of even harmonics.

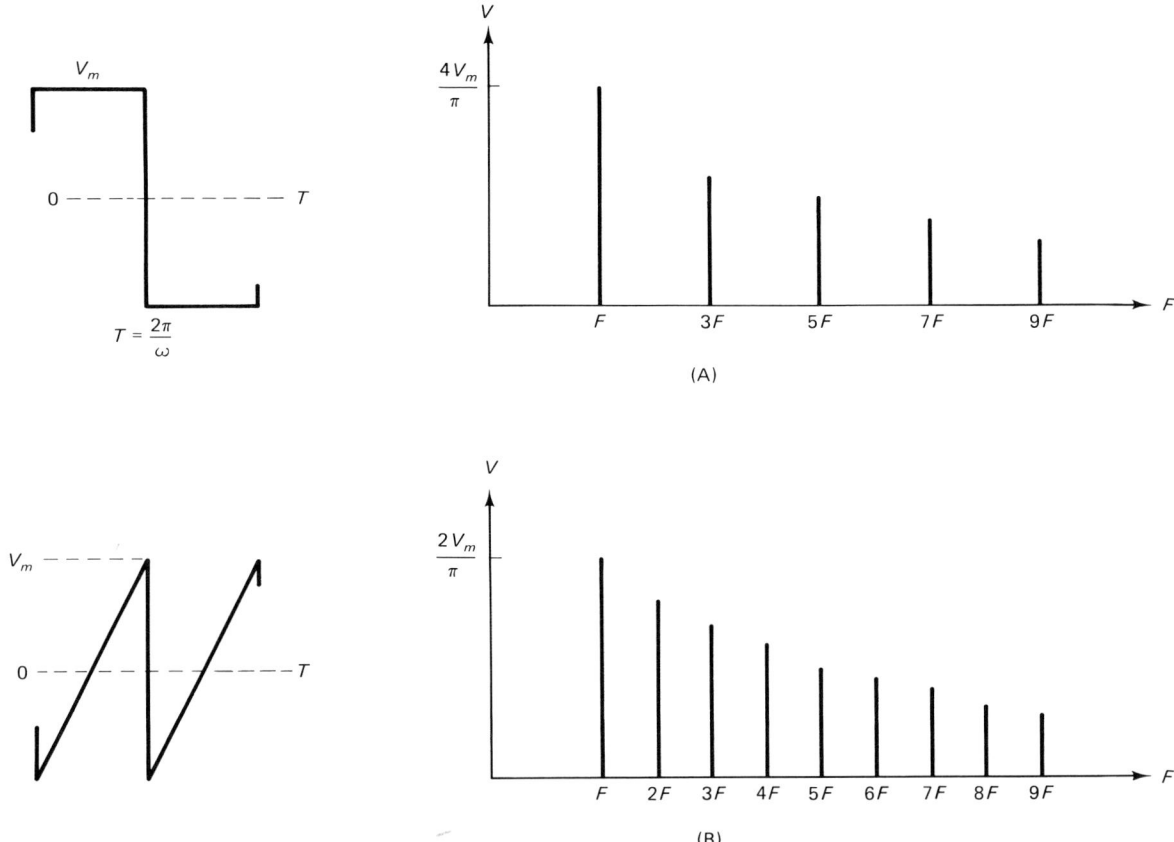

Fig. 2-7 Fourier series for two waveforms: (A) square wave and (B) sawtooth.

Quarterwave symmetry (Fig. 2-6D) exists when the left half and right half sides of the waveforms are mirror images of each other on the same side of the zero axis. Note in Fig. 2-6D, that above the zero axis the waveform is like a square wave, and indeed the left- and right-hand sides are mirror images of each other. Similarly, below the zero axis, the rounded waveform has a mirror image relationship between left and right sides. In this case, there is a full set of even harmonics, and any odd harmonics that are present are in-phase with the fundamental sine wave.

TRANSIENT SIGNALS

A transient signal is an event that occurs only once, randomly over a long period of time, or is periodic, but has a very short duration compared with its period (i.e., it is a very short duty cycle event). Many pulse signals fit the latter criterion, even though mathematically they are actually periodic.

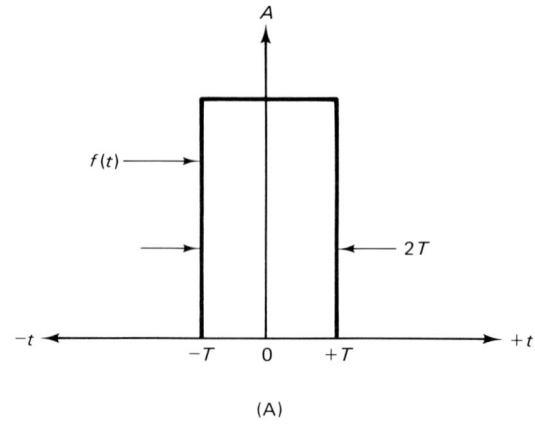

(A)

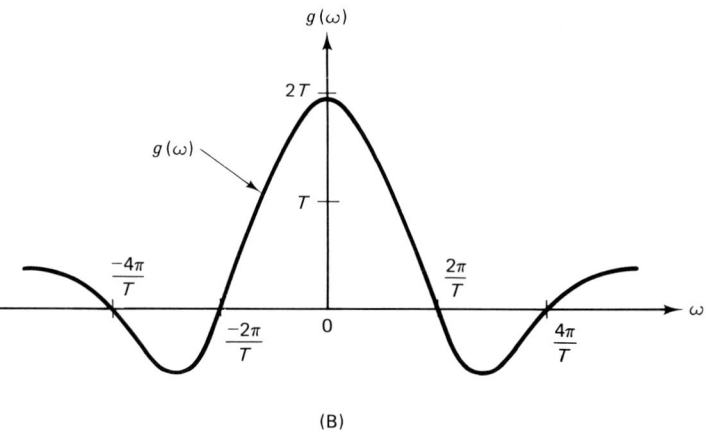

(B)

Fig. 2-8 (A) Pulse signal and (B) frequency domain distribution.

Transient signals are not represented properly by the Fourier series, but can nonetheless be represented by sine waves in a spectrum. The difference is that the spectrum of the transient signal is continuous rather than discrete (as in a periodic signal). Consider a transient signal of period $2T$, such as Fig. 2-8A. The spectral density, $g(\omega)$, is

$$g(\omega) = \int_{-\infty}^{+\infty} f(t)e^{-j\omega t}\, dt \qquad (2\text{-}6)$$

Equation (2-6) represents the spectral density $g(\omega)$. Given a spectral density, however, the original waveform can be reconstructed from

$$f(t) = \frac{1}{2\pi} \int_{-\infty}^{+\infty} g(\omega)e^{j\omega t}\, d\omega \qquad (2\text{-}7)$$

The shape of the spectral density region is shown in Fig. 2-8B. Note that the negative frequencies are a product of the mathematics and do not have physical reality. The shape of Fig. 2-8B is expressed by

$$g(\omega) = \frac{\sin \omega t}{\omega t} \qquad (2\text{-}8)$$

The general form $(\sin x)/x$ is also used for repetitive pulse signals, as well as the transient form shown in Fig. 2-8B.

CHAOTIC AND RANDOM SIGNALS

Chaotic and random signals are very similar to each other from a signals processing point of view, but are nonetheless profoundly different in subtle ways. Both signals are unpredictable in the sense that knowledge of the present state is not sufficient to either reveal a future state or reconstruct a past state of the signal. The philosopher might say that such signals are indeterminate, although today that designation only applies to truly random signals. The chaotic signal is now known to possess a pair of attributes that were previously considered impossible: the chaotic signal is both determinate and unpredictable.

Because the chaotic signal is relatively new in the context of instrumentation systems, indeed all of engineering, we will spend a little more time on this category than on the others (which are more widely known).

What characterizes the chaotic system is a hypersensitivity to initial conditions over its time history to such an extent that output results are either unpredictable or nearly so. Coupled with unpredictability is the loss of knowledge about the initial conditions in a system that is chaotic; information entropy increases with time in chaotic systems.

Information entropy arises from the fact that the initial conditions of any system, linear or nonlinear, can be known only to a certain extent. In a linear (or at least nonchaotic) system, errors in knowledge of initial conditions grow linearly with time, but in the chaotic system, errors grow exponentially. Therefore, unless the initial conditions are specified to infinite precision (an impossibility), the time history is impossible to predict for more than a short distance into the future. Beyond that point, trajectories diverge rapidly. Modeling of such a system using piecewise approximations becomes foolish: information entropy is increased, and no more knowledge about the fate of the system can be output than is input. The confidence level of the output information decays as rapidly as the errors grow. The term chaos is used to imply the loss of information and predictability in a system. It is possible to quantify the loss of information using a factor called the Lyapunov exponent of the system (of which, more later).

It was in the field of mechanics, especially dynamics, that chaos made some of its earliest and most significant advances [Moon 1987].

Looking beyond mechanics, one finds chaotic conditions in a large number of nonlinear systems. Chaos has been observed in a wide variety of situations including vibrations in buckled elastic planar and beam structures; systems with play, backlash, or friction; aeroelastic systems; magnetomechanical devices; three-dimensional large vibrations in structures; rotational systems; simple and harmonically forced electrical circuits; feedback control systems; hydraulic systems (especially when turbulence is present); and laser and optical systems. Almost any system containing either damping or nonlinearities (whether intended or accidental) is a candidate for chaotic behavior under the right conditions.

Chaos theory is not yet well known among working engineers (although this will surely soon change). In this section, we will examine some basic concepts, so that at least the jargon and word usage is understood.

Period and Period Doubling

The word period as used in chaos theory is, in some respects, similar to, but not quite the same as, its use in ordinary electrical and wave systems. Period refers to the number of iterations of an operation required to return the result to the same value. For example, in period-2 behavior, two iterations are required, that is, the result flips between two values. Consider the logistic equation of population growth: $Y = X_{n+1} = rX_n(1 - X_n)$. This equation shows several aspects of nonchaotic, prechaotic, and chaotic behavior and is easily modeled on a hand calculator or personal computer. The X term is the variable, while the r term is a fixed parameter. The initial conditions are the values of r and X.

The logistic equation is taken from the study of natural populations under conditions of resources being insufficient, sufficient, or abundant in varying degrees. Therefore, the value of X for this year's birth and death rates is essentially the result of last year's birth and mortality history. Therefore, in this equation, a feedback rule exists in which the next value of X is the immediate previous value of Y.

Let us examine the behavior of the logistic equation assuming a range of $0 < X < 1$. For parameter values of $0 < r < 1$, the equation runs to extinction after only a few iterations (Fig. 2-9A). This curve represents population death, and results from insufficient resources. But, for $1 < r < 3$, a different behavior is noted. The resources are sufficient to sustain a certain level of population, so the population rises to a certain value and then levels off in a stable equilibrium state (Fig. 2-9B). For values of $r > 3$, strange behavior is noted. Given initial conditions of $X = 0.1$ and $r = 3.2$, after a certain number of iterations X_{n+1} flip-flops between the values of 0.79994555 and 0.5130446 indefinitely; this is an example of period-2 behavior (Fig. 2-9C). Period-4 behavior (Fig. 2-9D) arises at $r = 3.44948973$, and the result oscillates through the sequence (again $X_0 = 0.1$): 0.8505646, 0.4384456,

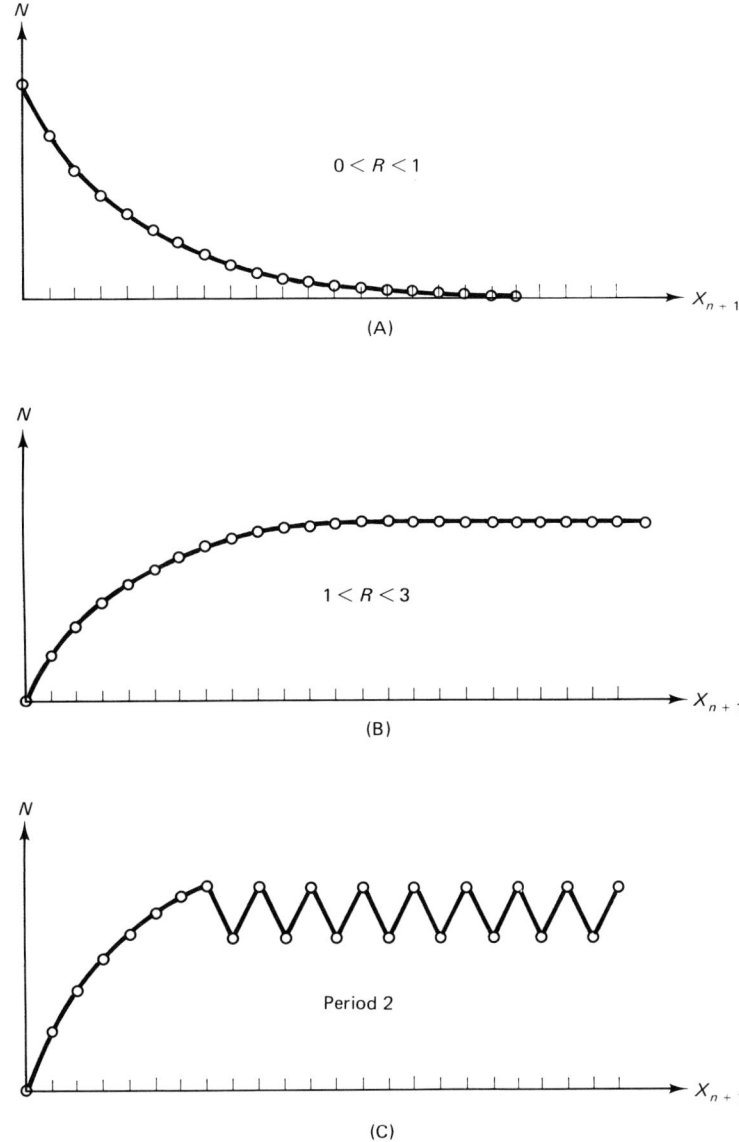

Fig. 2-9 Behavior of the logistic equation: (A) $R < 1$, (B) $1 < R < 3$, (C) period 2 ($R > 3$), (D) period 4, and (E) chaotic. (*Figure continues.*)

0.8493024, 0.4414927, 0.8505646, ... Similarly, for $3.8284 < r < 3.8415$, we see period-3 behavior. The concept of period is another way of saying that a system produces multivalued results. In period-2 behavior, for example, every value of X has two values of Y (and therefore X_{n+1}).

Period doubling occurs when a system jumps from lower order to higher order periodicity, for example, period 2 (Fig. 2-9C) to period 4 (Fig. 2-9D). In the logistic equation, period 2 commences at $r = 3$, period doubling to period 4 occurs at $r = 3.44948973$, and it doubles again to period 8 at $r = 3.545756694$. Period doubling is often a

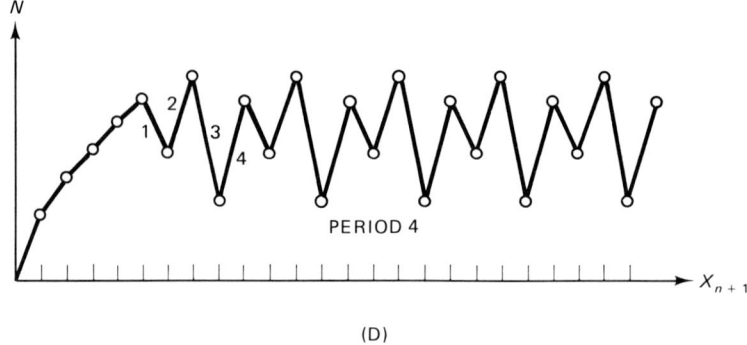

(D)

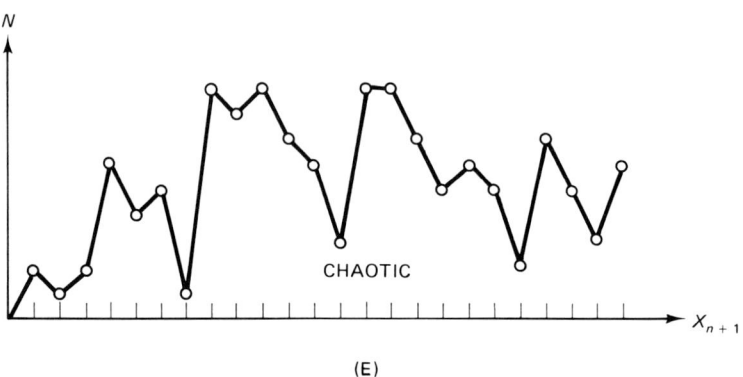

(E)

Fig. 2-9 (*continued*)

precursor to chaos, and in the sample equation, the onset of chaotic behavior (Fig. 2-9E) occurs at $r = 3.56994$. Instead of one, two, three, four, or eight values of Y for each X, there is an infinity of possible values, indeed a chaotic system.

Lyapunov Exponents

It is necessary to evaluate systems and data sets to determine whether or not they are chaotic, rather than random or some other category. A diagnostic test is the Lyapunov exponent (Γ). The defining characteristic of chaotic sets is that trajectories that are initially close together rapidly diverge (Fig. 2-10) at an exponential rate. For a time function, the divergence can be measured by

$$d(t) = d_0 2^{\Gamma t} \tag{2-9}$$

where $d(t)$ is the divergence at time t, d_0 the initial separation of the two trajectories, and Γ the Lyapunov exponent defined by

$$\Gamma = \frac{1}{t_n - t_0} \sum \log_2 \frac{dt_k}{d_0 T_{k-1}} \tag{2-10}$$

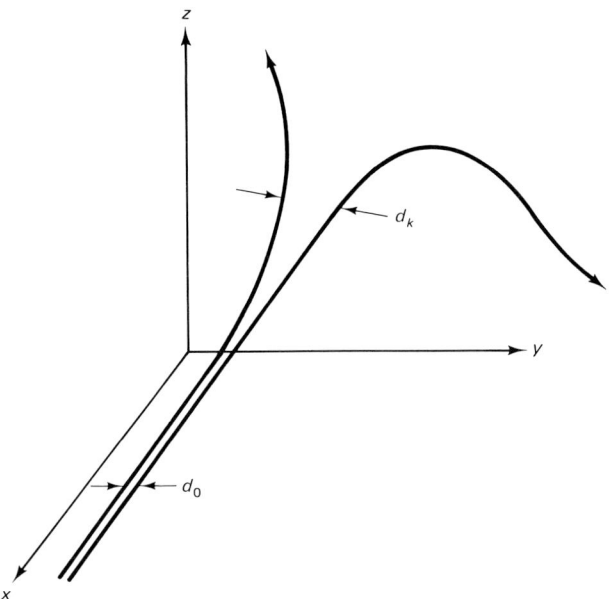

Fig. 2-10 Source of unpredictability: in chaotic systems the trajectories diverge rapidly because of tiny initial differences.

If $\Gamma > 0$, then the set is chaotic, but if $\Gamma = 0$ or $\Gamma < 0$, then the trajectory is nonchaotic. Within the space constraints imposed by the short format of this section, one is unable to properly explore the idea of Lyapunov exponents (this example is but one of several possible examples). The reader is encouraged to examine the literature cited at the end of this chapter for a deeper understanding of the subject.

DETECTING CHAOS

Chaotic behavior has been known in mathematics and physical systems for years, although it was only recognized and systematized since the 1970s. During the late 19th and early 20th centuries (approximately 1875–1925), there was an intellectual crisis in mathematics that was due, in part, to the inability of mathematicians to solve certain "demon" or "monster" curves often associated with chaotic phenomena. In physics and engineering, chaotic behavior was observed occasionally, but it was dismissed as either an experimental artifact or simply too hard to solve. An example is the behavior of a beam at the buckling point. Our equations are bounded to the region up to (but not including) the point of buckling and to the region beyond buckling. Behavior at the point of buckling is dauntingly complex to describe mathematically.

Moon asserted that chaotic conditions cannot exist in linear systems, but that "nonlinearity arises in a subtle way" in systems otherwise thought to be linear. A clue (albeit an imperfect clue) to chaos is nonperiodic variation where one would expect to find static, exponential, or periodic behavior. Another imperfect but useful clue is periodic behavior in a system that should not exhibit periodicity.

Moon provided nine points to address when determining whether a data system is chaotic or simply random. The following is a list of the nine points.

1. Look for one or more nonlinear elements in the system.

2. Look for random or unknown system inputs that could otherwise explain seemingly chaotic behavior.

3. Observe the history of the system over a sufficiently long time to detect aperiodicity.

4. Examine the phase map [(dy/dx) versus x] to identify nonclosed, nonrepeating wandering orbits.

5. Examine the Fourier spectrum of the system to look for broadband noise generated in response to a single-frequency, nonresonant

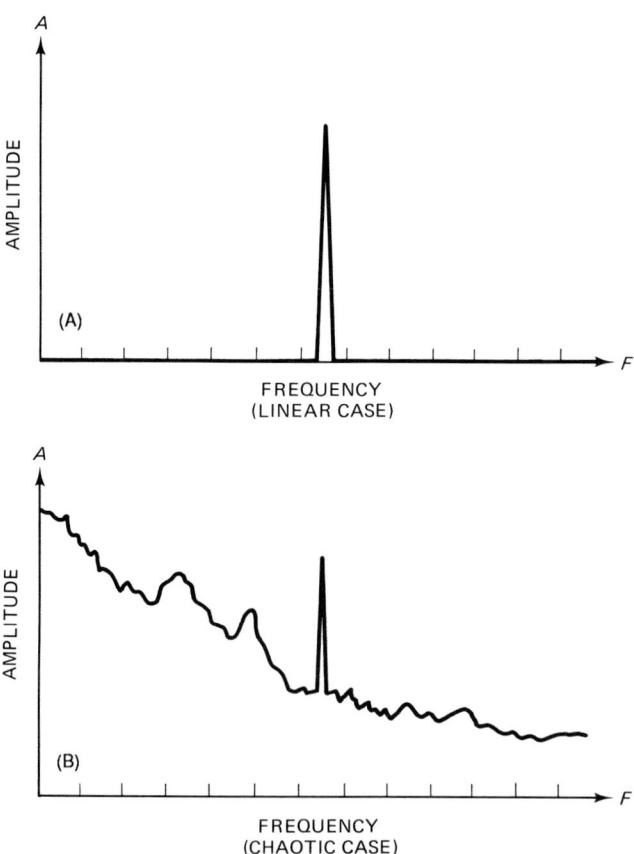

Fig. 2-11 Sign of potential chaos: (A) a normal spectrum expected of an ideal forced oscillator and (B) in prechaotic conditions, there is a large band of subharmonic energy.

forcing function. This phenomenon is obtained in systems with three or less degrees of freedom and often manifests itself in an unexplained prechaotic increase in incommensurate subharmonic noise generation (Fig. 2-11).

6. Take the Poincaré map [X_{n+1} versus X_n, where $X_{n+1} = f(X)$] of the data set. Look for period doubling bifurcations that lead to multivalued solutions to $f(X)$. Look for the existence of a strange attractor in the data.

7. Vary the system parameters to discern routes to chaos as indicated by either prechaotic or chaotic behavior. An exercise in this task is found in examining the logistic equation for which $0 < X_0 < 1$ and $0 < r < 5$.

8. Examine the Lyapunov exponent (a measure of sensitivity to initial conditions) for a positive value (which implies chaoticity).

9. Examine the dimension of the phase space map to discern any fractal (i.e., noninteger) dimension, which would imply chaoticity.

SAMPLED SIGNALS

The digital computer is incapable of accepting analog input signals, but rather requires a digitized representation of that signal. The analog-to-digital (A/D) converter will convert an input voltage (or current) to a representative binary word. If the A/D converter is either clocked or allowed to run asynchronously according to its own clock, then it will take a continuous string of samples of the signal as a function of time. When combined, these signals represent the original analog signal in binary form.

But, the sampled signal is not exactly the same as the original signal, and some effort must be expended to ensure that the representation is as good as possible. Consider Fig. 2-12. The waveform in Fig. 2-12A is a continuous voltage function of time, $V(t)$; in this case, a triangle waveform is seen. If the signal is sampled by another signal, $p(t)$, with frequency F_s and sampling period $T = 1/F_s$, as shown in Fig. 2-12B, and then later reconstructed, the waveform may look something like Fig. 2-12C. While this may be sufficiently representative of the waveform for many purposes, it would be reconstructed with greater fidelity if the sampling frequency (F_s) is increased.

Figure 2-13 shows another case in which a sine wave, $V(t)$ in Fig. 2-13A, is sampled by a pulse signal, $p(t)$ in Fig. 2-13B. The sampling signal, $p(t)$, consists of a train of equally spaced narrow pulses spaced in time by T. The sampling frequency F_s equals $1/T$. The resultant is shown in Fig. 2-13C and is another pulsed signal in which the amplitudes of the pulses represent a sampled version of the original sine wave signal.

The sampling rate, F_s, must, by Nyquist's theorem, be twice the maximum frequency (F_m) in the Fourier spectrum of the applied

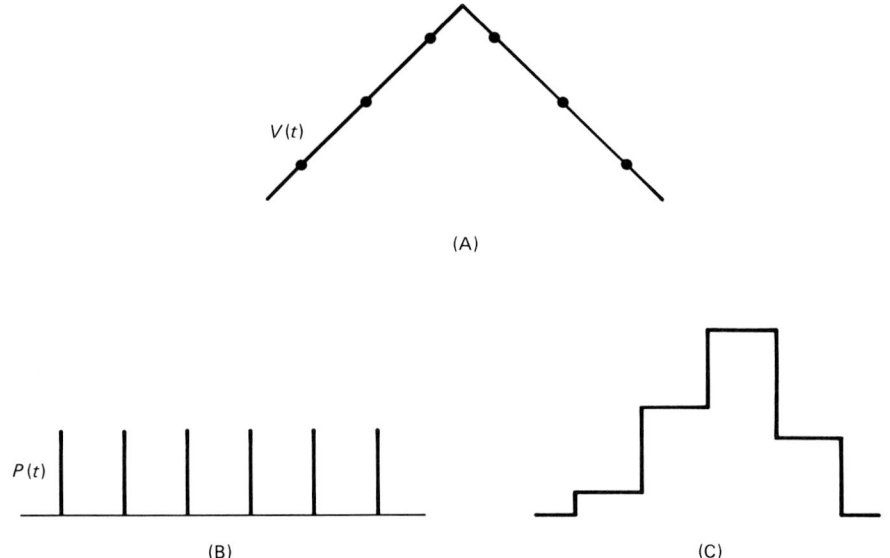

Fig. 2-12 Sampled signals: (A) original signal, (B) sampling pulses, and (C) reconstructed signal from too few samples.

analog signal, $V(t)$. In order to reconstruct the original signal after sampling, it is necessary to pass the sampled waveform through a low-pass filter that limits the bandpass to F_s.

The sampling process is analogous to a form of amplitude modulation (AM), in which $V(t)$ is the modulating signal, with a spectrum from DC to F_m, and $p(t)$ is the carrier frequency. The resultant spectrum is partially shown in Fig. 2-14 and resembles the spectrum of a double sideband with carrier AM signal. The spectrum of the modulating signal appears as sidebands around the carrier frequency, shown here as F_0. The actual spectrum is a bit more complex, as shown in Fig. 2-15. Like an unfiltered AM radio transmitter, the same spectral information appears not only around the fundamental frequency (F_s) of the carrier (shown at zero in Fig. 2-15), but also at the harmonics and subharmonics spaced at intervals of F_s up and down the spectrum.

Providing that the sampling frequency $F_s \geq 2F_m$, the original signal is recoverable from the sampled version by passing it through a low-pass filter with a cutoff frequency F_c, set to pass only the spectrum of the analog signal—but not the sampling frequency. This phenomenon is shown with the dotted line in Fig. 2-15.

When the sampling frequency $F_s < 2F_m$, then a problem occurs (see Fig. 2-16). The spectrum of the sampled signal looks similar to before, but the regions around each harmonic overlap such that the value of $-F_m$ for one spectral region is less than $+F_m$ for the next lower frequency region. This overlap results in a phenomenon called aliasing. That is, when the sampled signal is recovered by low-pass filtering, it will produce not the original sine wave frequency F_0 but a

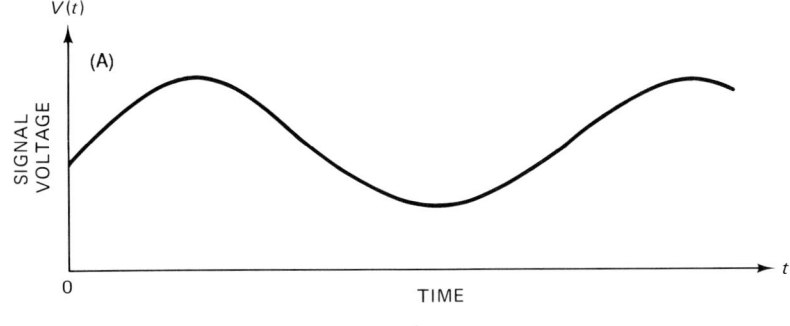

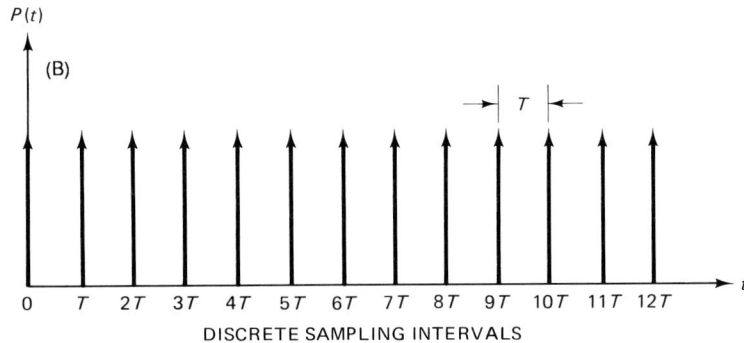

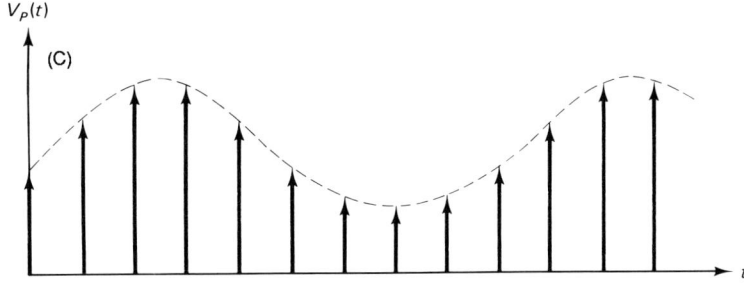

Fig. 2-13 (A) Analog signal $V = F(t)$, (B) sampling pulses, and (C) reconstructed signal.

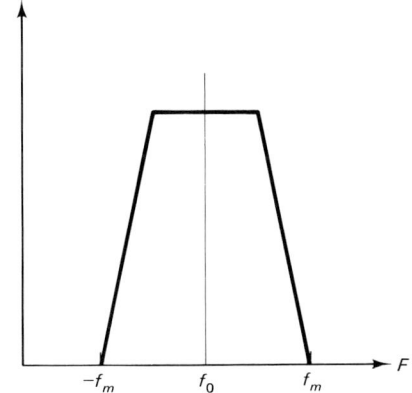

Fig. 2-14 Sampled signal simple spectrum.

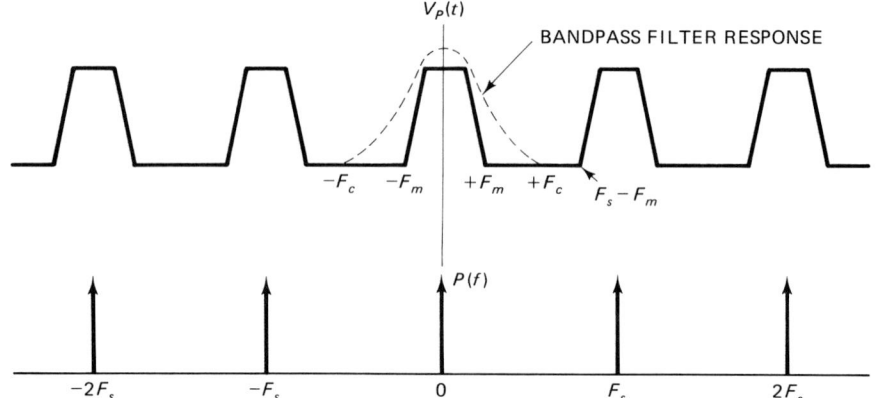

Fig. 2-15 Spectrum of a sampled signal when sampling frequency is above minimum required.

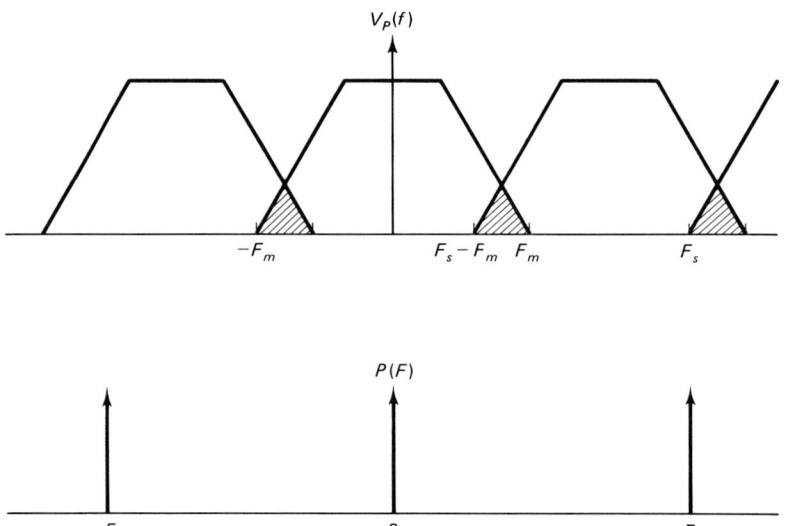

Fig. 2-16 Undersampling forces overlap of spectrum bands, creating aliasing phenomenon.

lower frequency equal to $(F_s - F_0)$, and the information carried in the waveform is thus lost or distorted.

The solution, for high fidelity sampling of the analog waveform for input to a computer, is to

1. bandwidth limit the signal at the input of the sampler or A/D converter with a low-pass filter with a cutoff frequency F_c selected to pass only the maximum frequency in the waveform (F_m) and not the sampling frequency (F_s), and

2. set the sampling frequency F_s at least twice the maximum frequency in the applied waveform's Fourier spectrum, that is, $F_s \geq 2F_m$.

Experience has shown that some users will not accept a reconstructed sampled waveform if the sample rate is $2F_m$. For example, medical ECG waveforms—in which $F_m = 100$ Hz—tend to look blocky when sampled at 200 Hz and then reconstructed. The user acceptance was much better when the waveform was sampled at 500 Hz, or $5F_m$. While that rate was expensive to accommodate in an eight-bit A/D converter at one time, it is now very low cost and should be used, Nyquist notwithstanding.

NOISE

An ideal electronic circuit produces no noise of its own, so the output signal from the ideal circuit contains only the noise that was in the original signal. But, real electronic circuits and components do produce a certain level of inherent noise of their own. Even the simple fixed-value resistor is noisy. Figure 2-17A shows the equivalent circuit for an ideal, noise-free resistor. The inherent noise is represented in Fig. 2-17B by a noise voltage source, V_n, in series with the ideal, noise-free resistance, R_i. At any temperature above absolute zero (0 K or about $-273°C$), electrons in any material are in constant random motion. Because of the inherent randomness of that motion, however, there is no detectable current in any one direction. In other words, electron drift in any single direction is canceled over short time periods by equal drift in the opposite direction. Electron motions are therefore statistically decorrelated. There is, however, a continuous series of random current pulses generated in the material, and those pulses are seen by the outside world as a noise signal. This signal is called by several names: thermal agitation noise, thermal noise, and Johnson noise.

Johnson noise is a so-called white noise because it has a very broadband (nearly gaussian) spectral density. The thermal noise spectrum is dominated by midfrequencies (10^4 to 10^5 Hz) and is essentially flat. The term white noise is a metaphor developed from white

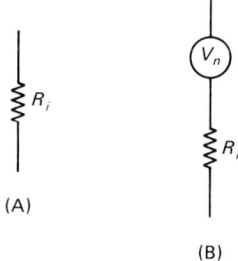

Fig. 2-17 (A) Model of the ideal input impedance. (B) Real input impedances also include a noise source.

light, which is composed of all visible color frequencies. The expression for Johnson noise is

$$V_n^2 = 4KTRB \ \text{V}^2/\text{Hz} \qquad (2\text{-}11)$$

where V_n is the noise voltage (V), K Boltzmann's constant (1.38×10^{23} J/K), T the temperature in degrees Kelvin (K), R the resistance in ohms (Ω), and B the bandwidth in hertz (Hz).

With the constants collected and the expression normalized to 1 kΩ, Eq. (2-11) reduces to

$$V_n = 4\sqrt{\frac{R}{1\ \text{k}\Omega}} \ \frac{\text{nV}}{\sqrt{\text{Hz}}} \qquad (2\text{-}12)$$

The evaluated solution of Eq. (2-12) is normally read nanovolts per square root hertz. In this equation, a 1-MΩ resistor will have a thermal noise of 126 nV$/\sqrt{\text{Hz}}$.

Several other forms of noise are present in linear integrated circuits (ICs) and other semiconductor amplifiers to one extent or another. For example, because current flow at the quantum level is not smooth and predictable, an intermittent burst phenomenon is sometimes seen. This noise is called popcorn noise and consists of pulses of many milliseconds duration. Another form of noise is shot noise (also called Schottky noise). The name shot is derived from the fact that the noise sounds like a handful of BB pellets thrown against a metal surface. Shot noise is a consequence of a DC current flowing in a conductor and is found from

$$I_n^2 = 2qIB \ \text{A}^2/\text{Hz} \qquad (2\text{-}13)$$

where I_n is the noise current in amperes (A), q the elementary electric charge (1.6×10^{-19} C), I the current in amperes (A), and B the bandwidth in hertz (Hz).

Finally, there is flicker noise, also called pink noise or $1/f$ noise. The latter name applies because flicker noise is predominantly a low-frequency (< 1000 Hz) phenomenon. This type of noise is found in all conductors and becomes important in IC devices because of manufacturing defects.

The noise spectrum in any given instrumentation system will contain elements of several kinds of noise, although in some systems one form or another may dominate the others. It is common to characterize noise from a single source using the root mean square (rms) value of the voltage amplitudes:

$$V_{n(\text{rms})} = \sqrt{\frac{1}{T}\int_0^T [F(t)]^2 \, dt} \qquad (2\text{-}14)$$

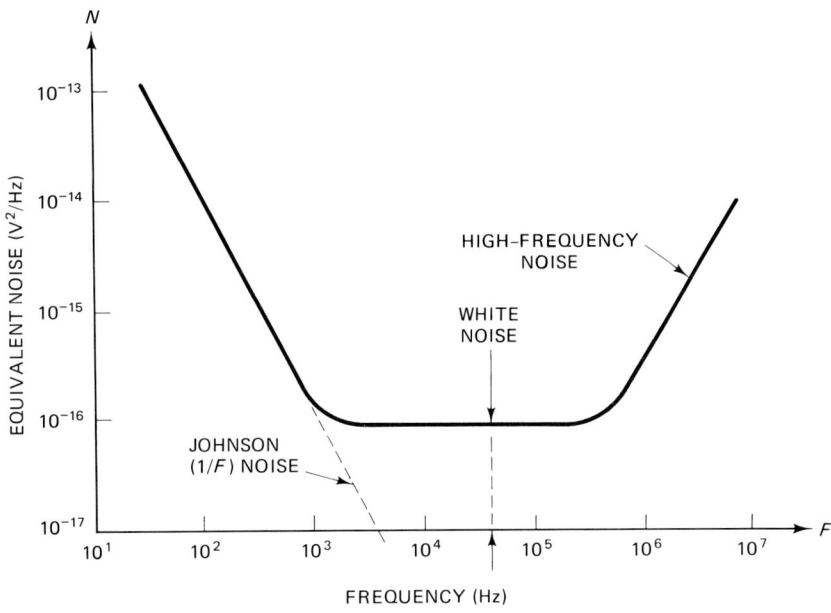

Fig. 2-18 Frequency spectrum showing classes of noise.

Figure 2-18 shows the noise spectrum profile for a typical system that contains $1/f$ noise, thermal or white noise, and some high-frequency noise.

Chaotic Noise

The classical description of noise in electronic systems has used the statistical language appropriate to random processes. But, more recent work indicates that noise is often either chaotic or contains a strongly chaotic component. In 1977, Benoit Mandelbrot, working on noise in digital communications systems at IBM's Watson Research Center in New York, demonstrated that such noise is not truly random, as engineers previously suggested, but rather that it matches a mathematical set called Cantor's dust. Although in the limit, Cantor's dust looks like a random noise signal, it is actually produced from a line segment iteratively divided and redivided according to a simple, highly ordered rule. Further, in 1988, Fote *et al.* discovered a chaotic component to $1/f$ noise in semiconductor electro-optical sensors.

SIGNAL-TO-NOISE RATIO

Amplifiers can be evaluated on the basis of the signal-to-noise ratio (S/N or SNR), denoted S_n. The goal of the circuit or instrument designer is to enhance the SNR as much as possible. Ultimately, the minimum signal level detectable at the output of an amplifier is that level which appears above the noise floor level. Therefore, the lower the system noise floor, the smaller the minimum allowable signal. Although often thought of as a radio receiver parameter, the SNR is applicable in other amplifiers where signal levels are low and gains are high. This situation occurs in scientific, medical, and engineering instrumentation, as well as other applications.

Noise resulting from thermal agitation of electrons is measured in terms of noise power (P_n) and carries the units of power (watts or its subunits). Noise power is found from

$$P_n = KTB \qquad (2\text{-}15)$$

where P_n is the noise power in watts (W), K Boltzmann's constant (1.38×10^{-23} J/K), and B the bandwidth in hertz (Hz).

Notice in Eq. (2-15) that there is no center frequency term, only the bandwidth (B). True thermal noise is gaussian (or near-gaussian), so the frequency content, phase, and amplitudes are equally distributed across the entire spectrum. Thus, in bandwidth limited systems, such as a practical amplifier or network, the total noise power is related to temperature and bandwidth. We can conclude that a 200-Hz bandwidth centered on 1-kHz produces the same thermal noise level as a 200-Hz bandwidth centered on 600-Hz or any other frequency.

Noise sources can be categorized as either internal or external. The internal noise sources are due to thermal currents in the semiconductor material resistances. It is the noise component contributed by the amplifier under consideration. If noise, or the S/N ratio, is measured at both the input and output of an amplifier, the output noise is greater. The internal noise of the device is the difference between the output noise level and input noise level.

External noise is the noise produced by the signal source, so it is often called source noise. This noise signal is due to thermal agitation currents in the signal source, and even a simple zero signal input termination resistance has some amount of thermal agitation noise. In fact, the simple terminated noise level might be higher than V_n because of component construction. For example, the noise signal produced by a carbon composition resistor has an additional noise source modeled as V_{na} in Fig. 2-19. This noise generator is a function of resistor construction and manufacturing defects.

Figure 2-20A is a circuit model showing that several voltage and current noise sources exist in an op-amp. The relative strengths of these noise sources, hence their overall contribution, varies with op-amp type. In an FET-input op-amp, for example, the current noise sources are tiny, but voltage noise sources are very large. On bipolar op-amps, the exactly opposite situation is obtained.

All of the noise sources in Fig. 2-20A are uncorrelated with respect to each other, so one cannot simply add noise voltages; only

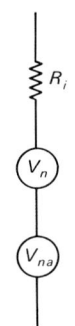

Fig. 2-19 Input impedance model showing noise sources.

noise power can be added. To characterize noise voltages and currents, they must be added in the root sum squares (RSS) manner.

Models such as Fig. 2-20A are too complex for most situations, so it is standard practice to lump all of the voltage noise sources into one source and all of the current noise sources into another source. The composite sources have a value equal to the RSS voltage (or current) of the individual sources. Figure 2-20B is such a model in which only a single current source and a single voltage source are used. The equivalent AC noise in Fig. 2-20B is the overall noise, given a specified value of source resistance, R_s, and is found from the RSS value of V_n and I_n:

$$V_{ne} = \sqrt{V_n^2 + (I_n R_s)^2} \tag{2-16}$$

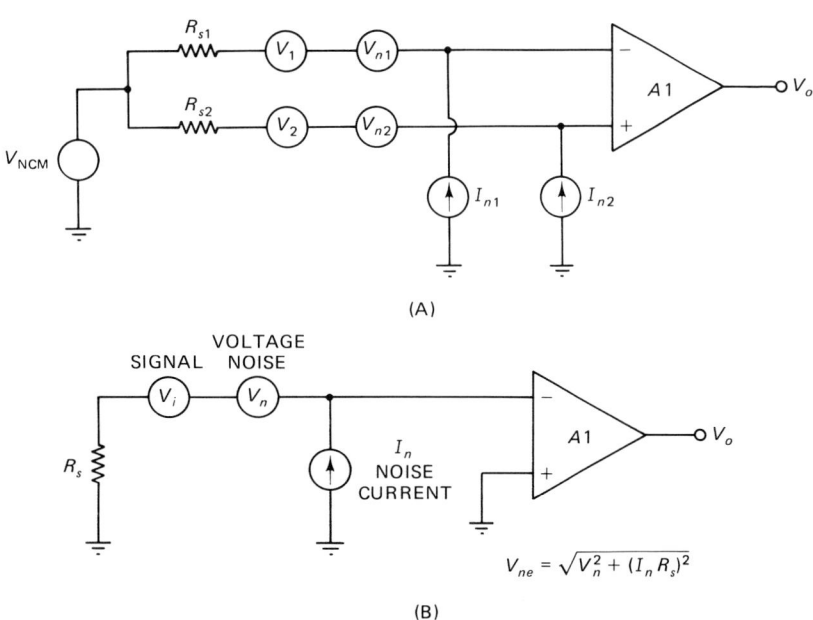

Fig. 2-20 (A) Operational amplifier showing various noise sources. (B) Combined noise sources representing the same circuit.

NOISE FACTOR, NOISE FIGURE, AND NOISE TEMPERATURE

The noise of a system or network can be defined in three different but related ways: noise factor (F_n), noise figure (NF), and equivalent noise temperature (T_e); these properties are definable as a simple ratio, decibel ratio, and temperature, respectively.

Noise Factor

For components such as resistors, the noise factor (F_n) is the ratio of the noise produced by a real resistor to the simple thermal noise of an ideal resistor. The noise factor of a system is the ratio of output noise power (P_{no}) to input noise power (P_{ni}):

$$F_n = \left. \frac{P_{no}}{P_{ni}} \right|_{T = 290 \text{ K}} \qquad (2\text{-}17)$$

In order to make comparisons easier, the noise factor is always measured at the standard temperature (T_0) of 290 K (standardized room temperature).

The input noise power, P_{ni}, is defined as the product of the source noise at standard temperature (T_0) and the amplifier gain (G):

$$P_{ni} = GKBT_0 \qquad (2\text{-}18)$$

It is also possible to define the noise factor F_n in terms of an output and input S/N ratio:

$$F_n = S_{ni}/S_{no} \qquad (2\text{-}19)$$

which is also

$$F_n = P_{no}/KT_0 BG \qquad (2\text{-}20)$$

where S_{ni} is the input signal-to-noise ratio, S_{no} the output signal-to-noise ratio, P_{no} the output noise power, K Boltzmann's constant (1.38×10^{-23} J/K), T_0 the standard temperature (290 K), B the network bandwidth in hertz, and G the amplifier gain.

The noise factor can be evaluated in a model that considers the amplifier ideal, and therefore only amplifies through gain G the noise produced by the input noise source:

$$F_n = \frac{KT_0 BG + \Delta N}{KT_0 BG} \qquad (2\text{-}21)$$

or

$$F_{n} = \Delta N / KT_0 BG \qquad (2\text{-}22)$$

where ΔN is the noise added by the network or amplifier, and all other terms are as defined above.

Noise Figure

The noise figure (NF) is a frequently used measure of an amplifier's goodness, or its departure from idealness. Thus, it is a figure of merit. The noise figure is the noise factor converted to decibel notation:

$$\text{NF} = 10 \log F_n \qquad (2\text{-}23)$$

where NF is the noise figure in decibels (dB), F_n the noise factor, and log refers to the system of base-10 logarithms.

Noise Temperature

The noise temperature (T_e) is a means for specifying noise in terms of an equivalent temperature. Evaluating Eq. (2-18) shows that the noise power is directly proportional to temperature in degrees Kelvin, and also that the noise power collapses to zero at the temperature of absolute zero (0 K).

Note that the equivalent noise temperature T_e is not the physical temperature of the amplifier, but rather a theoretical construct that is an equivalent temperature that produces that amount of noise power. The noise temperature is related to the noise factor by

$$T_e = (F_n - 1)T_0 \qquad (2\text{-}24)$$

and to noise figure by

$$T_e = \left[\text{antilog}\left(\frac{\text{NF}}{10}\right) - 1\right] KT_0 \qquad (2\text{-}25)$$

Now that we have the noise temperature T_e, we can also define the noise factor and noise figure in terms of noise temperature:

$$F_n = \frac{T_e}{T_0} + 1 \qquad (2\text{-}26)$$

and

$$\text{NF} = 10 \log\left(\frac{T_e}{T_0} + 1\right) \qquad (2\text{-}27)$$

The total noise in any amplifier or network is the sum of internally generated and externally generated noise. In terms of noise temperature,

$$P_{n(total)} = GKB(T_0 + T_e) \tag{2-28}$$

where $P_{n(total)}$ is the total noise power, and all other terms are as previously defined.

NOISE IN CASCADE AMPLIFIERS

A noise signal is seen by a following amplifier as a valid input signal. Thus, in a cascade amplifier, the final stage sees an input signal that consists of the original signal and noise amplified by each successive stage. Each stage in the cascade chain both amplifies signals and noise from previous stages and contributes some noise of its own. The overall noise factor for a cascade amplifier can be calculated from Friis' noise equation:

$$F_N = F_1 + \frac{F_2 - 1}{G1} + \frac{F_3 - 1}{G1G2} + \cdots + \frac{F_n - 1}{G1G2\ldots G_{n-1}}, \tag{2-29}$$

where F_N is the overall noise factor of N stages in the cascade, F_1 the noise factor of stage 1, F_2 the noise factor of stage 2, F_n the noise factor of the nth stage, $G1$ the gain of stage 1, $G2$ the gain of stage 2, and G_{n-1} the gain of stage $(n - 1)$.

As you can see from Eq. (2-29), the noise factor of the entire cascade chain is dominated by the noise contribution of the first stage or two. High-gain, low-noise amplifiers [such as electroencephalograph (EEG) preamplifiers] typically use a low-noise amplifier circuit for only the first stage or two in the cascade chain.

NOISE REDUCTION STRATEGIES

Although noise is a serious problem for the designer, especially where low signal levels are experienced, there are a number of common sense approaches to minimize the effects of noise on a system. In this section, we will examine several of these methods.

1. Keep the source resistance and the amplifier input resistance as low as possible. Using high-value resistances will increase thermal noise proportionally.

2. Total thermal noise is also a function of the bandwidth of the circuit. Therefore, reduce the bandwidth of the circuit to a minimum to also minimize noise. But, this job must be done mindfully because

signals have a Fourier spectrum that must be preserved for faithful reproduction or accurate measurement. The solution is to match the bandwidth to the frequency response required for the input signal.

3. Prevent external noise from affecting the performance of the system by appropriate use of grounding, shielding, and filtering.

4. Use a low-noise amplifier (LNA) in the input stage of the system.

5. For some semiconductor circuits, use the lowest DC power supply potentials that will do the job.

Using Feedback to Reduce Noise

Negative feedback is well known for reducing amplitude and phase errors, thereby reducing the distortion of an amplifier. Judicious use of feedback can also reduce the output noise of a signal conditioning amplifier.

Consider Fig. 2-21. This circuit model shows gain distributed into two blocks, $G1$ and $G2$. The total gain of the circuit (G) is the product $G1G2$. A noise source produces a noise signal, V_n, and injects it into a summation junction between $G1$ and $G2$. A feedback network with a transfer function, B, produces a signal BV_o that is summed with the input signal V_i. By inspection of Fig. 2-21, we know

$$V1 = V_i + BV_o \qquad\qquad (2\text{-}30)$$

$$V2 = V1G1 \qquad\qquad (2\text{-}31)$$

$$V2 = (V_i + BV_o)G1 + N \qquad\qquad (2\text{-}32)$$

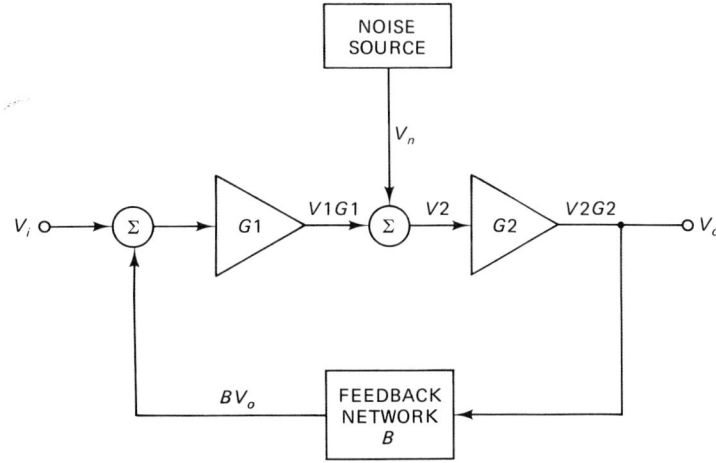

Fig. 2-21 Model of a feedback amplifier showing noise source.

and

$$V_o = V2G2 \tag{2-33}$$

Substituting Eq. (2-32) into Eq. (2-33),

$$V_o = \{[(V_i + BV_o)G1 + N]G2\} \tag{2-34}$$

and

$$V_o = G1G2V_i + BV_oG1G2 + NG2 \tag{2-35}$$

which, when rearranged, leads to

$$V_o = \frac{G1G2V_i + NG2}{1 - BG1G2} \tag{2-36}$$

and finally,

$$V_o = \frac{G1G2}{1 - BG1G2}\left(V_i + \frac{N}{G1}\right) \tag{2-37}$$

The result shown in Eq. (2-37) is consistent with Black's equation for feedback amplifiers, $[G_o = G/(1 - BG)]$, and demonstrates that the noise is reduced by the gain factor $(G1)$. This result is also consistent with the design philosophy inherent in Friis' equation.

Noise Reduction by Signal Averaging

If a signal is either periodic or repetitive, or can be made so, then it is possible to enhance the signal-to-noise ratio (S_n) by signal averaging. The basis for this simple signal processing technique is the assumption that noise meets the definition of either random or chaotic processes. If so, then noise tends to integrate to zero or near zero over time. If time-averaging integration is performed in a coherent manner, then a repetitive signal tends to build in value, while noise levels (being decorrelated) decrease. If we assume that the signal-to-noise ratio is

$$S_n = 20\log(V_i/V_n) \tag{2-38}$$

then for systems where $V_i < V_n$, the noise reduction by time averaging is

$$\bar{S}_n = S_n/(N)^{1/2} \tag{2-39}$$

where $\bar{S}_n$ is the time-averaged SNR, S_n the unprocessed SNR, and N the number of repetitions of the signal.

Example

An EEG system processes a 5-μV signal in the presence of a 500-μV random noise level. Calculate the unprocessed SNR, the processed SNR for 1000 repetitions of the signal, and the processing gain.

Solution

Unprocessed SNR:

$$S_{no} = 20 \log(V_i/V_n) = 20 \log(5 \ \mu V/500 \ \mu V) = -40 \ dB$$

Processed SNR:

$$\bar{S}_n = S_n/(N)^{1/2} = (-40 \ dB)/(1000)^{1/2} = -1.3 \ dB$$

Processing gain:

$$G_p = S_n - \bar{S}_{no} = (-1.3 \ dB) - (-40 \ dB) = +38 \ dB$$

The effect of time averaging is to increase the time required to collect data, so (by $F = 1/T$) time averaging is effectively a means of decreasing the bandwidth of the system.

Coherency is maintained in a system by ensuring that repetitive data points are processed in a consistent time relationship with respect to each other. The averager will be triggered by a repetitive event, and that action will start the process. Data points are always matched to other data points taken at the same elapsed time after the trigger for previous iterations. For example, the ith datum point following a current sweep is paired with all other ith points from previous sweeps and no others.

An example of signal averaging used to extract weak signals from larger noise signals is found in evoked potential studies of EEG waveforms. The EEG signal is chaotic in healthy brains (arising order—e.g., decreasing chaos—in the EEG signal often precedes the onset of an epileptic seizure) and tends to average around an amplitude of 50 μV. Features due to any given stimulus (such as a blinking light) average around 5 μV. The noise in this sense can be near-gaussian random noise, periodic noise from nearby electrical sources, or other EEG signals that are not due to the stimulus. These signals can range form 30 to 1000 μV (1 mV), so it is likely that the signal will be obscured. Neurologists and neurophysiologists use a repetitive stimulus, and then coherently signal average the EEG to extract the desired evoked signal potential. If sufficient repetitions are done, then the gaussian noise and other EEG components, being decorrelated, integrate toward zero. The evoked signal, however, tends to accumulate to larger values.

Figure 2-22 shows a typical EEG evoked potential data acquisition system. Although once limited to expensive dedicated systems, the

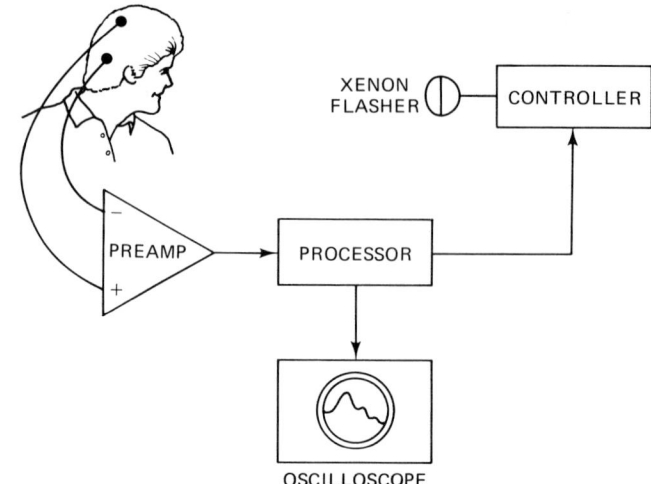

Fig. 2-22 Neurologist's evoked potentials recording setup.

advent of the microcomputer put evoked potential instrumentations within the reach of many more clinicians and researchers.

Although an analog integrator will perform the time-averaging function for a limited number of cases, the most usual method in modern equipment is to digitize the signal, perform the averaging function in software, and then reconvert the signal to analog form for display. Figure 2-23 shows the block diagram of a digital averager.

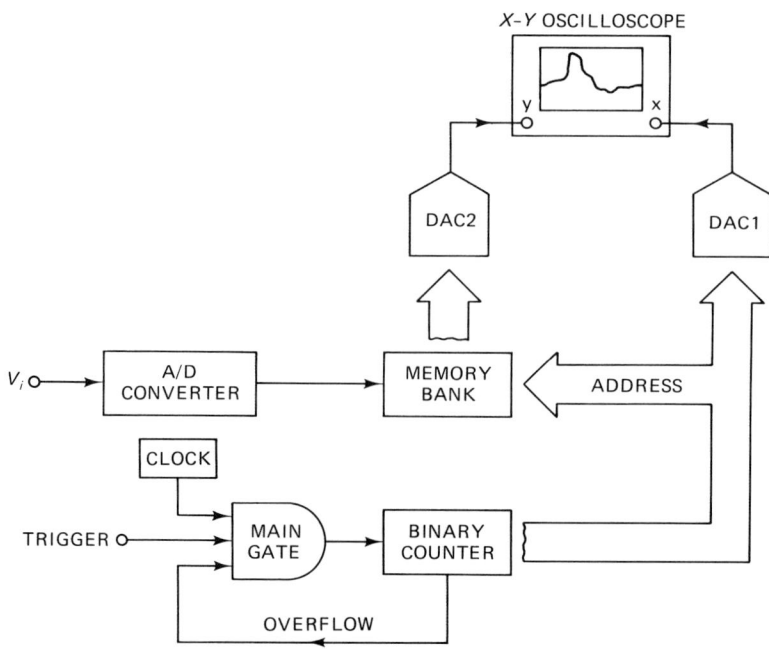

Fig. 2-23 Data acquisition system for evoked potentials recording.

The display device is an *X–Y* oscilloscope or paper chart recorder. The horizontal time base is derived from a digital-to-analog converter (DAC1) that is driven from the output of a binary counter. As the counter increments, it forces the DAC to output a linear sawtooth waveform. The same binary counter is also used as an address generator for a memory array. Thus, any given memory location always corresponds to the same point on the horizontal time base.

Input signal V_i is digitized by the A/D converter. The binary signals from the A/D are stored in the memory array. The data stored in the memory are summed with older data or a running arithmetic mean is calculated (both methods are popular).

Suppression of Periodic Noise and Interference

Sensor systems frequently encounter noise situations in which the noise signal contains both gaussian and periodic components. The gaussian components can be reduced to an acceptable level by simple linear filtering if the SNR is large (> 100) or by matched filtering if the SNR is low (< 100). The matched filter has a response that is the complex conjugate of the input signal frequency spectrum. By this means, it is able to maximize the signal power extracted, while at the same time minimizing the noise power.

A periodic noise signal, such as the 60-Hz sinusoidal noise from local AC power wiring, can also be suppressed using filtering methods. However, one must be aware in sampled data systems that an unfortunate choice of sampling frequency could result in a correlated relationship to the interfering signal and therefore an enhancement (rather than suppression) of the noise. In general, the interfering signal frequency and its harmonics and subharmonics should be avoided.

Figure 2-24 shows a sensor with an analog electronics subsystem. The sections include the sensor, a preamplifier, and a filter. The output signal (V_o) may be sent to a display (in purely analog systems), a signal sampler, or an A/D converter.

One of the most fundamental methods for suppressing noise is to shield the system (Fig. 2-25). Most periodic noise comes from electrical or magnetic fields in the vicinity of the sensor or its wiring, so

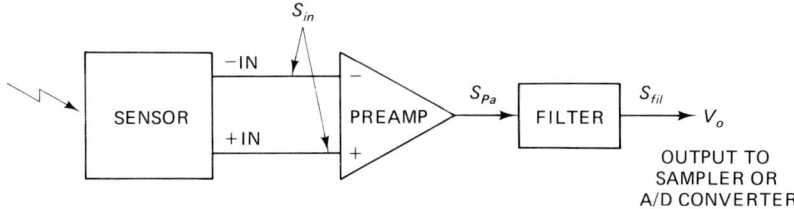

Fig. 2-24 Noise can be reduced by frequency selective filtering of the sensor data.

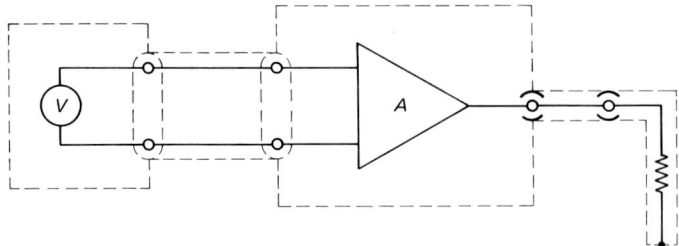

Fig. 2-25   Shielding is usually effective in reducing external radiated noise sources.

shielding is highly effective in preventing the noise. Thus, proper shielding will improve the SNR by reducing the interference signal level.

Another element to the noise suppression task is to design the sensor to preamplifier interface (again, see Fig. 2-25) as a differential signal. If the preamplifier is differential, then it will suppress common mode signals, that is those which affect both amplifier inputs equally. Because differential inputs are of opposite polarity, but see the same gain, they affect the output signal in an equal magnitude, but opposite manner. If the same signal is applied to both $-IN$ and $+IN$ simultaneously, the net output voltage is zero. Interfering signals created by external fields tend to be common mode, so are suppressed by a differential amplifier.

The measure of common mode rejection is given by the system's common mode rejection ratio (CMRR), especially that of the preamplifier. The CMRR is defined as

$$CMRR = A_d/A_{cm} \qquad (2\text{-}40)$$

where CMRR is the common mode rejection ratio, A_d the differential voltage gain of the preamplifier, and A_{cm} the common mode voltage gain of the preamplifier. When expressed in decibels, the CMRR is

$$CMRR_{dB} = 20\log(CMRR) \qquad (2\text{-}41)$$

or

$$CMRR_{dB} = 20\log(A_d/A_{cm}) \qquad (2\text{-}42)$$

The common mode rejection ratio can also be defined in terms of the signal to noise ratios at the input (S_{in}) and output (S_{pa}) ports of the preamplifier:

$$CMRR = \sqrt{S_{pa}/S_{in}} \qquad (2\text{-}43)$$

or

$$S_{pa} = S_{in}CMRR \qquad (2\text{-}44)$$

The filter shown in Fig. 2-24 can be used to improve the output signal-to-noise ratio both for gaussian noise and periodic noise. The filter input SNR is the same as the preamplifier output. The processing gain (G_p) defines the change in the SNR between the input and output of the filter:

$$G_p = S_{fil}/S_{pa} \tag{2-45}$$

and because the action of the filter is to limit the bandwidth,

$$G_p = BW_{pa}/2(BW_{fil}) \tag{2-46}$$

CONCLUSION

Now that we have discussed signals and noise, let us turn our attention to electrodes, sensors, and transducers. These are the devices that create the electrical input signals that are used in the instrument.

SELECTED READINGS

John D. Barrow, *The World within the World*, Clarendon Press, Oxford, England, 1988.

M. F. Barnsley, R. L. Devaney, B. B. Mandelbrot, H.-O. Peitgen, D. Saupe, and R. F. Voss, *The Science of Fractal Images*, Springer-Verlag, New York, 1988.

Pierre Berge, Yves Pomeau, and Christian Vidal, *Order within Chaos*, Wiley-Interscience, New York, 1986.

Paul Davies, *The Cosmic Blueprint*, Simon & Schuster, New York, 1988.

Robert L. Devaney, *An Introduction to Chaotic Dynamical Systems*, Addison-Wesley, New York, 1987.

A. Fote, J. McDonough, S. Kohn, and E. Fletcher, "Application of Chaos Theory to 1/F Noise," *NTIS Technical Report No. SD-TR-88-29*, El Segundo, California, 1988.

James Gleick, *Chaos: Making a New Science*, Viking Penguin, New York, 1987.

Benoit B. Mandelbrot, *The Fractal Geometry of Nature*, Freeman, New York, 1983; Revision of *Fractals*, 1977.

Francis C. Moon, *Chaotic Vibrations*, Wiley-Interscience, New York, 1987.

Ilya Prigogine and Isabelle Stengers, *Order Out of Chaos*, Bantam Books, New York, 1984.

3

Transducers, Sensors, and Signal Processing

The types of electronic instruments that are the subject of this book require some sort of analog input signal that is, often as not, either generated in some form of transducer device or acquired by a special-purpose electrode. In the next several chapters, we will examine specific forms of transducers, electrodes, and sensors. Before beginning that discussion, however, it is useful to settle on some basic definitions and concepts regarding these devices.

A sensor is a device that either acquires or generates an electrical signal that represents some physical phenomenon. The general class of sensors includes electrodes (such as the electrocardiograph electrodes used in medicine) and transducers (such as a pressure or temperature sensor).

As mentioned in Chapter 1, there is often an ambiguity regarding the use of the words *sensor* and *transducer*. A transducer, in our present context, is a data acquisition device that converts energy from one form to another for the purposes of measurement, data collection, monitoring, or control. It serves to provide input to the electronic instrument, so the form of energy "converted to" is electrical. Thus, a transducer is a device that converts energy derived from some physical parameter (e.g., force, pressure, temperature, or flow) to an electrical potential or current that is either proportional to the applied physical parameter or somehow otherwise correlated to it.

The word *transducer* also properly applies to certain output devices, such as the loudspeaker used in audio equipment, radios, and televisions. In that case, the transducer converts electrical energy to acoustical energy. But, in the context of this book, the output transducer is not applicable, so we limit the definition somewhat artificially to those devices that provide input signals.

Because the definition of *transducer* is limited to those devices that provide input signals to electronic instruments, we can state that transducers are all sensors. Also included in the general category of sensors, however, are metallic electrodes (the ECG example) and any other device that picks up electrical signals for input to the instrument. For purposes of consistency, the word sensor will be used here, unless some reason to use another word exists. In general, a sensor is any device that acts as a "sense organ . . . for electronic processing" (Geddes and Baker, 1968). This function is indicated by the choice of its masthead title for a technical journal: *Sensors: The Journal of Machine Perception*. The last word in the title indicates the purpose of sensors.

TRANSDUCTION

In the class of sensors that are also transducers, it is necessary to understand the concepts of transduction and transducible events or properties. A transducible property is a characteristic of the physical event[1] that is singularly able to represent that event and can be transformed into an electrical signal by some device or process. For instance, in an example given by Geddes and Baker,[2] it is noted that carbon dioxide (CO_2) absorbs electromagnetic wavelengths of 2.7, 4.3, and 14.7 μm. Although water also absorbs 2.7 μm of radiation to some extent, it is possible to make an infrared (IR) sensor that will respond to either 4.3 or 14.7 μm, or all three wavelengths, to measure the CO_2 content of a gas such as air. Transduction is the process of converting the transducible property into an electrical signal that can be input to an instrument.

PASSIVE VERSUS ACTIVE SENSORS

Another ambiguity found in discussions of sensors is the distinction between passive sensors and active sensors. Unfortunately, competing texts use exactly opposite definitions of these terms! This text adopts the form that is used by most people in the field, which is also consistent with usage in other areas of electronic engineering.

An active sensor is one that requires an external AC or DC electrical source to power the device. An example of the active sensor is the resistive strain gage pressure sensor that requires a $+7.5$-V DC regulated power supply to operate.

---

[1] The term *event* is used here in a broad sense that also includes physical phenomena of all types, not necessarily some happening at a point in time, as might be interpreted by the denotation of the word.

[2] L. A. Geddes and L. E. Baker, *Principles of Applied Biomedical Instrumentation*, Wiley, New York, 1968.

A passive sensor is one that provides its own energy, which, in order to not violate the conservation of energy law, means that it derives its energy from the phenomenon being measured. An example of a passive sensor is the thermocouple, which is used to measure temperature.

It is unfortunate that some authors invert these definitions, but if the above definitions are accepted, you will be consistent with the most common usage.

SENSOR ERROR SOURCES

Sensors, like all other devices, suffer from certain errors. In order to maintain consistency, an error is defined as the "difference between the measured value and the true value."[3] While the full range of possible errors is beyond the scope of this book, it is possible to break them into five basic categories: insertion errors, application errors, characteristic errors, dynamic errors, and environmental errors.

Insertion Errors

This class of error occurs during the act of inserting the sensor into the system being measured. This problem is a general problem with electronic measurements, indeed all measurements. For example, when measuring the voltage in a circuit, one must be certain that the inherent impedance of the voltmeter is very much larger than the circuit impedance, otherwise loading will occur, and that leads to error. Examples of this form of error include using a transducer that is too large for the system to measure pressures, one that is too sluggish for the dynamics of the system, or one that self-heats to the extent that excessive thermal energy is added to the system. Nineteenth-century British physicist Lord Kelvin formulated a "first rule of instrumentation," to the effect that "the measuring instrument must not alter the event being measured."[4]

Application Errors

These errors are operator caused, that is, the proverbial "cockpit trouble" referred to by airplane mechanics. Again, there are far too many of these errors that are possible, so we must settle on a couple of illustrative examples. One error seen in temperature measurements is either incorrect placement of the probe or erroneous insulation of the probe from the measurement site. Other examples seen in fluid pressure sensor applications include failure to purge the system of air and

---

[3]L. A. Geddes and L. E. Baker, *Principles of Applied Biomedical Instrumentation*, Wiley, New York, 1968.
[4]Ibid.

other gases ("bubbles in the line") and incorrect physical placement of the transducer so that a positive or negative pressure head is erroneously added to the correct reading.

Characteristic Errors

This category is that which is most often meant when discussing errors without otherwise qualifying the term. These errors are those that are inherent in the device itself, that is, the difference between the ideal published transfer function of the device and the actual characteristic. This form of error may include a DC offset value, an incorrect sensitivity slope, or a slope that is not perfectly linear.

Dynamic Errors

Many sensors are characterized and calibrated in a static condition, that is, with an input parameter that is either static or quasistatic in nature. Many sensors are heavily damped, so they will not respond to rapid changes in the input parameter. For example, thermistors tend to require many seconds to respond to a step-function change in temperature. That is, a thermistor in equilibrium will not jump immediately to the new resistance on an abrupt change in temperature. Rather, the device will change slowly toward the new value. Thus, if an attempt is made to follow a rapidly changing temperature with a sluggish sensor, the output waveform will be distorted and therefore will contain error. The issues to confront in respect to dynamic errors include response time, amplitude distortion, and phase distortion.

Environmental Errors

These errors are those that are derived from the environment in which the sensor is used. They most often include temperature, but may also include vibration, shock, altitude, chemical exposure, and other factors. These factors most often affect the characteristic errors of the sensor, so they are often lumped together with that category in practical application.

SENSOR TERMINOLOGY

Sensors, like other areas of technology, have their own terminology that must be understood before they can be properly applied. In this section are some of the most common terms.

Sensitivity

The sensitivity of the sensor is defined as the slope of the output characteristic curve ($\Delta Y / \Delta X$ in Fig. 3-1), or more generally, it is the minimum input of physical parameter that will create a detectable output change. In some sensors, the sensitivity is defined as the input parameter change required to produce a standardized output change. In

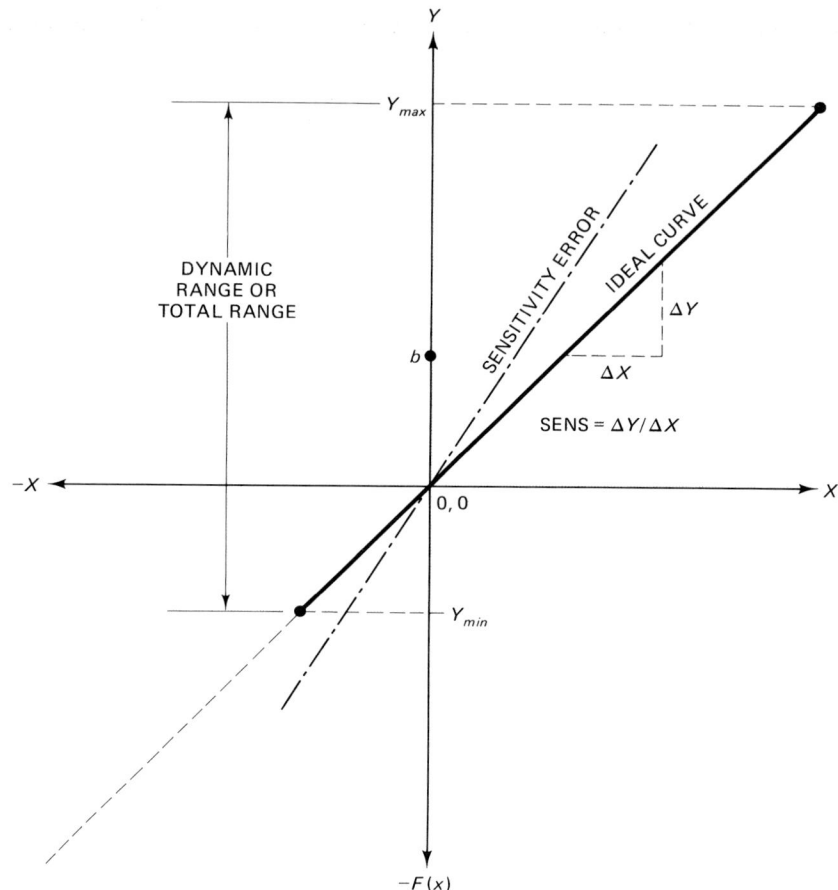

Fig. 3-1 Sensor characteristic curve.

still others, it is defined as an output voltage change for a given change in input parameter. For example, a typical pressure transducer may have a sensitivity rating of 10 μV/V/Torr; that is, there will be a 10-μV output voltage for each volt of excitation potential and each Torr of applied pressure.

Sensitivity Error

The sensitivity error (shown as a dotted curve in Fig. 3-1) is a departure from the ideal slope of the characteristic curve. For example, the pressure transducer discussed previously may have an actual sensitivity of 7.8 μV/V/Torr instead of 10 μV/V/Torr.

Range

The range of the sensor is the maximum and minimum values of the applied parameter that can be measured. For example, a given pressure sensor may have a range of -5 to $+5$ psi. Alternatively, it is often the case that the positive and negative ranges are unequal. For example, a certain medical blood pressure transducer is specified to have a low-end limit of -50 mm Hg (Y_{min} in Fig. 3-1) and a maximum high-end limit of $+450$ mm Hg (Y_{max} in Fig. 3-1).

Dynamic Range

The dynamic range is the total range of the sensor from minimum to maximum. That is, in terms of Fig. 3-1, $R_{dyn} = Y_{max} - |-Y_{min}|$.

Precision

The concept of precision refers to the degree of reproducibility of a measurement. In other words, if exactly the same value were measured a number of times, an ideal sensor would output exactly the same value every time. But, real sensors output a range of values distributed in some manner relative to the actual correct value. For example, suppose a pressure of 250 Torr, exactly, is applied to a sensor. Even if the applied pressure never changes, the output values from the sensor will vary considerably. Some subtle problems pop up in the matter of precision when the true value and the sensor's mean value are not within a certain distance of each other (e.g., the 1-σ range of the normal distribution curve).

Resolution

This specification is the smallest detectable incremental change of input parameter that can be detected in the output signal. Resolution can be expressed as either a proportion of the reading (or the full-scale reading) or in absolute terms.

Accuracy

The accuracy of the sensor is the maximum difference that will exist between the actual value (which must be measured by a primary standard or good secondary standard) and the indicated value at the output of the sensor. Again, the accuracy can be expressed as either a percentage of the full-scale reading or in absolute terms.

Offset

The offset error of a transducer is defined as the output that will exist when it should be zero, or alternatively, the difference between the actual output value and the specified output value under some particular set of conditions. An example of the first situation in terms of Fig. 3-1 would exist if the characteristic curve had the same sensitivity slope as the ideal, but crossed the Y axis (i.e., output) at b instead of zero. An example of the other form of offset is seen in the characteristic curve of a pH electrode shown in Fig. 3-2. The ideal curve will exist only at one temperature (usually 25°C), while the actual curve will between the minimum and maximum temperature, limits depending on the temperature of the sample and electrode.

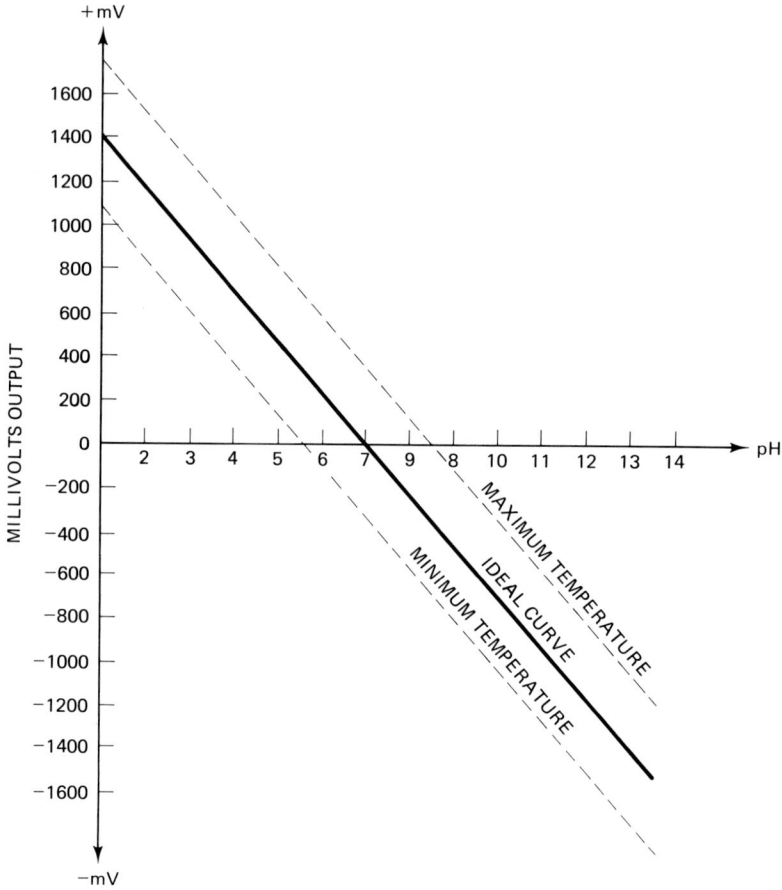

Fig. 3-2 Typical pH probe characteristic curve.

Linearity

The linearity of the transducer is an expression of the extent to which the actual measured curve of a sensor departs from the dial curve. Figure 3-3 shows a somewhat exaggerated relationship between the ideal, or least squares fit,[5] line and the actual measured, or calibration, line. (In most cases, it is the static curve that is used to determine linearity, and this may deviate somewhat from a dynamic linearity.) Linearity is often specified in terms of percentage of nonlin-

---

[5]The least squares method of linear regression is most often used to make calibration curves from points of data taken on the system. The basic assumption of the conventional least squares method is that the line is $Y = mx + b \pm e$, where e is an error term that represents the sum of all errors in the system. When there are two sources of error that cannot be conveniently summed, it is preferable to use the orthogonal least squares method of linear regression. In that method, the assumption is that the line is $Y \pm e1 = mX + b \pm e2$, in which $e1$ and $e2$ are decorrelated errors.

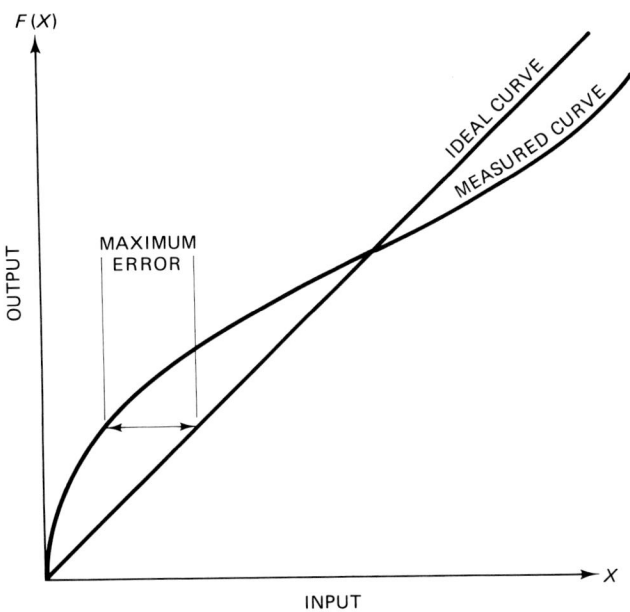

Fig. 3-3 Ideal calibration curve versus measured curve.

earity, which is defined as

$$\text{Nonlinearity (\%)} = \frac{D_{\text{in(max)}}}{\text{IN}_{\text{fs}}} \times 100 \qquad (3\text{-}1)$$

where nonlinearity (%) is the percentage of nonlinearity, $D_{\text{in(max)}}$ the maximum input deviation, and IN_{fs} the maximum, full-scale input.

The static nonlinearity defined by Eq. (3-1) is often subject to environmental factors, including temperature, vibration, acoustical noise level, and humidity. It is important to know under what conditions the specification is valid and to know that departures from those conditions may not yield linear changes of linearity.

Hysteresis

A transducer should be capable of following the changes of the input parameter regardless of which direction the change is made; hysteresis is the measure of this property. Figure 3-4 shows a typical hysteresis curve. Note that it matters from which direction the change is made. Approaching a fixed input value (e.g., point B in Fig. 3-4) from a higher value (e.g., point P) will result in a different indication than approaching the same value from a lesser value (point Q or zero). Note that input value B can be represented by $F(X)_1$, $F(X)_2$, or $F(X)_3$, depending on the immediate previous value; clearly an error due to hysteresis.

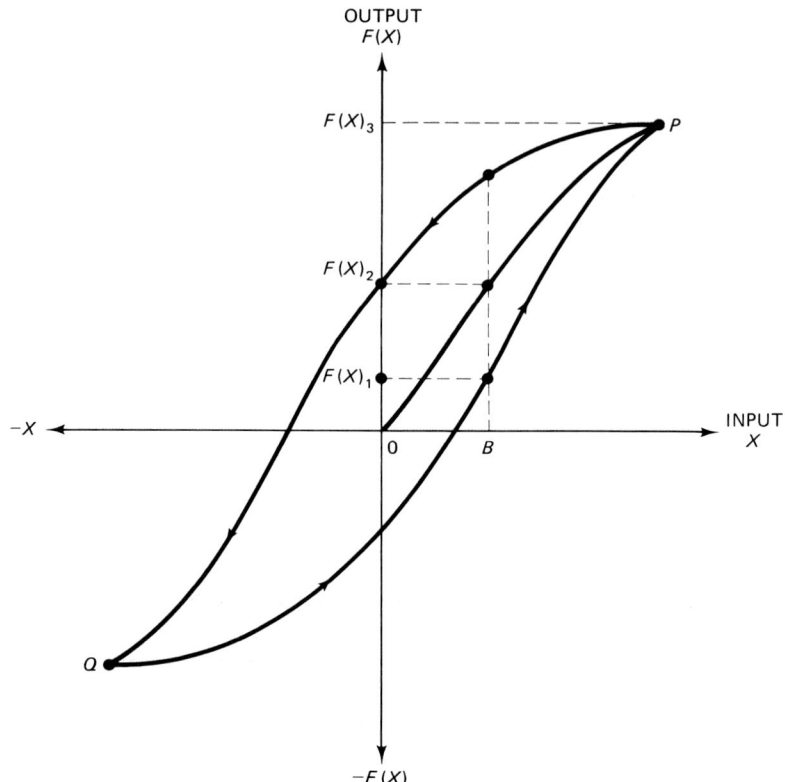

Fig. 3-4 Typical hysteresis curve.

Response Time

Sensors do not immediately change the output state when an input parameter change occurs. Rather, the output state will change to the new state over a period of time, called the response time (T_r in Fig. 3-5). The response time is defined as the time required for a sensor output to change from its previous state to a final settled value within a tolerance band of the correct new value. This concept is somewhat different from the notion of the time constant (T) of the system. This term can be defined in manner similar to a capacitor charging through a resistance, and it is usually less than the response time.

The curves in Fig. 3-5 show two types of response times. In Fig. 3-5A, the curve represents the response time following an abrupt, positive-going step-function change of the input parameter. The form shown in Fig. 3-5B is a decay time (T_d to distinguish from T_r, because they are not always the same) in response to a negative-going step-function change of the input parameter.

Dynamic Linearity

The dynamic linearity of the sensor is a measure of its ability to follow rapid changes in the input parameter. Amplitude distortion

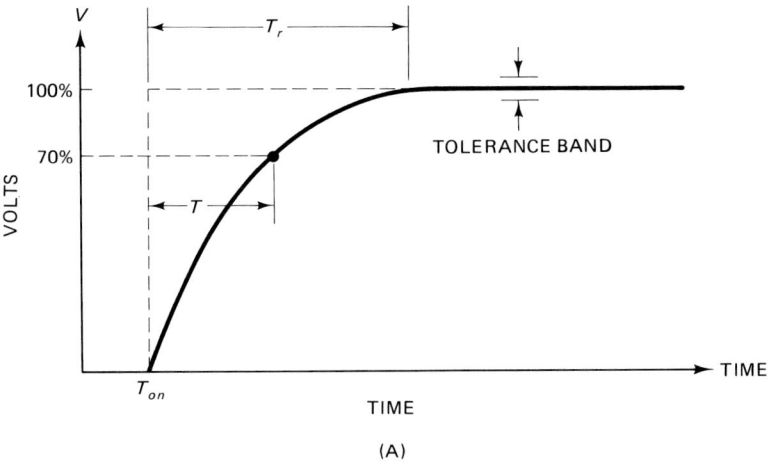

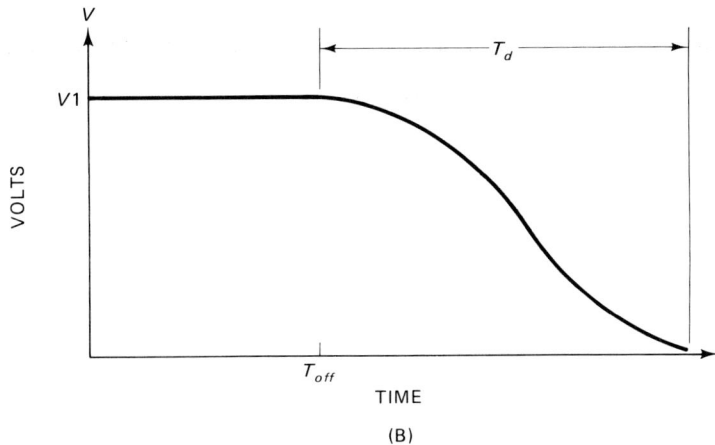

Fig. 3-5 Response times of a sensor: (A) rise time and (B) decay time.

characteristics, phase distortion characteristics, and response time are important in determining dynamic linearity. Given a system of low hysteresis (always desirable), the amplitude response is represented by

$$F(X) = aX + bX^2 + cX^3 + dX^4 + eX^5 + \cdots + K \qquad (3\text{-}2)$$

In Eq. (3-2), the term $F(X)$ is the output signal, the X terms represent the input parameter and its harmonics, and K is an offset constant (if any). The harmonics become especially important when the error harmonics generated by the sensor action fall into the same frequency bands as the natural harmonics produced by the dynamic action of the input parameter. Recall from Chapter 2 that all continuous waveforms are represented by a Fourier series of a fundamental sine wave and its harmonics. In the case of any nonsinusoidal wave-

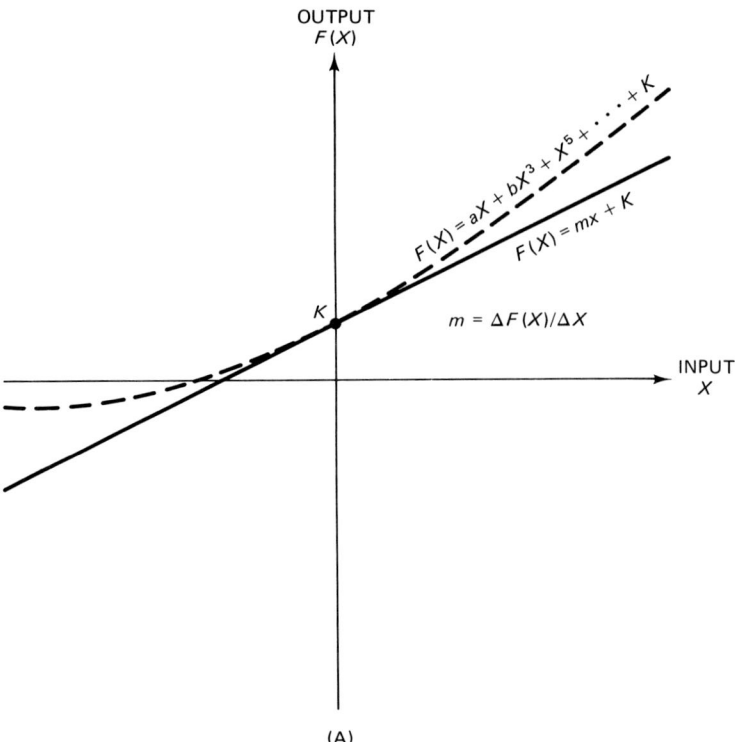

Fig. 3-6 (A) Symmetrical calibration curve; (B) asymmetrical calibration curve. (*Figure continues.*)

form (including time-varying changes of a physical parameter), there will be harmonics present that can be affected by the action of the sensor.

The nature of the nonlinearity of the calibration curve (Figs. 3-6A and 3-6B) tells something about which harmonics are present.[6] In the case of Fig. 3-6A, the calibration curve (shown as a dotted line) is asymmetrical, so only odd harmonic terms exist. Assuming a form for the ideal curve of $F(X) = mX + K$, Eq. (3-2) becomes, for the symmetrical case,

$$F(X) = aX + bX^2 + cX^4 + \cdots + K \qquad (3\text{-}3)$$

In the other type of calibration curve (Fig. 3-6B), the indicated values are symmetrical about the ideal $mX + K$ curve. In this case, $F(X) = -F(-X)$, and the form of Eq. (3-2) is

$$F(X) = aX + bX^3 + cX^5 + \cdots + K \qquad (3\text{-}4)$$

In the next section, we will take a look at some of the tactics and signal processing criteria that can be adapted to improve the nature of the data collected from the sensor.

[6]Richard S. C. Cobbold, *Transducers for Biomedical Applications*, Wiley, New York, 1974.

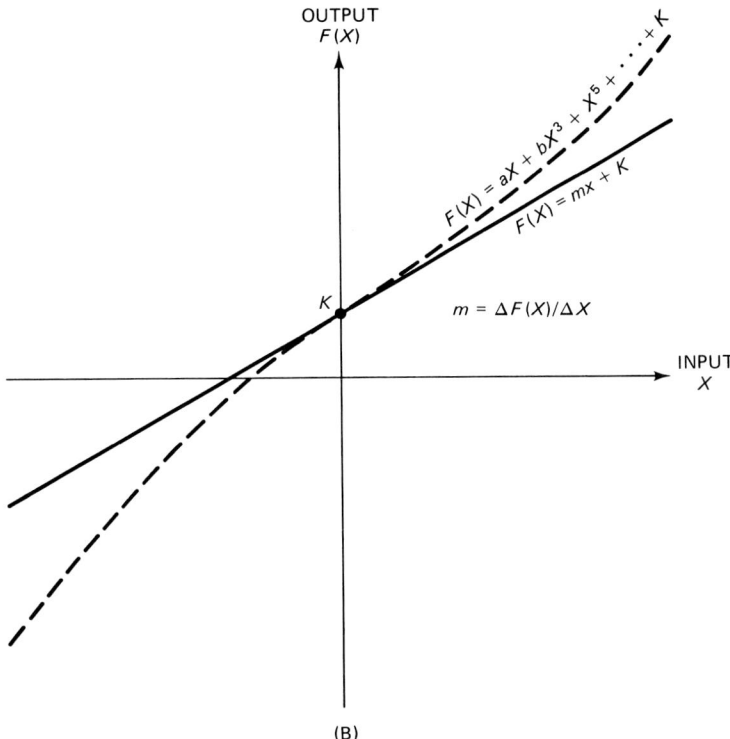

Fig. 3-6 (*continued*)

TACTICS AND SIGNAL PROCESSING FOR IMPROVED SENSING

The selection of sensors and the circuits that are connected to them can go a long way toward ensuring that the data acquired will accurately represent the physical phenomenon or event being detected.

For proper operation in a dynamic input environment, the sensor selected should have a flat response curve, that is, one that is free of amplitude distortion, phase distortion (which almost invariably causes amplitude distortion), or resonances.

An implication of these problems concerns the frequency response of the sensor and its signal processing system. Figure 3-7A shows a perfectly linear system in which the gain[7] is constant over the entire spectrum of frequencies, that is, in an ideal theoretical system from DC to daylight and beyond. But, real systems do not have such characteristics. Figure 3-7B shows the type of frequency response that might be found on real systems. In this example, the gain is flat between two frequencies, and over this region the performance is

---

[7]Gain is defined as the ratio of the output function to the input function. Because voltage gain (A_v) is used here for illustration purposes, the gain is V_o/V_{in}.

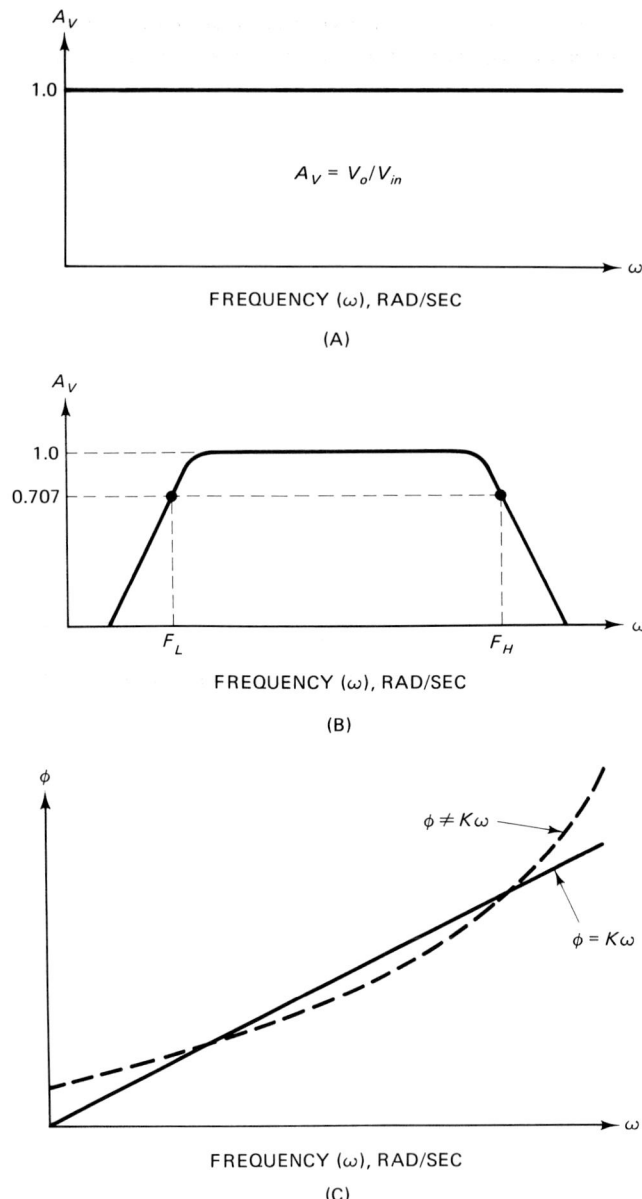

Fig. 3-7 (A) Broadbanded frequency response, (B) bandpass frequency response, and (C) sensor phase response.

similar to the ideal case. But, beyond these points, the gain falls off at a given slope. The breakpoint that defines the flat region is, by convention, taken to be the frequencies (F_L and F_H), at which the gain falls of to 70.7% of its gain in the flat region. These points are known as the -6-dB points in voltage systems and the -3-dB points in power systems.

When the frequency response is not entirely flat, one can expect to find phase distortion. Figure 3-7C shows the situations where the

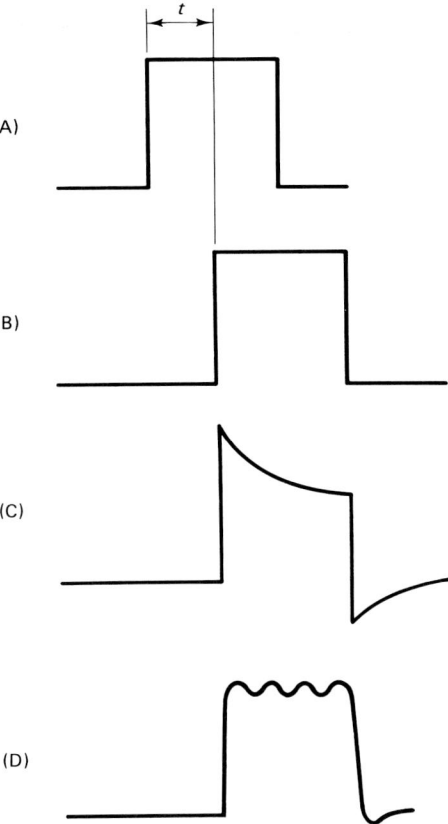

Fig. 3-8 (A) Square wave or pulse signal at input; (B) phase shifted in time, but not distorted at output; (C) severe low-frequency loss; and (D) phase distorted signal.

phase shift (ϕ) of the system is a linear function of frequency (solid line) and where it is a nonlinear function of frequency (dotted line).

We can see the effects of phase distortion in a somewhat simplistic sense in Fig. 3-8. Figure 3-8A is the applied signal, for example, the output of an ideal sensor in response to step-function changes of the measured input parameter. If the signal processing electronics and the sensor mechanism itself are perfectly ideal, then the only effect of the change will be displacement in time (t), as shown in Fig. 3-8B. There will be no distortion of the shape of the wave. In the presence of phase distortion, however, the wave will not only be time displaced, but also distorted. Figures 3-8C and 3-8D show two forms of distortion that can occur with phase nonlinearity.

A slightly different view of the same phenomenon is shown in Figs. 3-9 and 3-10. Consider a system in which the bandwidth can be varied across several limits, represented by curves a, b, and c in Fig. 3-9. Curve c represents the most restrictive of the three possibilities because it sharply limits both the low- and high-frequency responses, while curve a is the least restrictive. Note in Fig. 3-10 the various responses to the three bandwidths represented in Fig. 3-9.

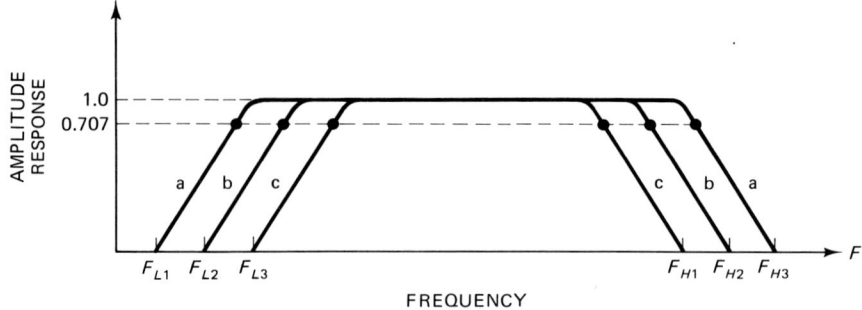

Fig. 3-9 Frequency responses of three different amplifiers.

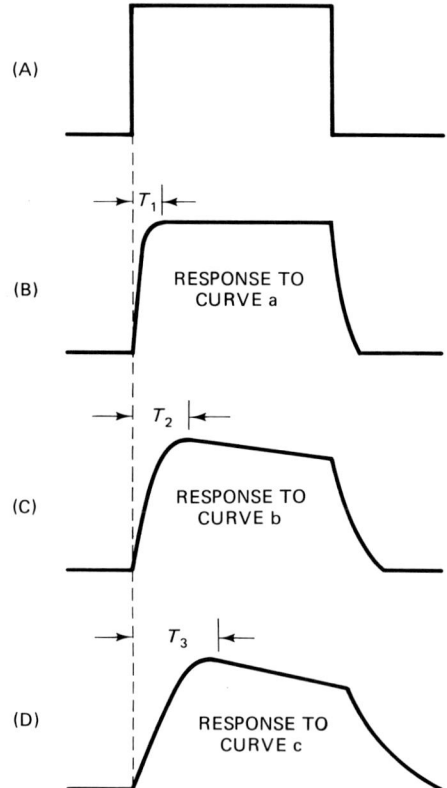

Fig. 3-10 Effects of the response of Fig. 3-9 on an input square wave.

These curves can be simulated by examining the response to square waves in resistor–capacitor (RC) filter networks. In fact, one of the problems that one must consider when using electronic filters is the effects of the -6-dB points on the applied waveform.

One might erroneously assume from the previous discussion that the instrument designer should select amplifiers with as wide a bandwidth as possible. That is not the case, however, because bandwidth can cause other problems at least as severe as those that are solved. Noise, for example, is proportional to bandwidth. It is possible to eliminate the problems of noise, plus certain input signal problems, such as ringing or resonances, by proper selection of the frequency response cutoff points. Thus, the selection of amplifier bandwidth and phase distortion characteristics is a trade-off between the need to make a high-fidelity recording of the input event and the other problems that can occur in the system.

In some systems, it is possible to reduce distortion of the output signal in dynamic systems by using two sensors in a differential configuration. By taking the difference between two sensors, the arithmetic shows that the even order terms will drop out, leaving less total harmonic distortion than in the single-ended case.

CONCLUSION

Now that we have discussed some of the basic problems of sensors, let us get down to examples of actual forms of sensors that might be employed in instrumentation systems. The treatment here is not exhaustive—indeed such would fill a number of very large books—but it is intended, instead, to be representative.

4

Temperature Sensors

Temperature sensors are, perhaps, most illustrative of some of the basic principles of transduction, so they are often the first discussed in a book such as this one. Thus shall it be here also, because there are several different transducible phenomena that can be used to measure temperature. There are a number of different types of temperature sensors available to the instrument designer, each based on different phenomena. In this chapter, we will take a look at some of the more popular types, with an eye toward being representative of the overall class of temperature sensors. First, let us review the different measurement scales used to define temperature.

DIFFERENT MEASUREMENT SYSTEMS

There are several different temperature scales used in the measurement of heat. The familiar Fahrenheit and Celsius (a.k.a. centigrade) scales, along with the less familiar Kelvin and Rankine scales, are frequently used in medical, scientific, and industrial measurements. The Celsius and Fahrenheit scales are arranged such that 0°C is the same temperature as 32°F. Both points are defined by the freezing point of water at standard temperature and pressure. The two scales can be converted to each other by the equation:

$$F - 32 = 1.8C \qquad (4\text{-}1)$$

where F is degrees Fahrenheit (°F) and C is degrees Celsius (°C).

The Kelvin scale uses the same size degree steps as the Celsius scale, but defines the zero degree point differently. In the Celsius scale, 0°C is the freezing point of water, while in the Kelvin scale it is absolute zero (the point where molecular activity ceases). Thus,

0 K = $-276.16°C$. The Celsius and Kelvin scales are converted by

$$K = °C - 273.16 \qquad (4\text{-}2)$$

and

$$°C = K + 273.16 \qquad (4\text{-}3)$$

The Rankine scale is to the Fahrenheit scale as the Kelvin scale is to the Celsius, that is, the size of Rankine degree steps are the same as Fahrenheit degrees, but the zero point is at absolute zero on the Fahrenheit scale. Thus, 0°R is approximately equal to $-459.7°F$.

TEMPERATURE TRANSDUCERS

Several different temperature sensors are commonly used: thermistors, thermocouples, and PN semiconductor junctions. Although applications for these different forms of transducer overlap, there are key parameters and other factors that often favor one or the other. Let us examine each of these.

THERMISTORS

Thermistors (i.e., thermal resistors) are resistors that are designed to change resistance value in a predictable manner with changes in applied temperature. The amount of change is designated by the temperature coefficient (a) of the material, which is measured in ohms of resistance change per ohm of resistance per degree Celsius ($\Delta\Omega/\Omega/°C$). A positive temperature coefficient (PTC) device (see Fig. 4-1) increases resistance with increases in temperature. Alternatively, a negative temperature coefficient (NTC) device decreases resistance with increases in temperature. The usual circuit symbols for thermistors are shown in Fig. 4-2. The indirectly heated variety uses an internal heating element. Several popular packaging styles are shown in Fig. 4-3.

Most thermistors have a nonlinear curve when plotted over a wide temperature range, but when limited to narrow temperature ranges, the linearity is considerably better. When such thermistors are used, however, it is necessary to ensure that the temperature will not go on excursions outside of the permissible linear range. There are methods for linearizing the thermistor, and these will be discussed in a later section.

Thermistors are among the oldest temperature sensors available. The temperature sensitivity of electrical resistance in silver sulfide was noted by physicist Michael Faraday in 1833. There are several different

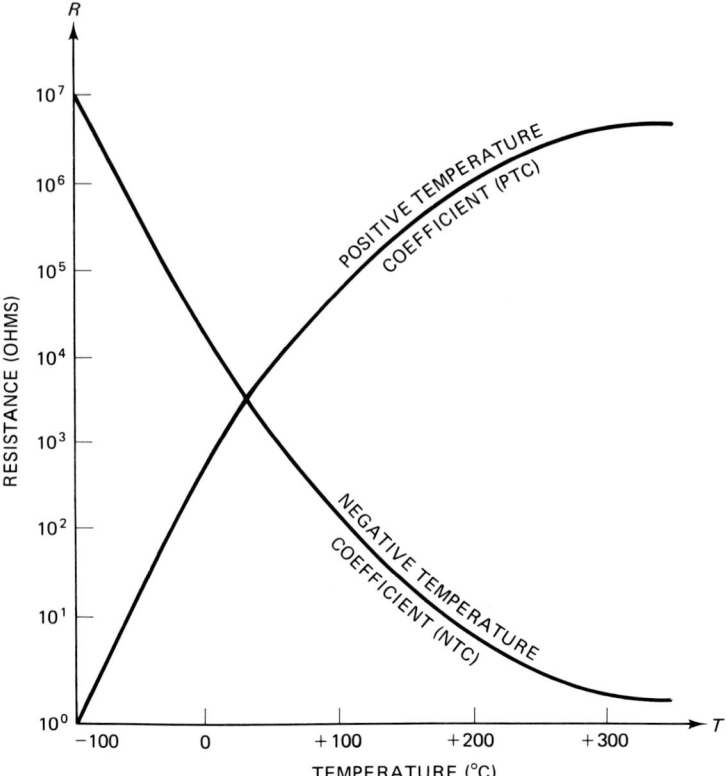

Fig. 4-1 Temperature versus resistance curves for positive and negative temperature coefficient thermistors.

types of thermistor, but the simplest is the wire element. Simple wire thermistor elements are based on two physical phenomena. First, materials tend to change physical dimensions with changes in temperature. Metals, for example, tend to expand when heated. Second, the resistance of a material is directly proportional to the length of the sample. Thus, when a metal is heated, it tends to expand, so its electrical resistance increases. Most metals have a positive temperature coefficient ($a > 0$). Copper, for example, has a temperature coefficient of $+0.004$.

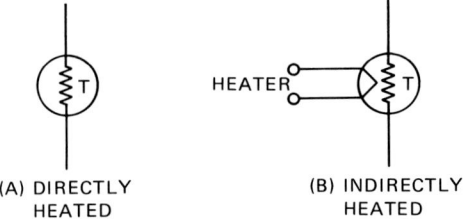

Fig. 4-2 Thermistor symbols: (A) directly heated and (B) indirectly heated.

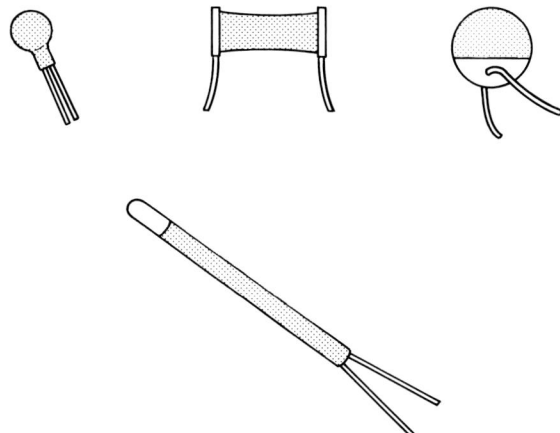

Fig. 4-3 Types of thermistor packages.

Not all materials have positive temperature coefficients, however. Some materials, such as carbon and some ceramics, have a negative temperature coefficient ($a = -0.0003$). Other materials, including certain metal alloys, have temperature coefficients that are nearly zero. For example, in manganin and constantan, the temperature coefficient is approximately $+0.00002$, and for nichrome, it is $+0.00017$. At one time, radio designers used nichrome wire to wind tuning inductors because of this property. Inductors wound with a low temperature coefficient wire do not change size when heated, so the drift due to the coil's inductance temperature sensitivity is reduced.

The change in resistance caused by changes in temperature is a function of a and the value of temperature change. For a wire element, the new resistance is found from

$$R_{T2} = R_{T1}[1 + a(T2 - T1)] \tag{4-4}$$

where R_{T1} is the starting temperature resistance, R_{T2} the final temperature resistance, a the temperature coefficient, $T1$ the starting temperature, and $T2$ the final temperature.

Wire elements are sometimes used as thermistors. Taut platinum wire elements, for example, are sometimes used in respiration sensors. Other thermistors are made of evaporated films, carbon or carbon compositions, or oxides of cobalt, manganese, magnesium, nickel, or uranium.

Thermistor Parameters

Before one can successfully use thermistors, it is necessary to first understand some of the basic properties of the thermistor. These are expressed in the form of certain standard parameters. The following are among the most commonly needed.

Cold (Zero-Power) Resistance

This parameter is the resistance of the thermistor at a standard reference temperature (usually either room temperature, 25°C, or the ice point of water, 0°C), under conditions of no self-heating power dissipation. This parameter is the cold resistance that is listed in the specifications sheet as the nominal resistance. For example, a device listed as a 1000-Ω thermistor has a resistance of 1000 Ω at the standard reference temperature (25°C unless otherwise specified). The conditions under which the thermistor is operated for measurement of the cold resistance include a requirement that the current through the device be sufficiently low to avoid self-heating.

Hot Resistance

The hot resistance of the thermistor is measured when the device is operated at a higher temperature than the cold resistance temperature. The higher temperature is due to ambient temperature, the current flow through the thermistor, the applied heater current (indirectly heated types only), or a combination of all of these factors. Equation (4-4) can be modified to find the hot resistance of wire elements:

$$R_T = R_0\big[1 + a(T - T_0)\big] \qquad (4\text{-}5)$$

For other forms of thermistor, the expression is

$$R_T = R_0 e^{B[(1/T)-(1/T_0)]} \qquad (4\text{-}6)$$

where T_0 is the reference temperature (25°C), T the new temperature, R_0 the thermistor resistance at the reference temperature, R_T the resistance at temperature T, a the coefficient of resistance, and B is a factor with units of temperature (usually between 1500 and 7000 K).

Resistance versus Temperature

This parameter is an expression of the characteristic shown in Fig. 4-1. The exact shape of the curve is a function of the thermistor in question, but it will be of the form shown in Fig. 4-1 and is quite nonlinear.

Resistance Ratio

The resistance ratio (R_T/R_0) is essentially a simplified expression of the R versus T curve. It states the ratio of the thermistor resistance at a specified resistance (50°C, 100°C, or 125°C) to the cold temperature (25°C) resistance.

Voltage versus Current

Directly heated thermistors have an unusual voltage versus current curve (Fig. 4-4) that includes both ohmic and negative resistance regions. Assuming a constant ambient temperature, an increase in the current through the thermistor will cause a linear increase in the voltage drop across the thermistor. Because this behavior is in accordance with Ohm's law, $V = IR$, that portion of the curve is called the ohmic region. At a certain point, however, internal self-heating becomes dominant and begins to alter the resistance of the thermistor. At this point, the voltage drop begins to decrease with increasing current flow. In other words, in this region the thermistor is a negative resistance device.

Current versus Time

A typical thermistor current versus time curve is shown in Fig. 4-5. The thermistor current would ideally snap to the level V/R_T when a step function voltage is applied (or the applied level is changed). However, because there is always a small amount of self-heating involved in any thermistor, this response is not linear. There is always a time lag between a change in applied voltage and the current in the thermistor reaching the level mandated by that voltage for the thermistor resistance.

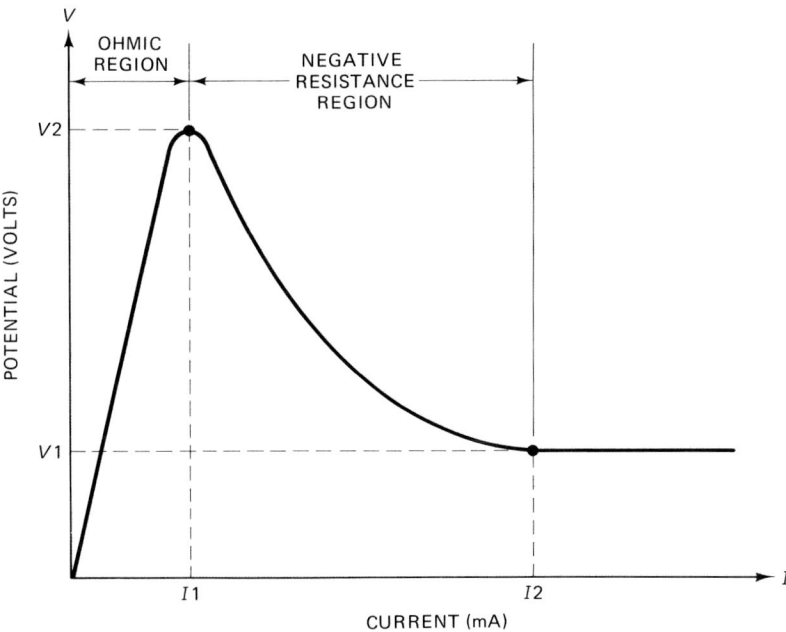

Fig. 4-4 V versus I curve for thermistor.

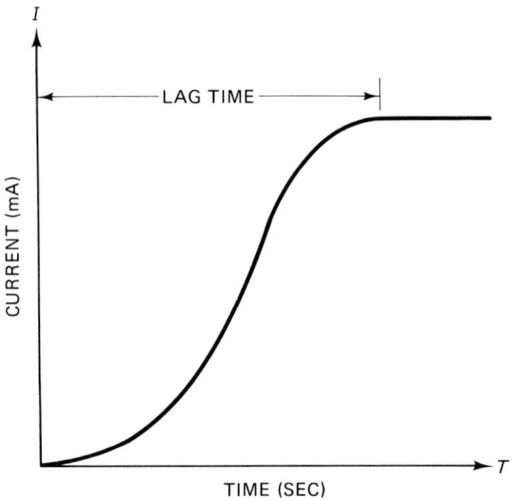

Fig. 4-5 Response time for current in thermistor when voltage is applied.

Maximum Power

This parameter is the maximum allowable constant power level (P_{max}) that the thermistor will handle without destruction, permanent alteration of characteristics, or degradation of its performance.

Dissipation Constant

This factor, symbolized in the specifications sheet by a lowercase Greek delta (δ), is the ratio of the change in power dissipation for small changes in the body temperature of the thermistor ($\delta = dP_d/dT_B$).

Sensitivity

The sensitivity (σ) of a thermistor is the ratio of resistance change to temperature change (dR/dT), expressed as a percent change per degree of temperature. Because the R versus T curve (Fig. 4-1) is nonlinear over most of its range, sensitivity factor numbers are valid only over a limited range. Typical values for σ run from 0.5%/°C to 4%/°C.

Temperature Range

The thermistor's characteristics are only specified over a limited temperature range, T_{min} to T_{max}. The value of T_{min} is typically -200°C, while T_{max} is typically $+650$°C (although there are devices with a narrower range).

Thermal Time Constant

The body temperature of a thermistor does not change instantaneously in response to a step-function change in ambient temperature. If T_i is the initial temperature and T_f is the final temperature, then the thermal time constant (t) is the time required for the body temperature of the thermistor to change 63.2% of the range between these two temperatures. The term 63.2 is derived from $[1 - e^{-t}]$ when $t = 1$ sec.

LINEARIZING THERMISTORS

The resistance versus time curve seen earlier in Fig. 4-1 is nonlinear over most of its range. For some measurements, therefore, it is necessary to either restrict the use of the device to a limited range of temperatures or to actually linearize the R versus T curve. There are several ways to linearize the curve. Some of them involve electronic circuits, so they will be discussed in detail after we have discussed the circuits involved. There are, however, two methods that only involve simple resistors or other thermistors. Figure 4-6A shows a linearization network used by a thermistor manufacturer. Although the network functions (to an outside observer) like a single, two-terminal thermistor, it actually consists of a network of resistors and thermistors.

A relatively easy method for linearizing a thermistor is shown in Fig. 4-6B. This method involves shunting a low temperature coefficient resistor, R_s, across the thermistor, R_t. The total value of the network is the parallel resistance of the two elements:

$$R_{total} = \frac{R_s R_t}{R_s + R_t} \qquad (4\text{-}7)$$

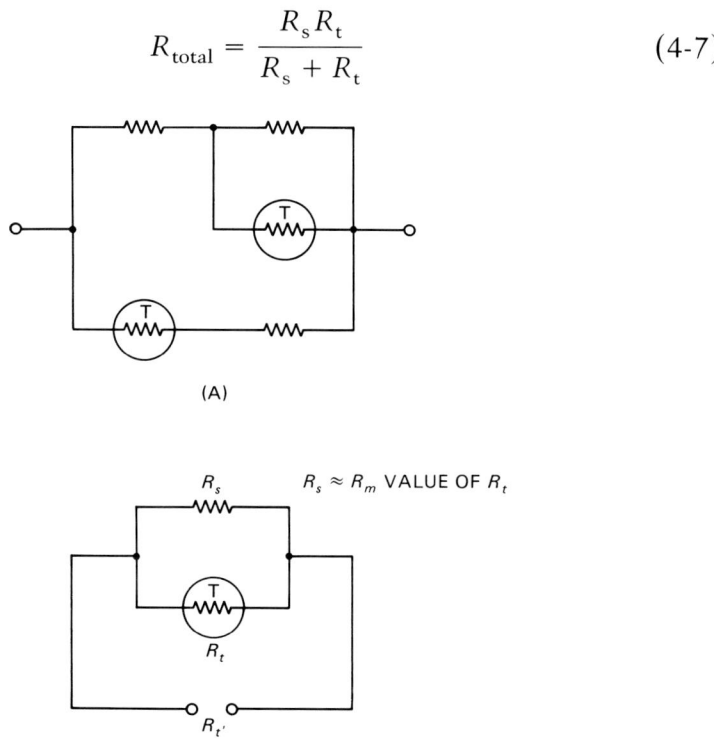

Fig. 4-6 Methods of thermistor linearization: (A) series–parallel and (B) parallel.

The value of R_s is the mean value (R_m) of R_t over the temperature range of interest. Suppose, for example, that you want to linearize a thermistor over the physiological temperature range (e.g., 30 to 45°C). The value of R_m in this case, hence the value of R_s, is the thermistor resistance at a temperature of $[30 + (45 - 30)/2]°C$, or 37.5°C. Modifying Eq. (4-4) gives us the expression for the value of the total resistance $R_{t'}$:

$$R_{t'} = \frac{R_m}{2}\left[1 + \frac{a}{2}(T - T_m)\right]. \qquad (4\text{-}8)$$

The thermistor is easy to use and is reasonably well behaved within the temperature range for which it is rated. But, when a wider temperature range is needed, especially when the temperature measurement is in a very hot environment, then the sensor of choice may well be the thermocouple.

THERMOCOUPLES

An example of a thermocouple is shown in Fig. 4-7. This type of transducer consists of two dissimilar metals or other materials (some ceramics and semiconductors are used) that are fused together at one end. Because the work functions of the two materials differ, there will be a potential difference generated across the open ends whenever the junction is heated. The potential is approximately linear with changes of temperature over relatively large ranges, although over very large ranges of temperature (for any given pair of materials) nonlinearity increases markedly.

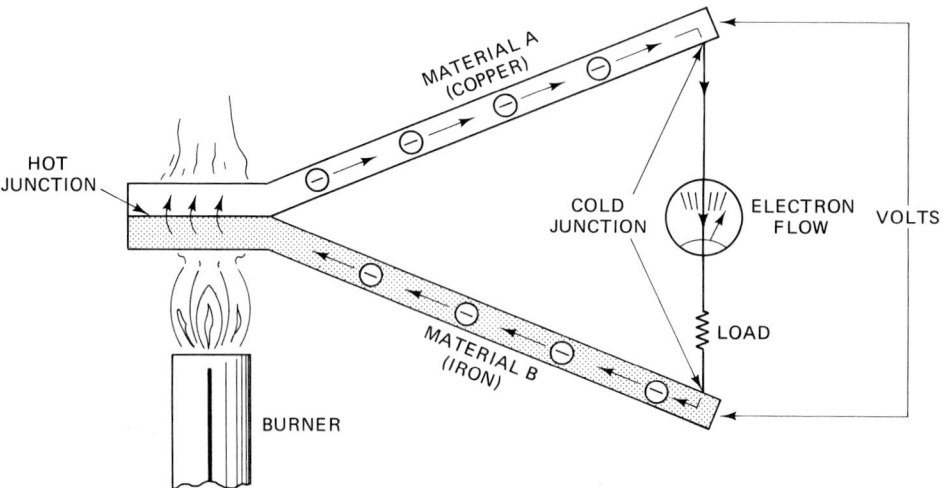

Fig. 4-7 Thermocouple temperature sensor depends on different work functions in different metals formed into an electrical junction.

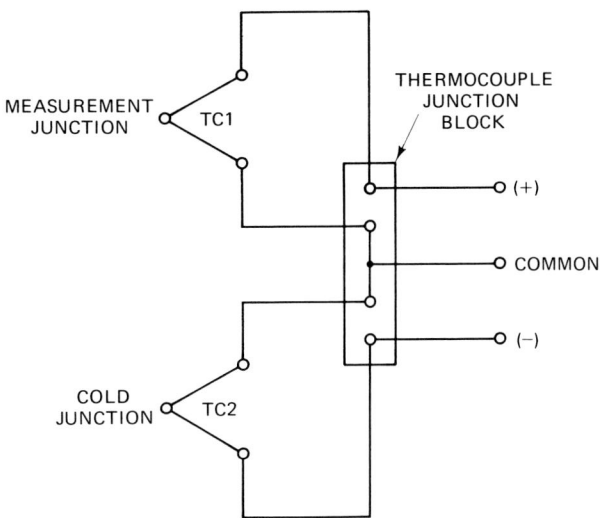

Fig. 4-8 Connection of thermistors in measurement and cold junctions.

Thermocouples are typically used in pairs (Fig. 4-8) or even threes. One junction will be used as the measurement thermocouple, while the other is the cold junction. Its name derives from the fact that some early systems required this junction to be bathed in an ice-water bath. In modern systems a synthetic cold junction, or just room temperature, may be used for the cold junction.

There is also a third junction in the circuit: the connection of the thermocouple wires to the brass or copper junction block. These junctions should be located at a distance from the cold junction and measurement junction in order to prevent heating from the same source.

The differential voltage between the two thermocouple junctions is proportional to the temperature difference and is used as the output voltage. This potential is found from the cubic equation:

$$E = a + bT + cT^2 + dT^3 \qquad (4\text{-}9)$$

where E is the output potential in volts; T the temperature of the measurement junction; and a, b, c, and d are constants that are a function of the materials used in the thermocouple.

Linearizing a Thermocouple

The equation governing the thermocouple demonstrates a strong non-linear, in fact cubic, dependence of the output voltage on temperature. In some cases, an approximation of the temperature is made using just the quadratic version of the equation (cubic term deleted or approximated with an additional constant). This practice was only reasonable in days when better linearization methods were not easily available. Analog circuits to solve the quadratic equation are, after all, somewhat easier than circuits for the cubic equation. But, today there is no reason to opt for the lesser of the two linearization methods.

As with other systems, there is more than one way to linearize the thermocouple. One could, for example, design a diode breakpoint generator with a cubic response and then sum its output with the thermocouple signal. But, that system is both cumbersome and subject to, of all things, thermal drift in the breakpoint generator diode circuits (this same phenomenon forms the basis for our next category of sensors, semiconductors). It is also possible to use a computer or computer-like circuit for linearization. The two computer methods involve (1) a lookup table to correct the value of output voltage for any given temperature and (2) an algorithm that will solve Eq. (4-9) for T given an output voltage. In both cases, the computer can be programmed with information on the specific type of thermocouple being used so that either the correct lookup table or the correct values of the coefficients of Eq. (4-9) are selected.

SOLID-STATE TEMPERATURE SENSORS

The last class of temperature sensor is the solid-state PN junction. If you take an ordinary solid-state rectifier diode (Fig. 4-9) and connect it across an ohmmeter, then you can see the temperature effect on diodes. Note the forward biased diode resistance at room temperature.

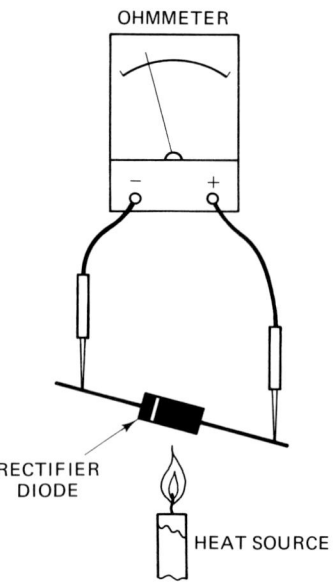

Fig. 4-9 Even ordinary rectifier diodes show leakage resistance as a function of temperature.

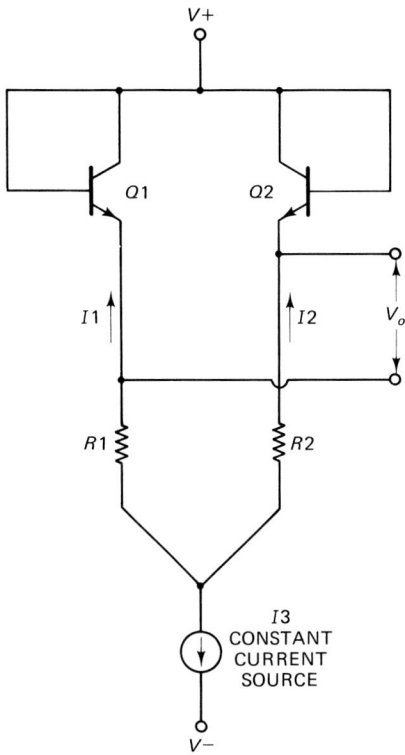

Fig. 4-10 Paired PN junction transistors used as temperature sensor.

Next, heat the diode temporarily with a lamp or soldering iron. The diode resistance drops dramatically as the temperature increases.

Most temperature transducers, however, use the diode-connected bipolar transistor, such as that shown in Fig. 4-10. We know that the base-emitter voltage (V_{be}) of a bipolar transistor is proportional to temperature. For a differential pair, as in Fig. 4-10, the transducer output voltage is given by

$$\Delta V_{be} = \frac{KT \ln(I_{C1}/I_{C2})}{q} \qquad (4\text{-}10)$$

where K is Boltzmann's constant (1.38×10^{-23} J/K), T the temperature in degrees Kelvin (K), and q the electronic charge (1.6×10^{-19} C per electron).

The K/q ratio is constant under all circumstances. The ratio I_{C1}/I_{C2} can be held constant artificially by making $I3$ a constant current source. The only variable in the equation, therefore, is temperature. In the following sections, we take a look at some commercial integrated circuit temperature devices based on this physical principle.

COMMERCIAL IC TEMPERATURE MEASUREMENT DEVICES

Several semiconductor device manufacturers offer temperature measurement/control integrated circuits (TMCIC). These devices are almost all based on the PN junction properties discussed earlier in this chapter, although at least one, by Analog Devices, Inc., uses an external thermocouple. In this section, we will look at the semiconductor TMCIC devices offered by National Semiconductor and Analog Devices, Inc.

The LM-335

The National Semiconductor LM-335 device shown in Fig. 4-11 is a three-terminal temperature sensor. The two main terminals are for power (and output), while the third terminal, shown coming out the body of the diode symbol is for adjustment and calibration. The LM-335 device is basically a special zener diode in which the breakdown voltage is directly proportional to the temperature, with a transfer function of close to 10 mV/K.

The LM-335 device and its wider range cousins the LM-135 and LM-235 devices operate with a bias current set by the designer. This current is not critical, but must be within the range of 0.4 to 5 mA. For most applications, designers seem to prefer currents in the 1-mA range.

The accuracy of the device is relatively decent and is more than sufficient for most control applications. The LM-135 version offers uncalibrated errors of 0.5 to 1°C, while the less costly LM-335 device offers errors of < 3°C. Of course, clever design can reduce these errors even further if they are out of tolerance for some particular application.

One difference between the three devices is the operating temperature ranges, which are as follows:

Device	Temperature range (°C)
LM-135	−55 to +150
LM-235	−40 to +125
LM-335	−10 to +100

There are two packages used for the LM-135 through LM-335 family of devices. The TO-92 is a small, plastic transistor case and is denoted by a Z suffix to the part number (e.g., LM-335Z), while the TO-46 is the small metal can transistor package (smaller than the

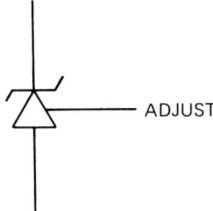

Fig. 4-11 LM-335 temperature sensor.

familiar TO-5 case). This case is identified with the suffix H or AH (e.g., LM-335H or LM-335AH).

The simplest, although least accurate, method of using the LM-335 device is shown in Fig. 4-12A. The LM-335 is essentially a temperature sensitive zener diode, and here it is connected as a zener diode. The series current-limiting resistor limits the current to around 1 mA. This value of $R1$ (4700 Ω) is appropriate for +5-V power supplies, as might be found in digital electronic instruments. The resistor value can be scaled upward for higher values of DC potential

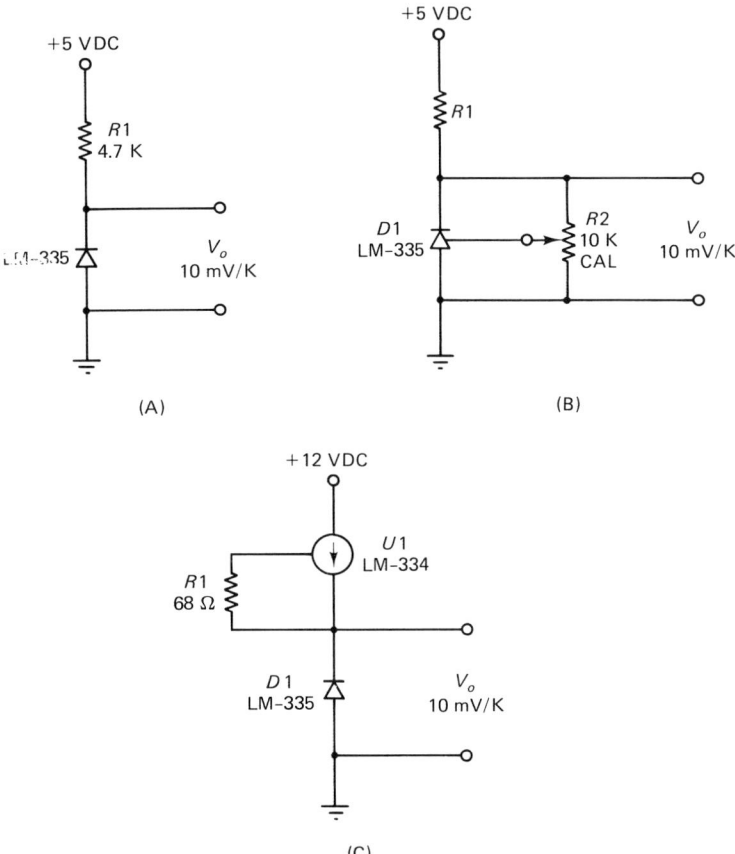

Fig. 4-12 (A) Simplest circuit for LM-335, (B) single-point calibration of the LM-335, and (C) use of LM-335 with LM-334 constant current source.

according to Ohm's law (keeping $I = 0.001$ A):

$$R_\Omega = (V+) \times 1000 \qquad (4\text{-}11)$$

For example, when the power supply voltage is $+12$ V DC, the value of the resistor in series with the LM-335 is

$$R_\Omega = (V+) \times 1000$$

$$R_\Omega = (12 \text{ V}) \times 1000 = 12{,}000 \ \Omega$$

The output signal in the circuit in Fig. 4-12(A) is taken across the LM-335 device. This voltage has an approximate rate of 10 mV/K. Recall that degrees Kelvin is similar to degrees centigrade, except that the zero point is at absolute zero (close to $-273.16°$C) rather than the freezing point of water. Using ordinary arithmetic shows us how much voltage to expect at any given temperature. For example, suppose we want to know the output voltage at 78°C. The first thing to do is convert the temperature to degrees Kelvin. This neat little trick is done by adding 273.16 to the centigrade temperature:

$$K = °C + 273.16 = 78°C + 273.16 = 351.16$$

Next, we convert the temperature to the equivalent voltage:

$$V = \frac{10 \text{ mV}}{K} \times 351 \text{ K} = (10 \text{ mV})(351) = 3510 \text{ mV} = 3.51 \text{ V}$$

One problem with the circuit of Fig. 4-12A is that it is not calibrated. While that circuit works well for many applications, especially those where precision is not needed, for other cases it might be better to consider the circuit of Fig. 4-12B. This circuit allows single-point calibration of the temperature. The calibration control is obtained from the 10-kΩ potentiometer in parallel with the sensor. The wiper of the potentiometer is applied to the adjustment input of the LM-335 device.

Calibration of the device is relatively simple. One only needs to know the output voltage (a DC voltmeter will suffice) and the environmental temperature in which the LM-335 exists. In some less than critical cases, one might take a regular glass mercury thermometer and measure the air temperature. Wait long enough after turning on the equipment for both the mercury thermometer and the LM-335 device to come to equilibrium. After that, adjust the potentiometer ($R2$) for the correct output voltage. For example, if the room temperature is 25°C (i.e., 298 K), then the output voltage will be 2.98 V. Adjust the potentiometer for 2.98 V under these conditions.

Another tactic is to use an ice-water bath as the calibrating source. The temperature 0°C (273.16 K) is defined as the point where

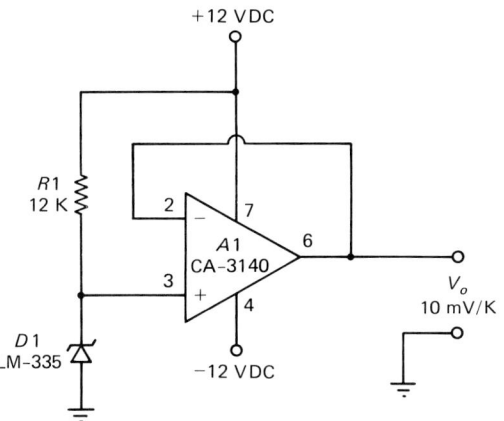

Fig. 4-13 Operational amplifier buffered LM-335 temperature sensor.

water freezes and is recognized by the fact that ice and water coexist (the ice neither melts nor freezes, it is in equilibrium). A mercury thermometer will show the actual temperature of the bath. The potentiometer is adjusted until the output voltage is 2.73 V.

Another tactic is to use a warmed oil bath for the calibration. The oil is heated to somewhat higher than room temperature (e.g., 40°C) and stirred slowly. Again, a mercury thermometer is used to read the actual temperature, and the potentiometer is adjusted to read the correct value. The advantage of this method is that the oil bath can be a constant temperature situation. There are numerous laboratory vessels on the market that will keep water or oil at a constant preset temperature.

Another connection scheme for the LM-335 is shown in Fig. 4-12(C). In this variation on the theme, we use at National Semiconductor LM-334 three-terminal adjustable current source for the bias of the LM-335 device. Again, the output voltage will be 10 mV/K.

All applications where the sensor is operated directly into its load suffer a potential problem or two, especially if the load impedance either changes or is lower than some limit. As a result, the buffered circuit of Fig. 4-13 is sometimes justified.

A buffer amplifier is one that is used for one or both of two purposes: (1) impedance transformation or (2) isolation of the circuit from its load. The impedance transformation factor is used when the source impedance is high (not true of the LM-335). The isolation factor is of somewhat more concern to us here. The operational amplifier in Fig. 4-13 places an amplifier between the sensor and its load. The gain of the amplifier in this case is unity, but a higher gain could be used if desired. In that case, simply substitute one of the gain amplifier circuits shown later in this book.

The operational amplifier shown here is a GE/RCA CA-3140 device. The reason for this is simply the freedom from bias currents exhibited by the BiMOS GE\RCA operational amplifiers. The bias currents found on many other operational amplifiers could conceivably introduce error. The CA-3140 is not the only operational amplifier that will work, however. Any low-input bias current model will work nicely.

The noninverting input of the operational amplifier is connected across the LM-335. In this respect, this circuit looks somewhat like the voltage reference circuits using zener diodes seen elsewhere. The bias for the LM-335 is derived from a 12-kΩ resistor. Because there is no voltage gain in this circuit, the output voltage factor is the same as in previous designs, 10 mV/K.

A circuit, such as in Fig. 4-13, sometimes proves useful in monitoring remote temperatures. If the operational amplifier is powered, a four-wire line is needed ($V-$, $V+$, ground, and temperature). The advantage is that the line losses are overcome by the higher output power of the operational amplifier. The LM-335 is a rugged, little, low-impedance device, however, and in many cases, such measures would not be needed.

Analog Devices AD-590

The Analog Devices, Inc., AD-590 (Fig. 4-14A) is another form of solid-state temperature sensor. This particular device is a two-electrode sensor that operates as a current source with a characteristic of 1 μA/K. The AD-590 will operate over the temperature range of -55 to $+150°C$. It is capable of a wide range of power supply voltages, being happy with anything in the range $+4$ to $+30$ V DC (this range

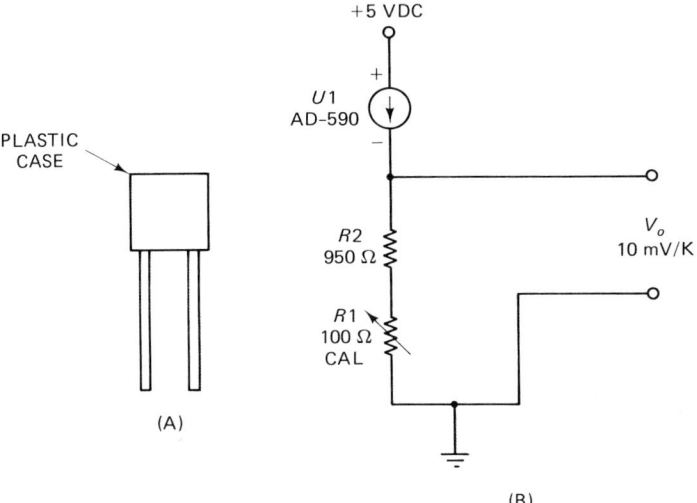

Fig. 4-14 (A) AD-590 temperature sensor; (B) simple single-point calibration of AD-590.

is more than sufficient for most solid-state circuit applications). Selected versions are available with a linearity of $\pm 0.3°C$ and a calibration accuracy of $\pm 0.5°C$.

The AD-590 comes in two different packages. There is a metal can (TO-52) that is recognized as the small size transistor package (smaller than the TO-5), and there is also a plastic flat pack available.

Being essentially a two-terminal current source, the AD-590 is simplicity itself in actual circuit operation. Figure 4-14B shows the most elementary circuit for the AD-590 that is able to be calibrated. Since it is a current source that produces a current proportional to temperature, we can convert the output to a voltage by passing it through a resistor. In Fig. 4-14B, the total resistance is approximately 1000 Ω and consists of the the resistance of $R2$ (950 Ω) and $R1$ (a 100-Ω potentiometer). From Ohm's law, we know that 1 $\mu A/K$ converts to 1 mV/K when passed through a 1000-Ω resistance. We can calculate the voltage output at any given temperature from this simple relationship:

$$V_o = \frac{1\ mV}{K} \times T \tag{4-12}$$

Thus, if we have a temperature of 37°C, which is (37 + 273) or 310 K, then the output voltage will be

$$V_o = \frac{1\ mV}{K} \times T = \frac{1\ mV}{K} \times 310\ K = 310\ mV$$

Potentiometer $R1$ is used to calibrate this system. You can make a "quick and dirty" calibration with an accurate mercury thermometer (laboratory grade recommended) at room temperature. Connect a digital voltmeter across the output and allow the system to come to equilibrium (should take about 10 min). Once the system is stable, adjust the potentiometer for the correct output voltage. For example, assume that the room temperature is 25°C, which is 77°F. This temperature converts to 298 K (272 + 25). The output voltage will be (1 mV × 298), or 298 mV (0.298 V). Using a $3\frac{1}{2}$ digit voltmeter is sufficient to make this measurement.

In some cases, it might be wise to delete the potentiometer and use a single 1000-Ω resistor in place of the network shown. There might be several reasons for this action. First, the calibration accuracy is not critical for the application at hand. Second, potentiometers are points of weakness in any circuit. Being mechanical devices, they are subject to stress under vibration conditions and may fail prematurely. If the temperature accuracy is not crucial, and reliability is, then consider the use of a single, fixed, 1% tolerance resistor in place of the network shown in Fig. 4-14B.

The circuit of Fig. 4-14B is sometimes used to make a temperature alarm. By using a voltage comparator to follow the network and biasing the comparator to the voltage that corresponds to the alarm temperature, we can create a transistor–transistor logic (TTL) level that indicates when the temperature is over the limit. A window comparator will allow us to have an alarm of either under- or overtemperature conditions. Some electronic equipment designers use this tactic to provide an overtemperature alarm. In one application, a commercial minicomputer generated a large amount of heat (it used a 65-A, +5-V DC power supply). The specification called for an air-conditioned room for housing the computer. An AD-590 device was placed inside at a critical point. If the temperature reached a certain level (45°C), then the comparator output snapped LOW and created an interrupt request to the computer. The computer would then sound an alarm and display an overtemperature warning message on the operator's CRT screen.

The circuit of Fig. 4-14B suffers from a problem: it allows calibration at only one temperature, which does not allow for optimization of the circuit. We can, however, improve the situation using the two-point calibration circuit of Fig. 4-15. In this case, we see an operational amplifier in the inverting follower configuration (see Chapter 12).

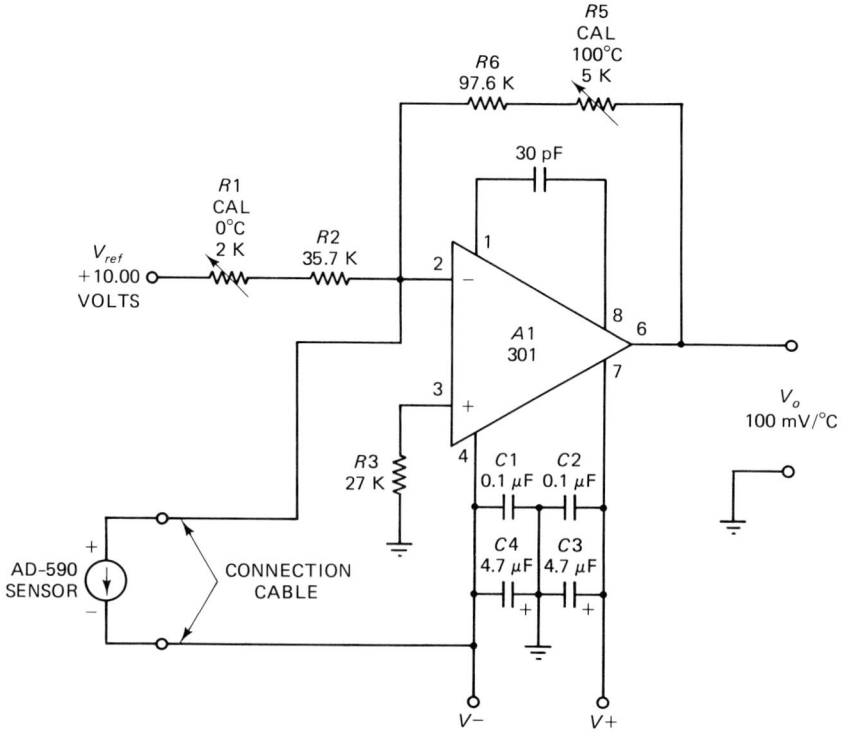

Fig. 4-15 Two-point calibration of the AD-590 sensor.

The summing junction of the amplifier (inverting input) receives two different currents. One current is the output of the AD-590 (i.e., 1 μA/K), while the other current is derived from the reference voltage V_{ref} (10.000 V). Adjustment of this current provides our zero-reference adjustment, while the overall gain of the amplifier provides the full-scale adjustment.

The operational amplifier selected is the LM-301 device, although almost any premium operational amplifier will suffice. The RCA CA-3140 BiMOS device, or some of those by either Analog Devices or National Semiconductor, will also work nicely. If the LM-301 or similar device is used, then be sure to use the 30-pF frequency compensation capacitor.

The $V-$ and $V+$ power supply lines are bypassed with 0.1- and 4.7-μF capacitors. The 0.1-μF capacitors are used for high-frequency decoupling and must be mounted as close as possible to the body of the operational amplifier. The values of these capacitors are approximate, and they may be anything from 0.1 to 1 μF.

Calibration of the device is simple, although two different temperature environments are required. The zero-degree centigrade adjustment ($R1$) can be made with the the sensor in an ice-water bath (as described previously). The upper temperature can be room temperature, provided that some means is available to measure the actual room temperature for comparison.

BIMETALLIC STRIPS

The bimetallic strip is an on–off temperature sensor that will allow the construction of a temperature-sensitive switch. An example of this form of temperature sensor is the water temperature sensor that plugs into the engine block of an automobile. When the water temperature reaches a certain level, the bimetallic strip closes a switch that lights up the TEMP or HOT alarm lamp on the dashboard.

Figure 4-16A shows the construction of a bimetallic strip thermoswitch. The two metals are selected to have radically different thermal coefficients of expansion and are bonded together. When they are heated, the two pieces of metal try to expand at differing rates, so the strip is forced into a radius of curvature, R. The value of this deflection radius is

$$R = \frac{(t1 + t2)^3}{6\delta(T2 - T1)t1t2} \qquad (4\text{-}13)$$

where R is the radius of curvature, $t1$ and $t2$ are the thicknesses of the two metal elements in the thermal strip, $T1$ is the resting temperature

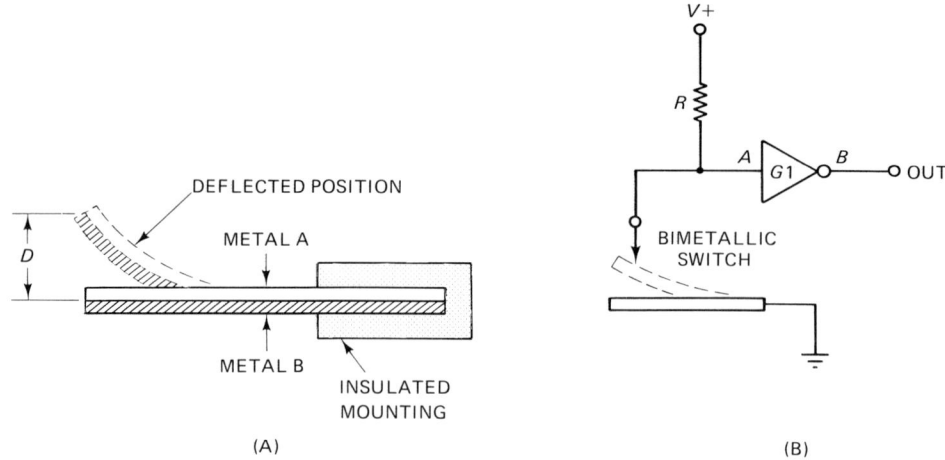

Fig. 4-16 (A) Bimetallic strip temperature sensor; (B) use of bimetallic strip as alarm sensor.

before curvature begins, $T2$ the final temperature (both $T1$ and $T2$ are in °C), and δ the difference in the thermal coefficients of expansion for the two metals.

The deflection of the end, D, is found from

$$D = \frac{KTL^2}{t} \tag{4-14}$$

where D is the deflection in inches; L the length of the strip in inches; T the temperature difference, $T2 - T1$; t the thickness of the strip in inches; and K a constant, typically 3×10^{-6} to 7×10^{-5}.

Figure 4-16B shows a typical electronic alarm circuit based on the bimetallic strip. A digital inverter, $G1$, is used as the sensor electronics. The rules of this device are simple: when the input (point A) is HIGH (near $V+$), then the output (point B) is LOW (near ground), and when the input is LOW, then the output is HIGH. Under normal conditions, below the alarm threshold, the bimetallic switch is open so the input of the gate is held HIGH by resistor R connected to the $V+$ source. Under this condition, the output of the gate is LOW. But, when the temperature passes a critical threshold, the bimetallic switch closes and the gate input is shorted to ground, so it is forced to the LOW level, and the output snaps HIGH to indicate an overtemperature condition.

5

Position, Displacement, Force, and Pressure Sensors

One of the largest classes of physical sensors on the market is used to measure position, displacement, force, and pressure. Several different techniques are used for these measurements, and we will touch on the most commonly encountered types. Specifications and examples of these devices can be found in various manufacturers' catalogs. Before discussing the sensors, however, we should first discuss some basic techniques and circuits used to form the sensors.

PIEZORESISTIVITY

All electrical conductors possess electrical resistance, which is opposition to the flow of current; resistance is measured in ohms. The resistance of any specific conductor is directly proportional to its length (see Fig. 5-1), and it is inversely proportional to its cross-sectional area. Resistance is also directly proportional to a property of the conductor material called resistivity. The relationship between length, area, and resistivity (ρ) is shown in Fig. 5-1. The equation in Fig. 5-1 clearly shows that resistance is related to length and cross-sectional area.

Piezoresistivity denotes the resistance change that takes place when either the length or area of a conductor is changed. Figure 5-1A shows a cylindrical conductor with an initial length (L_0) and a cross-sectional area (A_0). When a compression force is applied, as in Fig. 5-1B, the length reduces and the cross-sectional area increases. This situation results in a decrease in the electrical resistance. Similarly, when a tension force is applied (Fig. 5-1C), the length increases and the cross-sectional area decreases, so the electrical resistance will increase. Provided that the physical change is small, the change of

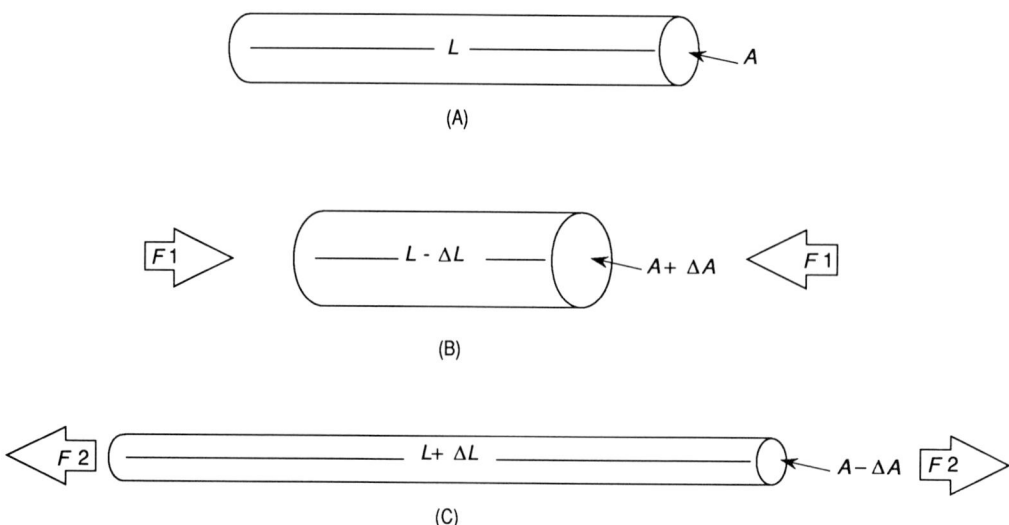

Fig. 5-1 Piezoresistivity is the change of resistance in a conductor due to mechanical deformation: (A) unstrained sample, (B) compression force applied, and (C) tension force applied. $F1$ is the compression force, and $F2$ is the tension force. $R = \rho L/A$.

electrical resistance is a nearly linear function of the applied force, so it can be used to make measurements of that force.

STRAIN GAGE ELEMENTS

A strain gage is a piezoresistive element, which may be a wire, metal foil, or semiconductor element, designed to create a resistance change when a force is applied. Strain gages can be classified as either bonded or unbonded. Figure 5-2 shows both methods of construction.

The unbonded strain gage is shown in Fig. 5-2A and consists of a wire resistance element stretched taut between two flexible supports. These supports are configured in such a way as to place a tension or compression force on the taut wire when external forces are applied. In the particular example shown, the supports are mounted on a thin metal diaphragm that flexes when a force is applied. Force $F1$ will cause the flexible supports to spread apart, placing an increased tension force on the wire and thereby increasing its resistance. Alternatively, when force $F2$ is applied, the ends of the supports tend to move closer together, effectively placing a compression force on the wire element and thereby reducing its resistance. In actuality, the wire's resting condition is tautness, which implies a tension force. So $F1$ increases the tension force from normal, and $F2$ decreases the normal tension.

The bonded form of strain gage is shown in Fig. 5-2B. In this type of device, a wire, foil, or semiconductor element is cemented to a thin

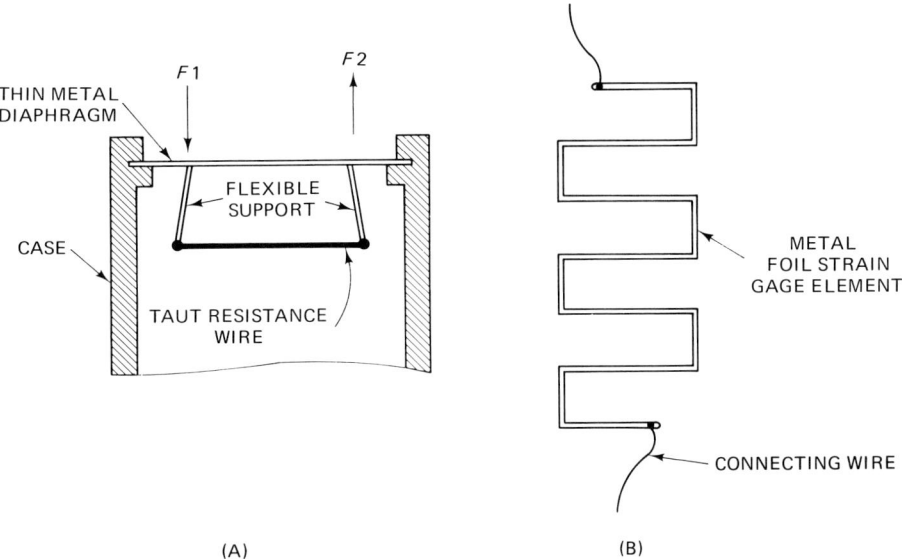

Fig. 5-2 Piezoresistive strain gages: (A) unbonded and (B) bonded.

metal diaphragm. When the diaphragm is flexed, the element deforms to produce a resistance change.

The linearity of both types of strain gage can be quite good, provided that the elastic limits of the diaphragm and element are not exceeded. It is also necessary to ensure that the change of length is only a small percentage of the resting length.

In the past, the standard wisdom held that bonded strain gages are more rugged, but less linear than unbonded models. Although this may have been the situation at one time, recent experience has shown that modern manufacturing techniques can produce rugged, linear, reliable units of both types of construction.

THE WHEATSTONE BRIDGE CIRCUIT

The Wheatstone bridge is a nineteenth-century holdover that finds a home in many modern electronic circuits. The classic form of Wheatstone bridge is shown in Fig. 5-3. There are four resistive arms to the bridge, each arm being labeled $R1$, $R2$, $R3$, and $R4$. The excitation voltage (V) is applied across two of the bridge nodes, while the signal is taken from the alternate two nodes (labeled C and D). We can consider this circuit as two series voltage dividers in parallel, one consisting of $R1$ and $R4$ and the other of $R2$ and $R3$ (see Fig. 5-4).

The output voltage from a Wheatstone bridge is the difference between the voltages at points C and D. When all of the arithmetic is finished, we find that the output voltage will be zero when the ratio $R4/R1$ is equal to the ratio $R3/R2$. If these ratios are not kept equal,

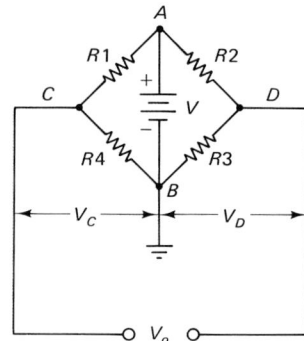

Fig. 5-3   Wheatstone bridge circuit.

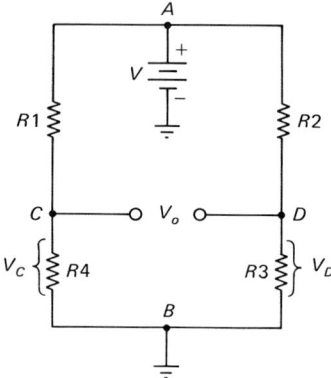

Fig. 5-4   Redrawn Wheatstone bridge circuit.

as is the case when one or more of the elements is a strain gage not at rest, then an output voltage is produced that is proportional to both the applied voltage and the change of resistance.

STRAIN GAGE CIRCUITRY

Before the resistive strain gage (or other form of resistive sensor) can be useful, it must be connected into a circuit that will convert its resistance changes into a current or voltage output. Most applications are voltage output circuits.

Figure 5-5 shows several popular forms of circuit. The circuit in Fig. 5-5A is both the simplest and least useful (although not useless); it is sometimes called the half-bridge circuit, or voltage divider circuit. The strain gage (SG) element of resistance R is placed in series with a fixed resistor, $R1$, across a stable DC voltage, V. The output voltage V_o is found from the simple voltage divider equation

$$V_o = \frac{VR}{R + R1}$$

(5-1)

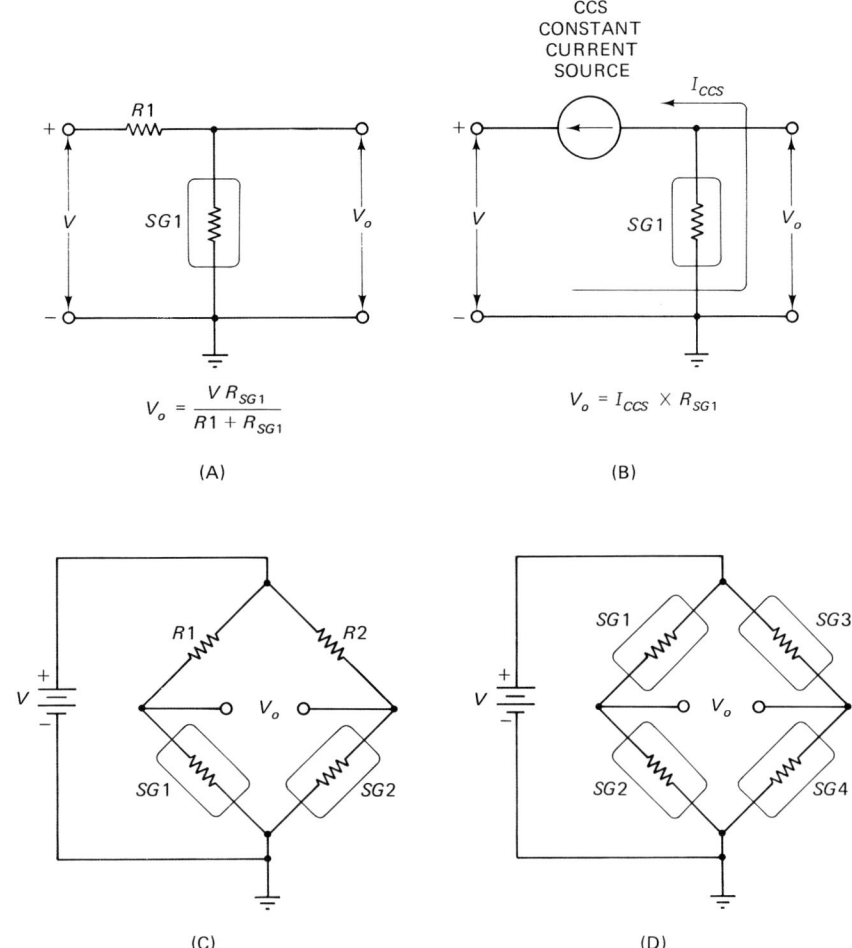

Fig. 5-5 Strain gage circuits: (A) voltage excited half-bridge, (B) current excited half-bridge, (C) two-element Wheatstone bridge, and (D) four-element Wheatstone bridge.

Equation (5-1) describes the output voltage V_0 when the sensor is at rest (i.e., nothing is stimulating the resistive element). When the element is stimulated, however, its resistance changes a small amount, h. The output voltage in that case is

$$V_0 = \frac{V(R + h)}{(R \pm h) + R1} \qquad (5\text{-}2)$$

Another form of half-bridge circuit is shown in Fig. 5-5B, but in this case, the strain gage is connected in series with a constant current source (CCS), which will maintain the current I at a constant level, regardless of changes in the strain gage resistance. In this case, $V_0 = I(R \pm h)$.

Both of the half-bridge circuits suffer from one major defect: output voltage V_o will always be present regardless of the value of the stimulus applied to the sensor. Ideally, in any sensor system, the output voltage should be zero when the applied stimulus is zero. For example, when a gas pressure sensor is open to atmosphere, then the gage pressure is zero, so the output voltage should also be zero. Also, the output voltage should be proportional to the value of the stimulus when the stimulus is not zero. A Wheatstone bridge circuit can have these properties. We can use strain gage elements for one, two, three, or all four arms of the Wheatstone bridge.

Figure 5-5C shows a circuit in which two strain gages (SG1 and SG2) are used in two arms of a Wheatstone bridge, with fixed resistors $R1$ and $R2$ forming the alternative arms of the bridge. It is usually the case that SG1 and SG2 are configured so that their actions oppose each other; that is, under stimulus, SG1 will have resistance $R + h$, and SG2 will have resistance $R - h$, or vice versa.

One of the most linear forms of sensor bridge is the circuit of Fig. 5-5D in which all four bridge arms contain strain gage elements. In most such sensors, all four strain gage elements have the same resistance (R), which will usually be a value between 50 and 1000 Ω.

Recall that the output from a Wheatstone bridge is the difference between the voltages across the two half-bridges. We can calculate the output voltage for any of the standard configurations from the equations given here (assuming all four bridges have nominally the same resistance, R):

One active element:

$$V_o = Vh/4R \qquad \text{(accurate to } \pm 5\% \text{ if } h < 0.1)$$

Two active elements:

$$V_o = Vh/2R$$

Four active elements:

$$V_o = Vh/R$$

where V_o is the output potential in volts (V), V the excitation potential in volts (V), R the resistance of all bridge arms, and h the quantity ΔR, which is the change in resistance in response to the applied stimulus.

SENSOR SENSITIVITY

The sensitivity factor (P) relates the output voltage (V) to the applied stimulus value (Q) and the excitation voltage. In most cases, the sensor maker will specify a number of microvolts (or millivolts) output potential per volt of excitation potential per unit of applied stimulus. In other words,

$$P = V_o/V/Q_0 \qquad (5\text{-}3)$$

or, written another way,

$$P = V_o/VQ \qquad (5\text{-}4)$$

where V_o is the output potential, V the excitation potential, and Q one unit of applied stimulus.

If we know the sensitivity factor, then we can calculate the output potential as follows:

$$V_o = PVQ. \qquad (5\text{-}5)$$

Equation (5-5) is the one that is most often used in circuit design.

Example

A certain fluid pressure sensor is often used for measuring human and animal blood pressures through an indwelling catheter. It has a sensitivity (P) of 5 $\mu V/V/Torr$, which means 5 μV output potential is generated per volt of excitation potential per Torr of pressure. Find the output potential when the excitation potential is $+7.5$ V DC and the pressure is 400 Torr (the usual high-end limit for such sensors):

$$V_o = PVQ$$

$$= \frac{5\mu V}{V \cdot Torr}(7.5\text{ V})(400\text{ Torr})$$

$$= (5 \times 7.5 \times 400)\ \mu V$$

$$= 15{,}000\ \mu V\ (\text{which is 15 mV or 0.015 V.})$$

BALANCING AND CALIBRATING A BRIDGE SENSOR

Few, if any, Wheatstone bridge sensors meet the ideal condition in which all four bridge arms have exactly equal resistances. In fact, the bridge resistance specified by the manufacturer is only a nominal value, and the actual value may vary quite a bit from the specified value. There will inevitably be an offset voltage (i.e., V_o is not zero when Q is zero). Figure 5-6 shows two circuits that will balance the bridge when the stimulus is zero.

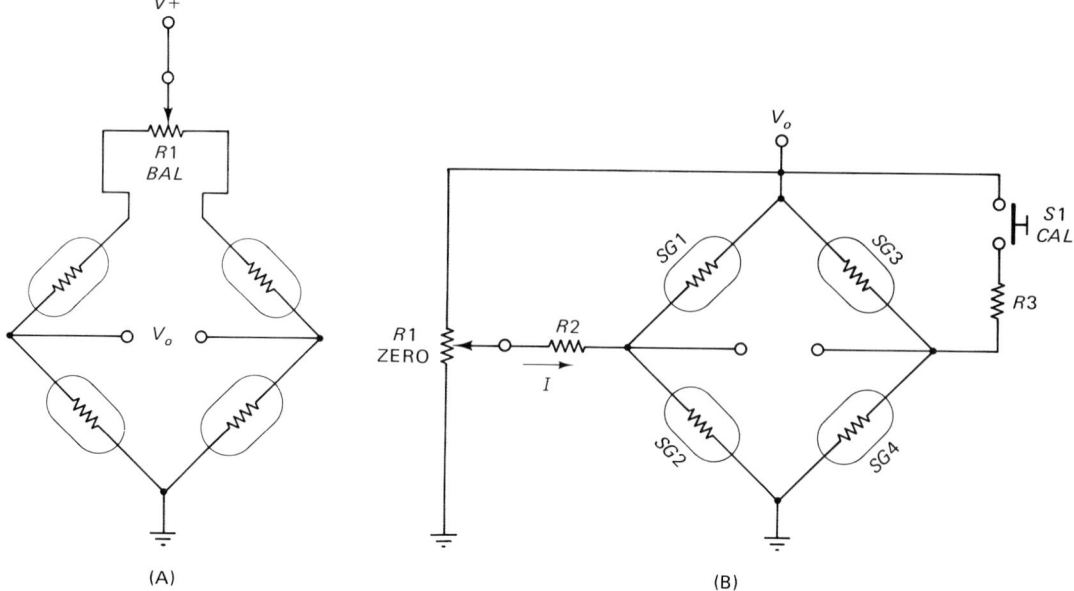

Fig. 5-6 (A) Crude bridge balancing scheme; (B) bridge balancing by injecting current into a bridge node.

In Fig. 5-6A, the balancing potentiometer is placed between the excitation potential and the excitation node. The resistance balance of the potentiometer is varied between the two legs of the bridge, nullifying any differences between them. The potentiometer is usually a precision type with 5 to 15 turns to cover the entire range.

In Fig. 5-6B, the purpose of the potentiometer is to inject a balancing current (I) into the bridge circuit at one of its nodes. $R1$ is adjusted, with the stimulus at zero, for zero output voltage.

Another application for this type of circuit is injecting an intentional offset potential. For example, on a digital scale, such a circuit is used to adjust for the tare weight of the scale, which is the sum of the platform and all other weights acting on the sensor when nobody is standing on the scale. This is also sometimes called empty weight compensation.

Calibration can be accomplished either the hard way or the easy (and less accurate) way. The easy way is to connect a calibration resistor across one arm of the circuit. The hard way is to set up the sensor in a system and apply the stimulus. The stimulus is measured, and the result is compared with the sensor output. For example, if you are testing a pressure sensor, connect a manometer (pressure measuring device containing a column of mercury) and measure the pressure directly. The result is compared with the sensor output. All sensors should be tested in this manner initially when placed in service and then periodically thereafter.

SENSOR CONSTRUCTION

Although many forms of construction are used in sensor manufacture, we can examine a generic force/pressure sensor in order to get a general idea of how it is done. Figure 5-7 shows a cutaway view of a typical bonded strain gage force/pressure sensor. This particular model uses a pair of strain gage elements ($R2$ and $R4$) and two fixed resistors in a Wheatstone bridge configuration. The case is a rigid structure that provides support for the thin metallic diaphragm and protection for the internal circuitry. The piezoresistive elements, $R2$ and $R4$, are cemented to the thin metallic diaphragm. When a force or pressure is applied to the diaphragm it distends and thereby applies strain to the strain gage elements.

In addition to the components shown in Fig. 5-7, there may also be other components. In some models, for example, scaling resistors are used. These resistors normalize the output voltage to account for

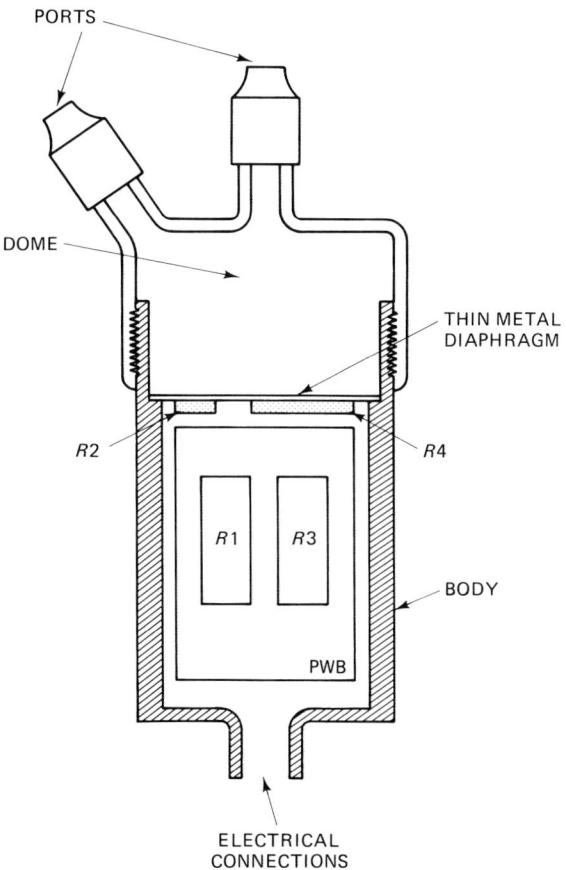

Fig. 5-7 Cutaway view of a pressure transducer based on bonded strain gages. (PWB ≡ printed wiring board.)

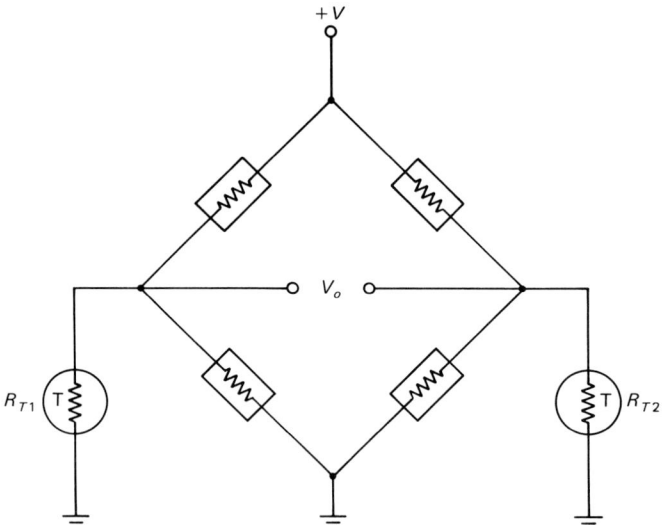

Fig. 5-8 Temperature compensated strain gage bridge.

variations in strain gage sensitivity. One commercial -100 to $+450$ Torr sensor, for example, advertises a sensitivity of 50 μV/V/Torr. In actual practice, however, the manufacturer packs a calibration sheet with each model, and in one lot of 10 sensors (all the same model), I found calibrations ranging from 30 to 60 μV/V/Torr.

Temperature compensation components are also part of the internal circuitry. Uncompensated sensors tend to drift with changes in temperature. In one case, I recall a pressure measuring servomechanism used in a medical research application in which the drift was so bad that the experimenter had to arrive up to four hours early in order to turn on the equipment to allow it to equilibrate. In that case, the temperature drift was due in large part to the electronics and to the sensor. The sensor portion could be compensated using a method similar to that shown in Fig. 5-8. The values and temperature coefficients of R_{T1} and R_{T2} depend upon the dV_o/dT experienced and the values of the sensor arm resistors (R).

SENSOR LINEARIZATION

Sensors are not perfect devices. Although the output function should be linear in a perfect world, real sensors are often highly nonlinear. For Wheatstone bridge strain gages, the constraints on linearity include making ΔR (called h in some equations) very small (5% or less) compared with the at-rest resistance.

There are several forms of linearization techniques used in sensor systems. An analog method is shown in Fig. 5-9. Here, we modify the circuit of the usual single strain gage Wheatstone bridge circuit. The

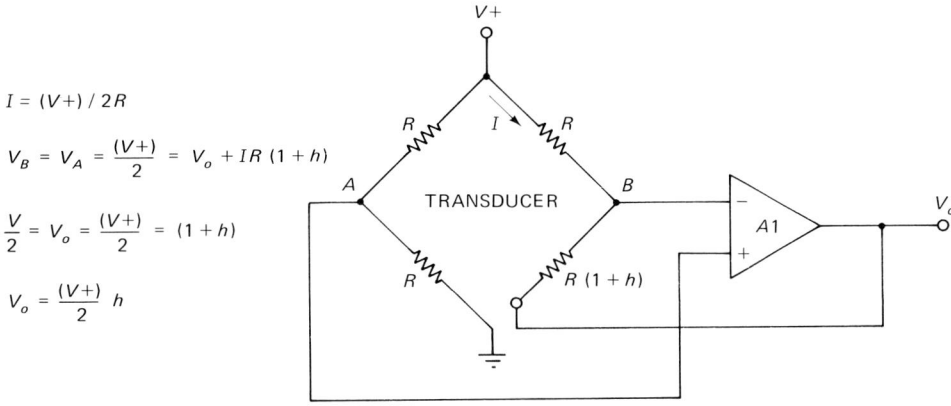

$$I = (V+) / 2R$$

$$V_B = V_A = \frac{(V+)}{2} = V_o + IR\,(1 + h)$$

$$\frac{V}{2} = V_o = \frac{(V+)}{2} = (1 + h)$$

$$V_o = \frac{(V+)}{2}\,h$$

Fig. 5-9 Partially linearized strain gage bridge.

ground end of one bridge resistor is lifted and applied to the output of a null-forcing amplifier, $A1$ (which is not the normal bridge amplifier as shown elsewhere). In this case, the resistor element $R(1 + h)$ is in the feedback network of operational amplifier $A1$. Small amounts of non-linearity are canceled with this circuit.

For larger nonlinearities, we must resort to other methods. Figure 5-10 shows a hypothetical sensor transfer function in which a voltage,

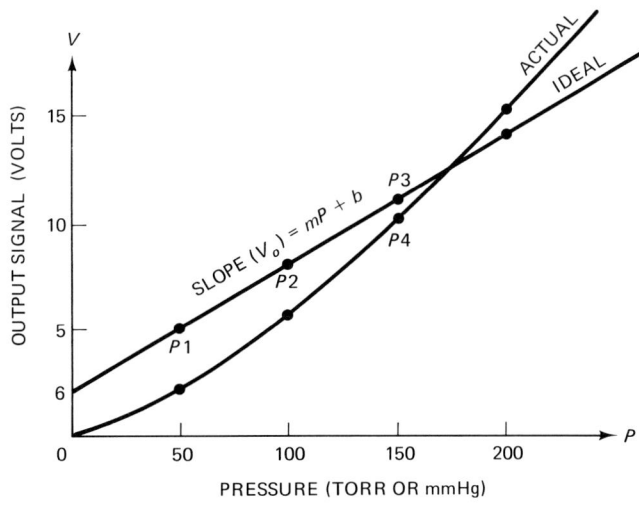

POINT	PRESSURE (TORR)	IDEAL	ACTUAL
P1	50	5	2
P2	100	8	6
P3	150	11	10
P4	200	14	16

Fig. 5-10 Calibration curves for ideal and actual pressure sensor.

V_o, is a function of applied pressure, P. The perfect sensor will obey the usual equation for a straight line, $V_o = mp + b$, in which m is the slope of the line and b is the offset. The actual curve may be a lot less straight.

Before digital computers were routinely used in instrumentation applications, we often used either special function circuits or diode breakpoint generators for linearization of the sensor. In cases where a special function circuit was used, the assumption was that the equation of the actual curve was known. The special function circuit generated the inverse of that function and summed it with the input voltage. In the case of the diode breakpoint generator, an offset voltage was added to or subtracted from the actual input signal to normalize it to the ideal. This method is piecewise linear, and its validity is dependent on the number of segments (one per breakpoint circuit)

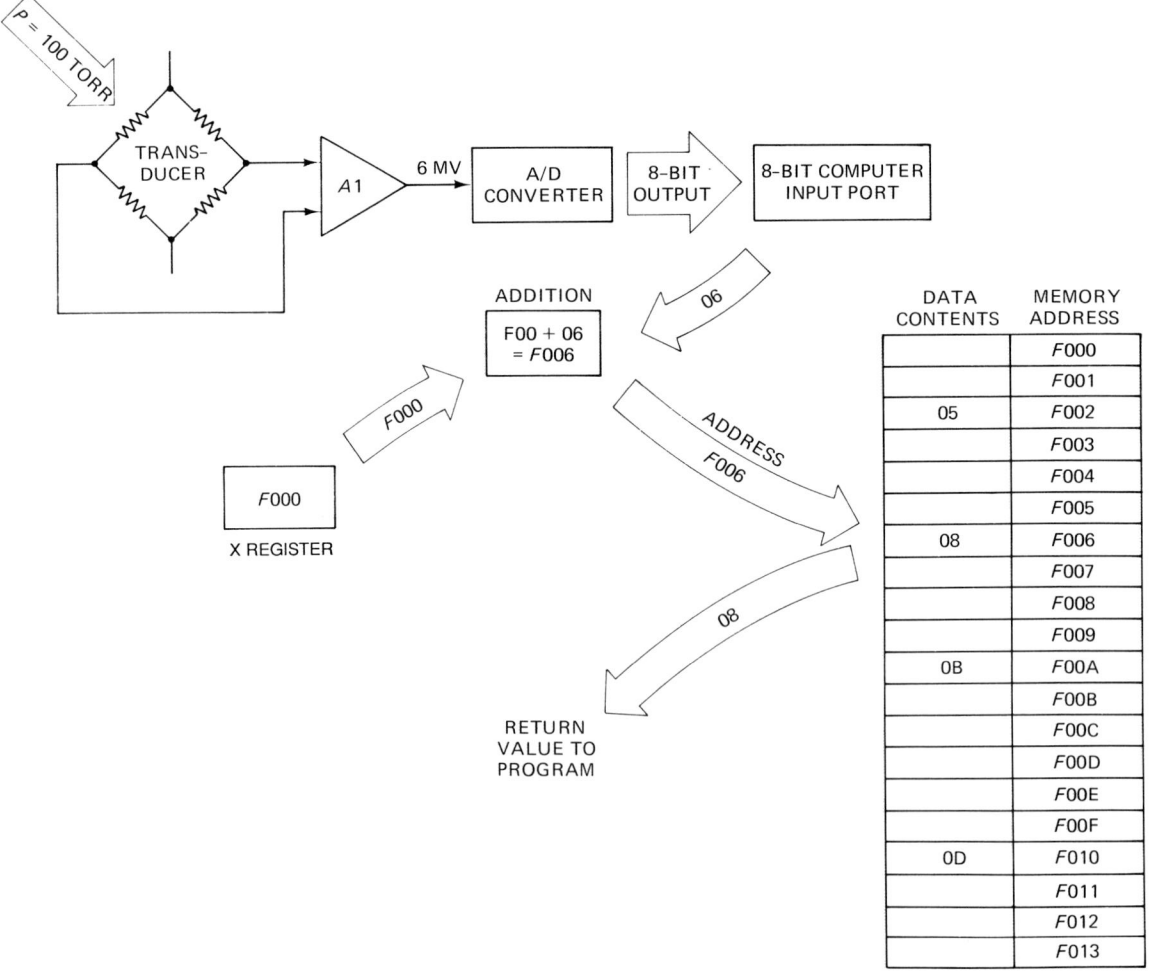

Fig. 5-11 Computer lookup table method for linearization.

used. Neither of these methods was particularly good—actually, both were often quite poor (thermal drift is a particular problem in breakpoint circuits).

Now that microprocessors are routinely used in instruments, a software method can be used to correct the error. If the equation defining the actual curve is known, then it is possible to write a software program that cancels the error. Alternatively, one could use the lookup table method of Fig. 5-11. This example shows only a limited number of data points for simplicity sake, but the actual number would depend upon the bit length of the A/D converter.

The values for the ideal transfer function are stored in a lookup table that begins at location HF000 in memory. The value HF000 is stored in the X register. When the A/D binary word is input to the computer, it is added to the contents of the X register. This value becomes the indexed address in the lookup table, where the correct value is found. Although a pressure sensor example is shown here, it is useful for almost any form of sensor.

SENSOR EXCITATION SOURCES

The DC Wheatstone bridge sensor requires a source of either AC or DC excitation voltage, with DC being the most common. Most sensors require an excitation voltage of 10 V DC or less. This voltage is critical, and exceeding it will create a very short life expectancy for the sensor. A typical fluid pressure sensor requires +7.5 V DC and operates best (least thermal drift that still produces usable output signal) at +5 V DC. A source of DC excitation that is stable, within specifications, and precise (in some cases) must be provided.

Non-DC Excitation

Although most Wheatstone bridge sensors are DC-excited, there are cases where non-DC sources are used. In Fig. 5-12, we see a sensor with pulsed excitation. A short duty cycle pulse train [typically 1000 to 5000 pulses per second (PPS)] is applied to the sensor in lieu of the DC source. The amplifier output is also a pulse train and is usually applied to an operational amplifier integrator. The voltage output of the integrator is a function of the repetition rate of the pulses (which is fixed) and the amplitude of the pulses.

AC excitation is shown in Fig. 5-13. The principal advantage of this system is that AC amplifiers can be made a lot more stable at the signal levels delivered by sensors than DC amplifiers. It is often the case that amplifier drift is of the same magnitude as the stimulus signal, which obscures the reading with considerable error. The AC

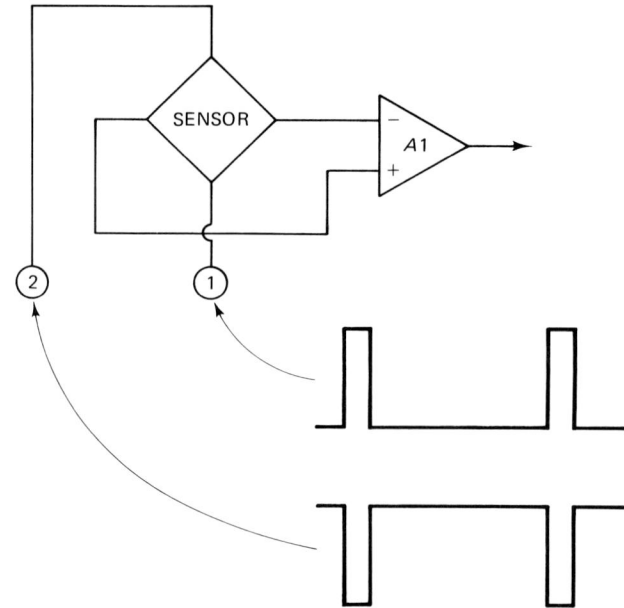

Fig. 5-12 Pulse excitation of strain gage bridge.

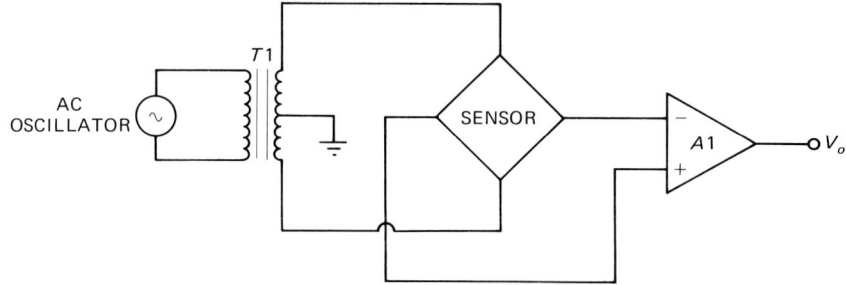

Fig. 5-13 AC carrier oscillator excited strain gage bridge.

amplifier eliminates some of that error by using negative feedback to improve stability.

The AC excitation method of Fig. 5-13 typically requires a synchronous, phase-sensitive detector driven by the same AC signal used to excite the sensor bridge. The filtered output of the phase-sensitive detector is a DC voltage that is proportional to the applied stimulus.

DC Excitation Sources

The simplest form of sensor excitation is the zener diode circuit in Fig. 5-14. A zener diode will regulate the voltage to a close tolerance that is sufficient for many applications. There are two problems with this circuit, however. First is the fact that the zener potential does not have a nice even value such as 5.0 V, but will have a value such as 4.7, 5.6,

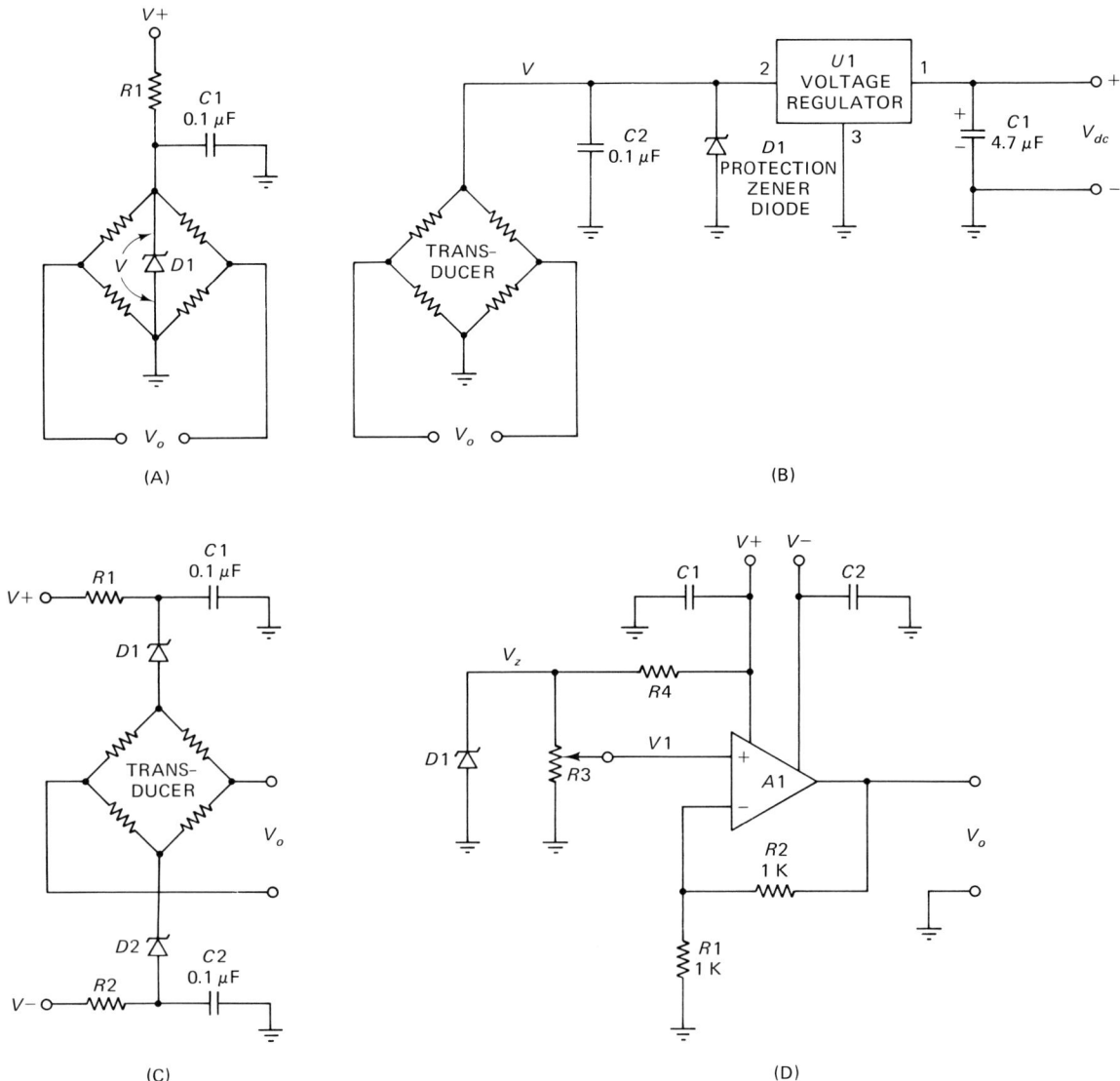

Fig. 5-14 DC excitation of strain gage bridges: (A) simple zener diode, (B) three-terminal voltage regulator, (C) bipolar zener diodes, and (D) variable precision supply.

6.2, or 6.8 V. The second defect is thermal drift. The zener voltage may vary somewhat with temperature in all but certain reference grade (band gap) zener diodes. Unless the application is not critical or the diode can be kept at a constant temperature, the method of Fig. 5-14A is not generally suitable.

Figure 5-14B shows a second method. In this case, the regulator is a three-terminal IC voltage regulator ($U1$) of the LM-309, LM-340, 78xx, or similar families. In many cases, the H version of the regulator (100 mA) can be used, although in others the 750-mA T-package devices must be specified. The selection depends upon the current normally drawn by the sensor, which is (V/R), where V is the regulator output voltage and R is the resistance of any one sensor element. In a typical case, the sensor will use a $+5$-V excitation potential. If the resistance of the R elements is $> 50\ \Omega$, then the current will be less than 100 mA. In that case, we can use a 100-mA LM-309H, LM-340H, and so forth.

The zener diode $(D1)$ in Fig. 5-14B is not used for voltage regulation, but rather for protection of the sensor. If the regulator $(U1)$ fails, then $+8$ to $+16$ V from the V + line will be applied to the sensor—which is fatal! The purpose of $D1$ is to clamp the voltage to a value that is greater than the excitation voltage, but less than the maximum allowable voltage rating of the sensor. In some cases, a small fuse is inserted in series with the input (pin number 1) of $U1$. The value of this fuse is set to roughly twice the current requirements (V/R) of the sensor and will blow if the zener diode voltage is exceeded. The fuse will add a certain amount of protection.

Some applications require a dual-polarity power supply. Figure 5-14C shows a version in which two zener diodes are used, one each for positive and negative polarities.

Neither the zener circuit nor the three-terminal regulator circuit will deliver precise output voltages. The voltage will be stable (that is, constant), but probably not precise. A typical three-terminal IC voltage regulator output voltage, for example, may vary several percent from sample to sample. If we need a precise voltage, then a circuit such as Fig. 5-14D might be used. This circuit is basically a standard operational amplifier voltage reference circuit in which the op-amp is a high current model, the National Semiconductor LM-13080.

The output voltage from Fig. 5-14D is determined by $R1$, $R2$, the setting of $R3$, and the value of zener diode, $D1$. The voltage V will be

$$V = (V1)(R2/R1) + 1$$

or since $R1 = R2 = 10\ \mathrm{k}\Omega$,

$$V = 2(V1)$$

The voltage $V1$, at the noninverting input of $IC1$, is a fraction of the zener voltage that depends upon the setting of potentiometer $R3$. We can adjust $V1$ from 1.13 to 6.8 V DC, so the sensor voltage can be set at any value from 2.26 to 13.6 V DC. In most cases, set V at 5.00, 7.50, or 10.00 V, depending upon the nature of the sensor.

SENSOR AMPLIFIERS

The basic DC differential amplifier is the most commonly used circuit for amplifying sensor signals. Fortunately, such amplifiers are easily

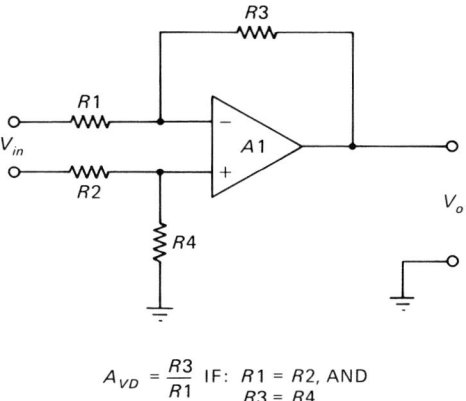

$$A_{VD} = \frac{R3}{R1} \quad \text{IF:} \quad \begin{array}{l} R1 = R2, \text{ AND} \\ R3 = R4 \end{array}$$

Fig. 5-15 DC differential amplifier for strain gage bridges.

constructed from simple operational amplifiers; Fig. 5-15 shows such a circuit (see Chapter 14 for more information on these circuits). Assuming that $R1 = R2$ and $R3 = R4$, the gain of the amplifier will be $R4/R2$, or $R3/R1$. The amplifier output voltage will be found from

$$V_o = (V_{in})(R3/R1)$$

where V_o is the amplifier output voltage, V_{in} the sensor output voltage, and $R1$ and $R3$ are the resistors in the amplifier circuit.

The amount of gain required from the amplifier is determined from a scale factor (SF), which is the ratio between the voltage representing full scale at the output of the amplifier and the voltage representing full scale at the output of the sensor:

$$SF = \frac{\text{voltage } V_o \text{ representing full scale}}{\text{sensor output voltage } V_{in} \text{ representing full scale}}$$

The amplifier output voltage required will depend upon the desired display method. For example, a strip-chart recorder might have a full-scale voltage range of 0.5 V, 1.0 V, or some such value. Alternatively, a digital panel meter (DPM) can be used for the output display. Most low-cost DPMs have a 0 to 1999 mV range, so one gains a great deal of utility by making the output voltage at full scale numerically the same at the DPM reading, for example, 1000 Torr being represented by 1000 mV. In that case, the DPM scale factor would be 1 mV/Torr, which is easy for humans to read. In that case, the "voltage V_o representing full scale" in the preceding equation is 1000 mV. The gain of the amplifier is the SF described earlier:

$$SF = \frac{\text{voltage } V_o \text{ representing full scale}}{\text{sensor output voltage } V_{in} \text{ representing full scale}}$$

$$= \frac{1000 \text{ mV}}{347.5 \text{ mV}} = 2.878$$

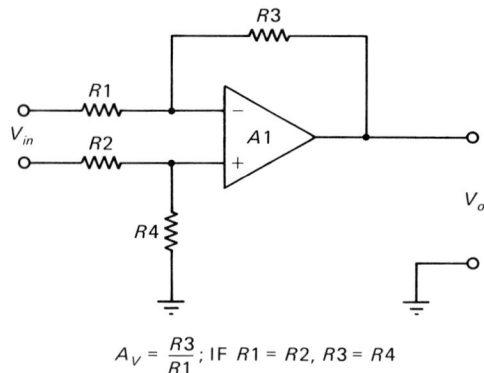

$$A_V = \frac{R3}{R1} ; \text{ IF } R1 = R2, R3 = R4$$

Fig. 5-16 DC differential amplifier.

Thus, a gain (SF) of 2.878 will provide the needed gain, so the ratio $R3/R1$ in Fig. 5-16 must be 2.878.

INDUCTIVE SENSORS

The other sensors covered in this chapter are resistive devices based on piezoresistive strain gages. In this section, we will look at two forms of inductive sensors.

In both examples, the sensor consists of a pair of inductors wound around a movable, permeable core. The core helps determine the inductance of each coil. When the core is evenly spaced between the two coils, the respective inductances are equal.

Figure 5-17 shows an inductive Wheatstone bridge sensor. The arms of the bridge comprise fixed resistors $R1$ and $R2$, plus inductors $L1$ and $L2$. The inductive reactances (X_L) of $L1$ and $L2$ are a function of the applied AC excitation frequency and the inductance of $L1$ and $L2$. When $L1 = L2$, for example, when the sensor is at rest, then the

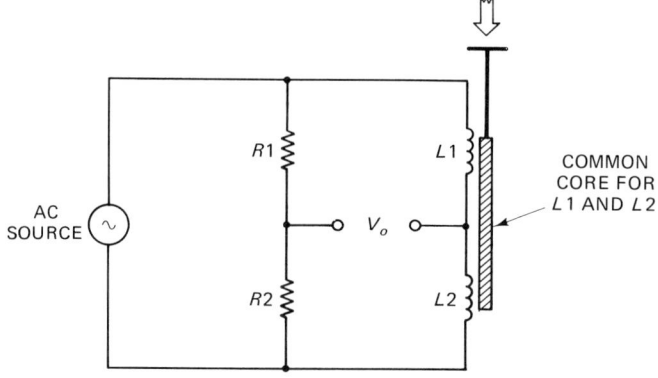

Fig. 5-17 Inductive bridge sensor.

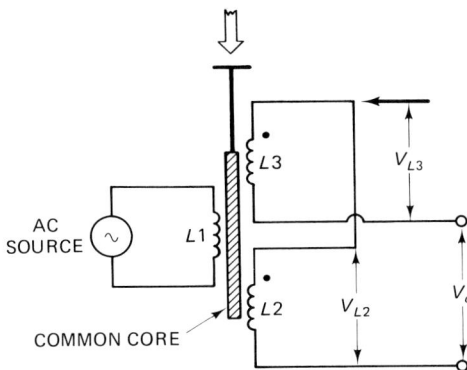

Fig. 5-18 Linear voltage differential transformer.

bridge is at null and V_o is zero. When the coil core is moved, as when a pressure or force is applied, the relative inductances of $L1$ and $L2$ change, so the reactances are no longer equal, and the bridge is unbalanced an amount proportional to the applied stimulus.

A linear variable differential transformer (LVDT) sensor is shown in Fig. 5-18. In this case, there are three coils because AC excitation is applied to the system via $L1$. Coils $L2$ and $L3$ are equal, and when the core is equally placed between the two, their inductances are also equal. But, when the core moves, the inductances are not equal. The operation of the LVDT depends upon the fact that the two output coils are connected in series opposing fashion such that the total output voltage is the algebraic difference. When $V_{L2} = V_{L3}$, the sum total voltage output is zero. Only when a stimulus is applied, when these voltages are not equal, will there be an output voltage.

Inductive sensors can be used as displacement sensors, but they have a very small range that is limited to the range of travel of the internal core. This distance can be improved, however, by use of a lever arm, reducing gears (angular displacement cases), or mechanical linkage.

PRESSURE AND FORCE SENSORS

The measurement of pressures is widespread in medicine, science, and industry. Although many of the specifics vary from application to application, many of the underlying principles are the same. Indeed, measuring fluid pressures, the difference between a chemical pipeline in a plant, a water main under the street, or the blood in the artery of a surgical patient is less a matter of approach and basic theory than of specific hardware selections.

Gas pressures are treated similar to fluid pressures, so they are also included in this discussion. In fact, one often uses air pressure over atmosphere to calibrate fluid pressure measurement instruments.

Given a rigorous definition of fluid, we must consider both liquids and gases. The difference between liquids and gases is that gases are compressible while the liquids are not, and compressibility affects the measurement technique.

WHAT IS PRESSURE?

Most people have some idea of how pressure is defined, but all too often, we find even practicing engineers have a poor idea of what they are measuring in their pressure monitors. For example, a group of engineering students was asked to define pressure only one year after taking Physics I and II. Several of them gave the correct answer, but most wrote hazy, ambiguous statements that indicated to the professor that they did not really understand what is meant by the concept. Some students came close, indicating that pressure is a force. This definition is still not correct: the proper definition is that pressure is force per unit area:

$$P = F/A$$

where P is the pressure in newtons per square meter (N/m^2) or pascals (Pa; $1\ N/m^2 = 1$ Pa), F the force in newtons, and A the area in square meters.

Pressure can be increased by either increasing the applied force or reducing the cross-sectional area over which the force operates. Alternate units for pressure are, using the CGS system, dynes per square centimeter (dyn/cm^2) and, using the British engineering system, pounds per square inch ($1b/in.^2$ or psi).

When the force in any system is constant or static (that is, nonvarying), then that pressure is said to be hydrostatic. If the force is varying, on the other hand, the force is said to be dynamic or hydrodynamic. Pressures in a fluid pipeline and physiological pressures (e.g., human arterial blood pressure) are examples of hydrodynamic pressures; the pressure head in a stoppered keg of beer is a hydrostatic pressure, at least until the bung is popped.

Pascal's principle (after French scientist and theologian Blaise Pascal, 1623–1662) governs pressures in closed systems. This physical law states that pressure applied to an enclosed fluid is transmitted undiminished to every portion of the fluid and to the walls of the containing vessel.

If a pressure is applied to the stoppered system (e.g. the syringe in Fig. 5-19), then, the same pressure is felt throughout the interior of the syringe. Changing the applied pressure at the rear of the plunger causes the same change to be reflected at every point inside the syringe.

Pascal's principle always holds true in hydrostatic systems. In hydrodynamic systems, it holds true only for quasistatic changes, that

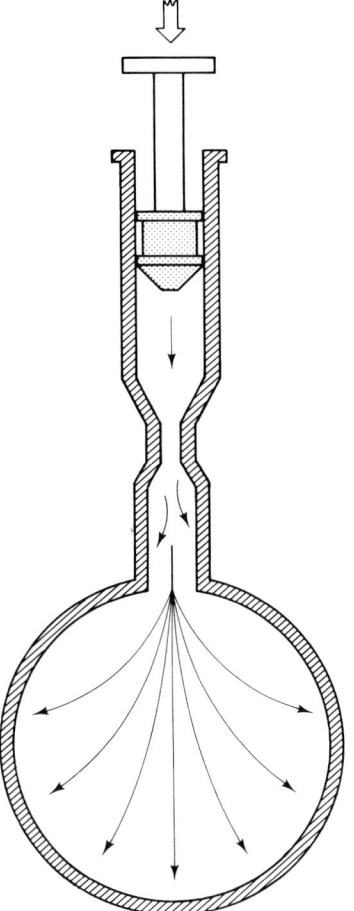

Fig. 5-19 Pressure applied to one end of a closed system is felt throughout the system.

is, when a very small change is made and the turbulence is allowed to die down before subsequent measurements are made. Pascal's principle holds approximately true for those hydrodynamic systems where the flow is reasonably nonturbulent and the pipe lumen is small compared with its length. The simple model holds true in those cases, however, only in the center of the flow mass, but not at the pipe wall boundaries. The study of pressure in turbulent or large lumen systems or in the boundary area close to the pipe or vessel wall is the subject of engineering mechanics and physics courses. We will assume that Pascal's principle either holds true absolutely or that the system can be made quasistatic for measurement or analysis purposes.

Pulsatile pressure systems result from a pumping action that is not constant (which includes most mechanical pumps). A piston pump or bellows pump, for example, places a pulsatile pressure waveform on the system. In physiological systems, the heart of the subject animal (or human) beats in a manner that produces a pulse flow

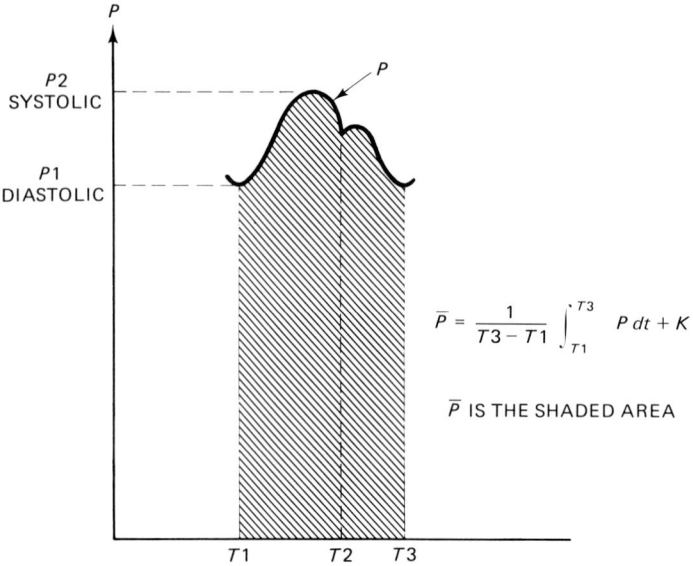

Fig. 5-20 Human arterial pressure waveform. The systolic is the maxima, the diastolic is the minima, and the mean arterial pressure (MAP) is the area under the curve (shaded portion).

(which can be felt with the finger tips where arteries run close to the surface, in the wrist, for example). Figure 5-20 shows the human arterial blood pressure waveform, here used as an example of pulsatile systems. There are several values that can be measured in this system:

1. peak pressure (called systolic in medical jargon),
2. minimum pressure (called diastolic in medical jargon),
3. dynamic average (one-half peak minus minimum), and
4. average pressure (i.e. time integral of P).

When one discusses pressure in a pulsatile system, one must also specify what pressure is intended! In a later section, we will discuss the methods used for electronically measuring these pressures.

An engineer designing a pressure measurement system for nonengineering personnel will want to consider the point of view of the client. Biomedical engineering personnel have a constant problem with both clinicians and researchers regarding the average pressure readings on the electronic blood pressure monitors. The problem was in the operational medical definition of mean arterial pressure (i.e., the time average of pressure). The correct definition, which is used in the design of the typical instrument, is written as

$$\overline{P} = \frac{1}{T3 - T1} \int_{T1}^{T3} P \, dt$$

In medical and nursing schools, and in typical intensive care unit nursing courses, however, a synthetic (and sometimes incorrect) definition is used (referring to Fig. 5-20):

$$\bar{P}' = \frac{(P2 - P1)}{3} + P1$$

In medical terminology, this definition states that the mean arterial pressure (or MAP) is equal to the diastolic pressure ($P1$) plus one-third the difference between the systolic and diastolic pressures ($P2 - P1$).

The problem faced by the biomedical engineer is that the synthetic definition is merely an approximation of the functional definition (the integral) for healthy people! In many sick people, however, the portion of the waveform between the dicrotic notch (time $T2$ in Fig. 5-20) and time $T3$ is very heavily damped, so the actual MAP is considerably less than the functional MAP actually measured by the electronic instrument. The simple test, which is revealed by plugging values into both equations, is to place a constant pressure on the system and see what happens to the readings. In that case, $P1 = P2 =$ MAP, so all three digital readouts should be the same.

In a later section, we will deal with the electronic measurement of pressures, so we will return to Fig. 5-20 to see the relationships of the various pressures.

BASIC PRESSURE MEASUREMENTS

The air forming our atmosphere exerts a pressure on the surface of the earth and all objects on the surface (or above it). This pressure is usually expressed in atmospheres (atm), pounds per square inch (lb/in.^{2} or psi), and other pressure units. The magnitude of 1 atm is approximately 14.7 psi at mean sea level.

If pressure is measured with respect to a perfect vacuum (defined as 0 atm), then it is called absolute pressure; and if against 1 atm (open air), it is called a gage pressure. Two gage pressures, or a gage pressure and an absolute pressure, can be measured relative to each other to form a single measurement called relative or differential pressure. Pressures in fluid pipelines, storage tanks, and the human circulatory system are usually gage pressures if measured at a point or differential pressures if measured between two points along a length.

Figure 5-21 shows the Torricelli manometer, named after Evangelista Torricelli (Italian scientist, 1608–1647), which is used to measure atmospheric pressure. An evacuated, small lumen glass tube stands vertically in a pool of mercury. The end that is inside the mercury (Hg) pool is open, while the other end is closed. The pressure exerted by the atmosphere on the surface of the mercury pool forces

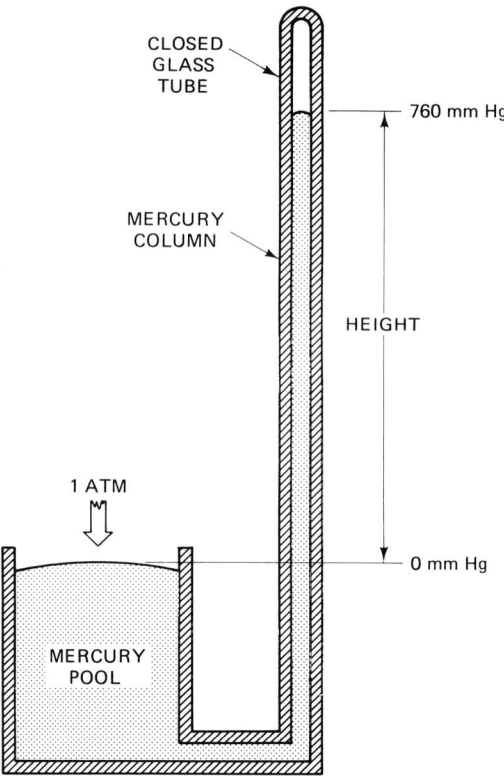

Fig. 5-21   Torricelli manometer.

mercury into the tube, forming a column. The mercury column rises in the tube until its weight (i.e., gravitational force) exactly balances the force of the atmospheric pressure. Torricelli found that a 760-mm column of mercury could be supported by atmospheric pressure at sea level. Thus, 1 atm is 760 mm Hg (also sometimes given in weather reports and aviation in inches, i.e., 1 atm = 760 mm Hg = 29.92 in. Hg).

The standard units of pressure, as established by scientists in international agreement and adopted in the United States by the National Bureau of Standards is the Torr (named after Torricelli), where 1 Torr is equal to 1 millimeter of mercury (i.e., 1 Torr = 1 mm Hg). In medicine and medical science (e.g., physiology), they still use mm Hg instead of the correct Torr.

Gage pressures are usually given in mm Hg (or inches) above or below atmospheric pressure. A manometer is any device that measures gage pressure, positive or negative. By convention, pressures above atmospheric pressure are signed positive, and those below atmospheric pressure are signed negative. Also by convention, negative gage pressures are called vacuums and negative-reading manometers are called vacuum gages (positive reading manometers are also called pressure gages). Both instruments are nonetheless properly called manometers.

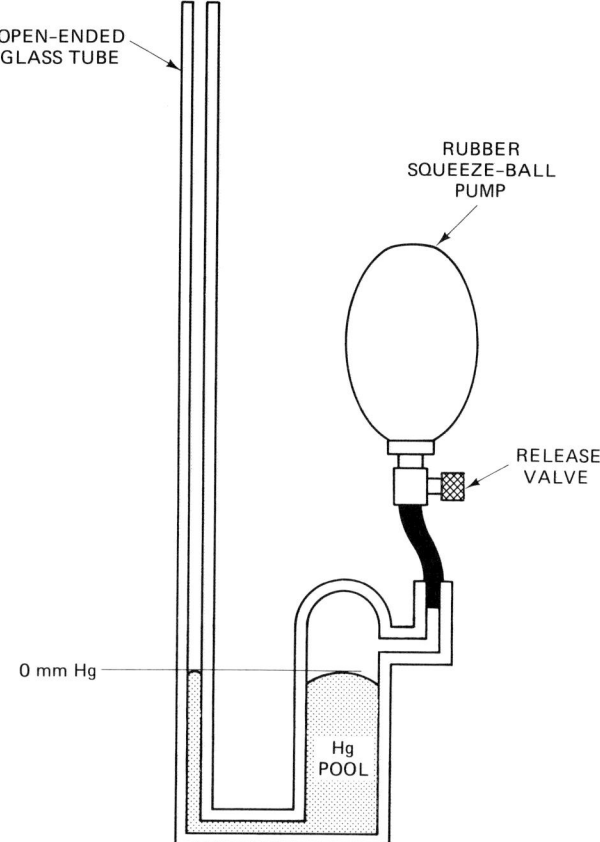

Fig. 5-22 Pressure pump manometer, as used in blood pressure gages.

All measurements require some form of reference point, and for gage pressures, the zero reference point is a pressure of 1 atm. Even though the absolute value of the atmospheric pressure varies from one place to another, and in the same location over the space of a few hours, the zero point can be established by setting the zero scale on the indicator by opening the manometer to the atmosphere.

Figure 5-22 shows a mercury manometer that is similar to those used to measure pressures and calibrate electronic pressure manometers. The open tube is connected to a mercury reservoir that is fitted with a rubber squeeze ball pump that can be used to increase pressure in the system. A valve is used to either open the chamber to the atmosphere or close it.

If the valve is open to the atmosphere, then the pressure on the chamber is equal to the pressure on the column, that is, 1 atm. Under this condition, the mercury column in the tube is at the same height as the mercury in the chamber. This point is defined as 0 mm Hg. If the valve is closed and the pressure inside the chamber is increased by operating the pump, then the mercury in the column will rise to a level proportional to the new pressure above atmospheric pressure.

If the rubber ball in Fig. 5-22 is replaced with a connection to a closed pressure system other than the squeeze ball, then the mercury will rise to a level proportional to the pressure in that system. We can use this manometer as a calibrating device by adding a T connector in the line between the rubber squeeze ball and the chamber. One port of the T goes to the rubber ball, one port goes to the chamber, and the third port goes to the sensor or other instrument being calibrated.

Gage pressure is used for measurement purposes, because it is a lot easier to be referenced at the zero point (open the manometer to the atmosphere) and can be easily recalibrated for each use (no matter where in the world the measurement is made). In addition, for most practical applications, the absolute pressure conveys no additional information content over gage pressure. Should absolute pressure be needed, then it becomes a relatively simple matter to measure the atmospheric pressure (with a device such as that in Fig. 5-21) and add that value to the pressure measured on the gage pressure manometer of Fig. 5-22.

PRESSURE PROCESSING

Only rarely is a simple pressure amplifier needed for dynamic measurements (the same is not true where static pressures are involved). We will want a system such as the one shown in Fig. 5-23. The pressure waveform, P, is the analog output of a pressure amplifier, and it is fed to four different circuits: a peak detector (maximum pressure), an

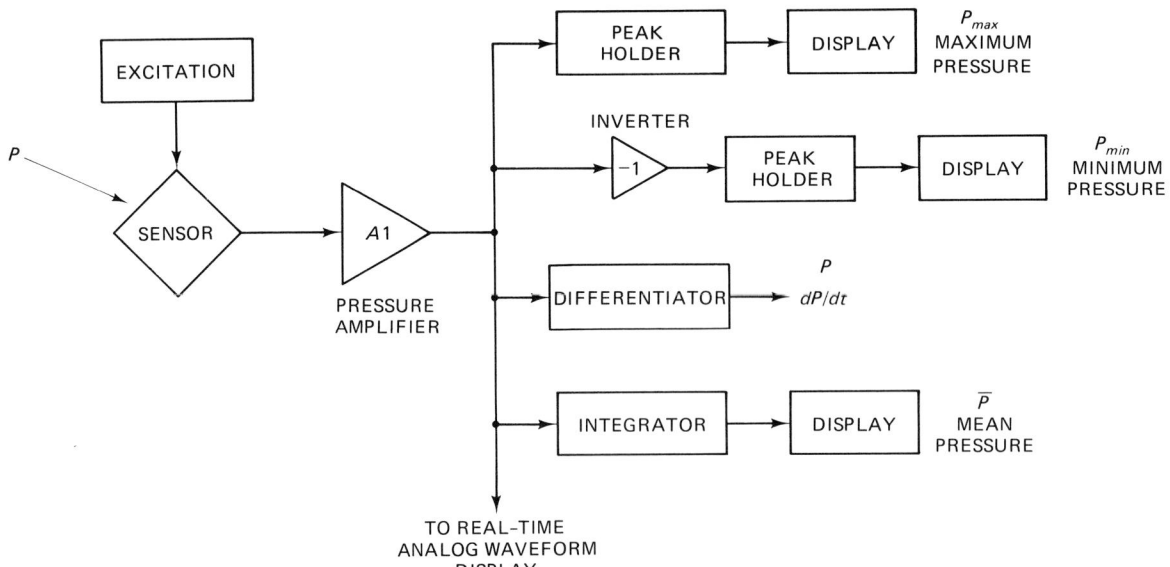

Fig. 5-23 Pressure processing system to display real-time waveform while calculating and displaying maxima, minima, dP/dt, and mean pressure.

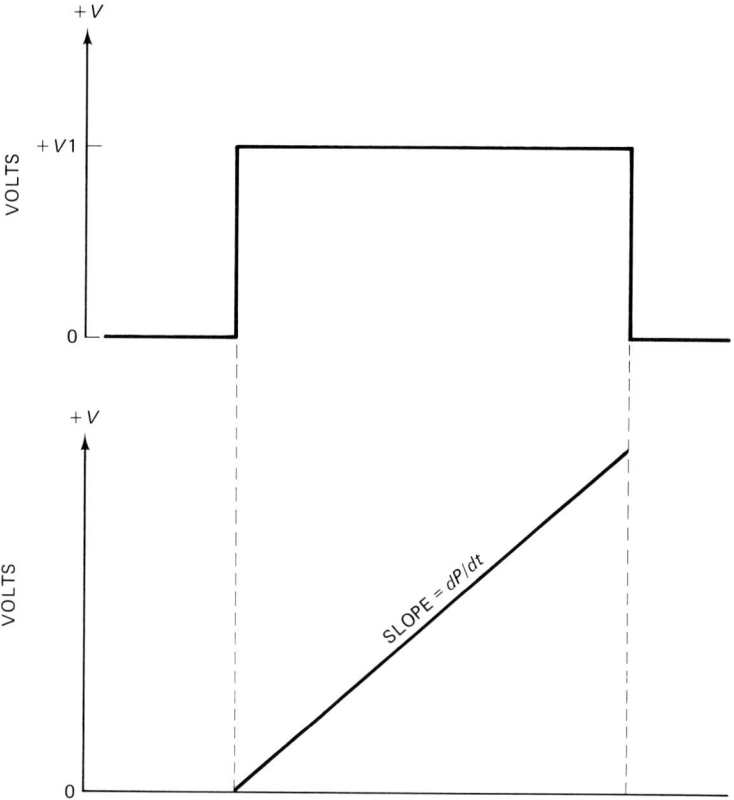

Fig. 5-24 The dP/dt calibration using a ramp function.

inverted peak detector (minimum pressure), a time integrator, and a differentiator (dP/dt).

The peak detectors and integrator can be calibrated by applying a constant value of pressure, P. But, since the differentiator measures dP/dt, a varying signal is needed. Typically, a square wave of the same amplitude as a standard value of the P signal is applied, as in Fig. 5-24. This signal produces an output from the differentiator (as shown), which is a linearly rising slope that can be measured.

POSITION, DISPLACEMENT, VELOCITY, AND ACCELERATION SENSORS

The parameters of position, displacement, velocity, and acceleration are related according to some very simple equations. Thus, when we measure one of these parameters, we can measure all of them by indirect implication. Let us consider each of these. Displacement is merely a change in position from one point to another, without regard for time or other variables. Figure 5-25 shows such one-dimensional

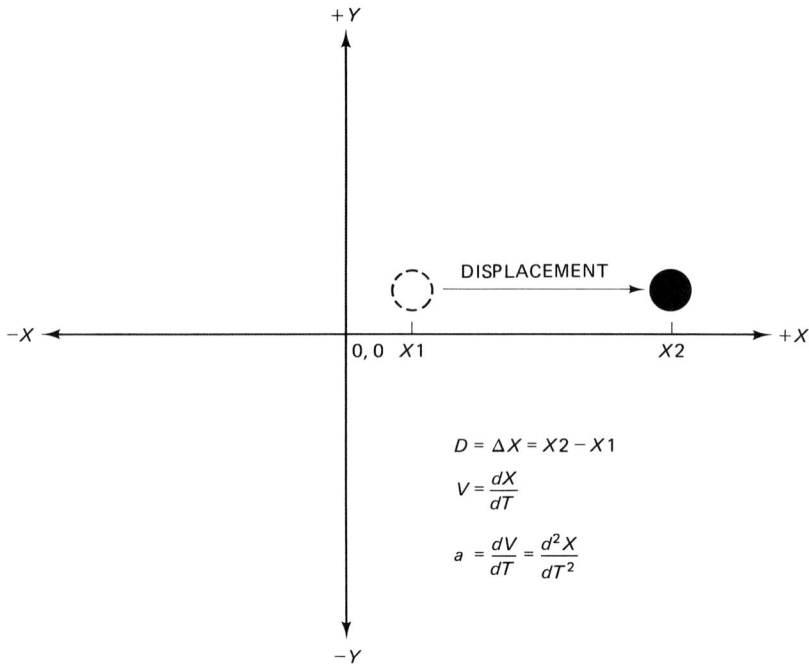

Fig. 5-25 Displacement measurement can lead to velocity and acceleration calculations.

displacement. The system defined in Fig. 5-25 is a Cartesian coordinate system, even though it could have easily been a single-dimensional line. The black ball in Fig. 5-25 was initially at position $X1$, but a hidden, invisible, adabiatic fairy (as professors like to say) came along and pushed it to position $X2$. The difference, $X2 - X1$, is its displacement.

Velocity is the measure of displacement per unit of time. In other words,

$$V = \frac{\Delta X}{\Delta T} = \frac{X2 - X1}{T2 - T1}$$

In the notation of calculus, when ΔX and ΔT are very tiny, the above expression is written

$$V = dX/dT$$

In other words, velocity is the first derivative of displacement, with respect to time. We can then differentiate the velocity expression and find the acceleration:

$$a = dV/dT$$

If acceleration is the first derivative of velocity, and velocity is the first derivative of displacement, then we can conclude that acceleration is also the second derivative of displacement:

$$a = d^2X/dt^2$$

Thus, we can compute acceleration and velocity from positional displacement by using either a differentiator circuit or a computer to differentiate the collected positional data.

POSITION SENSORS

A position sensor will create an output that is proportional to the position of some object along a given axis. For very small position ranges, we could use a simple strain gage or LVDT, although, once again, it must be realized that the range is extremely small. Some

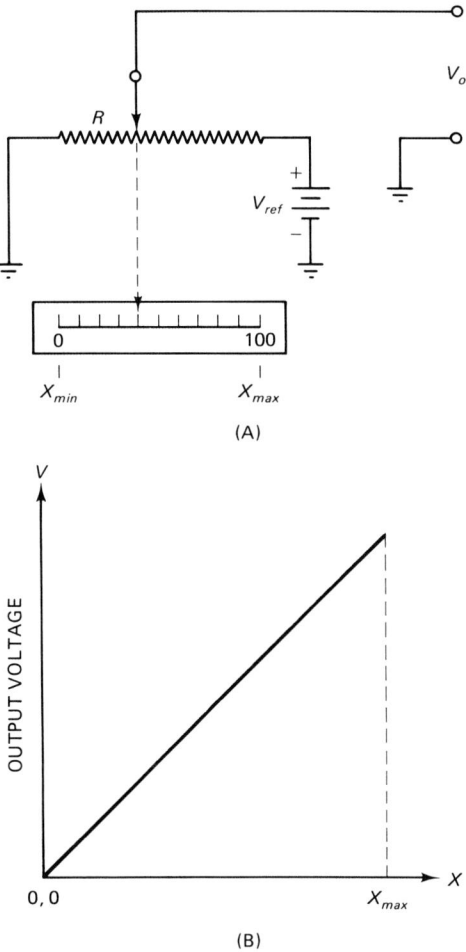

Fig. 5-26 (A) Simple one-quadrant resistive displacement sensor; (B) output characteristic.

amplification of the available range is obtainable by placing the strain gage at one end of a long lever arm and measuring the displacement of the opposite end by actually measuring the forces on the strain gage end. Keep in mind that strain gage position sensors are sometimes permanently damaged by overrange displacements; care is in order.

Perhaps the most common form of displacement sensor is the potentiometer. For applications tolerant of error, almost any ordinary linear-taper potentiometer can be pressed into service. For other applications, either a precision potentiometer or special potentiometers designed as position sensors can be used instead. For rotary (curvilinear) displacement, a rotary potentiometer is used, while for rectilinear displacement a slide potentiometer is used.

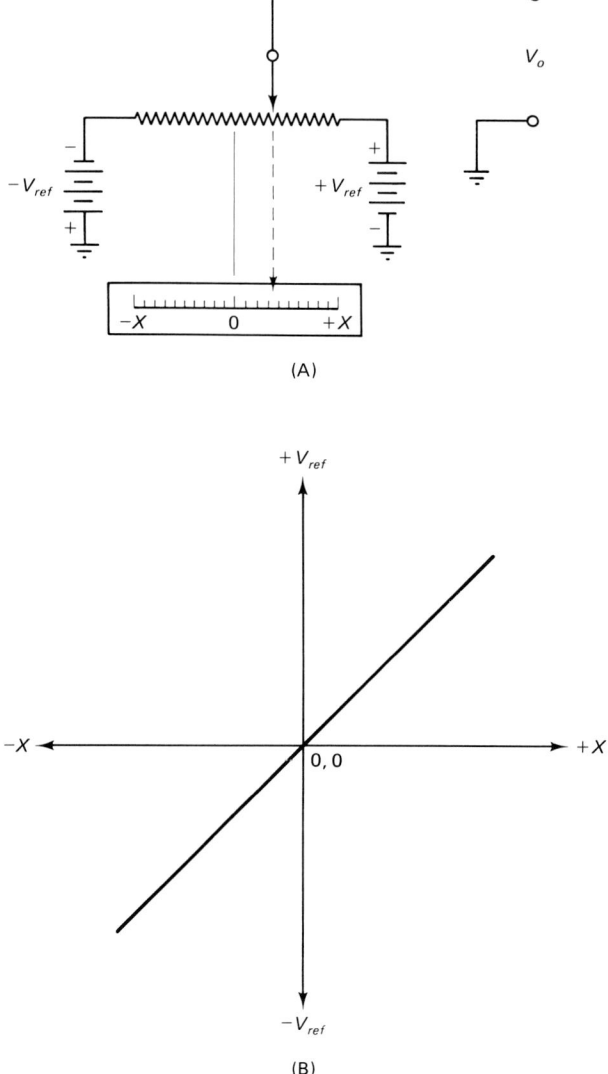

Fig. 5-27 (A) Two-quadrant resistive displacement sensor; (B) output characteristic.

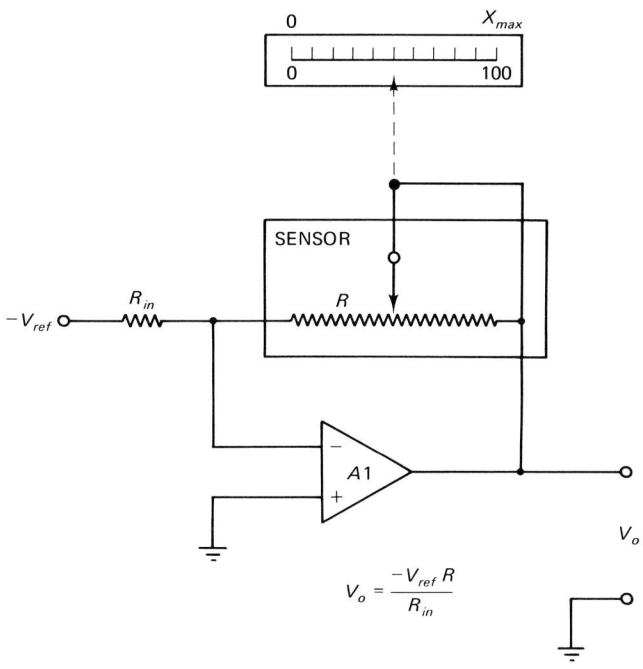

Fig. 5-28 Resistive displacement sensor in feedback loop of amplifier.

Figures 5-26 and 5-27 show two possible circuits for potentiometer displacement sensors. In Fig. 5-26A, we see a single-quadrant system. In this case, we have a single potentiometer with one end grounded and the other end connected to a reference voltage, V_{ref}. It is terribly important to make this voltage both stable and precisely known or error will creep into the system. The output voltage V_o is a function of the displacement X, as measured by the position of the potentiometer wiper along its element. The graph in Fig. 5-26B shows the voltage relationship to displacement.

A two-quadrant system is shown in Fig. 5-27A. It is functionally the same as the single-quadrant version, except that it uses two reference voltage supplies, one positive to ground ($+V_{ref}$) and the other negative to ground ($-V_{ref}$). The output voltage will be negative for negative displacements and positive for positive displacements about a zero reference point.

It is possible to make a four-quadrant system by using either four single-quadrant devices or two two-quadrant devices mechanically arranged at 90° angles to each other. Perhaps the most common form of these types of sensors is the so-called joystick, simple examples of which are found in computer game applications.

A circuit for mating a position or displacement sensor (potentiometer type) to an operational amplifier is shown in Fig. 5-28. In this circuit, the potentiometer sensor forms the feedback network of an operational amplifier. The input resistor ($R1$) is a precision type. Alternatively, it could be another (trimmer) potentiometer, and it could be used to trim errors in the main potentiometer. A $-V_{ref}$ source is applied in lieu of an input signal. The gain of an inverting follower operational amplifier circuit is the ratio of the feedback and input resistors:

$$A = -R2/R1$$

The transfer equation of the system is

$$V_o = (-V_{ref})(-R2/R1)$$

When the potentiometer is at zero displacement, $R2 = 0$, so $V_o = 0$, also. The output voltage is proportional to the displacement.

6

Electro-Optical Transducers

Many sensors use either light, infrared (IR) or ultraviolet (UV) as the transducible property. Various instrumentation techniques depend on the properties of light and light sensors. In some cases, only the existence or nonexistence of the light beam is important. Examples of this use are the PAPER OUT sensor on a computer printer and the light beam that counts entrances and exists from a building or controlled space. In other cases, it is the color of the light that is important. In still other cases, the absorption of particular colors from a wider spectrum (e.g., white or pink light) is the important factor. Whichever is the case, the light beam must be sensed before it can be used, and that is the subject of this chapter.

LIGHT

Light is a form of electromagnetic radiation, and in the ultimate sense, it is the same as radio waves, infrared (heat) waves, ultraviolet waves, and X-rays. The principal differences among these types of electromagnetic radiation are the frequency and wavelength (Fig. 6-1). The wavelength of visible light is 400 to 800 nm (1 nm = 10^{-9} m); infrared waves have longer wavelengths than visible light, and ultraviolet light has wavelengths shorter than visible light; X-radiation has wavelengths even shorter than most ultraviolet light, and it overlaps the upper UV and lower gamma radiation regions. Frequency and wavelength are related in electromagnetic radiation by the equation

$$\lambda = c/f, \qquad (6\text{-}1)$$

where c is the velocity of light (approximately 3×10^8 m/sec), λ the wavelength in meters, and f the frequency in hertz.

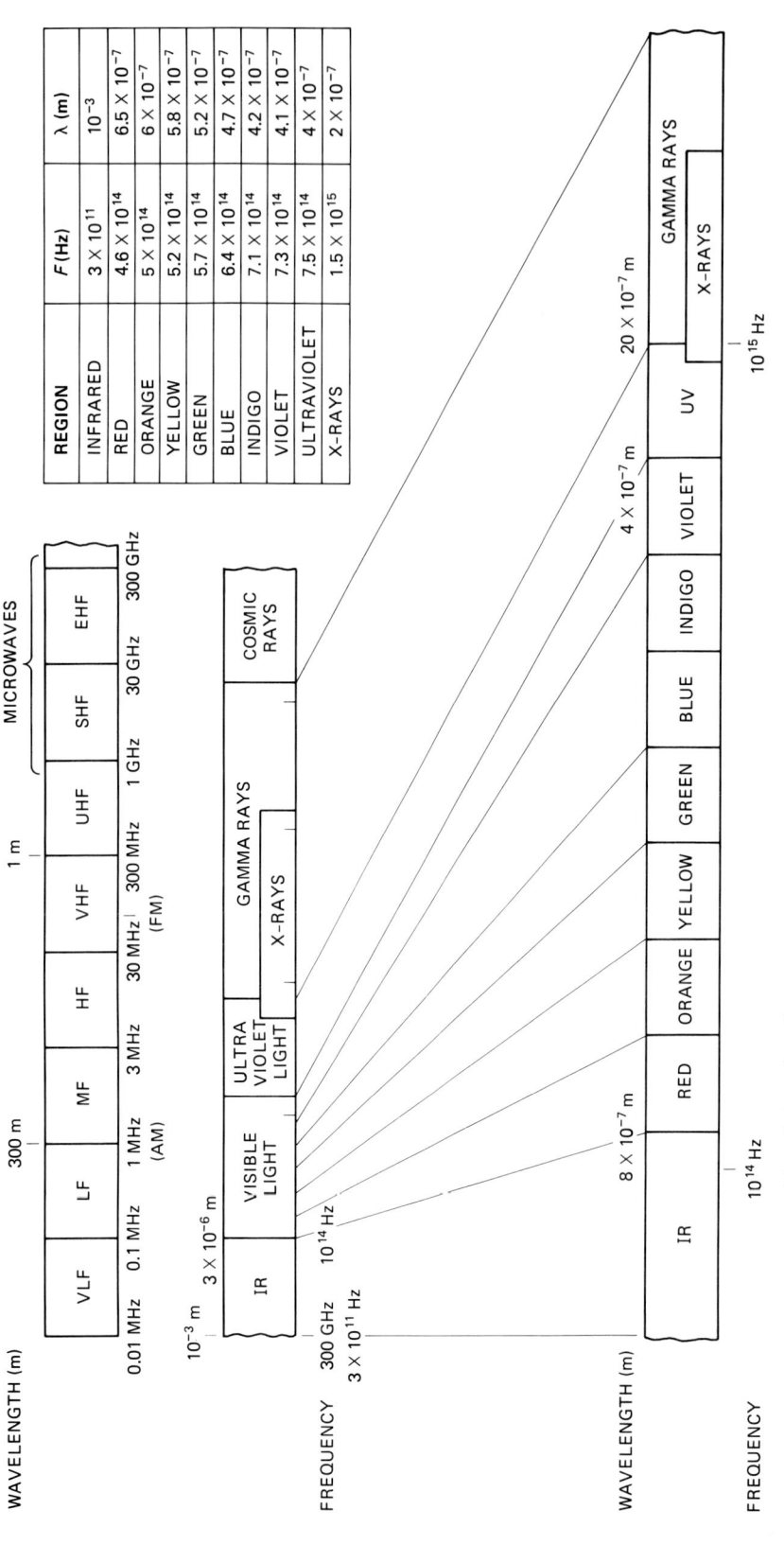

Fig. 6-1 The electromagnetic spectrum, highlighting the visible light and adjacent regions.

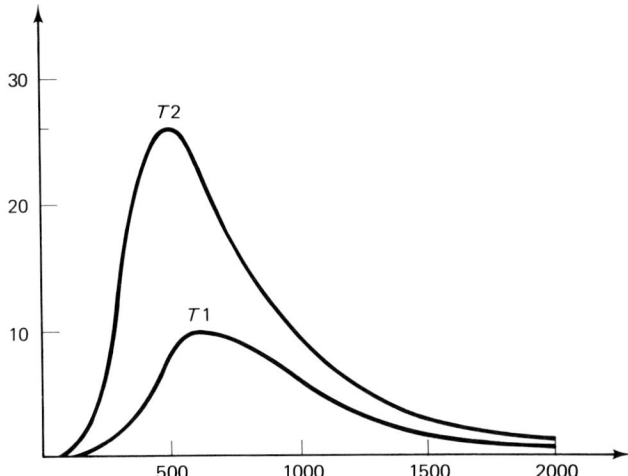

Fig. 6-2 Blackbody radiation curves.

From the above equation, you can see that light has a frequency on the order of 10^{14} Hz (compare, for interest, with the frequencies of AM and FM broadcast bands in the radio portion of the spectrum shown in Fig. 6-1).

Because IR light, UV light and X-radiation are similar in nature and wavelength to visible light, many of the sensors used with visible light also work to one extent or another in the adjacent regions of the electromagnetic spectrum. Although performance usually varies considerably, and some devices are not useful at all in certain areas, it is nonetheless true that workers dealing with those spectra may find these devices useful.

The photosensors described in this section depend upon quantum effects for their operation. Quantum mechanics arose as a new idea in physics in December 1900, the very dawn of the twentieth century, with a now-famous paper by German physicist Max Planck. He had been working on thermodynamics blackbody radiation (Fig. 6-2) problems and found the experimental results reported in nineteenth-century physics laboratories could not be explained by classical Newtonian mechanics, the then-prevailing world view of physics. The solution to the problem turned out to be a simple, but revolutionary, idea: energy exists in discreet bundles, not as a continuum. In other words, energy come in packets of specific energy levels; other energy levels are excluded. The name eventually given to these energy levels was quanta, thus was born quantum mechanics. The name given to energy bundles that operated in the visible light range was photons.

The energy level of each photon is expressed by the equation

$$U = ch/\lambda, \qquad (6\text{-}2)$$

or alternatively,

$$U = hv, \qquad (6\text{-}3)$$

where U is the energy in electronvolts (eV), c the velocity of light (3×10^8 m/sec), λ the wavelength, h Planck's constant (6.63×10^{-27} J erg sec), and v the frequency of light. (The constant ch is sometimes combined and expressed as 1240 eV/nm.)

The basis of light sensors is to construct a device that allows at least one electron to be freed from its associated atom by one photon of light. Materials in which the electrons are too tightly bound for light photons to do this work will not work well as light sensors.

Equation (6-3) shows the reason why certain forms of radiation cause cancer, while longer wavelengths (radio waves and microwaves) do not.[1] The high-frequency X-rays and UV light have tremendous energy that is not present in radio waves. Thus, these wavelengths are able to cause cellular damage that is (or once was) generally considered impossible with lower frequency radiations.

PHOTOTUBES AND THE PHOTOELECTRIC EFFECT

Interestingly, physicist Albert Einstein won the Nobel Prize in physics for his explanation of the photoelectric effect, not for either special or general theories of relativity, as is commonly assumed. Einstein wrote three seminal papers for the 1905 volume of *Annalen der Physik*, any of which could have qualified him for the Nobel: explanation of Brownian motion as a molecular effect, explanation of the photoelectric effect, and special relativity. (Because of the immensely important nature of these discoveries, souvenier hunters have stolen most of the 1905 volumes of that journal from the university libraries of the world. Although the 1905 volume is available in reproduction form, it is extremely rare in the original.)

The effect that Einstein explained in 1905 had long perplexed physicists. As early as 1887, radio research pioneer Heinrich Hertz was using a spark-gap receiving apparatus in his experiments. He noted that the spark-gap receiver was more sensitive when the gap region was illuminated by ultraviolet light. William Hallwachs explained this phenomenon in 1888 by describing the photoelectric effect also sometimes called the Hallwachs effect.

---

[1]Scientists and physicians have long held that only ionizing radiation, that is the radiation with wavelengths shorter than light, and microwave radiation from which thermal affects are created, are dangerous. Recent information on nonionizing, low-level radiation; low-frequency radiation (including 60 Hz from the AC power lines); and magnetic fields indicates that these may be capable of causing dangerous medical problems, such as leukemia and non-Hodgkins lymphoma. The reader is cautioned to examine the latest literature before accepting that only ionizing radiation is dangerous.

The photoelectric effect is the emission of electrons from a metallic surface in an evacuated space when light is shined onto that surface. Oddly, increasing the brightness of the light does not increase the energy level of the emitted electrons. If this were purely a mechanical kinetic event, then one would assume from the classical point of view that increasing the intensity of the light would increase the energy of the current emitted from the surface. It turned out, however, that changing the color of the light affected the electron energy level! Red light produced lower energy electrons than violet light; the energy level of the current flow is color sensitive. Once Planck's principle was known, however, Einstein was able to explain this effect by quantum principles. From the previous equations, you can see that the higher frequency (shorter wavelength), violet-colored light has significantly more energy than the red light. The energy expression for the photoelectric effect is

$$mV^2/2 = hv - \phi, \tag{6-4}$$

where m is the mass of the electron, v the velocity of the fastest emitted electrons, h Planck's constant (6.63×10^{-27} erg sec), v the frequency of the incident light, and ϕ the energy in ergs required to permit an electron to escape from the photoemissive surface.

Figure 6-3 shows a phototube based on the photoelectric effect. The cathode is a silver-oxide-coated metallic plate that is exposed to light; the metal plate is a material (such as silver-oxide) that will easily emit electrons when light is applied. The anode is positively charged (with respect to the cathode) by an external DC power supply. Electrons emitted from the photocathode are collected by this plate and can be read on an external meter (or other circuit).

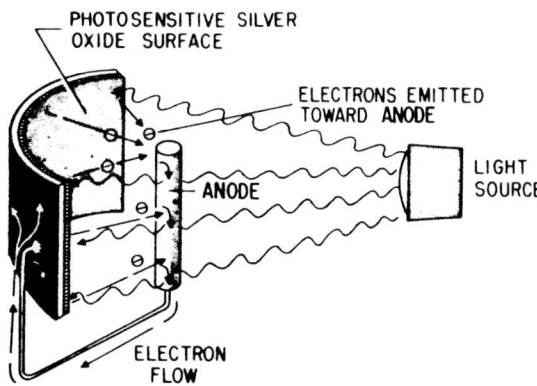

Fig. 6-3 Phototube based on the photoelectric effect uses the emission of electrons from a treated metallic surface when light impinges the surface.

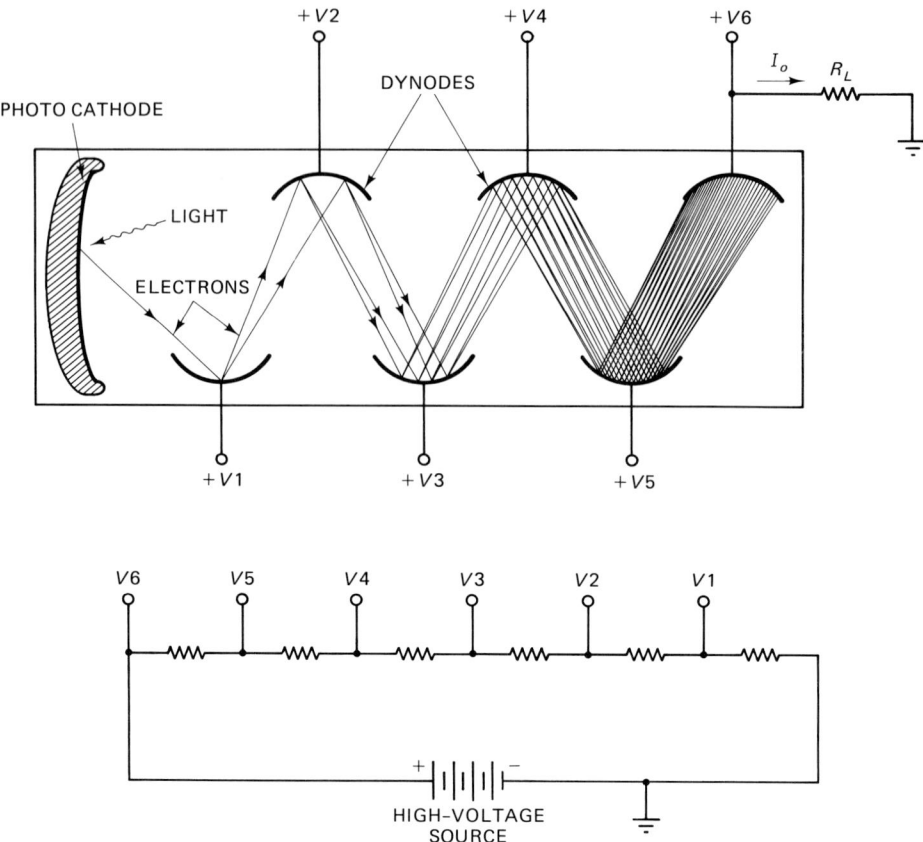

Fig. 6-4 Structure of the photomultiplier tube.

PHOTOMULTIPLIER TUBES

The photoemission process is less efficient than is needed in many cases, especially under low light-level conditions. The photoelectric tube system can be made more efficient by using a photomultiplier (PM) tube (Fig. 6-4). In this type of photosensor, there are a number of positively charged anodes, called dynodes, that intercept the electrons. When light impinges on the cathode, electrons are emitted because of the photoelectric effect. They are accelerated through a positive high-voltage potential ($V1$) to the first dynode. They acquire substantial kinetic energy during this transition, so when each electron strikes the metal, it gives up its kinetic energy. In the giving uo of this kinetic energy, some is converted to heat, while some is converted to an increased current flow by dislodging additional electrons from the dynode surface. Thus, a single electron caused two or more additional electrons to be dislodged. These electrons are accellerated by a high-voltage potential ($V2$) and reproduce the same effect at the second dynode. The process is repeated several times, and each time several more electrons join the cascade for each previously accelerated electron. Finally, the electron stream is collected by the last dynode or a separate anode and can be used in an external circuit.

PHOTOVOLTAIC CELLS

A photovoltaic cell is a device in which an electrical potential difference is generated, and thus a current made to flow in an external circuit, by shining light onto its surface. The common solar cell is an example of the photovoltaic cell.

Figure 6-5 shows two forms of photovoltaic cells. In Fig. 6-5A, a copper disk is coated with a layer of coper oxide, which is in turn covered with a semitransparent layer that passes light and collects emitted electrons. The copper-oxide cell was invented prior to World War I by Bruno Lange, and it was eventually marketed by Westinghouse under the trade name Photox cell.

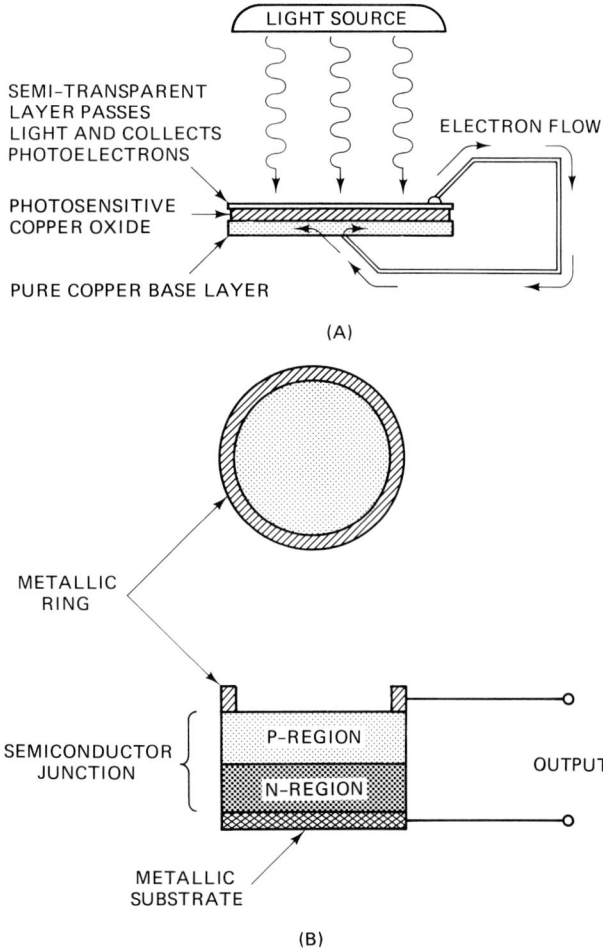

Fig. 6-5 Photocells: (A) copper oxide and (B) PN junction diode.

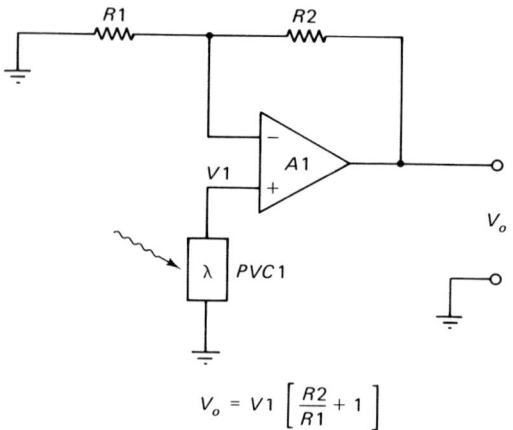

$$V_o = V1 \left[\frac{R2}{R1} + 1 \right]$$

Fig. 6-6   Photovoltaic cell buffered by an operational amplifier stage.

A similar photovoltaic cell is made of selenium. This cell was invented in the 1930s, and it was marketed by Weston Instruments under the trade name Photronic cell. In selenium cells, a layer of photosensitive selenium is coated onto an aluminum plate. Selenium cells produce an output of 0.45 VDC under 2000 fc (foot-candles) of illumination.

Figure 6-5(B) shows the structure of the silicon photovoltaic cell, discovered in 1958 by scientists at Bell Telephone Laboratories. The silicon cell consists of a PN junction of P- and N-type silicon. In the form shown in Fig. 6-5B, the N-type silicon is deposited onto a metallic substrate, which also forms the negative terminal of the cell. The P-type layer is diffused over the N-type and forms the surface exposed to light. The positive electrode is an annular ring deposited onto the exposed surface of the P-type silicon region. These cells output a potential of 0.27 to 0.6 V under illumination of 2000 fc.

Figure 6-6 shows a typical circuit for instrumentation applications of the photovoltaic cell (PVC1). The cell is connected across the input of a high-impedance amplifier, such as the noninverting operational amplifier shown. The output voltage is found from

$$V_o = V1 \left(\frac{R2}{R1} + 1 \right) \tag{6-5}$$

PHOTORESISTORS

A photoresistor is a device that changes electrical ohmic resistance when light is applied. Figure 6-7A shows the usual circuit symbol for photoresistors; it is the normal resistor symbol enclosed within a circle and given the Greek lambda (λ) symbol to denote that it is a resistor that responds to light. The structure of the photoresistor is shown in

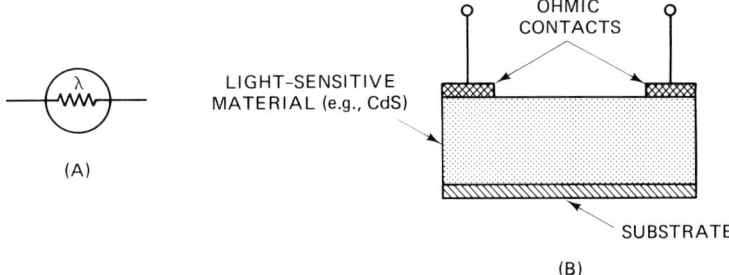

(C)

Fig. 6-7 (A) Photoresistor circuit symbol, (B) structure of a photoresistor, and (C) example of a photoresistor.

Fig. 6-7B. It consists of a layer of photoresistive material deposited onto a substrate. Metallic ohmic contacts are deposited onto the surface of the photoresistive material. Figure 6-7C shows a typical commercial photoresistive cell.

When photoresistors are specified, it is typical that a dark resistance and a light/dark ratio are given. In most common varieties, the electrical resistance is very high when dark and drops very low under intense light. The intensity of the light affects the resistance, so photoresistors can be used in photographic light meters, densitometers, colorimeters, and other devices.

Figure 6-8 shows three circuits in which photoresistors can be used. The half-bridge circuit is shown in Fig. 6-8A. In this circuit, the photoresistor is connected across the output of a voltage divider made up of the $R1$ and $PC1$. The output voltage is given by

$$V_o = \frac{V \times PC1}{R1 + PC1},\qquad (6\text{-}6)$$

where V_o is the output potential, V the applied excitation potential, and $R1$ and $PC1$ are in ohms.

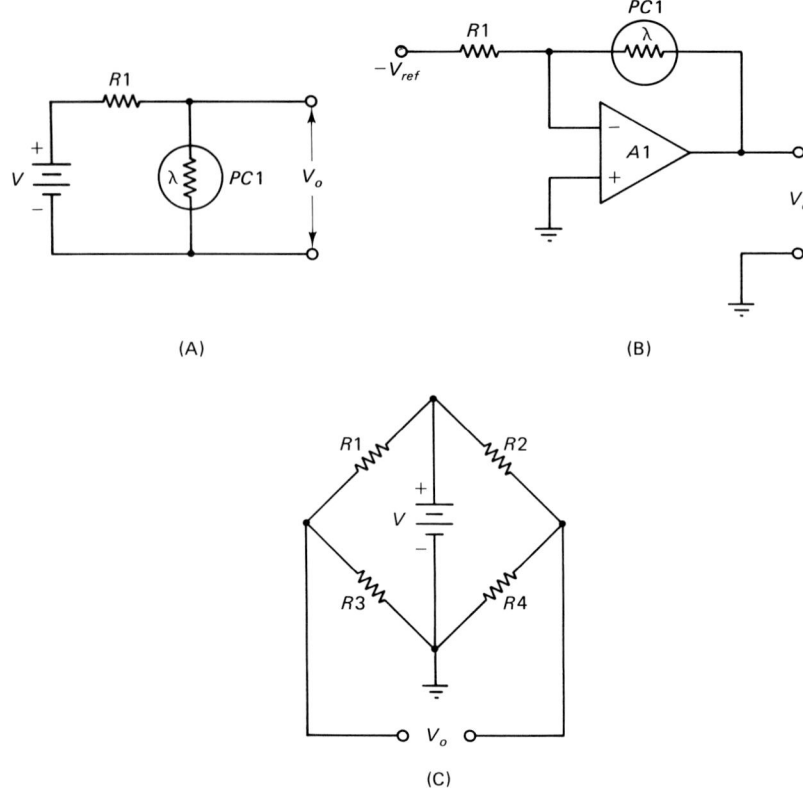

(A)   (B)

(C)

Fig. 6-8 (A) Half-bridge circuit for using photoresistor, (B) photoresistor used in feedback loop of an operational amplifier, and (C) Wheatstone bridge circuit.

A problem with this circuit is that the output potential does not drop to zero, but always has an offset value.

A second way to use the photoresistor is shown in Fig. 6-8B. Here, the photoresistor is the feedback resistor in an operational amplifier inverting follower circuit. The output voltage, V_o, is found from

$$V_o = -(-V_{ref})(PC1/R1). \tag{6-7}$$

The circuit of Fig. 6-8B provides a low-impedance output but, like the other half-bridge circuit (Fig. 6-8A), the output voltages does not drop to zero. In addition, with some photoresistors, the dynamic range of the operational amplifier may not match the dark/light ratio of the photoresistor at practical values of the $-V_{ref}$ potential.

The last photoresistor configuration is the Wheatstone bridge shown in Fig. 6-8C. This circuit allows the output voltage to be zero under the right circumstances, and it is the circuit favored by most designers. If a low-impedance output is required, or additional amplification is needed, then a differential DC amplifier can be connected across the output potential, V_o.

PHOTODIODES AND PHOTOTRANSISTORS

Perhaps the most modern light sensor is the PN junction in the form of special photodiodes. Certain types of diode, when the PN junction is illuminated, will increase the level of reverse leakage current available. Figure 6-9A shows the basic circuit used for these sensors. The diode is normally reverse biased, with a current limiting resistance (R_L) in series. Microammeter $M1$ measures the reverse leakage current that crosses the PN junction during this type of operation. When light strikes the Pn junction, the reading on $M1$ will increase.

The same principle applies to a class of NPN or PNP transistors called phototransistors (Fig. 6-9B). In these devices, the collector to emitter current flows when the base region is illuminated. These devices are the heart of optoisolator and optocoupler integrated circuits, and they are also used in various instrumentation sensor applications.

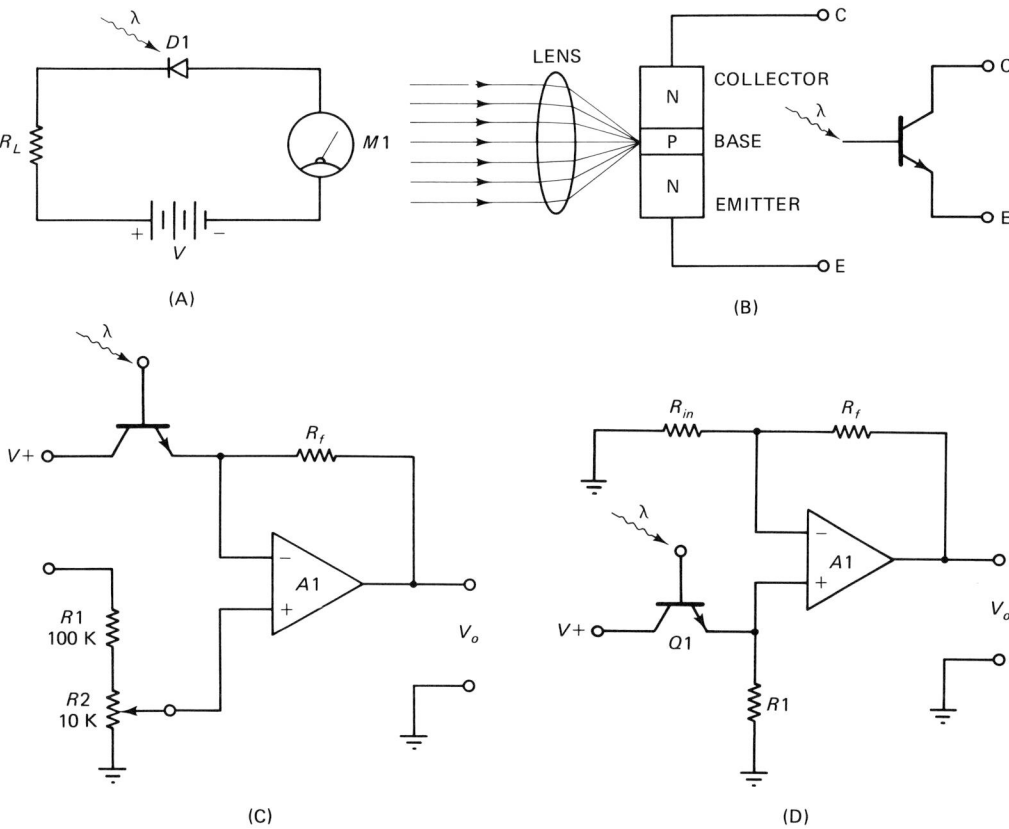

Fig. 6-9 (A) Basic PN junction photodiode circuit, (B) phototransistor, (C) use of phototransistor in inverting follower operational amplifier circuit, and (D) phototransistor used in noninverting follower circuit.

Figures 6-9C and 6-9D show ways in which photodiodes and phototransistors are used. The current flow through the device under light conditions is the transducible event, so we must connect them into a circuit that makes use of that property. The inverting follower operational amplifier circuit (Fig. 6-9C) will do that job. In all such cases, the output voltage, V_o, is equal to the product of the input current and the feedback resistance ($I_{in}R_f$). In the case of Fig. 6-9C, a zero control is added, and it can also be added to the other circuit as well.

A variation on the theme is the circuit of Fig. 6-9D. In this circuit, the phototransistor is applied to the noninverting input of the operational amplifier. The noninverting configuration (see Chapter 12) offers a very high input impedance, although without a zero control, there may well be a DC offset in V_o.

RADIATION AND X-RAY SENSORS

X-rays are similar to light waves in that they are electromagnetic waves, but with wavelengths shorter than either visible light or most ultraviolet radiation. X-rays are familiar to most readers because they have widespread use in medicine, science, and industry and at airport security checkpoints. X-rays are usually generated in a phenomenon called bremsstrahlung (see Fig. 6-10). When an incident electron, with energy E_i is smashed into a target containing heavy nuclei, then a strange thing happens. As the electron is deflected around the nucleus, it loses some energy and assumes a new energy level, E_D. The difference between incident and deflected electron energy levels ($E_i - E_D$) must, according to the law of conservation of energy, go somewhere, so it becomes a photon of X-ray energy.

Figure. 6-11 shows a simple X-ray generator tube based on the bremasstrahlung effect. It is a vacuum tube containing an electron emitting cathode (a heated thermionic filament) and a target anode. The materials of the target anode are selected to make X-ray generation easier. Besides the types of materials, the applied high-voltage (HV) potential ($V+$) determines the kinetic energy of the accelerated electrons, hence the frequency and wavelength of the emitted X-rays. In a few, very old TV sets, there was once a scare of X-ray emission caused by a new type of HV regulator tube that operated at higher than usual potentials. Medical, scientific, and industrial bremsstrahlung generators are somewhat more tightly designed than TV tubes, however.

A variant on the basic tube uses a rotating anode. The heat created by the kinetic energy of electrons smashing into the target is tremendous, and overheating is a major cause of lost X-ray tubes. The rotating anode spreads the heat energy over a larger volume of metal and incidentally produces a more narrowly focused beam.

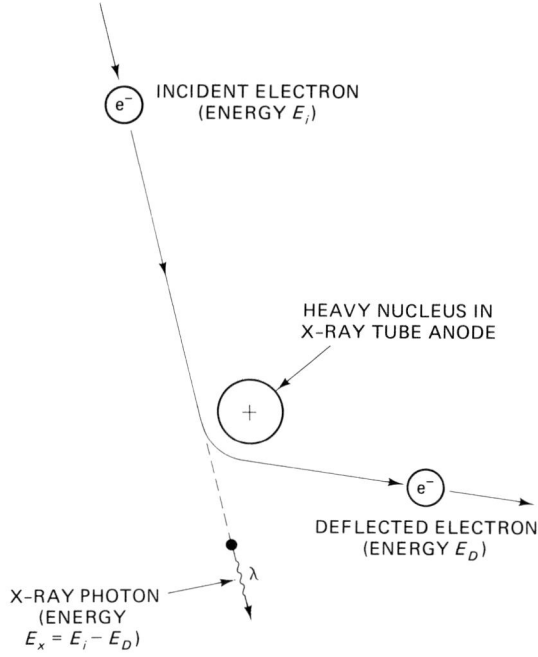

Fig. 6-10 Bremsstrahlung phenomenon generates X-ray photons by deflection of electron around a heavy nucleus. The wavelength, hence the energy, of the emitted photon is the difference between the incident and deflected electron energies.

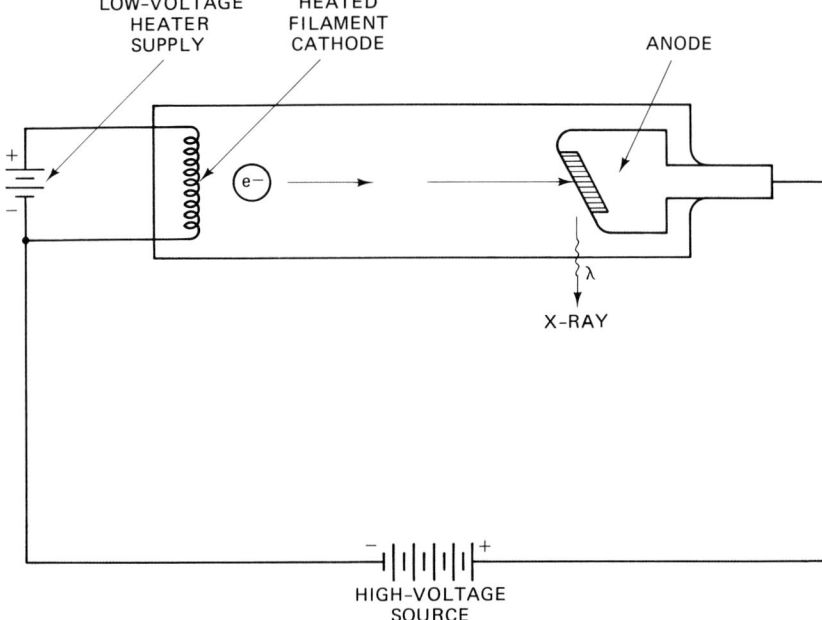

Fig. 6-11 Simple X-ray generator tube.

Many of the sensors used in light-operated devices will also work at X-ray wavelengths. Certain phototubes, photomultipliers, and photodiodes (and transistors) will also work in X-ray measurements. In this section, we will look at two sensors that are unique to radiation measurements: the Geiger–Mueller tube and the scintillation cell.

Geiger–Mueller Tubes

The Geiger–Mueller (G–M) tube (Fig. 6-12) is a glass or metallic cylinder that has been evacuated of air and then refilled to a less than atmospheric pressure with an ionizing gas (usually argon with a touch of bromine, at 100-Torr pressure). When radiation impinges such a gas, its energy forces the gas into ionization, thereby altering the electrical characteristics of the G–M tube. Those electrical characteristics are our transducible event. The general circuit for the G–M tube is also shown in Fig. 6-12. In most applications, the external circuitry consists of a power supply and a series current-limiting or load resistors.

There are three modes of operation for the G–M tube, and these modes affect both the subsequent circuitry and the voltage level required. These regions are shown in the curves in Fig. 6-13 and are designated as follows: ionization chamber mode (region B), proportional counter mode (region C), and Geiger counter mode (regions D and E). Region F is called the glow-discharge region.

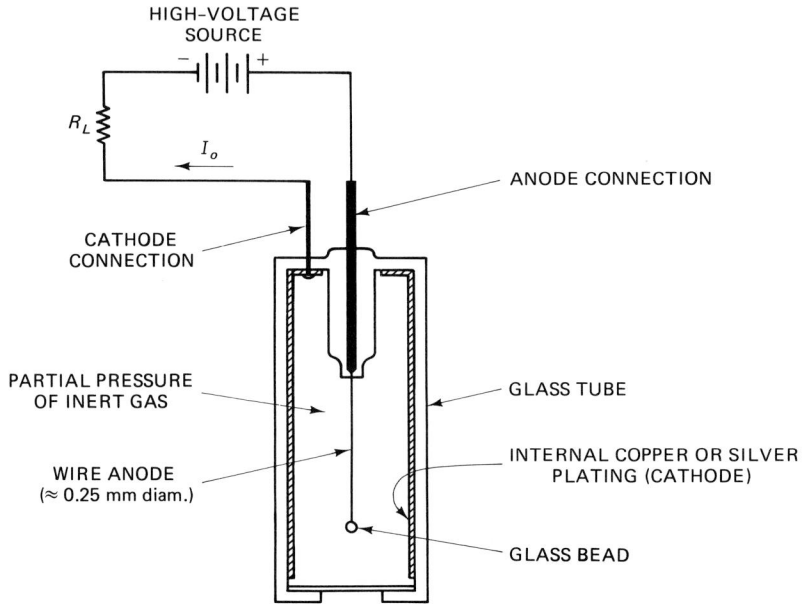

Fig. 6-12 Geiger–Mueller tube.

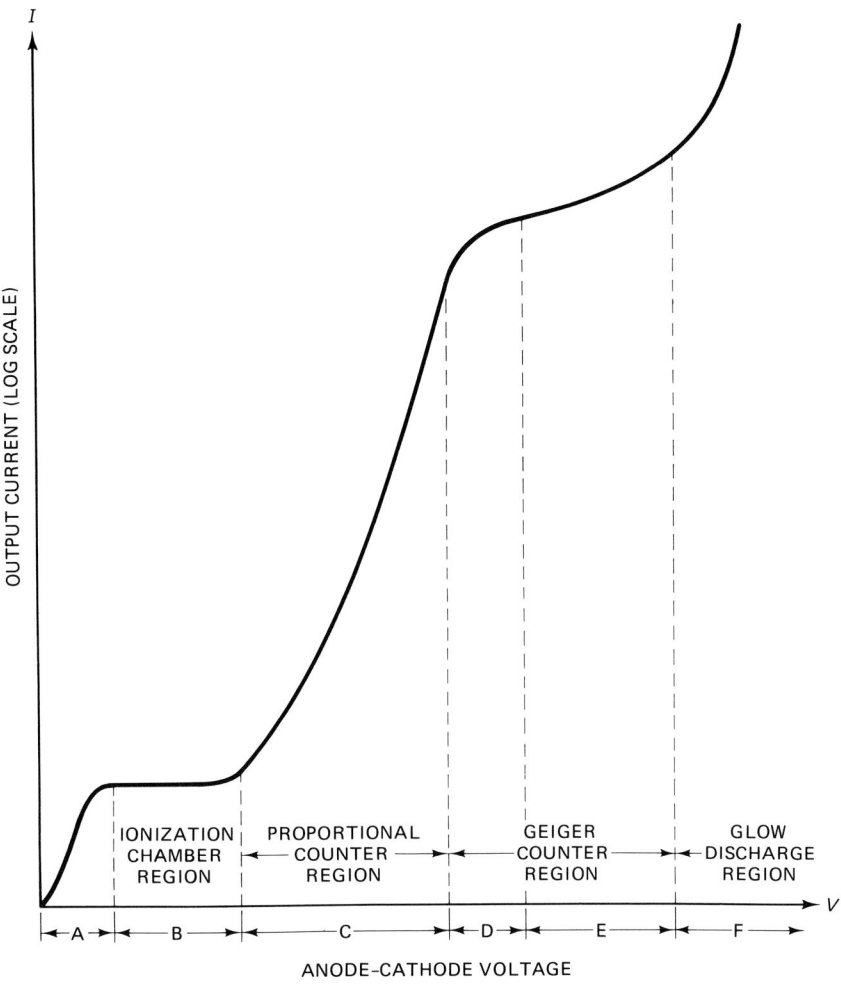

Fig. 6-13 *I* versus *V* characteristics of the Geiger–Mueller tube showing regions of operation.

In the ionization chamber mode, a weak electric potential is applied across the G–M tube, so only a few electrons are generated by the impinging radiation. Nearly all of these electrons are collected by the electrode. Therefore, we can assume that the current in the external load resistor is a measure of the amount of radiation.

In the proportional counter mode, the electric field is stronger, so the electrons generated by the ionizing gas reach sufficient kinetic energy to release additional electrons by kinetic collisions with gas molecules. The output across the load resistor will be a spike with an amplitude that is proportional to the kinetic energy of the ionizing radiation particle. In this mode, the G–M tube can count ionizing particles on a one-for-one basis.

The Geiger counter mode uses very high potentials, on the order of 800 to 2000 V. The operation is similar to that in the proportional counter mode, except that kinetic energies are so high that a multiplication effect takes place; the tube operates in an avalanche condition. The output pulses are of approximately the same amplitude ("all or nothing" operation) every time the tube fires. An external counter circuit will tell us the number of pulses per unit of time, hence the level of radiation.

Scintillation Counters

An example of a scintillation cell or counter is shown in Fig. 6-14. Scintillation is a word used to denote a process similar to that which generates light on the screen of a cathode ray tube. When a radiation particle strikes an atom of certain phosphorous materials, its kinetic energy may be added to the energy of the orbital electrons. When the electrons are thus excited, they jump to a higher energy state, which is unstable. When they fall back to their ground state, the energy absorbed goes off in the form of a photon of light. Certain crystal materials possess this property.

Figure 6-14 shows a scintillation device in which a scintillation crystal window is attached to the light input window of a photomultiplier tube. Radiation causes the crystal to scintillate, and the light thus produced is picked up and amplified by the PM tube. The current at the output of the PM tube is proportional to the light, hence the radiation level.

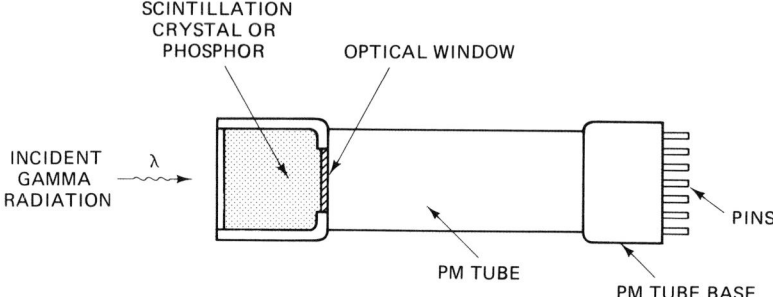

Fig. 6-14   The scintillation counter is made by marrying a photomultiplier tube with a scintillating crystal.

7

Flow and Flow Rate Sensors

The measurement of fluid or gas flow is the process of determining how much material is passing a given point along its pathway. The material measured might be blood in an artery, respiratory gases into and out of a ventilator tube, the amount of water passing though a pipe, or any of a number of other similar situations.

Flow can be either turbulent or smooth. The measurement of turbulent flow is dauntingly difficult, so we will concern ourselves mostly with the measurement of laminar flow situations. Fortunately, most mechanical flow sensors have inertia, so they tend to integrate out small variations due to turbulence (inertia of the sensor tends to act like a low-pass filter, or time averager). Ultrasonic sensors, which are covered later in this chapter, do not possess the inertia of the mechanical sensors, so sometimes they tend to be a bit sensitive to turbulence in the system. In ultrasonic systems, the sensor and its connecting plumbing should be designed to minimize introduced turbulence.

FLOW VOLUME VERSUS FLOW RATE

Many flow sensors actually measure flow rate, that is, the amount of material passing a point per unit of time. For example, a popular medical respiratory flow meter measures the patient's inspiration or expiration in liters per minute (l/min). Similarly, a certain fluid sensor measures fluid flow in cubic centimeters per second (cm^3/sec or cc/sec).

The flow volume can be derived from flow rate data by the simple expediant of integrating the flow signal, as in Fig. 7-1. The output of the flow sensor is amplified (and possibly further processed) in amplifier $A1$. The output of $A1$ is proportional to the flow rate. This same signal is applied to an integrator.

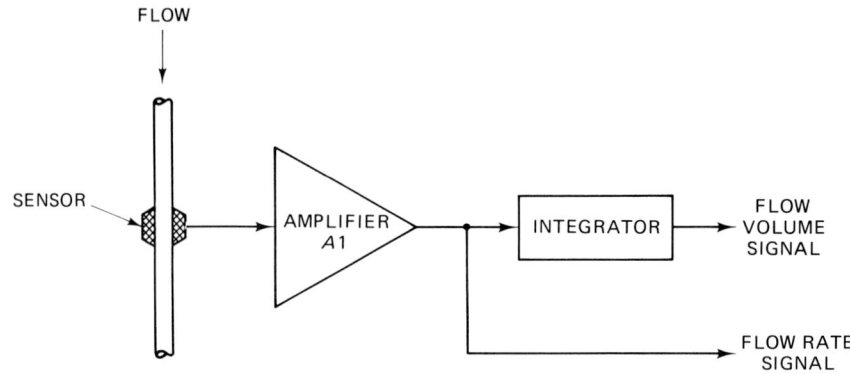

Fig. 7-1   Flow sensor instrumentation to produce flow rate and flow volume signals.

FLOW DETECTORS

A flow detector is a circuit that informs us about the presence of flow, but it does not quantify either the flow rate or the flow volume. Two such systems are shown in Fig. 7-2. In Fig. 7-2A, a thermistor bridge version is shown. Two thermistors, $RT1$ and $RT2$, are placed inside the flow container (in this case a "T" plumbing connector). The idea here is to detect whether or not air is getting through the line—how much is less important because this sensor is a simplified alarm circuit. Thermistor $RT1$ is placed in the air stream, while $RT2$ is in a dead space off the main flow (it makes the ambient measurement).

The two thermistors in Fig. 7-2A are connected in an ordinary Wheatstone bridge circuit and are excited by a DC source to the point of self-heating. When $R1/RT2 = R2/RT1$, the output voltage V_0 is zero. The voltage source (V) biases the thermistors only to the point of self-heating, but not greater. At this point, the thermistor resistance is the most sensitive to changes in air flow, which cools the surface of the device. When air flow changes the device resistance, the output voltage will also change.

The output voltage is proportional to the temperature change caused by the flowing air. While it is not impossible to calibrate this temperature change over a narrow range of flow rates, it is too difficult to consider in most applications. There are better suited sensors on the market.

Figure 7-2B shows a photooptical device that is used to the detect flow of opaque fluids. Here, we see a light source (LED in this case) shining across the flow path to a detector (phototransitor in this case). The fluid or gas must be opaque to the light, or the light frequency must be chosen for maximum attenuation. For example, ordinary visable red LEDs (or other light sources) can be used for an opaque liquid. For a gas such as carbon dioxide, however, we can appeal to the fact that CO_2 absorbs infrared energy. By making the light source and

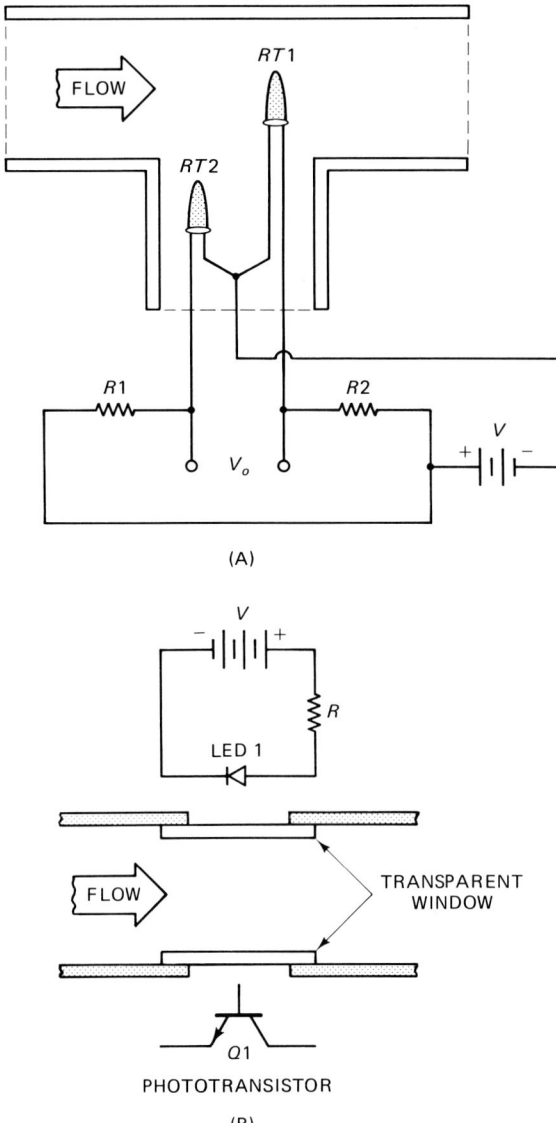

Fig. 7-2 (A) Thermistor gas or air flow detector, and (B) photoelectric flow sensor.

the detector IR-sensitive, we can detect the presence of CO_2. Other gases and liquids may absorb other wavelengths, so each system must be developed for its own capabilities.

POTENTIOMETER SYSTEMS

Figure 7-3A shows a system that uses a potentiometer displacement sensor to measure the flow volume (not the flow rate). This system is used on certain syringe pumps in medical devices. An ordinary glass or plastic syringe is placed in a saddle with a worm-gear pump and motor

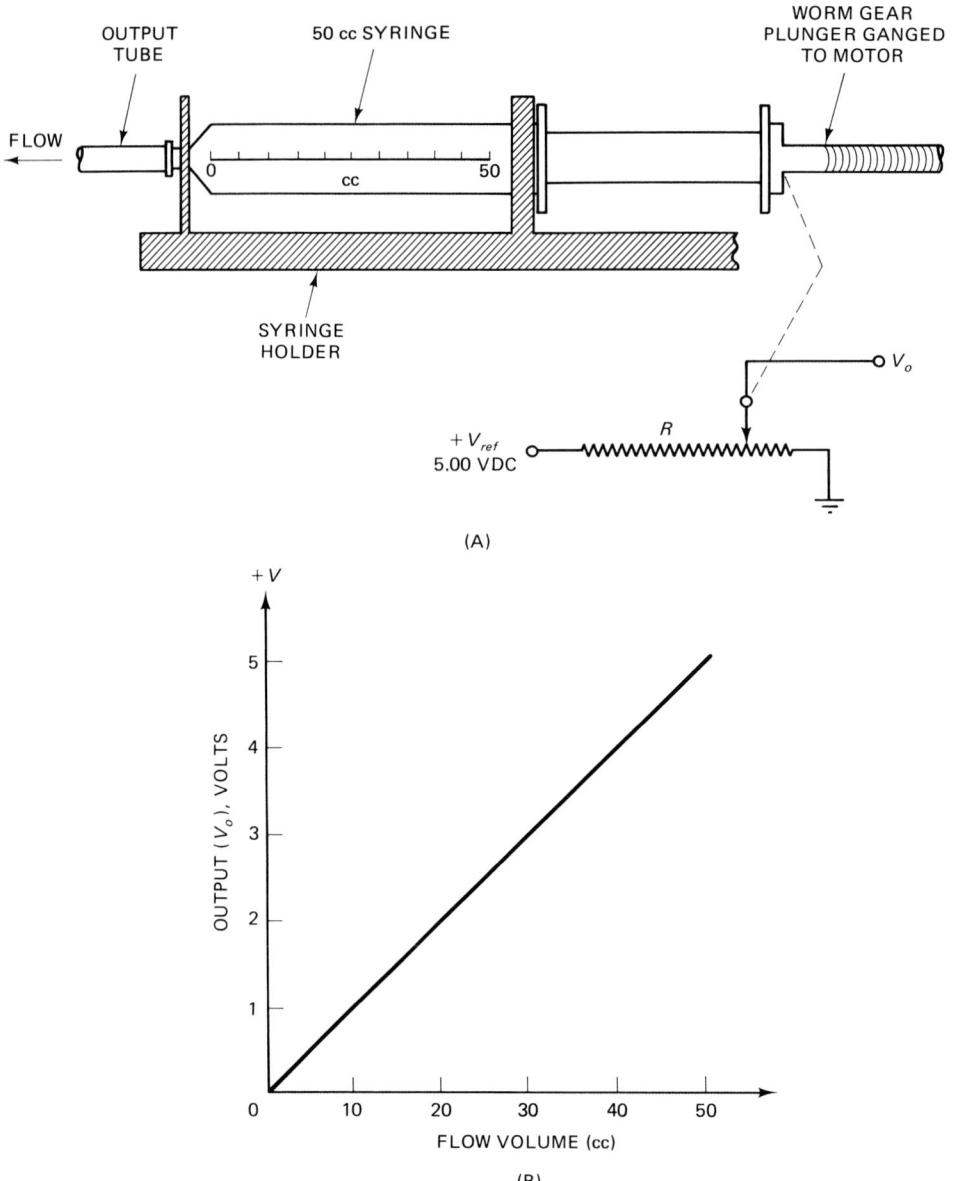

Fig. 7-3 (A) Syringe pump uses rectilinear potentiometer displacement sensor to measure flow volume; (B) output characteristic of displacement or volume sensor.

system. As the worm gear advances, it pushes the syringe plunger, thereby expressing fluid out of the tubing. A rectilinear potentiometer is ganged to the pressure plate on the end of the worm gear. Thus, the displacement of this plate is a measure of volume expressed. Consider an example in which a 50 cm^3 syringe is connected to a potentiometer that is, in turn, connected to a $+V_{ref}$ voltage of $+5.00$ V. Figure 7-3B shows the output function of this system. The output voltage is 5 V at 50 cm^3, so the scaling factor is 5 V/50 cm^3, or 100 mV/cm^3.

PRESSURE DROP SYSTEMS

Figure 7-4 shows two examples of a common form of flow rate sensor. An obstruction placed in the gas path will create a pressure drop (P_d) that is proportional to the square root of the flow rate:

$$P_d = K\sqrt{F},$$

where P_d is the pressure drop, K a sensitivity constant, and F the flow rate.

In some sensors, the obstruction is a narrowing of the path at a point (Fig. 7-4A). In others, the obstruction is a wire or plastic cloth mesh stretched across a constant diameter pathway (Fig. 7-4B). The usual mesh for medical respiratory measurements is 400 grid/inch. The pressure drop is measured by a differential pressure sensor. These sensors have two ports, one on either side of the diaphragm. Most pressure sensors measure gage pressure, so they are open on one side to the atmosphere. These devices are not suitable for the type of measurement described here. Instead, a differential pressure sensor is needed (as shown in Fig. 7-4).

A cannula pressure drop sensor is shown in Fig. 7-4C. The cannula is a double lumen tube in which the distal openings of the lumen channels are spaced ΔX apart. These openings are able to measure the pressures at two points, so a ΔP related to distance ΔX can be determined. The proximal ends of the lumen channels are connected across a differential pressure sensor. The expression for the pressure drop is

$$\frac{\Delta P}{\Delta X} = \left(1.1\zeta/\pi ga^2\right)(dF/dt) + \left(12.8\mu F/\pi ga^4\right),$$

where $\Delta P/\Delta X$ is the change of pressure over distance ΔX in cm H_2O/cm, ζ the fluid density in g/cm^3, a the inner diameter of the vessel, F the fluid flow rate in cm^3/sec, μ the fluid viscosity, and g is 980 cm/sec^2.

OTHER SENSOR SYSTEMS

Another gas flow system is shown in Fig. 7-5. Two versions are presented; Fig. 7-5A is a magnetic system, while Fig. 7-5B is an optical system. In the magnetic sensor, Fig. 7-5A, a small magnet is introduced into the flow stream. This form is used for both liquids and gases, and it is also usable in closed systems where it is difficult to introduce other forms of sensors. A pair of coils, $L1$ and $L2$, are placed at right angles to the flow path, at the point where the magnetic rotor

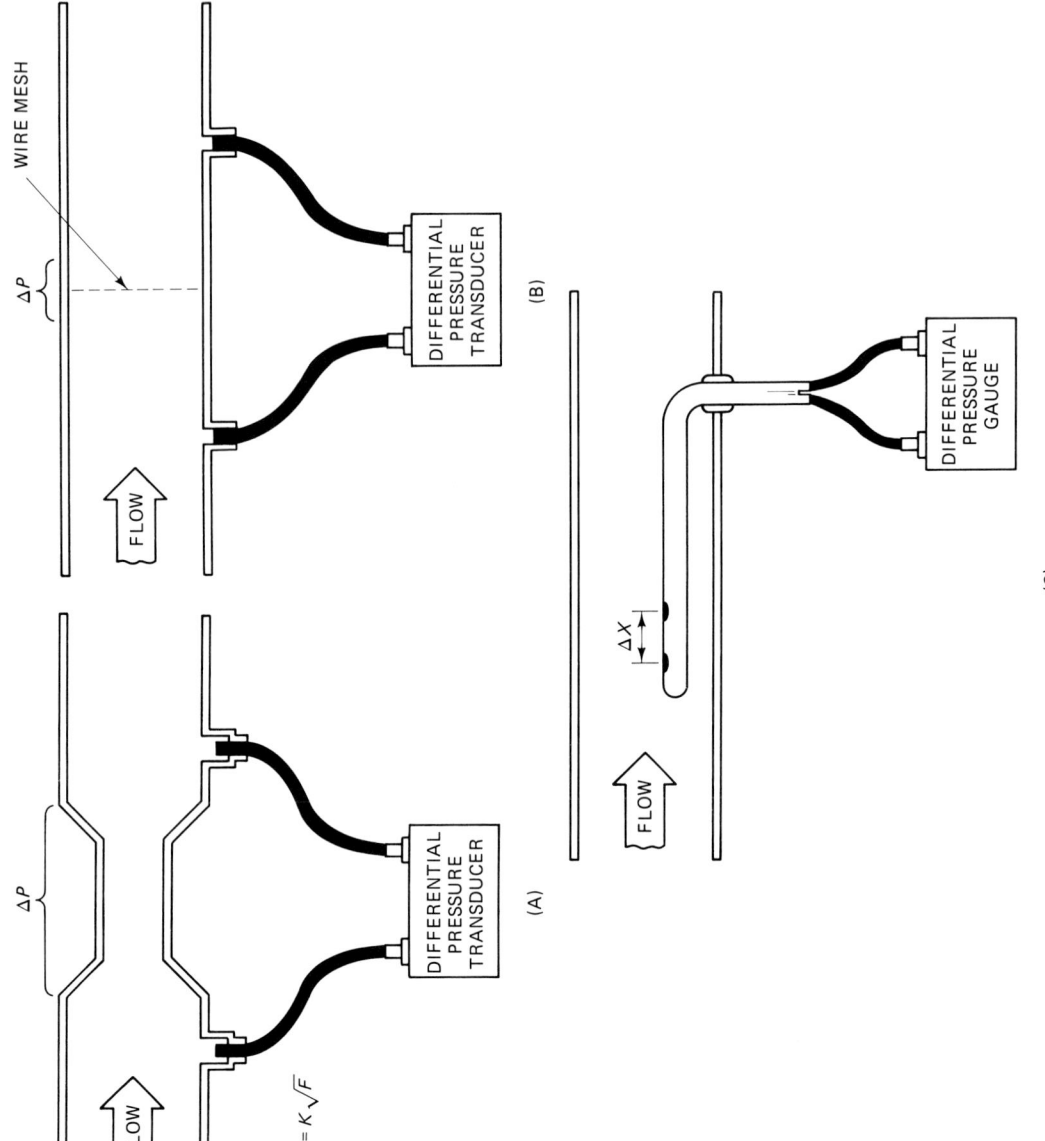

Fig. 7-4 Pressure flow sensors by restriction in the line: (A) reduced section version, (B) wire mesh screen version, and (C) dual lumen cannula version.

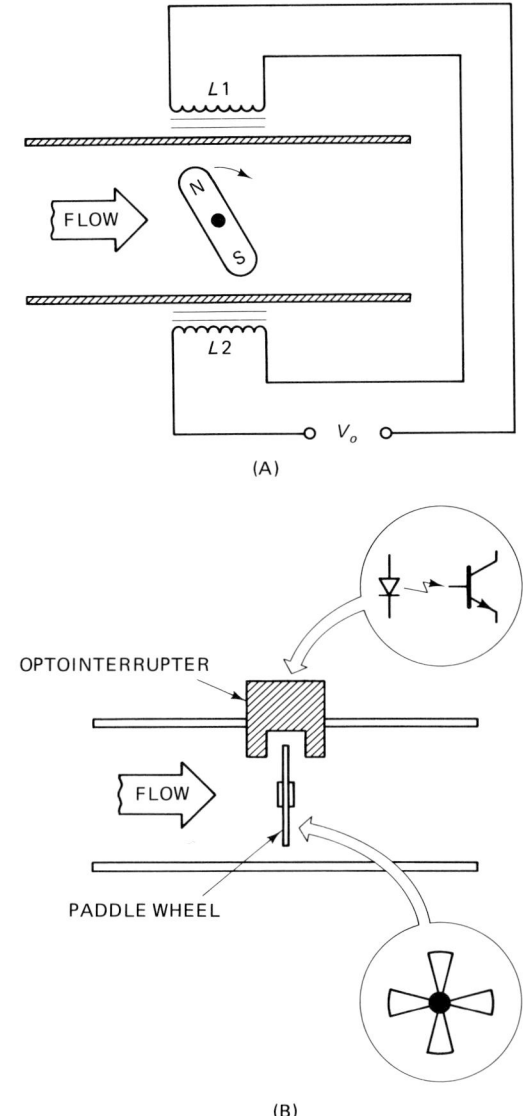

Fig. 7-5 (A) Magnetic flow sensor; (B) paddle wheel flow sensor.

is placed. When a moving magnetic field cuts across the turns of a coil, a current is introduced into the coil. Thus, a voltage, V_0, is found at the output of the series-aiding connected coils. The amplitude of the voltage is proportional to the magnetic field, while its frequency is proportional to the rotational frequency of the magnet. The rotational of the magnet is related to the flow rate. Sensors of this type are used in a wide variety of instrumentation applications.

Figure 7-5B shows an optical version. An optionterrupter is a device that places an LED and a phototransistor across an open path. When the path is blinded, the phototransistor is darkened; when the

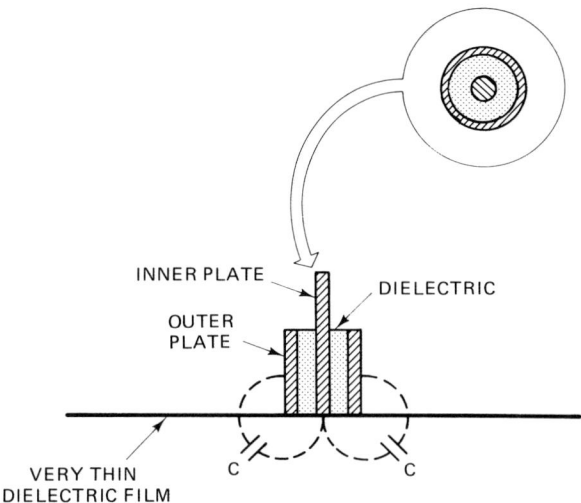

INNER PLATE

DIELECTRIC

OUTER
PLATE

VERY THIN
DIELECTRIC FILM

C   C

Fig. 7-6 The dielectric properties of very thin film make a capacitive sensor for film flow.

path is not blinded, the phototransistor is illuminated. Similar interrupters are used in other applications, such as the PAPER OUT sensor in computer printers and tape sensors in recorders. The light path in the sensor of Fig. 7-5B is interrupted by a multi-bladed fan placed in the flow path. Again, the frequency of the output signal is proportional to the flow rate.

With the right circuitry, a rotating flow rate sensor can also produce a flow volume signal. By using the AC signal to trigger a one-shot multivibrator, we obtain a train of pulses that are of constant duration and constant amplitude; the only variation is the pulse repetition rate (which is equal to the AC signal frequency from the sensor). Integrating these pulses produces a DC voltage that is proportional to the total area under all of the pulses per unit of time, in other words, the flow volume (i.e., integral of rate).

Figure 7-6 shows a method for measuring the flow of a thin liquid film. In this case, we have a sensor that is a coaxial, cylindrical capacitor. The nonconducting film flowing across the surface changes the dielectric constant of the capacitor, hence its capacitance. The actual capacitance can be measured in any of several ways, for example, either an electrometer or an FM oscillator can be used.

ULTRASONIC FLOW MEASUREMENT SYSTEMS

Ultrasonic waves are acoustical waves (i.e., vibrations in a media) that have a frequency above the range of human hearing. In measurement applications, ultrasonic can mean anything from about 20 kHz to more than 20 MHz. The important thing is that they are acoustical

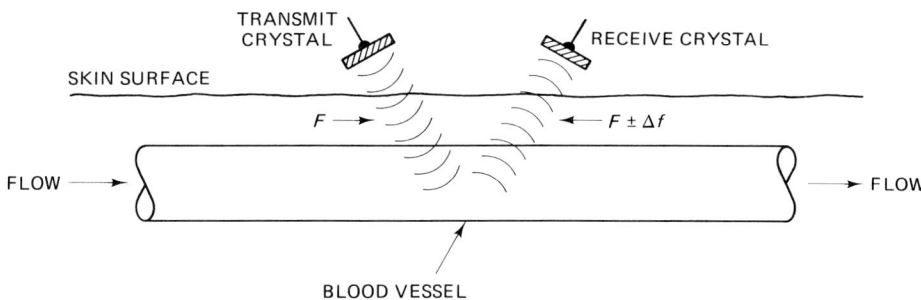

Fig. 7-7 Doppler flow detector will notify the presence of flow, but it is not easy to quantify.

waves, not electromagnetic (i.e., radio) waves. There are several types of ultrasonic sensors available, but most of them are either dynamic or piezoelectric. The dynamic forms are analogous to dynamic microphones: a thin diaphragm stretched over an electromagnet. These sensors are used for relatively low frequencies (< 100 kHz). The piezoelectric crystal forms of ultrasonic transducers are analogous to the crystal microphone. A piezoresistive crystal vibrates when an AC electrical signal of its resonant frequency is applied. This vibrational energy is applied either to a diaphragm or directly to the media being measured. Similarly, the same crystal will also produce an AC signal on its resonant frequency when vibrated by an acoustical wave of that frequency. Thus, the piezoelectric crystal can be used for both transmitting and receiving functions (this dual usage is why the term transducer is preferred here).

Figure 7-7 shows a Doppler flow detector based on piezoelectric crystals. Two crystals are used, one for transmitting and one for receiving. The frequency of the incident signal (F) is changed by the Doppler effect as fluid flows underneath the sensor. The reflected energy will contain frequency components of $F \pm \Delta F$, where ΔF is the Doppler shift.

In blood flow monitors based on this technique, it is nearly impossible to calibrate this sensor for the blood flow rate. The system is used to check for vessel patency, that is, whether or not blood is flowing, but it is not used to measure the blood flow rate. In other systems, the calibration might be possible, however, because the Doppler frequency shift is proportional to the fluid velocity. By filtering out those Doppler cells, we ought to be able to calibrate the system. The problems in medical systems are that blood vessels are distensible, that is, they flex as blood flows; blood flow is pulsatile; and blood vessels are not located at a constant distance beneath the surface.

Physicians use the Doppler flow meter to detect the presence of blood flowing in underlying arteries in arms and legs, especially after surgery or when treating a crushing injury. The sloshing sound created when the Doppler return signal is heterodyned with the incident signal

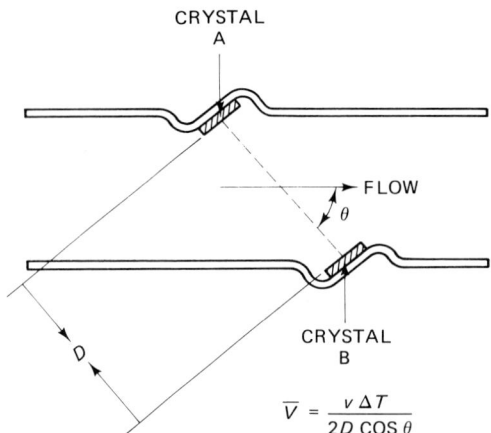

Fig. 7-8 Ultrasonic flow detector depends on the differential upstream and downstream transit times of the acoustical signal.

tells them that the vessel is patent. At one time, some physicians believed that the direction of the blood flow could be easily determined, provided that upstream and downstream receiving crystals were positioned on either side of the transmitting crystal. Not believing the theory, but eager to serve customers, one manufacturer made one of these so-called stereo flow meters, but by telephone informed inquirers that "it confers absolutely no advantage over the old monaural model, except for the fact that, after the patient is released, you can go back to the surgeon's locker room and brag: 'Yes, but mine is in stereo!'"

An example of a transit time flow sensor is shown in Fig. 7-8. This system depends upon the fact that the upstream and downstream transit times of acoustical pulses are different. A pair of piezoelectric crystal sensors are aimed at each other in an oblique path across the flow path. The angle between the crystal path and the flow path is θ. Both crystals are used for both transmitting and receiving functions. The system first fires a pulse from A to B and measures its transit time; next, it fires an upstream pulse from B to A and measures its transit time. The average flow velocity is proportional to the difference between the upstream and downstream transit times (ΔT):

$$\overline{V} = V\Delta T/2D \cos\theta,$$

where $\overline{V}$ is the average flow velocity, V the speed of the signal in the media, ΔT the difference between downstream and upstream transit times, and D and θ are as defined in Fig. 7-8.

A Doppler system is shown in Fig. 7-9. This system uses the frequency change of a wave scattered from particulate matter flowing in the fluid path, and it is particularly useful in blood flow meters. It has been shown that the change in frequency ΔF, is given by

$$\Delta F = VF_s(\cos\theta 1 + \cos\theta 2)/c,$$

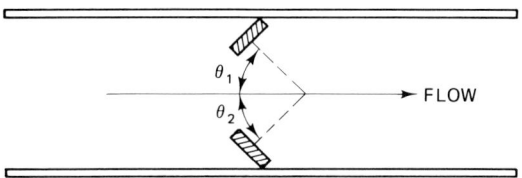

Fig. 7-9 Doppler flow meter.

where ΔF is the Doppler shift frequency, F_s the source frequency, V the fluid velocity, c the velocity of sound in the media, θ_1 the angle of the receiving crystal to the flow axis, and θ_2 the angle of the transmitting crystal to the flow axis.

ELECTROMAGNETIC FLOW METERS

An electromagnetic flow transducer uses as the transducible property the fact that a conductor moving in a magnetic field will create an electromotive force across its length. Figure 7-10 shows how this fact is used to measure the flow of conductive fluids. A vessel of radius a is carrying the fluid, with the flow coming out of the page, as shown by the vector arrow symbol inside the vessel. A magnetic field is set up by a horseshoe electromagnet, such that the magnetic field cuts across the fluid path. A pair of electrode contacts is placed in contact with the fluid orthogonal to the magnetic field. The electromotive force (emf) appearing across these electrodes is

$$V = FB/50\pi a,$$

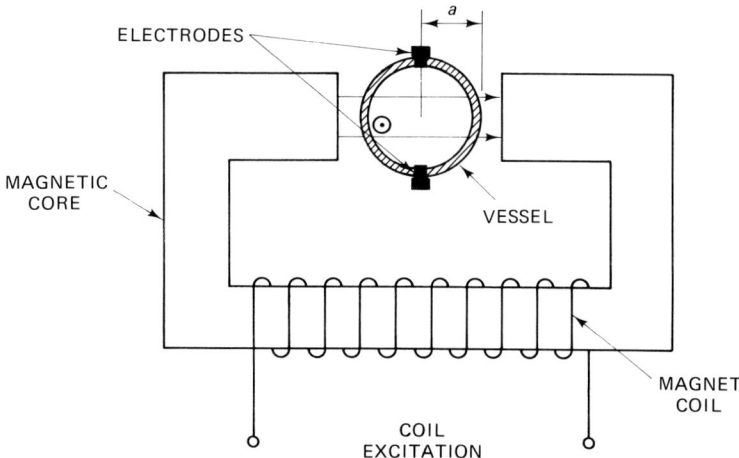

Fig. 7-10 Electromagnetic flow detector uses a magnetic field and orthogonal electrodes.

where V is the emf appearing across the electrodes in microvolts (μV), F the fluid volumetric flow in cm^3/sec, B the magnetic field in gauss (G), and a the radius of the vessel in centimeters.

Example

Calculate the potential appearing across the electrodes on a 1.2-cm-radius vessel, when a magnetic field of 350 G is present and the flow rate is 250 cm^3/sec.

Solution

$$V = FB/50\pi a = (250 \text{ cm}^3/\text{sec})(350 \text{ G})/(50)(3.14)(1.2 \text{ cm}) = 464 \ \mu V.$$

The potentials associated with the electromagnetic flow sensor are easily handled in electronic circuitry, making this method quite feasible for a broad range of applications.

8

Electrodes for Bioelectric Sensing

Bioelectrodes are a class of sensor that acquires the naturally occurring biophysical electrical signals, such as ECGs, EEGs, and electromyograms (EMG). Most such biopotentials are acquired from one of three forms of electrode: surface macroelectrodes, indwelling macroelectrodes, and microelectrodes. Of these, the first two are generally used *in vivo*, while the latter are used in *in vitro*. In this chapter, we will discuss the acquisition of biopotentials by dealing with the types of electrodes commonly used in biomedical instrumentation. Again, it should be recognized that the discussion is generic and representative, not exhaustive.

A problem with bipotential electrodes results from the fact that there is an exchange of ions between the metallic electrode and living tissue, which is effectively an electrolytic substance as far as electrical circuits are concerned. A half-cell potential exists across the metal–tissue interface, which ranges from -3 to $+3$ V, with -2 to $+2$ V being the most common. Although a range is given for these potentials in the various electrochemistry textbooks, the value for any one metallurgical mixture is relatively stable if the mechanical configuration of the electrode–tissue interface is stable.

Besides its initial material dependency, the actual potential exhibited by any given electrode may change slowly with time. Some candidate materials look good initially, but have such a large change with time and chemical environment that they are rendered almost useless in practical applications. For many materials, body fluids are terribly corrosive. For this reason, materials such as gold, some tungsten alloys, silver–silver chloride (Ag–AgCl), platinum, and platinum–platinum black are used to make practical biopotential electrodes. In general use for simple surface recording of bipotentials is the

Ag–AgCl electrode. Unless otherwise specified, you can generally assume this material is used.

By international scientific agreement, the zero reference point is the hydrogen–hydrogen (H–H) electrode, which is assigned a half-cell potential of 0 V by convention. All other electrode half-cell potentials are measured against the H–H zero reference. The half-cell potentials cited for any given electrode are the differential potential between the actual electrode and the H–H reference electrode.

The electrode half-cell potential becomes a tremendous problem in bioelectric signals acquisition because of the tremendous difference between these DC potentials and the biopotentials. A typical half-cell potential for a biomedical electrode is 1.5 V, while biopotentials are more than 1000 times less than the half-cell potential! The surface manifestation of the ECG signal is 1 to 2 mV, while EEG scalp potentials are on the order of 50 μV. Thus, the half-cell electrode voltage is 1500 times greater than the peak ECG potential, and 30,000 times greater than the EEG signal.

The instrument designer must provide a strategy for overcoming the effects of the massive half-cell potential offset. Because the half-cell potential forms a large DC component for the minute signal voltage, it is necessary to find an appropriate strategy that uses a combination of the following approaches.

First, we could use a differential DC amplifier to acquire the signal. If the electrodes are identical, then the half-cell potentials should be the same. Theoretically, at least, the equal potentials would be seen as a single, common-mode potential, thus they would cancel in the output. A limitation on this approach is that the gains required to process low-level signals also act on tiny differences between the two half-cell potentials. A difference of 1 mV—that is only 0.1% of the total—looks like any other 1 mV DC signal to the gain of a 1000-ECG amplifier.

Second, the signals acquisition circuit must be designed to provide a counter-offset voltage to cancel the half-cell potential of the electrode. While this approach has certain appeal, it is limited by the fact that the half-cell potential changes with time and the relative motion between the skin and electrode. Electrode motion can cause a widely varying baseline.

Third, we can AC-couple the input amplifier. This approach permits removal of the signal component from the DC offset. This option is, perhaps, the most appealing, especially where variations of the DC offset are of substantially lower frequency than the signal frequency components. In that case, the normal -3-dB frequency response limit can be of use to tailor the attenuation of variations in the DC offset. In some biomedical applications, however, signal components are near DC. For example, the frequency content of the ECG signal is 0.05 to 100 Hz. In medical ECG equipment, therefore, one can expect the baseline to shift every time the patient moves in bed.

In most cases, the first and third options are selected for biopotential amplifiers. The user will require an AC-coupled, differential input amplifier for signals acquisition.

SURFACE ELECTRODES

Surface electrodes are those that are placed in contact with the skin of the subject. Also in this category are certain needle electrodes of such a size as to prevent their being inserted inside a single cell (which is the criterion that defines a microelectrode). There is some case for including needle electrodes under the rubric indwelling electrodes, but that is not generally the practice in biomedical engineering.

Surface electrodes (other than needle electrodes) vary from 0.3 to 5 cm in diameter, with most being in the 1-cm range. Human skin tends to have a very high impedance compared with other voltage sources. Typically, normal skin impedance as seen by the electrode varies from 0.5 kΩ for sweaty skin surfaces to more than 20 kΩ for dry skin surfaces. Problem skin, especially dry, scaley, or diseased skin, may reach impedances in the 500-kΩ range. In any event, one must treat surface electrodes as a very high impedance voltage source—a fact that seriously influences the design of biopotential amplifier input circuitry. In most cases, the rule of thumb for a voltage amplifier is to make the input impedance of the amplifier at least 10 times the source impedance. For biopotential amplifiers, this requirement mean 5 MΩ or greater input impedance, a value easily achieved using premium bipolar, BiFET, or BiMOS operational amplifiers.

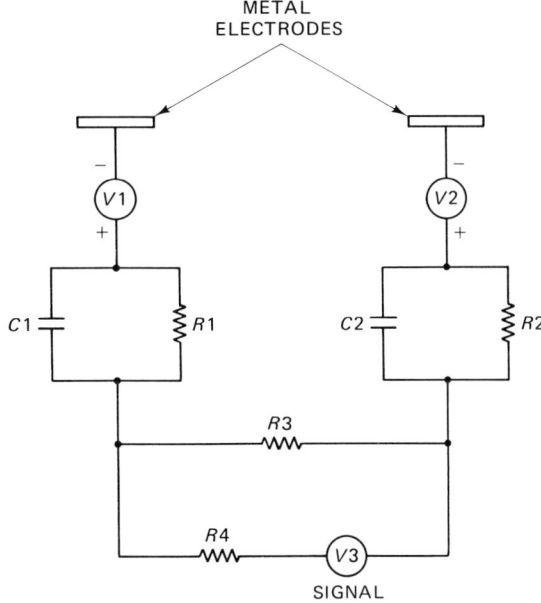

Fig. 8-1 Equivalent circuit models a pair of skin contact metallic electrodes.

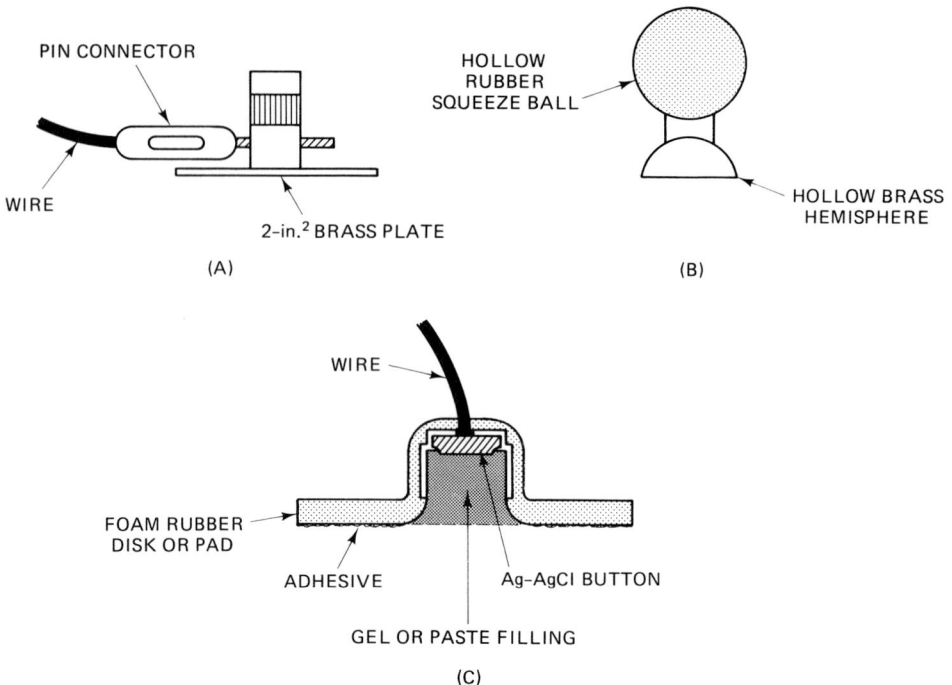

Fig. 8-2 (A) Brass plate ECG electrode, (B) suction cup ECG electrode, and (C) tower electrode.

ELECTRODE MODEL CIRCUIT

Figure 8-1 shows a circuit model of a biomedical surface electrode. This model more or less matches the equivalent circuit of ECG and EEG electrodes. In this circuit, a differential amplifier is used for signals processing, so it will cancel the effects of electrode half-cell potentials $V1$ and $V2$. Resistance $R3$ represents the internal resistances of the body, which are typically quite low. The biopotentials signal is represented as a differential voltage, $V3$. The other resistances in the circuit represent the resistances at the electrode–skin contact interface. The surprising aspect of Fig. 8-1 is the usual values associated with capacitors $C1$ and $C2$. While some capacitance is normally expected, it surprises people to learn that these contact capacitances can attain values of several microfarads.

TYPICAL SURFACE ELECTRODES

A variety of electrodes have been designed for surface acquisition of biomedical signals. Perhaps the oldest form of ECG electrodes in clinical use is the strap-on variety (Fig. 8-2A). These electrodes are 1- to 2-in.[2] brass plates that are held in place by rubber straps. A conductive gel or paste is used to reduce the impedance between the electrode and skin.

A related form of ECG electrode is the suction cup electrode shown in Fig. 8-2B. This device is used as a chest electrode in short-term ECG recording. For longer term recording or monitoring, such as continuous monitoring of a hospitalized patient in a coronary or intensive care unit, the paste-on column electrode is used instead.

A typical column electrode is shown schematically in Fig. 8-2C. The electrode consists of an Ag–AgCl metal contact button at the top of a hollow column that is filled with a conductive gel or paste. This assembly is held in place by an adhesive coated foam rubber disk.

The use of a gel- or paste-filled column that holds the actual metallic electrode off the surface reduces movement artifact. For this reason (among several others), the electrode in Fig. 8-2C is preferred for the monitoring of hospitalized patients.

There are, however, several problems associated with this type of electrode. One of the problems is the inability of the adhesive to stick on sweaty or "clammy" skin surfaces for long periods of time. The user also must avoid placing the electrode over boney prominences. Usually, the fleshy portions of the chest and abdomen are selected as electrode sites. Various hospitals have different protocols for changing the electrodes, but in general, the electrode is changed at least every 24 h, and often more frequently as few last as long as 24 h. In some hospitals, the electrode sites are moved, and the electrodes changed, once every 8-h nursing shift in order to avoid ischemia of the skin at the site.

The surface electrodes that we have discussed thus far are noninvasive types. That is, they adhere to the skin without puncturing it. In Fig. 8-3, we see the needle electrode. This type of ECG electrode is inserted into the tissue immediately beneath the skin by puncturing the skin at a large oblique angle (i.e., close to horizontal with respect to the skin surface). The needle electrode is used only for exceptionally poor skin, especially on anesthetized patients, and in veterinary situations. Of course, infection is an issue in these cases, so needle electrodes are either disposed of (one-time use) or resterilized in ethylene oxide gas.

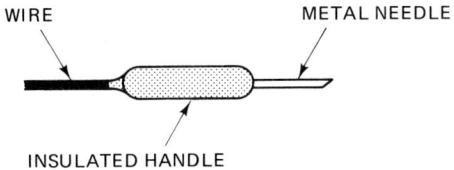

Fig. 8-3 Needle electrode.

INDWELLING ELECTRODES

Indwelling electrodes are those that are intended to be inserted into the body. These are not to be confused with needle electrodes, which are intended for insertion into the layers beneath the skin. The indwelling electrode is typically a tiny, exposed metallic contact at the end of a long, insulated catheter. In one application, the electrode is threaded through the patient's veins (usually in the right arm) to the right side of the heart in order to measure the intracardiac ECG waveform. Certain low-amplitude, high-frequency features (such as the bundle of His element) become visable only when an indwelling electrode is used.

MICROELECTRODES

The microelectrode is an ultrafine device that is used to measure biopotentials at the cellular level (Fig. 8-4). In practice, the microelectrode penetrates a cell that is immersed in an infinite fluid (such as physiological saline), which is in turn connected to a reference electrode. Although several types of microelectrode exist, most of them are of one of two basic forms: metallic contact or fluid filled. In both cases, an exposed contact surface of about 1–2 μm (1 μm = 10^{-6} m) is in contact with the cell. As might be expected, this fact makes microelectrodes very high impedance devices.

Figure 8-5 shows the construction of a typical glass–metal microelectrode. A very fine platinum or tungsten wire is slip-fit through a 1.5–2 mm glass pipette. The tip is etched, and then it is fire formed

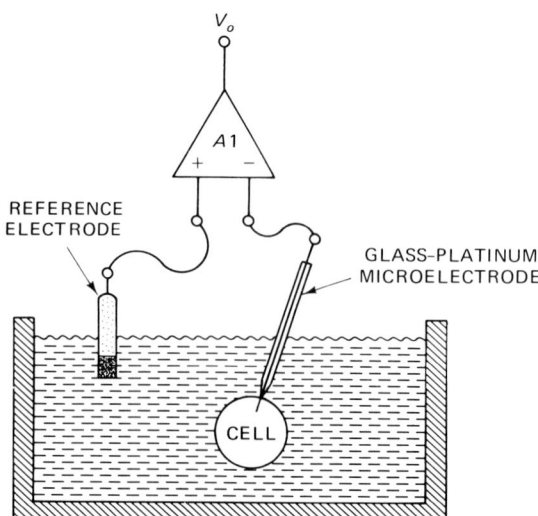

Fig. 8-4 Use of the microelectrode to measure cell potentials *in vitro*.

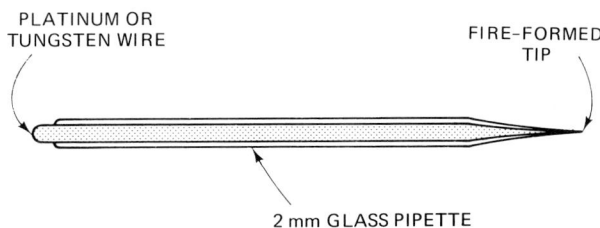

Fig. 8-5   Metallic wire microelectrode.

into the shallow angle taper shown. The electrode can then be connected to one input of the signal amplifier.

There are two subcategories of this type of electrode. In one type, the metallic tip is flush with the end of the pipette taper, while in the other there is a thin layer of glass covering the metal point. This glass layer is so thin it is measured in angstroms, and it drastically increases the impedance of the device.

The fluid-filled microelectrode is shown in Fig. 8-6. In this type, the glass pipette is filled with a 3-M solution of potassium chloride (KCl), and the large end is capped with an Ag–AgCl plug. The small end need not be capped because the 1-μm opening is small enough to contain the fluid.

The reference electrode is likewise filled with a 3-M solution of KCl, but it is very much larger than the microelectrode. A platinum plug contains fluid on the interface end, while an Ag–AgCl plug caps the other end.

Figure 8-7 shows a simplified equivalent circuit for the microelectrode (disregarding the contribution of the reference electrode). Analysis of this circuit reveals the signals acquisition problem due to the RC components. Resistor $R1$ and capacitor $C1$ are due to the effects at the electrode–cell interface and are (surprisingly) frequency dependent. These values fall off to a negligible point at a rate of $1/(2\ \pi F)^2$, and are generally considerably lower than R_s and $C2$.

Resistance R_s in Fig. 8-7 is the spreading resistance of the electrode and is a function of the tip diameter. The value of R_s in metallic microelectrodes without the glass coating is approximated by

$$R_s = P/4\pi r,$$

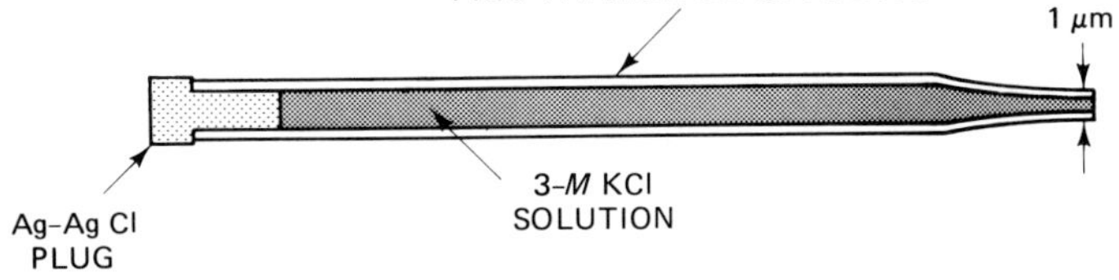

Fig. 8-6   KCl-filled liquid microelectrode.

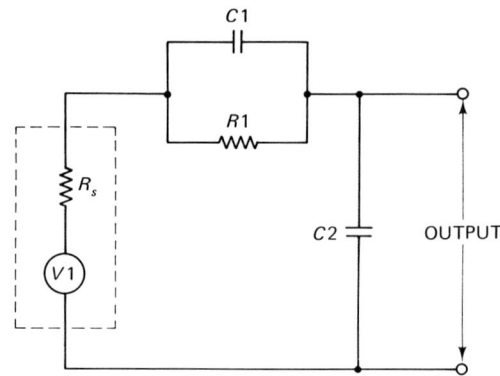

Fig. 8-7 Equivalent circuit of microelectrode.

where R_s is the spreading resistance in ohms, P the resistivity of the infinite solution outside the electrode (e.g., 70 Ω cm for physiological saline), and r the tip radius (typically 0.5 μm for a 1-μm electrode).

Assuming the typical values given in the previous equation, calculate the tip spreading resistance of a 1-μm microelectrode.

$$R_s = \frac{P}{4\pi r} = \frac{70 \ \Omega \ cm}{4\pi 0.5 \ \mu m \ (10^{-4}/\mu m)} = 111,999 \ \Omega.$$

The impedance of glass-coated metallic microelectrodes is at least 1 or 2 orders of magnitude higher than this figure.

For fluid-filled KCl microelectrodes with small taper angles ($\pi/180$ rad), the series resistance is approximated by

$$R_s = 2P/\pi ra,$$

where R_s is the resistance in ohms, P the resistivity (typically 3.7 Ω cm for $3M$ KCl), r the tip radius (typically 0.1 μm), and a the taper angle (typically $\pi/180$).

Example

Find the series impedance of a KCl microelectrode using the values shown in the previous equation.

$$R_s = \frac{2P}{\pi ra} = \frac{(2)(3.7 \ \Omega \ cm)}{(3.14)[0.1 \ \mu m \ (10^{-4}/\mu m)](3.14/180)} = 13.5 \ M\Omega.$$

The capacitance of the microelectrode is given by

$$C2 = \frac{0.55e}{\ln(R/r)} \ pF/cm,$$

where e is the dielectric constant of glass (typically 4), R the outside tip radius, r the inside tip radius (r and R are in the same units).

Example

Find the capacitance of a microelectrode if the pipette radius is 0.2 μm and the inside tip radius is 0.15 μm.

$$C2 = \frac{0.55e}{\ln(R/r)} \text{ pF/cm} = \frac{(0.55)(4)}{\ln(0.2\ \mu\text{m}/0.15\ \mu\text{m})} \text{ pF/cm} = 7.7 \text{ pF/cm}.$$

How do these values affect the performance of the microelectrode? Resistance R_s and capacitor $C2$ operate together as an RC low-pass filter. For example, a KCl microelectrode immersed in 3 cm of physiological saline has a capacitance of approximately 23 pF. Suppose it is connected to the amplifier input (15 pF) through 3 ft of small-diameter coaxial cable (27 pF/ft, or 81 pF). The total capacitance is 119 pF (23 + 15 + 81). Given a 13.5-MΩ resistance, the frequency response (at the -3-dB point) is

$$F = 1/2\pi RC,$$

where F is the -3-dB point in hertz, R the resistance in ohms, and C the capacitance in farads.

Example

For $C = 119$ pF (1.19×10^{-10} F) and $R = 1.35 \times 10^7$ Ω, find the frequency response of the upper -3-dB point.

$$F = 1/(2)(3.14)(1.35 \times 10^7)(1.19 \times 10^{-10}) = 99 \text{ Hz} \cong 100 \text{ Hz}.$$

Clearly, a 100-Hz frequency response, with a -6-dB/octave characteristic above 100 Hz, results in severe rounding of the fast rise-time action potentials. A strategy must be devised in the instrument design to overcome the effects of capacitance in high-impedance electrodes.

Neutralizing Microelectrode Capacitance

Figure 8-8 shows the standard method for neutralizing the capacitance of the microelectrode and associated circuitry. A neutralization capacitance, C_n, is in the positive feedback path, along with a potentiometer voltage divider. The value of this capacitance is

$$C_n = C/(A-1),$$

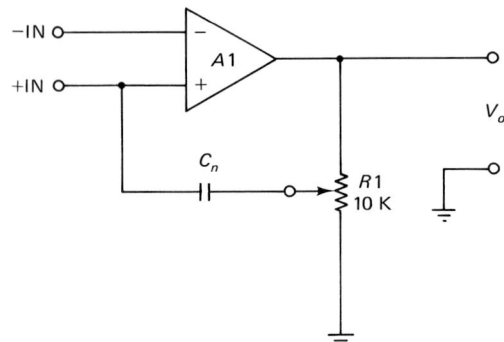

Fig. 8-8 Amplifier configuration to neutralize microelectrode capacitance.

where C_n is the neutralization capacitance, C the total input capacitance, and A the gain of the amplifier.

Example

A microelectrode and its cabling exhibit a total capacitance of 100 pF. Find the value of neutralization capacitance (Fig. 8-8) required for a gain-of-10 amplifier.

$$C_n = C/(A - 1) = (100\text{ pF})/(10 - 1) = 100\text{ pF}/9 = 11\text{ pF}.$$

9

Physical Measurements and Electronic Circuits

Sensors are used in electronic circuits to produce the information required. In this chapter, we will examine certain instrumentation methods in order to set the context for the types of electronic circuits that might be needed in other applications. Several different examples are selected to represent the larger universe of applications.

PHOTOCOLORIMETRY

One of the most basic forms of instrument circuits is also both the oldest and most commonly used: photocolorimetry. This device is used to measure the oxygen content trending in blood, the CO_2 content of air, the water vapor content in a gas, blood electrolyte (sodium, Na and potassium, K) levels, and a host of other similar measurements.

Photocolorimetry is basically a measurement technique in which light (or IR and UV) transmission over two paths is compared. Figure 9-1 shows the basic circuit of the most elementary form of colorimeter. Although the circuit is very basic, it is the actual circuit found in a widely used blood oxygen meter. The circuit is the Wheatstone bridge, and it uses a pair of photoresistor cells ($R2$ and $R4$) as the light sensors. Potentiometer $R5$ in Fig. 9-1 is used as a bridge balance control, and it is adjusted for zero output ($V_o = 0$) when the same light shines on both photoresistors. The output voltage from the bridge (V_o) will be zero when the two legs of the bridge are balanced. In other words, V_o is zero when $R1/R2 = R3/R4$. It is not necessary for the resistor elements to be equal (although that is often the case), only that their resistance ratios be equal. Thus, a 500 kΩ/50 kΩ ratio for

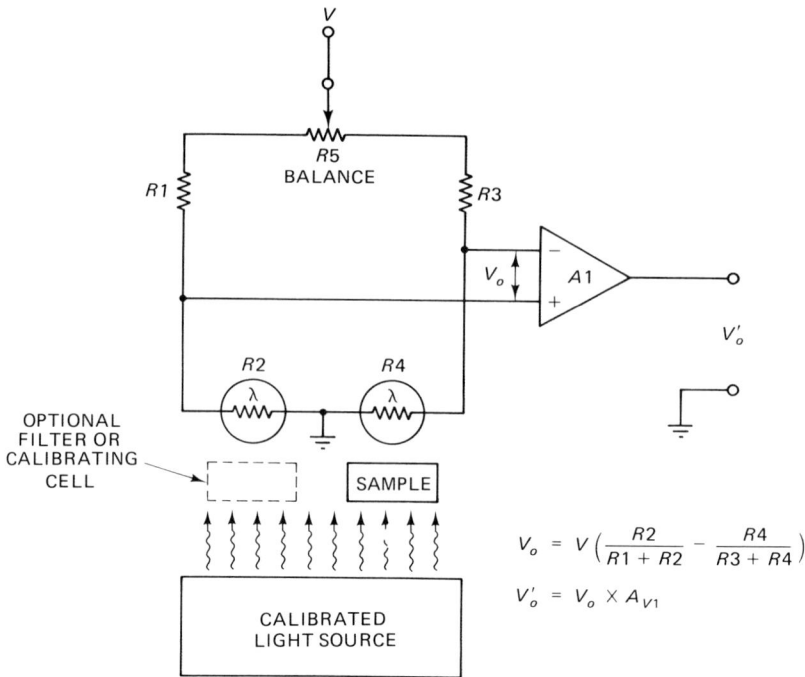

Fig. 9-1 Photocolorimeter places two photoresistors in a Wheatstone bridge circuit.

$R1/R2$ will produce zero output voltage when $R3/R4 = 100\ \text{k}\Omega/10\ \text{k}\Omega$.

The photoresistors are arranged so that light from a calibrated source illuminates both equally and fully, except when an intervening filter or sample is present in one or both pathways. Thus, the bridge can be nulled to zero using potentiometer $R5$ under this zero condition. In most instruments that are based on this principle, a translucent sample is placed between the light source and one of the photocells. The amount of transmission to light allowed by the sample is a measure of its density and is thus transducible. Let us look at a couple of different types of instruments to see how this principle is applied.

Blood O$_2$ Level

A classical (but still widely used) method for measuring blood oxygen level is the basic colorimeter of Fig. 9-1. It works because the redness of blood is a measure of its oxygenation. This instrument is nulled with neither a standard filter cell nor blood in the light path. A standard filter cell is introduced between the light source and $R2$, and a blood sample is placed in a standardized tube between the light source and $R4$. The degree of blood O$_2$ saturation in the sample is thus reflected by the difference in the bridge reading between the sample path and filter path. On one model, a separate resistor across the $R1/R2$ arm is used to bring the bridge back into null condition, and the dial for that resistor is calibrated in percent O_2. More modern instruments based on digital computer techniques provide the measurement in a more automatic manner.

Respiratory CO_2 Level

The exhaled air from humans is roughly 2 to 5% carbon dioxide (CO_2), while the percentage of CO_2 in normal room air is negligible. A popular form of "end CO_2 meter" is based on the fact that CO_2 absorbs IR waves at three discrete wavelengths. The light source in that type of photocolorimeter is actually either an IR LED or a cal rod (identical to the one that heats your coffee pot); the photocells are selected for good IR response. In this type of instrument, room air is passed through a glass cuvette placed between $R2$ and the heat source, while patient expiratory air is passed through the same type of cuvette placed between the heat source and $R4$. The difference in IR transmission across the two paths is a function of the percentage of CO_2 in the sample circuit.

The associated electronics (not shown) will allow zero and maximum span (i.e., gain) adjustments. The zero point is adjusted with room air in both cuvettes, while the maximum scale (usually 5% CO_2) is adjusted with the sample cuvette purged of room air and replaced with a calibration gas (usually 5% CO_2, 95% nitrogen). This calibration gas must be obtained from a local supplier, and it must be specified as a calibration gas. Otherwise, the quantities may be only approximate. Also, be sure of the type of measurement: calibration gases are available by either weight or volume.

Blood Electrolytes

Blood chemistry tests often include levels of sodium (Na) and potassium (K). An instrument commonly used for these forms of measurement is the flame photometer (Fig. 9-2). This form of colorimeter replaces the light source with a flame produced by a gas carburetor. The sample is injected into the carburetor and burned along with the gas–air mixture. The colors emitted on burning are proportional to the concentrations of Na and K ions in the sample. A special gas is used to burn cleanly with a blue flame when no sample is present. In medical applications, a specified size sample of a patient's blood is mixed with an indium or lithium calibrating solution (also a predetermined amount). The solution is well mixed and then applied to the carburetor. The instrument can infer the concentration of the two elements by comparing the intensities of the colors generated by burning Na and K ions with the intensity of the calibration color.

A warning is an order regarding these flame photometers. Flame photometers can give terribly flawed readings if improperly maintained. In all cases, it was not the fault of the instrument, but rather the

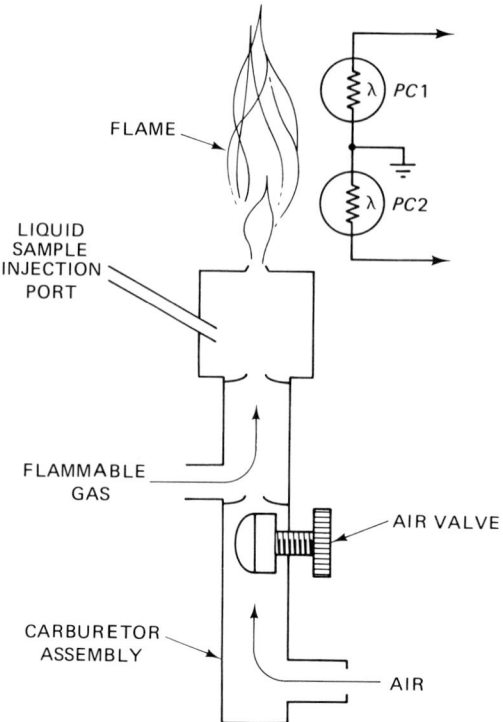

Fig. 9-2 Flame photometer colorimeter.

operator. The carburetor and surrounding glass structures must be cleaned frequently, or the buildup of material from past tests will unduly bias the results of the present test. In addition, pure carbonization of the associated glass windows will obscure the flame and may create both a lack of sensitivity and an erroneous reading (especially if the windows are not carboned uniformly).

The uses of colorimeters does not end with the medical laboratory. The clue to looking for a transducible event is the detection of either a density change or an absorption differential to one or two wavelengths, which is a transducible function of the parameter being measured; for example, IR light is absorbed by CO_2, and a certain O_2 saturation passes light at an 800-nm wavelength.

In the past, we had to rely only on those phenomena that were linear or changed in an easily discerned manner (e.g., logarithmically or exponentially). Today, however, the modern programmable digital microcomputer can be trained to unravel quite nonlinear phenomena and present us with an intelligible bit of information. We can either program the computer to solve the complicated mathematical equation that describes the phenomena or create a lookup table from purely empirical data. Alternatively, with certain types of curves, we can create a reverse or inverse curve to fit the data and output that to the human operator.

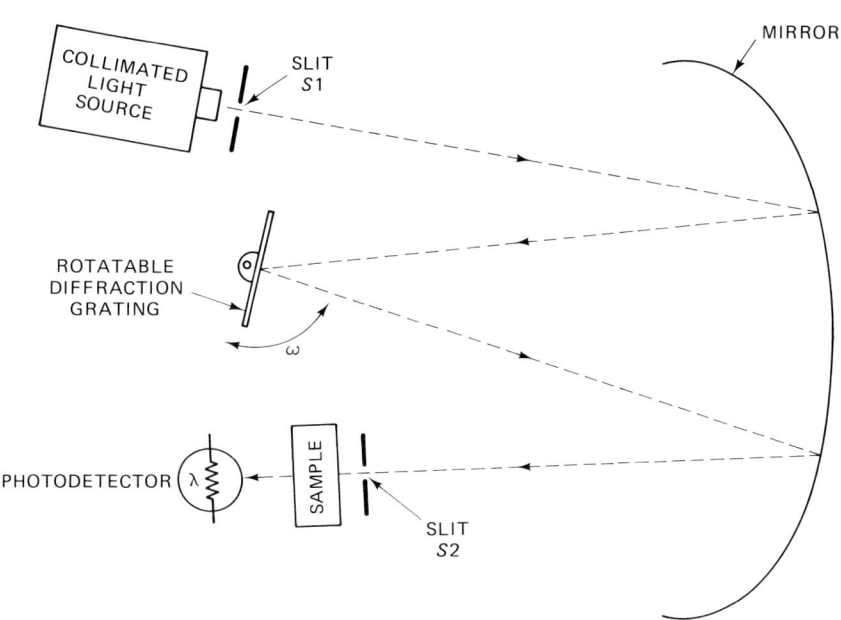

Fig. 9-3 Photospectrometer.

PHOTOSPECTROMETER

This class of instruments depends upon the fact that certain chemical compositions differentially absorb wavelengths of light. If you are familiar with the spectrum of sunlight created when light passes through a prism, or is reflected off of a diffraction grating, then you are familiar with the basis for the photospectrometer (see Fig. 9-3). Either the light beam or the photodetector must be moved to examine each of the light wavelengths in turn. In most cases, the sample and the photodetector are kept stable, while a diffraction grating or prism is rotated to permit all colors of the spectrum to fall on the same photodetector. By comparing the amplitudes of the light at different angles, we can infer the transmission of the respective colors, and create a chart that shows amplitude versus wavelength.

RATIOMETRIC MEASUREMENTS

A significant problem with photometric measurements is keeping the light source constant in both intensity and color. The simple fact is that all forms of light emitter change, and that change introduces artifact to the acquired data, especially if it occurs between calibration and measurement. The answer to many of these problems is the use of ratiometric measurement techniques (Fig. 9-4).

In a ratiometric system, the collimated light beam is passed through a 50% beam-splitting prism. This optical device splits the

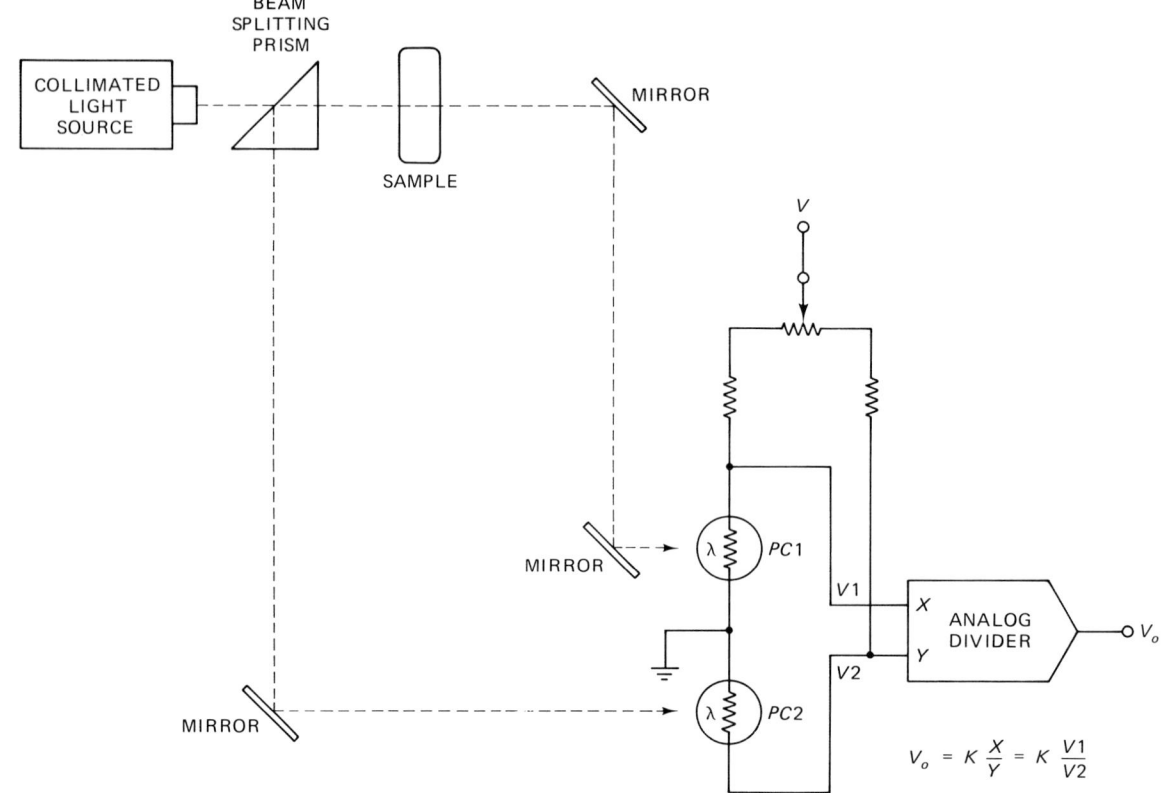

Fig. 9-4   Ratiometric photocolorimeter.

beam into two equal amplitude beams at right angles to each other. The incident beam is passed through an optical path of length L to photosensor $PC2$; the sample beam also passes through an optical path of length L, and it passes through the sample being measured as well. If the optical properties of the two paths are the same and they are of the same length, then the light intensities arriving at the two colorimeter photosensors will be the same, except for any lost in the sample. Thus, the Wheatstone concept can be used. First, however, some processing of the signal is needed. The instrument takes the ratio of $V1$ and $V2$, that is, the outputs of $PC1$ and $PC2$, respectively. Either in a computer or in an analog multiplier, it must take the ratio $V1/V2$. Thus, a change in the light level affects both photosensors equally, so the only difference between the two is the properties of the sample.

SENSOR / BRIDGE PREAMPLIFIER

Physiologists and other life scientists sometimes use a strain gage force displacement sensor such as the Grass FT-3 show in Fig. 9-5A

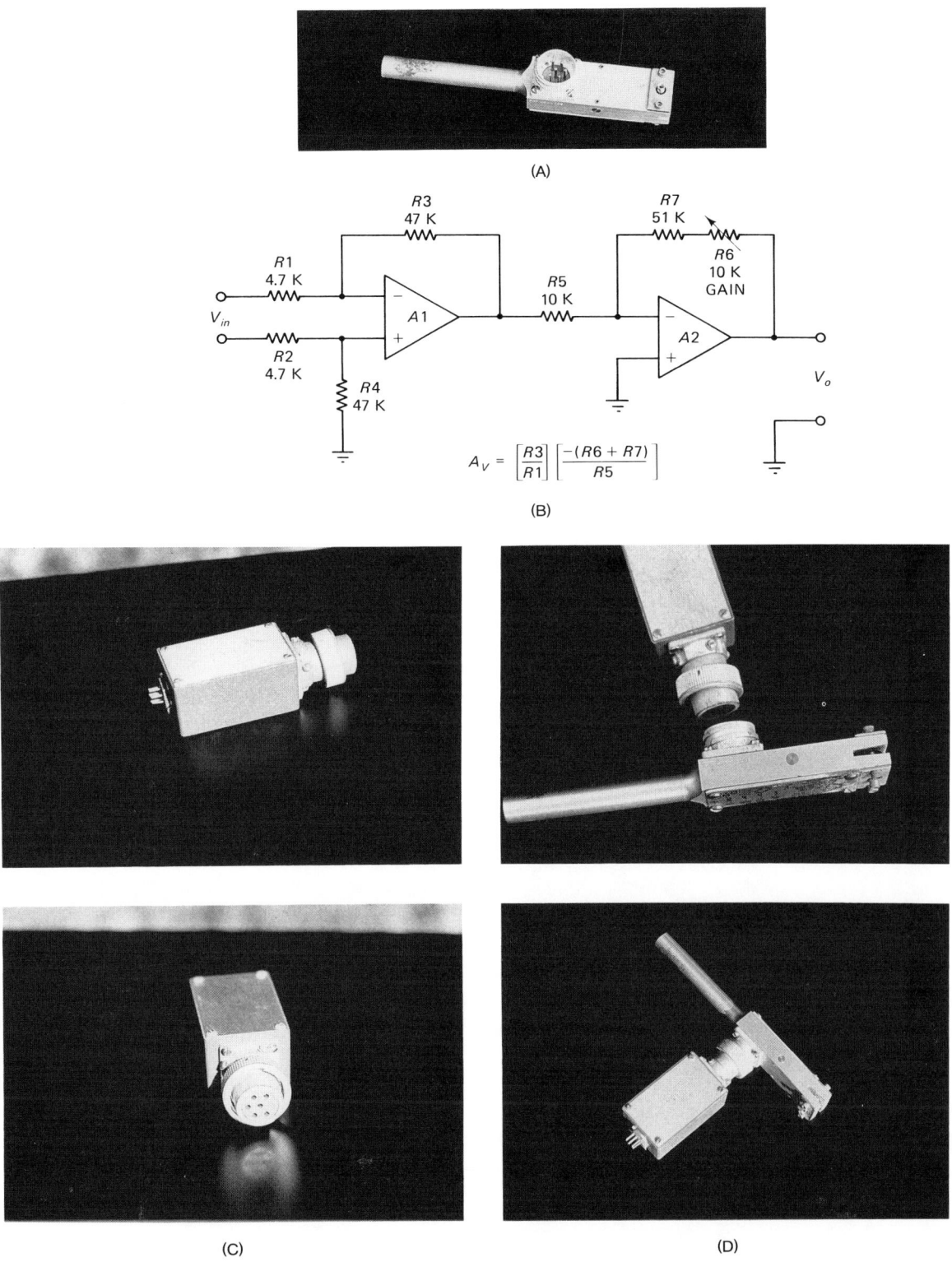

$$A_V = \left[\frac{R3}{R1}\right]\left[\frac{-(R6 + R7)}{R5}\right]$$

Fig. 9-5 (A) Grass FT-3 force–displacement sensor, (B) sensor amplifier, (C) mating connectors between the amplifier and sensor, and (D) amplifier box.

for their experiments. Small displacements of the operating arm at one end of the sensor housing produces changes in the internal Wheatstone bridge that, in turn, result in an output voltage, V_{ot}. The sensitivity factor (P) of the sensor is 35 μV/V/g-F: (the gram-force is a force equal to the force of gravity on a mass of 1g, i.e., 1g-F = 980 dyn. Forces to 10 g-F can be accommodated with excitation potentials up to +7.5 V DC.

A physiologist attempted to use an FT-3 sensor without a differential preamplifier because the manufacturer's offering was too expensive for the scientist's research budget. The experiment was to measure the minute contractions of a guinea pig heart *in vitro* in response to a certain stimulus. The sensor output, therefore, was a weak pulsatile voltage. The sensor was excited from a 6-V DC lantern battery. The balanced output of the sensor was used to drive the channel A and channel B inputs of an oscilloscope. When the oscilloscope was used in the A–B input mode, it poorly mimmicked a differential amplifier. Unfortunately, gain differences between the two channels resulted in a common mode rejection ratio (CMRR) problem. In addition, the lead wires were unshielded, which led to an excessive 60-Hz hum pickup from surrounding AC power wiring in the laboratory. The problem was exacerbated by the fact that the sensor output signals were very low compared with the 60-Hz induced signal.

It was decided to build a differential sensor preamplifier for the experimenter. The excitation voltage was standardized to +5 V DC because 5-V IC voltage regulators were easily available. For a maximum 4 g-F stimulus, which the scientist believed to be the absolute maximum that would be seen, the sensor output voltage was

$$V_{ot} = PVF = (35 \ \mu V/V \cdot g\text{-}F)(5.00 \ V \ DC)(4 \ g\text{-}F) = 700 \ \mu V,$$

or converting to volts,

$$V_{ot} = 0.0007 \ V.$$

Overcoming the 60-Hz interference problem required boosting the signal at the source to a higher level so that the induced 60-Hz signal was only a tiny fraction of the desired signal. In other words, the S/N needed to be improved. A strategy was adopted to boost V_{ot} to 10 mV/g-F, or 40 mV when the maximum 4 g-F was applied to the sensor. The required voltage gain was

$$A_{vt} = \frac{V_{ota}}{V_{ot}} = \frac{[40 \ mV(1 \ V/1000 \ mV)]}{0.0007 \ V} = \frac{0.04 \ V}{0.0007 \ V} = 57.14.$$

A two-stage preamplifier circuit (Fig. 9-5B) was designed and built. It used a DC differential input stage ($A1$) with a gain of 10 ($A_{vt} = 10$) to accommodate the balanced output of the Wheatstone bridge. The output signal of $A1$ (V_a) was boosted 10 times to 0.007 V (7 mV) when a 4 g-F was applied to the sensor.

Input resistors $R1$ and $R2$ needed to be at least 10 times the Thevanin's resistance of the sensor, which for a Wheatstone bridge with all elements equal is the resistance of any one element. For the Grass FT-3 sensor used here, the resistance elements were 200 Ω each, so any input resistor value > 2000 Ω was acceptable. A value of 4.7 Ω was selected for reasons of convenience. The gain A_{V1} of this stage is found from

$$A_{V1} = R3/R1,$$

so

$$R3 = A_{V1}R1.$$

Inserting the actual values,

$$R3 = (10)(4.7\ k\Omega) = 47\ k\Omega.$$

A variable gain stage ($A2$) was used to boost V_{ota} to 40 mV when $V_a = 7$ mV. The gain required of $A2$ was

$$A_{V2} = V_{ota\,(max)}/V_{a\,(max)},$$

or with the values inserted,

$$A_{V2} = 40\ mV/7\ mV = 5.714.$$

Another way to calculate A_{V2} was to divide the total required gain by A_{V1}:

$$A_{V2} = A_{vt}/A_{V1} = 57.14/10 = 5.714.$$

A value of 10 kΩ was selected for $R5$ for the simple reason that it is a standard value, and thus was easy to obtain. The value was, therefore, selected for reasons of economy and convenience. The value had to be > 1000 Ω in order to accommodate the normal < 100 Ω output impedance of the operational amplifier. For a gain of 5.714, the feedback resistance $R_f = (R6 + R7)$ needed to be

$$R6 + R7 = A_{V2}R5 = (5.714)(10\ k\Omega) = 57.14.$$

Although a single precision resistor can be obtained to accommodate the required resistance, the solution adopted was to break up the total resistance between two standard values. For example, $R6$ was set to 10 kΩ (and was a potentiometer), while $R7$ was a 51-kΩ fixed resistor.

The circuit of Fig. 9-5B was built inside a small aluminum instrument box (see Fig. 9-5C) that was fitted with a mate to the sensor output connector on one end. This configuration allowed the preamplifier to be mounted directly to the sensor housing.

A control box was connected to the preamplifier through $P2/J2$. This external box provided additional gain, as required. A gain-of-10 amplifier boosted the signal to a healthy 400 mV. The external box also provided an offset null (or ZERO) circuit and gain control (SPAN) to calibrate the system.

PRACTICAL DESIGN EXAMPLE: THE pH METER

Chemists use pH as a measure of the relative acidity or alkalinity of a fluid. If a solution is neither acidic nor basic it is said to be neutral and has a pH of 7.00. Acidic solutions have a pH < 7, while basic solutions have a pH > 7. Human blood has a normal pH of 7.4 ± 0.04 (i.e., 7.36 to 7.44), and pH values outside this range are considered pathological. A pH electrode is a special high-impedance chemical glass electrode that produces an output voltage that is a function of fluid pH (Fig. 9-6A). A typical pH electrode will output 0 V at pH = 7, plus or minus the offsets represented by a tolerance band and the electrode temperature.

In order to design a simple digital pH meter, it is necessary to create a circuit that produces a 1-V DC output at pH = 1, a 7-V DC output when pH = 7, and 14-V DC output when pH = 14.

The input impedance of the circuit must be > 100 mΩ because of the glass electrode source impedance. This requirement can be met by using a BiFET of BiMOS operational amplifier or an electrometer amplifier. The unity gain configuration is used in the input stage shown in Fig. 9-6B. The slope of the characteristic curve in Fig. 9-6A is negative, while the slope required by the design is positive (Fig. 9-6C). Therefore, an inverting stage ($A2$) is used to follow the input stage ($A1$). Accommodating electrode tolerances require both an offset control and a variable voltage gain.

The offset control is adjusted to set V_o to +7.00 V when the input voltage is zero ($V_{in} = 0$). The amplifier gain (A_{V2}) is set such that $V_o = +14.00$ V when $V_{in} = -1.4$ V or

$$A_{V2} = \left. \frac{V_{o(max)}}{V_{in(max)}} \right|_{pH=14}$$

$$= \frac{14 \text{ V}}{-1.4 \text{ V}} = -10.$$

Because the tolerance band is ± 80 mV, the gain has to have a range of

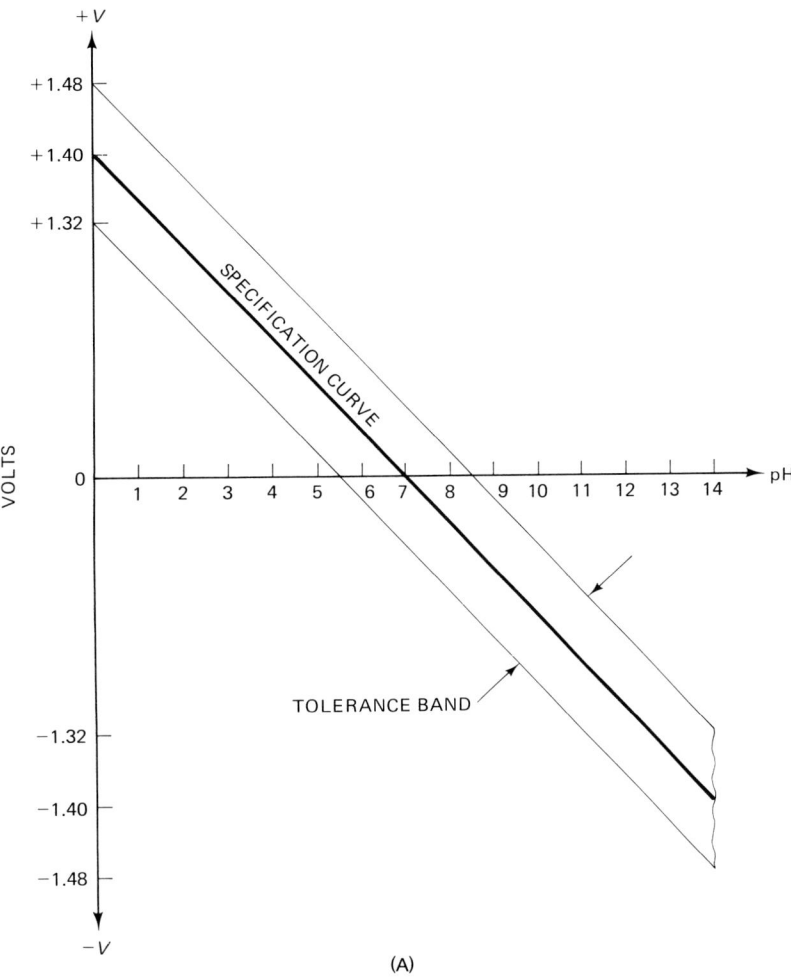

Fig. 9-6 (A) Characteristic curve of the typical pH curve, (B) simple pH meter, and (C) output characteristic. (*Figure continues.*)

operation. The maximum gain is

$$A_{V2(\text{max})} = 14\,\text{V}/-1.32\,\text{V} = -10.61,$$

and the minimum gain is

$$A_{V2(\text{min})} = 14\,\text{V}/-1.48\,\text{V} = -9.45.$$

The amplifier must have a gain that can be varied between -9.45 and -10.61. If $R1 = 10\,\text{k}\Omega$, then $R2 + R3$ must be variable between

$$R2 + R3 = (10\,\text{k}\Omega)(9.45) = 94.5\,\text{k}\Omega,$$

and

$$R2 + R3 = (10\,\text{k}\Omega)(10.61) = 106.1\,\text{k}\Omega.$$

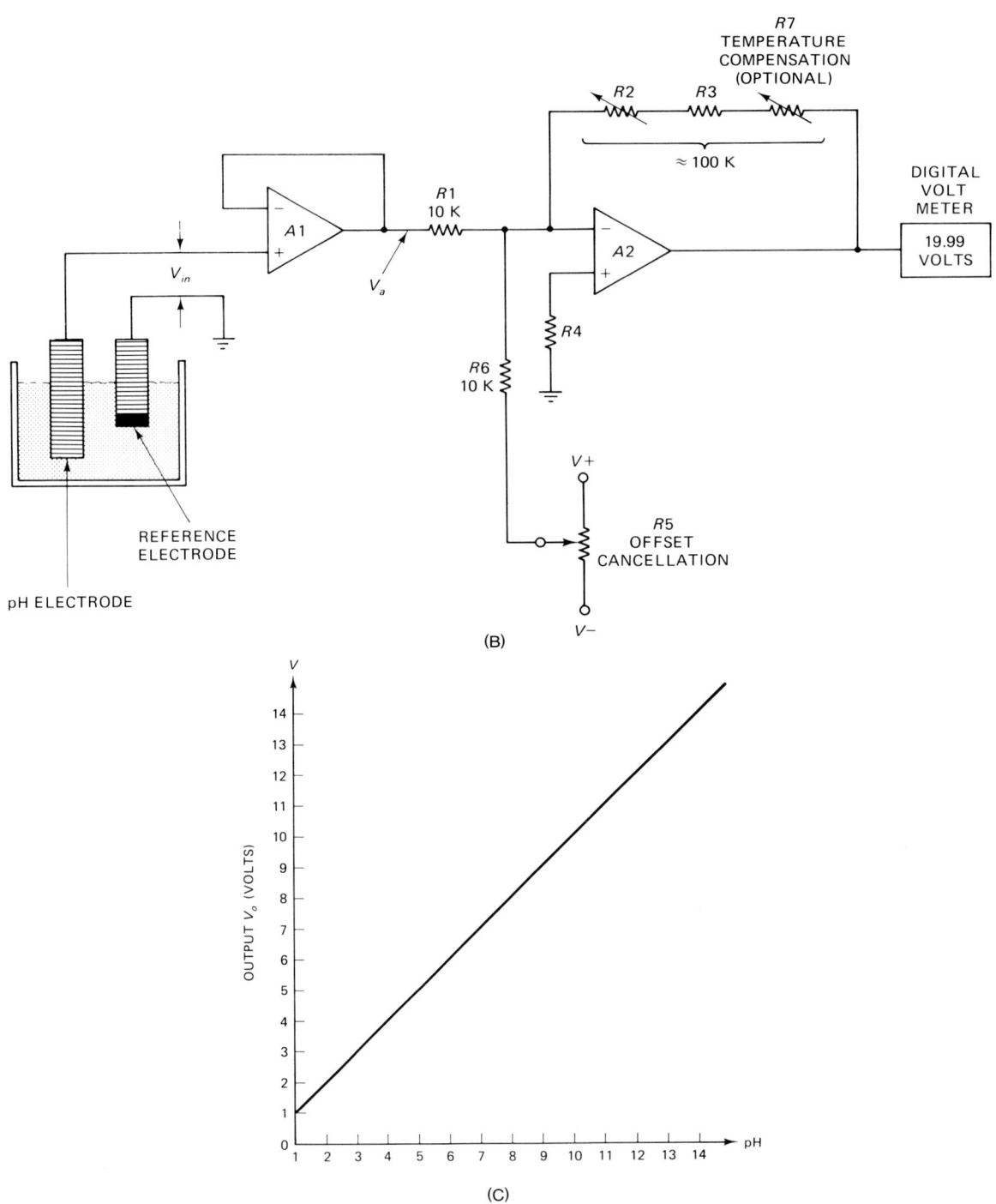

(B)

(C)

Fig. 9-6 (*continued*)

A combination of a 91-kΩ fixed resistor in series with a 20-kΩ multiturn potentiometer will solve the problem.

The offset control must set the voltage V_b to 1.00 when V_{in} = +1.4 V. Assuming a nominal gain of −10 for stage $A2$ and +1 for $A1$, voltage V_b would normally be $[(V_{in})(A_v)] = [(+1.4 \text{ V DC})(−10)]$ = −14.00 V DC. The potential at the wiper of the potentiometer (V_c) normally sees a norminal gain of −10 [i.e., $(R2 + R3)/R6$]. The goal is to offset −14 V DC to +1 V DC, a total of +15 V DC change. Given the gain of −10,

$$V_c = V_{b \, (desired)}/A_v + 15 \text{ V DC}/-10 = -1.5 \text{ V DC}.$$

Under the condition where the value of V_{in} is +1.4 V (representing pH = 1), and V_c = +1.3 V DC, the total output voltage will be

$$V_b = -(R2 + R3)\left(\frac{V_a}{R1} + \frac{V_c}{R6}\right)$$

$$= -(100 \text{ k}\Omega)\left[\frac{(+1.4 \text{ V})}{10 \text{ k}\Omega} + \frac{(-1.5 \text{ V})}{10 \text{ k}\Omega}\right]$$

$$= (-100)(0.14 - 0.15) = +1.00 \text{ V}.$$

If a temperature compensation control is needed, then another offset control is added to the circuit at $A2$. This control is operable from the instrument front panel. In some instruments, the temperature offset potential is provided by measuring the temperature of the solution being measured. In at least one case, a thermistor or PN junction temperature sensor is embedded into the pH electrode housing. The reader is left to think about the design of an automatic temperature compensation circuit when a 10-mV/K sensor is used, and the temperature range will be +20 to +45°C 0°C = 273 K).

CONCLUSION

In the previous sections, several different forms of instrumentation circuit were discussed: photocolorimeters, bridge amplifiers, and pH meters. In Chapter 10, we will take a look at what it takes to design reasonably reliable equipment, and in Chapter 11, we get down to the business of discussing the linear integrated circuit devices that form the basis for most viable analog instrument circuit designs.

10

Designing and Building Safe, Reliable Instruments

Reliability and safety are two very important considerations for the designer of electronic instruments. While we cannot even begin to address the full range of these engineering disciplines within a single, brief chapter, the highlights can nonetheless be touched upon. One of the principal factors affecting the reliability of electronic systems is heat, and that is an area that we can easily address.

KEEP IT COOL

Certain pieces of electronic equipment seem to constantly rotate in and out of the service shop for seemingly endless repairs, often repeats of the same defect. Some equipment just seems naturally failure prone. Be sure to consider overheating as a possible cause when you face such a situation. Ideally, the time to consider this possibility is at the initial design time. It is well recognized that heat is the great killer of electronic equipment. For decades, reliability engineers have identified heat as a principal cause of electronic component failure.

Many electronic device ratings are predicated on maintaining certain operating temperatures. One replacement audio power transistor, for example, offers a seemingly tremendous collector power dissipation (which is prominently advertised), but the full power is only available at room temperature (25–30°C). At temperatures above 30°C, the transistor must be derated substantially. Almost anyplace that the transistor is used inside a chassis, cabinet, or box the ambient temperature will exceed 30°C!

Similarly, the radio frequency (RF) power transistors used in radio communications transmitters fail as often from overheating as from the much advertised VSWR problems, but the heat problem is less well recognized in many service shops. In one case, there were constant failures of the RF power transistors in a small solid-state 100-W VHF power amplifier. Technicians repaired the rig several times before they drew the connection with heat. The trunk of a car during the summer months is extremely hot, while the passenger cabin cools off with air-conditioning in only a few minutes. Moving the amplifier to behind the dashboard cured the problem.

Reliability experts measure equipment performance in terms of mean time between failures (MTBF), which is usually expressed in hours. For example, an MTBF of 1000 implies that, for a large (statistically significant) number of samples of the equipment, the average will be 1 failure per 1000 h of operation. A general rule is that a 10°C rise in operating temperature will cut the MTBF in half.

How important is cooling in electronic equipment? Let us consider some examples. The author once worked in a university hospital repairing patient monitoring equipment. The slave ECG oscilloscopes at the nurses central station were a reliability nightmare. About once a week, usually at 3 a.m., the staff would call me to come repair one of the four oscilloscopes. Yet the same model oscilloscopes at bedside operated reliably. The problem was overheating of the central station oscilloscopes, which were mounted inside a completely closed console. After cutting 10 one-inch ventilation holes, and installing a pair of 100-cubic-feet-per-minute whisper fans, the central station oscilloscopes became as reliable as the beside oscilloscopes.

No one with any professional electronics experience will deny that heat is the great killer of electronic devices. Equipment that passes or delivers large amounts of either current or power must be kept cool for proper operation. The methods given in this chapter are simple and sufficient for most applications. While reliability engineers and the thermodynamicists will flinch at the lack of mathematical elegance, the methods are nonetheless effective for many practical applications.

There is only one simple rule: where there is excessive heat, remove it. What does excessive mean? If the equipment feels too hot to the touch or has a history of unexplained failures and/or repairs, then it is probably running too hot. An engineering thermal analysis will reveal the specific details, but the empirical "skin of the thumb" rule suffices for our present needs.

There are three basic tactics that can be used either singly or in combination: (1) radiate more of the heat, (2) improve natural ventilation, and (3) add or increase forced-air cooling. Water cooling is not an issue for most readers.

PROTECTING TRANSISTORS AND IC VOLTAGE REGULATORS

On small equipment, it is not always practical (or possible) to use forced air cooling, so you may have to provide substantial heat sinking for the semiconductors. In fact, even in most forced-air cooled equipments, the semiconductors will need these metal radiators. Figure 10-1A shows the metal TO-5 transistor package. Most of these transistors are mounted on printed wiring boards and are low-signal (and low-heat) devices. But some TO-5 transistors, such as the 2N3053 and certain 3- to 10-W RF power transistors, operate at moderate power levels. A finned "top-hat" heat sink, such as Fig. 10-1B, is mounted on the TO-5 package to radiate heat. There are also certain other spring clip versions of this same kind of heat sink.

Figure 10-2A shows two forms of plastic power device packages. These packages are typically used for audio power transistors (e.g., 2N5249), thyristors and three-terminal IC voltage regulators. In the regulator case, the devices are usually rated at 750 mA in free air and 1000 mA when heat sinked. Either vertical or horizontal finned sheet metal heat sinks, such as Fig. 10-2B, are used to provide heat dissipation. Be sure to specify a thin layer of silicone heat transfer grease between the metal tab surface on the transistor (or regulator) and the heat sink. Also, be certain that the mounting screw is properly torqued in order to facilitate heat transfer to the heat sink.

Sheet metal heat sinks for TO-3 transistors and three-terminal voltage regulators are mounted on a printed wiring board (PWB). As a rule of thumb (without making the correct thermal calculations) the bent sheet metal heat sinks are good up to about 10 W of power, and voltage regulators are good up to 1.5 amp (check the ratings and the conditions under which they are valid in the manufacturer's data sheets). For the 3-, 5-, and 10-A voltage regulators that also use a TO-3 package, it would be better to use a larger, finned heat sink. In those cases, go through the formal calculations to determine the size of heat sink required.

In many pieces of equipment, the metal chassis is used for heat sinking. In those cases, the transistors are bolted either directly to the metal chassis or mounted via mica insulators, if insulation is required.

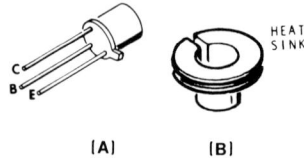

Fig. 10-1 (A) Small transistor package; (B) small transistor package heat sink.

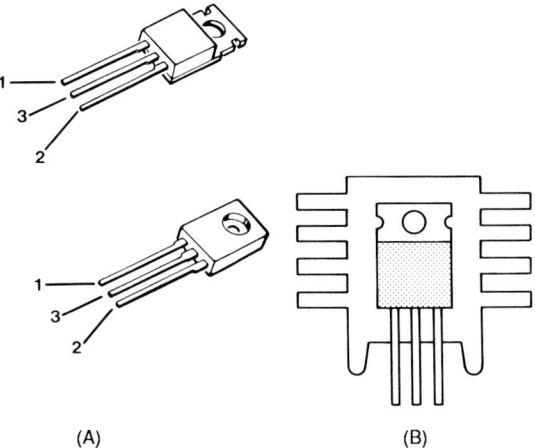

(A)   (B)

Fig. 10-2 (A) Plastic power transistor/regulator packages; (B) plastic power transistor/regulator in heat sink.

In both cases, silicone heat transfer grease is used between the semiconductor device and the chassis. This method is especially successful when the chassis is large or when it is particularly thick (i.e., has a high thermal mass).

Some PWBs use large areas of unetched copper foil and large metal ridges or blocks to provide better heat sinking. This method is used especially where there are no single devices that can be individually heat sinked (e.g., a TO-220 transistor), but rather when there are a large number of heat-producing devices, such as TTL ICs.

There are many different forms of large, finned heat sinks used for TO-3 (and other) transistors, high-current voltage regulators, high-current diodes, and SCRs; Fig. 10-3A shows a side view of one of these heat sinks. In this case, the TO-3 transistor is mounted on the flat central surface of the heat sink with screws. In most situations, it is wise to use a thin smear of silicone heat transfer grease between the device and the heat sink. This grease is especially needed when a mica insulator is placed between the semiconductor device and the heat sink. Again, it is necessary to make sure that the mounting screws are cinched down tight enough to allow maximum heat transfer (but not enough to distort the device package). The big issue in selecting a heat sink is the surface area.

When forced air is used to cool a heat sink—a good idea when the power or current is high—the orientation of the heat sink with respect to the airflow is sometimes important. Figure 10-3B shows the right and wrong ways to force air over the finned surfaces. Keep in mind, however, that orientation is not always critical, especially when air from the wrong direction is sufficient or blows over the entire surface. The designations right and wrong are merely general considerations for some critical applications.

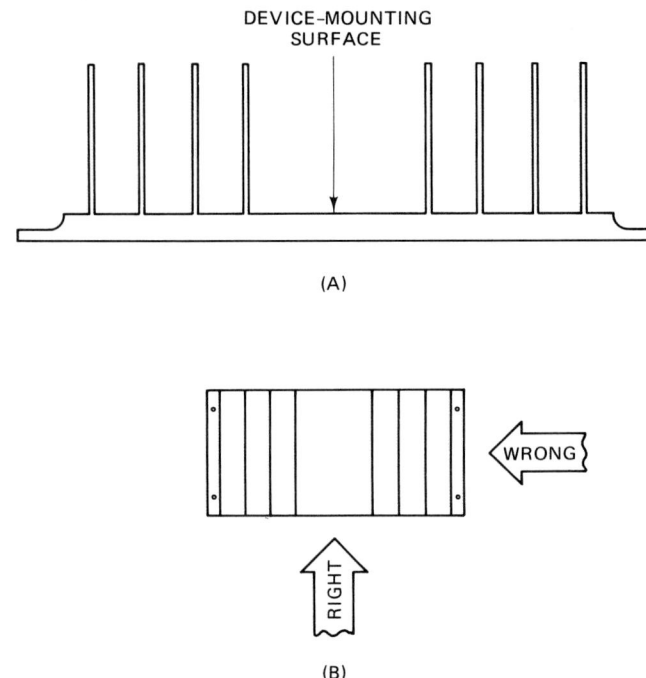

(A)

(B)

Fig. 10-3 (A) Finned heat sink (side view); (B) top view of right and wrong ways to pass air over the finned heat sink.

OTHER COMPONENTS

Certain components other than power transistors generate heat. Rectifier diodes and power resistors should be mounted with their bodies 0.125 to 0.250 in. off the PWB (see Fig. 10-4). This procedure allows the heat to dissipate into the air instead of the PWB material. If vibration is expected, then a Teflon spacer can be inserted between the component and the PWB. Many phenolic and some fiberglass printed wiring boards are badly damaged from the effects of a 10-W power resistor mounted flush to the surface.

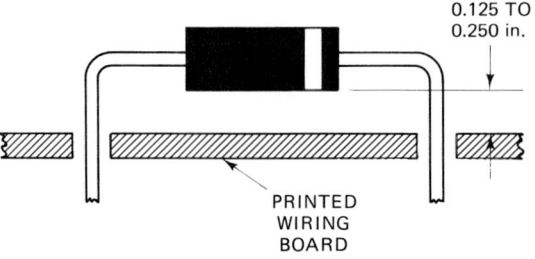

Fig. 10-4 Normally hot components, such as rectifiers and power resistors, should be mounted off the board 0.125 to 0.250 in. unless vibration is a problem. If vibration is present, then use a Teflon spacer between the part and the printed wiring board.

Besides reducing the operating life or limiting the power output of circuits, overheating can also decrease performance in other ways. Certain circuits, oscillators, for example, are inherently sensitive to heat. There was once a popular two-way radio transceiver that suffered terrible frequency drift because the master oscillator was located right next to the RF/IF (intermediate frequency) strip vacuum tubes. Although that was such a bad design error that nothing would really fix the situation, service technicians can improve the frequency stability markedly by adding some thermal insulating material between the RF/IF PWB and the aluminum oscillator shielded housing.

LARGE, MULTIBOARD EQUIPMENT

Figure 10-5A shows a typical large-scale multiboard device, such as a microcomputer or transceiver, in which plug-in printed wiring boards are installed on a socketed motherboard. Usually, these PWBs will be mounted in a closed cabinet for both electromagnetic interference (EMI) and aesthetic reasons. If we apply air broadside to the PWBs, then only the first one in the lineup will benefit. Figure 10-5B shows a top view that permits you to see right and wrong airflow directions. Obviously, air coming in from the sides is able to remove heat from the PWBs more effectively.

Figure 10-6 shows a method once used in a minicomputer. A large metal chassis with a motherboard mounted on it to hold the PWBs was used. There were 0.75-in. holes cut in both the chassis top and the motherboard to admit air between the boards. Although only one hole is shown between each board in this side view, there were

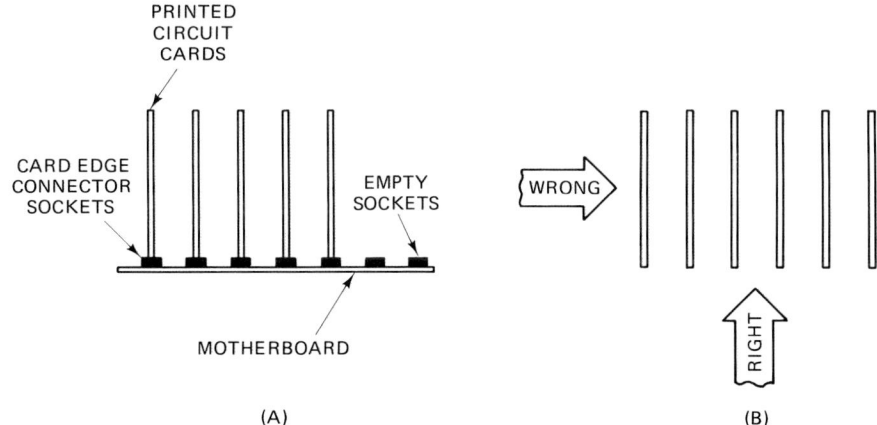

Fig. 10-5 (A) Multicard system in which plug-in printed wiring assemblies are mounted in sockets on a motherboard; (B) right and wrong ways to air-cool the multicard system.

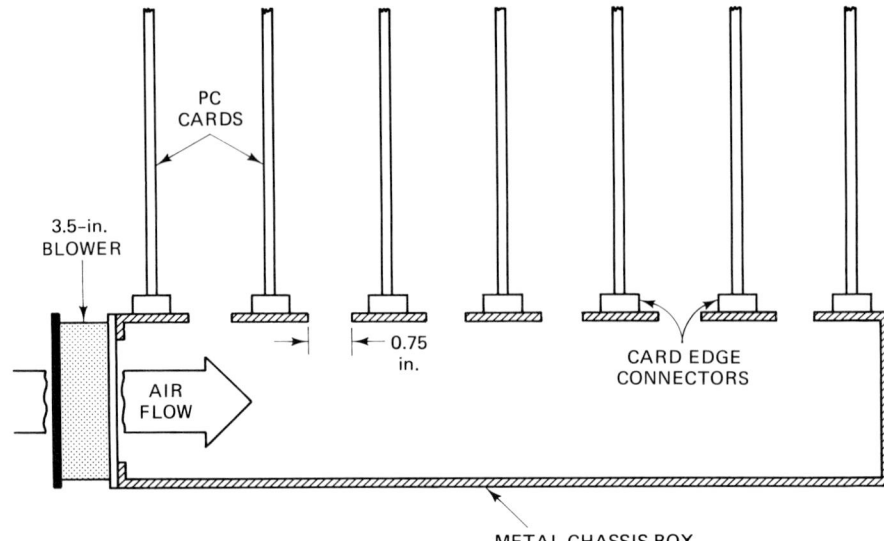

Fig. 10-6   Air system method for cooling the multicard system.

four per row in the actual computer. Air from the blower flowed up through the holes and across the electronics components on the PWBs.

Most RF power amplifier tubes used in medical diathermy and electrosurgery equipment, industrial inductive heaters, and radio transmitters must be forced-air cooled in order to realize their full ratings (some are absolutely dependent on cooling). Figure 10-7 shows two methods for providing the needed cooling air. In Fig. 10-7A, we see the situation in which a blower is mounted so that the air flow is directly over the glass envelope. The fan may be mounted either exterior to the RF compartment (as shown) or inside.

The other method, shown in Fig. 10-7B, assumes the use of air system tube sockets. A blower or fan supplies air to the bottom side of the socket, and the air is directed upward through holes in the socket and around the glass envelope. A chimney aids in keeping the airflow against the glass. Some air system sockets have plumbing connections for the air hose, while others are dependent upon pressurization of the lower compartment. In either case, the reason this socket is better is that the electrical connection seals in the glass are kept cooler. The plate cap lead seal should also be kept cool, if possible. Toward this end, some builders use a finned heat dissipating plate cap to make electrical connection to the anode.

IC PRINTED CIRCUIT BOARDS

The component density possible on modern PWBs makes it possible to make very small, high-density products, such as modern radio communications equipment and digital computers. Unfortunately, as the

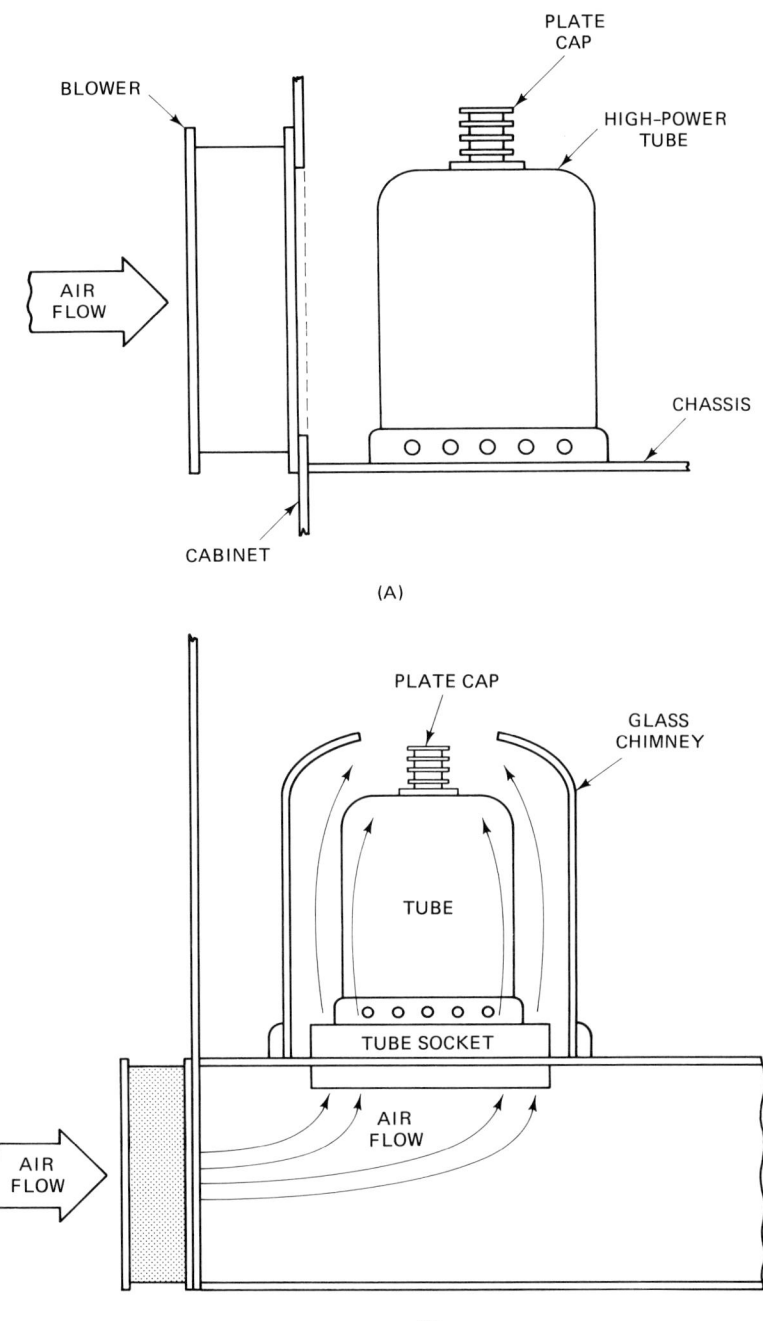

Fig. 10-7 (A) Method for direct cooling of large power devices, such as RF power tubes; (B) air system method for cooling power devices.

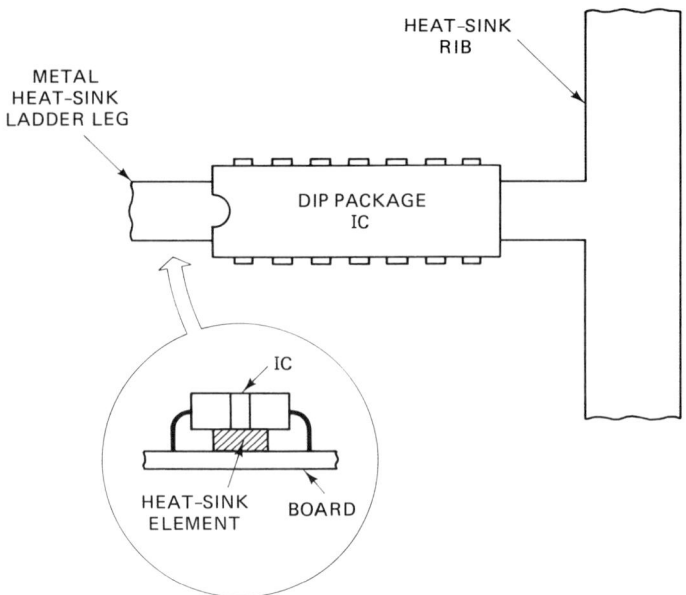

Fig. 10-8 Metallic ladder and cold plate method for cooling large multichip printed circuit boards.

number of IC devices on a card increases, so does the problem of cooling them. In some cases, impingement airflow, as discussed earlier, is neither feasible nor desirable, but we still have to remove the heat. One solution is shown in Fig. 10-8. This method uses a ladder heat sink built onto the board and a cold plate at the ends.

In Fig. 10-8, a heavy metal ladder is run underneath each IC device (see inset) and is joined to a large heat sink bar on the card edge. Heat is removed from the IC area by conduction. In some cases, airflow can be directed across the card edge heat sinks. In this type of construction, we usually want to place the most heat producing components as close as possible to the edges of the PWB, where the heat sink bar is located. The purpose of the metal ladder in many equipments is to conduct the heat from the printed wiring board to a cold plate, which is in turn cooled by forced air, radiation, or flowing water.

ELECTRICAL SAFETY CONSIDERATIONS IN EQUIPMENT DESIGN

When working on or using electronic equipment, it is possible to come into close proximity with voltage and current sources that are high enough to cause some damage, perhaps even death. In this section, we will point out the dangers and give some general advice on how to avoid electrical shock.

How Much Current Is Fatal?

Several years ago, a technician in a hospital electronics laboratory overheard an intern claim that 110 V AC from the wall socket is not dangerous because he was told in medical school that it is not the voltage that kills, it is the current. A bystander asked: "Doctor, have you ever heard of Ohm's law?" According to Ohm's law, the current is the quotient of voltage and resistance: $I = E/R$. A little statistic that the doctor apparently did not know was that 110 V AC from residential and business wall sockets is the most common cause of electrocution in the United States. In addition, medical studies reveal that the 50–60 Hz frequencies used almost worldwide in AC power distribution are in the most dangerous range of frequencies. Higher and lower AC frequencies are less dangerous (but not safer) than 50–60 Hz AC.

According to medical experts who have studied electrical shock, the killing factor is the current density in a certain section in the upper right-hand chamber of the heart called the sinoatrial node. Any current flow geometry through the body that causes a high current density in that section of the heart can induce fatal ventricular fibrillation. In general, for limb-contact electrical shock through intact skin (medical instrument designers have stricter guidelines), the following rules of thumb are accepted.

- 1–5 mA   Level of perception
- 10 mA   Level of pain
- 100 mA   Severe muscular contraction
- 100–300 mA   Electrocution

Keep in mind that these figures are approximate and are not to be taken as guidelines to approximate assumed risk. Death can occur under certain circumstances with considerably lower levels of current. For example, when you are sweating and are standing in salt water, then risks escalate tremendously. In medical situations, the level of current that can kill is in the 20–150 μA level because the current is induced directly into the body, which is essentially a saltwater environment (human skin has a resistance of 1000 to 20,000 Ω, internal tissue has a resistance of 50 Ω or so).

Is High Current Dangerous?

The author once attended a design review meeting for a high-power mobile transmitter. The specification called for insulation of low-voltage (28 V DC), high-current (50 A) DC power supply terminals. One of the engineers present sneered that that was something like asking him to insulate the battery terminals of his car. Implied in his comments was that low voltage can never hurt you. There are two false premises inherent in that opinion.

First, although low-voltage, high-current points rarely cause electrical shock, it is possible for dangerous shock to occur when the person has a very low electrical skin resistance (very sweaty) or has an open wound. Although the case did not result in electrocution, one electronics technician recently injured himself severely when he cut himself on a +5-V DC, 30-A computer power supply terminal. A large amount of current flowed in his arm and caused severe pain and some physical damage.

Second, the high current is extremely dangerous if you happen to be wearing a watch or jewelry! A two-way radio shop used 12-V automobile batteries and battery chargers for the troubleshooting bench supply for mobile service. A technician servicing the battery rack dropped a wrench, and it fell onto the battery making contact from (−) to (+) through his watchband. The large current turned the watchband red hot and gave him serious second and third degree burns.

Do not assume that low-voltage, high current power supplies are harmless—they are not!

Mechanisms of Electrical Shock

In order to raise our consciousness about how shock can occur let us take a look at certain scenarios of electrical shock that might occur in using electronic instruments. Figure 10-9 shows the direct approach to fatal electrical shock: you are grounded through conductive shoes and touch a hot point. You need not be outdoors to be affected by this scenario. A concrete garage, shop, or basement floor is a reasonably good conductor, as are wet leather and some forms of rubber shoes.

Figure 10-10 shows an indirect scenario that especially affects electronics workers. Consider the grounded instrument probe (in this case an oscilloscope). When you grasp that probe, you may be grounded

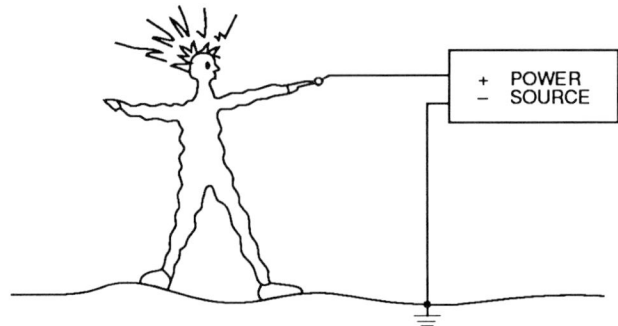

Fig. 10-9   Scenario for direct electrical shock: touching a hot line while grounded.

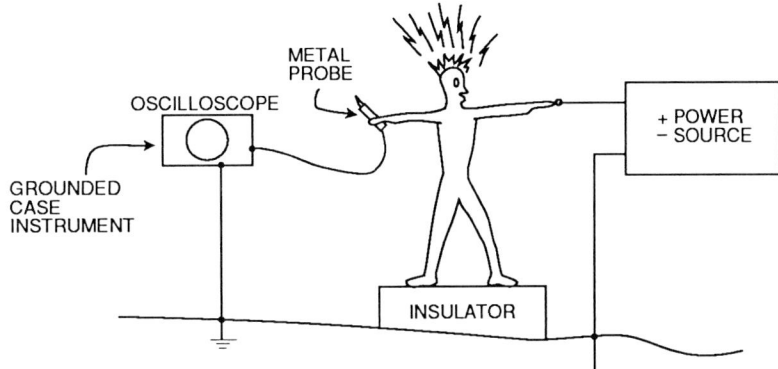

Fig. 10-10 Indirect scenario: one portion of body is connected to a power line, while the other is grounded through an electronic instrument.

through the oscilloscope shield and the power cord ground conductor. If you touch a hot point, then you will be shocked—and maybe killed.

A related scenario is shown in Fig. 10-11. Here, we have an AC/DC appliance, such as a low-cost radio or TV set. Note that the oscilloscope probe ground is connected to the set ground, which also happens to be one side of the AC power line. Everything is fine as long as the AC plug is oriented correctly in the wall and the wall socket is wired correctly. But, if you plug it into the wall receptacle backwards, then there will be an explosive short circuit and possible electrocution.

Another scenario is the fatal antenna erection job. It is never good practice to erect an antenna near a power line—**NEVER!** Every year we hear stories of people electrocuted because (1) an antenna they were working on fell across the power lines, (2) they tried to toss a wire antenna over the power line in order to raise the antenna above the lines, or (3) a ladder they were using fell across the power lines. These tactics will kill you.

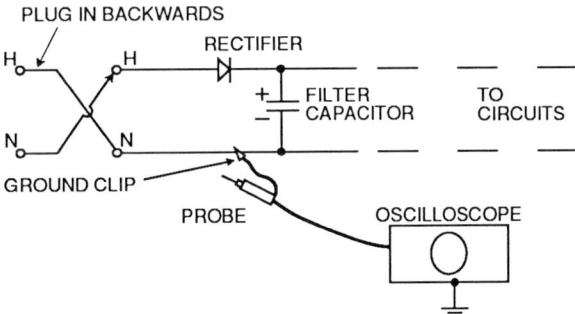

Fig. 10-11 AC/DC power supply represents a SEVERE HAZARD when the power line is reversed.

Some Cures for the Problem

Figure 10-12 shows the schematic for the usual U.S. residential AC electrical system. The power company distributes energy through high-voltage lines. When it arrives at a point a short distance from your home, office, or shop, it is stepped down in a "pole pig" transformer to 220 V AC (center tapped). The center tap of the transformer secondary is grounded: therein lies the root of the problem: the power system is ground referenced. The two ends of the 220-V AC secondary are brought into the house as a pair of 110-V AC hot lines. If you tap across the two lines, then you have a 220-V AC outlet; if you tap from the ground line to either hot line, then you have a 110-V AC outlet.

The root problem is that the electrical system in your home or shop is ground referenced. The solution is to make the little local electrical system on your workbench nonground referenced. Figure 10-13 shows the wiring on a bench. Transformer $T1$ is one of two forms of isolation transformer: a 1:1 transformer gives a 110-V AC isolated (nonground referenced) AC line from a 110-V AC standard line, a 2:1 transformer does the same thing from a 220-V AC line.

The second transformer in the circuit of Fig. 10-13 is an autotransformer, commonly but erroneously called a Variac (which is actually a trademark of one particular maker of these devices, not the generic term). These devices allow you to raise and lower the AC voltage for the equipment being tested. It is usually wisest to route the autotransformer output to selected outlets only and to monitor the outlet with an AC voltmeter.

If you either work with radio transmitters (or other high-power RF producing devices) or work near such generators, then you might want to place an EMI filter in the line at the points marked X. The EMI filter is an LC section that attenuates the RF, but not the 60-Hz power.

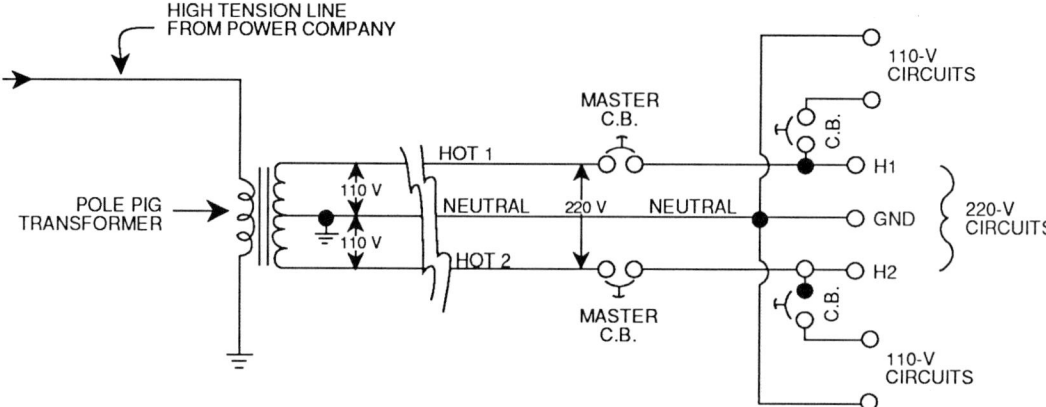

Fig. 10-12 Standard residential AC power system.

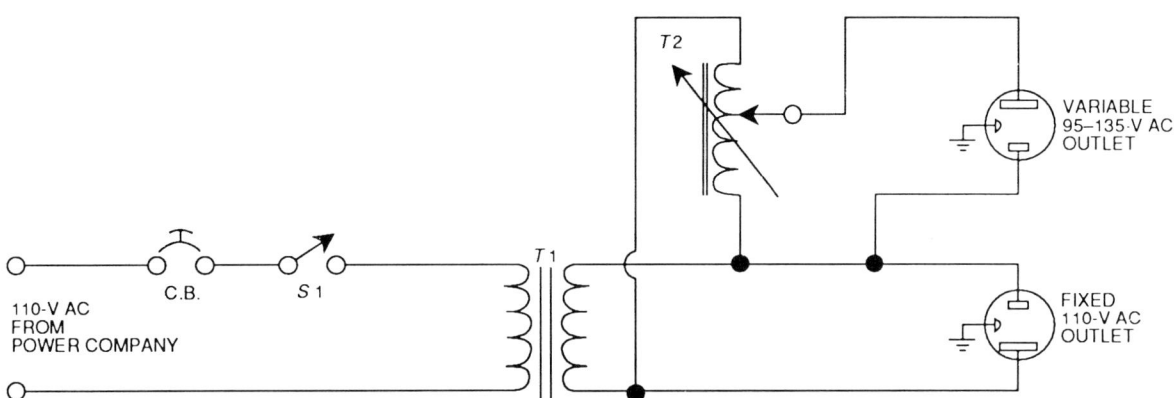

Fig. 10-13   Workbench power system for a safer environment.

The metal oxide varistor (MOV) is used to clip the amplitude of high-voltage line transients (100 μsec or so) that could either damage or interfere with the operation of the equipment on the bench.

The circuit breaker or fuse is used to protect equipment on the bench, as well as the transformer. It is always placed in the hot line, or in both lines; fuses and circuit breakers are never placed in the neutral line only. The switching shown in Fig. 10-13 breaks both lines. This approach is preferred on the theory that hot and neutral lines can be reversed accidentally and can leave you in the position of breaking a neutral line, and leaving the hot line alive as a hissing cobra.

Some General Advice for Safety

There is only one way to ensure that the AC line will not shock you: disconnect it. Make it your practice to NEVER work on equipment that has the plug inserted into the power outlet—NEVER. Do not trust switches, fuses, circuit breakers, or other people. If someone were to hand you a pistol, claiming that it was unloaded, the first thing you would do is to check it yourself—the same advice also holds true for the electrical connection (which can kill you just as a loaded and cocked pistol).

It is often advised that you work on high-voltage devices with your left hand in your pants pocket. That advice is based on the theory that the "left hand to either leg path" is supposedly the most deadly. Even if the physiology is correct, placing one hand in your pocket leaves you in an awkward position and unable to safely work on the circuit. It is probably better to use both hands, arrange it so that the work environment is inherently safe by design, and use safe working techniques.

What is a safe work environment? The power system should be isolated (as discussed). The floor should be insulated by a carpet, treated masonite, a plastic cover, a rubber mat, wooden planking, or some other good insulating material. The floor should always be well insulated and kept dry.

When working on high-voltage DC circuits, keep in mind that capacitors store energy. All filter capacitors must be discharged manually after the power is turned off. Also, *the capacitor must be discharged multiple times*. Even when a short circuit is placed across the capacitor terminals, not all of the energy is removed the first time. Some energy is stored in the dielectric even after the main charge is discharged.

Medical Device Safety

Electronic devices used in clinical and research medicine are subject to particularly strict rules regarding safety. The reason for this situation is that patients are at special risk of electrical shock because their bodies are often invaded for medical purposes. The necessary medical invasion breeches the skin, so it reduces the electrical resistance that protects humans in normal electrical environments. Because of low resistance, and a possibility of direct current paths to the heart, there is also a possibility of microshock, that is, shock from very small currents.

In general, a medical device must have a high degree of isolation between the patient and the AC power lines. This requirement often means that isolation amplifiers must be used at the input circuit. It also means that normal leakage between the AC power line carrying the current into the instrument and the instrument chassis be limited to 10 μA. This requirement means that special power transformers and construction techniques must be followed.

If you are designing instruments for medical use, then be aware that various government and industry bodies, as well as your malpractice or product liability insurer, publish standards for electrical safety. Pay particular attention to the U.S. Food and Drug Administration (medical devices) and a trade and professional group called the Association for the Advancement of Medical Devices. In addition, the various standards groups (ANSI) and those associated with the insurance industry may publish their own standards. Because there are legal, ethical, and regulatory issues, in addition to the engineering and scientific issues, one is advised to seek the counsel of a professional in the medical device industry for guidance.

11

Introduction to Linear Integrated Circuit Devices

Linear ICs are the common staple of analog circuit design today. Whether the ubiquitous operational amplifier, or one of many newly available special-purpose chips, the linear IC is a favorite among circuit designers. Early in the era of op-amps, one textbook proclaimed these devices made "...the contriving of contrivances a game for all." In this chapter, we will take a first look at linear IC devices and follow on with their application to analog instrumentation, data acquisition, and control circuits.

HISTORY OF INTEGRATED CIRCUIT DEVICES

The modern solid-state electronics era began in the late 1940s with the invention of the bipolar transistor. Only a decade later, engineers were working to build a device containing multiple transistors and resistors formed onto a single semiconductor substrate. In the early 1960s, the IC was born at Fairchild in California, and the modern chip revolution began. Now, literally hundreds of millions of IC devices are sold annually, and the IC semiconductor industry is a major contributor to the U.S. economy.

Two of the earliest commercially successful IC devices were the μA-703 RF/IF gain block and the μA-709 operational amplifier. The μA-703 device was a simple, high-frequency gain block and was frequently used as such in the mid-1960s. These devices were generally used as the frequency modulation (FM) intermediate frequency (IF) amplifier in broadcast and communications receivers. Typically, three to five μA-703 devices were used to provide the 80-dB or so minimum gain required of an IF amplifier in consumer FM broadcast receivers.

The μA-709 was an operational amplifier, or op-amp. Invented in 1948 (in vacuum tube form) by George Philbrick, the operational amplifier was named from the fact that it was originally intended to perform mathematical operations in analog computers. The unique property of the op-amp that makes it so useful and so versatile is that the circuit transfer function (V_0/V_{in}) is controlled entirely by the feedback network between output and input. Although the reasons may not be readily apparent, this attribute of the operational amplifier makes it one of the most powerful IC devices on the market.

There is a well-known production learning curve in the semiconductor industry under which costs drop dramatically after an initial high price period. In 1964, for example, electronic distributor catalogs listed the μA-709 operational amplifier for \$110.00, and two years later it was still \$79.00. Today, the μA-709 costs less than \$1.00

(A)

Fig. 11-1 (A) Microphotograph of integrated circuit; (B) internal circuit of a linear IC amplifier. (*Figure continues.*)

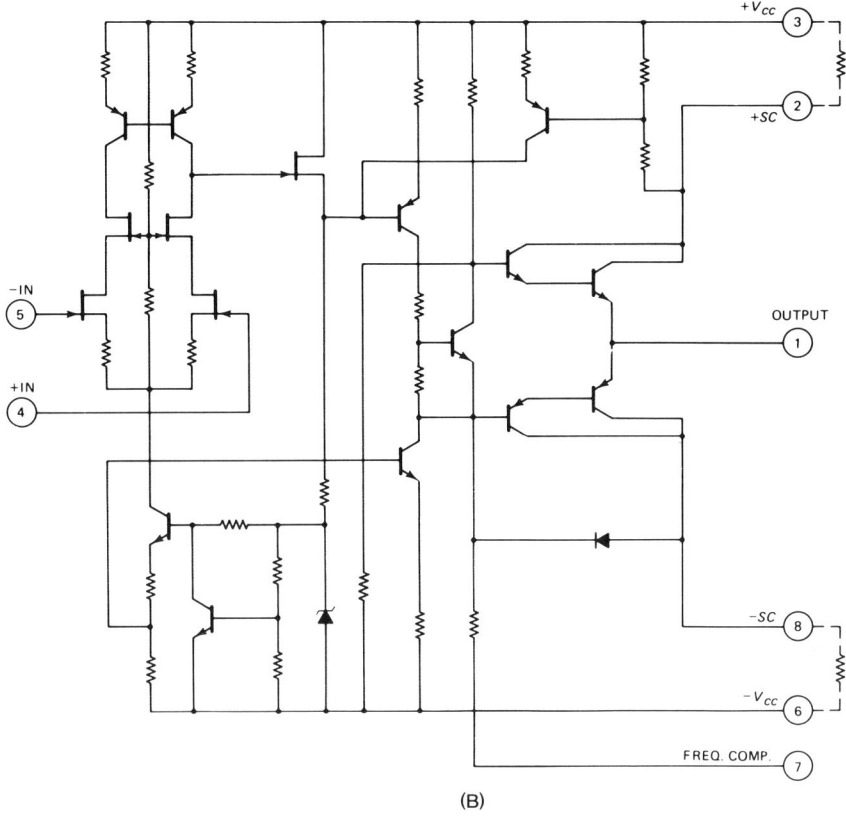

(B)

Fig. 11-1 (*continued*)

where it is still available at all. Devices with greatly superior performance over the μA-709 now cost only a few dollars and are easily available. Production yields and device availability have been considerably improved since the early years. Where designers in 1963 had to wait for rationed parts, operational amplifiers are available today from distributor stock inventories in almost any quantity that is needed.

The field of integrated circuit electronics became successful quickly because it allowed immense circuit density, permitting many circuits to be implemented with fewer overall discrete components. A very large number of devices can be packaged in a very small volume. Figure 11-1A shows a photomicrograph of a typical integrated circuit, while Fig. 11-1B shows the circuitry contained within a common IC operational amplifier.

Performance can be dramatically better in IC devices compared with the discrete component version of the same circuit. Consider thermal drift of DC amplifiers, for example. All DC amplifiers suffer an undesired change in output DC offset voltage that is a function of temperature: $\Delta V_{off} = F(T)$. Even low cost modern IC operational amplifiers are several orders of magnitude better with respect to drift than either traditional vacuum tube or discrete transistor models because all of the semiconductors and internal resistors (the main sources of drift) share the same thermal environment on a common substrate in IC devices. In discrete circuits, on the other hand, those components are spread out over several square inches of printed circuit board and thus do not share the same thermal environment, so drift will be more pronounced than in the IC version of the same basic circuit. Also, consider issues such as stray capacitance, inductance, and inadvertent series resistances due to conductor placement and length. These stray parameters are less of a factor in IC devices than in discrete circuits because of the small sizes and short distances involved in IC design.

Cost is another great advantage of the integrated circuit. Early transistor and vacuum tube operational amplifiers were not only larger and ran hotter than their modern IC counterparts, bu they were more costly as well. In addition, those early amplifiers only poorly approximated the performance of the ideal textbook version of the amplifier. Modern IC operational amplifiers come very close to the ideal, especially as regards input impedance and open-loop gain.

Most of the spectacular benefits of modern electronics derive from integrated circuit electronics technology. From consumer devices to commercial instruments to military equipment, the IC either improves the product considerably or makes it possible in the first place. Many of the marvels of modern electronics simply could not exist without the integrated circuit. The remainder of this book is designed to give you an overview of the field, as well as detailed technical information on the workings and applications of the most commonly available IC devices.

INTEGRATED CIRCUIT SYMBOLS

Manufacturers and users of modern integrated electronics devices have adopted several standard symbols for use in schematic diagrams. Figure 11-2 shows the most common versions of these symbols as they apply to linear IC devices.

The standard single-ended amplifier symbol is shown in Fig. 11-2(A). In this type of amplifier, the input signals are applied between the single input terminal and common (or "ground" if less rigorous terminology is allowed). The symbol shown in Fig. 11-2A has a V + DC power supply terminal and a ground or common terminal. In some actual schematic circuit diagrams, these terminals are not shown for the sake of simplicity. Do not assume, however, that they are not used in such cases.

A differential amplifier symbol is shown in Fig. 11-2B. This type of amplifier produces an output signal that is proportional to the difference between the two input signals. If A_v is the voltage gain of

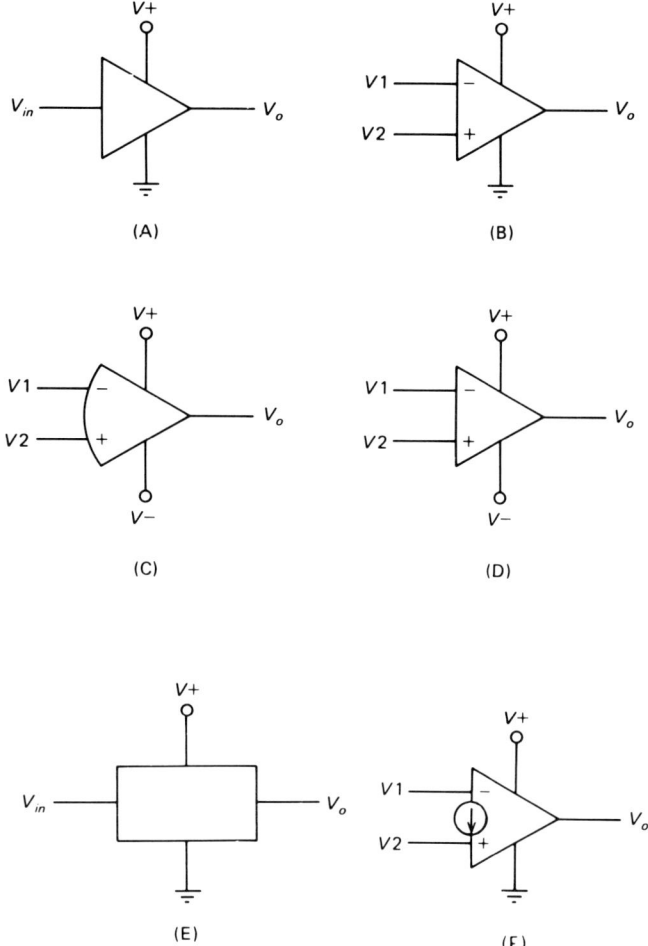

Fig. 11-2 Standard circuit symbols for linear integrated circuits: (A) generic single-ended linear amplifier, (B) generic differential linear amplifier, (C) traditional operational amplifier, (D) currently popular operational amplifier, (E) generic linear function module, and (F) current difference amplifier.

the amplifier, and $V1$ and $V2$ are the two input signals, then the output voltage is $A_v(V2 - V1)$.

There are always two inputs on a differential amplifier. The inverting input ($-$IN) produces an output signal that is 180° out of phase with the input signal. The noninverting input ($+$IN) produces an output that is in phase with the input signal.

Figures 11-2C and 11-2D show two alternate symbols used for the operational amplifier. The supposedly most proper version is that of Fig. 11-2C, although by common industry practice the more generic linear amplifier symbol of Fig. 11-2D is the de-facto standard.

The standard operational amplifier symbol has several features. First, there are two inputs (inverting and noninverting), because most op-amps are differential amplifier devices. In fact, only the differential amplifier gives true, full-range, operational amplifier performance. Second, there are two DC power supply terminals. The V + terminal requires a potential that is positive with respect to common, while the V − terminal requires a potential that is negative with respect to common. Third, the output is single ended, so the output signal is taken between the output terminal and the power supply common. Notice what is missing on this symbol? There is no ground or common terminal on the standard operational amplifier. The common connection is established by connecting together the cold ends of the V − and V + DC power supplies.

Some manufacturers use the universal or generic IC symbol shown in Fig. 11-2E for their products. Although these devices tend to be special function products, or limited use amplifiers, the same generic symbol is sometimes also used for other purposes.

Figure 11-2F is the symbol used for a special type of amplifier called a current difference amplifier (CDA), also called the Norton amplifier. The CDA produces an output voltage that is proportional to the difference between two input currents. The operation of the CDA is not exactly analogous to the op-amp (i.e., with the input voltages replaced by input currents), but it is at least similar. The symbol for the CDA differs from the normal op-amp symbol in order to distinguish its unique operation. The CDA symbol shown in Fig. 11-2F is the regular differential amplifier symbol with a current source symbol added along one edge to let the reader know that current mode operation is intended. Another form of linear IC amplifier, different from either the op-amp or CDA, is the operational transconductance amplifier, or OTA. This type of amplifier has a transfer function that relates output current to input voltage ($\Delta I_0/\Delta V_{in}$). Since the transfer function expression has the units A/V (or subunits thereof), the transfer function gain can be expressed in the units of conductance ("mhos" or the subunits millimhos or micromhos; siemens is the modern unit—1 S = 1 mho). Since these are units of conductance, the amplifier is called a transconductance amplifier. The name operational conveys the idea that some of the functions are similar to those of the operational amplifier. The OTA symbol is the same as the op-amp shown in Fig. 11-2D (sometimes with the letters OTA superimposed inside the triangle).

Although there is some variation in schematic symbols used by some manufacturers, the versions shown in Fig. 11-2 represent the vast majority of modern IC device applications.

LINEAR VERSUS DIGITAL IC DEVICES

The integrated circuit revolution brought us both linear and digital devices. It is, therefore, appropriate for us to consider the differences between the two types of device, especially because there are digital applications of linear devices and linear applications of digital devices.

Analog versus Digital Signals

It is instructive to examine the differences between analog and digital signals. Although the formal mathematical difference is more rigorously defined, the diagram of Fig. 11-3 shows the concept intuitively. An analog signal (Fig. 11-3A) is one that is continuous in both range and domain.

The range of an analog signal will be either a current (I) or voltage (V), while the domain is typically time (t). Thus, an analog signal is a continuous function of the form $V = F(t)$. The actual waveshape of the analog signal is not significant for sake of this definition, even though it is often critical for application purposes. A digital signal is discrete (i.e., noncontinuous in range and possibly in domain, also.

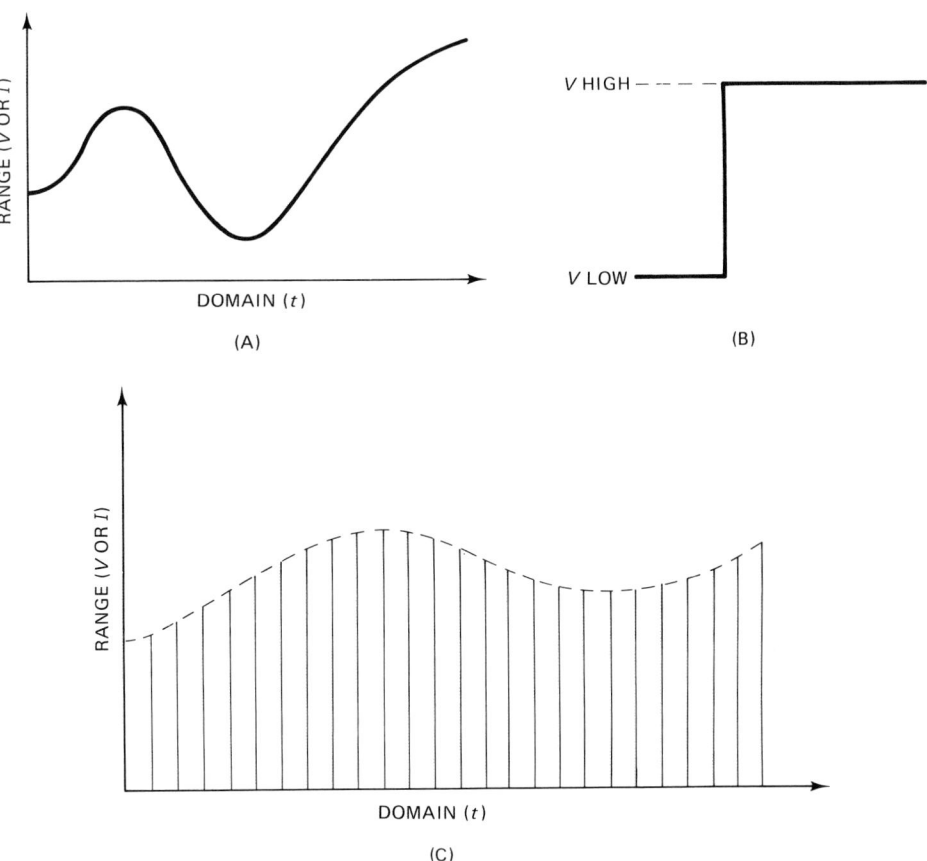

Fig. 11-3 (A) Analog signal, (B) digital signal, and (C) sampled analog signal.

In a digital signal, the range is a voltage or current that can take on only specified values (called VLOW and VHIGH in Fig. 11-3B). Although a digital signal may be continuous in the domain (t), many digital signals are discrete in the domain as well as the range. The key indentifying feature of the digital signal is the fact that the amplitude (range) can only take on specific discrete values. The overwhelming majority of digital signals are binary signals in which there are only two permissible states (e.g., VLOW and VHIGH in Fig. 11-3B). Modern digital computers use this type of signal.

The actual values of VLOW and VHIGH depend on the particular family of devices selected. In transistor–transistor logic (TTL) devices, VLOW nominally ranges from 0 to 0.80 V, and VHIGH is $+2.4$ to $+5.2$ V. In complementary metal oxide semiconductor (CMOS) devices, VLOW can be any voltage from 0 to -15 V, while VHIGH can be any voltage from 0 to $+15$ V. A constraint in the CMOS system is that the condition VLOW = VHIGH = 0 is not permitted. Furthermore, the absolute values of VLOW and VHIGH need not be equal. In other words, while VLOW = -10 V and VHIGH = $+10$ V is a permissible condition, VLOW = -5 V and VHIGH = $+10$ V (for example) is also permissible. The transition point from VLOW to VHIGH, or from VHIGH to VLOW, is one-half the difference between the two voltage levels.

A third category of signal is the sampled analog signal shown in Fig. 11-3C. This type of signal is continuous in range, but discrete in domain. The sampled signal is found extensively in instrumentation applications, especially in computerized analog instruments. A linear circuit is one that shows a simple scaler relationship between output and input signals. For example, a gain of 10 analog amplifier is linear because the output has the same waveshape as the input, but it has an amplitude that is 10 times larger. In contrast, both input and output signals of a digital circuit can take on only one of two values (VLOW or VHIGH); these are nonlinear circuits. For purposes of our discussion, a linear IC device is one that processes analog signals (some linear circuits may also process sampled signals as well as analog signals). Also included in this admittedly loose definition are certain nonlinear circuits discussed in a later chapter.

Applications of Linear versus Digital Devices

The inherent differences between linear and digital IC devices makes most applications clear-cut and unambiguous. There are areas, however, in which the situation is not so easily determined. For example, some CMOS digital gates can be biased in a way that makes them operate as linear amplifiers from DC up to AC frequencies exceeding 5 MHz. Similarly, operational amplifiers can work as clock oscillators in digital circuits, comparators, and coincidence detectors (i.e., AND gates) in a manner that suggests the binary situation that is normally expected in digital circuits. While it is true that the linear device output voltage must match the system voltage levels (VLOW and VHIGH) of the particular digital logic family that it works with, it is not difficult to interface such devices.

We will also see cases in which an inherently linear IC device (such as the operational amplifier) is used to produce some output signals that are decidedly nonlinear in that the output waveshape in no way resembles the input waveshape. Examples of this type of circuit are integrators, differentiators, and logarithmic amplifiers. These are, nonetheless, considered to be applications of linear IC devices in modern terminology.

COMMON PACKAGE TYPES

The integrated circuit is formed on a tiny chip of silicon material by a photolithographic process. Typical chip die sizes range around 100 mils (0.100 in.), with some being larger and others being smaller. The die is typically mounted inside a package and connected to the package pins by fine wires.

Figure 11-4A shows a die with a wire attached, while Fig. 11-4B shows a packaged die with a see-through window for illustration purposes. The connecting wires between package pins and solder pads on the die are around 10 mils (0.010 in.) diameter and in most cases are made of either gold or aluminum. Either an electric pulsed current or a thermosonic process is used to melt the end of the wire onto (and bond it with) the connecting pad on the semiconductor die.

The particular package style selected for any given IC device depends in part on the intended application and the number of pins required. For many IC devices, several different packages are available. The earliest IC packages were the 6-, 8-, 10-, and 12-lead metal can devices (Fig. 11-5A). These packages were redesigns of (and similar to) the TO-5 metal transistor package.

When viewed from the bottom the package, the keyway marks the highest number lead or pin (e.g., pin no. 8 in Fig. 11-5A), and pin no. 1 is the next pin clockwise from the keyway. One must be careful when looking at IC base diagrams to know whether a top or bottom view is depicted.

Perhaps the largest number of IC devices on the market today are sold in dual in-line packages (DIPs), examples of which are shown in Fig. 11-5B. DIPs are available in a wide variety of sizes, from 4 to more than 48 pins. Although many devices are available in other size DIPs, most linear devices are found in 8-, 14-, or 16-pin DIPs.

The DIP is symmetrical with respect to pin count regardless of the package size, so some means is needed to designate pin no. 1; Fig.

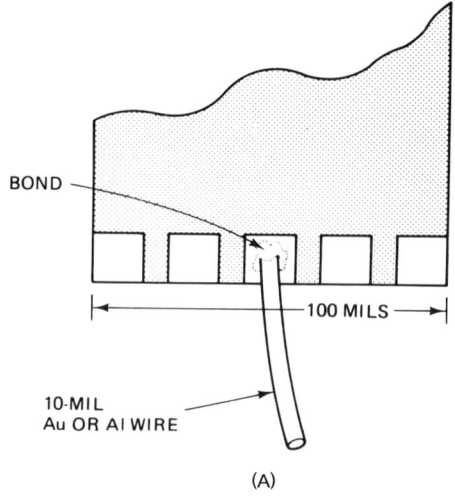

(A)

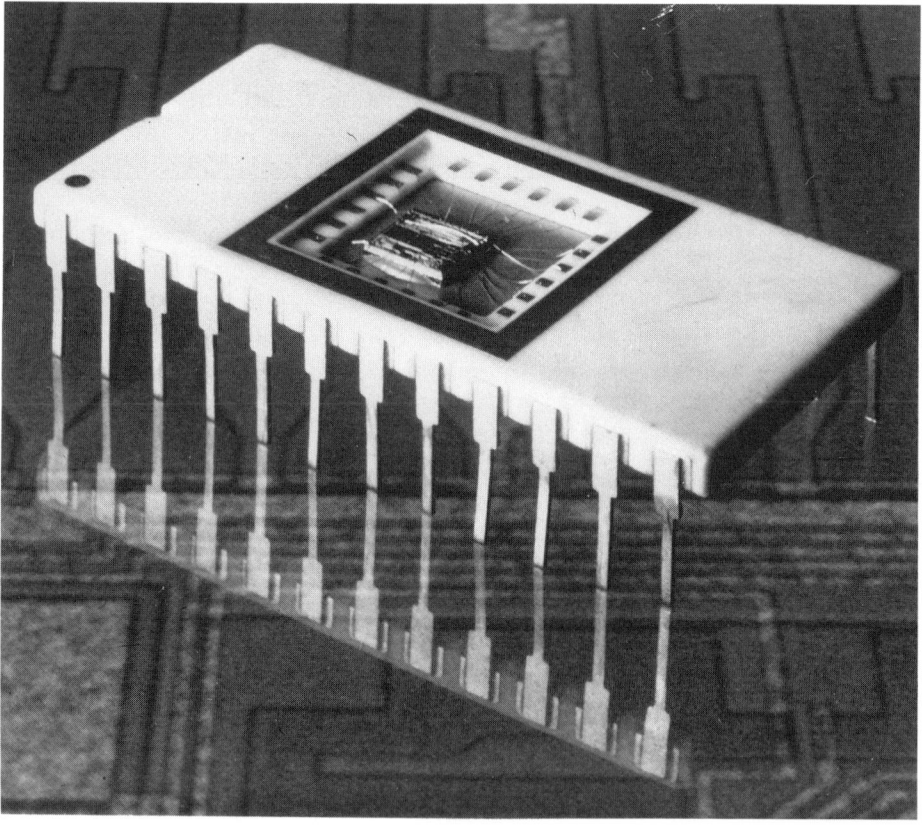

(B)

Fig. 11-4 Wire bonded to integrated circuit die.

11-5B shows several common methods. In all cases, the IC DIP is viewed from the top. On some devices, a painted dot, pimple, or dimple will mark pin no. 1. In other cases, a semicircular or square notch marks the end at which pin no. 1 is located. When viewed from the top, with the notch pointed away from you, pin no. 1 is to the left of the notch while the highest number pin is to the right of the notch.

Both plastic and ceramic materials are used in DIP construction. In general, the plastic packages are used in consumer and noncritical commercial (or industrial) equipment, while the ceramic packs are used in military and critical commercial equipment. The principal difference between the plastic and ceramic packages is the specified temperature range. Although exceptions exist, typical temperature range specifications are 0 to +70°C for commercial plastic devices and −10 to +80°C for ceramic. Military and some critical commercial ceramic devices are rated from −55 to +125°C. The principal differences between military chips and the highest grade commercial or industrial chips are the amount of testing, burn-in, and documentation that accompany each device.

An example of an IC flat pack is shown in Fig. 11-5C. This type of package is typically used where very high component density is required. Most flat-pack devices are digital ICs, although a few linear devices are also offered. DIP IC devices are mounted either in sockets or by insertion of the pins through holes drilled in a printed circuit

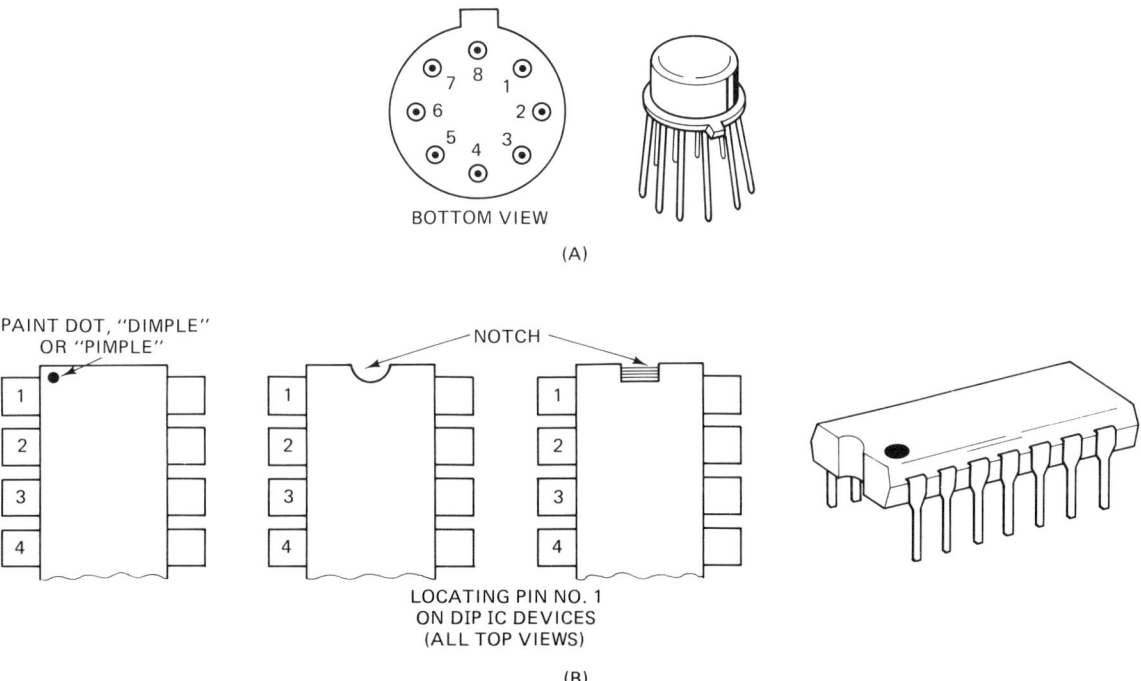

Fig. 11-5 Integrated circuit packages: (A) metal can, (B) dual in-line package (DIP), (C) flatpack, and (D) surface mounted device (SMD). (*Figure continues.*)

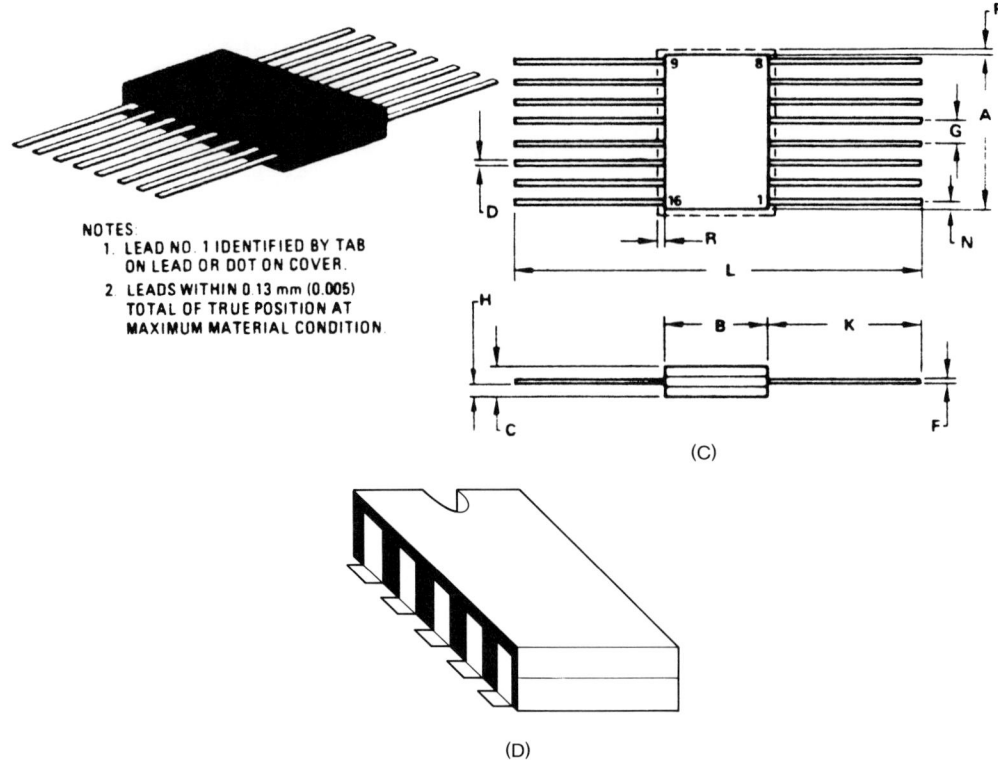

NOTES:
1. LEAD NO. 1 IDENTIFIED BY TAB ON LEAD OR DOT ON COVER.
2. LEADS WITHIN 0.13 mm (0.005) TOTAL OF TRUE POSITION AT MAXIMUM MATERIAL CONDITION.

(C)

(D)

Fig. 11-5 (*continued*)

board (PCB). Flat packs, on the other hand, are mounted on the surface by direct soldering to the conductive track of the PCB.

A relatively new style of package is the surface mounted device (SMD), an example of which is shown in Fig. 11-5D. The SMD technology represents a significant improvement in packaging density. SMD components can be mounted closer together than other types of packages and are amenable to automatic PCB production methods. It is expected that the SMD package will eventually overtake and replace other forms, especially in very large scale integration (VLSI) applications.

SCALES OF INTEGRATION

There are several different scales of integration found among IC devices. The ordinary small-scale integration (SSI) device consist of single gates, small amplifiers, and other smaller circuits. The number of components on each chip is on the order of 20 of less. Medium-scale integration (MSI) devices have a slightly higher degree of complexity and may have about 100 or so components on the chip. Devices such as operational amplifiers, shift registers, and counters, are usually classed as MSI devices. Large-scale integration (LSI) devices are mostly digital ICs and include functions such as calculators and microprocessors. Typical LSI devices contain from about 100 to 1000 components. The newer VLSI devices include some of the latest computer chips. The numbers and descriptions listed for SSI, MSI, and LSI devices are approximate only, but they do serve to provide guidelines. Most linear IC devices are either SSI or MSI, with the latter predominating.

DISCRETE VERSUS MONOLITHIC ICS VERSUS HYBRID CIRCUITS

The generic term microelectronics defines two distinctly different categories of solid-state devices: monolithic integrated circuits (which covers most devices discussed in this text) and hybrid circuits. A valid question is how are monolithic and hybrid circuits different from each other, and how are they both different from regular, discrete solid-state circuits?

The monolithic IC device is made using photolithographic and other processes on a single crystal of semiconductor material. The word monolithic applies because only a single piece of semiconductor material is used in the fabrication of the device (see Fig. 11-6A). A hybrid circuit, on the other hand, is a cross between the monolithic and printed circuit methods. Although larger than most monolithic devices, the tight packaging makes the hybrid very unlike the printed circuit board that it superficially resembles internally. The macroview of the hybrid makes it usable as if it were an integrated circuit, even though the internal structure is quite different from genuine IC devices.

The substrate diode shown in Fig. 11-6A is formed by the interface between the P-type substrate and the regions of the semiconductor crystal that are used for the other components. This diode is a natural (if undesired) feature of IC devices, and it is normally reverse biased. If circuit conditions force the diode into the forward biased condition, then damage to the device may result. Although some designers have been clever at using the substrate diode in certain applications, it is normally treated as if it were not there.

The typical hybrid consists of a multilayer (usually ceramic) substrate on which conductive tracks are printed by any of several deposition methods. The integrated circuits, transistors, and other semiconductor devices used in the hybrid circuit are either cemented or otherwise bonded to the ceramic substrate (Fig. 11-6B).

A critical difference between the hybrid and other printed circuit devices is that unpackaged semiconductors are used. The chips diodes and transistors are used in die form. In other words, the unpackaged dies are installed directly onto the ceramic substrate of the hybrid.

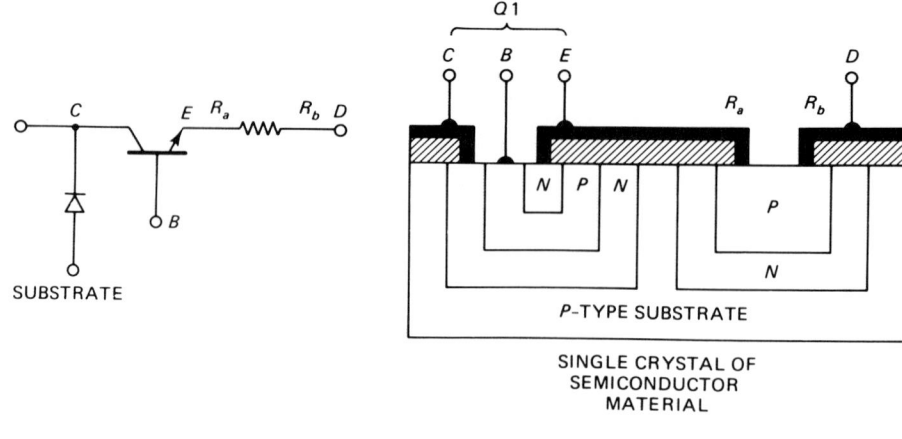

(A)

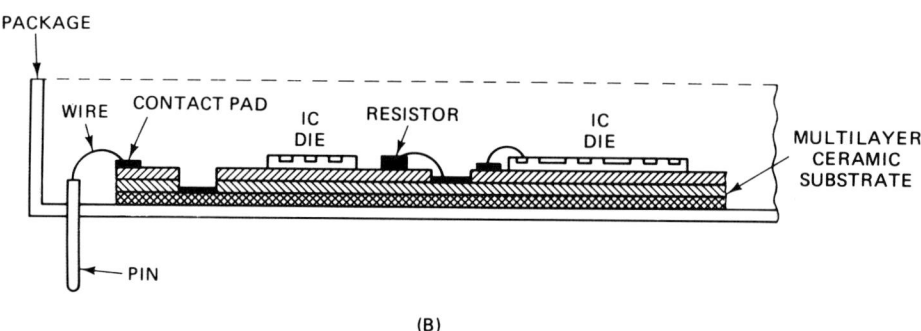

(B)

Fig. 11-6 (A) Circuit to be made into IC form and schematic of the equivalent IC, (B) hybrid microelectronic device, (C) delidded hybrid, and (D) discrete circuit and its implementation on PWB. (*Figure continues.*)

Tiny (about 10 mils) gold or aluminum wires connect the die electrical contacts either to the conductors of the hybrid substrate or to the hybrid package pinouts where appropriate.

The hybrid circuit may contain a variety of components or configurations that are difficult, expensive, or impossible to implement in monolithic IC form. Once the components are installed and the circuit is tested, the package lid is attached and the device is sealed, evacuated of air, and prepared for final testing and shipment. Once completed, the hybrid can then be treated as if it were a large integrated circuit. A photo of a delidded hybrid is shown in Fig. 11-6C.

A principal advantage of the hybrid is that it provides nearly IC-like packaging densities while being relatively easy to design and manufacture without a semiconductor foundry. The hybrid is used when either complexities or small quantities make monolithic IC implementation difficult or uneconomical.

Discrete electronic circuits are formed on PCBs of individually packaged active solid-state and passive components that are interconnected through photoetched copper tracks. The simple, discrete circuit

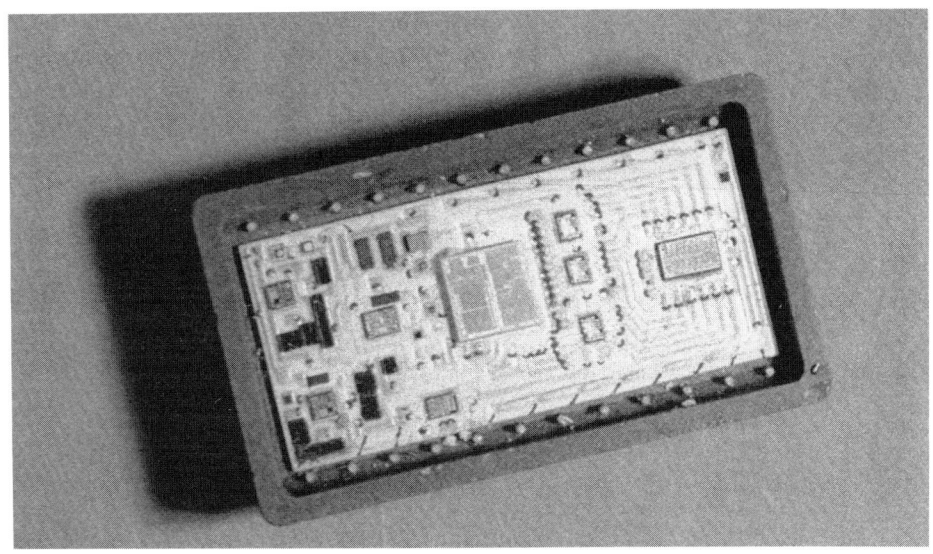

(C)

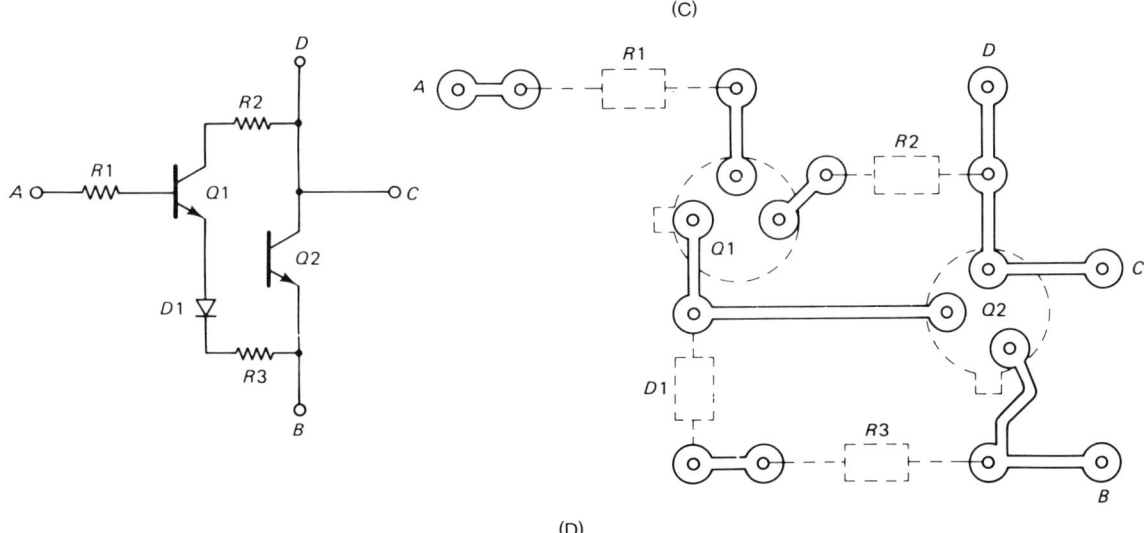

(D)

Fig. 11-6 (*continued*)

shown in Fig. 11-6D would easily fit onto a 100-mil (or less) IC die or a 250-mil hybrid substrate. At normal discrete component packaging densities, this same circuit would require on the order of a 7-in.^{2} (1000-mil) area on the PCB. By comparing the ratios of 1000 mil^2 to 250 mil^2 and 100 mil^2, we can see how IC and hybrid construction offers greatly reduced size.

LINEAR IC POWER SUPPLY REQUIREMENTS

Operational amplifiers and other linear IC devices normally operate from bipolar DC power supplies, such as shown in Fig. 11-7 (the pin

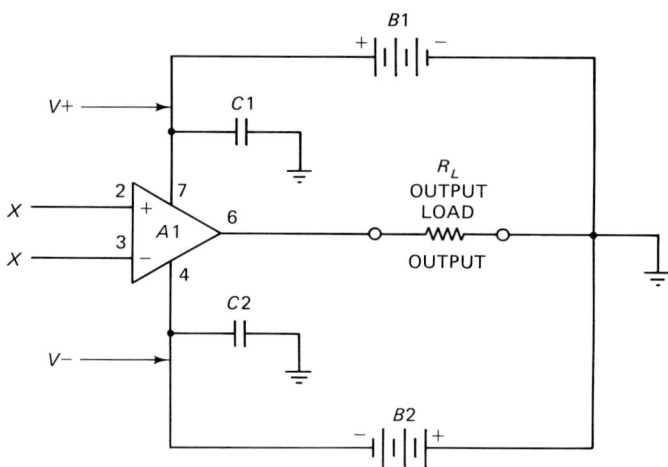

Fig. 11-7 Power for the operational amplifier is bipolar. That is, it uses two DC sources, one positive to ground and the other negative to ground.

numbers in this figure are for the so-called industry standard type 741 operational amplifier). This circuit (Fig. 11-7) shows that the two DC power supplies are completely independent of each other. The $V+$ power supply is positive with respect to common, while the $V-$ supply is negative with respect to common. The operational amplifier manufacturer will specify minimum and maximum values for $V-$ and $V+$. Typically, the maximum voltages will be on the order of ± 18 V, with some able to handle up to ± 22 V, and in at least one case ± 40 V.

There may be certain specified limitations on the maximum permissible supply voltages for any given device. For example, one such limitation that is occasionally seen in data sheets relates the $V-$ and $V+$ potentials. The manufacturer will specify that the quantity $(V+) - (V-)$ does not exceed a certain critical voltage. On some older devices, the maximum $V-$ and $V+$ ratings are both specified at 18 V. But, that does not mean that the permissible maximum pin-to-pin power supply potential is the algebraic sum of the two potentials (36 V)! The pin-to-pin supply voltage was specified not to exceed 30 V. Since $(+18$ V DC$) - (-18$ V DC$) = 36$ V, operating both supply terminals at their maximum voltage was not permitted. Modern devices rarely have such a limitation. One should, however, be careful to check this point on manufacturer data sheets for specific products before assuming otherwise.

Because most practical circuits operate with equal bipolar DC power supplies, it is also true that operational amplifiers with the 30-V pin-to-pin limit also effectively limit the $V-$ and $V+$ DC power supplies to not more than 15 V each.

Single DC Supply Operation

The operational amplifier (and many other linear IC devices) are designed for operation from split, dual, or bipolar power supplies (all three terms describe the same situation). There are, however, many applications in which only a single polarity DC power supply is available. In order to operate the op-amp in these cases, we must somehow supply the missing potential (e.g., with an on-board DC-to-DC converter) or devise a method for getting around the need for the missing potential.

It is reasonably easy to supply a missing potential. All that is needed is a DC-to-DC converter circuit that provides the needed voltage from the existing DC supply voltage. There are quite a few devices on the market that will produce either -15 V from $+15$ V (or $+12$ V, as the case may be) or will produce isolated ± 15-V potentials from an existing nonisolated $+12$- to $+15$-V potential. One can also design such circuits relatively easily.

A method for using a single DC power supply is shown in Fig. 11-8. A resistor voltage divider, $R1/R2$, is used to bias the noninverting input of the operational amplifier to some potential between ground and $V+$; the $V-$ terminal of the device is grounded. The bias voltage on the noninverting input also appears on the output terminal as a DC offset potential. Unless the circuit following the amplifier somehow does not care about this offset potential, the output terminal must be capacitor coupled to block the DC offset. The value of the bias voltage is found from

$$V1 = (V+)[R2/(R1 + R2)]. \qquad (11\text{-}1)$$

The capacitor shown in Fig. 11-8 is used to place the noninverting input at or near ground potential for AC signals while retaining the DC level produced by the resistor voltage divider. This capacitor sometimes causes noisy operation of the device when significant "ground loop" or "ground plane" noise is present, so it is often omitted in practical circuits. In most cases, however, it should be used. The value of the capacitor is such that its capacitive reactance is less than or equal to $R2/10$ at the lowest frequency of operation. For example, if the amplifier is designed to operate down to a frequency of 10 Hz, and the value of $R2$ is 2200 Ω (a typical value in real circuits), then the value of $C1$ must be such that it has a reactance of 220 Ω or less at 10 Hz. This requirement evaluates to

$$C1 = 1{,}000{,}000/(2\pi FX)$$

$$= 1{,}000{,}000/[(2)(3.14)(10\text{ Hz})(220\cdot\Omega)]$$

$$= 1{,}000{,}000/13{,}816 = 73\ \mu\text{F}.$$

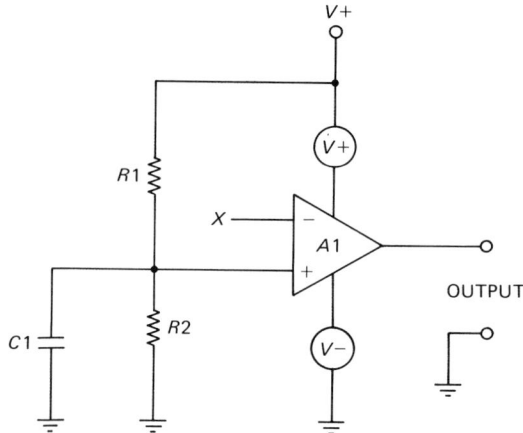

Fig. 11-8 Voltage divider $R1/R2$ is used to bias one input of the op-amp to allow single DC power supply operation.

Because 100 μF is the next higher standard value capacitor, most designers will select 100 μF for $C1$ instead of 73 μF.

PROTECTING LINEAR IC AMPLIFIERS

Operational and other linear IC amplifiers are sensitive to problems on the DC power supply lines. For example, the amplifier may oscillate if the DC lines are not properly decoupled. Furthermore, voltage variations, noise, and transients coupled to the DC power supply lines in one stage can affect the other stages in the same circuit, especially if power supply rejection is poor on the particular device.

We also sometimes find another problem, especially when breadboarding, troubleshooting, and using portable (battery operated) equipment: reversed polarity DC power supplies. The results can be catastrophic if the DC power supply potentials are reversed. An operational amplifier with reversed DC power supplies will probably be destroyed instantly. There are certain remedies available for each of the defects.

The problems of noise and oscillation due to cross-coupling between stages can be cured by using decoupling capacitors on the amplifier power supply terminals. Capacitors $C1$ and $C2$ (each 0.1 μF) in Fig. 11-9A are used to decouple high frequencies, while the low-frequency decoupling is provided by $C3$ and $C4$ (each typically 7 μF or higher). Why are two forms of capacitor needed at each op-amp power supply terminal? The higher value capacitors ($C3$ and $C4$) are typically aluminum or tantalum electrolytics. The performance of these capacitors drops drastically as frequency increases and may be very poor at higher frequencies that are nonetheless within the range of the device. At those high frequencies, the typical electrolytic capacitor is ineffec-

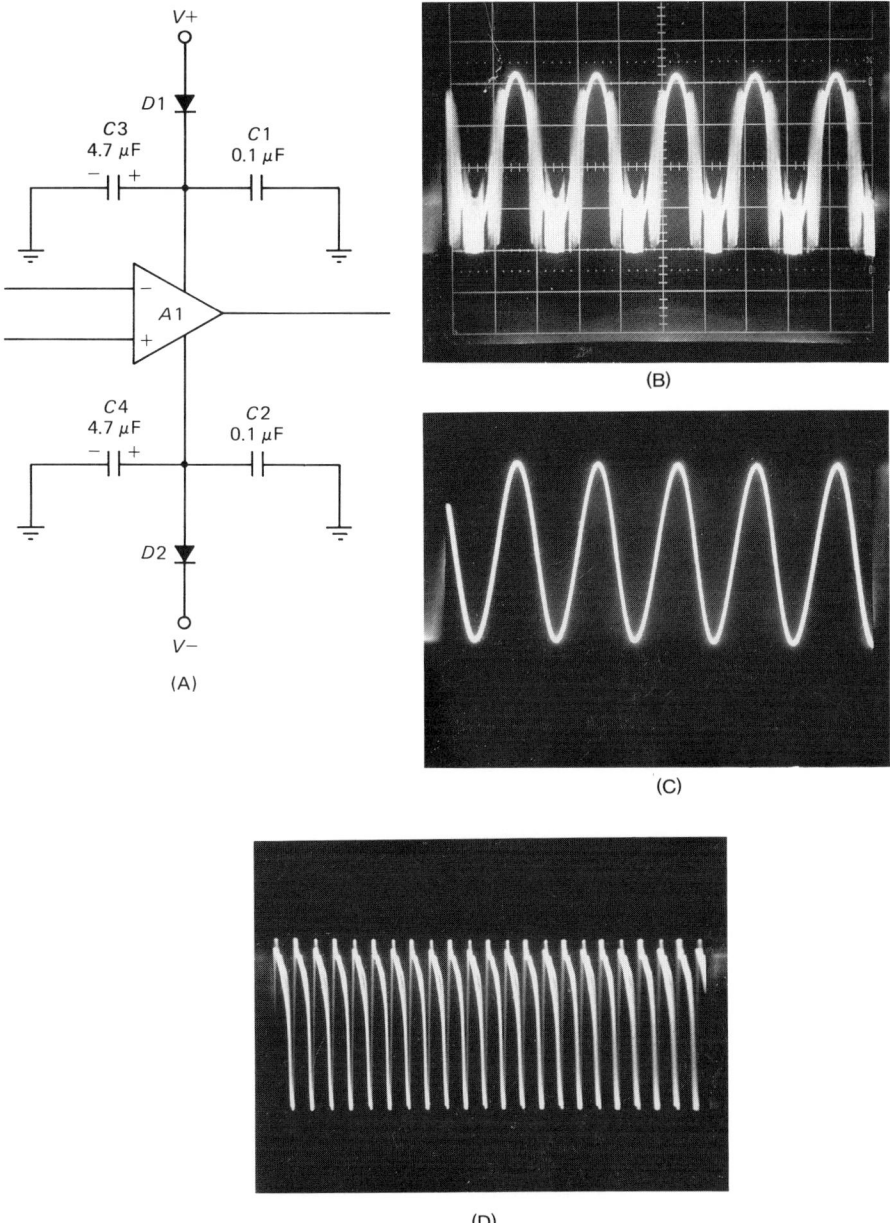

Fig. 11-9 Series power supply diodes used to protect the op-amp against reverse polarity.

tive. For this reason, we also sometimes use a smaller value capacitor, but one that is of a type that will work at higher frequencies (e.g., mylar, mica, and ceramic). This situation is changing, however, because certain new forms of capacitors are now available that offer high-frequency operation as well as high capacitance. The two-capacitor decoupling scheme is typically used on nonfrequency compensated op-amps or any linear IC with a high-gain–bandwidth product (e.g., > 5 MHz).

Figures 11-9B and 11-9C show the output waveform of a high-frequency operational amplifier in both nondecoupled (Fig. 11-9B) and decoupled (Fig. 11-9C) configurations. In Fig. 11-9B, we see a 400-Hz sinusoidal waveform that has high-frequency oscillation superimposed on it. In Fig. 11-9C, however, we see the same waveform after the decoupling capacitors were added to the $V-$ and $V+$ pins of the operational amplifier used in this experiment. Note that the oscillation is removed. In Fig. 11-9D, we see the actual oscillation (with the 400-Hz sine wave turned off) in an expanded oscilloscope time base range.

One practical rule of thumb ensures the success of the circuit in regard to noise and oscillation: place those capacitors as close as physically possible to the body of the amplifier. The 0.1-μF capacitors ($C1$ and $C2$) are more important than the higher value capacitors, so they should be closest to the IC amplifier body.

Protection against reverse polarity conditions is also shown in Fig. 11-9. Diodes $D1$ and $D2$ are placed in series with each DC power supply line. Under normal operation these diodes are forward biased, so they will conduct current to the amplifier. If someone accidentally connects one or both DC power supplies backward, then these diodes are reverse biased, so they will not conduct current. Thus, the series diodes protect the amplifier IC from incorrect power supply connection. Typical diodes for this application are any of the 1N400x series (1N4001 through 1N4007).

Another method for protecting the device is shown in Fig. 11-10. In this case, a zener diode is placed across the two DC power supply terminals of the amplifier. The zener potential must be greater than the maximum actual value, but less than the maximum permissible

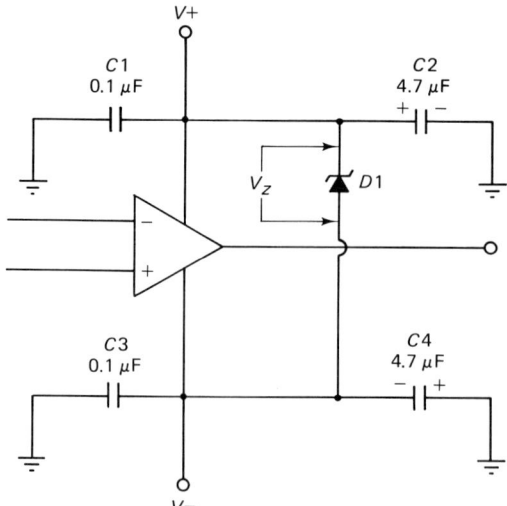

Fig. 11-10 Zener diode overvoltage and reverse polarity protection.

value, of the quantity $(V+) - (V-)$. For the case where DC power supplies of ± 12 V are used, this value is 24 V. A 28-V zener diode would be adequate if the maximum permissible value is specified as 30 V (provided that the power supply voltages are reasonably stable). Under these conditions, with V_z greater than the voltage between the terminals, zener diode $D1$ is reverse biased at a voltage lower than the zener potential. Thus, it is not conducting in normal operation.

In reverse polarity operation, diode $D1$ becomes forward biased in the normal nonzener mode. It will pass the current around the amplifier harmlessly. If an excessive $(V-)$ to $(V+)$ voltage is applied, then the zener diode will conduct and clamp the voltage to V_z.

A protection scheme for a multiple stage amplifier is shown in Fig. 11-11. There are two alternatives shown in this figure. In one case, 1N400x-series diodes in a reverse bias state are placed across the DC power supply lines and current-limiting resistors in series to prevent them from burning. The diodes are normally reverse biased, but when one or both DC power supplies are reversed, these diodes become forward biased and short the line to ground. The second alternative is to place the diodes in series with the line at the power supply terminals of the IC device (shown in dotted lines in Fig. 11-11). This method is analogous to the method of Fig. 11-9, except that it serves more than one amplifier.

POWER SUPPLIES FOR THE LABORATORY

Many readers will perform laboratory work to check out circuits. A power supply must be selected (or built) to work these experiments. Unless otherwise specified, the circuits in this book are designed to use either ± 12- or ± 14-V DC regulated power supplies. The DC power supply selected should offer either a single bipolar power supply or two independent 12-V DC supplies that are not ground referenced. The nongrounded feature allows the creation of a bipolar supply by connecting the positive output terminal of one supply to the negative output terminal of the other (see Fig. 11-12). Desirable features to have on a bench power supply include output voltages fixed at ± 12 V DC (or ± 15 V) output voltages adjustable from 0 to greater than 15 V DC, current available from each polarity not less than 100 mA, metered outputs (I and V), voltage regulation, current limiting for output short-circuit protection, and overvoltage protection.

Although not all good selections will have all of these features, those that do are clearly superior for most laboratory applications. In addition, it must be recognized that these recommendations fit the more or less standard laboratory bench and may vary considerably for certain highly specialized requirements.

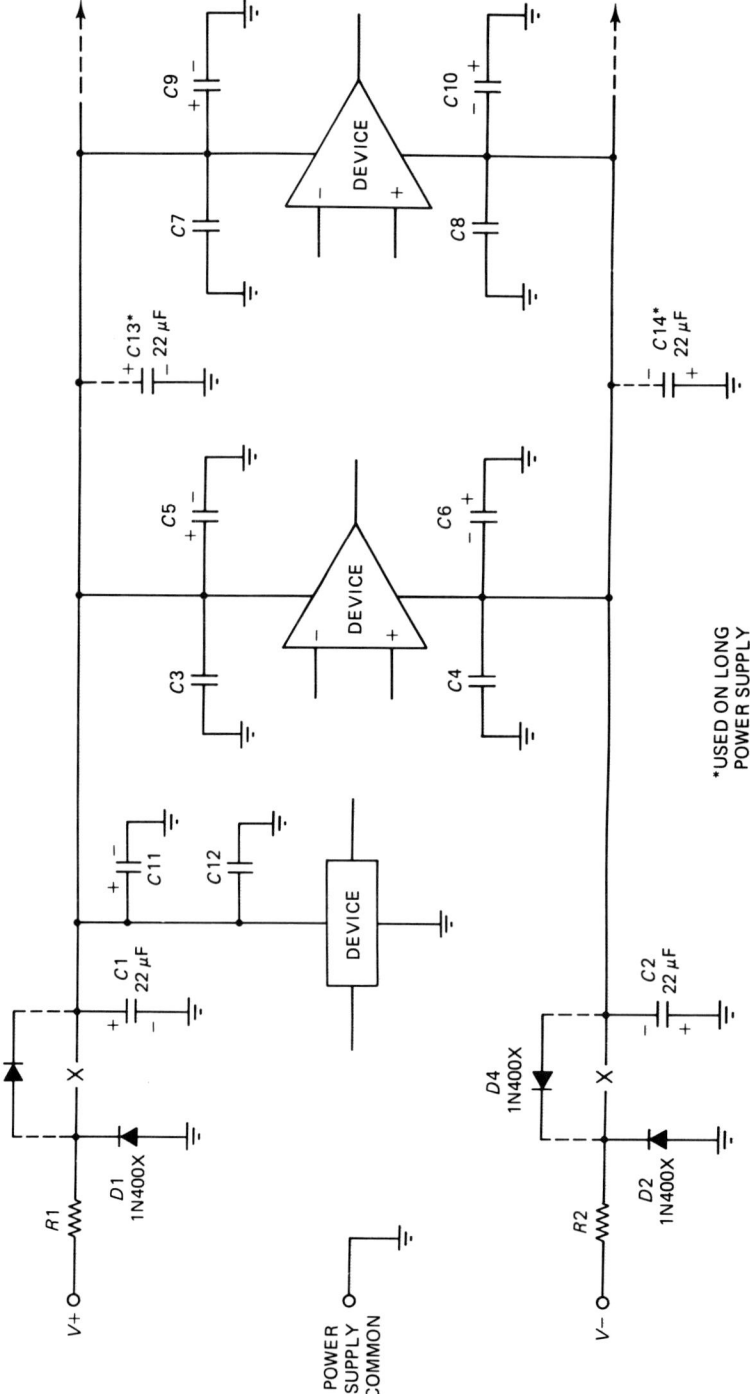

Fig. 11-11 Protection of a group of operational amplifiers and other devices.

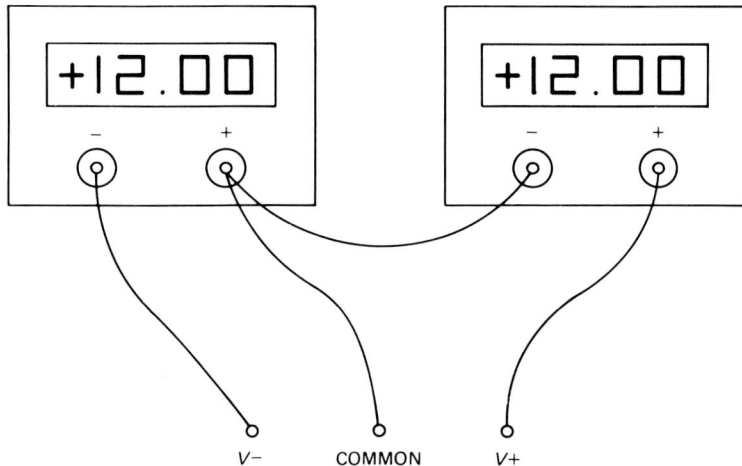

Fig. 11-12 Connection of external power supplies for bench experiments.

LABORATORY CONSTRUCTION PRACTICES

Successfully designing and debugging circuits on a professional level require certain skills and knowledge of lab construction practices. Note that these are not necessarily the same as production methods and, in fact, are probably not the same. In this section, we will take a look at some of these practical matters before proceeding with the rest of the text.

There are various levels of preproduction construction practice, of which four major categories emerge rather easily: (1) the partial breadboard, (2) the complete breadboard, (3) the brassboard, and (4) the full-scale development model.

The two breadboard categories involve using special construction methods to check circuit validity and make certain preliminary measurements. The breadboard is in no way usable in actual equipment, rather it is used in bench or laboratory levels of development.

The difference between partial and complete breadboards is merely one of scale. The partial breadboard is used to check out a partial circuit or a small group of circuits. The complete breadboard will be the full-up circuit and is much more extensive than the partial breadboard. There are fundamental differences in the type of breadboarding hardware used for both types, especially in large circuits.

The two commonly used types of hardware are shown in Figs. 11-13 and 11-14 . The device shown in Fig. 11-13 is an IC pin socket breadboard; this model is the Heath/Zenith Educational Systems ET-3300. Other, more extensive models are available from other sources. The ET-3300, like others of its type, offers three built-in DC power supplies: +12 V at 100 mA, −12 V at 100 mA, and +5 V at 1.5 A (for TTL digital logic IC devices). Located between the large

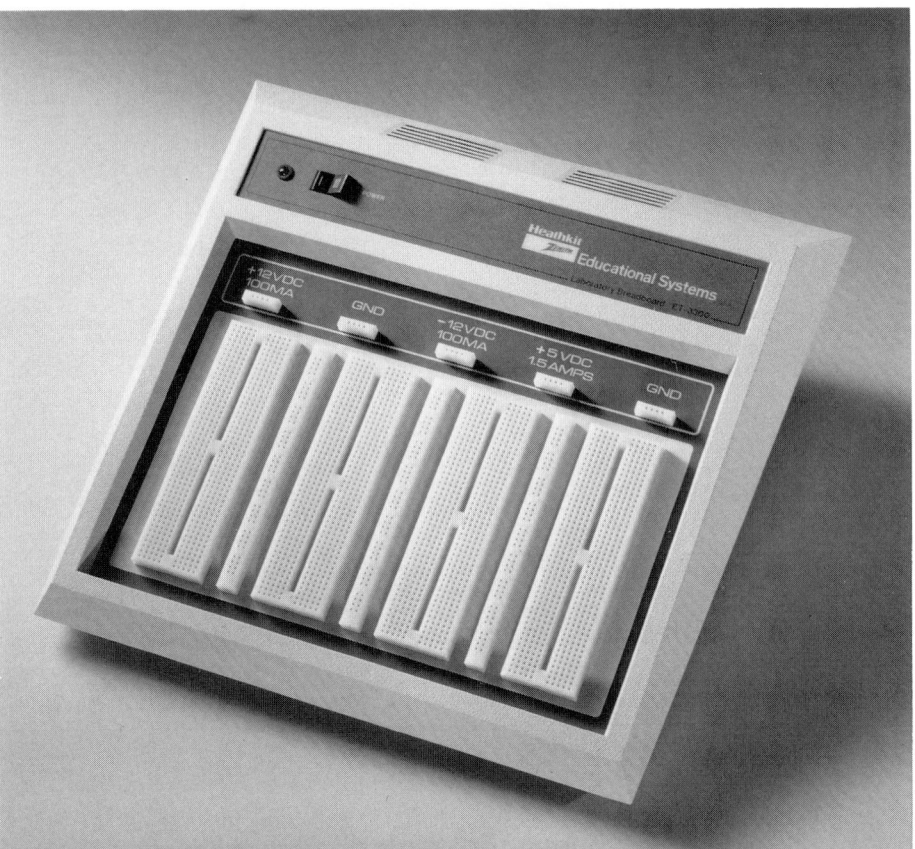

Fig. 11-13   An IC pin socket breadboard.

multipin IC sockets are bus sockets in which all pins in a given row are strapped together. These sockets are generally used for power distribution and common (ground) buses. Interconnections between sockets, power sources, and components are made using #22 insulated hookup wire. The bared ends of the wires are pressed into the socket holes.

The other method of breadboarding is wire wrapping (Fig. 11-14). Wire wrapping is the preferred method for full-up breadboards. Sockets with special square or rectangular wire-wrap pins are mounted on a piece of universal printed circuit board. These boards are perforated to accept wire-wrap IC sockets and usually have printed power distribution and ground buses. Using a special tool, the wire interconnects are strapped from point to point as needed by the circuit. The wire is wrapped around each contact tightly enough to break through the insulation to make the electrical connection.

In general, socketed breadboards are typically used for small-scale applications (for partial breadboarding) and for small design projects that do not justify the cost of the wire-wrap board. Wire wrapping is

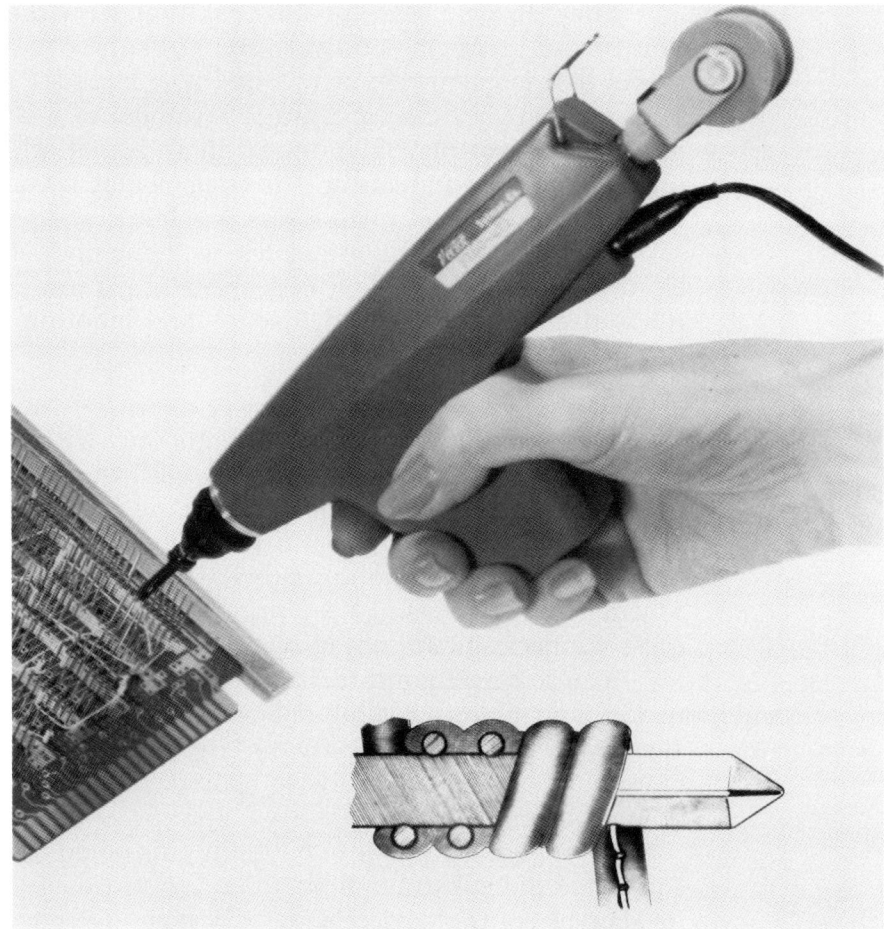

Fig. 11-14 The method of breadboarding by wire wrapping.

used for larger, more extensive, projects. Although the distinctions between types of breadboard, and the respective supporting hardware, are quite flexible, they hold true for a wide range of situations.

Brassboards are full-up model shop versions of the circuit that will plug into the actual equipment cabinet in which the final circuit will reside. Some brassboards are wire wrapped, but most are printed circuits. The important consideration is that the brassboard is near its final configuration and can be tested *in situ* in the actual equipment. In contrast, the breadboard is purely a laboratory model. The brassboard is usually made using model shop handwork methods and not routine automated production methods. The brassboard differs from the final version in that it might contain "dead bug" and "kluge card" modifications in which components are informally mounted on the board for purposes of testing circuit changes.

The full-scale development (FSD) model is the highest preproduction model. It might look very much like the first article production model that is eventually made, and it is intended for field tests of the final product. For example, an FSD model of a two-way land-mobile radio transceiver may be mounted in an actual vehicle and used for its intended purpose in tests or field trials. The FSD model is built as near to regular production methods as possible.

There are several principles to remember when breadboarding electronic circuits. While it is possible to get away with ignoring these rules, they constitute good practice and ignoring them is risky. The rules are

1. Insert and remove components only with the power turned off.
2. Wire and change wiring only with the power turned off.
3. Check all wiring prior to applying power for the first time.
4. Always do the power distribution and ground wiring first and then check these before any other wiring is done. (In many informal schematics the $V-$ and $V+$ wiring is omitted, but that does not mean these connections are not to be made.)
5. Use single-point (or star) grounding wherever possible. Do not use ground plane or ground bus unless signal frequencies are low, signal voltage levels are relatively large, and current drain from the DC power supply is low. Otherwise, ground noise and ground loop voltage drop problems will exist.
6. Applying a signal to IC input pins when there is no DC applied can cause the substrate diode to be forward biased, with the potential for damage to the device being very high. Therefore, (1) do not apply the signal until the DC power is turned on, (2) turn off the signal source prior to turning off the DC power supply, and (3) never apply a signal with a positive peak that exceeds $V+$ or a negative peak that exceeds $V-$.
7. All measurements are to be made with respect to common or ground, unless special test equipment is provided. For example, in Fig. 11-15, the differential voltage V_d is composed of two ground referenced

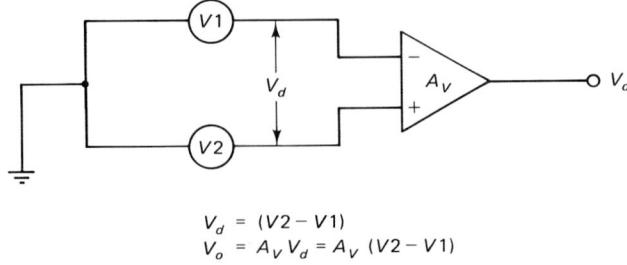

$$V_d = (V2 - V1)$$
$$V_o = A_V V_d = A_V (V2 - V1)$$

Fig. 11-15 Input signals to a differential amplifier.

voltages, $V1$ and $V2$. Measure these voltages separately and then take the difference between them (i.e., $V2 - V1$), unless a differential voltmeter is used.

In this chapter, we introduced the integrated circuit, discussed the power supply for the IC, and demonstrated some construction practices and techniques. In the next chapter, we introduce the ideal and practical operational amplifier, examine the applications of an op-amp with no feedback network, and discuss other important topics.

12

IC Operational Amplifiers

The operational amplifier is, by far, the most useful and most widely applied linear IC (integrated circuit) device on the market today. It was also one of the first IC devices on the market. No supposed instrumentation or linear circuit designer is properly educated until he or she can design with op-amps. These devices define the field of linear IC electronics. Because of their considerable importance in electronic instrumentation and in the chapters to follow, this chapter will discuss the IC operational amplifier in depth.

THE OPERATIONAL AMPLIFIER: AN OVERVIEW OF APPLICATIONS

The original operational amplifiers (1948) were designed to perform mathematical operations in analog computers; hence the name *operational amplifiers*. Their use in other applications follows from the fact that the op-amp is basically a very good DC differential amplifier with extremely high gain. The immense benefits and extreme flexibility of the op-amp derive from that one simple fact. Applications for the operational amplifier are found in instrumentation, process monitoring and control, servo control systems, signal processing, communications, measuring and testing circuits, alarm systems, medicine, science, and even in some digital computers. It is only somewhat exaggerated to call the operational amplifier a universal linear amplifier.

OPERATIONAL AMPLIFIERS

Operational amplifiers are probably the most widely used linear integrated circuits on the market today. They are, without a doubt, the most flexible linear IC devices available because one can manipulate the overall forward transfer function by manipulating the feedback network properties. Other "universal" linear IC amplifiers (e.g., CDA and OTA devices—see Chapter 13) have failed to overcome the op-amp in either sales or usefulness.

Figure 12-1 shows a simplified schematic diagram of the internal circuitry of the popular, and very low-cost, 741 operational amplifier device. There are three main sections to this circuit: input amplifier, gain stages, and output amplifier.

The output stage is a complementary symmetry push-pull DC power amplifier. Typical designs produce from 50 to 500 mW of output power. The output stage operates as a push-pull amplifier because transistors Q9 and Q10 have opposite polarities: Q9 is NPN, and Q10 is PNP. Output signal is taken from the junction of the Q9/Q10 emitters. If the two transistors conduct an equal amount of current, then the net output voltage is zero.

The intermediate gain stages are shown here in simplified block diagram form. These stages provide the high gain required, some level translation, and (in the case of the 741) some internal frequency compensation.

The input stage is a DC differential amplifier made from bipolar transistors. Although there are a few single-input devices on the market sold as "operational" amplifiers, they are in reality high-gain DC amplifiers, not true op-amps. The reason is that a true operational amplifier must be able to perform a wide range of mathematical operations, and that ability requires both inverting and noninverting input functions. This same reason is also used to explain why bipolar DC power supplies must be used in op-amp circuits: Results (i.e., output voltages) may be either zero, positive, or negative, and the device must accommodate all three of these possibilities. An amplifier must have differential inputs and bipolar output polarity in order to operate in all four quadrants of the Cartesian system.

DC Differential Amplifiers

Figure 12-2A shows a simplified DC differential amplifier circuit. Two transistors (Q1 and Q2) are connected at their emitters to a single constant current source (CCS), $I3$. Because current $I3$ cannot vary, a change in either $I1$ or $I2$ will also affect the other current. For example, an increase in current $I1$ means a necessary decrease in

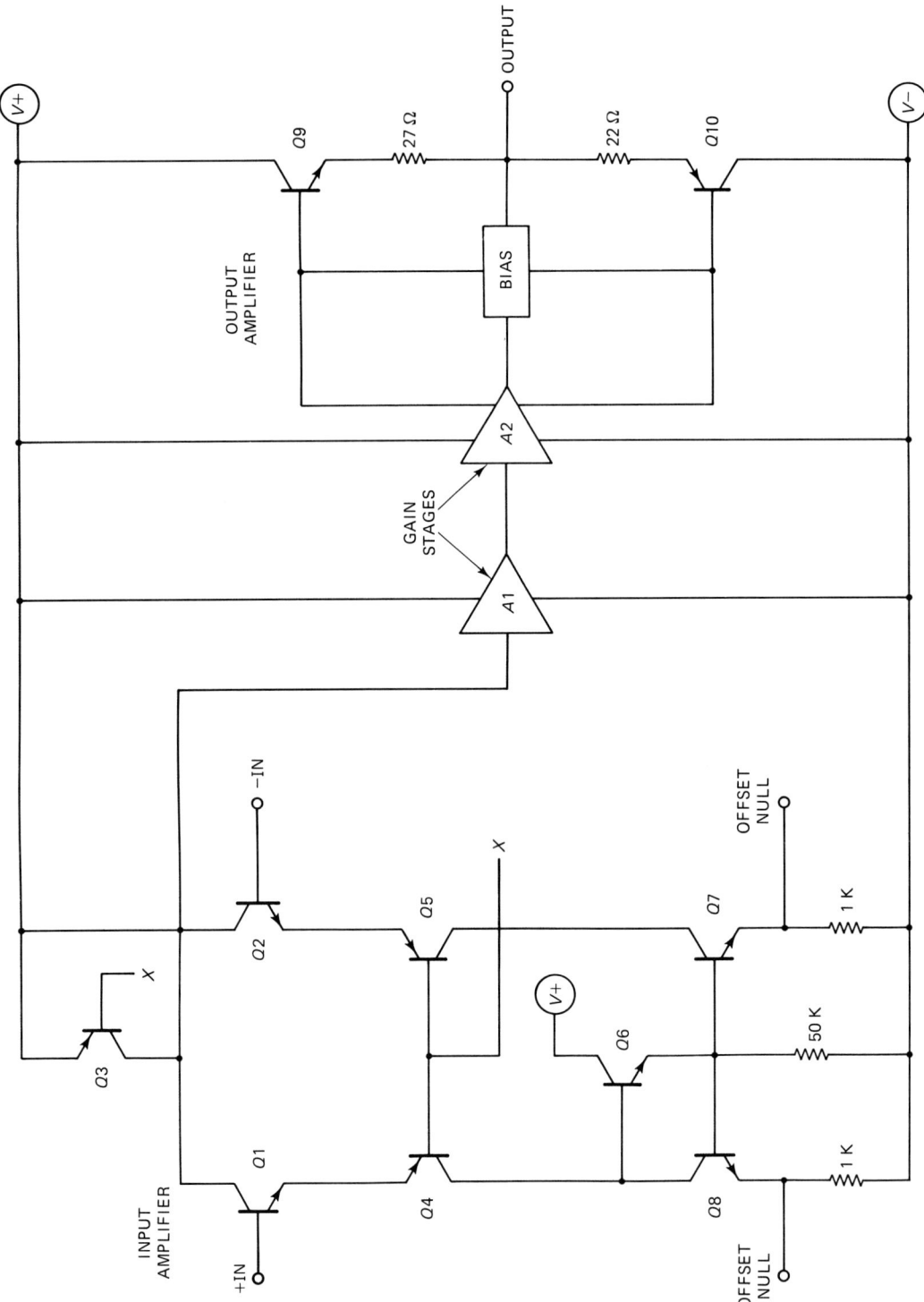

Fig. 12-1 Typical internal circuitry of operational amplifier.

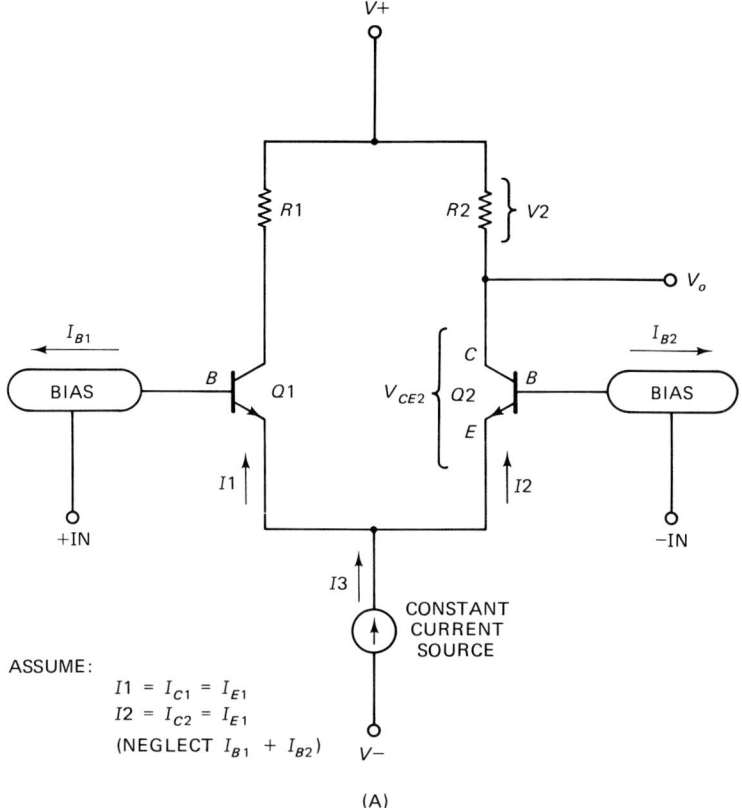

Fig. 12-2 Differential input circuitry: (A) actual circuitry and (B) redrawn equivalent circuit. (*Figure continues.*)

current $I2$ in order to satisfy Eq. (12-1):

$$I3 = I1 + I2 \qquad\qquad (12\text{-}1)$$

where $I3$ is the constant current, $I1$ the C-E current of transistor Q1, and $I2$ the C-E current of transistor Q2.

Although the collector and emitter currents for each transistor are not actually equal to each other (differing by the amount of the base current), we may conveniently neglect I_{B1} and I_{B2} for the time being. For purposes of this discussion we may assume that $I1 = I_{C1} = I_{E1}$, and $I2 = I_{C2} = I_{E2}$ (even though they are not, in fact, equal in real circuits).

Consider two voltage drops created by current $I2$ (which is the C-E current flowing in transistor Q2): V2 is the voltage drop $I2\,R2$, while V_{CE2} is the collector-to-emitter voltage drop across Q2. Voltage V_{CE2} depends on the conduction of Q2, which in turn is determined by the signal applied to the noninverting input ($+$IN).

If the voltages applied to both $-$IN and $+$IN inputs are equal, then $I1$ and $I2$ are equal. In that case, the quiescent values of $V2$ and

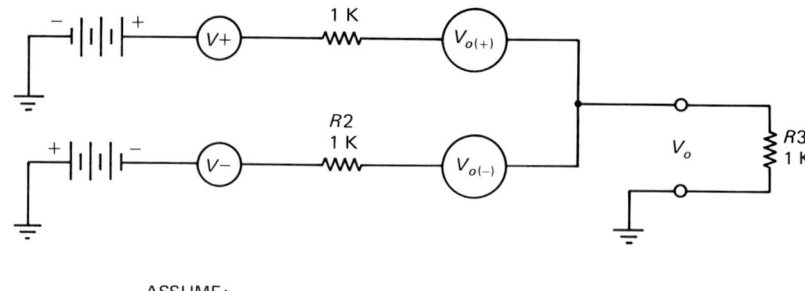

ASSUME:

$$V+ = +10 \ Vdc$$
$$V- = -10 \ Vdc$$
$$V_o = V_{o(+)} + V_{o(-)}$$
$$R1 = R2 = R3 = 1 \ K\Omega$$

A) $V_{O(-)} = \dfrac{(V-)(R3)}{(R2 + R3)} = \dfrac{(-10 \ Vdc)(1 \ K)}{(1 \ K + 1 \ K)} = \dfrac{(-10 \ Vdc)(1)}{2} = \quad -5 \ Vdc$

B) $V_{O(+)} = \dfrac{(V+)(R3)}{(R1 + R3)} = \dfrac{(+10 \ Vdc)(1 \ K)}{(1 \ K + 1 \ K)} = \dfrac{(+10 \ Vdc)(1)}{2} = \quad +5 \ Vdc$

C) $V_o = V_{o(-)} + V_{o(+)} = (-5 \ Vdc) + (+5 \ Vdc) = 0 \ Vdc$

(B)

Fig. 12-2 (*continued*)

V_{CE2} are approximately equal (as determined by internal bias networks), so the relative contributions of $V-$ and $V+$ to output potential V_o are equal. How does this work? A model circuit is shown in Fig. 12-2B to demonstrate how V_o is formed. The output voltage V_o is the sum of two contributors: $V_{o(-)}$ is the contribution from $V-$, and $V_{o(+)}$ is the contribution from $V+$. These voltages are derived from the voltage drops across $R1$ and $R2$, respectively, and in our model represent the voltage drops $V2$ and V_{CE2} above. As long as $R1$ and $R2$ are balanced, then the sum $V_o = V_{o(-)} + V_{o(+)}$ is equal to zero. But if either $R1$ or $R2$ change, then V_o will be nonzero. If you wish to prove this fact to yourself, repeat the arithmetic shown in Fig. 12-2B using a different value than 1 kΩ for either $R1$ or $R2$. For example, when $R1$ is changed to 2 kΩ, the output voltage V_o is -1.667 V DC instead of zero.

Now let us consider the operation of the inverting input ($-$IN), which is the base terminal of transistor Q2. Recall that an inverting input produces an output signal that is $180°$ out of phase with the input signal. In other words, as V_{-IN} goes positive, V_o goes negative; as V_{-IN} goes negative, V_o goes positive.

If a positive signal voltage is applied to the $-$IN input and $+$IN $= 0$, NPN transistor Q2 is turned on harder. The effect is an increase in current $I2$ because the collector–emitter resistance of Q2 drops. We now have an inequality between $V2$ and V_{CE2}: voltage $V2$ increases, while V_{CE2} decreases. The result is that the contribution of $V-$ to V_o is greater, so V_o goes negative. The base of Q2 is, therefore, the inverting input because a positive input voltage produces a negative output voltage.

If the signal voltage applied to the $-$IN input is negative instead of positive, the situation changes. In that case, Q2 starts to turn off, so $I2$ drops. Voltage V_{CE2} therefore increases and $V2$ decreases. The relative contribution of $V+$ to V_o is now greater than the contribution of $V-$, so V_o goes positive. Again inverting behavior is seen: A negative input voltage produces a positive output voltage.

Now consider the noninverting input, which is the base of transistor Q1. Recall that a noninverting input produces an output signal that is in-phase with the input signal. A positive-going input signal produces a positive-going output signal, and a negative-going input signal produces a negative-going output.

Suppose $-$IN $= 0$, and a positive signal voltage is applied to $+$IN. In this case $I1$ increases. Because Eq. (12-1) must be satisfied, an increase in $I1$ results in a decrease of $I2$ to keep $I3 = I1 + I2$ constant. Reducing $I2$ reduces $V2$ (which is $I2R2$, so the contribution of $V+$ to V_o goes up: V_o goes positive in response to a positive input voltage. This is the behavior expected of a noninverting input.

Now suppose that a negative signal voltage is applied to $+$IN. Now there is a decrease in the conduction of Q1, so $I1$ drops. Again, to satisfy Eq. (12-1) current $I2$ increases. This condition increases $V2$, reducing the contribution of $V+$ to V_o, forcing V_o negative. Because a negative input voltage produced a negative output voltage, we may again affirm that $+$IN is a noninverting input.

Categories of Operational Amplifiers

Now that we have discussed the basic operational amplifier, let us widen our discussion a bit to encompass a larger selection of devices. Table 12-1 shows a hierarchy of commonly available devices. Some of these devices have been on the market for a long time, while others are relatively new. This list is intended to be representative, rather than exhaustive. Let us take a brief look at each category.

General Purpose Op-Amps

These are "garden variety" operational amplifiers that are neither special purpose nor premium devices. Most of these devices are said to be frequency compensated, so designers trade off bandwidth for inherent high (but not absolute) stability. As such, the general purpose devices can be used in a very wide range of applications with very few external components. Op-amps are usually selected from this category, unless some property of another class brings unique advantage to some particular application.

Voltage Comparators

These devices are not strictly speaking operational amplifiers, but they are based on op-amp circuitry. While all op-amps can be used as

Table 12-1
Families of Operational Amplifiers

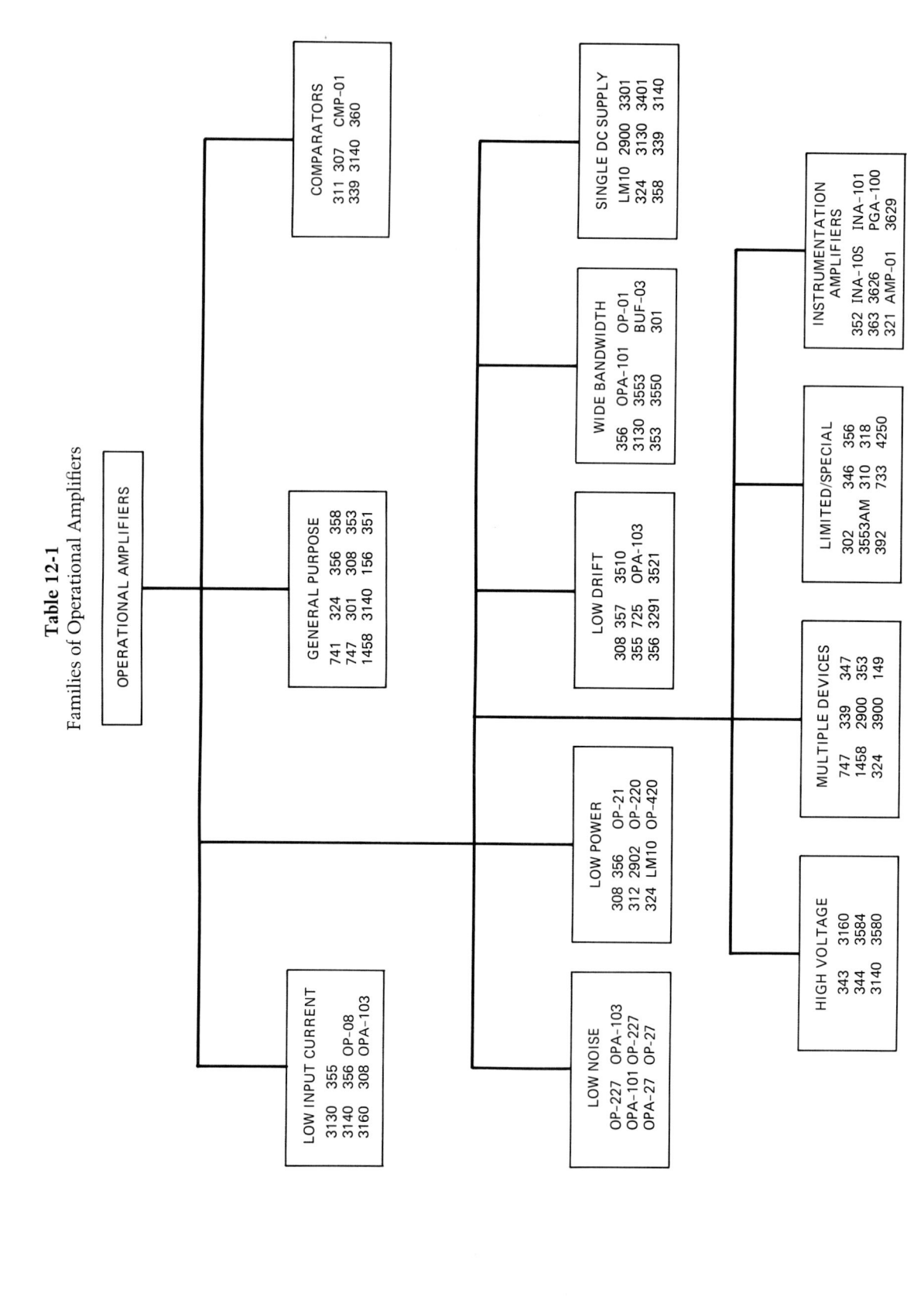

voltage comparators, the reverse is not true: IC comparators (e.g., LM-311) cannot usually be used as op-amps. More about comparator circuits will be said in a later section.

Low Input Current Op-Amps

Although ideal op-amps have zero input bias current, real devices have a small bias current due to input transistor biasing or leakage. This class of devices typically uses MOSFET, JFET, or superbeta (Darlington) transistors for the input stage instead of NPN/PNP bipolar devices. These transistors produce extremely low values of input current. The manufacturer may also elect to use a nulling method in-process that reduces input bias current. Low input current devices typically have picoampere-level currents, rather than microampere or milliampere input currents found in some other devices.

Low Noise Op-Amps

These devices are optimized to reduce internally generated noise. They are generally used in the first stage or two of a cascade chain to improve the overall noise performance of an amplifier circuit.

Low Power Op-Amps

This category of op-amp optimizes internal circuitry to reduce power consumption. Many of these devices also operate at very low DC power supply potentials (e.g., ± 1.5 V DC).

Low Drift Op-Amps

DC amplifier circuits can experience an erroneous change of output voltage as a function of temperature. Low-drift devices are internally compensated to minimize temperature drift. These devices are typically used in instrumentation circuits where drift is an important concern, especially when low-level input signals are handled.

Wide Bandwidth Op-Amps

Also called video op-amps in some literature, these devices have a very high gain–bandwidth product. One device, for example, has a G–B product of 100 MHz, compared with 300 kHz to 1.2 MHz for various 741-family devices.

Single DC Supply Op-Amps

These devices are able to provide op-amp-like behavior from a monopolar (typically $V +$) DC power supply. However, not all op-amp performance will be available from some of these devices, because the output voltage may not be able to assume negative values.

High-Voltage Op-Amps

Most op-amps operate at DC power supply potentials of ± 6 to ± 22 V DC. A few devices in the high-voltage category operate from

± 44 V DC supplies, and at least one proprietary hybrid model operates from ± 100 V DC power supplies.

Multiple Devices

This category requires only that more than one op-amp be included in the same package. Devices exist with two, three, or four operational amplifiers in a single package. The 1458 device, for example, contains two 741-family devices in either eight-pin metal cans or miniDIP packages.

Limited or Special Purpose Op-Amps

The devices in this category are designed for either a limited range of uses or for highly specialized uses. The type 302 buffer, for example, is an op-amp that is connected internally into the noninverting unity gain follower configuration. Some consumer audio devices also fit into this category. For example, several dual op-amp devices used for stereo audio are optimized for audio applications.

IC Instrumentation Amplifiers

Although the instrumentation amplifier (IA) is arguably a special-purpose device, it is sufficiently close to universal to warrant a class of its own. The IA is a DC differential amplifier made of either two or three internal op-amps. Voltage gain can be set by either one or two external resistors.

THE IDEAL OPERATIONAL AMPLIFIER

When you study any type of electronic device, be it transistor or integrated circuit, it is wise to start with an ideal representation of that device and then proceed to less practical devices. In some cases, the practical and ideal devices are so far apart that you might wonder at the wisdom of this approach. But IC operational amplifiers, even low-cost products, so nearly approximate the ideal op-amp of textbooks that the equations actually work. The ideal model analysis method thus becomes extremely useful for understanding the technology, learning to design new circuits, or figuring out how someone else's circuit works.

Later in this chapter we will discuss the inverting and the noninverting amplifier configurations of the op-amp. In those discussions we will derive the design equations that describe the operation of real circuits from both the ideal model and a feedback amplifier model. The usefulness of our simplified approach proceeds directly from the correspondence of the ideal and practical operational amplifier IC devices.

Properties of the Ideal Operational Amplifier

The ideal op-amp is characterized by seven properties. From this short list of properties we can deduce circuit operation and design equations. Also, the list gives us a basis for examining non-ideal operational amplifiers and their defects (plus solutions to the problems caused by those defects). The basic properties of the op-amp are

1. Infinite open-loop voltage gain
2. Infinite input impedance
3. Zero output impedance
4. Zero noise contribution
5. Zero DC output offset
6. Infinite bandwidth
7. Differential inputs that stick together

Let us take a look at these properties to determine what they mean in practical terms. You will find that some cheap op-amps only approximate some of these ideals, while for others on the list the approximation is extremely good.

Property No. 1—Infinite Open-Loop Gain (A_{vol})

The open-loop gain (A_{vol}) of any amplifier is its gain without either negative or positive feedback. By definition, negative feedback is a signal fed back to the input 180° out of phase. In operational amplifier terms this means feedback between the output and the inverting input.

Negative feedback has the effect of reducing the open loop gain (A_{vol}) by a factor (B) that depends on the transfer function and properties of the feedback network. Figure 12-3 shows the basic configuration for any negative feedback amplifier. The transfer equation for any circuit is the ratio of the output function and the input function. The transfer function of a voltage amplifier is, therefore, $A_{vol} = V_o/V_{in}$. In Fig. 12-3 the term A_{vol} represents the gain of the

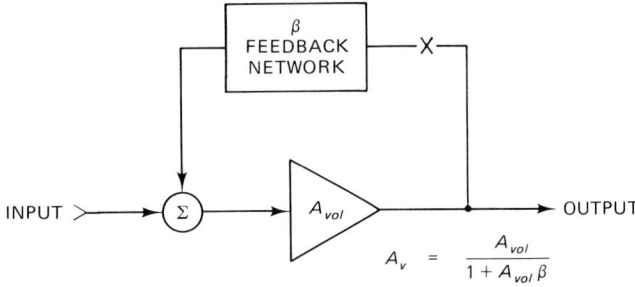

$$A_v = \frac{A_{vol}}{1 + A_{vol}\beta}$$

Fig. 12-3 Equivalent circuit for a feedback amplifier.

amplifier element only, that is, the gain with the feedback network disconnected. The overall transfer function of this circuit, with both amplifier element and feedback network (B) in the loop, is defined as

$$A_v = A_{vol}/(1 + A_{vol}B) \qquad (12\text{-}2)$$

where A_v is the closed-loop gain, A_{vol} the open-loop gain, and B the transfer equation for the isolated feedback network.

In the ideal op-amp A_{vol} is infinite, so the voltage gain is a function only of the feedback network. In real op-amps, the value of the open-loop gain is not infinite, but it is quite high. Typical values range from 20,000 in low-grade consumer audio models to more than 2,000,000 in premium units (typically it is 200,000 to 300,000).

Property No. 2—Infinite Input Impedance

This property implies that the op-amp input will not load the signal source. The input impedance of any amplifier is definable as the ratio of the input voltage to the input current: $Z_{in} = V_{in}/I_{in}$. When the input impedance is infinite, therefore, we must assume that the input current is zero. Thus, an important implication of this property is that the operational amplifier inputs neither sink nor source current. In other words, it will neither supply current to an external circuit nor accept current from an external circuit. We will depend on an implication of this property ($I_{in} = 0$) to perform the simplified circuit analysis used in this chapter to determine the gain equations.

Real operational amplifiers have some finite input current other than zero. In low-grade devices this current can be substantial (e.g., > 1 mA), and will cause a large output offset voltage error in high-gain circuits. The primary source of this current is the base bias currents from the NPN and PNP bipolar transistors used in the input circuits. Certain premium grade op-amps that feature bipolar inputs reduce this current to nanoamperes or picoamperes. In op-amps that use field effect transistors (FET) in the input circuits, on the other hand, the input impedance is quite high due to the very low leakage currents normally found in FET devices. The JFET input devices are typically called BiFET op-amps, while the MOSFET input models are called BiMOS devices. The RCA/GE CA-3140 device is a BiMOS op-amp in which the input impedance approaches $1.5~T\Omega$ (i.e., $1.5 \times 10^{12}~\Omega$)— which is near enough to infinite to make the input circuits of those devices approach the ideal.

Property No. 3—Zero Output Impedance

A voltage amplifier (of which class the op-amp is a member) ideally has a zero output impedance. All real voltage amplifiers, however, have a nonzero (but low) impedance. Figure 12-4 represents any voltage source (including amplifier outputs) and its load (external

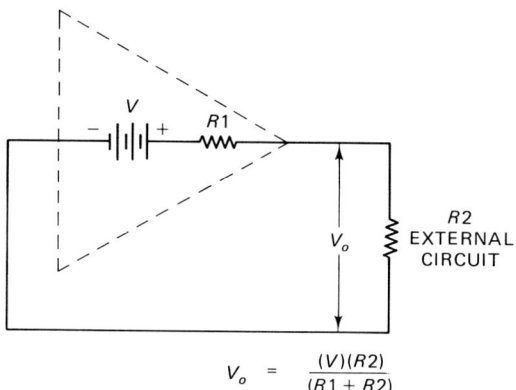

$$V_o = \frac{(V)(R2)}{(R1 + R2)}$$

Fig. 12-4   Equivalent output circuit showing internal resistance.

circuit). Potential V is a perfect internal voltage source with no internal resistance. Resistor $R1$ represents the internal resistance of the source, and $R2$ is the load. Because the internal resistance (which in amplifiers is usually called output resistance) is in series with the load resistance, the output voltage V_o that is available to the load is reduced by the voltage drop across $R1$. Thus, the output voltage is given by

$$V_o = V(R2)/(R1 + R2) \qquad (12\text{-}3)$$

It is clear from the above equation that the output voltage will equal the internal source voltage only when the output resistance of the amplifier ($R1$) is zero. In that case, $V_o = V(R2/R2) = V$. Thus, in the ideal voltage source, the maximum possible output voltage is obtained (and the least error) because no voltage is dropped across the internal resistance of the amplifier.

Real operational amplifiers do not have a zero output impedance. The actual value is typically less than 100 Ω, with many being in the neighborhood of 30 Ω. Thus, for typical devices, the operational amplifier output can be treated as if it were ideal.

A rule of thumb used by designers is to set the input resistance of any circuit that is given by a non-ideal voltage source output to be at least 10 times the output impedance of the previous stage. This situation is shown in Fig. 12-5. Amplifier $A1$ is a voltage source that drives the input of amplifier $A2$. Resistor $R1$ represents the output resistance of $A1$ and $R2$ represents the input resistance of amplifier $A2$. In practical terms, the circuit where $R2 > 10R1$ will yield results acceptably close to "ideal" for many purposes. In some cases, however, the rule $R2 > 100R1$ must be followed if greater precision is required.

Property No. 4—Zero Noise Contribution

All electronic circuits, even simple resistor networks, produce noise signals. A resistor creates noise due to the movement of elec-

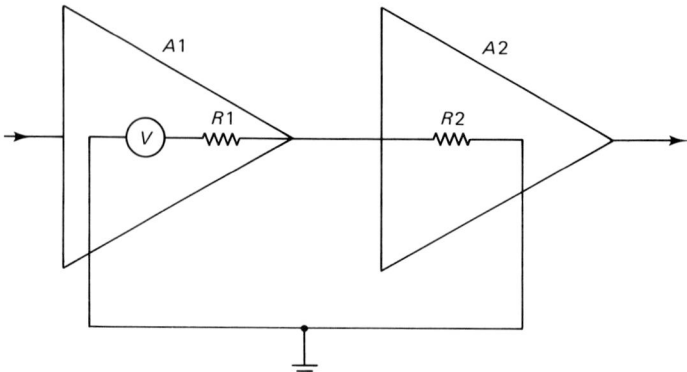

Fig. 12-5 Equivalent cascade circuit.

trons in its internal resistance element material. This phenomenon was discussed at length in Chapter 2. In the ideal operational amplifier, zero noise voltage is produced internally. Thus, any noise in the output signal must have been present in the input signal as well. Except for amplification, the output noise voltage will be exactly the same as the input noise voltage. In other words, the op-amp contributed nothing extra to the output noise. This is one area where practical devices depart quite a bit from the ideal. Practical op-amps do not approximate the ideal, except for certain higher-cost premium low-noise models.

Amplifiers use semiconductor devices that create not merely resistive noise (as described above), but they also create special noise of their own. There are a number of internal noise sources in semiconductor devices, and any good text on transistor theory will give you more information on them. For present purposes, however, assume that the noise contribution of the op-amp can be considerable in low signal-level situations. Premium op-amps are available in which the noise contribution is very low, and these devices are usually advertised as premium low-noise types. Others, such as the RCA/GE metal-can CA-3140 device, offer relatively low noise performance when the DC supply voltages are limited to ± 5 V DC and the metal package of the op-amp is fitted with a flexible "TO-5" style heatsink.

Property No. 5—Zero Output Offset

The output offset voltage of any amplifier is the output voltage that exists when it should be zero. The voltage amplifier sees a zero input voltage when the inputs are both grounded. This connection should produce a zero output voltage. If the output voltage is nonzero, then there is an output offset voltage present. In the ideal op-amp, this offset voltage is zero volts—but real op-amps exhibit at least some amount of output offset voltage. In the real IC operational amplifier the output offset voltage is nonzero, although it can be quite low in some cases.

Property No. 6—Infinite Bandwidth

The ideal op-amp will amplify all signals from DC to the highest AC frequencies. In real op-amps, however, the bandwidth is sharply limited. There is a specification called the gain–bandwidth (G–B) product, which is symbolized by F_t. This specification is the frequency at which the voltage gain drops to unity (1). The maximum available gain at any frequency is found from dividing the maximum required frequency into the gain–bandwidth (G–B) product. If the value of F_t is not sufficiently high, then the circuit will not behave in classical op-amp fashion at some higher frequencies within the range of interest.

Some op-amps have G–B products in the 10- to 20-MHz range. Others, on the other hand, are quite limited. The 741 family of devices is very limited, such that a device will perform as an op-amp only to frequencies of a few kilohertz. Above that range, the gain drops off considerably. But in return for this apparent limitation, we obtain nearly unconditional stability; such op-amps are said to be frequency compensated. It is the frequency compensation of those devices that both reduces the G–B product and provides the inherent stability. Noncompensated op-amps will yield wider frequency response, but only at the expense of a tendency to oscillate. Those op-amps may spontaneously oscillate without any special encouragement if certain precautions are not taken in the circuit design.

Property No. 7—Differential Inputs Stick Together

Most operational amplifiers have two inputs: an inverting (−IN) input and an noninverting (+IN) input. "Sticking together" means that a voltage applied to one of these inputs also appears at the other input. This voltage is real—it is not merely some theoretical device used to evaluate circuits. If you apply a voltage to, say, the inverting input, and then connect a voltmeter between the noninverting input and the power supply common, the voltmeter will read the same potential on the noninverting input as it did on the inverting input. The implication of this property is that both inputs must be treated the same, mathematically. This fact will make itself felt when we discuss the concept of virtual as opposed to actual grounds, and again when we deal with the noninverting follower circuit configuration.

The inverting follower circuit produces an output signal that is 180° out of phase with its input signal. The noninverting follower, as you might expect, produces an output signal that is in-phase with its input signal. Almost all other operational amplifier circuits are variations on either inverting or noninverting follower circuits. Understanding these two configurations will allow you to understand, and either design or modify, a wide variety of different circuits using IC operational amplifiers.

STANDARD OPERATIONAL AMPLIFIER PARAMETERS

Understanding operational amplifier circuits requires knowledge of the various parameters given in the specification sheets. The most commonly needed parameters are described here.

Open-Loop Voltage Gain Voltage gain (A_{vol}) is defined as the ratio of output voltage to input signal voltages (V_o/V_{in}), which is a dimensionless quantity. The open-loop voltage gain is the gain of the circuit without feedback (i.e., with the feedback loop open). In an ideal operational amplifier A_{vol} is infinite, but in practical devices it will range from about 20,000 for low-cost devices to over 1,000,000 in premium devices.

Large Signal Voltage Gain This gain figure is defined as the ratio of the maximum allowable output voltage swing (usually one to several volts less than $V-$ and $V+$) to the input signal required to produce a swing of ± 10 volts (or some other standard).

Slew Rate This parameter specifies the ability of the amplifier to make the transition from one output voltage extreme to the other extreme while delivering full-rated output current to the external load. The slew rate is measured in terms of voltage change per unit of time. The 741 operational amplifier, for example, is rated for a slew rate of 0.5 volts per microsecond (0.5 V/μ). Slew rate is usually measured in, and specified for, the unity gain noninverting follower configuration.

Common Mode Rejection Ratio A common mode voltage is one that is presented simultaneously to both inverting and noninverting inputs. In an ideal operational amplifier, the output signal resulting from the common mode voltage is zero, but in real devices it is nonzero. The common mode rejection ratio (CMRR) is the measure of the device's ability to reject common mode signals, and it is expressed as the ratio of the differential gain to the common mode gain. The CMRR is usually expressed in decibels, with common devices having ratings between 60 and 120 dB (the higher the number, the better the device).

Power Supply Rejection Ratio Also called power supply sensitivity, the PSRR is a measure of the operational amplifier's insensitivity to changes in the power supply potentials. The PSRR is defined as the change of the input offset voltage (see below) for a one-volt change in one power supply potential (while the other is held constant). Typical values are in microvolts or millivolts per volt of power supply potential change.

Input Offset Voltage The voltage required at the input to force the output voltage to zero when the input signal voltage is zero. The output voltage of an ideal operational amplifier is zero when V_{in} is zero.

Input Bias Current This current is the current flowing into or out of the operational amplifier inputs. In some sources, this current is defined as the average difference between currents flowing in the inverting and noninverting inputs.

Input Offset (Bias) Current The difference between inverting and noninverting input bias current when the output voltage is held at zero.

Input Signal Voltage Range The range of permissible input voltages as measured in the common mode configuration.

Input Impedance The resistance between the inverting and noninverting inputs. This value is typically very high: 1 MΩ in low-cost bipolar operational amplifiers and over 10^{12} Ω in premium BiMOS devices.

Output Impedance This parameter refers to the resistance looking back into the amplifier's output terminal, and is usually modeled as a resistance between output signal source and output terminal. Typically the output impedance is considerably less than 100 Ω.

Output Short Circuit Current The current that will flow in the output terminal when the output load resistance external to the amplifier is zero ohms (i.e., a short to common).

Channel Separation This parameter is used on multiple operational amplifier integrated circuits, that is, devices in which two or more operational amplifiers share the same package with common power supply terminals.

This separation specification tells us something of the isolation between the op-amps inside the same package; the separation is measured in decibels (dB). The 747 dual operational amplifier, for example, offers 120 dB of channel separation. From this specification we may imply that a one-microvolt change will occur in the output of one of the amplifiers when the other amplifier output changes by one volt (20 Log [1 μV] = 120 dB).

MINIMUM AND MAXIMUM PARAMETER RATINGS

Operational amplifiers, like all electronic components, are subject to certain maximum ratings. If these ratings are exceeded, the user can expect either premature—often immediate—failure, or at least unpredictable operation. The ratings mentioned here are the most commonly used.

Maximum Supply Voltage This potential is the maximum that can be applied to the operational amplifier without damaging the device. The operational amplifier uses $V+$ and $V-$ DC power supplies that are typically ± 18 V DC, although some exist with much higher maximum potentials.

Power Dissipation   This rating is the maximum power dissipation (Pd) of the operational amplifier in the normal ambient temperature range (80° Celsius in commercial devices and 125° Celsius in military-grade devices). A typical rating for op-amps is 500 mW (0.5 W).

Maximum Power Consumption   The maximum power dissipation, usually under output short circuit conditions, that the device will survive. This rating includes both internal power dissipation and device output power requirements.

Maximum Input Voltage   This potential is the maximum that can be applied simultaneously to both inputs. Thus, it is also the maximum common-mode voltage. In most bipolar operational amplifiers the maximum input voltage is very nearly equal to the power supply voltage. There is also a maximum input voltage that can be applied to either input when the other input is grounded.

Differential Input Voltage   This input voltage rating is the maximum differential mode voltage that can be applied across the inverting ($-$IN) and noninverting ($+$IN) inputs.

Maximum Operating Temperature   The maximum temperature is the highest ambient temperature at which the device will operate according to specifications with a specified level of reliability. The usual rating for commercial devices if 70° or 80° Celsius, while military components must operate at up to 125° Celsius.

Minimum Operating Temperature   There is a minimum operating temperature, that is, the lowest temperature at which the device operates within specifications. Commercial devices operate down to either 0 or $-10°$ Celsius, while military components operate down to $-55°$ Celsius.

Output Short-Circuit Duration   This rating is the length of time the operational amplifier will safely sustain a short circuit of the output terminal. Many modern operational amplifiers are rated for indefinite output short-circuit duration.

Maximum Output Voltage   The maximum output potential of the operational amplifier is related to the DC power supply voltages. Operational amplifiers have one or more bipolar PN junctions between the output terminal and either $V-$ or $V+$ terminals. The voltage drop across these junctions reduces the maximum achievable output voltage. For example, if there are three PN junctions between the output and power supply terminals, then the maximum output voltage is $[(V+) - (3 \times 0.7)]$, or $[(V+) - 2.1]$ V. If the maximum $V+$ voltage permitted is 15 V, the maximum allowable output voltage is $[(15 \text{ V}) - (2.1 \text{ V})]$, or 12.9 V. It is not always true, especially in older devices, that the maximum negative output voltage is equal to the maximum positive output voltage. A related rating is the maximum output voltage swing, which is the absolute value of the voltage swing from maximum negative to maximum positive.

PRACTICAL OPERATIONAL AMPLIFIERS

Now that we have examined the ideal operational amplifier and some typical device specifications, let us turn our attention to practical devices. Because of its popularity and low cost we will concentrate on the 741 device. The 741 family also includes the 747 and 1458 "dual 741" devices. Although there are many better operational amplifiers on the market, the 741 and the members of its close family are considered the industry standard generic op-amp devices.

Figure 12-6 shows the two most popular packages used for 741. Figure 12-6A is the eight-pin miniDIP package, while Fig. 12-6B is the eight-pin metal can package. The 741 is also available in flatpacks and 14-pin DIP packages, although these are becoming rare today. The miniDIP pin-outs for a 1458 dual op-amp are shown in Fig. 12-6C. The 741 has the pins described here.

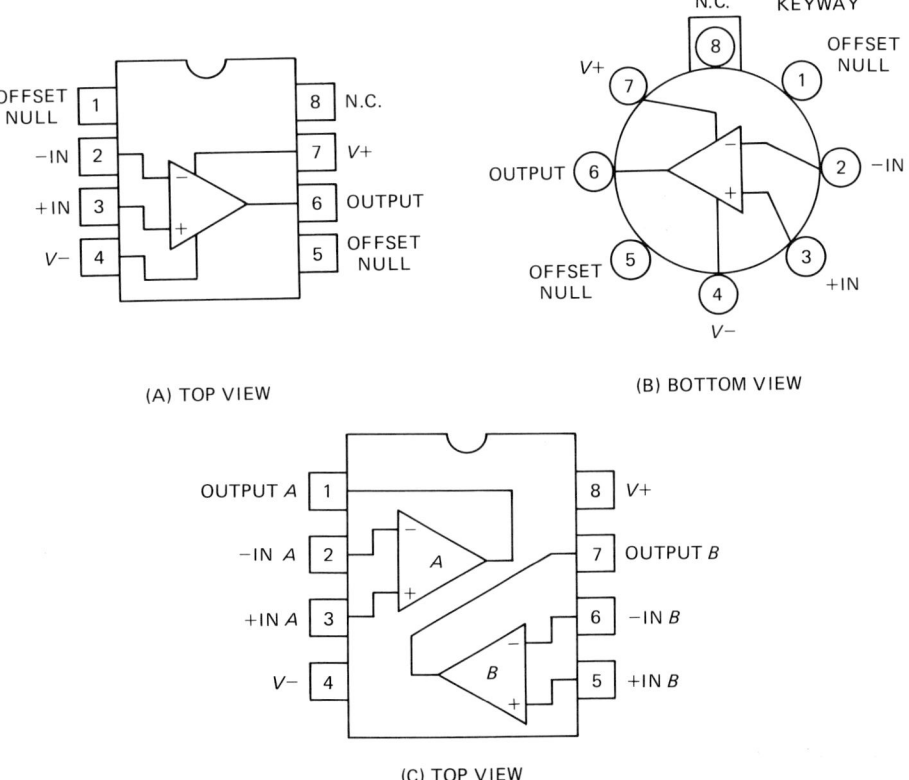

Fig. 12-6 Packaging for industry standard op-amp (741) in (A) DIP and (B) metal can packages; (C) dual op-amp such as 1458 device.

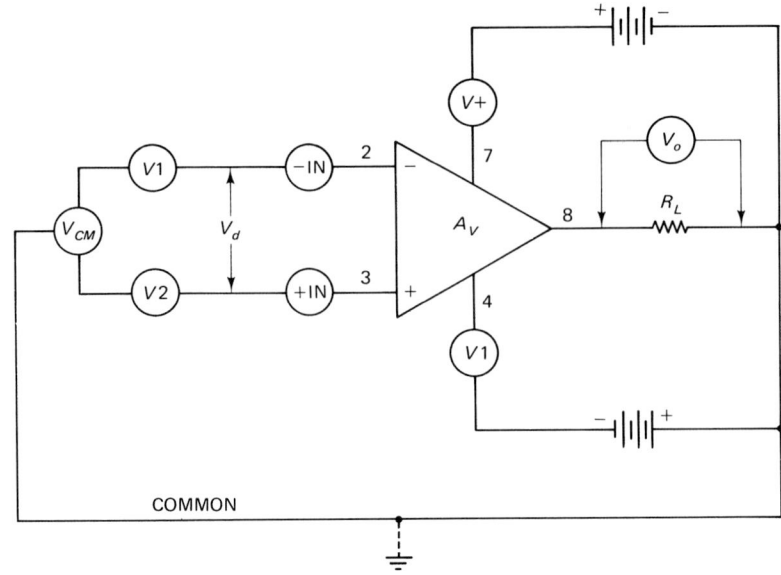

Fig. 12-7   Input signals applied to differential op-amp.

Inverting Input (−IN), Pin No. 2   The output signals produced from this input are 180° out of phase with the input signal applied to −IN.

Noninverting Input (+IN), Pin No. 3   Output signals are in-phase with signals applied to the +IN input terminal.

Output, Pin No. 6   On most op-amps, the 741 included, the output is single-ended. This term means that output signals are taken between this terminal and the power supply common (see Fig. 12-7). The output of the 741 is said to be short-circuit proof because it can be shorted to common indefinitely without damage to the IC.

V + DC Power Supply, Pin No. 7   The positive DC power supply terminal.

V − DC Power Supply, Pin No. 4   The negative DC power supply terminal.

Offset Null, Pins 1 and 5   These two terminals are used to accommodate external circuitry that compensates for offset error voltages.

The pin-out scheme shown in Fig. 12-6 is considered the *de-facto* "industry standard" for generic single operational amplifiers. Although there are numerous examples of amplifiers using different pin-outs than those shown in 12-6, a very large percentage of the available devices use this scheme.

STANDARD CIRCUIT CONFIGURATION

The standard circuit configuration for 741-family operational amplifiers is shown in Fig. 12-7. The pin-outs are industry standard. The

Table 12-2
Operational Amplifier Parameters

	Minimum	Typical	Maximum	Units
Input offset voltage V_{io}	—	1	5	mV
Input bias current	—	80	500	nA
Input resistance	0.3	2.0	—	MΩ
Input voltage range	± 12	± 13	—	V
Large signal voltage gain	50	200	—	V/mV[a]
Output voltage swing				
$R_L = 10\ k\Omega$	± 12	± 14	—	V
$R_L = 2\ k\Omega$	± 10	± 13	—	V
Output short circuit	—	25	—	MA
CMRR	80	90	—	dB
PSRR	77	96	—	dB[b]
Rise time ($A_v = 1$)	—	0.3	—	μs
BW	—	1.16	—	MHz
Power consumption	—	50	85	mW
Power supply	—	± 22		V DC
Power dissipation	—	500	—	mW
Differential input voltage	—	± 30	—	V
Input voltage	—	± 15	—	V

[a]$V_s = \pm 15$ V DC, $V_o = \pm 10$ V.
[b]$R_L \le 10\ k\Omega$.

output signal voltage is impressed across load resistor R_L connected between the output terminal (pin no. 6) and the power supply common. Most manufacturers recommend a 2 kΩ minimum value for R_L. Also, note that some operational amplifier parameters shown in Table 12-2 are based on a 10 kΩ load resistance. Because it is referenced to common, the output is single-ended.

The ground symbol shown in Fig. 12-7 indicates that it is optional. The point of reference for all measurements is the common connection between the two DC power supplies ($V-$ and $V+$). Whether or not this point is physically connected to a ground, the equipment chassis, or a dedicated system ground bus is purely optional in some cases and is required in others. Whether it is required or not, however, the determination is made on the basis of circuit factors other than the basic nature of the op-amp.

The $V-$ and $V+$ DC power supplies are independent of each other. Do not make the mistake of assuming that these terminals are merely different ends of the same DC power supply. In fact, $V-$ is negative with respect to common, while $V+$ is positive with respect to common.

The two input signals in Fig. 12-7 are labeled $V1$ and $V2$. Signal voltage $V1$ is the single-ended potential between common and the inverting input ($-$IN), while $V2$ is the single-ended potential between common and the noninverting input ($+$IN). The 741 operational amplifier is differential, as indicated by the fact that both $-$IN and

+IN are present. Any differential amplifier produces an output that is proportional to the difference between the two input potentials. In Fig. 12-7 the differential input potential V_d is the difference between $V1$ and $V2$:

$$V_d = V2 - V1 \qquad (12\text{-}4)$$

Signal voltage V_{cm} in Fig. 12-7 is the common mode signal, that is, a potential that is common to both $-IN$ and $+IN$ inputs. This potential is equivalent to the situation $V1 = V2$. In an ideal operational amplifier there will be no output response at all to a common-mode signal. In real devices, however, there is some small response to V_{cm}. The freedom from such responses is called the common-mode rejection ratio (CMRR).

OPEN-LOOP VOLTAGE GAIN

The flexibility of the operational amplifier is due in large part to the extremely high open-loop DC voltage gain of the device. By definition, the open-loop voltage gain (A_{vol}) is the gain of the amplifier without feedback. If the feedback network (shown earlier in Fig. 12-3) had been interrupted at point X, then the gain of the circuit becomes A_{vol}. The effect of negative feedback is to reduce overall circuit gain to something less than A_{vol}.

The open-loop voltage gain of operational amplifiers is always very high. Some audio amplifier devices intended for consumer electronics applications offer A_{vol} of 20,000, while certain premium operational amplifiers offer gains to 1,000,000 and more. Depending on the specific device surveyed, the 741 op-amp will typically exhibit A_{vol} values in the 200,000 to 300,000 range.

A consequence of such high values of A_{vol} is that very small differential input signal voltages will cause the output to saturate. On the 741 device the value of the maximum permissible output voltage, $\pm V_{sat}$, is typically about 1 V (or a little less) below the power supply potential of the same polarity (certain BiMOS devices—such as the CA-3130 or CA-3140—operate to within a few tenths of a volt of the supply rail). For ± 15 V DC supplies, the maximum 741 output potential is ± 14 V DC. Let us consider the maximum input potential that will not cause saturation of the op-amp at four popular values of power supply potential, assuming the one-volt less rule. The calculation is

$$V_{in(max)} = \pm V_{sat}/A_{vol} \qquad (12\text{-}5)$$

Power supply	$A_{vol} = 300,000$ $\pm V_{sat}$	$\pm V_{in(max)}$
$\pm$ 6 V DC	$\pm$ 5 V DC	17 μV
$\pm$ 10 V DC	$\pm$ 9 V DC	30 μV
$\pm$ 12 V DC	$\pm$ 11 V DC	37 μV
$\pm$ 15 V DC	$\pm$ 14 V DC	47 μV

One of the consequences of high A_{vol} is that op-amps usually saturate at either the $V-$ or $V+$ power supply rails when the input lines are either shorted to common or floating open. This phenomena is due to tiny imbalances in the input bias conditions internal to the device, which are random in nature. Accordingly, one might expect half of a group of op-amps to saturate at $+V_{sat}$ and half to saturate at $-V_{sat}$. This situation is probably true for a very large number of devices procured from various lots and various manufacturers. However, a collection of, say, one hundred devices purchased at the same time from the same manufacturer will show a marked tendency toward either $-V_{sat}$ or $+V_{sat}$, not the expected random distribution.

The reason is that the input bias current imbalances tend to be design and process related, so are generally uniform from device to device within a given lot from the same source. For example, in a lot of twenty 741 devices of the same brand tested, 19 flipped to $+V_{sat}$ and only one flipped to $-V_{sat}$ at turn-on. According to some authorities, we should have expected 10 to fall into each group; but that's not what usually happens in real situations.

The behavior of operational amplifiers in the open-loop configuration leads to one category of applications that takes advantage of the very high values of A_{vol}: voltage comparators.

VOLTAGE COMPARATORS

A voltage comparator is basically an operational amplifier that has no negative feedback network (Fig. 12-8A). The open-loop gain of the operational amplifier is very large, on the order of 200,000 to 300,000 for many common devices and higher for premium devices. Thus, with no negative feedback the operational amplifier functions as a very high gain DC amplifier with an output that saturates at a very tiny input potential.

So what use is an amplifier that saturates with only a few microvolts of input signal voltage? Such an amplifier can be used as a voltage comparator. The voltage comparator is used to compare two input voltages and issue an output signal that indicates their relationship ($V1 = V2$, $V1 > V2$, or $V1 < V2$). In Fig. 12-8A potential $V1$ is applied to the inverting input and $V2$ is applied to the noninverting input. If $V1 = V2$, then $V_o = 0$. Otherwise, the output voltage obeys

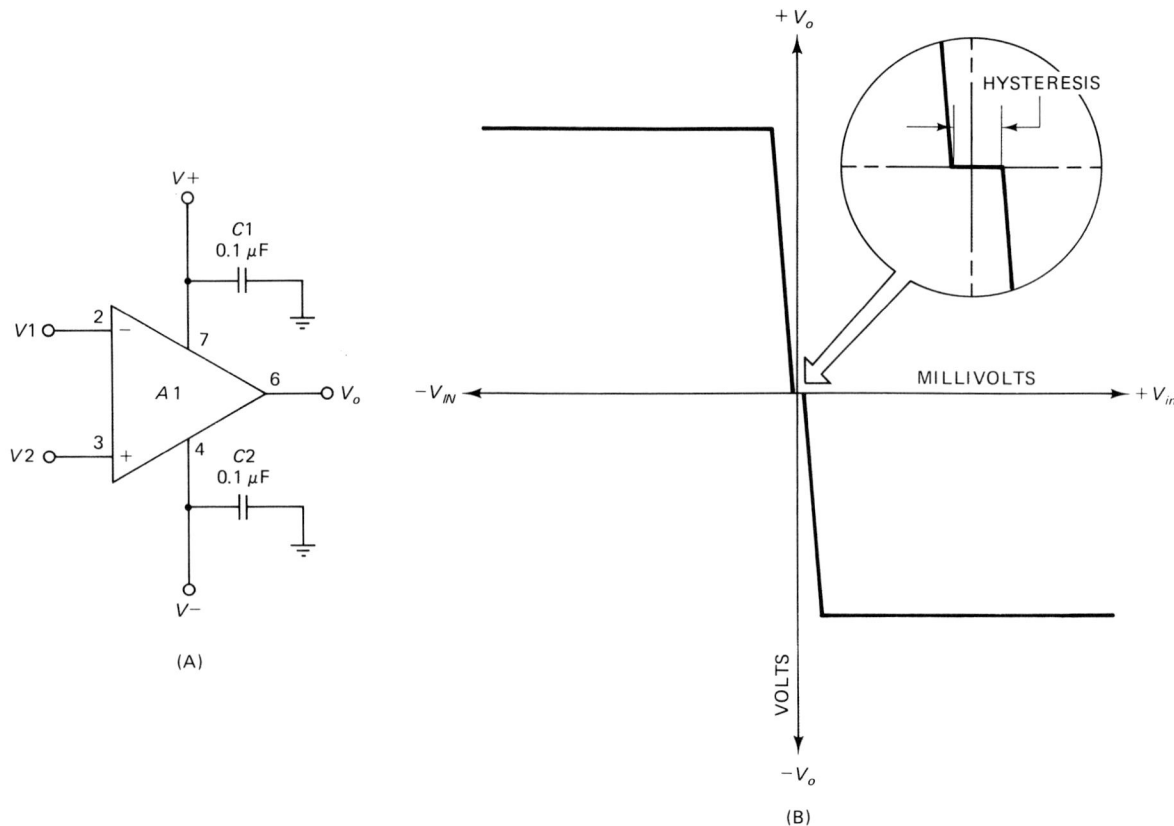

Fig. 12-8   (A) Operational amplifier comparator and (B) hysteresis.

the relationship shown in Fig. 12-8B, which is the transfer function of the comparator. According to the normal rules for operational amplifiers, when $V1$ is larger than $V2$ (see Fig. 12-8A) it looks like a positive input to the inverting input; so the output potential is saturated just below $V-$. Alternatively, when $V1$ is smaller than $V2$ it looks like a negative input potential; so the output is saturated just below $V+$.

In Fig. 12-8B there is a small hysteresis band around zero where no output changes occur. This is an unfortunate defect in practical operational amplifiers. It is possible to measure the hysteresis of operational amplifiers and IC comparators on the laboratory bench. In one such experiment involving operational amplifier and IC comparator (LM-311) devices, hysteresis bands of 1 to 25 mV were found. Not surprisingly, the low-cost 741 family of operational amplifiers had the highest hysteresis levels (on the order of 25 mV). The LM-311 devices had 8–10 mV of hysteresis. Certain other devices had 10–20 mV of hysteresis. The overall best device in the experiment was the RCA/GE CA-3140, a BiMOS operational amplifier. The CA-3140 device uses the industry standard 741 pinouts, which are shown in Fig. 12-8A.

The LM-311 device (Fig. 12-9A) is a low-cost voltage comparator in IC form. Although based internally on op-amp circuitry, this device

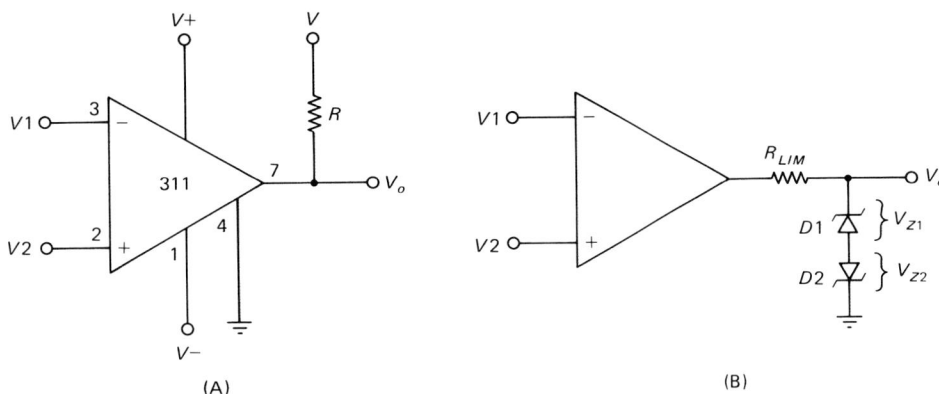

Fig. 12-9 (A) LM-311 comparator and (B) zener diode output limiting.

is specifically designed as a voltage comparator. It has a ground terminal (pin 1), contrary to op-amp practice, and requires an output pull-up resistor (R) to a positive power supply voltage. The output terminal can drive loads such as low-current relay coils, lamps, and LEDs operated at potentials up to 40–50 V (depending on the category of the 311 device) and current loads of 50 mA. If the LM-311 is operated so as to be compatible with TTL digital logic, the pull-up resistor is terminated in a +5 V DC potential; it usually has a value of 1.5 to 3.3 kΩ.

A means for limiting the output level, improving the sharpness of the transfer function corners (see Fig. 12-8B), and improving speed by reducing latch-up problems, is shown in Fig. 12-9B. In this circuit a pair of back-to-back zener diodes are connected across the output line. When the output voltage is HIGH it is limited to $V_{Z1} + 0.7$ V; and when LOW it is limited to $V_{Z2} + 0.7$ V. These potentials represent the reverse bias zener voltages of D1 and D2, plus the normal forward bias voltage drop of the other diode (which is forward biased).

High Drive Capacity Comparators

Figure 12-10 shows a means for increasing the drive capacity of the comparator. In this circuit a bipolar transistor (2N3704, 2N2222, etc.) is used to control a larger load than the device could normally handle (such as the relay coil shown here). The output voltage (V_o) of the comparator is used to set up the DC bias for the NPN transistor. When the comparator output is HIGH, then the transistor is biased hard-on and the load is grounded through the transistor's collector-to-emitter path. Alternatively, when the comparator output is LOW, then the transistor is reverse biased and the load remains ungrounded.

The diode across the relay coil is essential for any inductive load. When the magnetic field surrounding an inductor (such as a relay coil)

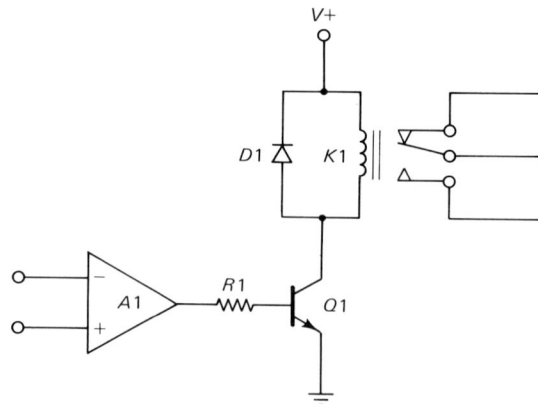

Fig. 12-10   Circuit for driving external load such as a relay.

collapses, the counter-electromotive force (CEMF) generates a high voltage spike that is capable of either damaging components or interrupting circuit operation (especially digital circuits). The diode is normally reverse biased, but for the CEMF spike it is forward biased. The diode therefore clamps the high voltage spike to about 0.7 V.

Current Mode Comparators

Figure 12-11 shows two comparator circuit techniques applied to the same circuit. One technique is a zero offset control used to reduce the effects of the hysteresis band, while the other technique is the current mode configuration. The offset control (R4) biases one input to a small but nonzero level so that it is ready to trip when the other input is also nonzero. In this particular case the noninverting input is grounded ($V2 = 0$), but it could as easily be connected to a nonzero voltage.

Current mode operation is usually faster and less prone to latch-up than voltage mode operation. For this reason, current mode comparators are sometimes used in high-speed analog-to-digital converters (A/D). Assume that the noninverting input is grounded. In this case, the output potential V_o will reflect the relationship of the two currents. If $I1 = I2$, then $V_o = 0$. This circuit is, to the outside observer, a voltage comparator in that $I1 = V1/R1$ and $I2 = V2/R2$. Of course, the circuit is also useful for accommodating current output devices such as the LM-334 temperature monitor IC as well as voltages, as shown in Fig. 12-11.

Zero-Crossing Detectors

Figure 12-12A shows a zero-crossing detector circuit. In this case a comparator is connected with its noninverting input grounded. When

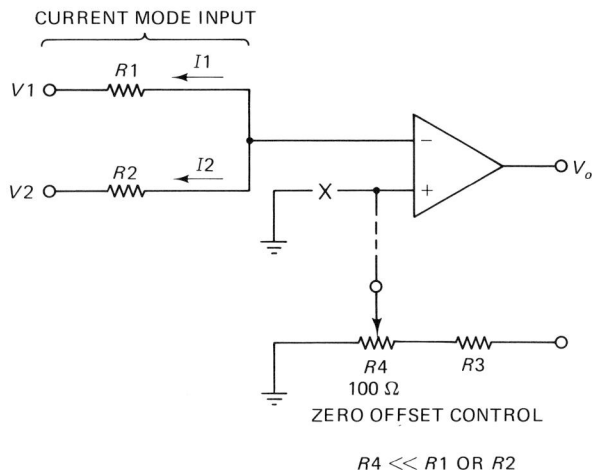

Fig. 12-11 Current mode input comparator with zero offset control.

V_{in} is nonzero, the output will also be nonzero. But when the input voltage crosses zero, the output briefly goes to zero, producing the differentiated output pulse shown. These relationships are shown in Fig. 12-12B.

Window Comparators

A window comparator is shown in Fig. 12-13. This circuit consists of two voltage comparators connected such that one or the other input is activated when the input voltage (V_{in}) exceeds either positive or negative limits. The limits are set by setting $V1$ or $V2$ reference voltages. A possible application for this circuit is alarm systems (for example, over- and under-temperature alarms), and other applications where a range of permissible values exists between two forbidden regions.

Pre-Biased Comparator (Voltage Level Detector)

Figure 12-14A shows a method for biasing either comparator input to a specific reference voltage. This circuit is called a voltage level detector. Although in this case the noninverting input is biased and the inverting input is active, the roles can just as easily be reversed. Two methods of biasing are used: resistor voltage divider and zener diode. If $R2$ is replaced with a zener diode, then the reference potential is the zener potential. In that case, $R1$ is the normal current-limiting resistor needed to protect the zener from self-destruction. In the case where a resistor voltage divider is used, the bias voltage $V1$ is set by the voltage

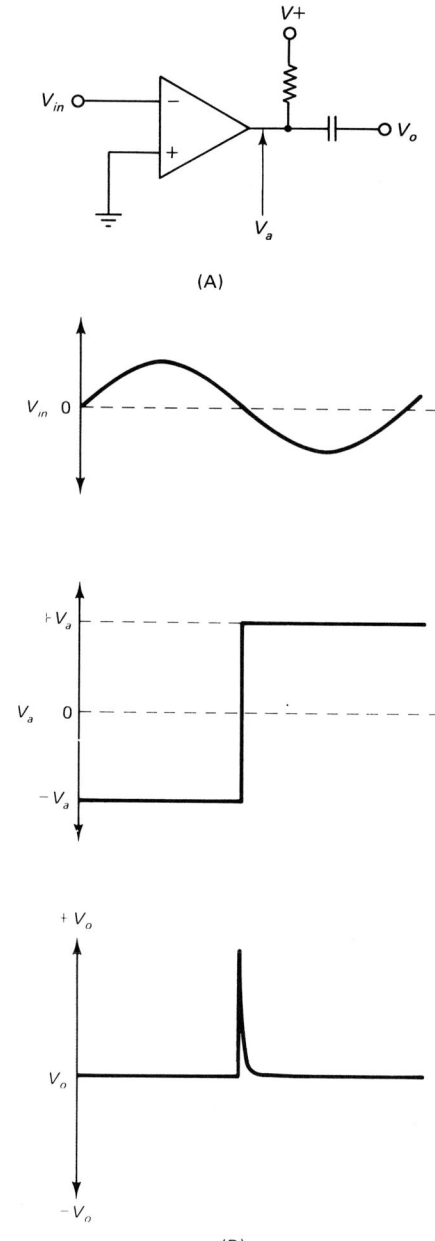

Fig. 12-12   (A) Zero crossing circuit and (B) input and output circuits.

divider equation:

$$V1 = R2(V+)/(R1 + R2) \qquad (12\text{-}6)$$

Temperature Alarm

Figure 12-14B shows an over-temperature circuit based on Fig. 12-14A. In this circuit the inverting input is biased by voltage divider $R1/R2$,

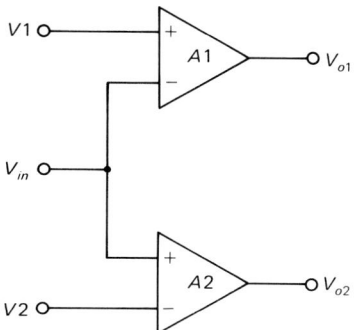

Fig. 12-13   Window comparator circuit.

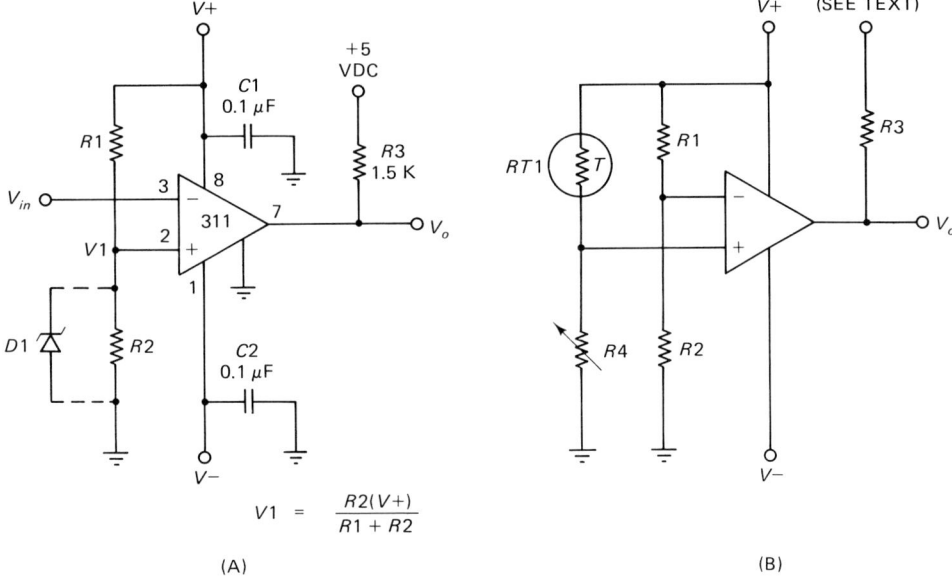

$$V1 = \frac{R2(V+)}{R1 + R2}$$

(A)   (B)

Fig. 12-14   (A) Set voltage comparator and (B) thermistor-controlled comparator.

while the noninverting input is set by another voltage divider, $R4/RT1$. Resistance $RT1$ is a thermistor, which has a resistance proportional (or inversely proportional in some types) to the temperature. Potentiometer $R4$ is used to set the "trip-point" temperature. The values of the resistors depend on the set trip-point desired and the resistance of the thermistor over the range of temperatures being monitored.

Pulse Width Controller

Pulse width modulation is used in many communications systems, motor and load controllers, switching DC power supplies, and other applications. A pulse width modulator will vary the width of an output pulse proportionally to an applied input voltage. We can use a voltage

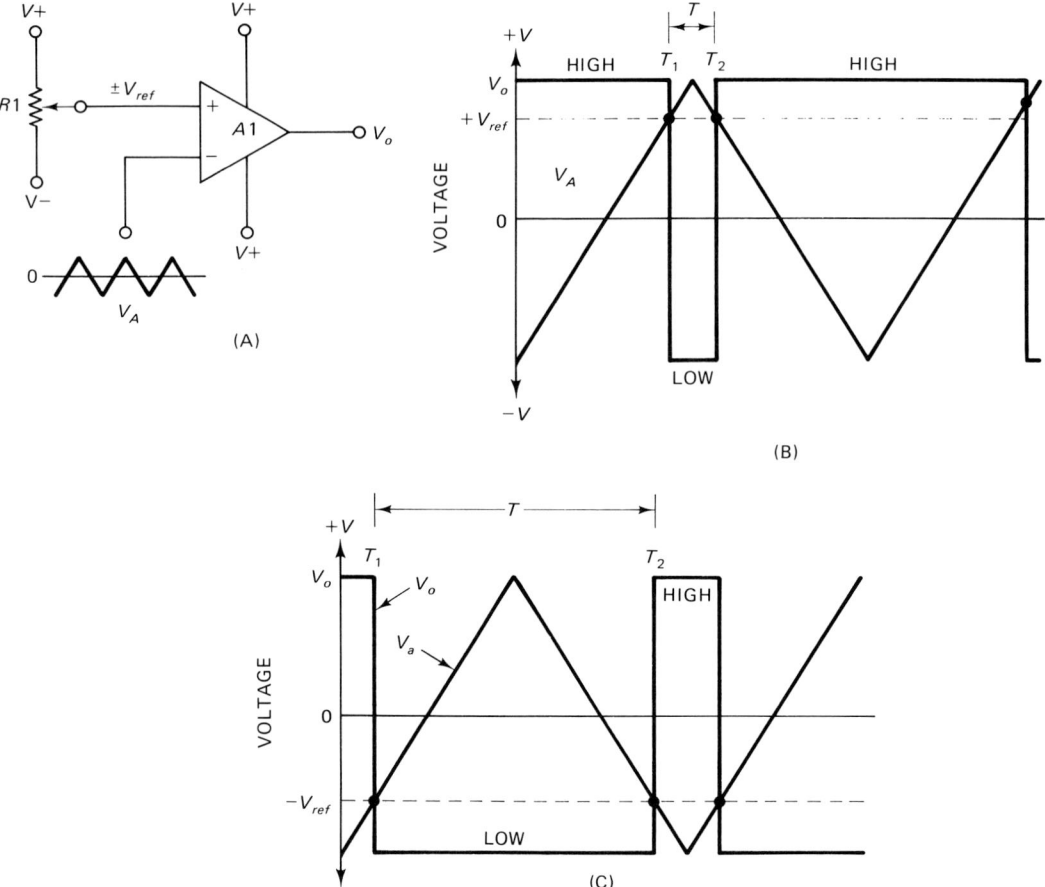

Fig. 12-15 Pulse width modulator based on comparator circuit.

comparator to form a basic pulse width modulator, as shown in Fig. 12-15A. In this circuit a triangular waveform is applied to the inverting input of the comparator, while the DC reference potential (V_{ref}) output of a potentiometer is applied to the noninverting input.

Figures 12-15B and 12-15C show the relationship between the output terminal and the two input signals, V_a and V_{ref}. Examine Fig. 12-15B. Note that V_{ref} is a positive DC potential. As long as the applied triangle waveform is less than $+V_{ref}$, the output of the comparator V_o is HIGH. But at time $T1$ the two voltages become equal, so the output flips as the triangle waveform increases in value above $+V_{ref}$. During the period $T2-T1$, signal V_a is greater than $+V_{ref}$; so V_o is LOW. At time $T2$, however, the situation again reverts to the situation in which V_a is less than $+V_{ref}$; so V_o drops LOW again.

Now consider a slightly different situation in Fig. 12-15C. In this case the value of V_{ref} is readjusted to a negative value, $-V_{ref}$. As in the previous case, the output V_o is LOW during the period $T2-T1$; but note that this segment is much longer than it was in Fig. 12-15B. This difference is caused by the relationship between V_{ref} and V_a under the two different situations.

INVERTING AND NONINVERTING FOLLOWER CONFIGURATIONS

There are two basic configurations for operational amplifier voltage amplifier circuits: inverting and noninverting. For some reason lost in antiquity, these are usually called "follower" circuits. So in this section we will examine the basic inverting follower and noninverting follower circuits.

Inverting Follower Circuits

The inverting follower is an operational amplifier circuit configuration in which the output signal is $180°$ out of phase with the input signal. Figure 12-16A is a cathode ray oscilloscope (CRO) presentation that shows the relationship between input and output signals for an inverting follower with a gain of -2. Note the phase reversal present in the output signal with respect to the input signals. To achieve this inversion, the inverting input ($-IN$) of the operational amplifier is active and the noninverting input ($+IN$) is grounded.

Figure 12-16B shows the basic configuration for the inverting follower (also called inverting amplifier) circuits. The noninverting input is not used, so it is set to ground potential. There are two resistors in this circuit: resistor R_f is the negative feedback path from the output to the inverting input, while R_{in} is the input resistor. We will examine the R_f/R_{in} relationship to determine how gain is fixed in this type of circuit. But first, let us take a look at the implications of grounding the noninverting input in this type of circuit.

What Is "Virtual" Ground?

A *virtual ground* is a connection or circuit point that acts like a ground, even though it is not physically connected to either a truly grounded point or the circuit common point. While this definition sounds strange at first, it is not an unreasonable description of a virtual ground. Unfortunately, that terminology is confusing and therefore leads to an erroneous implication that the virtual ground somehow doesn't really function as a ground. Let us examine the concept of a virtual ground.

Earlier you learned the properties of the ideal operational amplifier. One of those properties tells us that differential inputs "stick together." Put another way, this property means that a voltage applied to one input appears on the other input also.

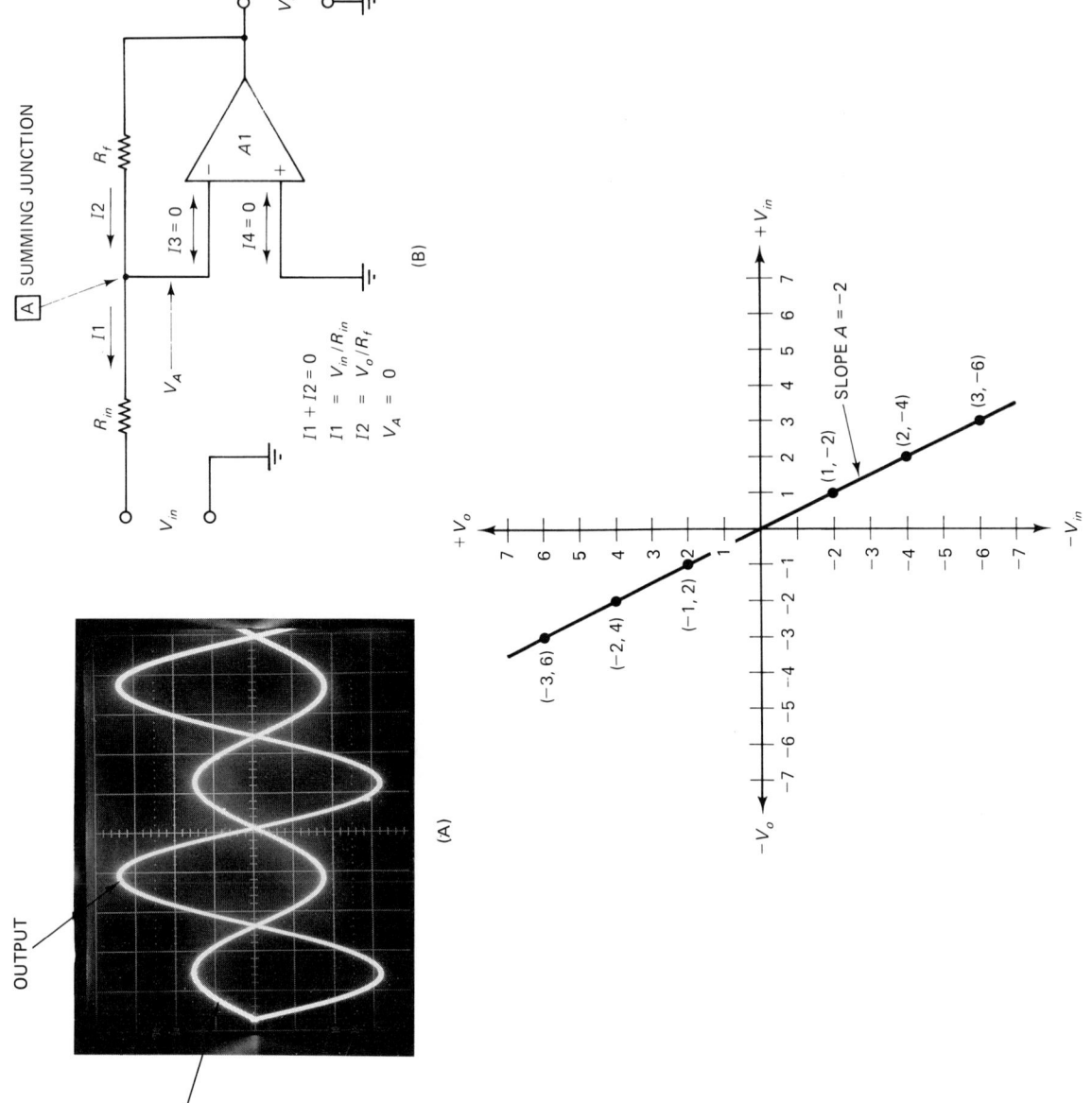

Fig. 12-16 Inverting amplifier: (A) input and output waveforms, (B) circuit, and (C) transfer characteristic.

In the arithmetic of op-amps, therefore, we must treat both inputs as if they are at the same potential. This fact is not merely a theoretical device, either; if you actually apply a potential, say 1 V DC, to the noninverting input, the same 1 V DC potential can also be measured at the inverting input.

In Fig. 12-16B, the noninverting input is grounded, so it is at 0 V DC potential. This fact, by the properties of the ideal op-amp, means that the inverting input of the op-amp is also at the same 0 V DC ground potential. Since the inverting input is at ground potential, but has no physical ground connection, it is said to be at virtual (as opposed to physical) ground. A virtual ground is, therefore, a point that is fixed at ground potential (0 V DC), even though it is not physically connected to the actual ground or common of the circuit. The choice of the term *virtual ground* is unfortunate, for the concept is actually quite simple—the terminology makes it sound abstract.

Developing the Transfer Equation
for the Inverting Follower Circuit

The transfer equation of any circuit is the output function divided by the input function. For an operational amplifier used as a voltage amplifier, therefore, the transfer function describes the voltage gain.

$$A_v = V_o/V_{in} \tag{12-7}$$

where A_v is the voltage gain (dimensionless), V_o the output signal potential, and V_{in} the output signal potential (V_o and V_{in} are in the same units).

In the inverting follower circuit (Fig. 12-16B) the gain is set by the ratio of two resistors, R_f and R_{in}. Let us make a step-by-step analysis to see if we can find this relationship. Consider the currents flowing in Fig. 12-16B. The input bias currents, $I3$ and $I4$, are assumed to be zero for purposes of analysis. This is a reasonable assumption because our model is an ideal operational amplifier. In a real op-amp these currents are nonzero and have to be accounted for, but in the analysis case we use the ideal model. Thus, in the following analysis we can ignore bias currents (assume that $I3 = I4 = 0$).

Remember that the summing junction (point A) is a virtual ground and is at ground potential because the noninverting input is grounded. Current $I1$ is a function of the applied input voltage V_{in} and the input resistance R_{in}. By Ohm's law, then, the value of $I1$ is

$$I1 = V_{in}/R_{in} \tag{12-8}$$

Further, we know that current $I2$ is also related by Ohm's law to the output voltage V_o and to the feedback resistor R_f (again, because the summing junction is at 0 V DC).

$$I2 = V_o/R_f \tag{12-9}$$

How are $I1$ and $I2$ related? These two currents are the only currents entering or leaving the summing junction (recall that $I3 = 0$), so by Kirchoff's current law (KCL) we know that

$$I1 + I2 = 0 \tag{12-10}$$

so

$$I2 = -I1 \tag{12-11}$$

We can arrive at the transfer function by substituting Eqs. (12-8) and (12-9) into Eq. (12-11).

$$I2 = -I1$$

$$\frac{V_o}{R_f} = \frac{-V_o}{R_{in}} \tag{12-12}$$

Algebraically rearranging Eq. (12-12) yields the transfer equation in standard format.

$$V_o/V_{in} = -R_f/R_{in} \tag{12-13}$$

According to Eq. (12-7), the gain (A_v) of the circuit is V_o/V_{in}, so we may also write Eq. (12-13) in the form:

$$A_v = -R_f/R_{in} \tag{12-14}$$

We have shown above that the voltage gain of an op-amp inverting follower is merely the ratio of the feedback resistance to the input resistance ($-R_f/R_{in}$). The minus sign indicates that a 180° phase reversal takes place. Thus, a negative input voltage produces a positive output voltage, and vice versa.

We often see the transfer equation, Eq. (12-13), written to express output voltage in terms of gain and input signal voltage. The two expressions are

$$V_o = -A_v V_{in} \tag{12-15}$$

and

$$V_o = -V_{in}(R_f/R1) \tag{12-16}$$

The transfer function ($A_v = V_o/V_{in}$) can be plotted on graph paper in terms of input and output voltage. Figure 12-16C shows the plot V_o versus V_{in} for an inverting amplifier with a gain of -2. In the case of a perfect amplifier the Y-intercept is 0 V. Given the nature of Fig. 12-16C the basic form for our purposes becomes $V_o = A_v V_{in} \pm V_{offset}$.

Inverting Amplifier Transfer Equation by Feedback Analysis

In the section above, we developed the inverting amplifier transfer equation from the ideal model of the operational amplifier. Now let us consider the same matter from the point of view of the generic feedback amplifier to see if Eq. (12-14) is valid. When used in a closed-loop circuit, the operational amplifier is a feedback amplifier; so feedback analysis will result in the same transfer equation as the ideal model analysis.

Figure 12-17 shows an operational amplifier with its feedback network. The overall gain of this type of amplifier is defined by the following expression:

$$A_v = A_{vol}C/(1 + A_{vol}B) \qquad (12\text{-}17)$$

where A_v is the closed-loop voltage gain, A_{vol} the open-loop voltage gain, C the transfer equation of the input network, and B the transfer equation of the feedback network.

Two networks must be considered in this analysis: the input network (C) and the feedback network (B); both networks are resistor voltage-divider attenuators, so we can expect B and C to be fractions. The expression for the input network in Fig. 12-17 is

$$C = R_f/(R_f + R_{in}) \qquad (12\text{-}18)$$

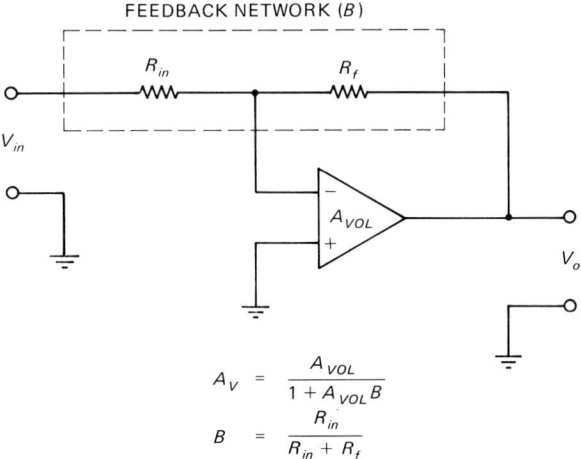

Fig. 12-17 Inverting amplifier with feedback network.

The C term is needed because the input signal is attenuated by the R_{in}/R_f voltage divider network. If the signal is applied directly to the inverting input, as it might be in certain other feedback amplifiers, then this input attenuation term is unity, so it disappears from Eq. (12-17).

The feedback transfer equation is defined by the feedback voltage divider R_f/R_{in}:

$$B = R_{in}/(R_f + R_{in}) \tag{12-19}$$

We can now substitute the expressions for B [Eq. (12-19)] and C [Eq. (12-18)] into the equation for the standard feedback amplifier, Eq. (12-20).

$$A_v = A_{vol}C/(1 + A_{vol}B) \tag{12-20}$$

$$A_v = \frac{A_{vol}[R_f/(R_f + R_{in})]}{1 + A_{vol}[R_{in}/(R_f + R_{in})]} \tag{12-21}$$

$$A_v = \frac{R_f/(R_f + R_{in})}{(1/A_{vol}) + [R_{in}/(R_f + R_{in})]} \tag{12-22}$$

Because A_{vol} is infinite in ideal devices (and very high in practical devices), in term $1/A_{vol} \rightarrow 0$, so we may write Eq. (12-22) in the form:

$$A_v = \frac{R_f/(R_f + R_{in})}{R_{in}/(R_f + R_{in})} \tag{12-23}$$

Earlier we discovered that $A_v = R_f/R_{in}$. If the feedback analysis is correct, then Eq. (12-23) will be equal to R_f/R_{in}. Solving this relationship we invert and multiply:

$$\left[\frac{R_f + R_{in}}{R_{in}}\right] \times \left[\frac{R_f}{R_f + R_{in}}\right] = \frac{R_f}{R_{in}} \tag{12-24}$$

$$\frac{R_f}{R_{in}} = \frac{R_f}{R_{in}} \tag{12-25}$$

Equation [12-25] demonstrates the equality of the two methods, proving that the transfer equation, Eq. (12-15), derived earlier is valid.

The following equations apply to inverting followers:

$$A_v = -R_f/R_{in} \tag{12-26}$$

$$V_o = -A_v V_{in} \tag{12-27}$$

$$V_o = -V_{in}(R_f/R_{in}) \tag{12-28}$$

Multiple Input Inverting Followers

We can accommodate multiple signal inputs on an inverting follower by using a circuit such as Fig. 12-18. There are a number of applications of such circuits: summers, audio mixers, instrumentation, and so forth. The multiple input inverter of Fig. 12-18 can be evaluated exactly like Fig. 12-16B, except that we have to account for more than one input. Again appealing to KCL, we know that

$$I1 + I2 + I3 + \cdots + I_n = I_f \tag{12-29}$$

Also by Ohm's law, considering that summing junction A is virtually grounded, we know that

$$I1 = V1/R1 \tag{12-30}$$

$$I2 = V2/R2 \tag{12-31}$$

$$I3 = V3/R3 \tag{12-32}$$

$$I_n = V_n/R_n \tag{12-33}$$

$$I_f = V_o/R_f \tag{12-34}$$

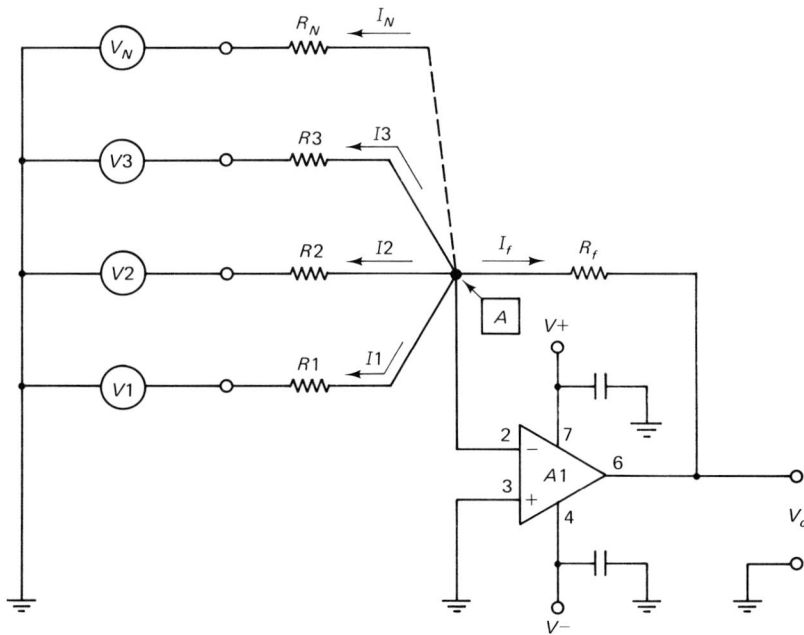

Fig. 12-18 Multiple-input inverting amplifier.

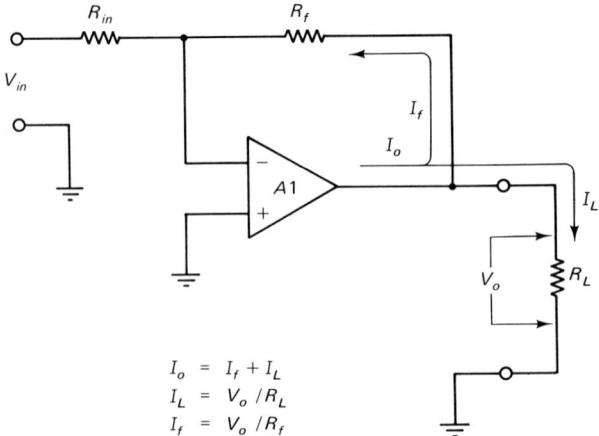

$$I_o = I_f + I_L$$
$$I_L = V_o / R_L$$
$$I_f = V_o / R_f$$

Fig. 12-19   Output currents on inverting follower.

Substituting Eqs. (12-30) through (12-34) into Eq. (12-29):

$$\frac{V1}{R1} + \frac{V2}{R2} + \frac{V3}{R3} + \cdots + \frac{V_n}{R_n} = \frac{V_o}{R_f} \qquad (12\text{-}35)$$

or, algebraically rearranging Eq. (12-36) to solve for V_o:

$$V_o = R_f \left[\frac{V1}{R1} + \frac{V2}{R2} + \frac{V3}{R3} + \cdots + \frac{V_n}{R_n} \right] \qquad (12\text{-}36)$$

Equation (12-36) is the transfer equation for the multiple input inverting follower.

Output Current

The output current I_o must be supplied by the output terminal of the op-amp. Typically small-signal op-amps supply 5 to 25 mA of current depending on the device, while power op-amps such as the Burr-Brown OPA-511 supply up to 5 A at potentials of ± 30 V. The output current (I_o) splits into two paths (Fig. 12-19); a portion of the output current flows into the feedback path (I_f) and a portion flows into the load (I_L). The total current is

$$I_o = I_f + I_L \qquad (12\text{-}37)$$

where I_o is the output current, I_f the feedback current (V_o/R_f), and I_L the load current (V_o/R_L).

In normal voltage amplifier service, both I_f and I_L tend to be very small compared with the available output current. But in applications where load and feedback resistances are low, the output currents may approach the maximum specified value. To determine whether this limit is exceeded, divide output potential V_o by the parallel combination of R_f and R_L, or

$$\frac{(V_{o(max)})(R_f + R_L)}{R_f R_L} < I_{o(max)} \qquad (12\text{-}38)$$

where $V_{o(max)}$ is the maximum expected output voltage, $I_{o(max)}$ the maximum allowable output current, R_f the feedback resistance in ohms, and R_L the load resistance in ohms.

In general, the output current limit is not approached on ordinary devices unless load and/or feedback resistances are less than 1000 Ω. Power devices, of course, can drive a lower load and/or feedback resistance combination.

Response to AC Signals

Thus far our discussion of inverting amplifiers has assumed a DC input signal voltage. The behavior of the circuit in response to AC signals (e.g., sine waves, square waves and triangle waves) is similar. Recall the rules for the inverter: positive input signals produce negative output signals, and negative input signals produce positive output signals. These relationships mean that a 180° phase shift occurs between input and output. The relationship is shown in Fig. 12-20A.

Although the DC-coupled op-amp will respond to AC signals, there is a limit that must be recognized. If the peak value of the input signal gets too great, then output clipping (Fig. 12-20B) will occur. The peak output voltage will be

$$V_{o(peak)} = -A_v V_{in(peak)} \qquad (12\text{-}39)$$

where $V_{o(peak)}$ is the peak output voltage, $V_{in(peak)}$ the peak input voltage, and A_v the voltage gain.

For every value of $V-$ and $V+$ power supply potentials there is a maximum attainable output voltage $V_{o(max)}$. As long as the peak voltage is less than this maximum allowable output potential, then the input waveform will be faithfully reproduced in the output (except that it will be amplified and inverted). But if the value of $V_{o(peak)}$ determined by Eq. (12-39) is greater than $V_{o(max)}$, clipping will occur.

In a linear voltage amplifier, clipping is undesirable. The maximum output voltage can be used to calculate the maximum input signal voltage:

$$V_{in(max)} = \frac{V_{o(max)}}{A_v} \qquad (12\text{-}40)$$

There are occasions when clipping is desired. For example, in a radio transmitter, circuits called modulation limiters are often simple

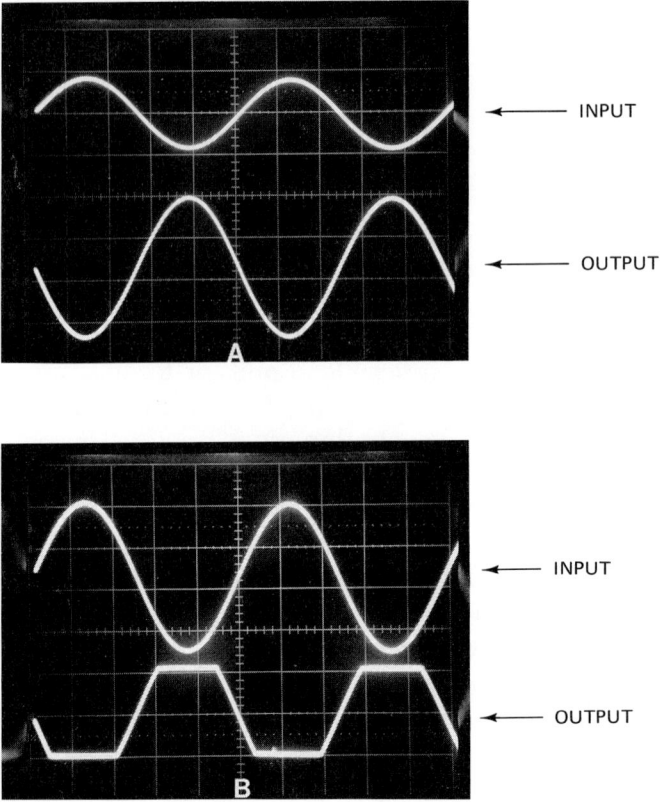

Fig. 12-20   (A) Unclipped output waveform and (B) clipped output waveform.

clippers followed by an audio low-pass filter that removes the harmonic distortion created by clipping. Another case where clipping is desired is in generating square waves from sine waves. The goal in that case is to drive the input so hard that sharp clipping occurs. Although there are better ways to realize this goal, the overdriven clipper square-wave generator does work.

Response to AC Input Signals with DC Offset

The case considered in the previous section assumed a waveform that is symmetrical about the zero volts baseline. In this section we will examine the case where an AC waveform is superimposed on a DC voltage. Figure 12-21 shows an inverting amplifier circuit with an AC signal source in series with a DC source. In Fig. 12-22A, a 4-V peak-to-peak square wave is superimposed on a 1-V DC fixed potential. Thus, the nonsymmetrical signal will swing between $+3$ V and -1 V.

The output waveform is shown in Fig. 12-22B. With the 180° phase inversion and the gain of -2 depicted in Fig. 12-21, the waveform will be a nonsymmetrical oscillation between -6 V and

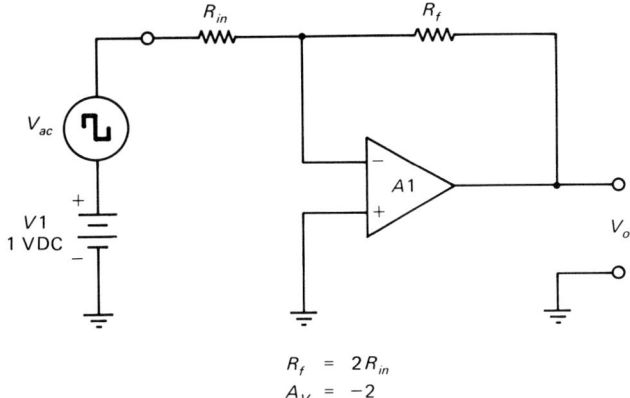

$$R_f = 2R_{in}$$
$$A_V = -2$$

Fig. 12-21 Inverting amplifier with DC component superimposed on AC input signal.

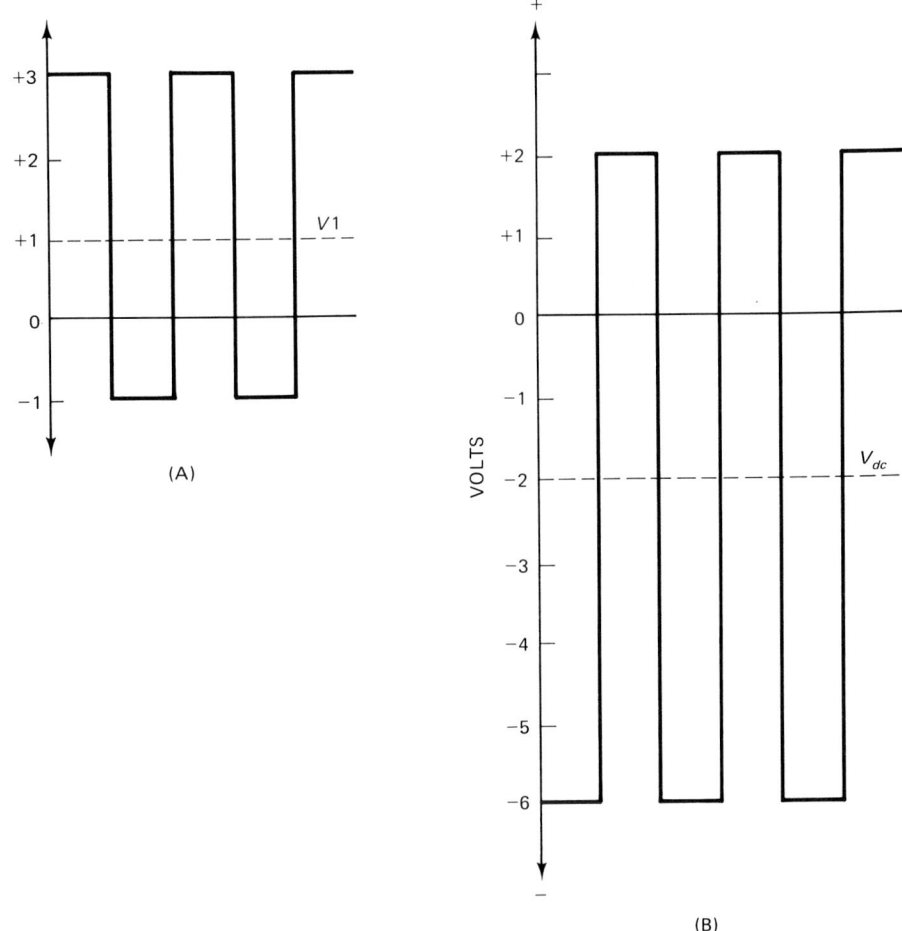

Fig. 12-22 Output waveforms for two different conditions.

+2 V. Because of gain ($A_v = -2$) the degree of asymmetry has also doubled to 2 V DC.

Dealing with AC signals that have a DC component can lead to problems at high gain and/or high input signal levels. As was true in the case of the high amplitude symmetrical signal, the output may saturate at either $V-$ or $V+$ power supply rails. If this limit is reached, then clipping will result. The DC component is seen as a valid input signal, so it will drive the output to one power supply limit or the other. For example, if the op-amp in Fig. 12-21 has a $V_{o(max)}$ value of ± 10 V (with $A_v = -2$), a +4-V positive input signal will saturate the output, while the negative excursion can reach -7 V before causing output saturation.

Response to Square Waves and Pulses

Most amplifiers respond in a congenial manner to sinusoidal and triangular waveforms. Some amplifiers, however, will exhibit problems dealing with fast rise-time waveforms such as square waves and pulses. The source of these problems is the high-frequency content of these waveforms.

All continuous mathematical functions (including electronic waveforms) are made of a series of harmonically related sine and cosine constituent waves (and possibly also a DC component). The sine wave consists of a single frequency, or fundamental sinusoidal wave. All nonsinusoidal waveforms, however, are made up of a fundamental sine wave plus its harmonics. The actual wave shape is determined by the number of harmonics present, which particular harmonics are present (i.e., odd or even), the relative amplitudes of those harmonics, and their phase relationship with respect to the fundamental. These factors can be deduced from the quarterwave and/or halfwave symmetry of the wave. The listing of the constituent frequencies forms a Fourier series and determines the bandwidth of the system required to process the signal. For example, the symmetrical square wave is made up of a fundamental-frequency sine wave (F), plus odd harmonics (3F, 5F, 7F, ...) up to (theoretically) infinity (as a practical matter, most square waves are "square" if the first 100 harmonics are present). Furthermore, if the square wave is truly symmetrical, then all of the harmonics are in-phase with the fundamental. Other wave shapes have different Fourier spectrums.

In general, the rise time of a pulse is related to the highest significant frequency in the Fourier spectrum by the rule-of-thumb approximation:

$$F = \frac{0.35}{T_r} \qquad (12\text{-}41)$$

where F is the highest Fourier frequency in hertz, and T_r is the pulse rise time in seconds.

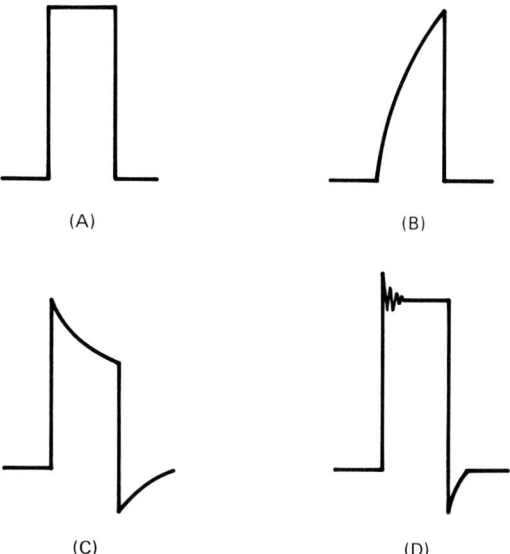

Fig. 12-23 (A) Square-wave signal, (B) effect of high-frequency loss, (C) effect of low-frequency loss, and (D) ringing.

Because pulse shape is a function of the Fourier spectrum for that wave, the frequency response characteristic of the amplifier has an effect on the wave shape of the reproduced signal. Figure 12-23 shows an input pulse signal (Fig. 12-23A) and two possible responses. The response shown in Fig. 12-23B results from attenuation of the high frequencies. The rounding shown will be either moderate or severe depending on the -12-dB bandwidth of the amplifier—in other words, by how many harmonics are attenuated by the amplifier frequency response characteristic, and to what degree. This problem becomes especially severe when the fundamental frequency (or pulse repetition rate) is high, the rise time is very fast, and the amplifier bandwidth is low.

Frequency-compensated operational amplifiers achieve their claimed unconditional stability by rolling off the high-frequency response drastically above a few kilohertz. A type 741AE is a frequency-compensated op-amp with a gain–bandwidth product of 1.25 MHz and an open-loop gain of 250,000. Frequency response at maximum gain is 1.25 MHz/250,000, or 5 kHz. Thus, we can expect good square-wave response only at relatively low frequencies. A rule of thumb for square waves is to make the amplifier bandwidth at least 100 times the fundamental frequency. As with all such rules, however, this one should be applied with caution even though it is assimilated into your collection of standard engineering wisdom.

The other class of problems is shown in Figs. 12-23C and 12-23D. In this case we see peaking and ringing of the pulse. Three principal causes of these phenomena are found. First, a skewed bandpass characteristic in which either the low frequencies are attenuated (or amplified less) or the high frequencies are amplified more. Second, there are LC resonances in the circuit that give rise to ringing. Although not generally a problem at low frequencies, video operational amplifiers may see this problem. Third, there are significant harmonics present at frequencies where circuit phase shifts add up to 180° and where the loop gain is unity or greater. When combined with the 180° phase shift inherent in inverting followers, we have Barkhausen's criteria for oscillation. Under some conditions the device will break into sustained oscillation. In other cases, however, oscillation will occur only on fast rise-time signal peaks, as shown in Fig. 12-23D.

Some Basic Rules

We must consider several factors when designing inverting follower amplifiers. First, we obviously must consider the voltage gain required by the application. Second, we must consider the input impedance of the circuit. That specification is needed to prevent the amplifier input from loading down the driving circuit. In the case of the inverting follower, the input impedance is the value of the input resistor (R_{in}), and a simple design rule is in effect:

> The input resistor (hence the input impedance) should be equal to or greater than 10 times the source resistance of the previous circuit.

The implication of this rule is that we must determine the source resistance of the driving circuit and then make the input impedance of the operational amplifier inverting follower at least 10 times larger. When the driving source is another operational amplifier, we can assume that the source impedance (i.e., the output impedance of the driving op-amp) is 100 Ω or less (it is actually much less). For these cases, make the value of R_{in} at least 1000 Ω (i.e., 10 × 100 Ω = 1000 Ω). This value is based on consideration of available output load current. In other cases, however, we have a slightly different problem. Some sensors, for example a thermistor for measuring temperature, have a much higher source resistance. One thermistor has an advertised resistance that varies from 10 kΩ to 100 kΩ over the temperature range of interest; so a minimum input impedance of 1 MΩ (i.e., 10 × 100 Ω) is required. When the input impedance gets this high the designer might want to consider the noninverting follower rather than the inverting follower configuration.

In the inverting follower circuit the choice of input impedance drives the design, so it is part of the design procedure:

1. Determine the minimum allowable input resistance (i.e., 10 or more times the source impedance).

2. If the source resistance is Ω or less, try 10 Ω as an initial trial input resistance (R_{in}).

3. This value might be lowered if the feedback resistor (R_f) becomes too high for the required gain. The value of R_{in} is the input resistance, or 10 kΩ, whichever is higher.

4. Determine the amount of gain required. In general, the closed-loop gain of a single inverting follower should be less than 500. For gains higher than that figure, use a multiple op-amp cascade circuit. Some low-cost op-amps should not be operated at closed-loop gains greater than 200. The reason for this rule is the problems that are found in real (versus ideal) devices. In those cases the distributed gain of a cascade amplifier may prove easier to tame in practical situations.

5. Determine the frequency response (i.e., the frequency at which the gain drops to unity). From steps 3 and 4 we can calculate the minimum-gain bandwidth product of the op-amp required.

6. Select the operational amplifier. If the gain is high, for example more than 100, then you might want to select a BiMOS or BiFET operational amplifier to limit the output offset voltage caused by the input bias currents. Select a 741-family device if (a) you don't need more than a few kilohertz frequency response and (b) the unconditionally stable characteristics of the 741 is valuable for the application.

7. Also look at the package style. For most applications the eight-pin miniDIP package is probably the easiest to handle. The eight-pin metal can is also useful, and it can be made to fit eight-pin miniDIP positions by correct bending of the leads.

8. Select the value of the feedback resistor:

$$R_f = \text{ABS}(A_v R_{in}) \qquad (12\text{-}42)$$

9. If the value of the feedback resistor is too high, that is, beyond the range of standard values (about 20 MΩ or so), or too high for the input bias currents, try a lower input resistance.

Altering AC Frequency Response

The natural bandwidth of an amplifier is sometimes too great for certain specific applications. Noise power, for example, is a function of bandwidth as indicated by the expression $P_n = KTBR$, where K is the Boltzmann constant, T the temperature in Kelvin, B the bandwidth in hertz, and R the resistance in ohms. Thus, it is possible that the signal-to-noise ratio will suffer in some applications if the bandwidth is not limited to that which is actually needed to process the expected waveform. In other cases we find that the rejection of spurious signals suffers if we fail to tailor the bandwidth of an amplifier circuit to that which is required by the bandwidth of the applied input signal. Amplifier stability is improved if the loop gain of the circuit is reduced to less than one at the frequency at which the circuit phase shift (including internal amplifier phase shift) reaches 180°. When the distributed phase shift is added to the 180° phase shift seen normally on inverting amplifiers, Barkhausen's criteria for oscillation is satisfied and the amplifier will oscillate. Those criteria are

1. total phase shift of 360° at the frequency of oscillation;
2. output-to-input coupling (may be accidental); and
3. loop gain of unity or greater.

If these criteria are satisfied at any frequency, the operational amplifier will oscillate at that frequency. For the present we will discuss just one technique in case you need to know the method in performing laboratory exercises.

The design goal in tailoring the AC frequency response is to roll off the voltage gain at the frequencies above a certain critical frequency F_c. This frequency is determined by evaluating the application; it is defined as the frequency at which the gain of the circuit drops off

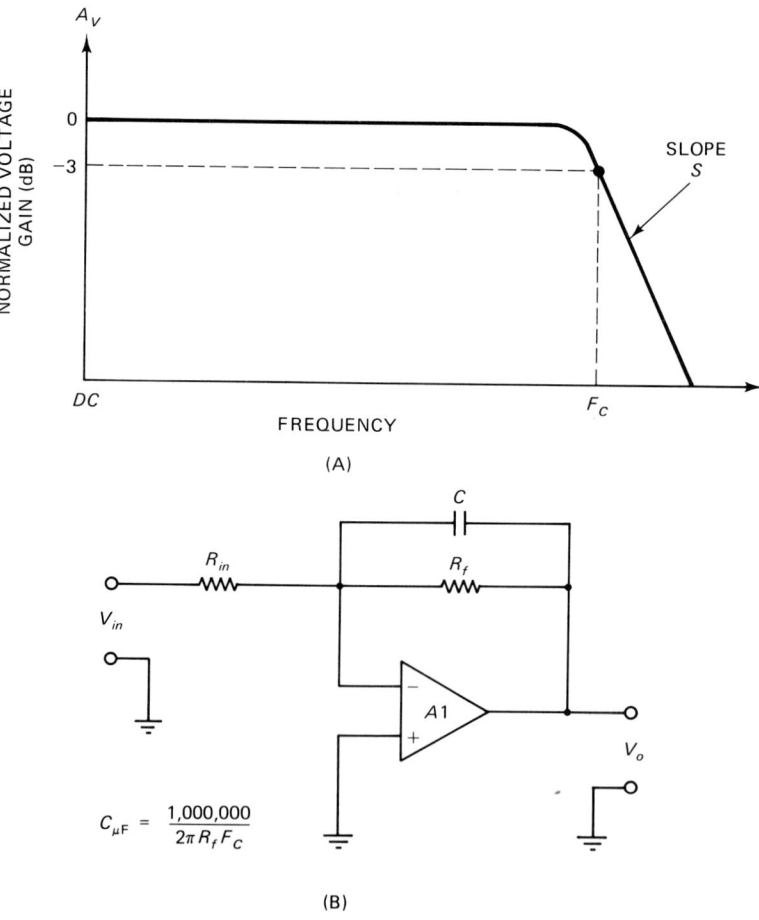

Fig. 12-24 (A) Frequency response tailoring drops off the high-frequency end by slope S and (B) inverting follower with capacitor C to tailor frequency response.

−3 db from its in-band voltage gain. The response of the amplifier should look like Fig. 12-24A; it is shown here in normalized form in which the maximum in-band gain is taken to be 0 dB. Above the critical frequency the gain drops off −6 dB/octave (an octave is a 2:1 change in frequency) by shunting a capacitor across the feedback resistor, as shown in Fig. 12-24B. The reactance of the capacitor is shunted across the resistance of R_f, so the gain is reduced. The low-pass filter characteristic is achieved because the capacitive reactance becomes lower as frequency increases. The value of the capacitor is found from

$$C = \frac{1}{2\pi R_f F_c} \tag{12-43}$$

where C is the capacitance in farads, R_f the feedback resistance in ohms, and F_f the −3-dB frequency in hertz.

Alternatively, to calculate the capacitance of C in microfarads (μF) we use Eq. (12-44).

$$C_{\mu F} = 1,000,000/2\pi R_f F_c \tag{12-44}$$

Noninverting Followers

The next standard op-amp circuit configuration is the noninverting follower. This type of amplifier uses the noninverting input of the operational amplifier to apply signal. In this configuration the output signal is in-phase with the input signal (Fig. 12-25). There are two basic noninverting configurations: unity gain and greater-than-unity gain.

Figure 12-26 shows the circuit for the unity gain noninverting follower. The output terminal is connected directly to the inverting

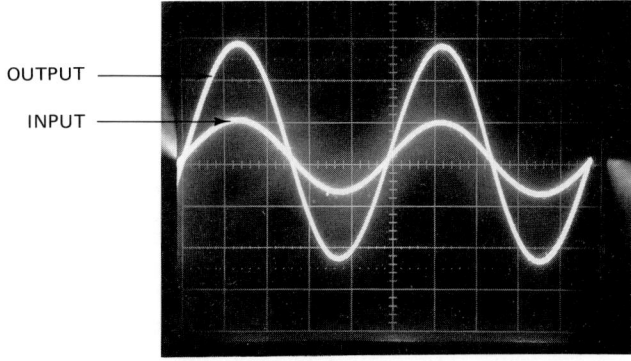

Fig. 12-25 Input and output waveforms of gain of ×2 noninverting amplifier.

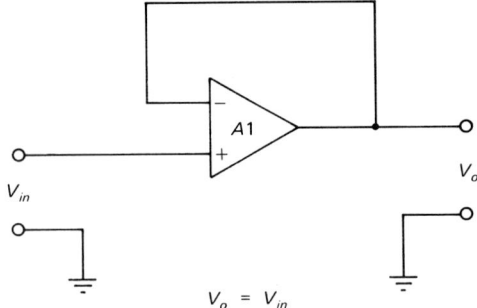

Fig. 12-26 Unity-gain noninverting follower circuit.

input, resulting in 100 percent negative feedback. Recall the voltage gain expression for all feedback amplifiers:

$$A_v = A_{vol}C/(1 + A_{vol}B) \qquad (12\text{-}45)$$

where A_v is the closed-loop voltage gain (i.e., gain with feedback), A_{vol} the open-loop voltage gain (i.e., gain without feedback), B the feedback factor, and C the input attenuation factor.

In this circuit the input signal is applied directly to $+$IN; so $C = 1$ and it can therefore be ignored. The feedback factor B represents the transfer function of the feedback network. When that network is a resistor voltage divider network, the value of B is a decimal fraction that represents the attenuation of the op-amp output voltage before it is applied to the op-amp inverting input. In the unity-gain follower circuit the value of B is also 1, so it too is ignored. The feedback amplifier equation therefore reduces to

$$A_v = A_{vol}/(1 + A_{vol}) \qquad (12\text{-}46)$$

Consider the implications of Eq. (12-46) for common operational amplifiers. With a gain of 300,000 (not unusual), Eq. (12-46) evaluates to 0.9999967. A gain of 0.9999967 is close enough to 1.0 to justify calling the circuit of Fig. 12-26 a *unity* gain follower.

Applications of Unity-Gain Followers

What use is an amplifier that does not amplify? First of all, it is not strictly true that the circuit does not amplify. It has a unity voltage gain, but the power gain is greater than unity. The three principal uses of the unity gain noninverting follower are buffering, power amplification, and impedance transformation.

A "buffer" amplifier is placed between a circuit and its load to improve the isolation between the two. An example is use of a buffer amplifier between an oscillator or waveform-generator circuit and its load. The buffer is especially useful where the load exhibits a varying impedance that could result in "pulling" of the oscillator frequency.

Such unintentional frequency modulation of the oscillator is very annoying because it makes some oscillator circuits unable to function and causes others to function poorly.

Another common use for buffer amplifiers is isolation of an output connection from the main circuitry of an instrument. An example might be an instrumentation circuit that uses multiple outputs, perhaps one to the A/D converter input of a digital computer and another to an analog oscilloscope or strip chart recorder. By buffering the analog output to the oscilloscope we prevent short circuits in the display wiring from affecting the signal to the computer, and vice versa.

A special case of buffering is represented by using the unity-gain follower as a power driver. A long cable run may attenuate low-power signals. To overcome this problem we sometimes use a low impedance power source to drive a long cable. This application points out the fact that a unity-gain follower actually does have power gain (the unity-gain feature refers only to the voltage gain). If the input impedance is typically much higher than the output impedance, yet $V_o = V_{in}$, then by V^2/R the delivered power output is much greater than the input power. Thus, the circuit of Fig. 12-26 is unity gain for voltage signals and greater-than-unity gain for power. It is therefore a power amplifier.

The impedance transformation capability is obtained from the fact that an op-amp has a very high input impedance and a very low output impedance. Let us illustrate this application by a practical example. Figure 12-27A is a generic equivalent of a voltage source driving a load ($R2$). The resistance $R1$ represents the internal impedance of the signal source impedance. The signal voltage V is reduced at the output (V_o) by whatever voltage is dropped across source resistance $R1$. The output voltage is found from

$$V_o = V(R2)/(R1 + R2) \qquad (12\text{-}47)$$

By way of example: If the ratio of $R1/R2$ is, say, 10:1, then a 1-V DC potential is reduced to 0.091 V DC across $R2$. Ninety percent of the signal amplitude is lost. With a unity-gain noninverting amplifier, as in Fig. 12-27B, the situation is entirely changed. If the amplifier input impedance is very much larger than the source resistance, and the amplifier output impedance is very much lower that the load impedance, then there is very little loss and V will closely approximate V_o.

Noninverting Followers with Gain

Figure 12-28A shows the circuit for the noninverting follower with gain. In this circuit, the signal (V_{in}) is applied to the noninverting input, while the feedback network (R_f/R_{in}) is almost the same as it was in the inverting follower circuit. The difference is that one end of R_{in} is grounded.

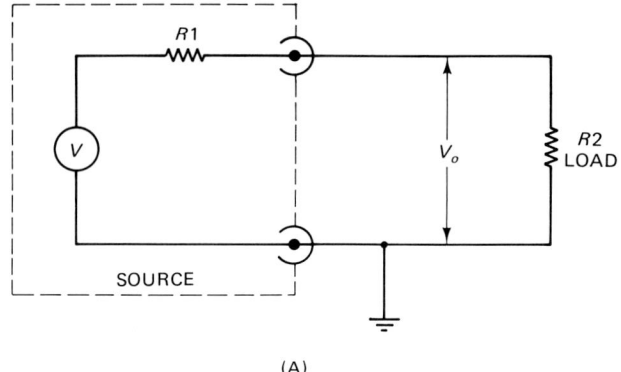

(A)

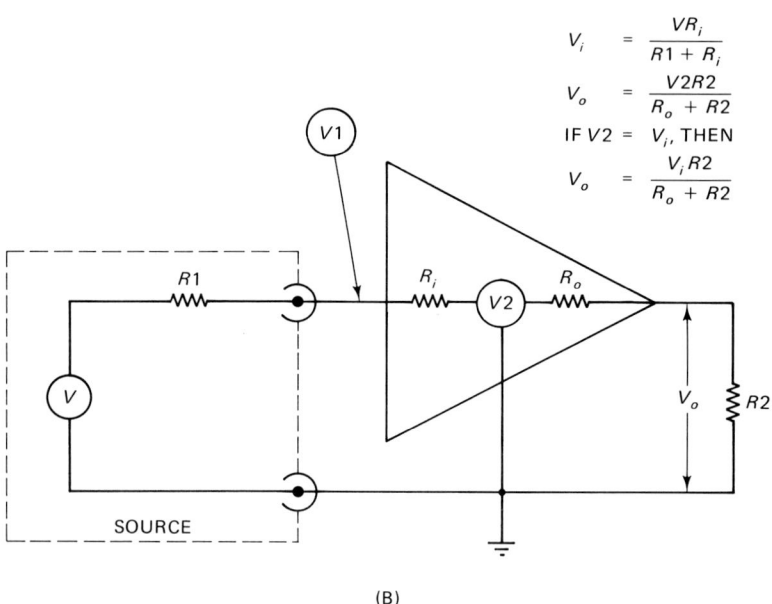

$$V_i = \frac{VR_i}{R1 + R_i}$$

$$V_o = \frac{V2R2}{R_o + R2}$$

IF $V2 = V_i$, THEN

$$V_o = \frac{V_i R2}{R_o + R2}$$

(B)

Fig. 12-27 (A) Output circuit and load and (B) input circuit and op-amp circuit model.

We can evaluate this circuit using the same general method as was used in the inverting follower case. We know from Kirchhoff's current law, and the fact that the op-amp inputs neither sink nor source current, that $I1$ and $I2$ are equal to each other. Thus, the Kirchhoff expression for these currents at the summing junction (point A) can be written as

$$I1 = I2 \tag{12-48}$$

We know from the properties of the ideal op-amp that any voltage applied to the noninverting input (V_{in}) also appears at the inverting

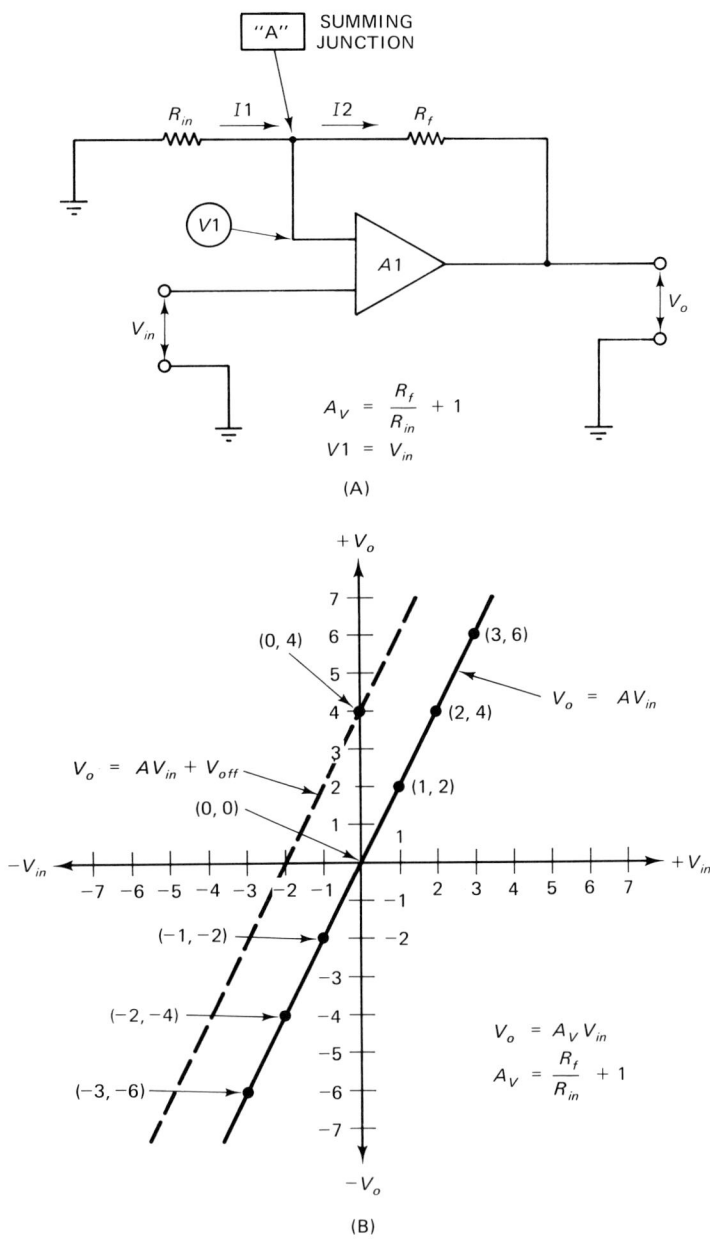

Fig. 12-28 Noninverting follower with gain: (A) circuit and (B) characteristic curve.

input. Therefore,

$$V1 = V_{in} \qquad (12\text{-}49)$$

From Ohm's law we know that the value of current $I1$ is

$$I1 = V1/R_f \qquad (12\text{-}50)$$

or, because $V1 = V_{in}$,

$$I1 = V_{in}/R_f \qquad (12\text{-}51)$$

Similarly, current $I2$ is equal to the voltage drop across resistor R_f, divided by the resistance of R_f. The voltage drop across resistor R_f is the difference between output voltage V_o and the voltage found at the inverting input, $V1$. By Ideal Property No. 7, $V1 = V_{in}$. Therefore,

$$I2 = (V_o - V_{in})/R_f \qquad (12\text{-}52)$$

We can derive the transfer equation of the noninverting follower by substituting Eqs. (12-51) and (12-52) into Eq. (12-48).

$$V_{in}/R_{in} = (V_o - V_{in})/R_f \qquad (12\text{-}53)$$

We must now solve Eq. (12-53) for output voltage V_o.

$$V_{in}/R_{in} = (V_o - V_{in})/R_f \qquad (12\text{-}54)$$

$$V_o = V_{in}\left(\frac{R_f}{R_{in}} + 1\right) \qquad (12\text{-}55)$$

Equation (12-55) is the transfer equation for the noninverting follower. The transfer function V_o/V_{in} for a gain of -2 noninverting amplifier is shown in Fig. 12-25B. The solid line assumes no output offset voltage is present (i.e., $V_0 = A_vV_{in} + 0$), while the dotted line represents a case where the offset voltage is nonzero.

Advantages of Noninverting Followers

The noninverting follower offers several advantages. In our discussion of the unity-gain configuration we mentioned that buffering, power amplification, and impedance transformation were advantages. Also, in the gain noninverting amplifier configuration we are able to provide voltage gain with no phase reversal.

The input impedance of the noninverting followers shown thus far is very high, being essentially the input impedance of the op-amp itself. In the ideal device, this impedance is considered infinite, while in practical devices it may range from 500,000 to more than 10^{12} Ω. Thus, the noninverting follower is useful for amplifying signals from any high-impedance source, regardless of whether or not impedance transformation is a circuit requirement.

When the gain required is known (as it usually is in practical situations), we select a trial value for R_{in} and then solve the gain equation to find R_f. This new version of the equation is

$$R_f = R_{in}(A_v - 1) \qquad (12\text{-}56)$$

Determine from evaluating the application whether or not the trial result obtained from this operation is acceptable. If the result is not acceptable, then work the problem again using a new trial value.

What does "acceptable" mean? If the value of R_f is exactly equal to a standard resistor value, then all is well. But, as in the case above, the value (118,800 Ω) is not a standard value. What we have to determine, therefore, is whether or not the nearest standard values result in an acceptable gain error (which is determined from the application). Both 118 Ω and 120 Ω are standard values, with 120 Ω being somewhat easier to obtain from distributor stock inventories. Both of these standard values are less than 1 percent from the calculated value, so this result is acceptable if a 1-percent gain error is within reasonable tolerance limits for the application.

The AC Response of Noninverting Followers

The noninverting amplifier circuits discussed in the preceding sections are DC amplifiers. Nonetheless, as with the inverting amplifiers considered earlier, the noninverting amplifier will also respond to AC signals up to the upper frequency response limit of the circuit.

Figure 12-29 shows the input signal situation for a noninverting follower. In this case there is an AC signal source in series with a DC potential ($V1$), which are applied to the noninverting input of the operational amplifier. A square-wave input signal (Fig. 12-30A) is applied to the input, but it is offset by a DC component (Fig. 12-30B). If the amplifier has a gain of $+2$, the output signal will be as shown in Fig. 12-30C. This signal swings from $+1$ V to $+5$ V. The offset of 1.5 V DC is amplified by two, and becomes a 3-V DC offset, with the AC signal swinging about this level.

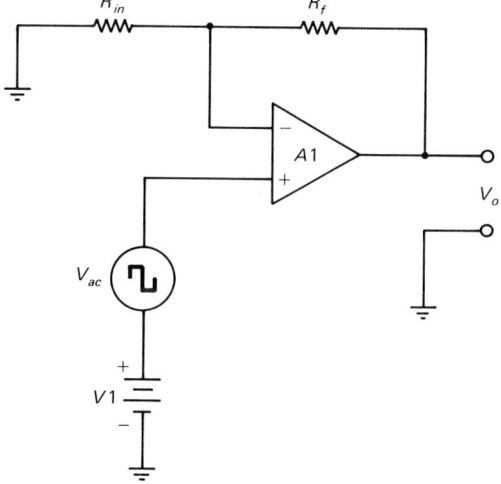

Fig. 12-29 Noninverting follower with DC offset to an AC signal.

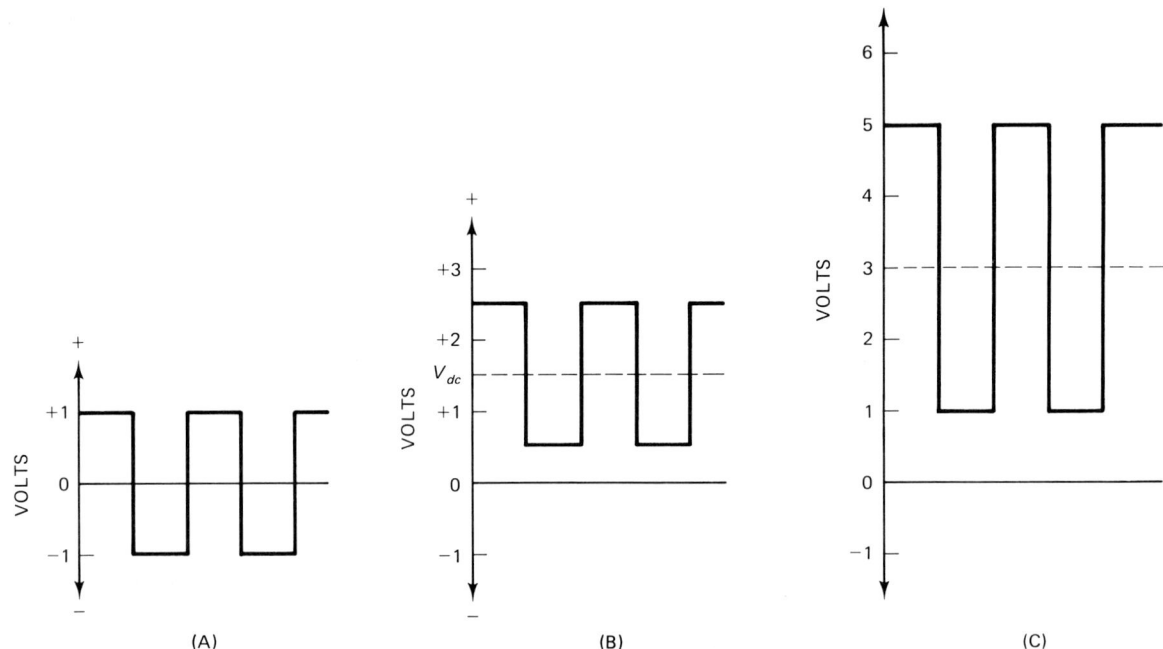

Fig. 12-30 Output waveforms from circuit of Fig. 12-29 under different conditions.

FREQUENCY RESPONSE TAILORING

It is possible to custom tailor the upper-end frequency response of the inverting-follower operational amplifier with a capacitor shunting the feedback resistor; the same method also works for the noninverting follower. In this section we will expand the subject and discuss not only tailoring of the upper -3-dB frequency response, but also a lower -3-dB limit as well. The capacitor across the feedback resistor (Fig. 12-31) sets the frequency at which the upper end frequency response

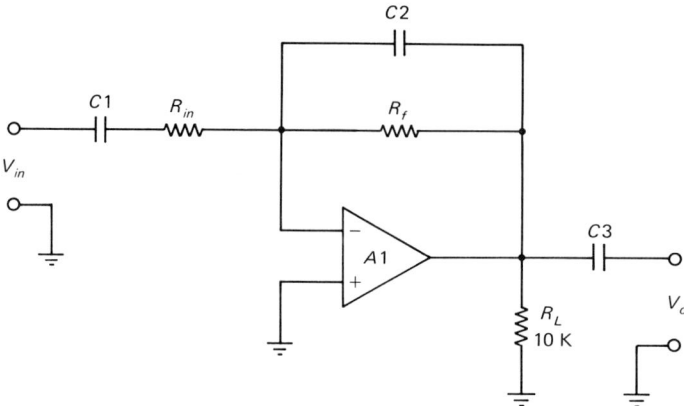

Fig. 12-31 AC-coupled and frequency-tailored inverting follower.

falls off -3-dB below the low end in-band gain. The gain at frequencies higher than this -3-dB frequency falls off at a rate of -6 dB/octave (an octave is a 2:1 frequency ratio), or -20 dB/decade (a decade is a 10:1 frequency ratio). The value of capacitor $C2$ is found by

$$C2_{\mu F} = 1{,}000{,}000/2\pi R_f F_c \qquad (12\text{-}57)$$

where $C2_{\mu F}$ is the capacitance of $C2$ in microfarads (μF), R_f the resistance in ohms (Ω), and F_c the upper -3-dB frequency in hertz (Hz).

The low-frequency response is controlled by placing a capacitor in series with the input resistor, which makes the inverting follower an AC-coupled amplifier. Figure 12-31 is the circuit for an inverting follower that uses AC-coupling at both input and output circuits. Capacitor $C2$ limits the upper -3-dB frequency response point. Its value is set by the method discussed above. The lower -3-dB point is set by the combination of R_{in} and input capacitor $C1$. This frequency is set by the equation

$$C1 = 1{,}000{,}000/2\pi R_{in} F \qquad (12\text{-}58)$$

where $C1$ is in microfarads (μF), R_{in} the resistance in ohms (Ω), and F the lower -3-dB point in hertz (Hz).

In some cases we will want to AC-couple the output circuit (although it is optional in most cases). Capacitor $C3$ is used to AC-couple the output, thus preventing any DC component that is present on the op-amp output from affecting the following stages. Resistor R_L is used to keep capacitor $C3$ from being charged by the offset voltage from op-amp $A1$. The value of capacitor $C3$ is set to retain the lower -3-dB point, using the resistance of the stage following as the R in the foregoing equations.

AC-Coupled Noninverting Amplifiers   The noninverting amplifiers discussed thus far have all been DC-coupled. They will respond to signals from either DC or near-DC up to the frequency limit of the amplifier selected. Sometimes, however, we do not want the amplifier to respond to DC or slowly varying near-DC signals. For these applications we select an AC-coupled noninverting follower circuit. In this section we will examine several AC-coupled noninverting amplifiers.

Figure 12-32A shows a capacitor input AC-coupled amplifier circuit. It is essentially the same as the previous circuits, except for the input coupling network, $C1/R3$. The capacitor in Fig. 12-32A serves to block DC and very low frequency AC signals. If the op-amp has zero (or, more realistically, very low) input bias currents, we can safely delete resistor $R3$. For all but a few commercially available devices, however, resistor $R3$ is required if closed-loop gain is high. Input bias currents will charge capacitor $C1$, creating a voltage offset that is seen

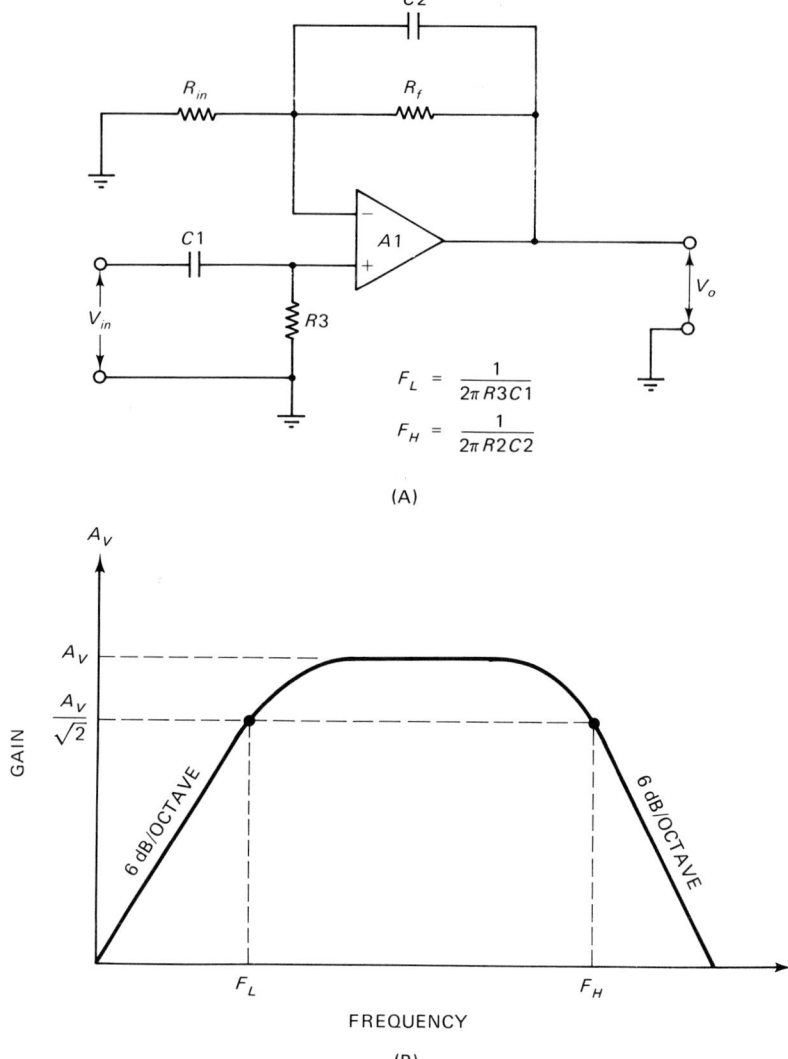

$$F_L = \frac{1}{2\pi R3 C1}$$

$$F_H = \frac{1}{2\pi R2 C2}$$

(A)

(B)

Fig. 12-32   AC-coupled noninverting follower.

by the op-amp as a valid DC signal and amplified to form an output offset voltage. In some devices the output saturates from the $C1$ charge shortly after turn-on; resistor $R3$ prevents such latch-ups because it keeps $C1$ discharged.

Resistor $R3$ also sets the input impedance of the amplifier. Previous circuits had a very high input impedance because that parameter was determined only by the (extremely high) op-amp input impedance. In Fig. 12-32A, however, the input impedance seen by the source is equal to $R3$.

Another effect of resistor $R3$ and capacitor $C1$ is to limit the low-frequency response of the circuit. Filtering occurs because $R3 C1$ forms a high-pass filter (see Fig. 12-32B). The -3-dB frequency is found from

$$F = 1,000,000/2\pi(R1)(C1) \qquad (12\text{-}59)$$

where F is the -3-dB frequency in hertz, $R1$ the resistance in ohms, and, $C1$ the capacitance in microfarads.

The form of Eq. (12-59) is backwards from the point of view of practical circuit-design problems. In most cases, we know the required frequency-response limit from the application. We also know from the application what the minimum value of $R3$ should be (derived from source impedance), and we often set it as high as possible as a practical matter (e.g., 10 MΩ). Thus, we want to solve the equation for C, as shown here:

$$C = 1,000,000/2\pi(R1)F \qquad (12\text{-}60)$$

All terms are as defined for Eq. (12-59).

The technique of Fig. 12-32A works well for dual polarity DC power supply circuits. In single polarity DC power supply circuits, however, the method falls down because of the large DC offset voltage present on the output. For these applications we use a circuit such as shown in Fig. 12-33.

The circuit in Fig. 12-33 is operated from a single $V+$ DC power supply (the $V-$ terminal of the op-amp is grounded). To compensate for the $V-$ supply being grounded, the noninverting input is biased to

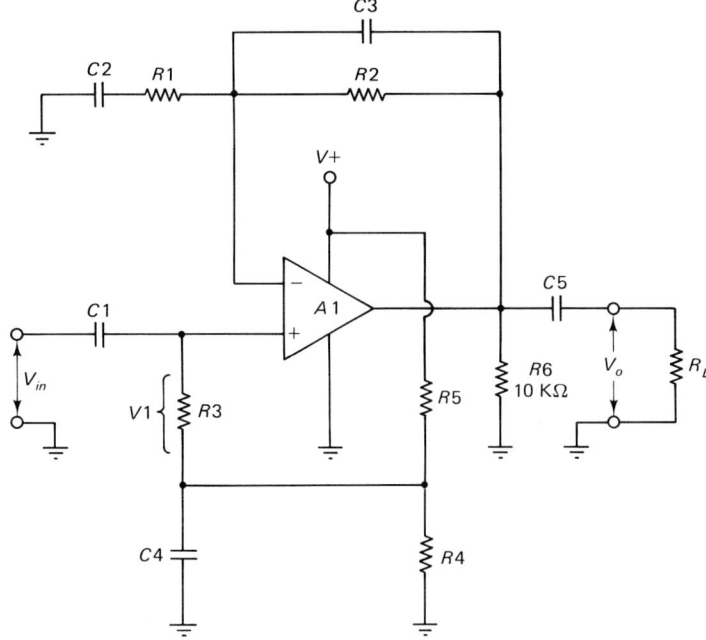

Fig. 12-33 AC-coupled single power supply noninverting amplifier.

a potential of

$$V1 = (V+)\left[\frac{R4}{R4 + R5}\right] \qquad (12\text{-}61)$$

If $R4 = R5$, then $V1$ will be $(V+)/2$. Because the noninverting input typically sinks very little current, the voltage at both ends of $R3$ is the same (i.e., $V1$). The circuit of Fig. 12-33 does not pass DC and some low AC frequencies because of the capacitor coupling. Also, because capacitor $C3$ shunts feedback resistor $R2$, there is also a roll-off of the higher frequencies. The high-frequency roll-off -3-dB point is found from

$$F = 1,000,000/2\pi(R2)(C3) \qquad (12\text{-}62)$$

where F is the -3-dB frequency in hertz, $R2$ the resistance in ohms, and $C3$ the capacitance in microfarads.

We can restate Eq. (12-62) into a more useful form that takes into account the fact that we usually know the value of $R2$ (from setting the gain), and the nature of the application sets the minimum value of frequency F. We can rewrite Eq. [12-62] in a form that yields the value of $C3$ from these data.

$$C3 = 1,000,000/2\pi(R3)F \qquad (12\text{-}63)$$

The lower -3-dB frequency is set by any or all of several RC combinations within the circuit.

1. $R1/C2$
2. $R3/C1$
3. $R3/C4$
4. $R_L/C5$

Resistor $R1$ is part of the gain-setting feedback network. Capacitor $C2$ is used to keep the "cold" end of $R1$ above ground at DC while keeping it grounded for AC signals.

Resistor $R3$ is the input resistor and serves the same purpose as the similar resistor in the previous circuit. At mid-band the input impedance is set by resistor $R3$, although at the extreme low end of the frequency range the reactance of $C4$ becomes a significant feature. In general, X_{C4} should be less than or equal to $R3/10$ at the lowest frequency of operation.

Capacitor $C1$ is in series with the input signal path and serves to block DC and certain very low frequency signals. The value of $C1$ should be

$$C1 = 1,000,000/2\pi F(R3) \qquad (12\text{-}64)$$

where $C1$ is in microfarads, F in hertz, and $R3$ in ohms.

Capacitor $C5$ is used to keep the DC output offset from affecting the succeeding stage. The 10-kΩ output load resistor ($R6$) keeps $C5$ from being charged by the DC offset voltage. The value of $C5$ should be

$$C5 > 10^6/2\pi F(R3) \tag{12-65}$$

where $C5$ is the capacitance in microfarads, F the low-end -3-dB frequency in hertz, and R_L the load resistance in ohms.

Transformer-Coupled Noninverting Amplifiers

Figure 12-34 shows the circuit for a transformer coupled noninverting follower. This type of circuit is often used in audio and broadcasting equipment. In those applications audio signals are passed over a 600-Ω balanced line. The point we will make here is that this circuit is an AC-only amplifier, with upper and lower -3-dB points determined by (a) the frequency response of the transformer ($T1$), the limitations of the operational amplifier, and any capacitances shunting feedback resistor $R2$.

The gain of the amplifier in Fig. 12-34 is given in Eq. (12-66):

$$A_v = (V_{in})\left(\frac{N_s}{N_p}\right)\left(\frac{R2}{R1} + 1\right) \tag{12-66}$$

where A_v is the voltage gain, N_s the number of turns in the secondary

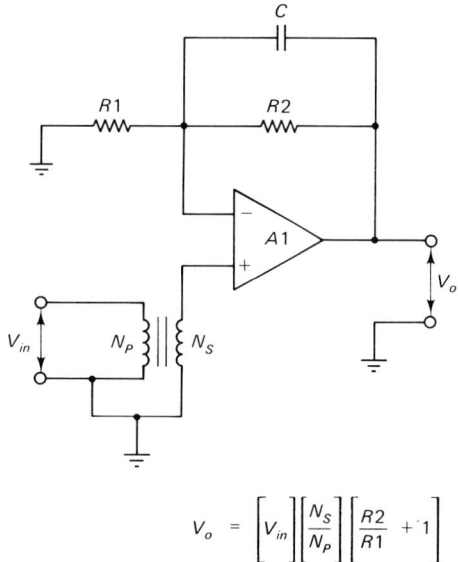

$$V_o = \left[V_{in}\right]\left[\frac{N_S}{N_P}\right]\left[\frac{R2}{R1} + 1\right]$$

Fig. 12-34 Transformer-coupled AC amplifier.

of $T1$, N_p the number of turns in the primary of $T1$, $R2$ the op-amp feedback resistor, and $R1$ the op-amp input resistor.

DEALING WITH PRACTICAL OPERATIONAL AMPLIFIERS

In the early part of this chapter we discussed the ideal operational amplifier. Such a hypothetical device is a good model-making tool, but it doesn't really exist. While it makes our analysis easier, it cannot actually be purchased and used in practical circuits. All real operational amplifiers depart somewhat—in some areas a great deal—from the ideal. We find, for example, that open-loop gain is not really infinite; but rather it has very high values in the range from about 20,000 to more than 1,000,000. Similarly, real operational amplifiers don't have infinite bandwidth; in fact, most are intentionally made severely bandwidth limited. Such amplifiers are said to be unconditionally stable or frequency-compensated devices (an example is the 741 device). Although stability is a highly desirable feature for many applications, it is obtained at the expense of frequency response. In this section we will deal with some of the more common problems found in real devices and the solutions to these problems.

Input Offset Current

Input offset current is measured in a test circuit such as shown in Fig. 12-35. Input offset current can be specified by the relationship between

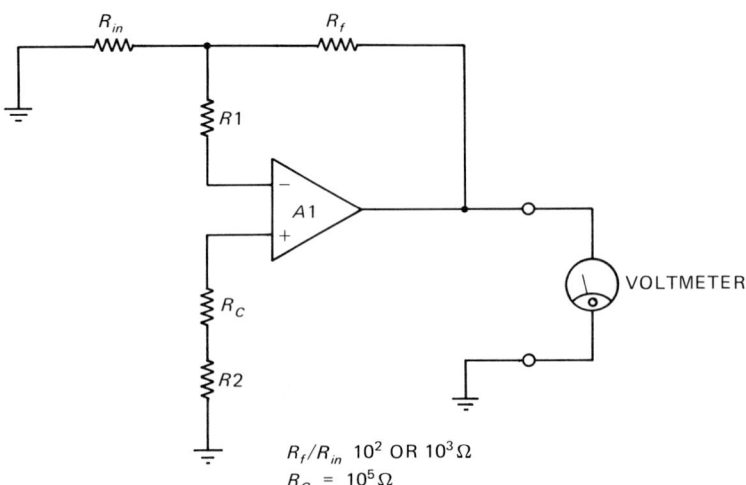

$$R_f/R_{in} \; 10^2 \text{ OR } 10^3 \, \Omega$$
$$R_C = 10^5 \, \Omega$$

Fig. 12-35 Op-amp measurement circuit for input offset.

two different output offset voltages that are taken under different input conditions.

$$I_{io} = \frac{R_{in}(V_{o1} - V_{o2})}{R2(R_f + R_{in})} \qquad (12\text{-}67)$$

The first output voltage, V_{o1}, is measured with $R1$ and $R2$ connected in the circuit. Voltage V_{o2} is then measured with resistors short-circuited, but with all other conditions remaining the same. The resultant output voltage can, along with V_{o1}, be used in Eq. (12-67) to determine input offset current.

Input Offset Voltage

The input offset voltage is the voltage required to force the output voltage (V_o) to zero when the input voltage is also zero. The operational amplifier is connected in an inverting amplifier configuration such as shown in Fig. 12-36. To make the measurement, the input terminal is connected to ground. The input offset voltage is found by measuring the output voltage (when $V_{in} = 0$) and then using the voltage divider equation:

$$V_{io} = V_o R_{in}/(R_f + R_{in}) \qquad (12\text{-}68)$$

Greater accuracy is achieved if the gain of the amplifier is 100, 1000 (or even higher), provided that such gains can be accommodated without saturating the amplifier.

Input Bias Current

This test requires a pair of closely matched resistors connected between the op-amp inputs ($-IN$ and $+IN$) and ground (Fig. 12-37). Power is applied to the operational amplifier, and the voltage is measured at each input. The value of resistors $R1$ and $R2$ must be high enough to create a measurable voltage drop at the level of current

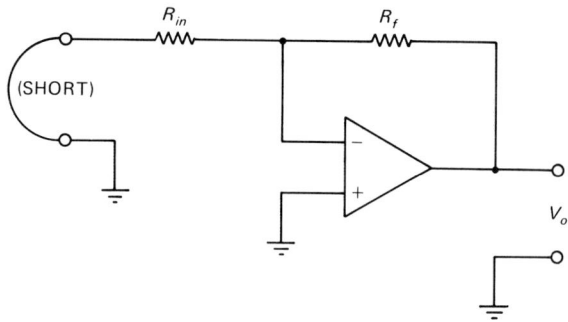

Fig. 12-36 Test circuit for output offset voltage.

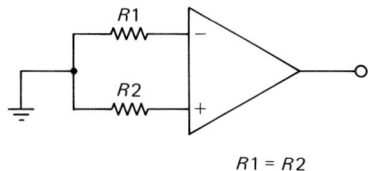

Fig. 12-37   Offset current test circuit.

anticipated. Although the actual value of these resistors is not too critical, the match between them ($R1 = R2$) is critical to the success of the measurement. The definition of "readable" voltage drop depends on the instrumentation available to do the job. For example, assume that an actual input bias current of 5 μA (microamperes) is flowing, and the resistors are each 10 kΩ. In this case the measured voltage will be (by Ohm's law)

$$V = IR$$

$$= (5 \times 10^{-6} \text{ A}) \times (10,000 \text{ }\Omega)$$

$$= 5 \times 10^{-2} \text{ V} = 50 \text{ mV}$$

If your voltage measuring equipment is not capable of measuring these levels, then higher resistor values would be required. If the two inputs had ideally equal input bias currents, only one measurement would be needed. Since real devices usually have unequal bias currents, however, it is sometimes necessary to measure both and use the higher input bias current measurement. Alternatively, the root sum squares (RSS) value can be used.

POWER SUPPLY SENSITIVITY

Power supply sensitivity (PSS) is the worst case change of input offset voltage for a 1.0 V DC change of one DC power supply voltage (either $V-$ or $V+$), with the other supply potential being held constant. The same test configuration is used to measure this parameter (ψ) as was used to measure input offset voltage (Fig. 12-36). First, the two power-supply voltages are set to equal levels and the input-offset voltage is measured. One of the power supply voltages is then changed by precisely 1.00 V DC and the input offset voltage is again measured. The power supply sensitivity (PSS) is given by

$$\psi = dV_{io}/dV_o \tag{12-69}$$

The actual power supply sensitivity is the worst case when this measurement is made under four conditions: (1) $V+$ increased

1 V DC, (2) $V+$ decreased 1 V DC, (3) $V-$ increased 1 V DC, and (4) $V-$ decreased 1 V DC. The worst case of these four measurements is taken to be the true power supply sensitivity.

SLEW RATE

Slew rate is a measure of the operational amplifier's ability to shift between the two possible opposite output-voltage extremes while supplying full output power to the load. This parameter is usually specified in terms of volts per unit of time (e.g., 30 V/μs).

A saturating square wave is usually used to measure the slew rate of an operational amplifier. The square wave must have a rise time that substantially exceeds the expected slew rate of the operational amplifier. The value of rise time is found from examination of the leading edge of the output waveform on an oscilloscope while the input is over-driven by the square wave. The time measured is that which is required for the output to slew from 10 percent of the final value to 90 percent of the final value. It must be noted that slew rate can be affected by gain, so the value at unity gain will not match either the slew rate under open-loop or very high gain closed-loop conditions. Once the switching time is known, the slew rate (S_r) is closely approximated by

$$S_r = \frac{(V+) + \text{ABS}(V-)}{T_s}   \tag{12-70}$$

where S_r is the slew rate in volts per microsecond (V/μs), $V+$ the positive supply voltage, $\text{ABS}(V-)$ the absolute value of the negative supply voltage, and T_s the switching time.

Since most manufacturers specify slew rate for the open-loop configuration in their data sheets, we can use this relationship to approximate the switching times of specific operational-amplifier digital circuits.

It is possible to improve the closed-loop slew rate at any given gain figure through the use of appropriate lag compensation techniques (Fig. 12-38). Keep the values of R_f and R_{in} low when trying to improve slew rates; values less than 10 kΩ will be best. The compensation capacitor will have a value of

$$C = \left[\frac{(R_f + R_{in})}{4\pi R_f R_{in}} \right]\left[\frac{F_{oi}}{10^m} \right]   \tag{12-71}$$

where F_{oi} is the -3-dB half-power point, R_f the feedback resistance, R_{in} the input resistance, and m the quantity $[A_{vol(dB)} - A_{v(dB)}]/20$.

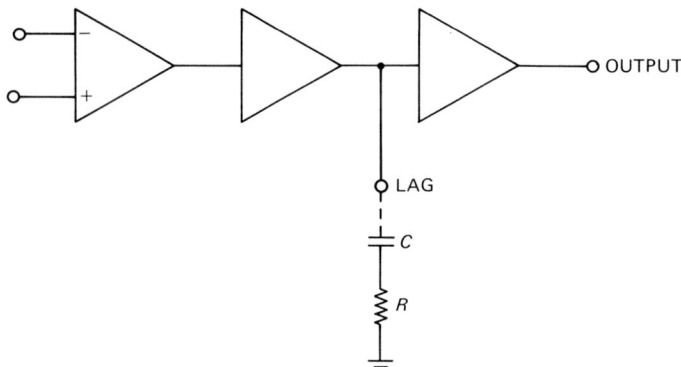

Fig. 12-38 Lag compensation.

The resistor value is found from

$$R = 1/2\pi F_{oi}C \qquad (12\text{-}72)$$

It is the usual practice to measure slew rate in the noninverting unity-gain voltage follower configuration because that circuit generally has the poorest slew rate in most op-amp devices. As in the previous test, the worst-case figure is used as a matter of standard practice.

The unity-gain follower is driven by a square wave of sufficient amplitude to drive the device well beyond the full saturation point. This criterion is necessary to eliminate the rounded curves that will exist at points just below full saturation. The output waveform can then be examined with a wide-band oscilloscope that has a time base fast enough to allow for a meaningful examination. The trace (Fig. 12-39) will be a straight line with a certain slope. It is standard practice to measure rise time as the time of transition from 10 percent of full

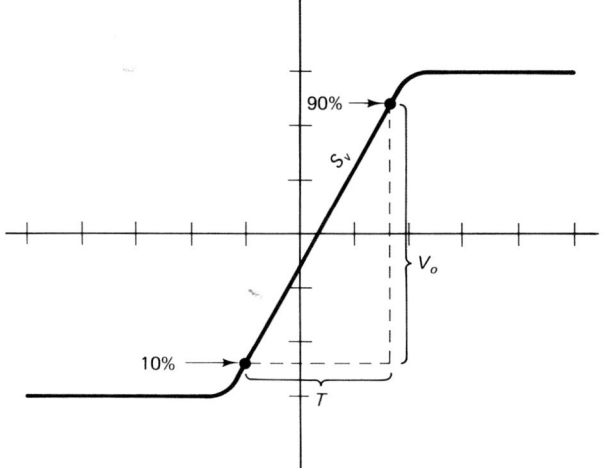

Fig. 12-39 Slew rate is the speed of transition from 10 to 90 percent of overdriven square-wave input.

amplitude to 90 percent of full amplitude and to adjust the time-base triggering of the oscilloscope so that the slope covers several horizontal scale divisions. The slew rate (S_r) is then found from the slope of the trace on the oscilloscope.

$$S_r = V_o/t \qquad (12\text{-}73)$$

Phase Shift

The phase shift (ϕ) of an operational amplifier can be measured using a sine wave and an oscilloscope. In one version of this test an X − Y oscilloscope (or a dual channel model with an X − Y capability) is

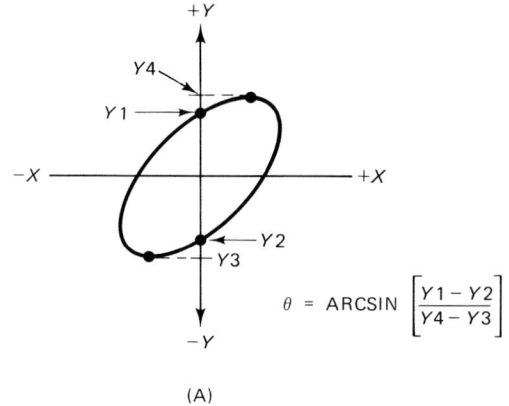

$$\theta = \text{ARCSIN} \left[\frac{Y1 - Y2}{Y4 - Y3} \right]$$

(A)

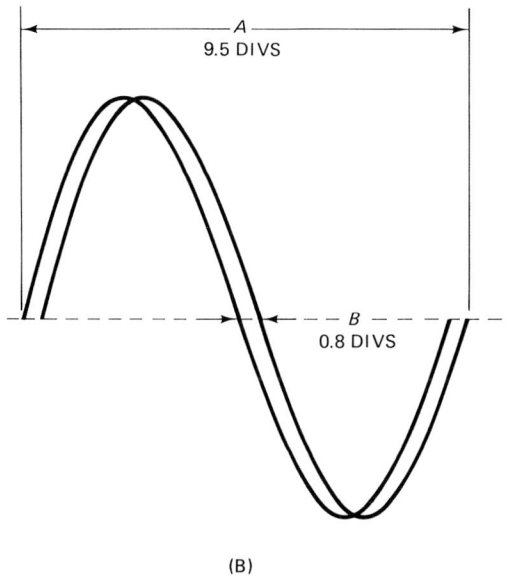

(B)

Fig. 12-40 Measurement of phase shift on oscilloscope: (A) Lissajous pattern method and (B) time-base method.

used. The input signal is applied to the vertical channel of the CRO, while the operational amplifier output is applied to the horizontal channel. The gains for the two channels are set to produce equal beam deflections.

The points marked $Y1$ through $Y4$ (Fig. 12-40A) are measured and the phase shift is calculated from

$$\phi = \sin^{-1}[(Y1 - Y2)/(Y4 - Y3)] \qquad (12\text{-}74)$$

An alternative approach uses a dual-trace CRO in which the input signal is applied to one channel and the output signal is applied to the other channel. The noninverting unity-gain operational amplifier configuration is used. The CRO channel gains are adjusted to be identical, and the traces are superimposed (Fig. 12-40B) on each other. The phase shift is found from

$$\phi = 360B/A \qquad (12\text{-}75)$$

This test only works properly if both channels of the CRO are identical to each other, and thus have the same internal phase shift. In addition, the CRO must have a high internal chopping frequency in the dual-beam mode. It must be recognized that instrument limitations constrain the usefulness of this test.

Common Mode Rejection Ratio

The common mode rejection ratio (CMRR) is defined as the ratio of the differential gain to the common mode gain.

$$\text{CMRR} = A_{vd}/A_{vcm} \qquad (12\text{-}76)$$

where CMRR is the common mode rejection ratio, A_{vd} the differential voltage gain, and A_{vcm} the common mode voltage gain.

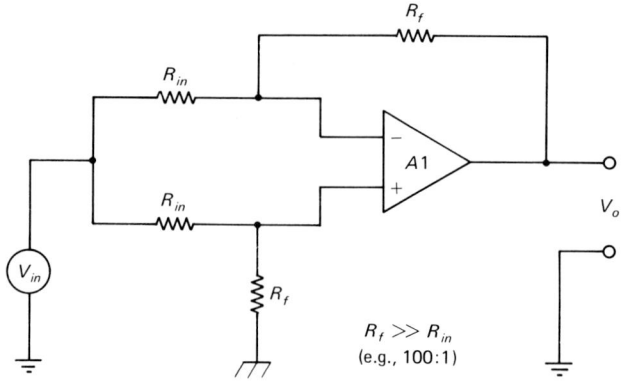

Fig. 12-41 Differential DC amplifier with common mode signal applied.

The common mode voltage gain is not always a linear function of common mode input voltage (V_{cm}), so it is usually specified at the maximum allowable value for V_{cm}. The amplifier is connected into a circuit such as shown in Fig. 12-41, and CMRR is determined from

$$\text{CMRR} = \frac{(R_f + R_{in})V_{in}}{R_{in}V_o} \qquad (12\text{-}77)$$

provided that $R_f \gg R_{in}$, such as 100:1.

DC ERRORS AND THEIR SOLUTIONS

In this section we will examine DC errors in operational amplifiers. These errors are in the form of current and voltage levels that conspire to force the output voltage to differ from the theoretical value in any given case. In many cases the error will be specified in terms of an actual output voltage that is nonzero at a time when it should be zero (e.g., when $V_{in} = 0$). We will also propose some circuit tactics that minimize or eliminate certain errors.

Output Offset Compensation

The DC error factors result in an output offset voltage V_{oo} that exists between the output terminal and ground at a time when V_o should be zero. This phenomenon helps explain discrepancies between those voltages that actually are found to exist and those that the equations say should exist. One method for classifying output offset voltages is by the causes: input offset voltage and input bias current.

Input offset voltage V_{io} (Fig. 12-42A) is defined as the differential input voltage required to force the output to zero when no other signal is present ($V_{in} = 0$). A reasonably good model for the input offset voltage phenomenon (Fig. 12-42B) is a voltage source with one end connected to ground and the other end connected to the noninverting input. Although voltage source polarity is shown here, the actual polarity found in any given situation may be either positive or negative to ground, depending on the device being tested. Values for input offset are typically from one to several millivolts. The popular type 741 operational amplifier is specified to have a 1 to 5 mV input offset voltage, with 2 mV being listed as typical.

The value of the output offset voltage (V_{oo}) caused by an input offset voltage is given by

$$V_{oo} = R_f V_{io} \qquad (12\text{-}78)$$

If the circuit gain is low and V_{in} remains at relatively high values, the input offset voltage may be of little practical consequence. It is

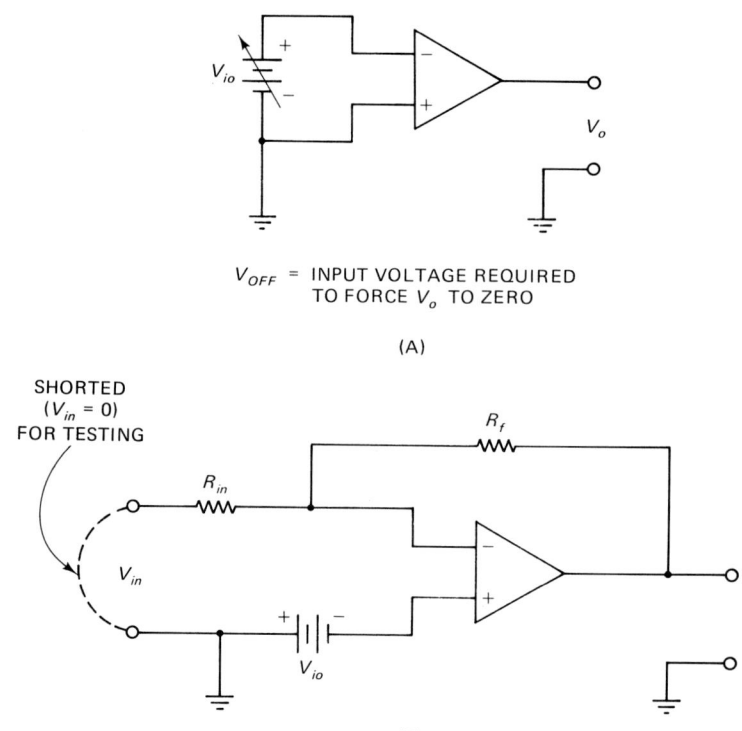

V_{OFF} = INPUT VOLTAGE REQUIRED
TO FORCE V_o TO ZERO

(A)

(B)

Fig. 12-42 Input offset voltage test set-up.

primarily where either high values of A_v or low values of V_{in} are encountered that the input offset voltage becomes a problem. (These conditions are often encountered together.)

The second major cause of output offset voltage is spurious input current. This can be further subdivided into two more classes: normal input bias and input offset current.

Figure 12-43A shows a typical operational amplifier differential input stage. Whenever bipolar (NPN or PNP) transistors are used in this stage, some small input bias current will be required for normal operation. This is one of those unavoidable conditions inherent in the nature of the transistor rather than any deficiency in the internal op-amp circuitry. The problem that input bias currents cause becomes acute when high values of input and feedback resistances are used (Fig. 12-43B). When these resistors are in the circuit the bias current causes a voltage drop across the resistances even when $V_{in} = 0$, causing an output voltage equal to

$$V_{oo} = I_b R_f \qquad (12\text{-}79)$$

Figure 12-43C shows the use of a compensation resistor R_c to reduce the offset potential due to input bias currents. This same resistor also improves thermal drift. This resistor has a value equal to

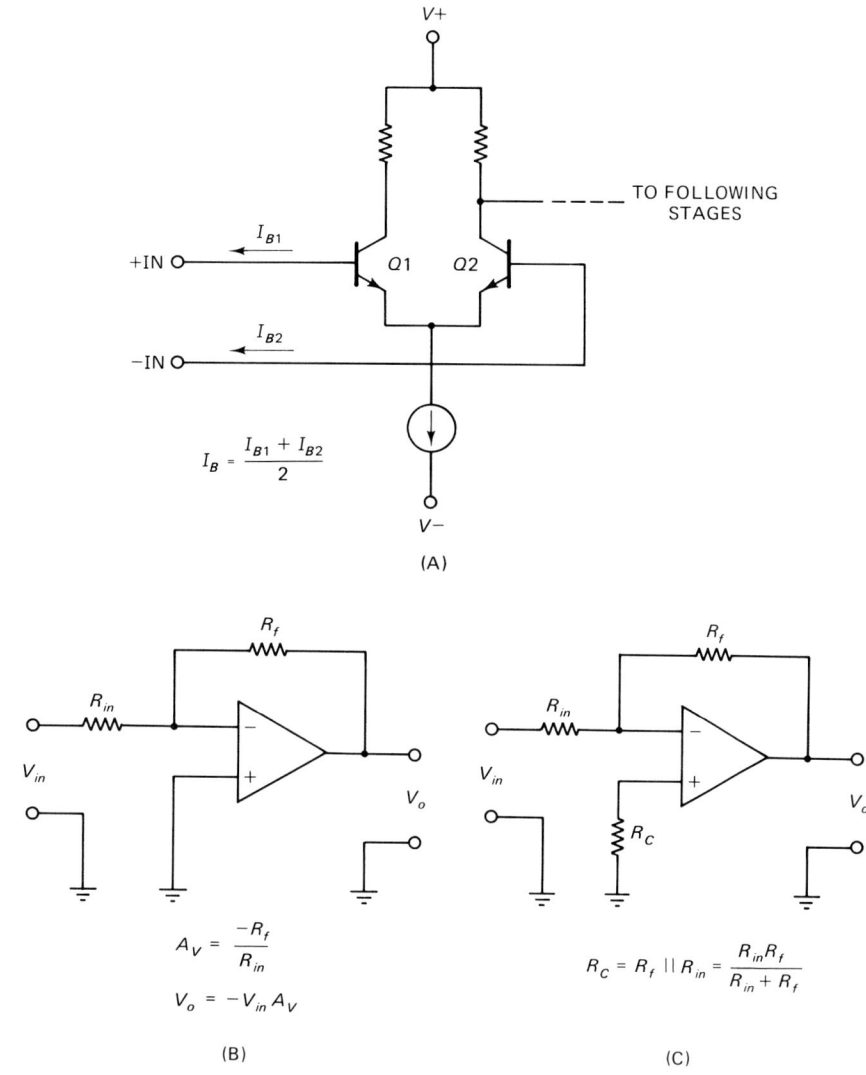

Fig. 12-43 (A) Differential input stage, (B) inverting amplifier, and (C) compensation resistor to overcome input offset current.

the parallel combination of the other two resistors.

$$R_\mathrm{c} = R_\mathrm{f} R_\mathrm{in}/(R_\mathrm{f} + R_\mathrm{in}) \tag{12-80}$$

Approximately the same bias current flows in both inputs, so the compensation resistor will produce the same voltage drop at the noninverting input as appears at the inverting input. Because the operational amplifier inputs are differential, the net output voltage is zero.

The method of Fig. 12-43C is used where the source of the DC offset potential in the output signal is due to the input bias currents.

If the input signal contains an undesired DC component, the DC component will also create an amplifier output error. Depending on

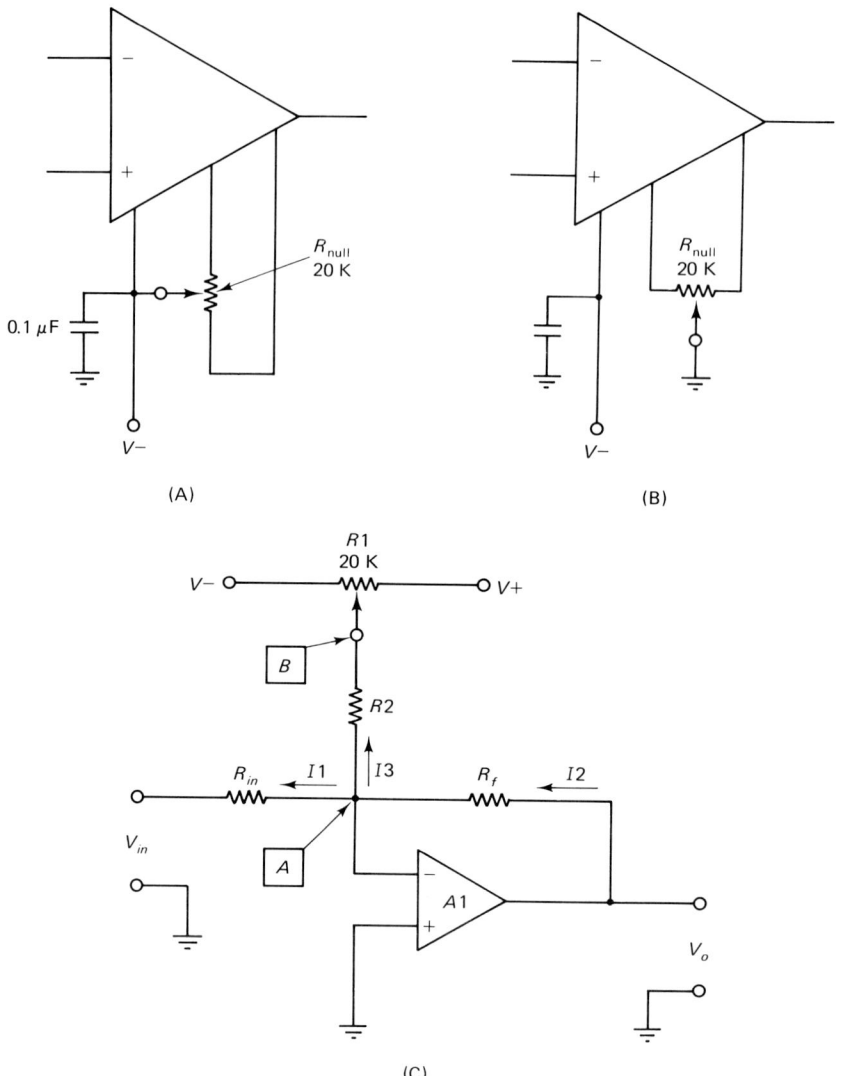

Fig. 12-44 Offset null circuits: (A) Using null terminals, (B) alternate offset null method, and (C) universal offset null.

the specific situation, we may need considerably greater range of control than is offered by Fig. 12-43C. For this purpose we turn to one of several external offset null circuit techniques.

Figure 12-44 shows two methods for nulling output offsets, regardless of whether the source is bias currents, other op-amp defects, or an input signal DC component. Figure 12-44A shows the use of the offset null terminals that are found on some operational amplifiers. A potentiometer is placed between the terminals, while the wiper is connected to the $V-$ power supply. This potentiometer is adjusted to produce the null required; the input terminals are shorted together at ground potential, and the potentiometer is adjusted to produce 0-V DC output. In the LM-101/201/301 family of devices the wiper terminal of the potentiometer is connected to common rather than to $V-$ (see Fig. 12-44B).

Figure 12-44C shows a nulling circuit that can be used on any operational amplifier, inverting or noninverting, except the unity-gain noninverting follower. A counter current ($I3$) is injected into the summing junction (point A) of a magnitude and polarity such as to cancel the output offset voltage. The voltage at point B is set to produce a null offset at the output. The output voltage component due to this voltage is

$$V_o' = -V_B(R_f/R2) \qquad (12\text{-}81)$$

If finer control over the output offset is required, then the potentiometer can be replaced with another resistor voltage divider consisting of a low-value potentiometer (e.g., 1 kΩ) in series with two fixed resistors, one connected to each end of the potentiometer. Thus, the potentiometer resistance is only a fraction of the total voltage-divider resistance. In some cases, narrower limits are set by using zener diodes or three-terminal IC voltage regulators at each end of the potentiometer to reduce the range of $V-$ and $V+$.

Thermal Drift

Another category of DC error is thermal drift. This error is usually given in the data sheets relative to input conditions. The specification is usually related to the change of input offset voltage per degree Celsius of temperature change. Typical figures for common operational amplifiers are in the range one to five microvolts per degree Celsius (1–5 μV/°C), with typical values around 3 μV/°C. Keep in mind, however, that this specification is usually an expression of drift in steady-state circuits and may not accurately represent drift under dynamic situations. These conditions may well be a lot higher than the drift given in the specification sheet for a device. There are actually two sources of drift in the input circuit; both current and voltage are at fault.

A potential source of increased thermal drift is in the use of resistor networks to null offset voltages (see previous section and Fig. 12-44C). The current passing through the network adds to the drift because, as the resistors change value in response to temperature changes, the op-amp sees a varying current as a valid input signal. The overall drift performance is thus made worse. Use of low temperature coefficient resistors helps this situation. In some cases, an op-amp manufacturer will recommend an offset null circuit that is different from those described in the previous section. Such circuits are preferred because they are generally designed to temperature-compensate the process for a specific device. Although the null methods shown previously will work well for nulling and are reasonable in stable temperature (or noncritical) situations, they may fall down in cases where temperature varies, because they unbalance input offset currents.

The input offset current of an op-amp can result in an offset voltage. A solution to the problem shown in that discussion is the use of a compensation resistor between the noninverting input and ground. That resistor has a value equal to the parallel combination of feedback and input resistances. Although couched in terms of the inverting follower circuit in the previous discussion, the actual goal is to ensure that both inputs ($+$IN and $-$IN) look out to the same resistance to ground. Thus, the differential nature of the op-amp inputs tends to cancel out the effects of the voltage drop created by bias currents flowing in external resistances.

The same method also works to improve thermal drift. The bias current is a function of temperature, hence thermal drift is due to bias current changes as a function of temperature change. These effects are smoothed out when both inputs see the same resistance. It is sometimes important to use identical, low-temperature-coefficient resistors in the op-amp input and feedback circuits. In other cases, thermistors and/or PN junction devices are used to temperature-compensate circuits.

Another means for reducing overall thermal drift is to reduce thermoelectric sources in the circuit. A phenomena called the Seebeck effect exists when dissimilar metals are joined together. If two metals with different quantum work functions are joined together in a junction, a voltage is generated between them that is proportional to the junction temperature. The Seebeck effect is used to form temperature sensors called thermocouples (see Chapter 4). Ordinary copper wire or printed wiring-board tracks produce thermoelectric voltages of 120 to 60 μV/ΔT when brought into contact with the Kovar^R leads used on integrated circuits. Lead or tin solder reduces the effect to 1–5 μV/ΔT, and cadmium-based solders to less than 0.5 μV/ΔT.

For very low level signals, especially where high gain is also present, several steps are taken to reduce thermal drift. First, operate the device at the lowest possible internal junction temperature. Two methods are useful in this respect. One is to use a heatsink (where possible) on the IC device. The other is to operate the temperature-sensitive stage at the lowest practical DC power supply voltages; for example, at ± 6 V DC instead of ± 15 V DC. Second, ensure that both amplifier inputs see the same resistance to ground (use a compensation resistor). Third, make all resistors in the circuit low-temperature-coefficient types. Fourth, eliminate thermoelectric sources. Fifth, stabilize the ambient temperature. Finally, if necessary use thermistor or PN junction devices to temperature-compensate the circuit.

Although drift may not be too important in some circuits, it can become critical where low-level signals are being processed. If an amplifier has a drift of 10 $\mu V/°C$ and the circuit is expected to maintain its performance over a large temperature range, a significant drift component may exist when low input signals are processed. For example, a temperature change of 20°C will result in a 200-μV error, or in other terms, an error of 20 percent of a 1-mV input signal.

AC Frequency Stability

Operational amplifiers are subject to spurious oscillations, especially those that are not internally frequency compensated. Figure 12-45 shows the plot of the open-loop phase shift versus frequency for a typical operational amplifier. From DC to a certain frequency there is essentially zero phase-shift error, but above that breakpoint the phase error increases rapidly. This change is due both to the internal resistances and capacitances of the amplifier acting as an RC phase-shift network and to the phase shift of the feedback network. This shift is called the propagation phase shift. At some frequency F the propagation phase-shift error reaches 180°, which when added to the 180° inversion that is normal in an inverting follower amplifier, adds up to the 360° phase shift that satisfies Barkhausen's criteria for oscillation. At that frequency the amplifier will become an oscillator.

In some cases, we may need to use a variant of the methods shown in Fig. 12-46. In Fig. 12-46A we see lead compensation. If the operational amplifier is equipped with compensation terminals (usually either pins 1 and 8 or 1 and 5, on standard packages), then connect a small-value capacitor (20–100 pF) as shown. An alternate

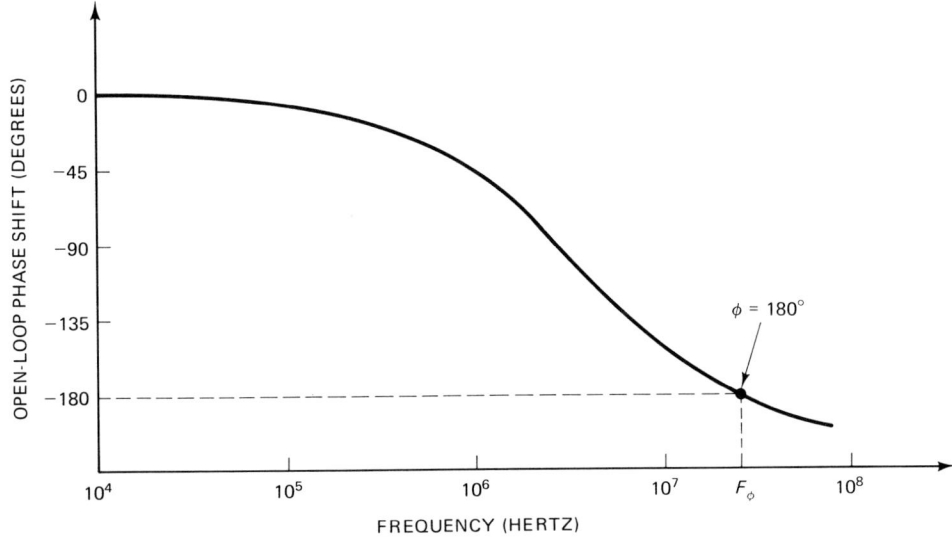

Fig. 12-45 Phase shift versus frequency.

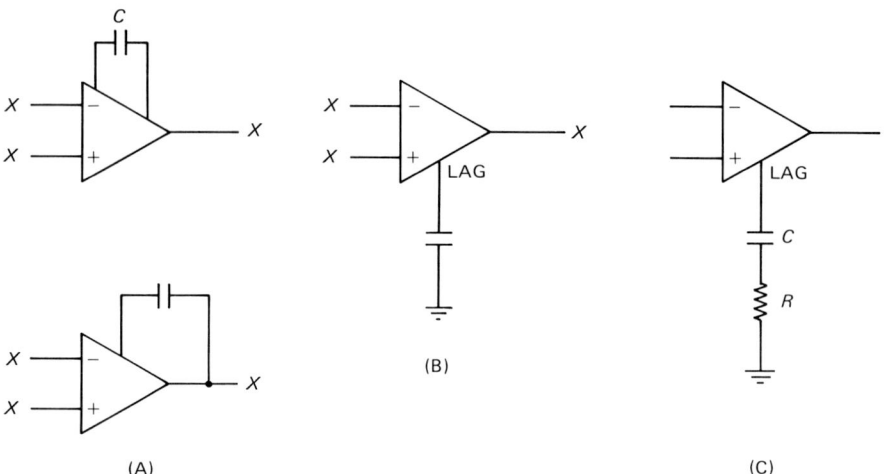

Fig. 12-46 Frequency compensation methods: (A) Lead, (B) capacitor lag, and (C) resistor-capacitor lag.

scheme is to connect the capacitor from a compensation terminal to the output terminal.

The recommended capacitance in manufacturers' specification sheets is for the unity-gain noninverting follower configuration. For a gain follower the capacitance is reduced by the feedback factor B.

$$C = C_m B \tag{12-82}$$

where C is the required capacitance, C_m the recommended unity-gain capacitance, and B is the feedback factor $R_{in}/(R_{in} + R_f)$.

Lag compensation is shown in Fig. 12-46B and 12-46C. In this case, either a single capacitor (Fig. 12-46B) or a resistor-capacitor network (Fig. 12-46C) is connected from the compensation terminal to ground. A related method places the resistor–capacitor series network between the inverting and noninverting input terminals.

The object of these methods (see Fig. 12-47) is to reduce the high-frequency loop gain of the circuit to a point where the total loop gain is less than unity at the frequency where the 180° phase shift occurs. The amount of compensation required to accomplish this goal determines the maximum amount of feedback that can be used without violating the stability requirement.

Several factors conspire toward allowing an operational amplifier to oscillate at times when this is highly undesirable. Quite often these oscillations occur at frequencies far in excess of the passband of the associated circuit. Two of these factors, both of which can be overcome, are positive feedback via the DC power supply and spurious internal phase shift.

Figure 12-48 shows the input and output waveforms from a unity-gain inverting follower amplifier ($R_{in} = R_f$) that used no decou-

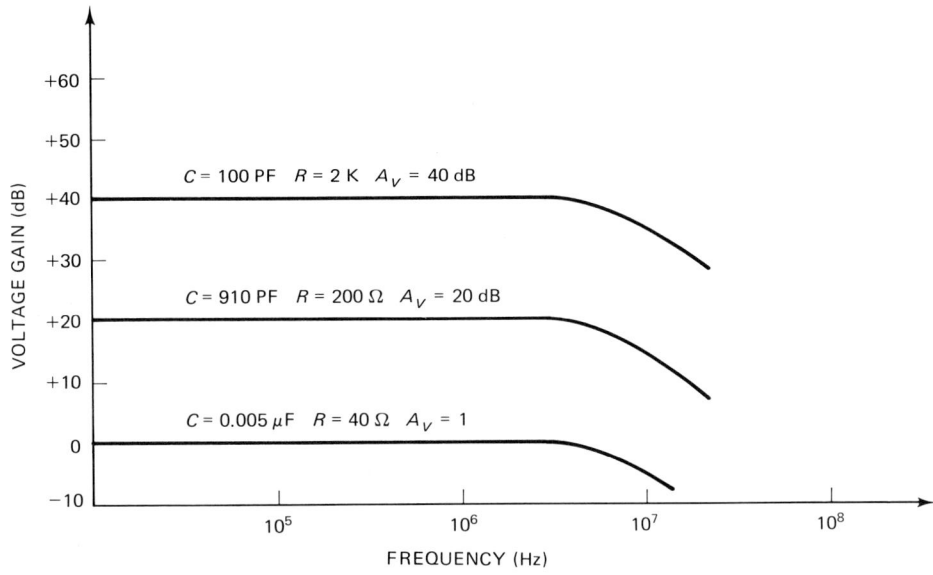

Fig. 12-47 Gain with various *RC* combinations.

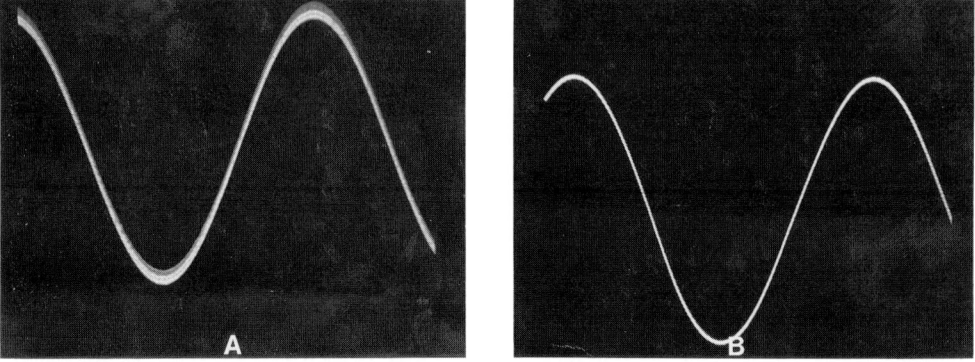

Fig. 12-48 Waveforms of unity-gain inverting follower without decoupling capacitors, showing high-frequency oscillation in output.

pling of $V-$ and $V+$ terminals. The output waveform shows a high-frequency oscillation superimposed on the output sine wave. The cause of this problem is a high AC impedance to ground through the DC power supply terminals. The use of power supply decoupling helps somewhat for this problem, and it is considered poor engineering practice to use an uncompensated operational amplifier without those decoupling capacitors.

Diagnosing and Fixing Instability

The process of finding and fixing instability sources in linear IC amplifier circuits is sometimes believed (erroneously) to rely a little on magic, but in reality the process is relatively straightforward. There may be, however, a certain amount of cut-and-try involved. But first, be sure that the problem you are trying to solve is indeed an instability in the circuit, and not an external problem. Three kinds of external problem can mimic op = amp instability. One is a defective source signal. For example, a low-frequency (sub-Hertzian) "motorboating" oscillation may be due to a high AC impedance to ground in the amplifier DC power supply, but it may also be due to an oscillation in the signal source. Another problem of this sort is 60-Hz AC hum caused by a broken shield or ground wire on the input line. In fact, don't look for any other cause if the oscillation is exactly 60 Hz (the AC power line frequency) until this possibility is checked.

The second external problem is noisy or defective power supply lines. Examine the $V-$ and $V+$ lines on an oscilloscope to be sure that the power is clean. A 120-Hz oscillation could indicate an excessive ripple condition in a DC power supply. Additional ripple filtering or a voltage regulator in each DC line ($V-$ and $V+$) are usually the successful solutions.

Finally, we sometimes see electromagnetic interference in IC amplifies. Strong local fields, such as RF interference from a local broadcasting station, may get into the circuit and appear to be high-frequency oscillations. Such interference may be direct radiation, or may be carried into the circuit on an input line, an output line, or a DC power supply line. Figure 12-49 shows an op-amp circuit containing several anti-interference techniques. The small value capacitors should be used with caution because they may attenuate signals at higher frequencies. Also, they may add to the phase shift of signals in the feedback network and thereby cause an oscillation condition even while eliminating the interference.

Once the obvious noninstabilities have been eliminated, it is necessary to home in on possible causes of actual instabilities. First, examine the circuit for any unintended common impedances between input and output circuits, or between stages in a cascade chain of amplifiers. Two areas of concern are often found: the DC power supply lines and the ground. The power supply circuits are usually handled by the bypass capacitors discussed earlier.

Common impedances in the ground system, usually called ground loops, are suppressed by using large conductive ground planes (usually on the component side of the printed circuit board) rather than thin point-to-point conductors. Diagnosis and resolving of ground loops is subject to design and a few rules, beyond which a cut-and-try approach is advised. The cut-and-try method involves moving ground connections and/or actual components around the board to find cold spots on the ground plane. Some rules to follow are

1. Use large ground-plane surfaces where possible.
2. Use single-point "star" grounding rather than either random grounding or daisy-chain grounding.

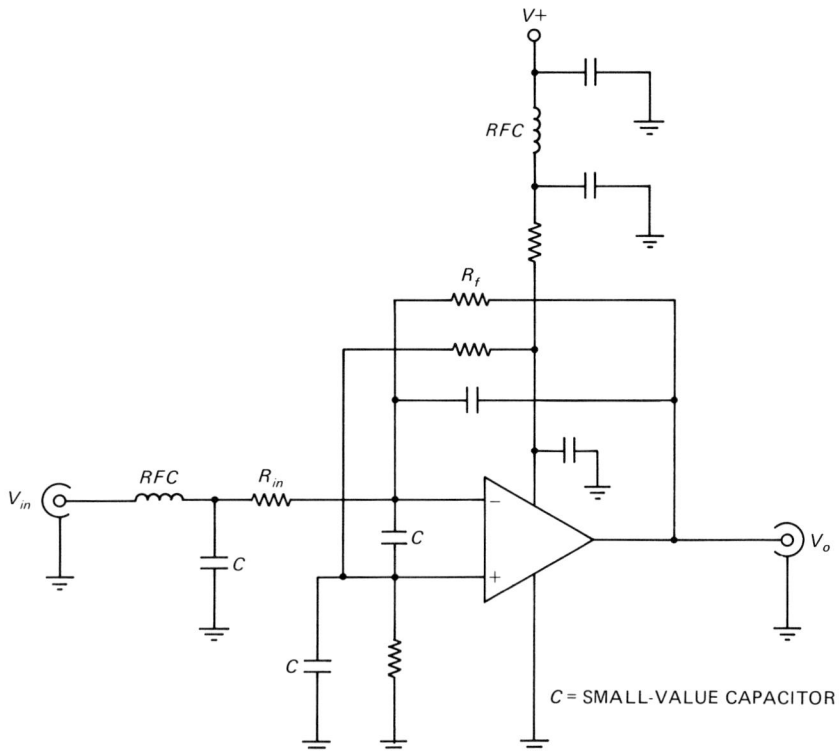

Fig. 12-49 Heavily compensated operational amplifier with EMI protection in place.

3. Keep the DC power supply and AC signal lines separate except at the single "star" point.

Another source of difficulty, especially (but not exclusively) in high-gain circuits, is erroneous circuit layout. Keep input and output circuits separated to the greatest extent possible. A source of unintended feedback is sometimes found when several stages are in cascade. In response to printed wiring-board size constraints some designers double a circuit back on itself on the board. That arrangement places the output circuit components next to either the input circuit components or the intermediate stages. Such a layout induces radiation feedback, and it is especially likely if the total propagation feedback is 360° (e.g., as in noninverting amplifiers, or circuits with an even number of inverting amplifiers).

Once layout problems, common impedances, and other causes are ruled out, it is time to evaluate the linear IC amplifier circuit itself for instability problems. Two distinct areas must be investigated: feedback network and the rest of the circuit. Although not an absolute indicator, the frequency of oscillation often tells you where the problem is located. Measure the oscillation frequency (F_o) and compare it to the frequency at which the amplifier gain drops to unity (F_t, which is found in the specifications sheet or data book). If F_o is close to F_t, it

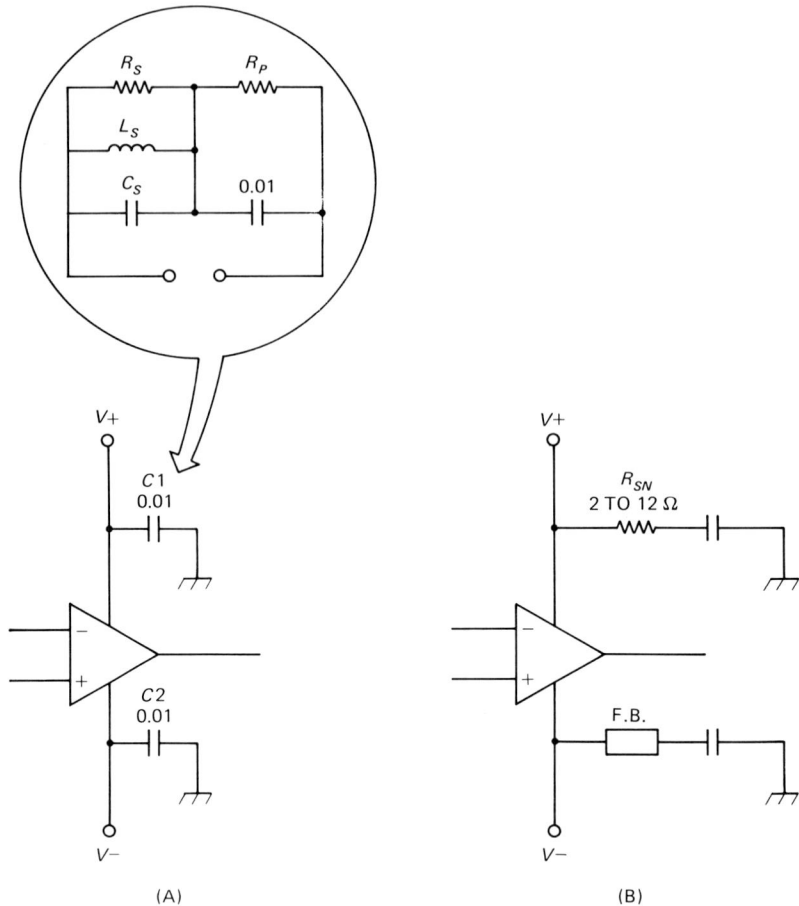

Fig. 12-50 (A) Op-amp decoupling capacitor and equivalent circuit at VHF, (B) ferrite bead and losser resistor decoupling.

is probable that the problem is in the feedback network. A diagnostic aid is to temporarily increase the amplifier closed-loop gain by a factor of 2–10, and then observe the effect on stability. If the oscillation ceases, or F_o drops appreciably, the problem is probably in the feedback network. If neither of these events occurs, the problem is most likely in another part of the circuit (the feedback loop is exonerated).

Only a few linear IC devices (and practically no op-amps) operate into the VHF region. Yet op-amps and other low-frequency devices sometimes oscillate in the 50- to 200-MHz region, far beyond the bandwidth of the device. The cause of these VHF parasitic oscillations is the output power amplifier in the IC device. The effect is especially likely when resonances are present.

One source of a stray resonance that leads to parasitics is the use of the wrong capacitor type on the DC power supply terminals. In Fig. 12-50A, for example, disk ceramic capacitors are used for $V-$ and $V+$ bypassing. Such capacitors have significant stray capacitance and inductance (see inset) that tend to resonate in the VHF region. Fixes

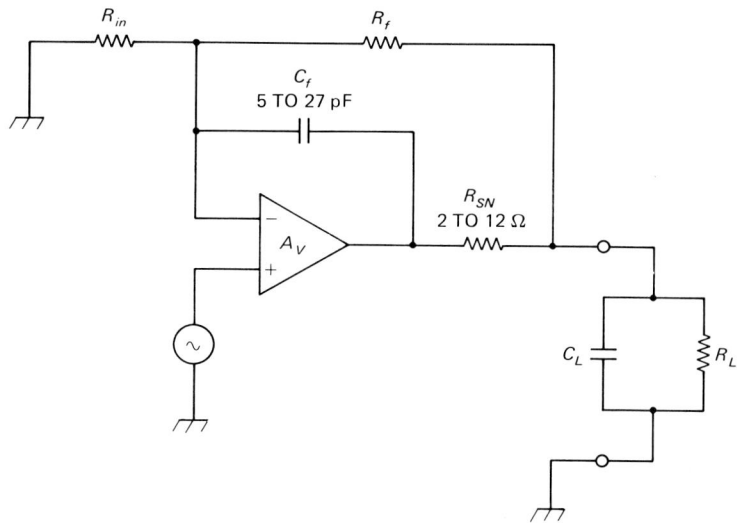

Fig. 12-51 Losser resistor and capacitor used to snuff out oscillations.

for test problems are shown in Fig. 12-50B. The $V+$ lead bypass capacitor uses a 2- to 12-Ω "snubber" resistor in series to lower the Q of the stray resonant LC circuit elements. If the Q is lowered sufficiently, the LC circuit elements will be unable to cause an oscillation. In the $V-$ lead of the same op-amp we see an alternate fix. A ferrite bead is slipped over the lead to the bypass capacitor. This bead acts like an RF choke at VHF frequencies, but it is practically transparent to low frequencies.

An often hidden source of feedback problems is capacitance in the load of the amplifier. Such capacitance adds to the propagation phase shift of the feedback network, possibly causing oscillation. If a load is known to be capacitive, then identification of the problem is easy. But it often happens that other sources of capacitance are found in a circuit. For example, shielded or coaxial cables have a high value of capacitance per unit of length. Similarly, some chassis or in-line connectors offer significant capacitances, and circuit stray capacitance may also be significant. A circuit fix that isolates a capacitance load from an IC amplifier output is shown in Fig. 12-51. A small feedback capacitor reduces the closed-loop gain at frequencies where oscillation is likely to occur while essentially not affecting lower frequencies. Isolation is obtained by the series snubber resistor R_{SN}.

13

Nonoperational IC Linear Amplifiers

The operational amplifier is a simple voltage amplifier with the simple transfer function $A_v = V_o/V_{in}$. While that type of device is the most commonly used form of IC linear amplifier, there are cases where another form of linear amplifier is needed. In this chapter we will take a look at two other popular forms of IC linear amplifiers: the operational transconductance amplifier (OTA) and the current difference amplifier (CDA), also sometimes called the Norton amplifier. These devices are not likely to replace the operational amplifier in the marketplace, but they do find their own niche in the integrated electronics marketplace.

OPERATIONAL TRANSCONDUCTANCE IC AMPLIFIERS

The operational transconductance amplifier (OTA) is based on a transfer function that relates an output current to an input voltage. In other words,

$$G_m = dI_o/dV_{in} \tag{13-1}$$

where G_m is the transconductance in mhos or micromhos, I_o the output current, and V_{in} the input voltage.

The operational transconductance amplifier equivalent circuit is shown in Fig. 13-1. The differential input circuit is similar to the input

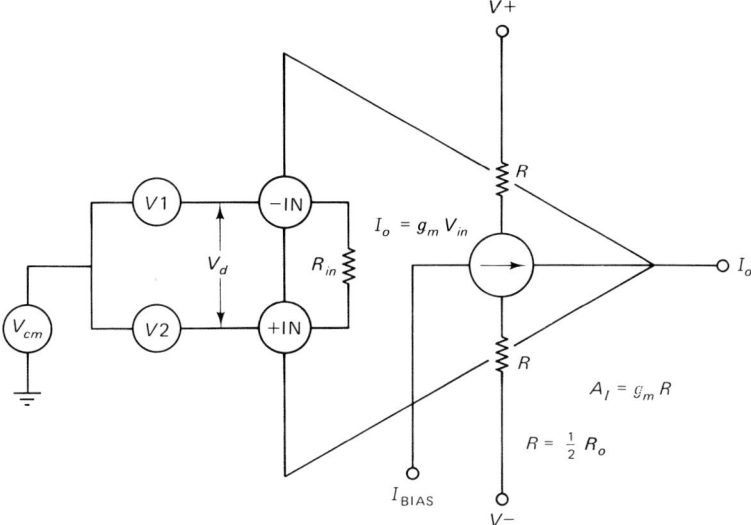

Fig. 13-1 Operational transconductance amplifier equivalent circuit.

circuit of the operational amplifier because each are differential voltage inputs ($-$IN and $+$IN). The input voltages are differential signal V_d (i.e. $V2 - V1$) and common mode signal V_{cm}. The output side of the amplifier, however, is a current source that produces an output current I_o that is proportional to the gain and the input voltage. The current gain (A_{gm}) of this circuit is a function of the transconductance (I_o/V_{in}) and the load resistance R:

$$A_{gm} = G_m R \qquad (13\text{-}2)$$

where A_{gm} is the gain, G_m is the transconductance (I_o/V_{in}), and R is the load resistance (one-half the output resistance R_o).

G_m and R must be expressed in equivalent reciprocal units. In other words, when G_m is in siemens (i.e., mhos) then R is in ohms. Millisiemens–milliohms and microsiemens–microohms are also paired.

Perhaps the most common commercial versions of the OTA are the RCA CA-3080, CA-3080A, and CA-3060 devices. The CA-3080 devices are available in the eight-pin metal IC package using the pinouts shown in Fig. 13-2. The CA-3080 devices will operate over DC power supply voltages from ± 2 to ± 15 V, with adjustable power consumption of 10 to 30 mW. The gain is from 0 to the product $G_m R$. The input voltage spread is ± 5 V. The bias current can be set to as high as 2 mA.

Note that the pinouts for the CA-3080 device are industry standard operational amplifier pinouts, except for the bias current applied

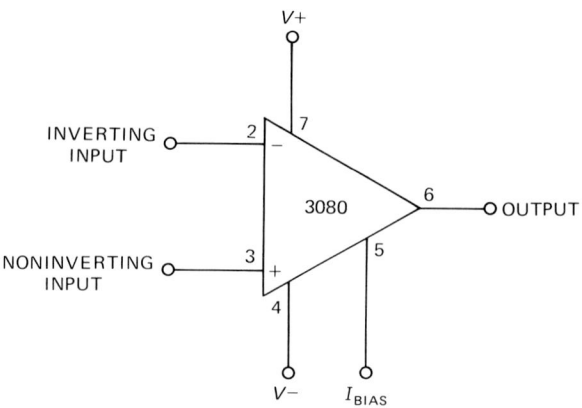

Fig. 13-2 CA-3080 OTA.

to pin number 5:

- $V-$ on pin 4
- $V+$ on pin 7
- Inverting input ($-$IN) on pin 2
- Noninverting input ($+$IN) on pin 3
- Output on pin 6

The operating parameters of the operational transconductance amplifier are set by the bias current (I_{bias}). For example, on the CA-3080 device the transconductance is 19.2 times higher than the bias current.

$$G_m = 19.2 I_{\text{bias}} \tag{13-3}$$

where G_m is in milliohms and I_{bias} is in milliamperes.

In many actual design cases you will know the required value of G_m from knowledge of I_o/V_{in}, so the G_m required can be set by adjusting the bias current. In those cases, the I_{bias} is found by rewriting expression (13-3).

$$I_{\text{bias}} = G_m/19.2 \tag{13-4}$$

The CA-3080 output resistance of the device is also a function of the bias current.

$$R_o = 7.5/I_{\text{bias}} \tag{13-5}$$

where R_o is the output resistance in megohms (MΩ) and I_{bias} is the bias current in milliamperes (mA).

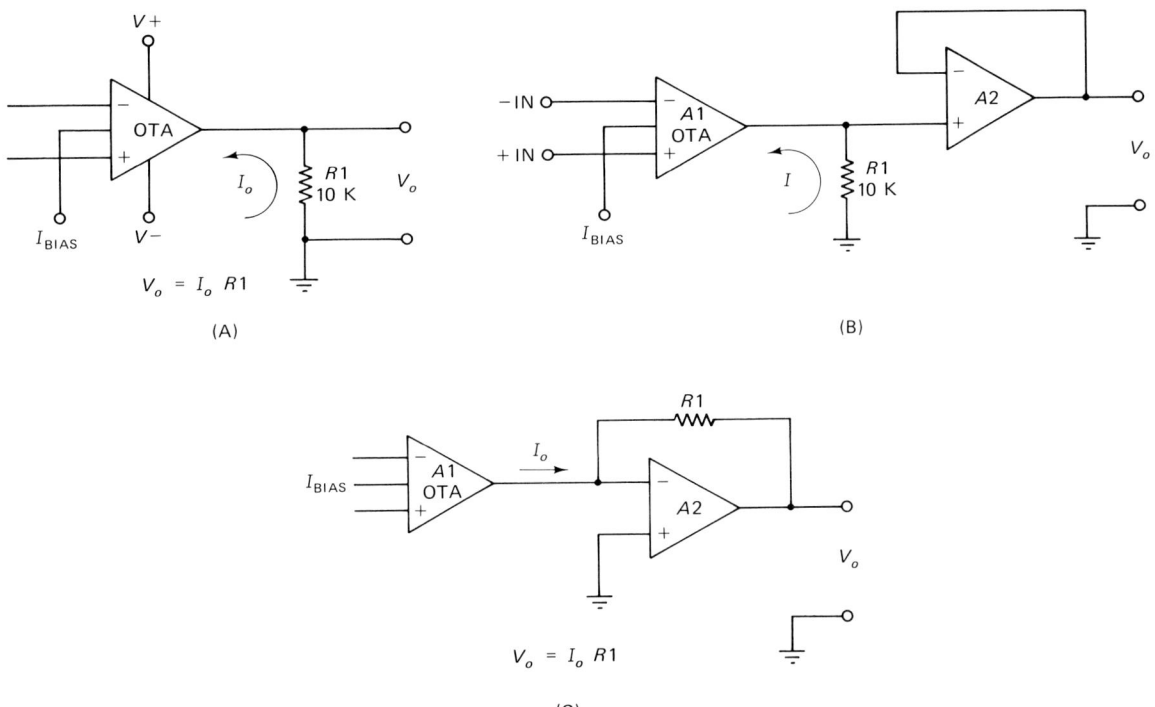

Fig. 13-3 OTA circuit configuration to produce an output voltage signal from the normal output current signal.

VOLTAGE AMPLIFIER FROM THE IC OTA

The OTA is a current-output device, but it can be used as a voltage amplifier when one of the circuit strategies of Fig. 13-3 is used. The simplest method is the resistor load shown in Fig. 13-3A. Because the output of the OTA is a current (I_o), we can pass this current through a resistor ($R1$) to create a voltage drop. The value of the voltage drop (and the output voltage V_o) is found from Ohm's law.

$$V_o = I_o R1 \tag{13-6}$$

A problem with this circuit is that the source impedance is very high, being equal to the value of $R1$. In the example shown in Fig. 13-3A, the output impedance is 10 Ω. This problem can be overcome by adding a unity-gain noninverting operational amplifier such as A2 in Fig. 13-3B. The output voltage in this case is the same as for the nonamplified version $I_o R1$, although the output impedance is very low.

Another form of low-impedance output circuit is shown in Fig. 13-3C. This form uses the inverting follower configuration of the operational amplifier ($A2$). The output voltage is the product of the OTA output current (I_o) and the op-amp feedback resistor ($R1$).

$$V_o = I_o R1 \tag{13-7}$$

In both Figs. 13-3B and 13-3C the output impedance is equal to the operational amplifier output impedance, which is typically something less than 100 Ω.

OTA APPLICATIONS

An analog multiplier is a circuit that produces a voltage that is the product of two input voltages.

$$V_o = rV_xV_y \qquad (13\text{-}8)$$

where V_o is the output voltage, V_x the voltage applied to the X-input, V_y the voltage applied to the Y-input, and r a proportionality constant.

There are many applications for the multiplier circuit, although some of them are now generally performed in a digital computer or processor. Immediately one can see instrumentation applications, even in this era of computers. Also possible are amplitude modulation and demodulation tasks for analog multipliers. A frequency-compensated OTA amplifier is shown in Fig. 13-4.

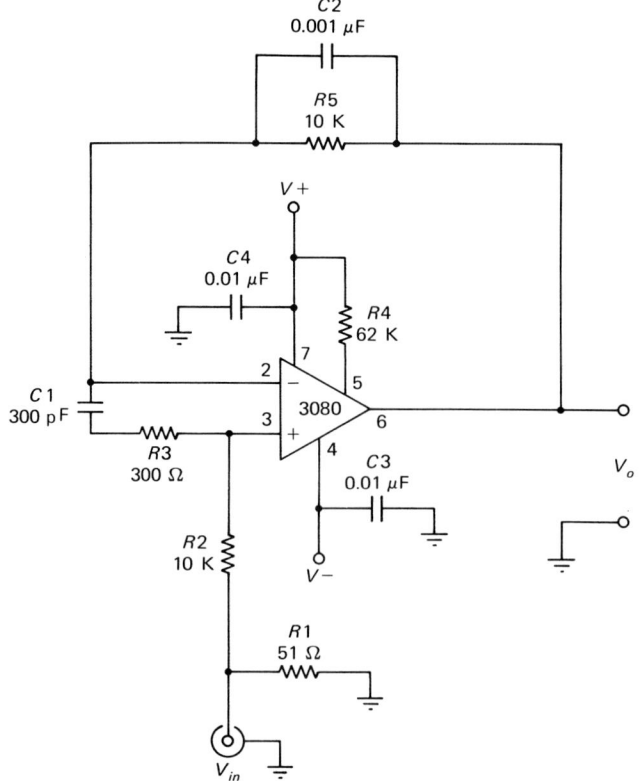

Fig. 13-4 Frequency-compensated OTA amplifier.

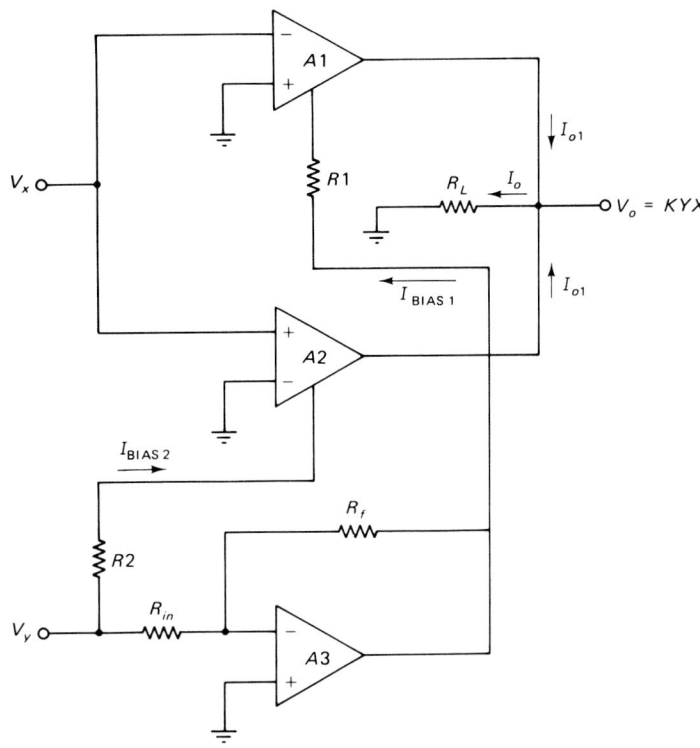

Fig. 13-5 OTA analog XY multiplier circuit.

Operational transconductance amplifiers can be used to make two-quadrant and four-quadrant analog multipliers. Consider Fig. 13-5, which is an analog XY multiplier based on the CA-3060 quad OTA. Recall from Eq. (13-1) that

$$G_{\mathrm{m}} = I_{\mathrm{o}}/V_{\mathrm{in}} \qquad (13\text{-}9)$$

Therefore,

$$I_{\mathrm{O}1} = -V_x G_{\mathrm{M}1} \qquad (13\text{-}10)$$

and

$$I_{\mathrm{O}2} = +V_x G_{\mathrm{M}2} \qquad (13\text{-}11)$$

Because the value of R_{o} for each OTA is very large compared with the load, we can simply sum the two output currents.

$$I_{\mathrm{o}} = I_{\mathrm{O}2} + I_{\mathrm{O}1} \qquad (13\text{-}12)$$

by $V_{\mathrm{o}} = I_{\mathrm{o}} R_{\mathrm{L}}$:

$$V_{\mathrm{o}} = (I_{\mathrm{O}2} + I_{\mathrm{O}1}) R_{\mathrm{L}} \qquad (13\text{-}13)$$

$$V_{\mathrm{o}} = [(+V_x)(G_{\mathrm{M}2}) + (-V_x)(G_{\mathrm{M}1})] R_{\mathrm{L}} \qquad (13\text{-}14)$$

$$V_{\mathrm{o}} = (G_{\mathrm{M}2} - G_{\mathrm{M}1})V_x R_{\mathrm{L}} \qquad (13\text{-}15)$$

Recall that $G_m = KI_{bias}$. For amplifier $A2$ we know that

$$I_{bias} = [(V-) + (V_x)]/R1 \qquad (13\text{-}16)$$

hence,

$$G_{M2} = K[(V-) + (V_x)] \qquad (13\text{-}17)$$

Using a similar line of reasoning we arrive at

$$G_{M1} = K[(V-) + (V_y)] \qquad (13\text{-}18)$$

Combining Eqs. (13-16), (13-17), and (13-18) yields

$$V_o = V_x KR_L([(V-) + V_x] - [(V-) - (V_y)]) \qquad (13\text{-}19)$$

or, after simplifying terms,

$$V_o = 2KR_L V_x V_y \qquad (13\text{-}20)$$

Equation (13-20) is the transfer equation for Fig. 13-5. It has the same form as Eq. (13-8) in which $r = 2KR_L$.

CURRENT DIFFERENCE AMPLIFIERS

The current difference amplifier (CDA), also called the Norton amplifier, is another nonoperational linear IC amplifier that performs similarly to the op-amp, but not exactly the same. The CDA has certain features that make it uniquely useful for certain applications. One place where the CDA is more useful than the operational amplifier is in circuits that process AC signals, but are limited to a single-polarity DC power supply. An example is automotive electronics equipment, limited to a single 12- to 14.4-V DC battery power supply that uses the car chassis for negative common return. There are other cases where the linear IC amplifier is but a minor feature of the circuit, most of which operates from a single DC power supply. It might be wasteful in such circuits to use the operational amplifier. We would have to either bias the operational amplifier with an external resistor network or provide a second DC power supply.

The normal symbol for the CDA is shown in Fig. 13-6. This symbol looks much like the regular op-amp symbol, except that a current source is placed along the side opposite the apex. This symbol is typically used for several products such as the LM-3900 device, which is a quad Norton amplifier. You may sometimes find schemat-

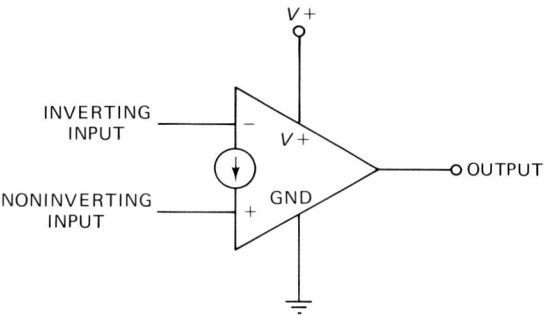

Fig. 13-6 Current difference amplifier symbol.

ics where the op-amp symbol is used for the CDA, but that is technically incorrect usage.

CDA CIRCUIT CONFIGURATION

The input circuit of the CDA differs radically from the operational amplifier. Recall that the op-amp used a differential input amplifier driven from a constant current source supplying the collector-emitter current. The CDA is quite different, however, as can be seen in Fig. 13-7. The overall circuit of a typical CDA is shown in Fig. 13-7A; an alternate form of the input circuit is shown in Fig. 13-7B. Transistor Q7 in Fig. 13-7A forms the output transistor and Q5 is the driver. Both the NPN output transistor and the PNP driver transistor operate in the emitter follower configuration. Transistors Q4, Q5, Q6, and Q8 are connected to serve as current sources. The input transistor is Q3, and it operates in the common emitter configuration. The base of Q3 forms the inverting (-IN) input for the CDA.

The noninverting input of the CDA is formed with a current-mirror transistor, Q1 (transistor Q1 in Fig. 13-7A is diode-connected and serves exactly the same function as diode D1 in Fig. 13-7B). The dynamic resistance offered by the current-mirror transistor (Q2) is given by

$$r = 26/I_b \qquad (13\text{-}21)$$

where r is the dynamic resistance of Q2 in ohms and I_b is the base bias current of Q3 in milliamperes.

Equation (13-21) is used only at or near normal room temperature because I_b will vary with wide temperature excursions. For most common applications, however, the room temperature version of the equation will suffice. Data sheets for specific current difference amplifiers give additional details for amplifiers that must operate outside the

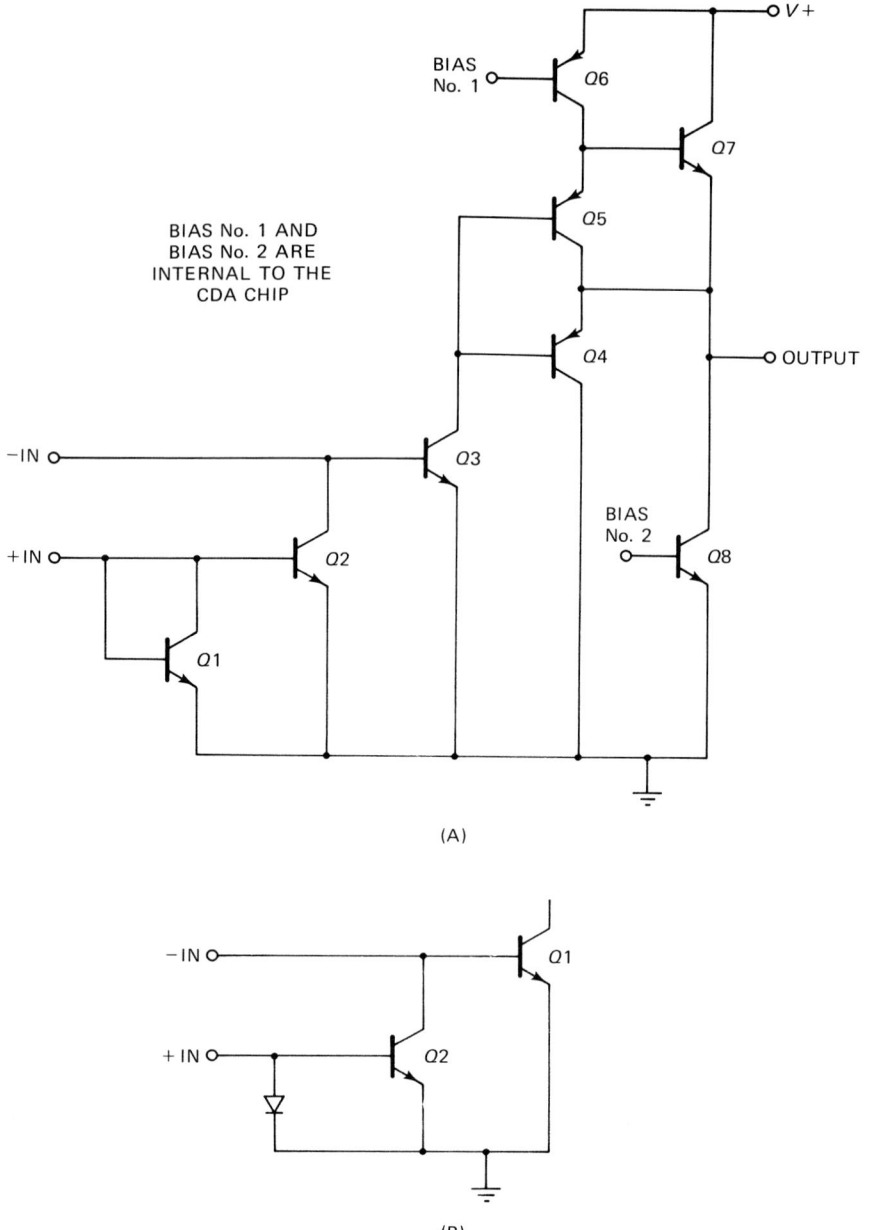

Fig. 13.7 (A) CDA internal circuit and (B) current-mirror input circuit.

relatively narrow temperature range of 25 to 30 °C specified for the
simplified equation.

CDA INVERTING FOLLOWER CIRCUITS

Like operational amplifiers, the CDA can be configured in either
inverting or noninverting follower configurations. The inverting fol-

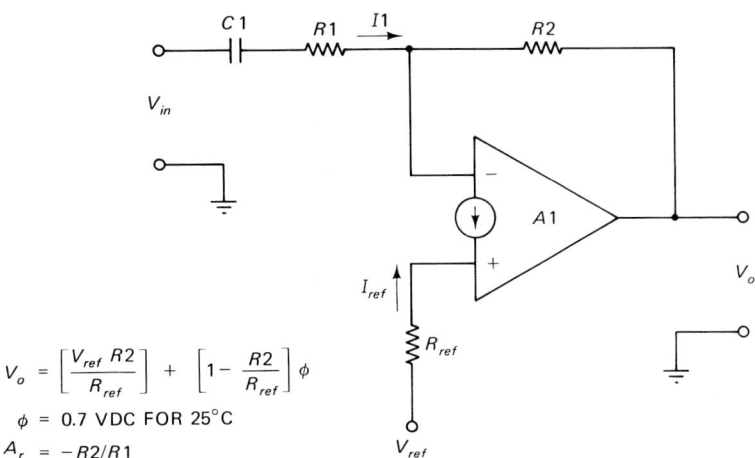

$$V_o = \left[\frac{V_{ref}\,R2}{R_{ref}}\right] + \left[1 - \frac{R2}{R_{ref}}\right]\phi$$

$$\phi = 0.7 \text{ VDC FOR } 25°C$$

$$A_r = -R2/R1$$

Fig. 13-8 AC-coupled CDA inverting amplifier.

lower is shown in Fig. 13-8. In many respects this circuit is very similar to that of operational amplifiers. The voltage gain of the circuit is set approximately by the ratio of the feedback to the input resistor.

$$A_v = -R2/R1 \tag{13-22}$$

where A_v is the voltage gain, $R2$ the feedback resistance, and $R1$ the input resistance.

(Note: $R1$ and $R2$ are in the same units. The minus sign indicates that a $180°$ phase reversal occurs between input and output signals.)

We must provide a bias to the current-mirror transistor (i.e., Q2 in Fig. 13-7A), so resistor R_{ref} is connected in series with the noninverting input of the CDA and a reference voltage source V_{ref}. In many practical circuits the reference voltage source is merely the $V +$ supply used for the CDA. In other cases, however, some other potential might be required, or alternatively the reference current is required to be regulated more tightly (or with less noise) than the supply voltage. Ordinarily the reference current is set to some convenient value between 5 and 100 μA. For $V +$ power supply values of $+ 12$ VDC, for example, it is common to find a 1-MΩ resistor used for R_{ref}. In that case, the input reference current is $I_{ref} = 12/1,000,000 = 12 \; \mu$A.

A constraint placed on CDAs is that the input resistor ($R1$) used to set gain must be high compared with the value of the current-mirror dynamic resistance [Eq. (13-21)]. The CDA becomes nonlinear (i.e., distorts the input signal) if the input resistor value approaches the current-mirror resistance r. In that case, the voltage gain is not $-R2/R1$, but rather

$$A_v = R2/(R1 + r) \tag{13-23}$$

where A_v is the voltage gain, $R2$ the feedback resistance, $R1$ the input resistance, and r the current mirror resistance.

Equation (13-23) essentially reduces to Eq. (13-22) when we can force $R1$ to be very much larger than r. This goal is easily achieved in most circuits because r is tiny. [Work a few examples with normal bias currents in Eq. (13-21).]

The output voltage of the CDA will exhibit an offset potential even when the AC input signal is zero. This potential is given by

$$V_o = \left[\frac{V_{ref}R2}{R_{ref}} + 1\right] - \left[\frac{R2}{R_{ref}}\right]\Phi \qquad (13\text{-}24)$$

where V_o is the output potential in volts, V_{ref} the reference potential in volts (usually V +), $R2$ the feedback resistance in ohms, R_{ref} the current-mirror bias resistance in ohms, and Φ a temperature-dependent factor (0.70 V for room temperature).

The capacitor in series with the input circuitry has the effect of limiting the low-end frequency response. The -3-dB cutoff frequency is a function of the value of this capacitor and the input resistance $R1$. This frequency, F, is given by

$$F = 1{,}000{,}000/2\pi R1 C1 \qquad (13\text{-}25)$$

where F is the lower-end -3-dB frequency in hertz (Hz), $R1$ the resistance in ohms (Ω), and $C1$ the capacitance in microfarads (μF).

In some CDA circuits there is also a capacitor in series with the output terminal. The purpose of that capacitor is to prevent the DC offset that is inherent in this type of circuit from affecting following circuits. The output capacitor will also limit the low-end frequency response. The same form of equation [i.e., Eq. (13-25)] is used to determine this frequency, but it uses the input resistance of the load as the R term.

As is often the case with equations presented in electronics books, Eq. (13-25) is not necessarily in the most useful form. In most cases, you will know the input resistance ($R1$) from the application. It is typically not less than 10 times the source impedance and forms part of the gain equation. Consideration of driving source impedance and voltage gain tends to determine the value of $R1$. The required low-end frequency response is usually determined from the application. You generally know (or can find out) that frequency spectrum of input signals. From the lower limit of the frequency spectrum you can determine the value of F. Thus, you will determine F and $R1$ from other considerations than the circuit. You therefore need a version of Eq. (13-25) that calculates capacitor $C1$ from knowledge of $R1$ and F. Using elementary algebra,

$$C1 = 1{,}000{,}000/2\pi R1 F \qquad (13\text{-}26)$$

All terms are as defined for Eq. (13-25).

Design Example

Design an AC amplifier with a gain of 100 based on a CDA in which the input impedance is at least 10 Ω and the -3-dB frequency response is 3 Hz or lower. Assume DC power supplies of ± 15 V DC (see Figure 13-9 for final circuit).

1. Set the reference current to the noninverting input to a value between 5 and 100 μA. Select 15 μA.

$$R3 = V/I_{\text{ref}}$$

$$R3 = (15 \text{ V DC})/(0.000015) = 1 \text{ M}\Omega$$

2. Set the gain resistors. $R1$ can be 10,000 Ω to meet the input impedance requirement.

(From $A_{\text{v}} = R2/R1$, we know that $R2 = A_{\text{v}}R1$.)

$$R2 = A_{\text{v}}R1$$

$$R2 = (100)(10,000) = 1 \text{ M}\Omega$$

3. Find the value of input capacitor $C1$ when $F_{-3 \text{ dB}}$ is 3 Hz.

$$C1 = 1,000,000/2\pi R1 F$$

$$C1 = 1,000,000/(2)(3.14)(10,000)(3) = 5.3 \ \mu\text{F}$$

Because 5.3 μF is a nonstandard value, select the next higher standard value (which is 6.8 μF).

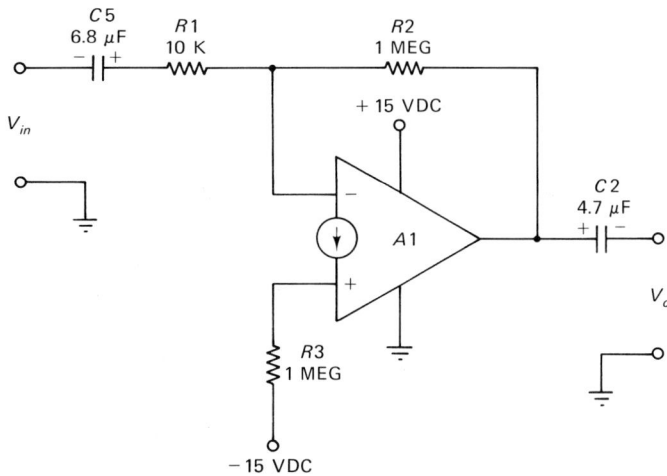

Fig. 13-9 Audio CDA amplifier circuit.

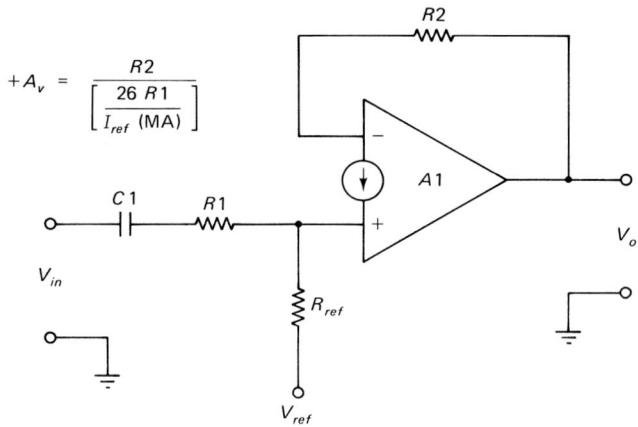

Fig. 13-10   AC-coupled noninverting CDA amplifier.

The value of output capacitor $C2$ can be arbitrarily set to 6.8 μF if the load impedance is 10,000 Ω. If the load impedance is higher than that value, use 4.7 μF or 6.8 μF. If the load impedance is very much higher than 10,000 Ω or is lower than 10,000 Ω, calculate the value using the same equation as for $C1$ but with the load resistance substituted for the input impedance.

NONINVERTING AMPLIFIER CIRCUITS

The noninverting amplifier CDA configuration is shown in Fig. 13-10. This circuit retains the reference current bias applied to the noninverting input, but rearranges some of the other components. As in the case of the inverting amplifier configuration, the noninverting amplifier uses $R2$ to provide negative feedback between the output terminal and the inverting input. Unlike the inverting CDA circuit, however, input resistor $R1$ is connected in series with the noninverting input. The gain of the noninverting CDA amplifier is given by

$$A_v = \frac{R2}{26R1/\left[I_{ref}\,(\mathrm{mA})\right]} \tag{13-27}$$

where A_v is the voltage gain, $R1$ and $R2$ the resistance in ohms (Ω), and I_{ref} the bias current in milliamperes (mA).

The reference current I_{ref} is set to a value between 5 and 100 μA. Unlike the situation in the inverting amplifier, the value of this current is partially responsible for setting the gain of the circuit. Some clever designers have used this current as a limited gain control for some CDA stages. The value of the resistor that provides the reference current (R_{ref}) is set by Ohm's law, considering the required value of reference current and the reference voltage (V_{ref}). In most common applications, the reference voltage is merely one of the supply voltages.

The value of R_{ref} is determined from

$$R_{ref} = V_{ref}/I_{ref} \qquad (13\text{-}28)$$

where R_{ref} is the reference resistor in ohms, V_{ref} the reference potential in volts, and I_{ref} the reference current in amperes. (Note: 1 μA = 0.000001 A.)

The value of input impedance is approximately equal to $R1$, provided that $R1$ is much higher than the dynamic resistance of the current mirror inside the CDA (which is typically the case). As was true in the inverting follower case, the input capacitor ($C1$) sets the low-end frequency response of the amplifier. The -3-dB frequency is given by exactly the same equation as for the inverting case [see Eqs. (13-25) and (13-26)].

Figure 13-11 shows a modification of the noninverting follower circuit that allows for a noisy reference source. This type of amplifier circuit might be used where the DC power supplies for the reference voltage are electrically noisy. Such noise could come from other stages in the circuit or from outside sources.

The purpose in Fig. 13-11 is to form a reference voltage from a resistor voltage-divider circuit consisting of $R3$ and $R4$. The value of V_{ref} will be

$$V_{ref} = (V+)R3/(R3+R4) \qquad (13\text{-}29)$$

where V_{ref} is the reference potential in volts, $V+$ the supply potential in volts, and $R3$ and $R4$ the resistance in ohms.

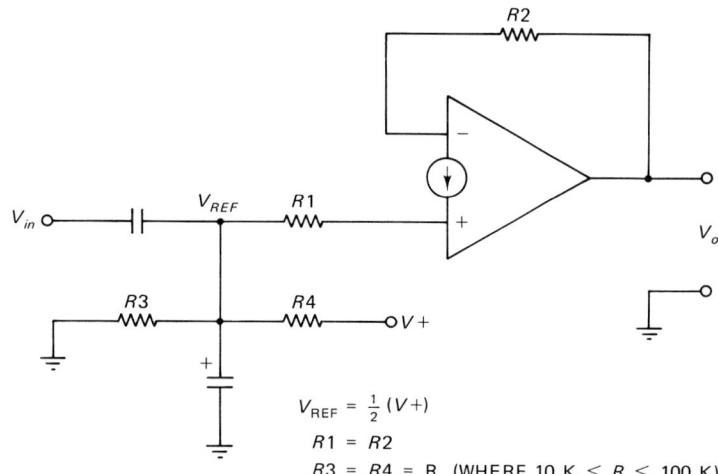

Fig. 13-11 CDA amplifier.

Inspection of Eq. (13-29) reveals that $V_{ref} = (V+)/2$ when $R3 = R4$, which is the usual case in practical circuits. The reference current is

$$I_{ref} = V_{ref}/R1 \qquad (13\text{-}30)$$

Supergain Amplifier

There is a practical limit to voltage gain using standard resistor values and standard circuit configurations (a similar problem also exists for operational amplifiers). In Figure 13-12 we see a means for overcoming the limitations. This supergain amplifier circuit forms a noninverting follower in a manner similar to the earlier circuit, except that feedback resistor R2 is driven from an output voltage divider network rather than directly from the output terminal of the CDA. The voltage gain of the circuit of Fig. 13-12 is given by

$$A_v = (R2/R1)[(R3 + R4)/R3] \qquad (13\text{-}31)$$

Capacitor $C1$ is set using the same Eq. (13-26) that was used previously, and $C2$ is set to have a capacitive reactance of $R4/10$ at the lowest frequency of operation (in other words, the low-end -3-dB point).

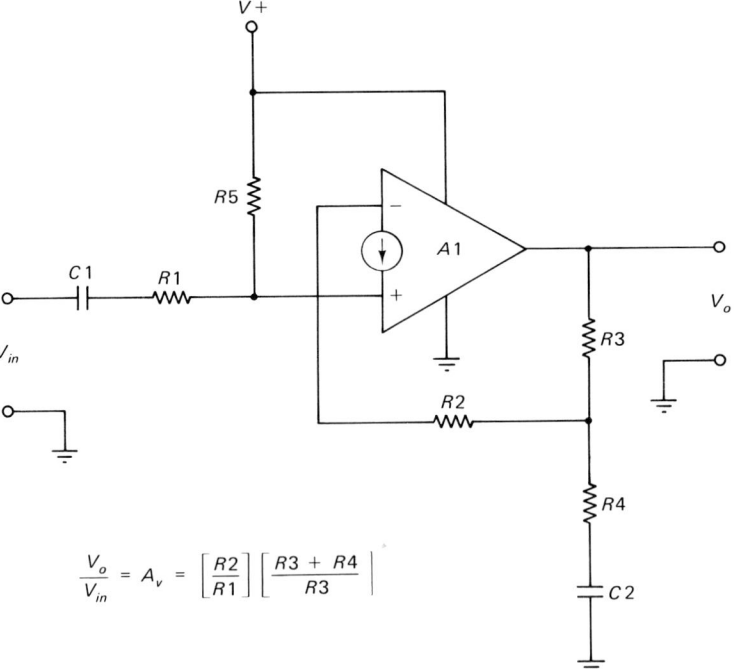

Fig. 13-12   Supergain CDA amplifier.

CDA DIFFERENTIAL AMPLIFIERS

A differential amplifier is one that will produce an output that is proportional to the gain and the difference between potentials applied to the inverting and noninverting inputs. Figure 13-13 shows the circuit of a CDA differential amplifier. It is similar to the operational amplifier version of this simple circuit in several respects. For example, the two input resistors are equal, and the differential voltage gain is the ratio of the negative feedback resistor ($R3$) and the input resistor.

$$A_v = R3/R2 \qquad (13\text{-}32)$$

If $R1 = R2$, then $R3 = R4 = R5$.

The input impedance (differential) of this circuit is twice the value of the input resistances: $R_{in} = R1 + R2$. The bias current is provided through the two series resistors $R4$ and $R5$.

Assuming that $R1 = R2 = R$ and $C1 = C2 = C$, we can calculate the low-end -3-dB frequency from the equation:

$$F = 1{,}000{,}000/2\pi RC \qquad (13\text{-}33)$$

where F is the -3-dB frequency in hertz (Hz), R the resistance in ohms (Ω), and C the capacitance in microfarads (μF).

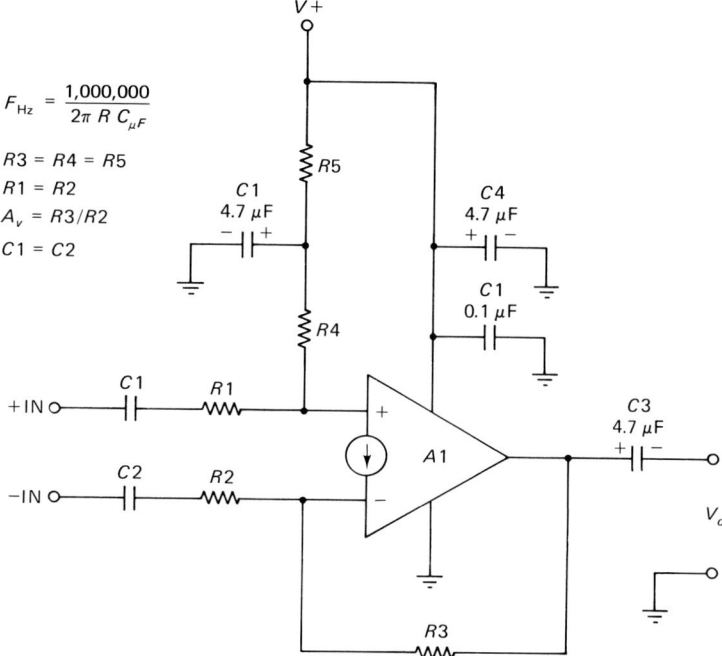

Fig. 13-13 Differential AC-coupled CDA amplifier.

When this circuit is used as a 600-Ω line receiver, we can make the two input resistors 330 Ω each (or 270 Ω) if a small mismatch can be tolerated. Ideally, the input resistors will be 300 Ω each (which may require two resistors for each input resistor).

AC Mixer/Summer Circuits

The CDA mixer or summer circuit is shown in Figure 13-14. This circuit is used to combine two or more inputs into one channel. The basic circuit is an inverting follower. Each input sees a gain that is the quotient of the feedback resistor to its input resistor:

$$A_{V1} = -R2/R3 \tag{13-34}$$

$$A_{V2} = -R2/R4 \tag{13-35}$$

$$A_{V3} = -R2/R5 \tag{13-36}$$

From Eqs. (13-34) through (13-36) we can deduce that the output voltage is found from

$$V_o = R2\left[\frac{V1}{R3} + \frac{V2}{R4} + \frac{V3}{R5}\right] \tag{13-37}$$

The frequency response of each channel is found from the usual equation for -3-dB frequency.

$$F = 1{,}000{,}000/2\pi RC \tag{13-38}$$

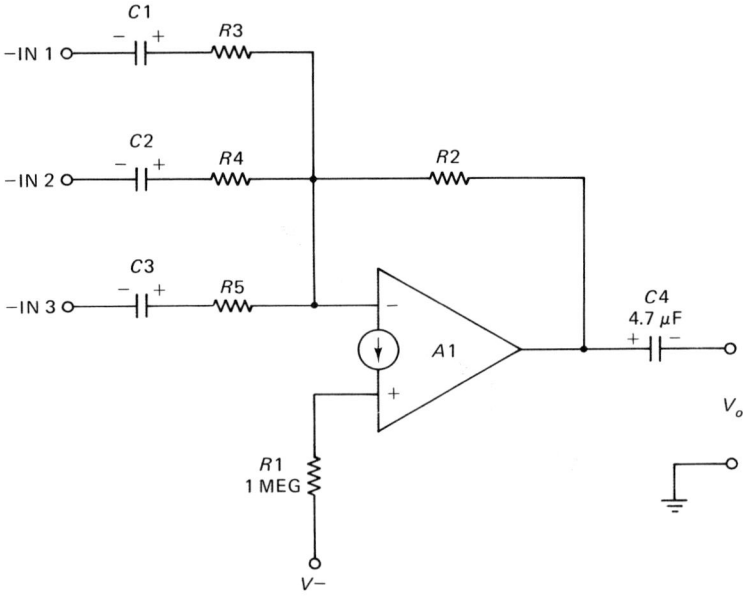

Fig. 13-14 CDA mixer.

where F is the -3-dB frequency in hertz (Hz), R the input resistance ($R3$, $R4$, or $R5$), and C is the input capacitance ($C1$, $C2$, or $C3$).

Differential Output 600-Ω Line Driver Amplifier

The 600-Ω line used in broadcasting electronics and professional audio recording requires either a center-tapped output transformer or a linear amplifier with a push-pull output to drive the line. Figure 13-15 shows the circuit of a CDA 600-Ω line driver amplifier. This circuit basically consists of two separate amplifiers, one an inverting follower and the other a noninverting follower.

The bias resistors ($R5$ and $R6$) are set to provide a small bias current in the range 5 to 100 μA. This current is found from Ohm's law, $I_{ref} = (V+)/R$. In the example shown in Fig. 13-15 the resistors are set to 2 MΩ for a supply voltage of $+15$ V DC.

The capacitors in the circuit set the low-end -3-dB point in the frequency response curve. These capacitor values are set as shown here.

Assuming that $R1 = R2 = R3 = R4$ and $R5 = R6$,

$$C1 = 1,000,000/2\pi FR1 \tag{13-39}$$

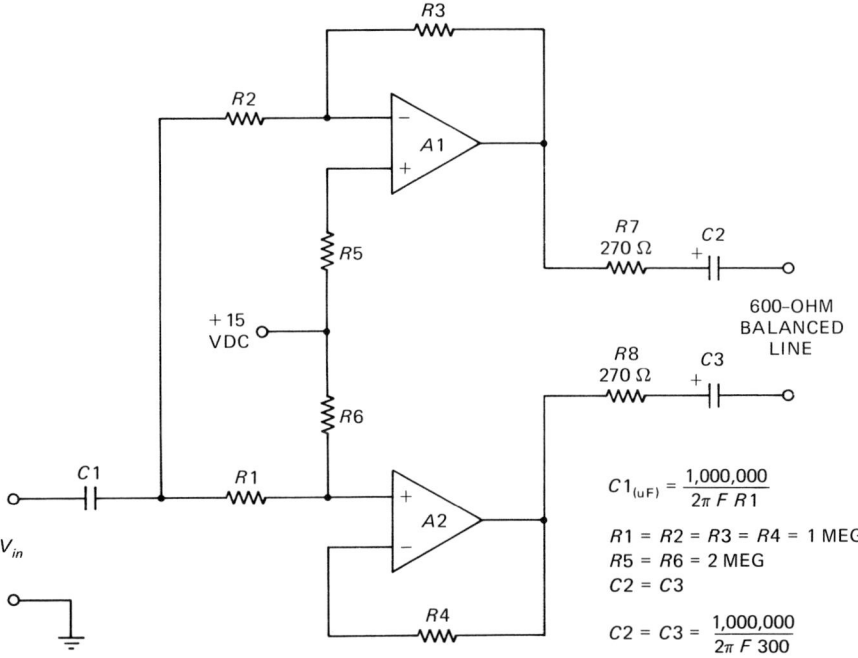

Fig. 13-15 600-Ω line driver amplifier.

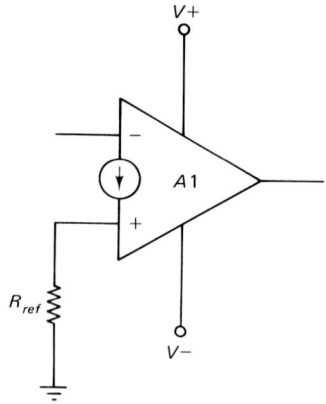

Fig. 13-16 Grounded reference resistor CDA amplifier.

Assuming $C2 = C3$,

$$C2 = 1{,}000{,}000/2\pi F(300) \tag{13-40}$$

where $C1$ and $C2$ are in microfarads, F in hertz, and $R1$ in ohms.

For best results a multiple CDA integrated circuit such as the LM-3900 should be used for this application. Such a circuit would allow the drift to be controlled because the two halves would both drift at the same rate (they share a common thermal environment).

Using Bipolar DC Power Supplies

The current difference amplifier is designed primarily for single polarity power supply circuits. In most cases, the CDA will operate with a $V +$ DC power supply in which one side is grounded. We can, however, operate the CDA in a circuit with a bipolar DC power supply using a circuit such as Fig. 13-16. The reference resistor R_{ref} is connected from the noninverting input to ground. The $V -$ and $V +$ power supplies are each ground referenced and of equal potential. Thus, the 5 to 100-μA bias current is found from $(V +)/R_{\text{ref}}$.

14

DC Differential Operational Amplifier Circuits

Most operational amplifiers have differential inputs. That is, there are two inputs ($-\text{IN}$ and $+\text{IN}$) that each provide the same amount of voltage gain but are of opposite polarity with respect to each other. The inverting input ($-\text{IN}$) of the operational amplifier provides an output signal that is $180°$ out of phase with the input signal. In other words, a positive-going input signal will provide a negative-going output signal, and vice versa. The noninverting input ($+\text{IN}$) produces an output signal that is in-phase with the input signal. For this type of input, a positive-going input signal will produce a positive-going output signal. We will use these properties to understand a class of amplifiers in which both inputs are used. But before going further with our discussion of differential amplifiers, let us revisit the basic differential input stage of an operational amplifier.

DIFFERENTIAL INPUT STAGE

Figure 14-1 shows a hypothetical input stage for a differential input operational amplifier. Transistors Q1 and Q2 are a matched pair that share common $V-$ and $V+$ DC power supplies through separate collector load resistors. The collector of transistor Q2 serves as the output terminal for the differential amplifier. Thus, collector voltage V_o is the output voltage of the stage. The exact value of the output voltage is the difference between supply voltage ($V+$) and the voltage drop across resistor $R2$ (i.e., V_{R2}).

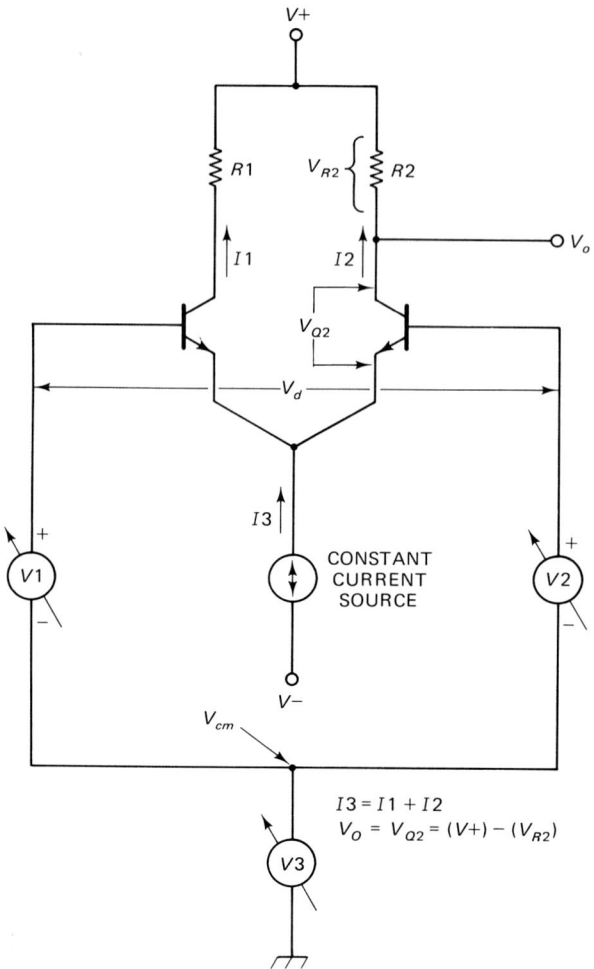

Fig. 14-1   Differential input amplifier.

The emitter terminals of the two transistors are connected together and are fed from a constant current source (CCS). For purposes of analysis, we can assume that the collector and emitter currents ($I1$ and $I2$) are equal to each other (even though these currents are not actually equal—they are very close to each other—this convention is nonetheless useful for our present purpose). Because of Kirchhoff's current law (KCL), we know that

$$I3 = I1 + I2 \tag{14-1}$$

Because current $I3$ is a constant current, a change in either $I1$ or $I2$ will change the other. For example, an increase in current $I1$ must force a decrease in current $I2$ to satisfy Eq. (14-1). Now let us consider how the stage works—first, the case where $V1 = V2$. In this case, both

Q1 and Q2 are biased equally, so the collector currents ($I1$ and $I2$) are equal to each other. Under the normal situation, this case will result in V_{R2} being equal to V_o, and also equal to $(V +)/2$.

Next, let us consider the case where $V1$ is greater than $V2$. In this case, transistor Q1 is biased on harder than Q2, so $I1$ will increase. Because $I1 + I2$ is always equal to $I3$, an increase in $I1$ will force a decrease in $I2$. A decrease of $I2$ will also reduce V_{R2} and increase V_o. Thus, an increase in $V1$ increases V_o, so the base of transistor Q1 is the noninverting input of the differential amplifier.

Now let us consider the case where $V2$ is greater than $V1$. In this case, transistor Q2 is biased on harder than Q1 so $I2$ will increase and $I1$ will decrease. An increase in $I2$ forces a larger voltage drop (V_{R2}) across resistor $R2$, so V_o will go down. Thus, an increase in $V2$ forces a decrease in V_o, so the base of transistor Q2 is the inverting input of the differential amplifier.

Voltage $V3$ is a common mode signal, so it will affect transistors Q1 and Q2 equally. For this kind of signal the output voltage (V_o) will not change.

The base bias currents required to keep transistors Q1 and Q2 operating become the input bias currents that make the practical op-amp non-ideal. To make the input impedance high, the device manufacturer makes these currents very low. Some manufacturers offer operational amplifiers that use MOSFET transistors (called BiMOS op-amps) or JFET transistors (BiFET op-amps) for the input stage. These types of transistors inherently have lower input bias currents than bipolar NPN or PNP transistors, and in fact the bias currents approach mere leakage currents.

INPUT SIGNAL CONVENTIONS

Figure 14-2 shows a generic differential amplifier with the standard signals applied. Signals $V1$ and $V2$ are single-ended potentials applied to the $-IN$ and $+IN$ inputs, respectively. The differential input signal V_d is the difference between the two single-ended signals: $V_d =$

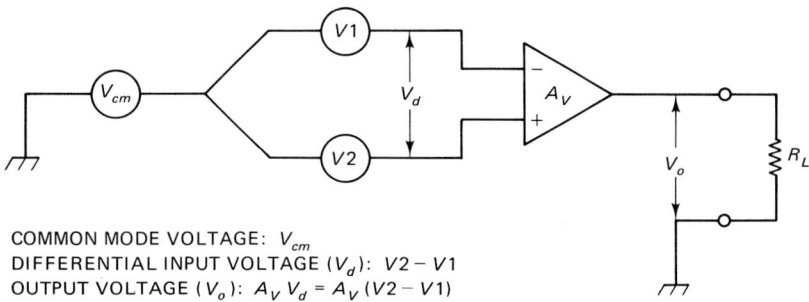

COMMON MODE VOLTAGE: V_{cm}
DIFFERENTIAL INPUT VOLTAGE (V_d): $V2 - V1$
OUTPUT VOLTAGE (V_o): $A_V V_d = A_V (V2 - V1)$

Fig. 14-2 Signals applied to differential amplifier.

$(V2 - V1)$. Signal V_{cm} is a common mode signal; that is, it is applied equally to both $-IN$ and $+IN$ inputs. These signals are described further here.

Common Mode Signals

First let us consider the common mode signal V_{cm}. A common mode signal is one that is applied to both inputs at the same time. Such a signal might be either a voltage such as V_{cm} or a case where voltages $V1$ and $V2$ are equal to each other and of the same polarity (i.e., $V1 = V2$). The implication of the common mode signal is that, being applied equally to inverting and noninverting inputs, the output voltage is zero. Because the two inputs have equal but opposite polarity gains for common mode signals, the net output signal in response to a common mode signal is zero.

The operational amplifier with differential inputs cancels common mode signals. An example of the usefulness of this property is in the performance of the differential amplifier with respect to pick-up of 60-Hz hum from local AC power lines. Almost all input signal cables for practical amplifiers will pick up 60-Hz radiated energy and convert it to a voltage that is seen by the amplifier as a genuine input signal. In a differential amplifier, however, the 60-Hz field will affect both inverting and noninverting input lines equally, so the 60-Hz artifact signal will disappear in the output.

The practical operation amplifier will not exhibit perfect rejection of common mode signals. A specification called the common mode rejection ratio (CMRR) tells us something of the ability of any given op-amp to reject such signals. The CMRR is usually specified in decibels (dB) and is defined as

$$CMRR = A_{vd}/A_{cm} \qquad (14\text{-}2)$$

or, in decibel form:

$$CMRR_{db} = 20 \log[A_{vd}/A_{cm}] \qquad (14\text{-}3)$$

where CMRR is the common mode rejection ratio, A_{vd} the voltage gain to differential signals, and A_{cm} the voltage gain to common mode signals.

In general, the higher the CMRR the better the operational amplifier. Typical low-cost devices have CMRR ratings of 60 dB or more, while better devices exhibit CMRR values up to 120 dB.

Differential Signals

Signals $V1$ and $V2$ in Fig. 14-2 are single-ended signals. The total differential signal seen by the operational amplifier is the difference between the single-ended signals.

$$V_d = V2 - V1 \tag{14-4}$$

The output signal from the differential operational amplifier is the product of the differential voltage gain and the difference between the two input signals (hence the term *differential* amplifier). Thus, the transfer equation for the operational amplifier is

$$V_o = A_v(V2 - V1) \tag{14-5}$$

DIFFERENTIAL AMPLIFIER TRANSFER EQUATION

The basic circuit for the DC differential amplifier is shown in Fig. 14-3A. This circuit uses only one operational amplifier, so it is the simplest possible configuration. Later you will see additional circuits based on two or three operational amplifier devices. In its most common form the circuit of Fig. 14-3A is balanced such that $R1 = R2$ and $R3 = R4$.

Consider the redrawn differential amplifier circuit shown in Fig. 14-3B. Assume that source resistances R_{S1} and R_{S2} are zero. Further assume that $R1 = R2 = R$ and that $R3 = R4 = kR$, where k is a multiplier of R.

1. Set $V2 = 0$. In this case V_o is found from

$$V_{O1} = (-kR/R)V1 \tag{14-6}$$

$$V_{O1} = -k(V1) \tag{14-7}$$

2. Now assume that $V1 = 0$ instead.

$$V_a = (V2)kR/(kR + R) \tag{14-8}$$

$$V_a = (V2)k/(k + 1) \tag{14-9}$$

$$V_{O2} = [(kR/R) + 1]V_a \tag{14-10}$$

$$V_{O2} = [(kR/R) + 1][(V2)k/(k + 1)] \tag{14-11}$$

$$V_{O2} = (k + 1)[(V2)k/(k + 1)] \tag{14-12}$$

$$V_{O2} = (V2)k \tag{14-13}$$

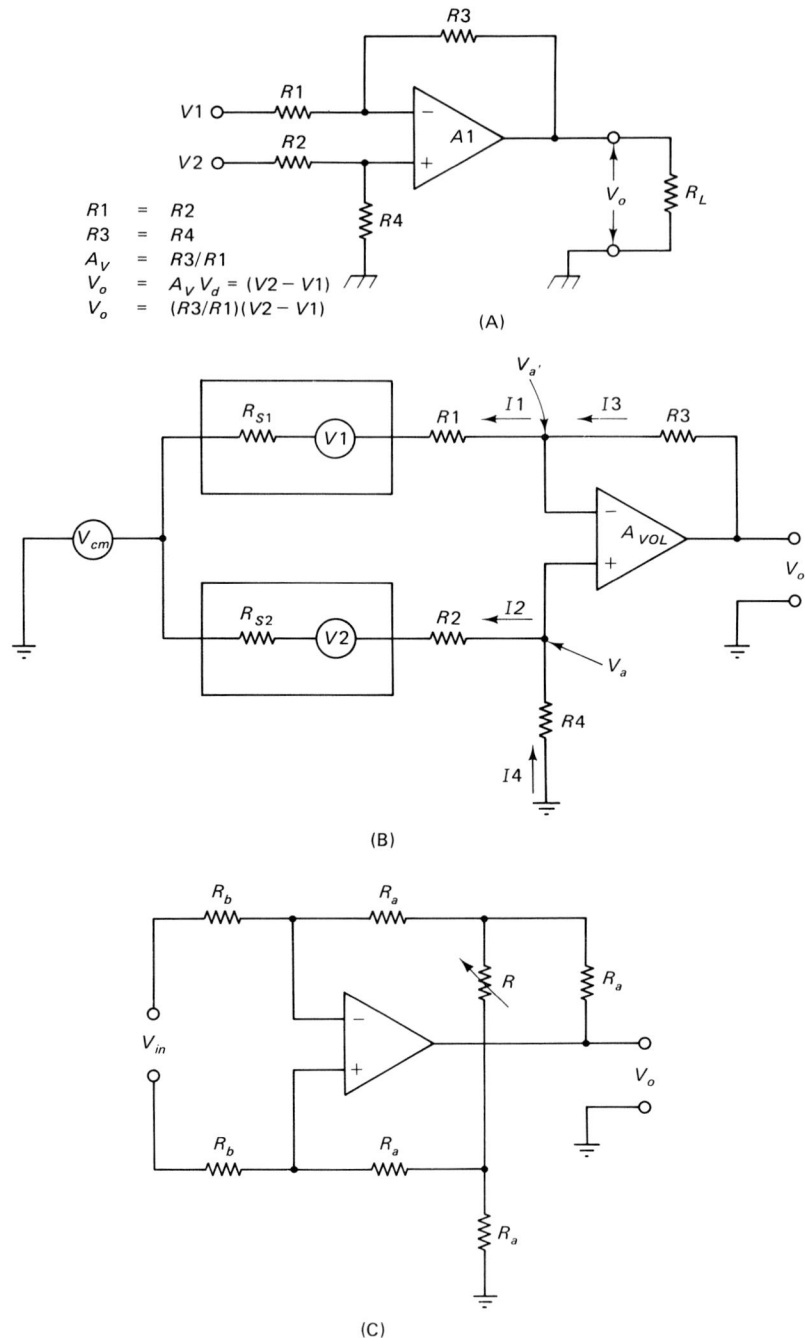

R1 = R2
R3 = R4
A_V = R3/R1
V_o = $A_V V_d$ = (V2 − V1)
V_o = (R3/R1)(V2 − V1)

Fig. 14-3 DC differential amplifier: (A) circuit, (B) equivalent circuit, and (C) attempt at producing gain control.

3. Now, by superimposing the two expressions for V_o:

$$V_o = V_{O2} + V_{O1} \tag{14-14}$$

$$V_o = (V2)k - (V1)k \tag{14-15}$$

$$V_o = k(V2 - V1) \tag{14-16}$$

According to Eq. (14-16), therefore, the output voltage is the product of the difference between single-ended input potentials $V1$ and $V2$ and a factor k. The differential input voltage (V_d) is

$$V_d = V2 - V1 \tag{14-17}$$

while factor k is the differential voltage gain, A_{vd}. Thus, the output voltage is

$$V_o = V_d A_{vd} \tag{14-18}$$

We may also use a less parametric analysis by appealing to Fig. 14-3B. We know that the following relationships are true (assuming $R_{S1} = R_{S2} = 0$):

$$V_a = (V_{cm} + V2)R^4/(R2 + R4) \tag{14-19}$$

$$I1 = -I3 \tag{14-20}$$

$$I1 = (V_{cm} + V1 - V_{a'})/R1 \tag{14-21}$$

$$I3 = (V_o - V_{a'})/R3 \tag{14-22}$$

From the properties of the ideal op-amp we know that voltage $V_{a'} = V_{a'}$ so

$$I1 = (V_{cm} + V1 - V_a)/R1 \tag{14-23}$$

$$I3 = (V_o - V_a)/R3 \tag{14-24}$$

Combining equations and solving for V_o:

$$V_o = V_{cm}\left[\frac{R3R4 + R1R4 - R2R3 - R3R4}{R1(R2 + R4)}\right]$$

$$-[R3V1/R1] + \left[\left(\frac{R4}{R2}\right)\left(\frac{1 + R3/R1}{1 + R4/R2}\right)\right]V2 \tag{14-25}$$

Assuming that $R1 = R2$ and $R3 = R4$, Eq. (14-25) resolves to

$$V_o = (R3/R1)(V2 - V1) \qquad (14\text{-}26)$$

Equation (14-26) is similar to Eqs. (14-5) and (14-16). In this case, A_{vd} is $R3/R1$, and V_d is $(V2 - V1)$. The standard transfer equation for the single op-amp DC differential amplifier is

$$V_o = V_d A_{vd} \qquad (14\text{-}27)$$

$$V_o = V_d(R3/R1) \qquad (14\text{-}28)$$

It is difficult to build a DC differential amplifier with variable gain control. It is, for example, very difficult to get two ganged potentiometers (used to replace $R3$ and $R4$ in Fig. 14-3A) to track well enough to vary gain while maintaining required balance. Figure 14-3C shows one attempt to solve the problem. In this case, a potentiometer is connected between the midpoints of the two feedback resistances. This circuit works, but the gain control is a nonlinear function of the potentiometer setting. It is generally assumed to be a better practice to use a post-amplifier stage following the differential amplifier and to perform gain control in that stage. Alternatively, use one of the differential amplifier circuits shown later in this chapter that uses more than one operational amplifier.

COMMON MODE REJECTION

Figure 14-4 shows two different situations; Fig. 14-4A shows a single-ended amplifier while Fig. 14-4B shows a differential amplifier in a similar situation. In these circuits a noise signal V_n is placed between

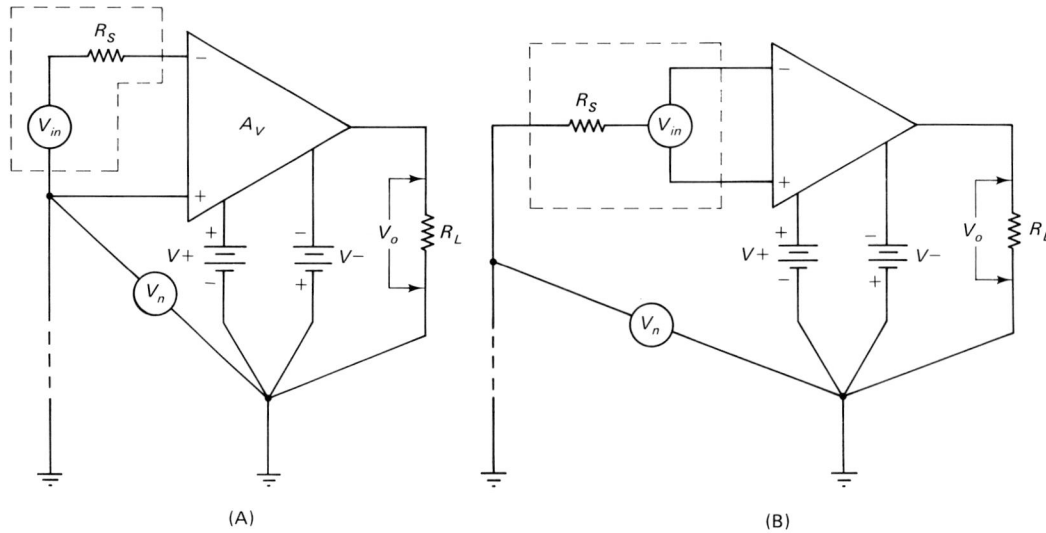

(A)   (B)

Fig. 14-4 (A) Signals applied to a single-ended amplifier and (B) signals applied to a differential amplifier.

the input ground and the output ground. This noise signal might be either AC or DC noise. In Fig. 14-4A we see the case where the noise signal is applied to a single-ended input amplifier. The input signal seen by the amplifier is the algebraic sum of the two independent signals: $V_{in} + V_n$. Because of this fact, the amplifier output signal will see a noise artifact equal to the product of the noise signal amplitude and the amplifier gain, $-A_v V_n$. Now consider the situation of a differential amplifier depicted in Fig. 14-4B. The noise signal in this case is common mode, so it is essentially canceled by the common mode rejection ratio. Of course, in non-ideal amplifiers the actual situation is that the input signal V_{in} is subject to the differential gain, while the noise signal V_n is subject to the common mode gain. If an amplifier has a CMRR of 90 dB, for example, the gain seen by the noise signal will be 90 dB down from the differential gain.

The common mode rejection ratio (CMRR) of the operational amplifier DC differential circuit is dependent principally on two factors. The first is the natural CMRR of the operational amplifier used as the active device. The second, the balance of the resistors, $R1 = R2$ and $R3 = R4$. Unfortunately, the balance is typically difficult to obtain with common fixed resistors (not considering precision types). We can use a circuit such as shown in Fig. 14-5 to compensate for the mismatch. In this circuit, $R1$ through $R3$ are exactly the same as in previous circuits. The fourth resistor, however, is a potentiometer (see $R4$). The potentiometer will "adjust out" the CMRR errors caused by resistor and other circuit mismatches.

A version of the circuit with greater resolution is shown in the inset to Fig. 14-6. In this version the single potentiometer is replaced by a fixed resistor and a potentiometer in series, the sum resistance

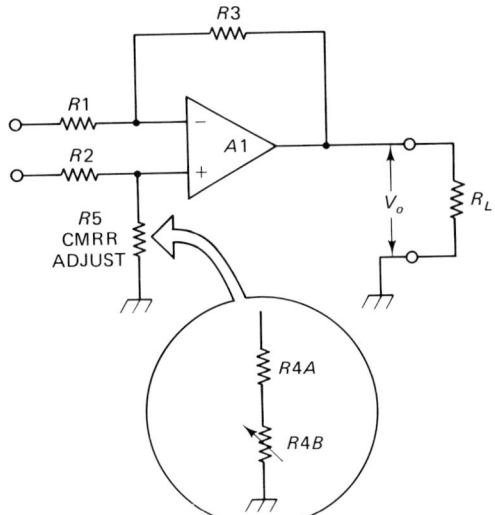

Fig. 14-5 Simple method for common mode rejection adjustment.

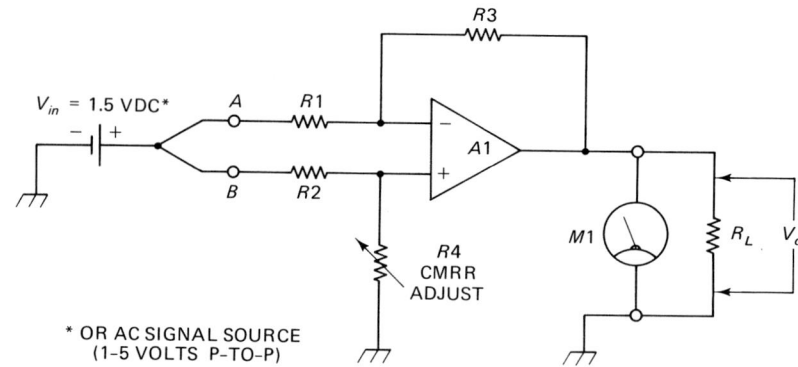

Fig. 14-6 Adjustment of CMRR.

($R4A + R4B$) is equal to approximately 20 percent more than the normal value of $R3$. Ordinarily, the maximum value of the potentiometer is 10 to 20 percent of the overall resistance.

The adjustment procedure for either version is the same.

1. Connect a zero-center DC voltmeter to the output terminal (M1).

2. Short together inputs A and B, and then connect them to either a signal voltage source or ground.

3. Adjust potentiometer $R4$ (CMRR ADJ) for zero volts output.

4. If the output indicator (meter M1) has several ranges, then switch to a lower range and repeat step 3 until no further improvement is possible.

Alternatively, connect the output to an audio voltmeter or oscilloscope, and connect the input to a 1-V to 14-V peak-to-peak AC signal that is within the frequency range of the particular amplifier. For audio amplifiers, a 400- to 1000-hertz, 1-V signal is typically used.

PRACTICAL DIFFERENTIAL AMPLIFIER CIRCUITS

Figure 14-7 shows the circuit of the simple DC differential amplifier based on the operational amplifier device. The gain of the circuit is set by the ratio of two resistors.

$$A_v = R3/R1 \tag{14-29}$$

or

$$A_v = R4/R2 \tag{14-30}$$

provided that $R1 = R2$ and $R3 = R4$.

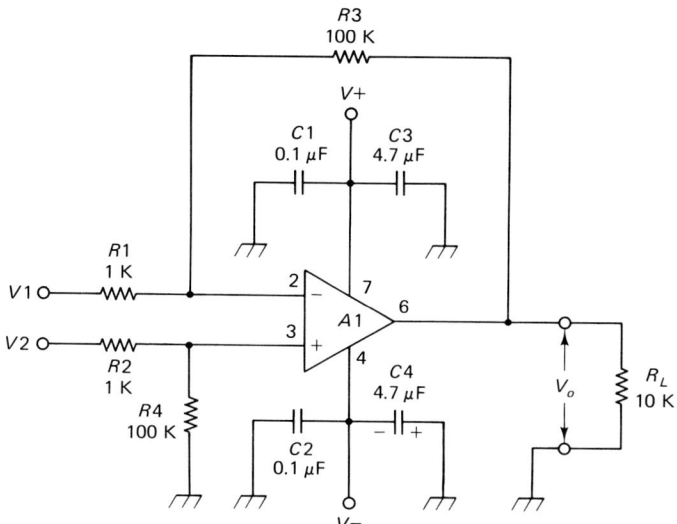

Fig. 14-7 DC differential amplifier with gain of 100.

The output voltage of the DC differential amplifier is given by

$$V_o = (R3/R1)(V2 - V1) \qquad (14\text{-}31)$$

or

$$V_o = A_v(V2 - V1) \qquad (14\text{-}32)$$

where V_o is the output voltage, $R3$ the feedback resistor, $R1$ the input resistor, A_v the differential gain, $V1$ the signal voltage applied to the inverting input, and $V2$ the signal voltage applied to the noninverting input.

Again, the constraint is that $R1 = R2$ and $R3 = R4$. These two balances must be maintained or the common mode rejection ratio (CMRR) will deteriorate rapidly. In many common applications, the CMRR can be maintained within reason by specifying 10% tolerance resistors for $R1$ through $R4$. But where superior CMRR is required, especially where the differential voltage gain is high, closer tolerance resistors (1% or better) are required.

Design Example

Design a DC differential amplifier with a gain of 100. Assume that the source impedance of the preceding stage is about 100 Ω (refer to Fig. 14-7).

Because the source impedance is 100-Ω, we need to make the input resistors of the DC differential amplifier 10 times larger (or more). Thus, the input resistors ($R1$ and $R2$) must be at least 1000 kΩ: $R1 = R2 = 1000\ \Omega$.

Set the value of the two feedback resistors (keep in mind that $R3 = R4$). The values of these resistors is found from rewriting Eq. (14-29):

$$R3 = A_v R1 \qquad\qquad (14\text{-}33)$$

$$R3 = (100)(1000) \qquad\qquad (14\text{-}34)$$

$$R3 = 100{,}000\ \Omega \qquad\qquad (14\text{-}35)$$

Figure 14-7 shows the finished circuit of this amplifier. The values of the resistors are $R1 = 1$ kΩ, $R2 = 1$ kΩ, $R3 = 100$ kΩ, and $R4 = 100$ kΩ.

For best results, make $R1$, $R2$, $R3$, and $R4$ 1% precision resistors.

The pin-outs shown in Fig. 14-7 are for the industry standard 741 family of operational amplifiers. These same pinouts are found on many other different op-amp products as well.

The DC power supply voltages are usually either ± 12 V DC or ± 15 V DC. Lower power supply potentials can be accommodated, however, if a corresponding reduction in the output voltage swing is tolerable. A typical lower grade op-amp produce a maximum output voltage that is approximately 3 V lower than than the supply voltage. For example, when $V+$ is 12 V DC the maximum positive output voltage permitted is $(12 - 3)$ V or 9 V.

The decoupling bypass capacitors shown in Fig. 14-7 are used to keep the circuit stable, especially in those cases where the same DC power supplies are used for several stages. The low-value 0.1-μF capacitors ($C1$ and $C2$) are used to decouple high-frequency signals. These capacitors should be physically placed as close as possible to the body of the operational amplifier. The high value capacitors ($C3$ and $C4$) are needed to decouple low-frequency signals. Two values of capacitors are needed because on some op-amps the high-value capacitors needed for low-frequency decoupling are typically electrolytics (tantalum or aluminum), some of which are ineffective at high frequencies. Thus, we must provide smaller-value capacitors that (1) have a low enough capactive reactance to do the job and (2) are effective at high frequencies. The extra capacitor is used on uncompensated op-amps that have a high gain–bandwidth product.

A different situation is shown in Fig. 14-8. Most differential amplifiers have relatively low input impedances, which is a function of factors such as input bias currents and so forth. The amplifier in Fig. 14-8 uses a high input impedance by virtue of the high values of resistors $R1$ and $R2$. To attain this input impedance, however, we need to specify an operational amplifier for $A1$ that has a very low input bias current (i.e., a very high natural input impedance). The BiMOS devices, which use MOSFET input transistors, and the BiFET, which uses JFET input transistors, are good selections for DC differ-

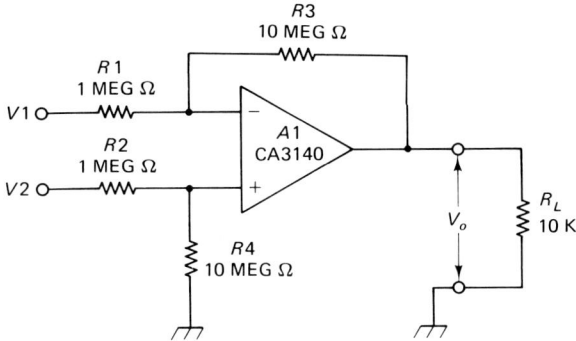

Fig. 14-8 DC differential amplifier with high-input impedance, gain of 10.

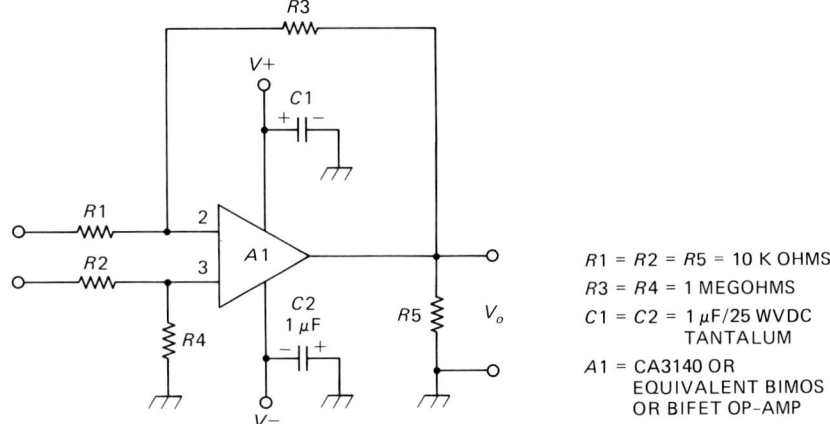

$R1 = R2 = R5 = 10$ K OHMS
$R3 = R4 = 1$ MEGOHMS
$C1 = C2 = 1 \mu F/25$ WVDC
TANTALUM

$A1 =$ CA3140 OR
EQUIVALENT BIMOS
OR BIFET OP-AMP

Fig. 14-9 DC differential amplifier.

ential amplifier circuits when very high input impedances are needed. Figure 14-9 shows the complete circuit, including required decoupling capacitors.

ADDITIONAL DIFFERENTIAL AMPLIFIERS

The simple DC differential amplifier circuits shown in this chapter are useful for low-gain applications and for those applications where a low to moderate input impedance is permissible (e.g., 300 to 200,000 Ω). Where a higher gain is required, we must resort to a more complex circuit called the operational amplifier instrumentation amplifier, or IA. In the next chapter we will examine the classical three-device IA circuit, as well as several integrated circuit instrumentation amplifiers (IC IA) that offer the advantages of the IA in a single small IC package. Some of those devices are now among the most commonly used in many instrumentation applications; they will be considered below.

INSTRUMENTATION AMPLIFIERS

The simple DC differential amplifier discussed earlier suffers from several important drawbacks. First, there is a limit to the input impedance (Z_{in} is approximately equal to the sum of the two input resistors).

Second, there is a practical limitation on the gain available from the simple single-device DC differential amplifier. If high gain is attempted, then either the input bias current tends to cause large output offset voltages, or the input impedance becomes too low. In this section we will demonstrate a solution to these problems in the form of the instrumentation amplifer (IA). All of these amplifiers are differential amplifiers, but they offer superior performance over the simple DC differential amplifiers of the last section. The instrumentation amplifier can offer higher input impedance, higher gain, and better common mode rejection than the single-device DC differential amplifier.

Simple IA Circuit

The simplest form of instrumentation amplifier circuit is shown in Fig. 14-10. In this circuit the input impedance is improved by connecting inputs of a simple DC differential amplifier ($A3$) to two input amplifiers ($A1$ and $A2$) that are each of the unity-gain, noninverting follower configuration (their use here is as buffer amplifiers). The input amplifiers offer an extremely large input impedance (a result of the noninverting configuration) while driving the input resistors of the actual amplifying stage ($A3$). The overall gain of this circuit is the

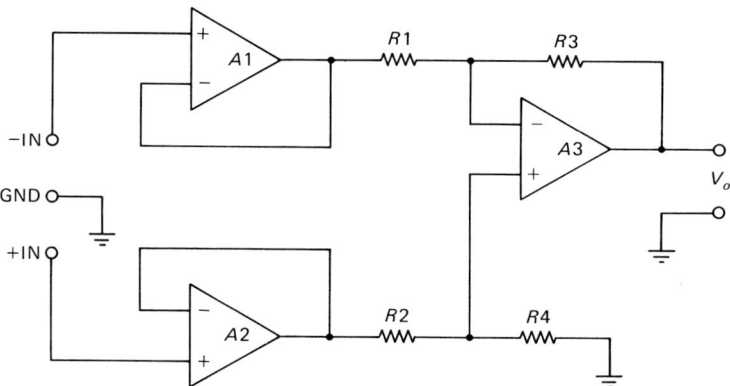

Fig. 14-10   Simple three-device instrumentation amplifier.

same as for any simple DC differential amplifier.

$$A_v = R3/R1 \qquad (14\text{-}36)$$

where A_v is the voltage gain, assuming $R1 = R2$ and $R3 = R4$.

It is considered best practice if $A1$ and $A2$ are identical operational amplifiers. In fact, it is advisable to use a dual operational amplifier for both $A1$ and $A2$ (i.e., two op-amps in a common IC package). The common thermal environment of the dual operational amplifier will reduce thermal drift problems. The very high input impedance of superbeta (Darlington), BiMOS, and BiFET operational amplifiers make them ideal for use as the input amplifiers in this type of circuit.

One of the biggest problems with the circuit of Fig. 14-10 is that it wastes two good operational amplifiers. The most common form of instrumentation amplifier circuit uses the input amplifiers to provide voltage gain in addition to providing a higher input impedance. Such amplifier circuits are discussed in the next section.

Standard Instrumentation Amplifiers

The standard instrumentation amplifier (IA) is shown in Fig. 14-11A. Like the simple circuit discussed previously, this circuit uses three operational amplifiers. The biggest difference is that the input amplifiers ($A1$ and $A2$) are now used in the "noninverting follower with gain" configuration. Like the circuit of Fig. 14-10, the input amplifiers are ideally BiMOS, BiFET, or superbeta input types to obtain the maximum possible input impedance. Again, for best thermal performance use a dual, triple, or quad operational amplifier for this application. The signal voltages shown in Fig. 14-11A follow the standard pattern: Voltages $V1$ and $V2$ form the differential input signal ($V2 - V1$), while voltage V_{cm} represents the common mode signal because it affects both inputs equally.

Let us evaluate Fig. 14-11A by first examining the behavior of the input stages $A1$ and $A2$. In the partial circuit of Fig. 14-12 the output voltage V_c is the difference between the output potential of $A1$ (i.e., $V3$) and the output potential of $A2$ (i.e., $V4$). Given that resistor $R1$ is shared by both $A1$ and $A2$, we count its value as $R1/2$ for each calculation. Our method is to calculate $V3$ when $V2 = 0$, and $V4$ when $V1 = 0$, and then superimpose the result to find $V_c = V4 - V3$.

$$1. \qquad V4 = V2\left[\frac{R3}{R1/2} + 1\right] \qquad (14\text{-}37)$$

$$2. \qquad V3 = V1\left[\frac{R2}{R1/2} + 1\right] \qquad (14\text{-}38)$$

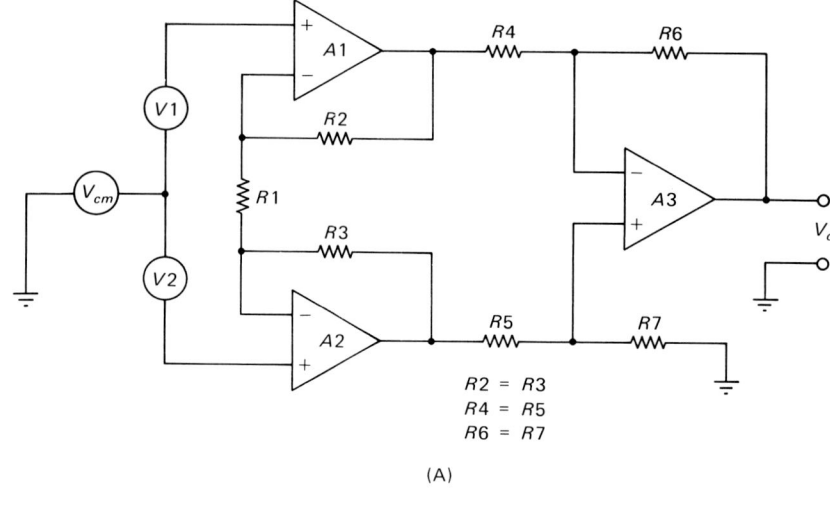

(A)

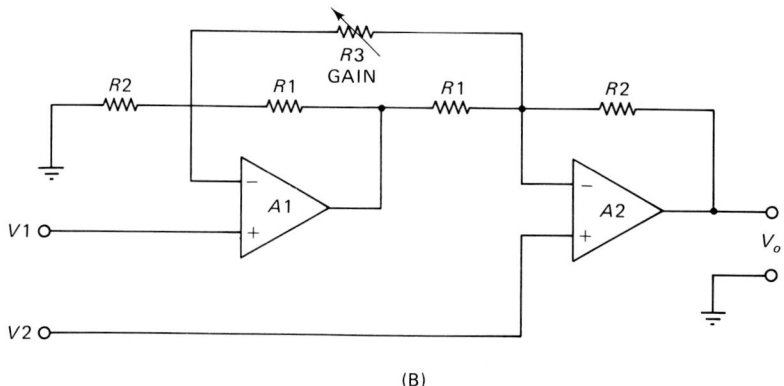

(B)

Fig. 14-11 (A) Standard instrumentation amplifier and (B) two-device instrumentation amplifier.

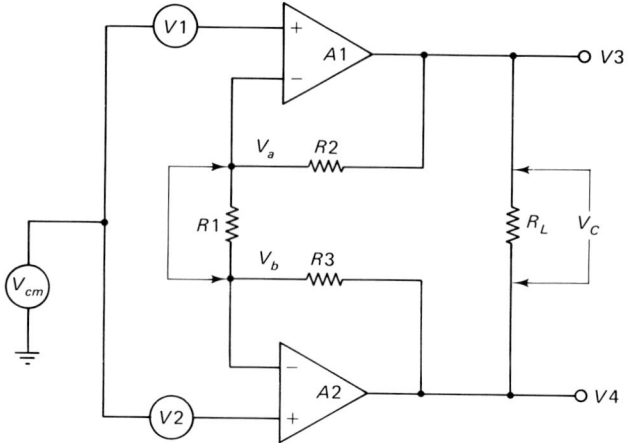

Fig. 14-12 Input circuit to the standard instrumentation amplifier.

Therefore,

3.
$$V4 - V3 = V2\left[\frac{R3}{R1/2} + 1\right]$$

$$- V1\left[\frac{R2}{R1/2} + 1\right] \quad (14\text{-}39)$$

Because $V_c = V4 - V3$ and $R2 = R3 = R$, we may rewrite Eq. (14-39) in the form:

4.
$$V_c = V2\left[\frac{R}{R1/2} + 1\right] - V1\left[\frac{R}{R1/2} + 1\right] \quad (14\text{-}40)$$

5.
$$V_c = (V2 - V1)\left[\frac{R}{R1/2} + 1\right] \quad (14\text{-}41)$$

6.
$$V_c = (V2 - V1)\left[\frac{2R}{R1} + 1\right] \quad (14\text{-}42)$$

7. Or, in the form that identifies specific circuit components,

$$V_c = (V2 - V1)\left[\frac{2(R2)}{R1} + 1\right] \quad (14\text{-}43)$$

In the form of a standard differential amplifier:

$$V_c = V_d A_{V12} \quad (14\text{-}44)$$

So, we may conclude by comparing Eqs. (14-43) and (14-44):

$$V_d = V2 - V1 \quad (14\text{-}45)$$

$$A_{V12} = \frac{2(R2)}{R1} + 1 \quad (14\text{-}46)$$

From our discussions earlier we know that the gain of $A3$ in Fig. 14-11A is

$$A_{V3} = R6/R4 \quad (14\text{-}47)$$

The overall gain of Fig. 14-11A is

$$A_{V13} = A_{V12}A_{V3} \quad (14\text{-}48)$$

By substituting Eqs. (14-46) and (14-47) into Eq. (14-48) we arrive at the transfer equation for the instrumentation amplifier:

$$A_v = \left[\frac{2(R2)}{R1} + 1\right]\left[\frac{R6}{R4}\right] \qquad (14\text{-}49)$$

provided $R = R3$, $R4 = R5$, $R6 = R7$.

An alternate instrumentation amplifier circuit is shown in Fig. 14-11B. This circuit also offers the advantage of high input impedance, but it uses only two operational amplifier devices rather than three (as were used in Fig. 14-11A). The gain of this circuit is given by

$$A_{vd} = \frac{R2(2R1 + R3)}{R1R3} + 1 \qquad (14\text{-}50)$$

In most practical situations the problem will be to select a value for the gain-ranging resistor that is consistent with the required differential voltage gain and the values of the remaining resistors ($R1$ and $R2$). For this application we use the following equation:

$$R3 = \frac{2(R2)}{A_v - 1 - (R2/R1)} \qquad (14\text{-}51)$$

A problem with the simple DC differential amplifier is providing variable gain control. That problem is easily solved in instrumentation amplifier circuits.

GAIN CONTROL FOR THE IA

It is difficult to provide a gain control for a simple DC differential amplifier without adding an extra amplifier stage (for example, an inverting follower postamplifier with a gain of 0 to -1). For the instrumentation amplifier, however, resistor $R1$ can be used as a gain control provided that the resistance does not go to a value near zero ohms. Figure 14-13 shows a revised circuit with resistor $R1$ replaced by a series circuit consisting of fixed resistor $R1A$ and potentiometer $R1B$. This circuit prevents the gain from rising above the level set by $R1A$. Do not use a potentiometer alone in this circuit because it can have a disastrous effect on the gain. Note in Eq. (14-49) that the term $R1$ appears in the denominator. If the value of $R1$ gets close to zero, the gain goes very high (in fact, supposedly to infinity if $R1 = 0$). The maximum gain of the circuit is controlled by using the fixed resistor in series with the potentiometer. The gain of the circuit in Fig. 14-13 varies from a minimum of 167 (when R1B is set to 2000 Ω) to a

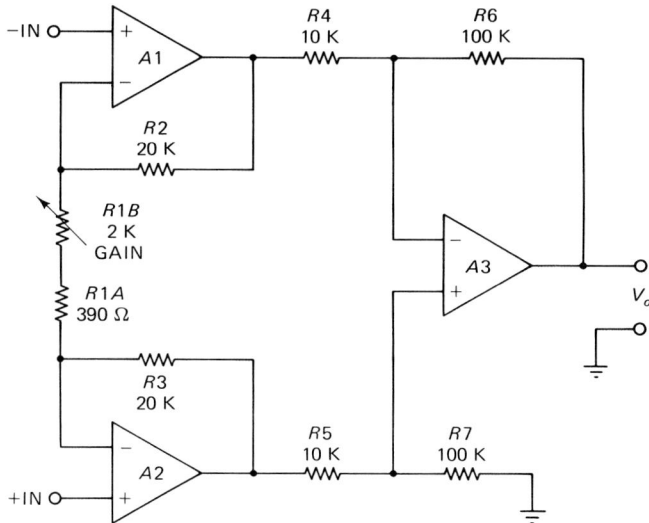

Fig. 14-13 Gain controlled instrumentation amplifier.

maximum of 1025 (when R1B is zero). The gain expression for Fig. 14-13 is

$$A_\text{v} = \left[\frac{2(R2)}{R1A + R1B} + 1 \right]\left[\frac{R6}{R4} \right] \qquad (14\text{-}52)$$

or, rewriting Eq. (14-52) to take into account that R1A is fixed,

$$A_\text{v} = \left[\frac{2(R2)}{390 + R1B} + 1 \right]\left[\frac{R6}{R4} \right] \qquad (14\text{-}53)$$

where $R1B$ varies from 0 to 2000 Ω.

COMMON MODE REJECTION RATIO ADJUSTMENT

The instrumentation amplifier is no different from any other practical DC differential amplifier in that there will be imperfect balance for common mode signals. The operational amplifiers are not ideally matched, so there will be a gain imbalance. This gain imbalance is further deteriorated by the mismatch of the resistors. The result is that the instrumentation amplifier will respond at least to some extent to common mode signals. As in the simple DC differential amplifier we can provide a common mode rejection ratio adjustment by making resistor $R7$ variable (see Fig. 14-14).

One configuration of Fig. 14-14 uses a single potentiometer ($R7$) that has a value that is 10 to 20 percent larger than the required

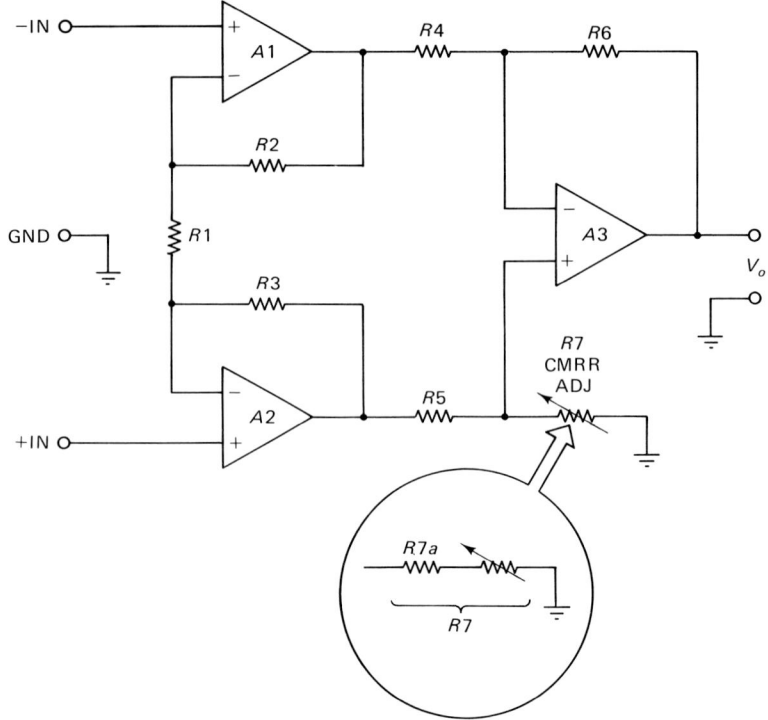

Fig. 14-14   CMRR adjust on instrumentation amplifier.

resistance of $R6$. For example, if $R14$ is 100 kΩ, R7 should be 110 to 120 kΩ. Unfortunately, these values are somewhat difficult to obtain, so we would pick a standard value for $R7$ (e.g., 100 kΩ) and then select a value for $R6$ that is somewhat lower (e.g., 82 kΩ of 91 kΩ).

The second configuration of Fig. 14-14 uses a fixed resistor in series with a potentiometer. The general rule is to make $R7A$ approximately 80 percent of the total required value and $R7B$ 40 percent of the required value. As was true in the other configuration, the sum of $R7A$ and $R7B$ is approximately 110 to 120 percent of the value of resistor $R6$. Resistance values of this sort permit the total resistance to vary from less than to greater than the nominally required value.

AC INSTRUMENTATION AMPLIFIERS

What is the principal difference between DC amplifiers and AC amplifiers? A DC amplifier will amplify both AC and DC signals up to the frequency limit of the particular circuit being used. The AC amplifier, on the other hand, will not pass or amplify DC signals. In fact, AC amplifiers will not pass AC signals of frequencies from close to DC to some lower -3-dB bandpass limit. The gain in the region between near-DC and the full-gain frequencies within the passband rises at a rate determined by the design, usually $+6$ dB/octave (an octave is a

2 : 1 frequency change). The standard low-end point in the frequency response curve is defined as the frequency at which the gain drops off -3 dB from the full gain.

Figure 14-15A shows a modified version of the instrumentation amplifier that is designed as an AC amplifier. The input circuitry of $A1$ and $A2$ is modified by placing a capacitor in series with each op-amp's noninverting input. Resistors $R8$ and $R9$ are used to keep the input bias currents of $A1$ and $A2$ from charging capacitors $C1$ and $C2$. In some modern low input current operational amplifiers these resistors are optional because of the extremely low levels of bias current that are normally present.

The -3-dB frequency of the amplifier in Fig. 14-15A is a function of the input capacitors and resistors (assuming that $R8 = R9 = R$ and $C1 = C2 = C$).

$$F = 1{,}000{,}000/2\pi RC_{\mu F} \qquad (14\text{-}54)$$

where F is the -3-dB frequency in hertz, R is in ohms, and $C_{\mu F}$ is in microfarads.

The equation given in Eq. (14-54) for frequency response is not the most useful form. In most practical cases you will know the required frequency response from evaluation of the application. Furthermore, you will know the value of the input resistors ($R9$ and $R10$) because they are selected for high input impedance, and they are by convention either $> 10 \times$ or $> 100 \times$ the source impedance (depending on the application). Typically, these resistors are selected to be 10 MΩ. You will therefore need to select the capacitor values from Eq. (14-55).

$$C_{\mu F} = 1{,}000{,}000/2\pi RF \qquad (14\text{-}55)$$

where $C_{\mu F}$ is the capacitance of $C1$ and $C2$ in microfarads, R the resistance of $R9$ and $R10$ in ohms, and F the -3-dB frequency in hertz.

The AC instrumentation amplifier can be adapted to all the other modifications of the basic circuit discussed earlier in this chapter. We may, for example, use a gain control (replace $R1$ with a fixed resistor and a potentiometer) or add a CMRR ADJ control. In fact, these adaptations are probably necessary in most practical AC IA circuits.

In many instrumentation amplifier applications it is desirable to provide selectable AC or DC coupling, as well as the ability to ground the input of the amplifiers. This latter feature is especially desirable in circuits where an oscilloscope, strip-chart paper recorder, digital data logger, or computer is used to receive the data. By grounding the input of the amplifier (without also grounding the source, which could be dangerous), it is possible to set (or at least determine the $V_d = 0$ baseline. Figure 14-15B shows a modified input circuit that uses a

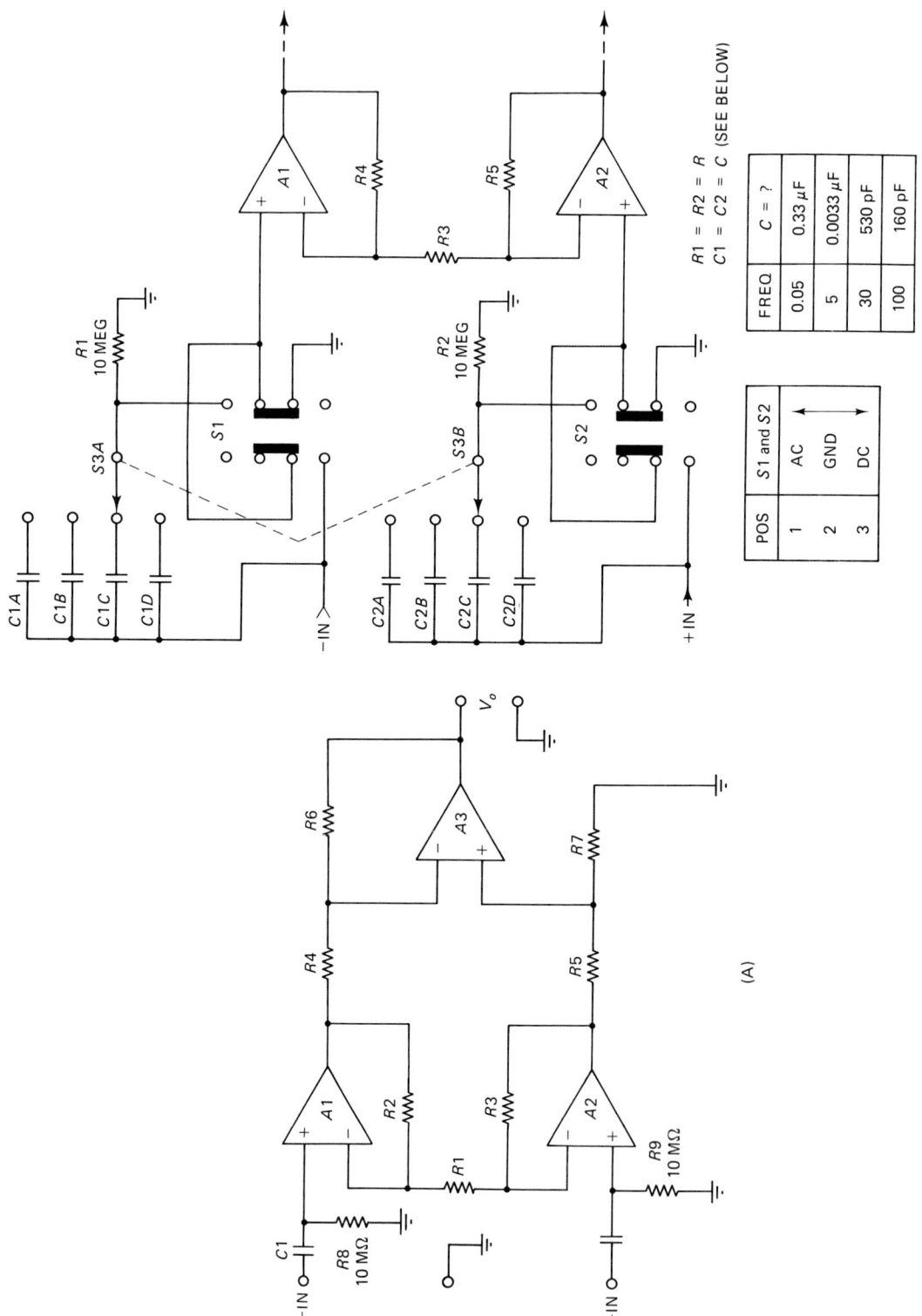

Fig. 14-15 (A) AC-coupled instrumentation amplifier and (B) AC-coupled instrumentation amplifier with input control and frequency tailoring.

POS	S1 and S2
1	AC
2	GND
3	DC

$R1 = R2 = R$
$C1 = C2 = C$ (SEE BELOW)

FREQ	$C = ?$
0.05	0.33 μF
5	0.0033 μF
30	530 pF
100	160 pF

switch to select AC–GND–DC coupling. In addition, a second switch is provided that sets the low-end -3-dB frequency response point according to Eq. (14-56). A table of popular frequency response limits is shown inset to Fig. 14-15B.

Design Example: An ECG Amplifier

The heart in humans and animals produces a small electrical signal that can be recorded through skin-surface electrodes and displayed on an oscilloscope or paper strip-chart recorder. This electrical signal is called the electrocardiograph or ECG signal. The peak value of the ECG signal at the skin-surface electrodes is on the order of one millivolt (1 mV). To produce a 1-V signal to apply to a recorder or oscilloscope, therefore, a gain of 1000 is required. The ECG amplifier must provide a gain of 1000 or more. Furthermore, because skin has a relatively high electrical resistance (1 to 20 kΩ with 1-cm Ag–AgCl electrodes), the ECG amplifier must also have a very high input impedance.

Another requirement for the ECG amplifier is that it be an AC amplifier. The reason for this requirement is that metallic electrodes (such as Ag–AgCl) applied to the electrolytic skin produces a DC halfcell potential. This potential tends to be on the order of 1 to 2 V, so it is more than 1000 times higher than the signal voltage. By making the amplifier respond only to AC, we eliminate the artifact caused by the DC halfcell potential.

The frequency selected for the -3-dB point of the ECG amplifier must be very low, close to DC, because the standard ECG waveform contains very low frequency components. The typical ECG signal has significant frequency component in the range 0.05 to 100 Hz, which is the industry standard frequency response for diagnostic ECG amplifiers (some clinical monitoring ECG amplifiers use 0.05 to 40 Hz to eliminate muscle artifact due to patient movements).

The typical ECG amplifier has differential inputs because the most useful ECG signals are differential in nature, and to suppress 60-Hz hum picked-up on the leads and the patient's body. In the most simple case, the right arm (RA) and left arm (LA) electrodes form the inputs to the amplifier, with the right leg (RL) defined as the common (see Fig. 14-16). The basic configuration of the amplifier in Fig. 14-16 is the AC-coupled instrumentation amplifier discussed earlier. The gain for this amplifier is set to slightly more than $\times 1000$, so a one-millivolt ECG peak signal will produce a one-volt output from this amplifier. Because of the high gain it is essential that the amplifier be well balanced. This requirement suggests the use of a dual amplifier for $A1$ and $A2$. An example might be the CA-3240 device, which is a dual BiMOS device that is essentially two CA-3140s in a single eight-pin miniDIP package. Also, in the interest of balance, resistors having tolerance of 1 percent or less should be used for the equal pairs.

The lower end -3-dB frequency response point is set by the input resistors and capacitors. In this case, the particular combination of values shown produces a response of 0.048 Hz.

The CMRR ADJUST control in Fig. 14-16 is usually a 10- to 20-turn trimmer potentiometer. It is adjusted in the following manner.

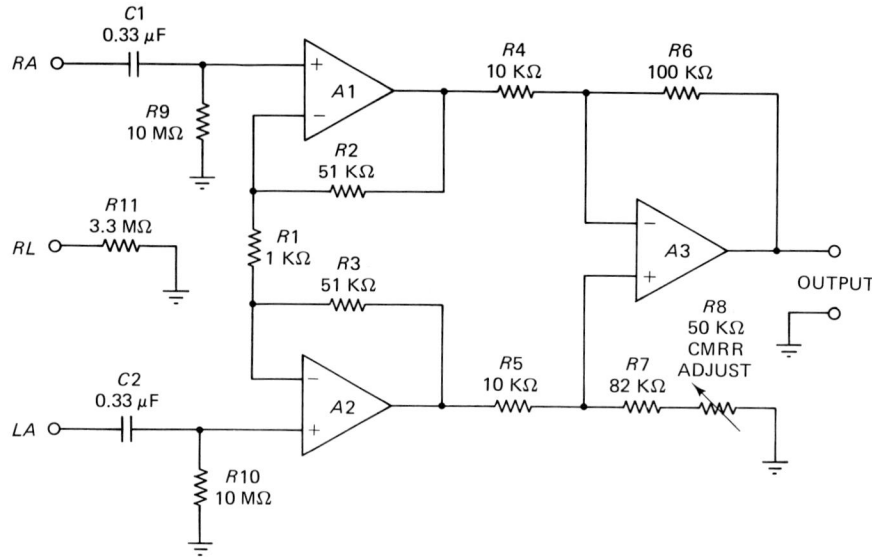

Fig. 14-16   Non-isolated ECG preamplifier.

1. Short together the RA, LA, and RL inputs.

2. Connect a DC voltmeter to the output (either a digital voltmeter or an analog meter with a 1.5-V DC scale (alternatively, a DC-coupled oscilloscope can be used, but be sure to identify the zero baseline).

3. Adjust CMRR ADJUST control ($R7$) for zero volts output. Disconnect the RL input and connect a signal generator between RL and the still-connected RA/LA terminal.

4. Adjust the output of the signal generator for a sine-wave frequency in the range 10 to 40 Hz and a peak-to-peak potential of 1 to 3 V.

5. Using an AC scale on the voltmeter (or an oscilloscope), again adjust CMRR ADJUST ($R7$) for the smallest possible output signal. It may be necessary to read just the voltmeter or oscilloscope input range control to greater sensitivity to observe the best null.

6. Remove the RA/LA short. The ECG amplifier is ready to use.

A suitable post-amplifier for the ECG preamplifier is shown in Fig. 14-17. This amplifier is placed in the signal line between the output of Fig. 14-16 and the input of the oscilloscope or paper chart recorder used to display the waveform. The gain of the post-amplifier is variable from 0 to +2; it will produce a 2-V maximum output when a 1-mV ECG signal provides a 1-V output from the preamplifier. Because of the high-level signals used, this amplifier can use ordinary 741 operational amplifiers.

The frequency response of the amplifier is set to an upper −3-dB point of 100 Hz, with the response dropping off at −6 dB/octave above that frequency. This frequency-response point is determined by capacitor $C3$ operating with resistor $R12$.

$$F_{HZ} = 1,000,000/2\pi RC$$

$$F_{Hz} = \frac{1,000,000}{(2)(3.14)(100,000)(0.015\ \mu F)} = 106\ Hz$$

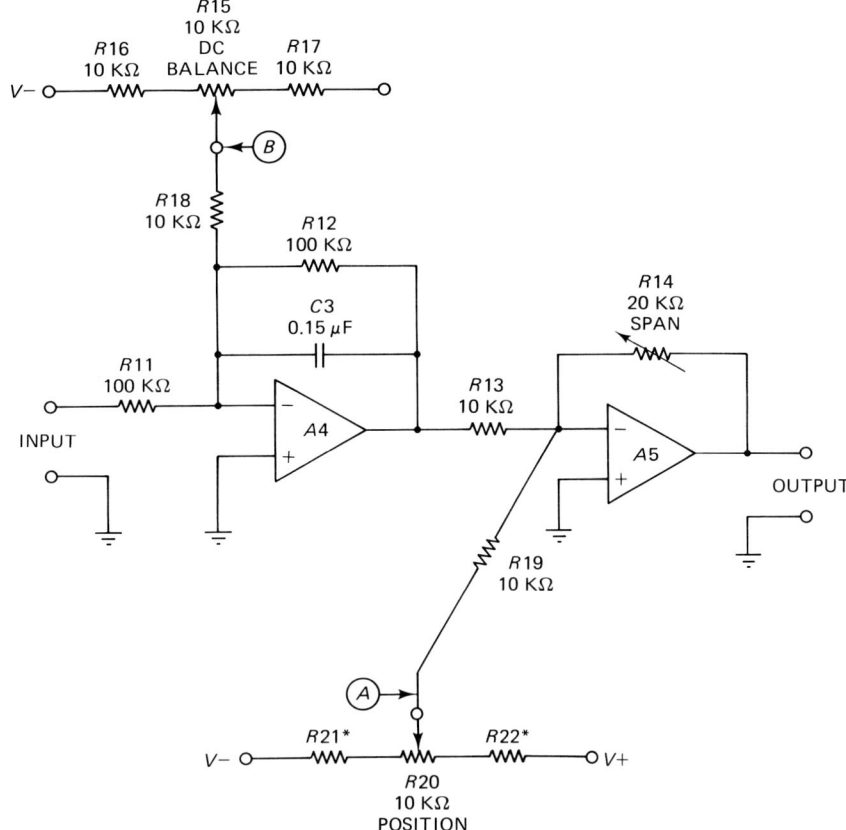

Fig. 14-17   Universal rear-end post amplifier.

There are three controls in the post-amplifier circuit: span, position, and DC balance. The span control is the 0 to 2 gain control, and the label *span* reflects instrumentation users' language, rather than electronics language. The position control sets the position of the output waveform on the display device. Resistors $R21$ and $R22$ are selected to limit the travel of the beam or pen to full scale. Set these resistors so that the maximum potential at the end terminals of $R20$ corresponds to full-scale deflection of the display device. Resistors $R21$ and $R22$ are selected to limit the travel of the beam or pen to full scale. Set these resistors so that the maximum potential at the end terminals of $R20$ corresponds to full-scale deflection of the display device.

The DC BALANCE control is used to cancel the collective effects of offset potentials created by the various stages of amplification. This control is adjusted as shown here.

Follow the CMRR ADJUST procedure, and then reconnect the short-circuit at RA, LA, and RL. The voltmeter is moved to the output of Fig. 14-17. Adjust the position control for zero volts at point A. Adjust the DC BALANCE control for zero volts at point B. Adjust the SPAN control ($R14$) through its entire range from zero to maximum while monitoring the output voltage. If the output voltage does not shift, then no further adjustment is needed.

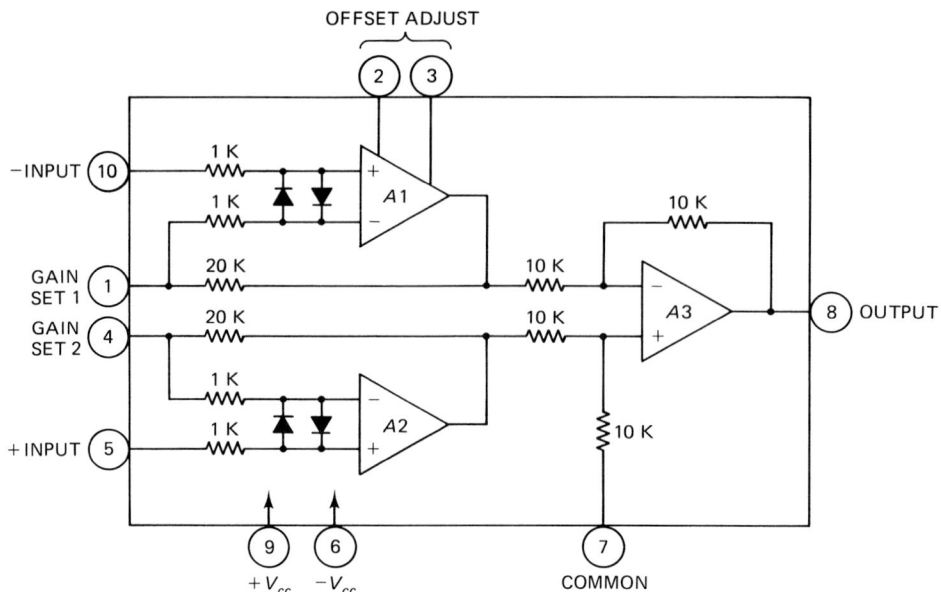

Fig. 14-18   Burr–Brown INA-101 instrumentation amplifier.

If the output voltage in the previous step varies as the span control is varied, adjust DC balance until varying the span control over its full range does not produce an output voltage shift. Repeat this step several times until no further improvement is possible.

Remove the RA, LA, RL short; the amplifier is ready for use.

IC INSTRUMENTATION AMPLIFIERS: SOME COMMERCIAL EXAMPLES

The operational amplifier truly revolutionized analog circuit design. For a long time, the only additional advances were that op-amps were vastly improved (they became nearer the ideal). While that was an exciting development, those devices were not truly new. The next big breakthrough came when the analog device designers made an IC version of the instrumentation amplifier shown in Fig. 14-11A, the integrated circuit instrumentation amplifier (IC IA). Today, several manufacturers offer substantially improved IC IA devices.

The Burr–Brown INA-101 (Fig. 14-18) is a popular IC IA device; a sample INA-101 circuit is shown in Fig. 14-19. This IC IA amplifier is simple to connect and use: There are only DC power connections, differential input connections, offset adjust connections, ground, and an output. The gain of the circuit is set by

$$A_{vd} = (40\ \Omega/R_g) + 1 \qquad (14\text{-}56)$$

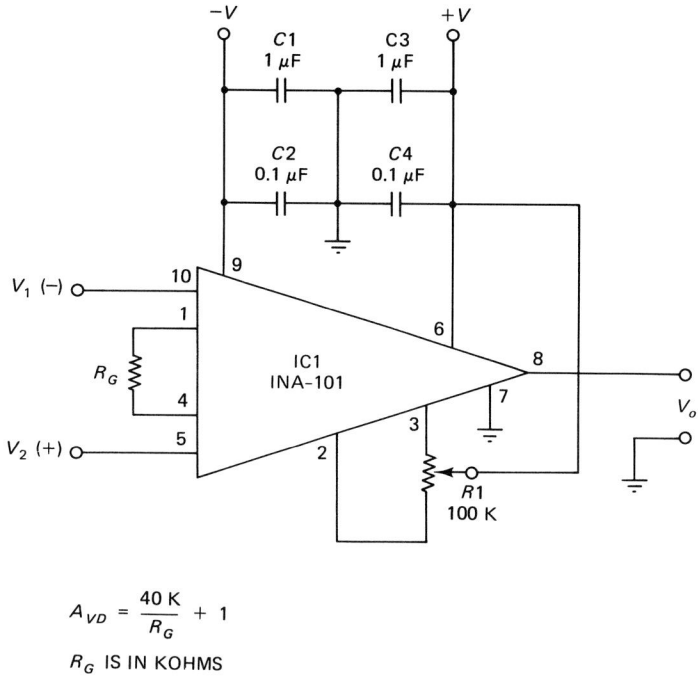

$$A_{VD} = \frac{40\text{ K}}{R_G} + 1$$

R_G IS IN KOHMS

Fig. 14-19 INA-101 circuit.

The INA-101 is a low-noise, low input bias current integrated-circuit version of the IA of Fig. 14-11A. The resistors labeled $R2$ and $R3$ in Fig. 14-11A are internal to the INA-101 and are 20 kΩ each, hence the 40-kΩ term in Eq. (14-56).

Potentiometer $R1$ in Fig. 14-19 is used to null the offset voltages appearing at the output. An offset voltage is a voltage that exists on the output at a time when it should be zero (i.e., when $V1 = V2$, so that $V1 - V2 = 0$). The offset voltage might be internal to the amplifier, or alternatively, a component of the input signal. DC offsets in signals are common, especially in biopotentials amplifiers such as ECG and EEG, and in chemical transducers such as pH, pO2, and pCO2.

Another IC IA is the LM-363-xx device shown in Figure 14-20; the miniDIP version is shown in Fig. 14-20A (an eight-pin metal can is also available), while a typical circuit is shown in Figure 14-20B. The LM-363-xx device is a fixed-gain IC IA. Three versions of the LM-363-xx are enumerated according to gain:

Designation	Gain (A_v)
LM-363-10	$\times 10$
LM-363-100	$\times 100$
LM-363-500	$\times 500$

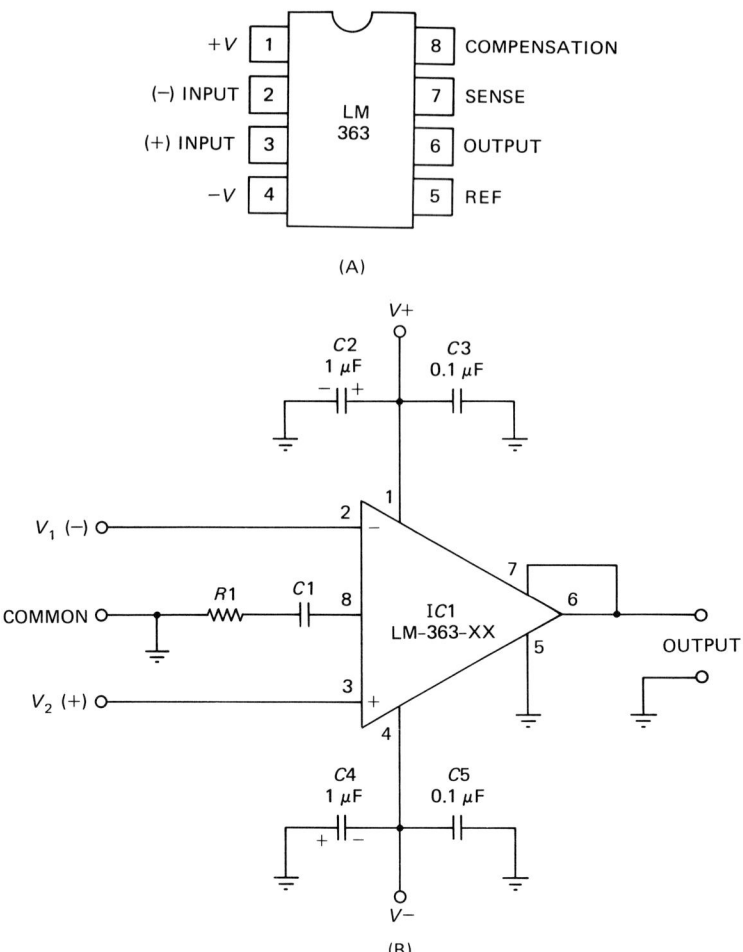

Fig. 14-20 (A)LM-363-xx instrumentation amplifier and (B) typical circuit.

The LM-363-xx is useful in places where one of the standard gains is required and minimum space is available. Two examples spring to mind. The LM-363-xx can be used as a transducer preamplifier, especially in noisy signal areas; the LM-363-xx can be built onto (or into) the transducer to build up its signal before sending it to the main instrument or signal acquisition computer. Another possible use is in biopotentials amplifiers. Biopotentials are typically very small, especially in lab animals. The LM-363-xx can be mounted on the subject and a higher-level signal sent to the main instrument.

A selectable gain version of the LM-363 device is shown in Figure 14-21A; the 16-pin DIP package is shown in Figure 14-21A, while a typical circuit is shown in Figure 14-21B. The type number of this device is LM-363-AD, which distinguishes it from the LM-363-xx devices. The gain can be ×10, ×100, or ×1000 depending on the

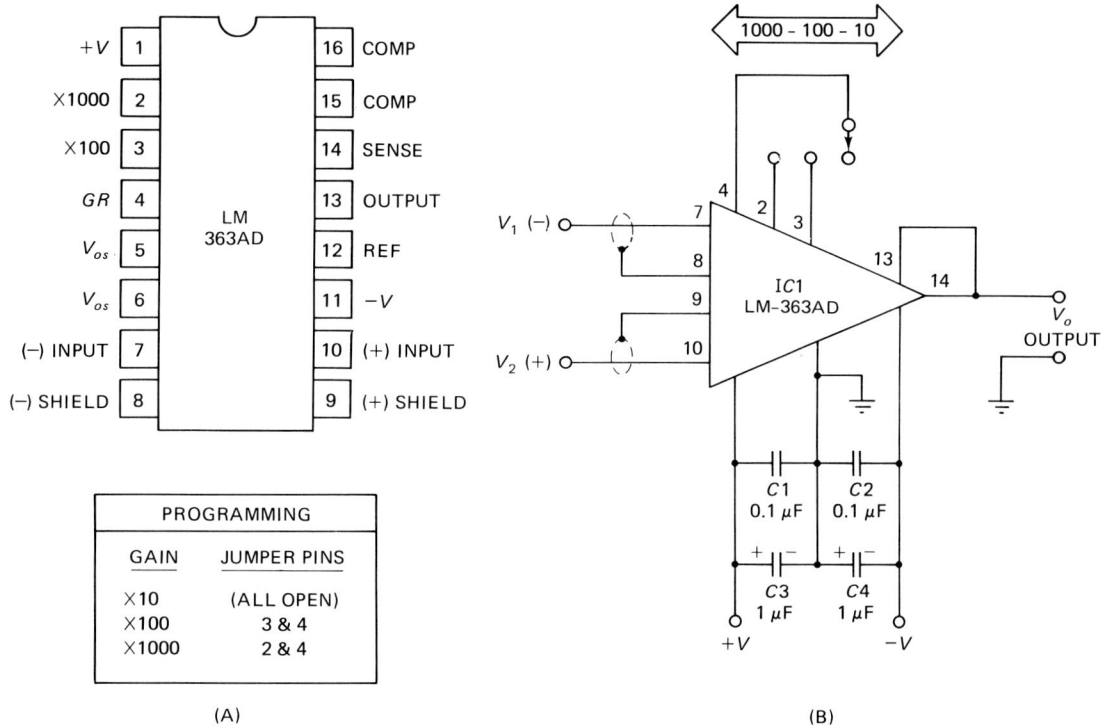

Fig. 14-21 (A) LM-363-AD instrumentation amplifier and (B) typical circuit.

programming of the gain setting pins (2, 3, and 4). The programming protocol is given here.

Gain Desired	Jumper Pins
× 10	(All Open)
×100	3 and 4
×1000	2 and 4

Switch S1 in Figure 14-21B is the GAIN SELECT switch. This switch should be mounted close to the IC device, but it is quite flexible in mechanical form. The switch could also be made from a combination of CMOS electronic switches (e.g., 4066). The DC power supply terminals are treated in a manner similar to the other amplifiers. Again, the 0.1-μF capacitors need to be mounted as close as possible to the body of the LM-363-AD. Pins 8 and 9 are guard shield outputs. These pins are a feature that makes the LM-363-AD more useful for many instrumentation problems than other models. By outputting a signal sample back to the shield of the input lines, we can increase the common-mode rejection ratio. This feature is used a lot in biopotentials amplifiers and in other applications where a low-level signal must pass through a strong interference (high noise) environment. Guard shield theory is discussed more fully here.

The LM-363 devices will operate with DC supply voltages of ± 5 to ± 18 V DC, with a common mode rejection ratio (CMRR) of 130 dB. The 7-nV/[SQR(Hz)] noise figure makes the device useful for low noise applications. (A 0.5-nV model is also available).

GUARD SHIELDING

One of the properties of the differential amplifier, including the instrumentation amplifier, is that it tends to suppress interfering signals from the environment. The common mode rejection process is at the root of this capability. When an amplifier is used in a situation where it is connected to an external signal source through wires, those wires are subjected to strong local 60-Hz AC fields from nearby power-line

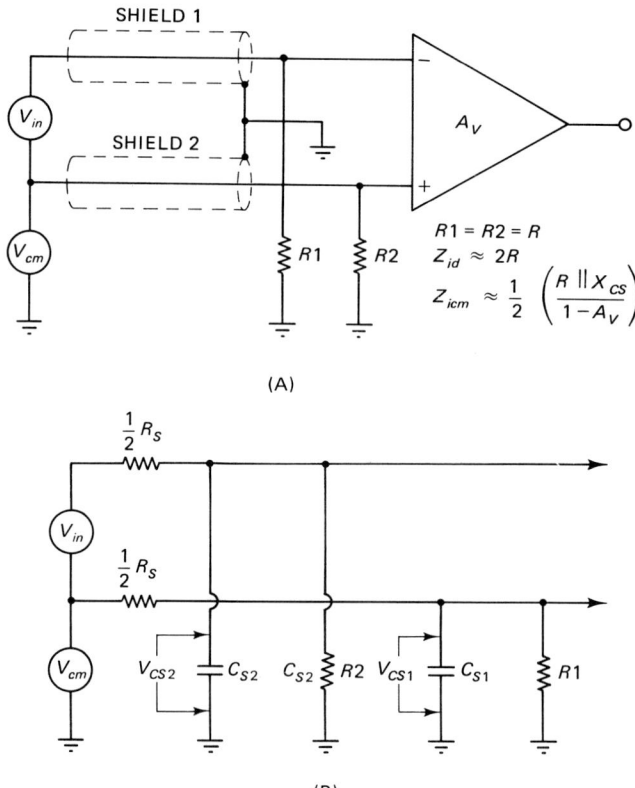

(A)

(B)

Fig. 14-22 (A) Shielded input leads on differential amplifier and (B) equivalent circuit.

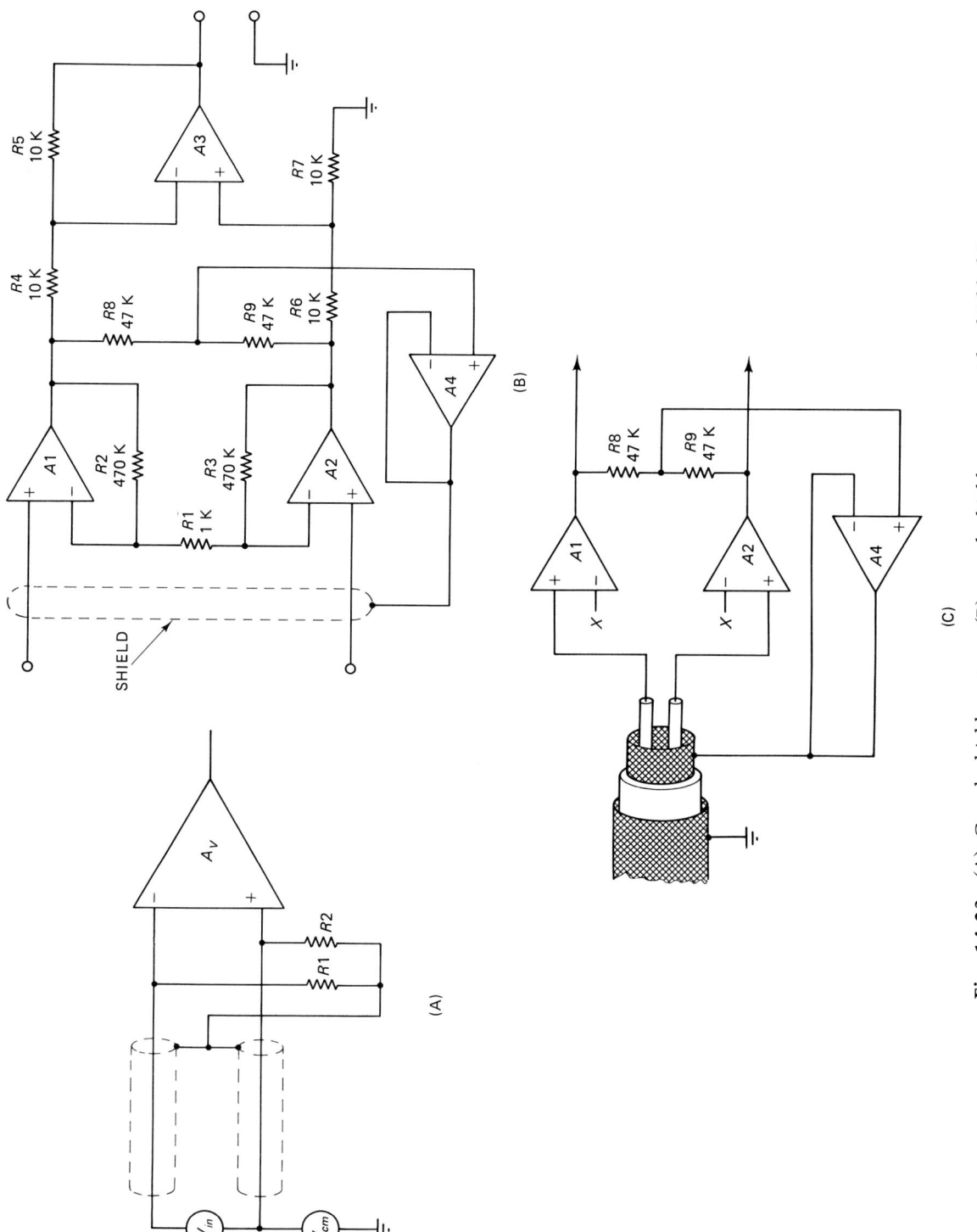

Fig. 14-23 (A) Guard shield system, (B) guard shield system with shield driver amplifier, and (C) practical guard shield system.

wiring. Fortunately, in the case of the differential amplifier, the field affects both lines equally, so the induced interfering signal is canceled out by the common mode rejection property of the amplifier.

Unfortunately, the cancellation of interfering signals is not total. There may be, for example, imbalances in the circuit that tend to deteriorate the CMRR of the amplifier. These imbalances may be either internal or external to the amplifier circuit. Figure 14-22 shows a common scenario. In this figure we see the differential amplifier connected to shielded leads from the signal source V_{in}. Shielded lead wires offer some protection from local fields, but there is a problem with the standard wisdom regarding shields: It is possible for shielded cables to manufacture a valid differential signal voltage from a common mode signal!

Figure 14-22B shows an equivalent circuit that demonstrates how a shielded cable pair can create a differential signal from a common mode signal. The cable has capacitance between the center conductor and the shield conductor surrounding it. In addition, input connectors and the amplifier equipment internal wiring also exhibits capacitance. These capacitances are lumped together in the model of Fig. 14-22B as C_{S1} and C_{S2}. As long as the source resistances and shunt resistances are equal and the two capacitances are equal, there is no problem with circuit balance. But inequalities in any of these factors (which are commonplace) creates an unbalanced circuit in which common mode signal V_{cm} can charge one capacitance more than the other. As a result, the difference between the capacitance voltages, V_{CS1} and V_{CS2}, is seen as a valid differential signal.

A low-cost solution to the problem of shield-induced artifact signals is shown in Fig. 14-23A. In this circuit a sample of the two input signals are fed back to the shield, which in this situation is not grounded. Alternatively, the amplifier output signal is used to drive the shield. This type of shield is called a guard shield. Either double shields (one on each input line) as shown or a common shield for the two inputs can be used.

An example of an improved guard shield for the instrumentation amplifier is shown in Fig. 14-23B. In this case a single shield covers both input lines, but it is possible to use separate shields. In this circuit a sample of the two input signals is taken from the junction of resistors $R8$ and $R9$ and fed to the input of a unity-gain buffer/driver guard amplifier ($A4$). The output of $A4$ is used to drive the guard shield.

Perhaps the most common approach to guard shielding is the arrangement shown in Fig. 14-23C. Here we see two shields used; the input cabling is double-shielded insulated wire. The guard amplifier drives the inner shield, which serves as the guard shield for the system. The outer shield is grounded at the input end in the normal manner, and serves as an electromagnetic-interference suppression shield.

CONCLUSION

In this chapter we have discussed both the generic DC differential amplifier circuit and the more complex instrumentation amplifier. In the next chapter (Chapter 15), we will take a look at a highly specialized form of amplifier that is also usually differential. The operant property of the next class of device however is isolation, that is, an extremely large impedance between the input terminals and the DC power supply that is derived from the AC power mains.

15

Isolation Amplifiers
and Their Applications

There are many applications in which ordinary solid-state differential amplifiers are either themselves in danger, because of a hostile electrical environment, or they present a danger to the users. An example of the former is an amplifier in a high-voltage experiment such as a biochemist's electrophoresis system, while the latter is represented by certain devices used in clinical medicine. Many of the commercial products of this type now available on the market are not, strictly speaking, integrated circuits, but rather they are hybrid circuits. Nonetheless, it is important to discuss these devices in any book on linear IC amplifiers because they are used in a manner that is similar to monolithic devices.

An isolation amplifier (Fig. 15-1) has an extremely high impedance between the signal inputs and those power supply terminals that are connected to a DC power supply that are, in turn, connected to the AC power mains. Thus, in isolation amplifiers there is an extremely high resistance (e.g., 10^{12} Ω) between the amplifier input terminals and the AC power line. In the case of medical equipment, the designer's goal is to prevent minute leakage currents (which can cause fatal "microshock") from the 60-Hz AC power lines from being applied to the patient. Current levels that are normally negligible to humans can theoretically be fatal to a hospital patient in situations where the body is invaded by medical devices that are electrical conductors. In other cases, the high impedance is used to prevent high voltages at the signal inputs from adversely affecting the rest of the circuitry. Modern isolation amplifiers can provide more than 10^{12} Ω of isolation between the AC power lines and the signal inputs, which is sufficient for both cases.

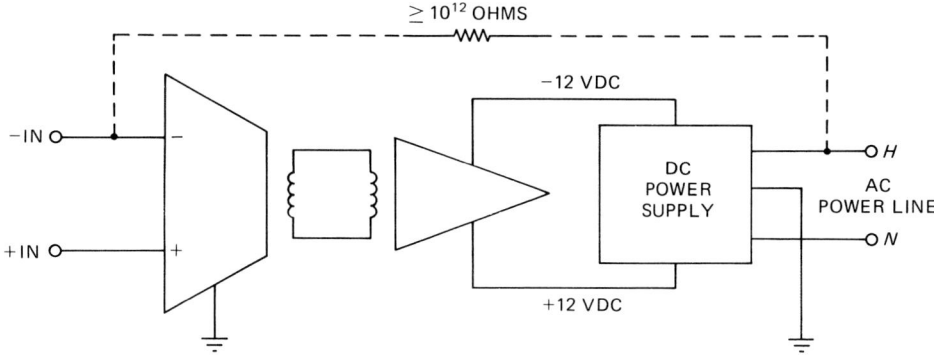

Fig. 15-1 Isolation differential amplifier.

Several different circuit symbols are used to denote the isolation amplifier in schematic diagrams, but the one that is the most common is shown in Fig. 15-2. It consists of the regular triangular amplifier symbol broken in the middle to indicate isolation between the "A" and "B" sections. The following connections are usually found on the isolation amplifier:

Nonisolated "A" Side V + and V − DC power supply lines (to be connected to a DC supply powered by the AC lines), output to the rest of the (nonisolated) circuitry, and (in some designs) a nonisolated

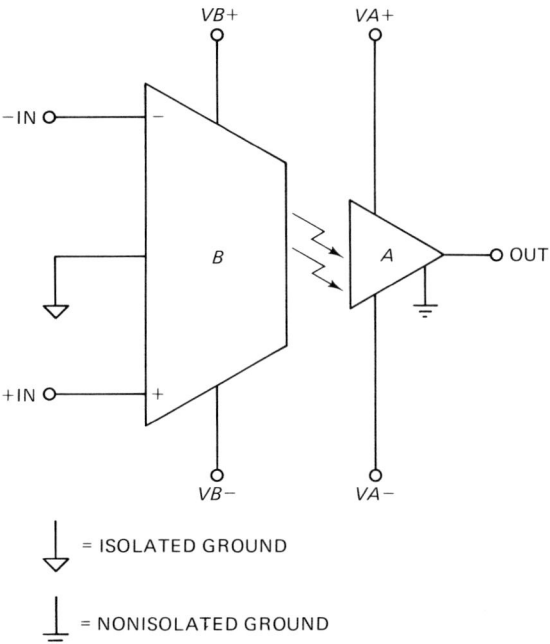

= ISOLATED GROUND

= NONISOLATED GROUND

Fig. 15-2 Isolation amplifier symbol.

ground or common. This ground is connected to the chassis or main system ground also served by the main DC power supplies.

Isolated "B" Side Isolated $V+$ and $V-$, isolated ground or common, and the signal inputs. The isolated power supply and ground are not connected to the main power supply or ground systems. Batteries are sometimes used for the isolated side, while in other cases special isolated DC power supplies derived from the main supplies are used.

APPROACHES TO ISOLATION AMPLIFIER DESIGN

Different manufacturers use different approaches to the design of isolation amplifiers. Common circuit approaches to isolation include battery powered, carrier operated, optically coupled, and current loading. These methods are discussed in detail in the next sections.

Battery-Powered Isolation Amplifiers

The battery approach to isolation amplifier design is perhaps the simplest to implement, but it is not always most suitable due to problems inherent in battery upkeep. A few products exist, however, that use a battery-powered front-end amplifier, even though the remainder of the equipment is powered from the 100-V AC power line. Other products are entirely battery powered. A battery-powered amplifier or instrument is isolated from the AC power mains only if the battery is disconnected from the charging circuit during use. In some battery-powered instruments used in medicine, mechanical interlocks and electrical logic circuitry prevent the instrument from being turned on if the AC power cord is still attached. Later in this chapter we will study a battery powered cardiac output computer as a design example of this type of isolation.

Carrier-Operated Isolation Amplifiers

Figure 15-3A shows an isolation amplifier that uses the carrier signal technique to provide isolation. The circuitry inside of the dashed line is isolated from the AC power lines (in other words, the "B" side of Fig. 15-2). The voltage gain of the isolated section is typically in the range $\times 10$ to $\times 500$.

The isolation is provided by separation of the ground, power supply, and signal paths into two mutually exclusive sections by high-frequency transformers T1 and T2. These transformers have a

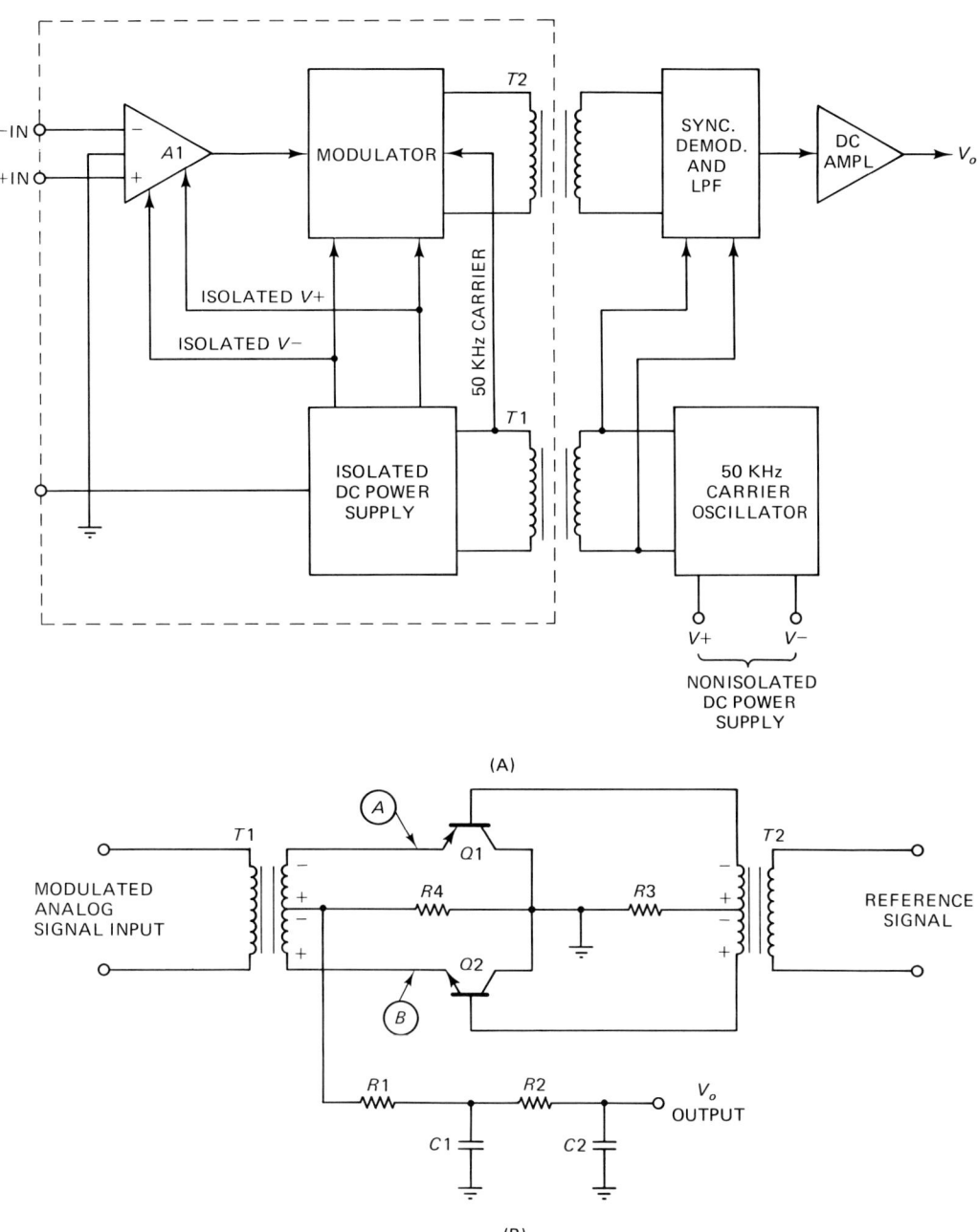

Fig. 15-3 (A) Carrier-operated isolation amplifier and (B) phase-sensitive detector.

design and core material that works very well in the ultrasonic (20 to 500 kHz) region, but that is very inefficient at the 60-Hz frequency used by the AC power lines. This design feature allows the transformers to easily pass the high-frequency carrier signal, while severely attenuating 60-Hz energy. Although most models use a carrier frequency in the 50- to 60-kHz range, examples of carrier amplifiers exist over the entire 20- to 500-kHz range.

The carrier oscillator signal is coupled through transformer T1 to the isolated stages. Part of the energy from the secondary of T1 is directed to the modulator stage; the remainder of the energy is rectified and filtered, and it is then used as an isolated DC power supply. The DC output of this power supply is used to power the input "B" amplifiers and the modulator stage.

An analog signal applied to the input is amplified by A1, and it is then applied to one input of the modulator stage. This stage amplitude modulates the signal onto the carrier. Transformer T2 couples the signal to the input of the demodulator stage on the nonisolated side of the circuit. Either envelope or synchronous demodulation may be used, although the latter is considered superior. Part of the demodulator stage is a low-pass filter that removes any residual carrier signal from the output signal. Ordinary DC amplifiers following the demodulator complete the signal processing chain.

An example of a synchronous demodulator circuit is shown in Fig. 15-3B. These circuits are based on switching action. Although the example shown uses bipolar PNP transistors as the electronic switches, other circuits use NPN transistors, FETs, or CMOS electronic switches (e.g., 4066 device).

The signal from the modulator has a fixed frequency in the range from 20 to 500 kHz, and it is amplitude modulated with the input signal from the isolated amplifier. This signal is applied to the emitters of transistors Q1 and Q2 (via T1) in push-pull. On one-half of the cycle, therefore, the emitter of Q1 will be positive with respect to the emitter of Q2. On alternate half-cycles, the opposite situation occurs: Q2 is positive with respect to Q1. The bases of Q1 and Q2 are also driven in push-pull, but by the carrier signal (called here the reference signal). This action causes transistors Q1 and Q2 to switch on and off out of phase with each other.

On one-half of the cycle, the polarities are as shown in Fig. 3B; transistor Q1 is turned on. In this condition point A on T1 is grounded. The voltage developed across load resistor R4 is positive with respect to ground.

On the alternate half-cycle, Q2 is turned on, so point B is grounded. But the polarities have reversed, so the polarity of the voltage developed across R4 is still positive. This causes a full-wave output waveform across R4, which when low-pass filtered becomes a DC voltage level proportional to the amplitude of the input signal. This same description of synchronous demodulators also applies to the circuits used in some carrier amplifiers (a specialized laboratory amplifier used for low-level signals).

A variation on this circuit replaces the modulator with a voltage-controlled oscillator (VCO) that allows the analog signal to frequency modulate a carrier signal generated by the VCO. The power supply carrier signal is still required, however. A phase detector, phase-locked

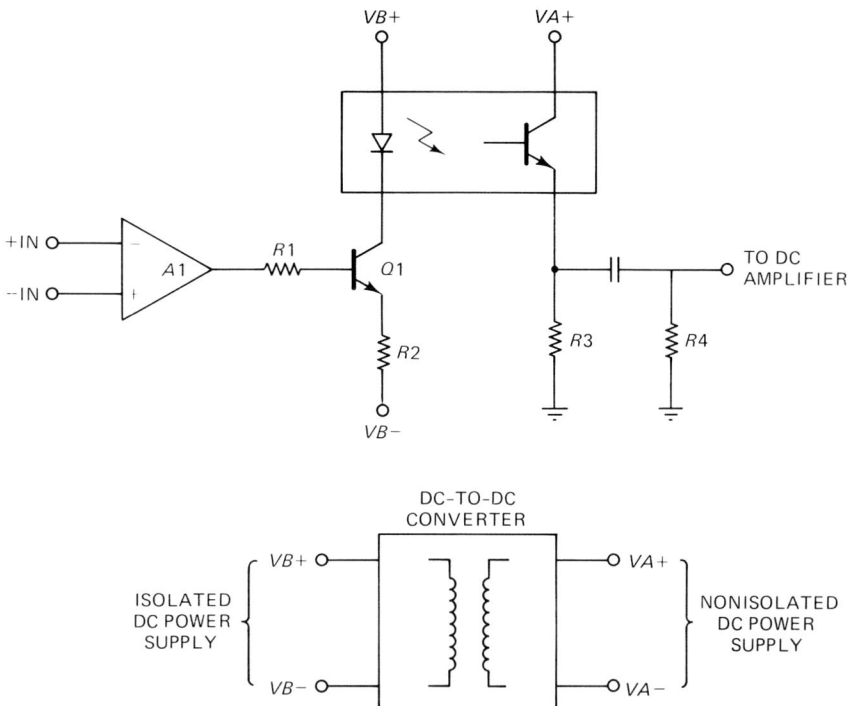

Fig. 15-4 Optocoupled isolation amplifier.

loop (PLL), or pulse-counting frequency modulation (FM) detector on the nonisolated side recovers the signal.

Optically Coupled Isolation Amplifier Circuits

Electronic optocouples (also called optoisolators) are sometimes used to provide the desired isolation. In early designs of this class, a light-emitting diode (LED) was mounted together with a photoresistor or phototransistor. Modern designs, however, use integrated circuit (IC) optoisolators that contain an LED and phototransistor inside of a single DIP IC package.

There are actually several approaches to optical coupling. Two common methods are the carrier and the direct methods. The carrier method is the same as discussed in the previous section, except that an optoisolator replaces transformer T2. The carrier method is not the most widespread in optically coupled isolation amplifiers because of the frequency response limitations of some IC optoisolators.

The more common direct approach is shown in Fig. 15-4. This circuit uses the same DC-to-DC converter to power the isolated stages as was used in other designs. It keeps A1 isolated from the AC power mains, but it is not used in the signal coupling process. In some designs, the high-frequency carrier power supply is actually a separate block from the isolation amplifier.

The LED in the optoisolator is driven by the output of isolated amplifier $A1$. Transistor Q1 serves as a series switch to vary the light output of the LED proportional to the analog signal from $A1$. Transistor Q1 normally passes sufficient collector current to bias the LED into a linear portion of its operating curve. The output of the phototransistor is AC coupled to the remaining amplifiers on the nonisolated side of the circuit, so that the offset condition created by the LED bias is eliminated.

Although not strictly speaking an isolated amplifier by the definition used herein, another category of optical isolation is especially attractive for applications where the environment is too hostile for ordinary electronics. It is possible to use LED and phototransistor transmitter and receiver modules in a fiber optic system to provide isolation. A battery-powered (or otherwise isolated) amplifier will sense the desired signal, convert to an AM or FM light signal, and transmit it down a length of fiber optic cable to a phototransistor receiver module. At that point the signal will be recovered and processed by the nonisolated electronics.

Current-Loading Isolation Amplifiers

A unique current-loading isolation amplifier was used in the front end of an electrocardiograph (ECG) medical monitor that found widespread acceptance. A simplified schematic is shown in Fig. 15-5. Notice that there is no obvious coupling path for the signal between the isolated and nonisolated sides of the circuit. The gain-of-24 isolated input preamplifier ($A1$) in Fig. 15-5 consists of a high input impedance operational amplifier.

This amplifier is needed to interface with the very high source impedance normal to electrodes in ECG systems. The output of $A1$ is connected to the isolated -10-V DC power supply through load resistor $R1$. This power supply is a DC-to-DC converter operating at 250 kHz. Transformer T1 provides isolation between the floating power supplies on the isolated "B" side of the circuit and the nonisolated "A" side of the circuit (which are powered by the AC line).

An input signal causes the output of $A1$ to vary the current loading of the floating -10-V DC power supply. Changing the current loading proportional to the analog input signal causes variation of the T1 primary current that is also proportional to the analog signal. This current variation is converted to a voltage variation by amplifier $A2$. An offset null control ($R3$) is provided in the $A1$ circuit to eliminate the offset at the output due to the quiescent current flowing when the analog input signal is zero. In that case, the current loading of T1 is constant—but still provides an offset to the $A2$ amplifier.

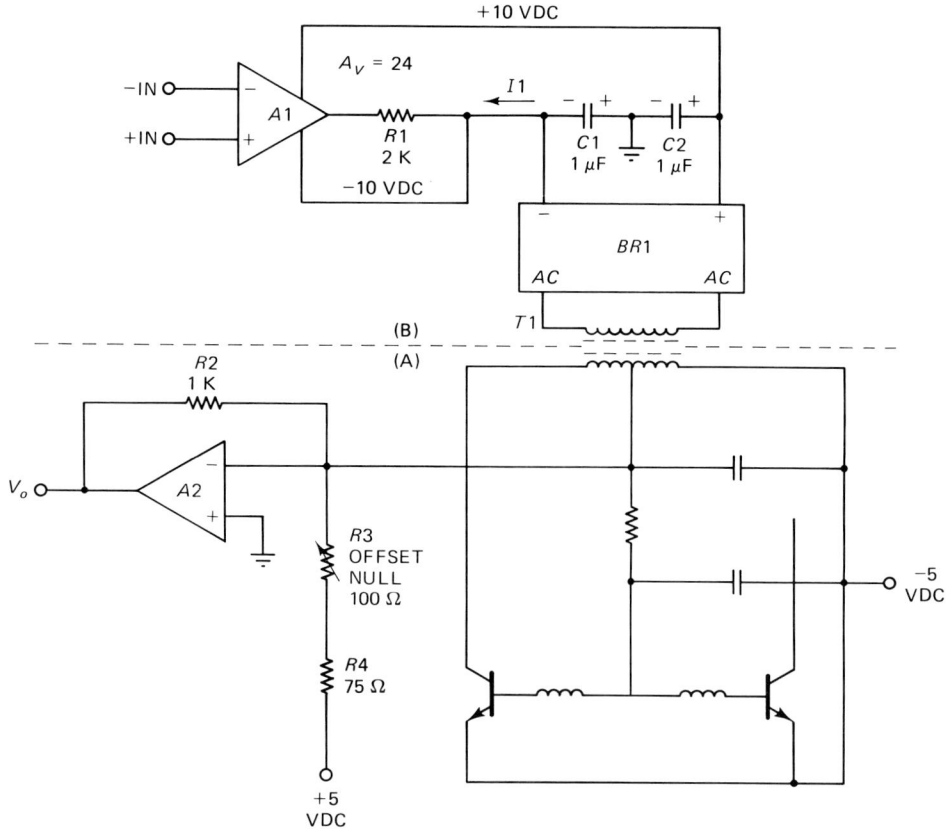

Fig. 15-5 Current-loading isolation amplifier.

Cardiac Output Computer

The problems presented by most electronic signal acquisition situations are simple compared with the problems presented in measuring human cardiac output. The principal difference is that cardiac output is usually measured using an invasive surgical technique on living humans. This type of measurement is presented here to demonstrate a data acquisition technique that for the sake of safety almost absolutely requires an isolation amplifier to interface with the signal source.

Cardiac output (CO) is defined as the rate of blood volume pumped by the heart. The question being asked of the CO measurement is "How much blood is this person pumping per unit of time." Cardiac output is measured in units of liters of blood per minute of time (l/min). In healthy adults CO typically reaches a value between 3 and 5 l/min.

A quantitative measure of cardiac output is the product of the stroke volume and the heart rate. The stroke volume is merely the volume of blood expelled from the heart ventricle (lower chamber) during a single contraction of the heart. Cardiac output is calculated from

$$CO = VR \qquad\qquad (15\text{-}1)$$

where CO is the cardiac output in liters per minute (l/min), V the stroke volume in liters per beat (l/beat), and R the heart rate in beats per minute (beat/min).

It is difficult, and usually impossible (except on animals in laboratory settings) to directly measure cardiac output using any technique based on the above equation. The main problem is obtaining good stroke volume data without excessive risk to the patient. There are, however, several related indirect methods that yield CO data for the physician.

The thermodilution method of cardiac output measurement has become the standard indirect method for measuring cardiac output in clinical settings, and it is also popular among laboratory scientists. Thermodilution technique forms the basis for most clinical and research cardiac output computers now on the market. One reason why thermodilution is preferred is that no poisonous injectates are used (as they are in radioopaque or optical dye dilution methods); only ordinary medical intravenous (IV) solutions such as normal saline or 5-percent dextrose in water (D$_5$W) are used. The thermodilution measurement of cardiac output is made using a special hollow catheter that is inserted into one of the patient's veins, usually on the right arm (the brachian vein is popular).

The catheter is multilumened, and one of the lumens has its output hole several centimeters from the catheter tip. This proximal lumen is situated so that it is outside the heart (close to the input valve on the right atrium) when the tip is all the way through the heart, resting in the pulmonary artery (Fig. 15-6). Other lumens in the catheter output at the tip, so they are used to measure pressures in the pulmonary artery in other procedures. A thermistor in the tip registers a resistance change with changes in blood temperature.

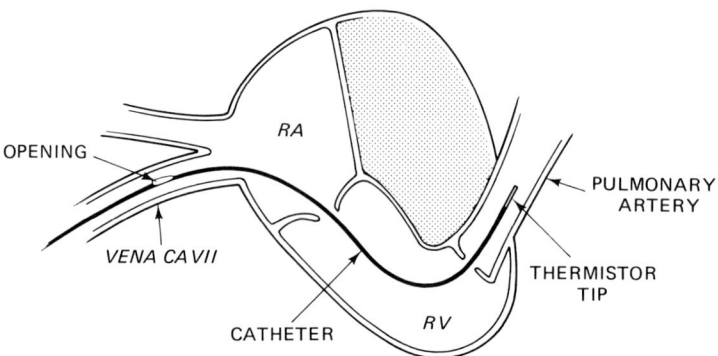

Fig. 15-6 Method for measuring cardiac output.

Most thermodilution cardiac output computers operate on a version of Eq. (15-2).

$$CO = \frac{64.8 C_t V_i [T_b - T_i]}{\int T_{b'} \, dt} \tag{15-2}$$

where CO is the cardiac output in liters per minute (l/min), 64.8 represents a collection of other constants and the conversion factor from seconds to minutes, C_t is a constant that is supplied with the injectate catheter that accounts for the temperature rise in the portion of the outside of the patient's body, T_b the blood temperature in degrees Celsius, T_i the temperature of the injectate in degrees Celsius, $T_{b'}$ the temperature of the blood as it changes due to mixing with the injectate, and V_i the injectate volume.

The mathematical symbol in the denominator of Eq. (15-2) tells us that the temperture of the blood at the output side of the heart is integrated; in other words, the computer finds the time-average of the temperature as it changes.

Example

A special cardiac output computer "dummy catheter" test fixture enters a temperature signal that simulates a temperature change of 10° Celsius for a period of 10 seconds. Find the expected reading during a test of the instrument if the following front panel settings are entered: $C_t = 49.6$, $T_b = 37°$, injectate temperature T_i is 25°, and injectate volume V_i is 10 ml.

Solution

$$CO = \frac{(64.8) C_t V_i [T_b - T_i]}{\int T_{b'} \, dt} \ (l/min)$$

$$CO = \frac{(64.8)(49.6)(10 \text{ ml} \times 1 \ 1/1000 \text{ ml})[(37) - (25)]}{(10 \text{ degrees})(10 \text{ sec})}$$

$$CO = 385.68/100 = 3.9 \ l/min$$

The thermistor in the end of the catheter is usually connected in a Wheatstone bridge circuit (see Fig. 15-7). The DC excitation of the bridge is critical. Either the short-term stability of this voltage must be very high, or a ratiometric method must be used to cancel excitation potential drift. In addition, it is necessary to limit the bridge excitation potential to about 200 mV for reasons of safety to the patient (electri-

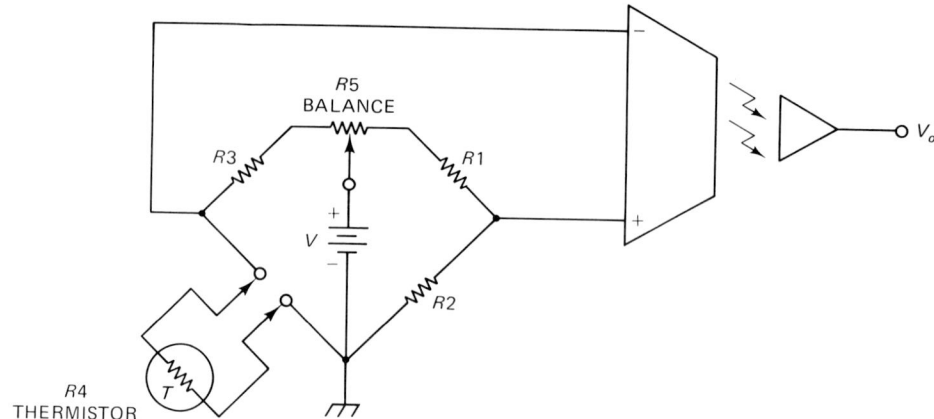

Fig. 15-7   Isolation amplifier used in thermal measurement system.

cal leakage is especially dangerous because the thermistor is inside the heat or pulmonary artery). This low value of excitation voltage promotes both patient safety and thermistor stability through freedom from self-heating induced drift, even though imposing a greater burden on the amplifier design.

Figure 15-7 shows a simplified schematic of a typical cardiac output computer front-end circuit. The thermistor ($R4$) is in a Wheatstone bridge circuit consisting also of $R1$ through $R4$, with potentiometer $R5$ serving to balance the bridge. An autobalancing or zeroing method is sometimes used for this function. Those circuits use a digital-to-analog converter (DAC) to inject a current into one node of the bridge, and that current nulls the bridge circuit to zero. The physician waits a few minutes for the thermistor to equilibrate with blood temperature (usually it is in this condition by the time it is threaded through the venous system to the pulmonary artery). The output of the bridge, depending on the design, is typically 1.2 to 2.5 millivolts per degree Celsius, with 1.8 mV/°C being quite common. This signal is amplified approximately 1000 times to 1 V/°C by the preamplifier. This preamplifier is an isolated amplifier for reasons of patient safety. The output of this circuit, V_o, is used in the denominator of an equation as in the foregoing example.

The block diagram for the sample analog cardiac output computer is shown in Fig. 15-8. The front-end circuitry from Fig. 15-7 is in the blocks marked "Bridge" and "Pre-amp." The isolator circuit is merely a buffer amplifier that permits V_0 to be output to an analog paper chart recorder. Analysis of the wave shape reveals errors of technique—and thus explains odd readings that are not supported by other clinical facts. For this reason the physician often demands an analog output, so that it can be visually inspected. The temperature signal (V_o) is integrated, and then it is sent to an analog divider where it is combined with the temperature difference signal ($T_b - T_i$) and the constants (all

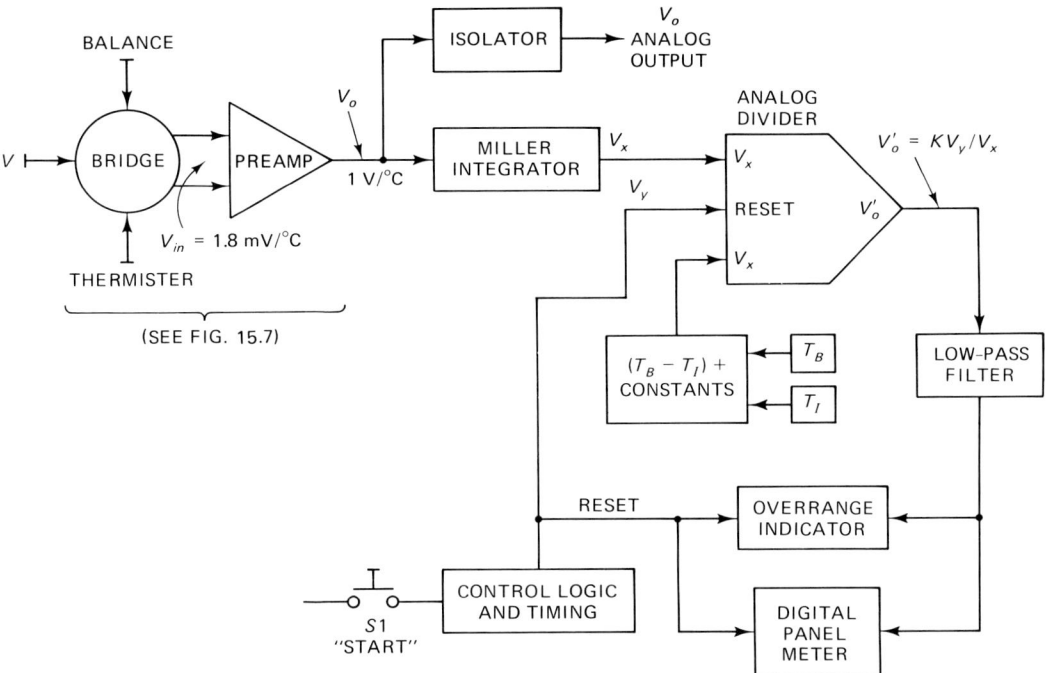

Fig. 15-8 Computer block diagram of analog cardiac output.

represented by a single voltage). The low-pass filtered output of the analog multiplier is a measure of the cardiac output; it is displayed on a digital voltmeter.

Sensor Isolation Applications

Medical applications of isolation amplifiers are perhaps the best known because they contribute much to patient safety. Nonetheless, there is also a substantial body of nonmedical applications for isolation amplifiers. These amplifiers are especially useful whenever the environment (e.g., high voltages) can adversely effect the device, or when a conventional amplifier can perturb the signal source (certain transducers fall into this category). Figure 15-9 shows two situations where an isolation amplifier is interfaced to different forms of sensor. In Fig. 15-9A the sensor is a photodiode used to detect the existence of an electrical arc across a spark gap (SG). In some such systems strong electric fields can destroy conventional ground-referenced systems.

Another sensor application is shown in Fig. 15-9B. A thermocouple (TC) is a temperature transducer consisting of two dissimilar metals forming a junction. If the metals have different work functions, an electric potential is generated across the ends of the wires that is proportional to both the work function difference and the junction temperature. Ordinarily, the TC can be connected to any amplifier. But

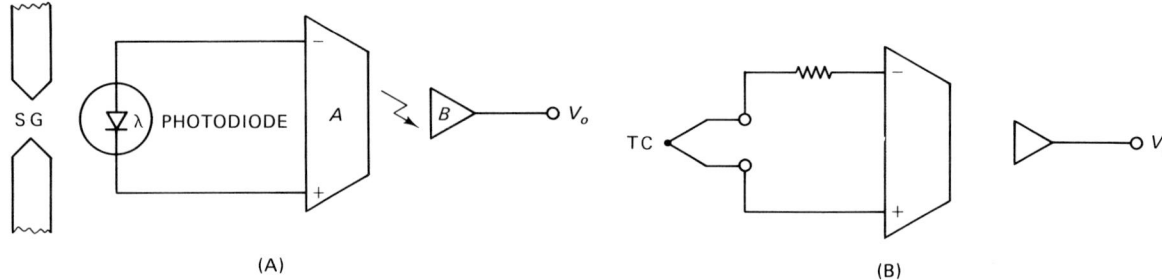

Fig. 15-9   Isolation amplifier used to interface (A) photodiode and (B) thermocouple.

in certain cases the local electrical environment is too adverse for the electronic components. An example is a high-voltage melting vessel for melting metals. It is necessary to know and control the temperature, but the vessel floats electrically at a high AC potential.

Electrophoresis Column

Electrophoresis is a process used in biochemistry laboratories to separate proteins from biological fluids by using an electrical field as the selection factor. In an electrophoresis column (Fig. 15-10), a pair of platinum electrodes connected to a constant-current, high-voltage DC power supply creates an electrical field between the top and bottom of the column. Various protein molecules within the fluid react very specifically to the field and will migrate to, and settle at, a specific electrical potential level within the field (hence, to a specific height within the column). The chemist can then extract the particular

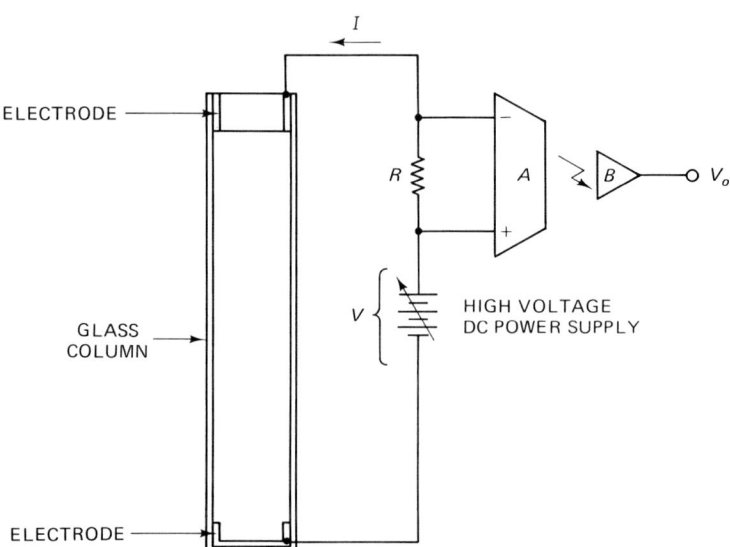

Fig. 15-10   Electrophoresis column.

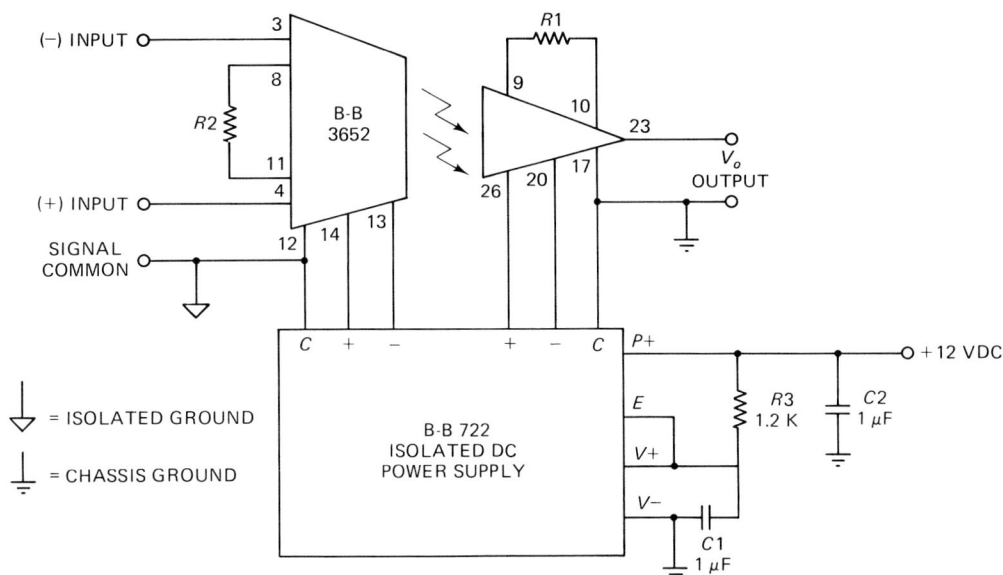

Fig. 15-11 Connections for the B–B 3652 isolation amplifier.

protein of interest if the height of the column and the potential gradient are known.

It is necessary to continuously monitor the current flowing in the column. If computer data logging is used, an isolation amplifier may be required to keep the high voltage of the circuit from the computer. A series resistor is inserted into the column circuit, and the voltage drop across it is proportional to the current flowing (Ohm's law). The differential inputs of the isolated side of the amplifier are bridged across this resistor to indirectly measure the current drain.

COMMERCIAL PRODUCT EXAMPLE

Figure 15-11 shows the circuit of an isolation amplifier based on the Burr–Brown 3652 device. The DC power for both the isolated and nonisolated sections of the 3652 is provided by the 722 dual DC-to-DC converter. This device produces two independent ±15 V DC supplies that are each isolated from the 60-Hz AC power mains and from each other. The 722 device is powered from a +12 V DC source that is derived from the AC power mains. In some cases, the nonisolated section (which is connected to the output terminal) is powered from a bipolar DC power supply that is derived from the 60-Hz AC mains, such as a –12 + V DC or ±15 V DC supply. In no instance, however, should the isolated DC power supplies be derived from the AC power mains.

There are two separate ground systems in this circuit, symbolized by the small triangle and the regular three-bar chassis ground symbol. The isolated ground is not connected to either the DC power supply ground or common, or the chassis ground. It is kept floating at all times, and it becomes the signal common for the input signal source.

The gain of the circuit is approximately

$$\text{GAIN} = 1,000,000/(R1 + R2 + 115) \qquad (15\text{-}3)$$

In most design cases, the issue is the unknown values of the gain setting resistors. We can rearrange Eq. (15-3) to solve for $(R1 + R2)$.

$$(R1 + R2) = \frac{1,000,000 - (115 \times \text{GAIN})}{\text{GAIN}} \qquad (15\text{-}4)$$

where $R1$ and $R2$ are in ohms and GAIN is the voltage gain desired.

For a common set of resistor values and a gain of 1000, the equation above evaluates to $R1 = R2 = 440\ \Omega$.

INPUT-PROTECTED DEVICES

In certain applications the isolation amplifier inputs are subjected to transient high-voltage spikes that last tens of milliseconds. These pulses are much longer than the high-voltage transients expected from the normal residential and commercial AC power systems. An example of a long-duration, high-voltage transient pulse is the medical defibrillator device. Ventricular fibrillation is a pathological, rapidly fatal, asynchronous arrhythmia of the heart. Instead of beating normally, the fibrillating heart merely quivers. Physicians can "jump start" the patient's heart by applying a high-voltage shock across the chest. The electric shock forces all fibers of the heart (a muscle) to contract simultaneously. When the shock wave passes, and the heart unclenches from the contracted condition, the normal synchronized rhythm will hopefully be restored. A typical medial defibrillator consists of a high-voltage capacitor charged to 400 J or more, a vacuum transfer relay, and associated control circuits. A typical capacitor in a Lown waveform defibrillator is 16 μF, and it is charged to more than 7000 volts. Because of circuit resistive losses from several sources, the potential delivered across the patient's chest is approximately 3000 volts.

An electrocardiograph (ECG) monitor is used during resuscitation attempts so that the physician or emergency medical crew can judge the effectiveness of the defibrillation effort. Unfortunately, because the ECG electrodes are attached to the patient's body close to the same points as the defibrillator electrodes, the discharging capacitor in the

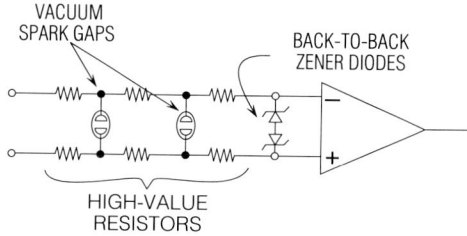

Fig. 15-12 High-voltage protection devices.

defibrillator presents an input potential of 1000 to 3000 V for a period up to 25 ms (although in most designs it is only 5 to 10 ms).

ECG preamplifiers are normally isolation amplifiers for patient safety reasons, and they typically employ one or more tactics to protect the amplifier from the high-voltage defibrillator blast. Figure 15-12 shows several methods. First, high resistance value resistors are connected in series with each input lead of the differential preamplifier. These resistors serve both a voltage drop and a current limiting function. Second, there may be one or more spark gap (SG) devices shunted across the inputs. In some earlier designs the spark gap was actually an NE-2 or NE-51 neon glow lamp. The glow lamp served to reduce the reaction time of the spark gap between the internal electrodes. Finally, there may be either a pair of back-to-back zener diodes or a metal oxide varistor (MOV) device shunted across the amplifier input lines. These devices limit the maximum differential voltage that can be applied across the amplifier inputs. Typical protection specifications, depending on model and intended application, range from about 1500 V to more than 3500 V.

CONCLUSION

Although the isolation amplifier is considerably more expensive than common IC linear amplifiers, there are applications where these amplifiers are absolutely critical. Wherever the instrument could cause injury to a human, or wherever the environment is such that the electronics must be isolated as far as possible, the isolation amplifier is the device of choice (at least in the front end).

16

Nonlinear (Diode) Applications
of Linear IC Devices

Although the operational amplifier is well known for its capabilities as a linear amplifier, there are also numerous nonlinear applications for which the op-amp is very well suited. In this chapter, techniques are presented that are commonly employed in fields as diverse as instrumentation, control, and communications. Of particular interest here are circuits in which PN junction diodes are used, such as precise rectifiers, bounded-value circuits, and clippers and clampers.

REVIEW OF THE PN JUNCTION DIODE

The PN junction diode is the oldest solid-state electronic component available. Indeed, naturally occurring diodes of galena crystals (lead sulfide, PbS) were used to prior to World War I as the demodulator (a.k.a. detector) in crystal set radio receivers. During World War II radar research led to the development of the 1N34, 1N60, and 1N63 germanium video detector diodes and the 1N21 and 1N23 microwave diodes.

The PN junction diode ideally has a transfer characteristic like Fig. 16-1A. When the anode is positive with respect to the cathode (Fig. 16-1B), the diode is forward biased, so it conducts current.

Alternatively, when the anode is negative with respect to the cathode (Fig. 16-1C), the diode is reverse biased and no current flows. Figure 16-1D shows the effect of this unidirectional current flow of a sine-wave input signal. Notice that in the halfwave rectified output only the positive peaks are present.

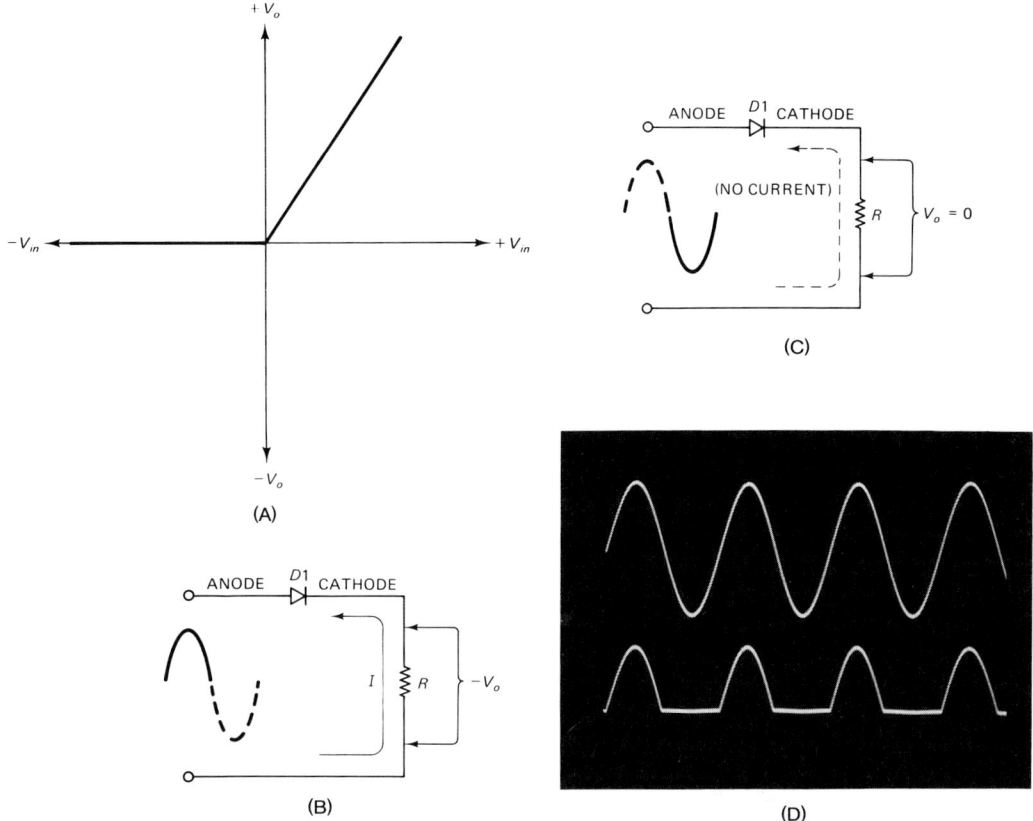

Fig. 16-1 (A) Ideal diode transfer characteristic, (B) halfwave rectifier with positive input, (C) halfwave rectifier with negative input, and (D) oscilloscope waveforms showing input and output of halfwave rectifier.

Real diodes fail to meet the ideal in several important respects. Figure 16-2 shows a transfer characteristic for a practical, non-ideal diode. For the ideal diode the reverse current is always zero, while in real diodes there is a minute leakage current (I_L) flowing backwards across the junction. A manifestation of this current can be seen by measuring the forward and reverse resistances of a PN junction diode. The forward resistance is very low, while the reverse resistance is very high—but not infinite, as one might expect from a supposed open circuit.

Another departure from the ideal in the reverse bias region is the avalanche point (V_z) at which reverse current flow increases sharply. At this point, the reverse bias voltage is too great and causes breakthrough. When carefully regulated, the breakdown potential is both sharply defined and reasonably stable (except for a slight temperature dependence). In such cases the device is called a zener diode; it is used as a voltage regulator.

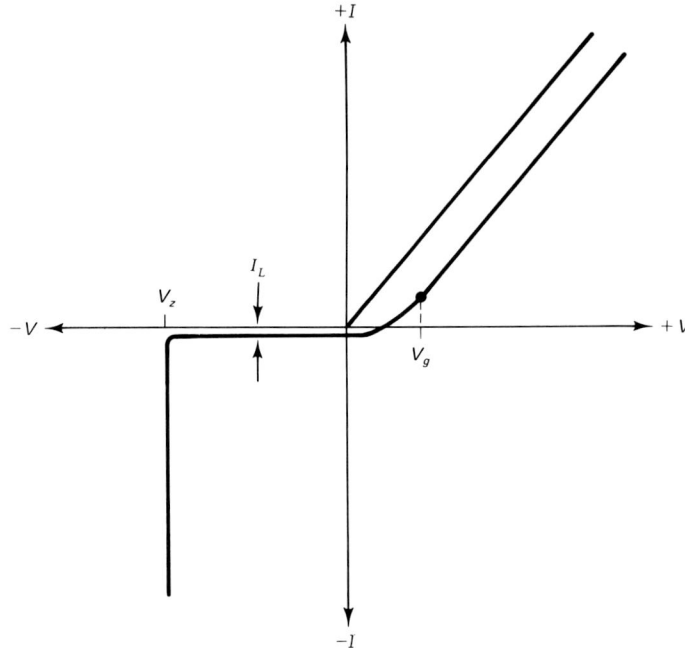

Fig. 16-2 Actual versus ideal diode transfer characteristic.

In the forward-biased region there are other anomalies that depart from the ideal. In the ideal case there is an ohmic relationship between current flow and applied forward voltage. Similarly, there is a linear relationship between applied forward voltage (V_f) and output voltage V_o. In real diodes, however, there is a significant departure from the ideal transfer characteristic. Between zero volts and a critical junction potential (V_g) the characteristic curves are quite nonlinear. The actual value of this potential is a function of both the type of semiconductor material used and the junction temperature. In general, V_g will be 0.2 to 0.3 V for germanium (Ge) diodes (1N34, 1N60, etc.), and 0.6 to 0.7 V for silicon (Si) diodes (1N400 ✕ , 1N914, 1N4148, etc.). In the 0 to V_g region the diode forward resistance is a variable function of V_f and T, and the I-vs-V_f characteristic is logarithmic. Above V_g the characteristic becomes more nearly linear.

PRECISE DIODE CIRCUITRY

A precise diode circuit combines an active device such as an operational amplifier with a pair of diodes to essentially "servo-out" the errors of the non-ideal diode. Two advantages obtain from this arrangement. First, the circuit will rectify very small AC signals between zero volts and V_g (i.e., $0 < V < V_g$, about 0.65 V). Second, the rectification

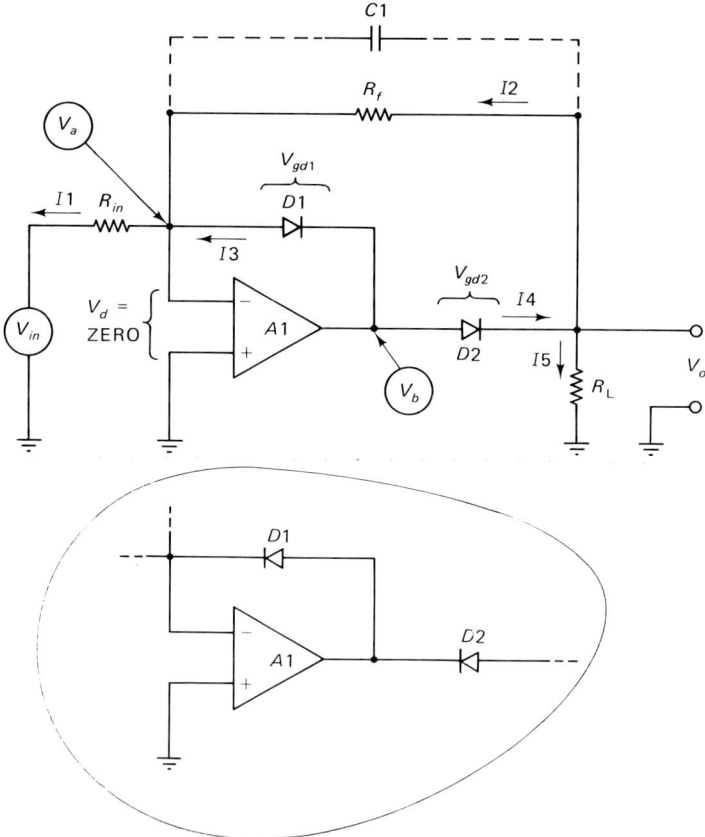

Fig. 16-3   Precise halfwave rectifier circuit.

will be more nearly linear than is the case with the diode alone, even in the diode's ohmic range.

Figure 16-3 shows the circuit for the inverting halfwave precise rectifier. A basic assumption in this circuit is that load impedance R_L is purely resistive and therefore contains no energy production or storage elements. The circuit is essentially an inverting follower amplifier with two PN junction diodes (D1 and D2) added. Halfwave rectification occurs because the circuits offers two different gains that are dependent on the polarity of the input signal. For positive values of V_{in} the gain (V_o/V_{in}) is zero, while for negative values of V_{in} the voltage gain is R_f/R_{in}.

Consider operation of the circuit for positive values of V_{in}. The noninverting input (+IN) is grounded, so it is at zero volts. By the properties of the ideal op-amp we must consider the inverting input (−IN) as if it were also grounded ($V_a = 0$). Recall that this concept is called a virtual ground. Thus, differential voltage V_d is zero.

When $V_{in} > 0$—that is, when it is positive—current $I1 = +V_{in}/R_{in}$. To maintain the equality $I1 + I3 = 0$, conserving Kirchhoff's current law (KCL), the op-amp output voltage V_b swings negative, but is limited by the D1 junction potential to V_g (about 0.6 to 0.7 V). With $V_b < 0$, even by only 0.6 to 0.7 V, diode D2 is reverse biased and therefore cannot conduct. Currents $I2$, $I4$, and $I5$ are zero. Thus, for positive values of V_{in} the output voltage V_o is zero.

Now consider operation for $V_{in} < 0$. Under this input condition op-amp output voltage V_b swings positive, forcing diode D1 to become reverse biased and D2 to conduct. To converse KCL (as before) for this case ($I1 + I2 = 0$), current $I2$ will have the same magnitude but opposite direction relative to $I1$. Because $V_{in}/R_{in} = -V_o/R_f$, the voltage gain ($A_v = V_o/V_{in}$) reduces to $-R_f/R_{in}$, as is appropriate for an inverting amplifier. Thus, the gain for negative input voltages ($V_{in} < 0$) is $-R_f/R_{in}$, while for positive input voltage ($V_{in} > 0$) it is zero. From that difference comes halfwave rectification.

The voltage drop across diode D2 is about $+0.6$ to $+0.7$ V; it is "servoed-out" because D2 is in the negative feedback loop around $A1$. Voltage V_b is correspondingly higher than V_o to null the effects of V_{gd2}.

The precise rectifier is capable of halfwave rectifying very low level input signals. The minimum signal allowed is given by

$$V_{in} > V_g/A_{vol} \tag{16-1}$$

where V_{in} is the input signal voltage, V_g the diode junction potential (0.6 to 0.7 V), and A_{vol} the open-loop gain of the amplifier.

In Eq. (16-1) the term A_{vol} the open-loop gain, which for DC and low-frequency AC signals is extremely high. But at some of the frequencies at which precise diodes operate, the input frequency is a substantial fraction of the gain–bandwidth product, so A_{vol} will be less than might otherwise be true. For example, if the gain–bandwidth product is 1.2 MHz, the gain at 100 Hz is 12,000. But at 1000 Hz, a typical frequency for precise rectifier operation, the gain is only 1200.

Circuit operation of the precise rectifier is shown by the waveforms in Fig. 16-4. If the sine wave is applied (Fig. 16-4A), the output voltage V_o will be zero from time $T1$ to $T2$ (positive input voltage), while V_b will rest at $-V_g$ (about -0.6 to -0.7 V). Between $T2$ and $T3$ the input is negative, so V_o will be a positive voltage with a halfwave sine shape (Fig. 16-4B). But note the behavior of V_b, the op-amp output (Fig. 16-4C). From $T1$ to $T2$ the output rests at $-V_g$, but at $T2$ it snaps to a value of $2V_g$ to the positive. The halfwave sine shape rests on top of the $+V_g$ offset caused by V_{gd2}. Figure 16-4D shows this same situation in the form of the transfer characteristic (V_o-vs-V_{in}).

The circuit of Fig. 16-3 as shown will rectify and invert negative peaks of the input signal. To accommodate the positive peaks one need only reverse the polarity of diodes D1 and D2.

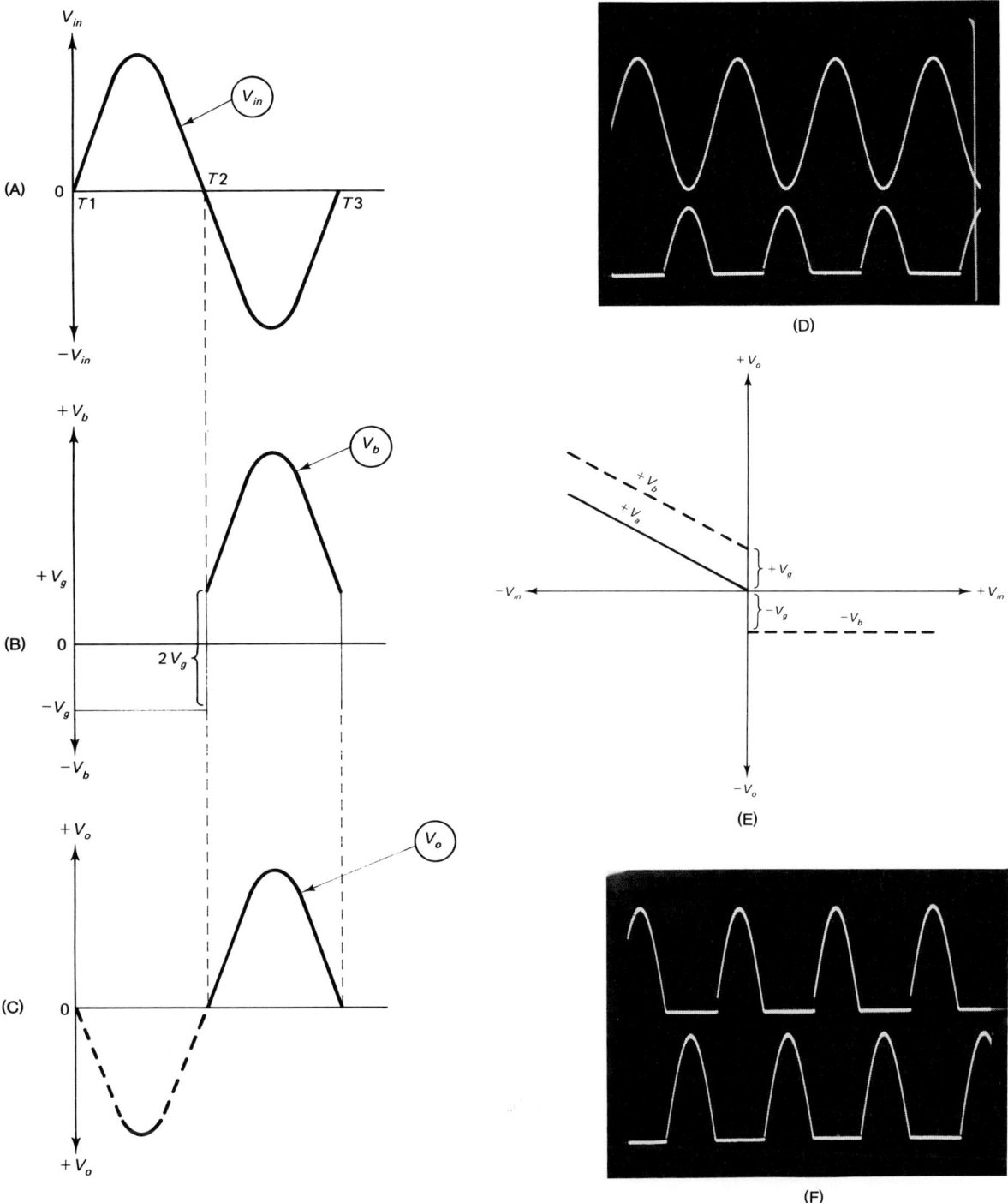

Fig. 16-4 Precise diode circuit waveforms: (A) input sine wave, (B) output terminal o op-amp, (C) output voltage V_o, (D) actual oscilloscope photo showing input and output voltages, (E) ideal rectifier transfer characteristic, and (F) output signals for both forward and reverse diode configurations.

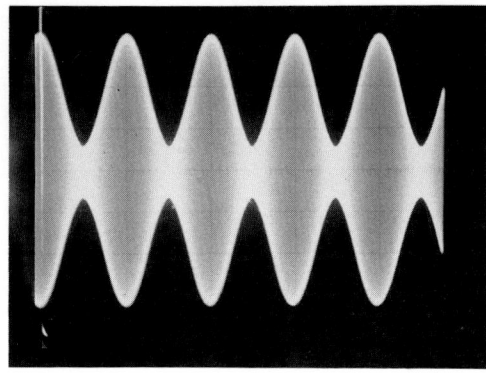

Fig. 16-5   Amplitude modulated carrier waveform.

Precise Diodes as AM Demodulators

One of the several applications of the precise diode circuit is demodulation or detection of amplitude modulated (AM) carrier signals. An AM signal (Fig. 16-5) is one in which a sinusoidal carrier signal of frequency F_c is varied in amplitude by a lower-frequency modulating frequency F_m. The modulating signal might be either sinusoidal or nonsinusoidal.

The most familiar use of AM is in radio broadcasting and communications. Not so familiar, perhaps, is the use of AM in nonradio applications such as instrumentation. One common use is in certain AC-excited Wheatstone bridge transducers. In one popular pressure transducer the excitation signal is a 10-V peak-to-peak, 2400-Hz sine wave. The output of the transducer is an AC signal that is proportional to the excitation voltage and the applied pressure. Thus, the pressure signal is used to modulate the 2400-Hz carrier signal.

Demodulation of an AM signal is usually done by envelope detection. This type of detector is basically a halfwave rectifier and a low-pass filter. The precise diode circuit of Fig. 16-3 can be used to demodulate AM signals; it offers an advantage because of its ability to accommodate weak signals. Low-pass filtering is obtained by shunting capacitor $C1$ across feedback resistor R_f.

For any given type of operational amplifier there is a maximum carrier frequency that can be accepted for AM demodulation. This frequency is related to the gain bandwidth product of the particular device selected:

$$F_{c(max)} = F_t / [100 \text{ ABS}(A_v)] \qquad (16\text{-}2)$$

where $F_{c(max)}$ is the maximum allowable carrier frequency, F_t the device gain–bandwidth product (i.e., the frequency at which $A_{vol} = 1$), and ABS(A_v) the absolute value of the closed-loop voltage gain.

The operational amplifier slew rate (S) must be sufficient to handle the input signal carrier frequency. In general,

$$S > \frac{\Delta V_{in}}{\Delta_t} \qquad (16\text{-}3)$$

or, for the specific case of sinusoidal carriers:

$$S > 2\pi F_c \text{ ABS}(A_v) V_{in(peak)} \qquad (16\text{-}4)$$

where S is the slew rate in volts per microsecond (μs), ΔV_{in} the change in input voltage over time Δt, and Δt the time over which V_{in} changes. The value of capacitor $C1$ (Fig. 16-3) is found from

$$C1 = \frac{(F_c/F_m)^{1/2}}{2\pi F_c R_f} \qquad (16\text{-}5)$$

Polarity Discriminator Circuits

A polarity discriminator is a circuit that will produce outputs that indicate whether the input voltage is zero, positive, or negative. Applications for this type of circuit include alarms, controls, and instrumentation.

Figure 16-6A shows a typical polarity discriminator circuit. The basic configuration is the inverting follower op-amp circuit, but with two negative feedback circuits. Each feedback path contains a diode, but since the diodes are connected in the opposite polarity sense, the polarity of the output potential will determine which one conducts and which is reverse biased.

Consider first the case where the input signal V_{in} is positive. In this case, current I_{in} flows away from the summing junction toward the source, and has a magnitude of $+V_{in}/R_{in}$. The output terminal of the op-amp will swing negative, causing diode D1 to be reverse biased and D2 to be forward biased. Current $I1$ is zero, and $I2$ is equal to V_{o2}/R_f. Output voltage V_{o2} is negative; it has a value of $V_{o2} = V_o - 0.6$ V. Output voltage V_{o1} is zero.

Now consider the opposite case, where V_{in} is negative. The current flows away from the source toward the summing junction. The output terminal of the op-amp swings positive, causing diode D1 to become forward biased while D2 is reverse biased. Current $I2$ is then zero and $I1$ is V_{o1}/R_f. In this case, V_{o1} is positive and V_o is zero.

The operation of this circuit can be seen in the waveforms shown in Fig. 16-6B and in the transfer characteristics shown in Fig. 16-6C.

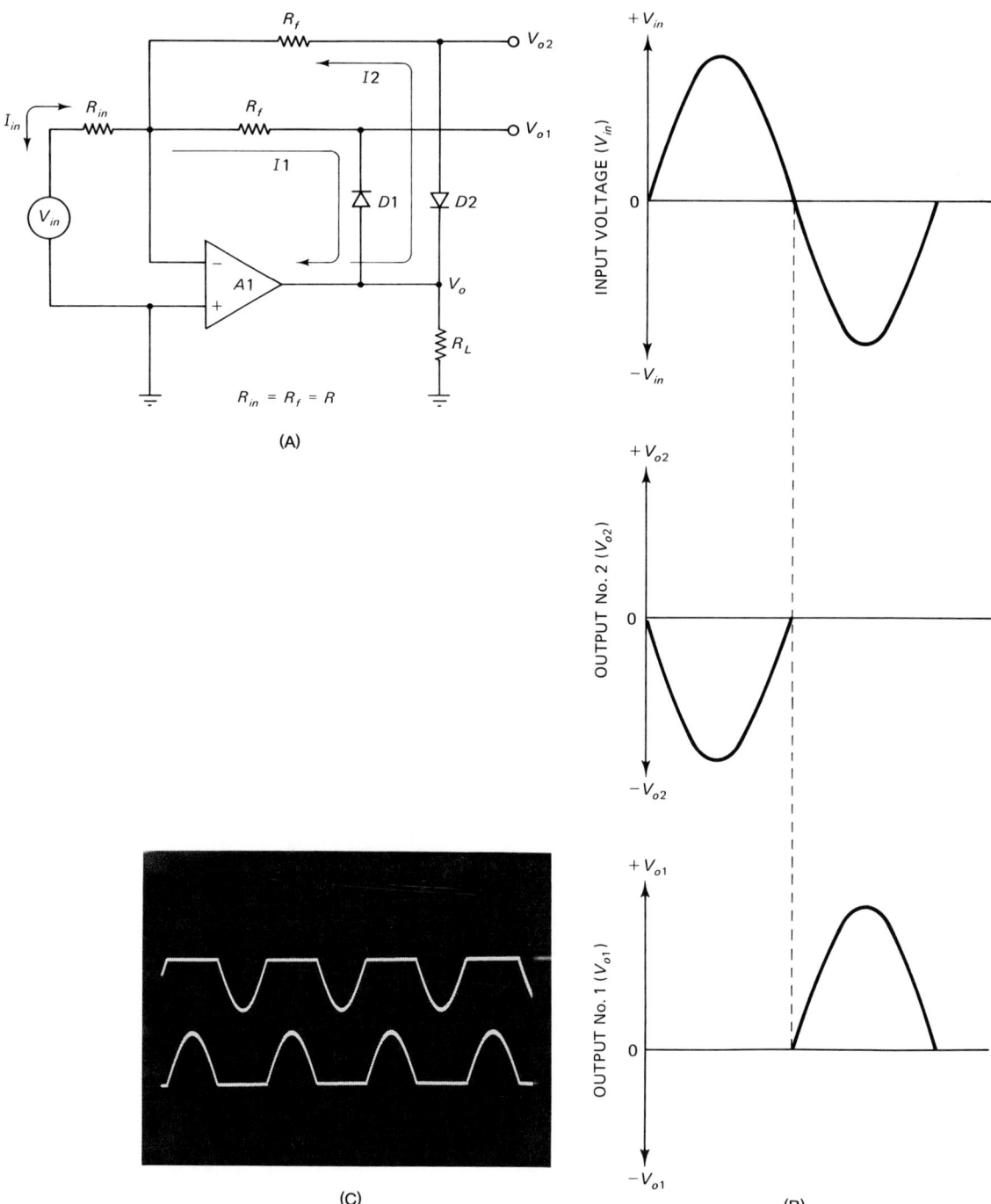

Fig. 16-6 (A) Polarity discriminator circuits, (B) circuit wave forms, (C) oscilloscope photo showing V_{o1} and V_{o2}, and (D) transfer characteristics. (*Figure continues*.)

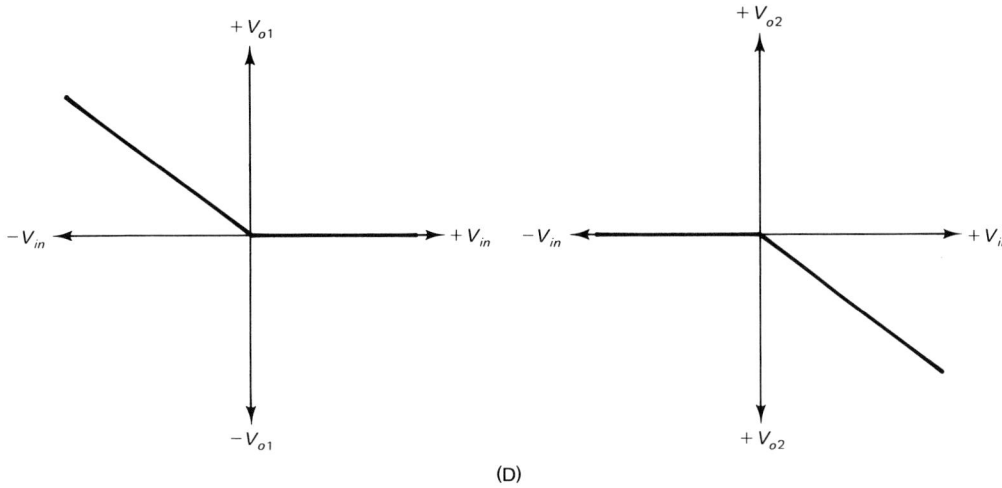

(D)

Fig. 16-6 (*continued*)

Fullwave Precise Rectifier

The fullwave rectifier uses both halves of the input sine wave. Recall that the halfwave rectifier removes one polarity of the sine wave; the fullwave rectifier preserves it. Figure 16-7 shows the relationships present in a fullwave rectifier circuit. In Fig. 16-7A we see the input

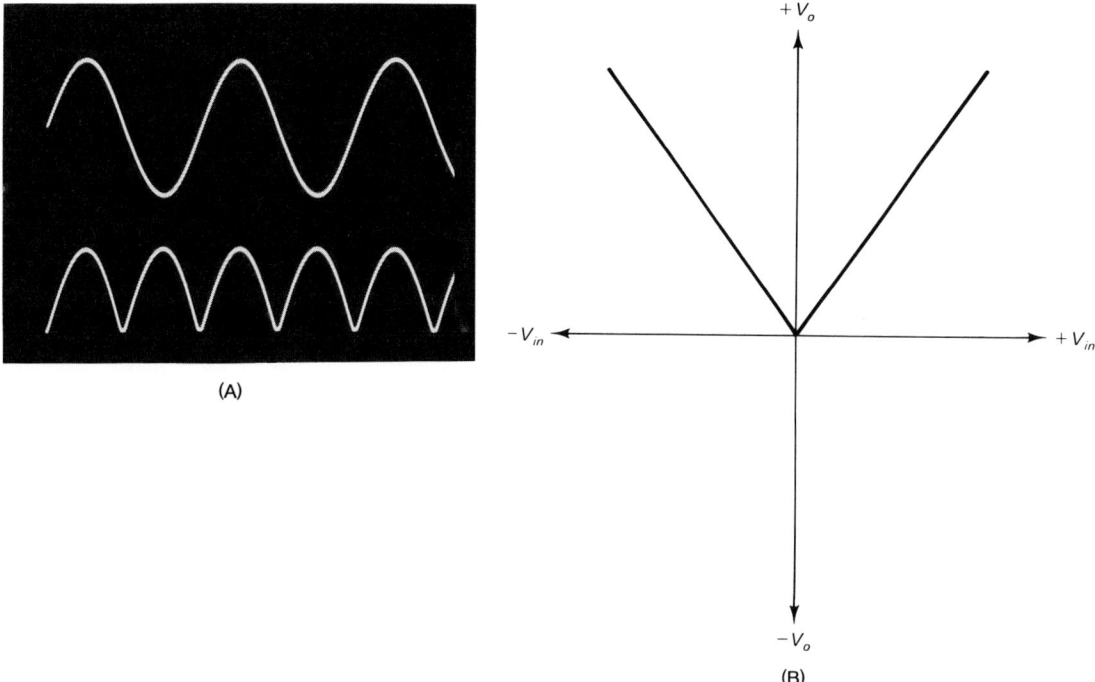

Fig. 16-7 (A) Sine-wave input (top) and fullwave rectified output (lower) and (B) transfer characteristic of the fullwave rectifier.

sine wave and the pulsating DC output of a fullwave rectifier. Note that the negative halves of the sine waves are flipped over and appear in the positive-going direction. The characteristic function for the full-wave rectifier is shown in Fig. 16-7B. Because the output voltage is always positive, regardless of whether the input signal is positive or negative, the fullwave rectifier can be called an absolute value circuit. The output voltage will be either

$$V_o = k|V_{in}| \qquad\qquad (16\text{-}6)$$

or

$$V_o = -k|V_{in}| \qquad\qquad (16\text{-}7)$$

depending on the direction of the diodes within the circuit.

While the fullwave rectifier has major applications in DC power supply, it is the absolute value feature that makes the fullwave rectifier important for instrumentation and other related purposes.

There are several methods for creating a precise fullwave rectifier, a.k.a. absolute value amplifier, and some of these are shown in Fig. 16-8 The first circuit (Fig. 16-8A) is based on the polarity discriminator circuit of Fig. 16-6. The two outputs, V_{o1} and V_{o2}, are applied to the inputs of a DC differential amplifier.

A second approach is shown in Fig. 16-7B. In this circuit a pair of oppositely connected diodes are applied to the inputs of a simple DC differential amplifier. This approach is not as well regarded because of the fact that the diodes in the input stage are not in the feedback loop, so their voltage drops (V_g) are not servoed out.

Figure 16-8C shows the usual absolute value amplifier circuit. It consists of a pair of precise halfwave rectifier circuits connected such that their respective outputs are summed at the input of an output buffer amplifier. Amplifier $A1$ is connected as an inverting precise rectifier, while amplifier $A2$ is connected as a noninverting precise rectifier. The waveform of the signal at the summing point, that is, the input of amplifier $A3$, is the absolute value of the input waveform.

There are two common variations of this circuit. As shown, the input circuit of $A3$ uses a resistor to ground that is very much larger than the summing resistors (R). This relationship prevents the loading of the signal by voltage divider action between R and R_i. In some modern BiFET and BiMOS amplifiers, no R_i is needed because the bias current is very nearly zero.

The other alternative is to make $R_i = R$, causing the input voltage to $A3$ to be reduced to one-half by $V(R)/(R + R)$. For this case, it is necessary to make $A3$ a gain of two noninverting follower amplifier (see inset to Fig. 16-8C).

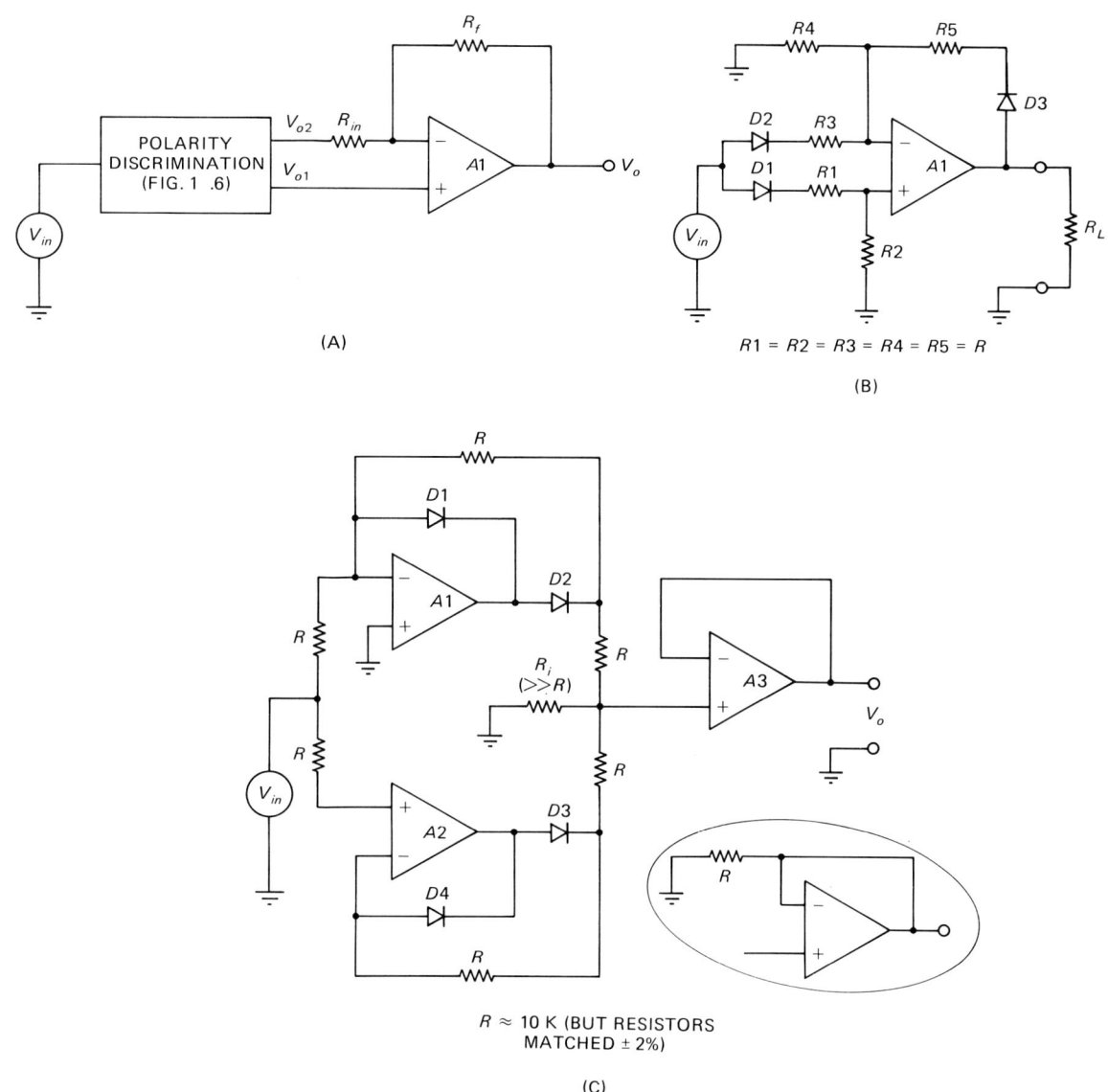

Fig. 16-8 (A) Absolute value circuit based on polarity discriminator, (B) a simple absolute value circuit, and (C) preferred active absolute value circuit.

Zero-Bound and Dead-Band Circuits

Zero-Bound Circuits

A zero-bound circuit is one in which the output voltage is limited such that it will be nonzero for certain values of input voltage and zero for all other input voltages. The term does not mean that the values of V_{in} are in any way constrained, but rather that there are constraints on allowable output voltages. The output of a zero-bound circuit indicates when the input signal exceeds a certain threshold, and by how much.

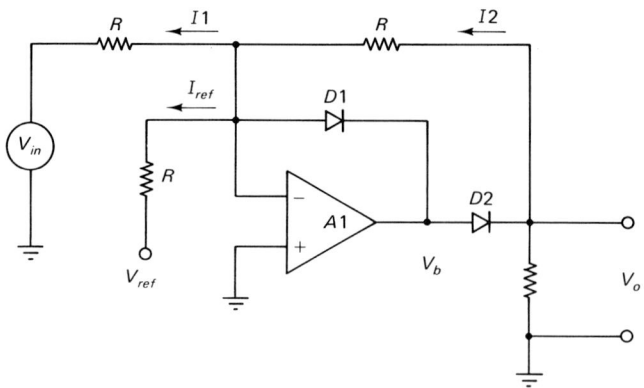

Fig. 16-9   Zero-bound circuit.

Figure 16-9 shows a zero-bound amplifier circuit. This circuit is based on the halfwave precise rectifier circuit of Fig. 16-3; it functions in exactly the same way except for the extra input reference current I_{ref}. The effect of I_{ref} is to offset the trip point at which the input voltage takes effect.

To understand this circuit we can use an analysis similar to the method used, earlier, based on the properties of the ideal operational amplifier. We know from Kirchhoff's current law (KCL) and the fact that op-amp inputs are neither sink nor source for current, that the following relationship is true:

$$I1 + I_{ref} + I2 = 0 \tag{16-8}$$

or

$$I1 + I_{ref} = -I2 \tag{16-9}$$

We also know that

$$I1 = V_{in}/R \tag{16-10}$$

$$I_{ref} = V_{ref}/R \tag{16-11}$$

$$I2 = V_o/R \tag{16-12}$$

Thus,

$$V_{in}/R + V_{ref}/R = -V_o/R \tag{16-13}$$

and after multiplying both sides by R:

$$V_{in} + V_{ref} = V_o \tag{16-14}$$

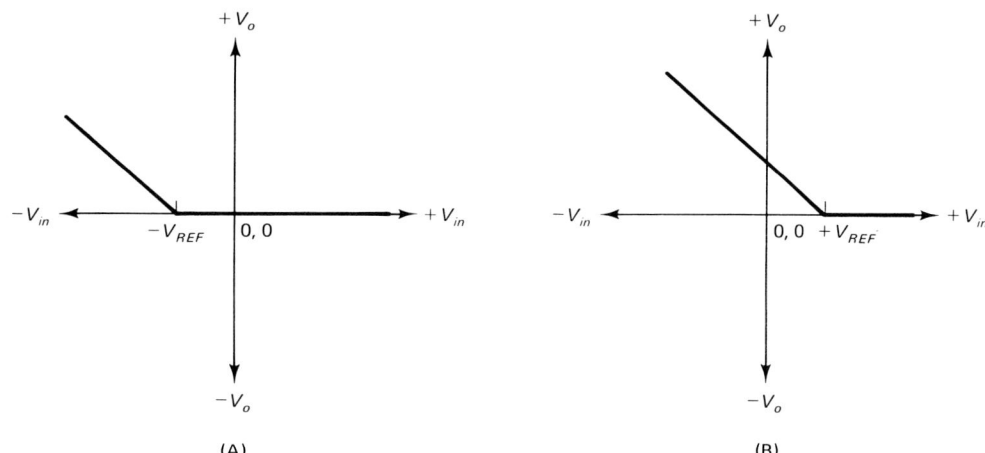

Fig. 16-10 Zero-bound transfer characteristic (A) for $-V_{\text{ref}}$, and (B) for $+V_{\text{ref}}$.

Thus, the output voltage is still proportional to the input voltage, but it is offset by the value of V_{ref}. The transfer characteristics for this circuit are shown in Fig. 16-10. In Fig. 16-10A the value of V_{ref} is negative, while in Fig. 16-10B the value of V_{ref} is positive. In both cases the transfer curve is offset by the reference signal potential.

Consider the operation of the circuit in Fig. 16-9 under two conditions: $V_{\text{in}} > 0$ and $V_{\text{in}} < 0$. First assume that $V_{\text{ref}} = 0$. For the positive input ($V_{\text{in}} > 0$), the output of the operational amplifier ($A1$) swings negative (the circuit is an inverter), causing diode D2 to be reverse biased and D1 to be forward biased. Output voltage V_{o} is zero in this case. The output voltage is bound to zero for all values of $V_{\text{in}} > 0$.

Now consider the two cases where V_{ref} is not zero: $V_{\text{ref}} > 0$ and $V_{\text{ref}} < 0$. Figure 16-11 shows several cases of a zero-bound circuit such as Fig. 16-9. In all of these examples a 741 operational amplifier was connected with a pair of 1N4148 silicon signal diodes; the value of R was selected as 22 kΩ. The excitation signal was an 8-V p-p sine wave at a frequency of 700 Hz. Figure 16-11A shows the action of the circuit without the reference voltage applied ($V_{\text{ref}} = 0$); the circuit operates as a normal precise rectifier. As will be true of all of the examples in Fig. 16-11, the input sine wave is shown in the upper trace and V_{o} is shown in the lower trace on the dual-beam oscilloscope. In Fig. 16-11B, $V_{\text{ref}} = -1.2$ V. Notice the clipping action. The amplitude of the waveform is 5.2 V base to peak. Because the input signal is 8 V p-p, the positive peak is 4 V peak. Thus, the baseline of the output signal voltage is at a level of [($+4$ V) $-$ (5.2 V)], or -1.2 V—which is the value of V_{ref}. In this circuit, the zero bounding occurs at all negative potentials greater than -1.2 V; only those signals more positive than this value can pass to the output of the circuit.

The waveforms in Fig. 16-11C are similar to those in Fig. 16-11B, except that the reference voltage has been increased to -3.4 V. In this

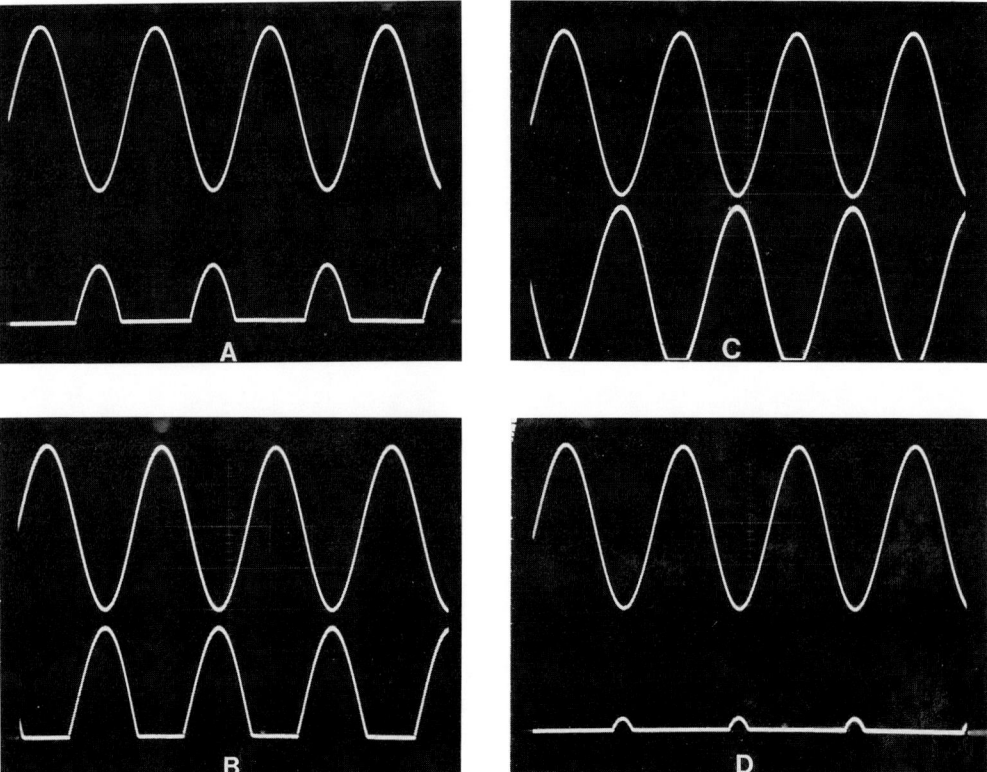

Fig. 16-11 Oscilloscope waveforms for various conditions in a zero-bound circuit.

case, only (4 − 3.4 V) or 0.6 V of the negative peak is unable to pass to the output. Exactly the opposite situation is shown in Fig. 16-11D. Here the reference voltage has the same magnitude, but it is reversed in polarity (+3.4 V DC). Note that only the top 0.6 V of the positive peak shows; all lower voltages are zero bounded.

Dead-Band Circuits

A dead-band circuit is one in which two zero-bound circuits work together to produce a summed output. Figure 16-12A shows the transfer characteristic of such a circuit. Two different threshold values are shown in this curve. The circuit will output signals only when the input signal is less than the lower threshold ($V_{in} < -V_{th}$), or greater than the upper threshold ($V_{in} > +V_{th}$). This behavior is shown relative to a sine wave input signal. The output will be zero for all values of input signal within the shaded zone.

Keep in mind that the output voltage will not suddenly snap to a high value above the threshold potential, but rather it will be equal to the difference between the peak voltage and the threshold voltage. Assuming unity gain for both reference voltage and input signal voltage, the output peaks will be $[(+V_p - (+V_{th})]$ and $[(-V_p - (-V_{th})]$. A

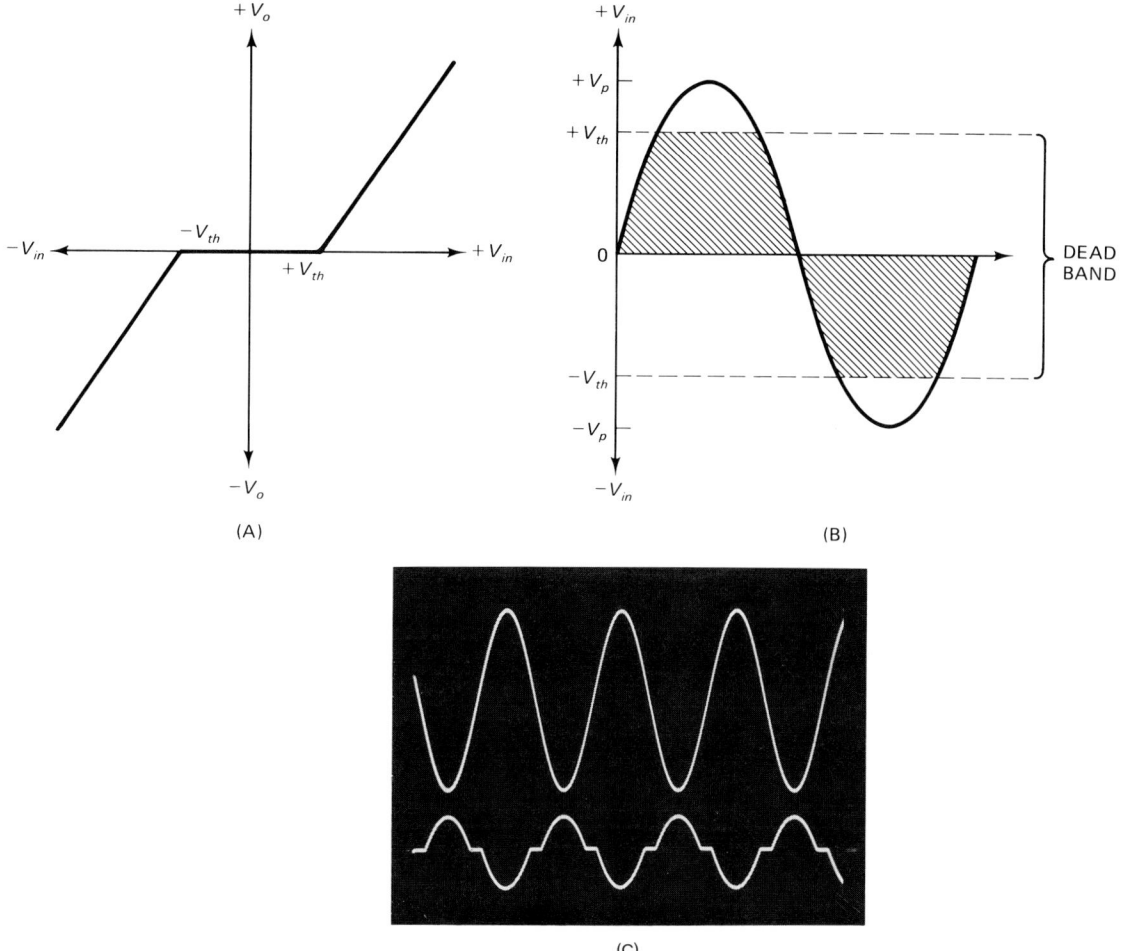

Fig. 16-12 (A) Dead-band transfer characteristic. (B) Shaded zone will not produce output signal. (C) Actual oscilloscope trace of dead-band circuit.

waveform such as this is shown in Fig. 16-12C. Note here that the two threshold voltages are not equal to each other, so they produce different peak values.

The dead-band amplifier circuit consists of a pair of zero-bound circuits summed together (Fig. 16-13). Both zero-bound circuits are similar to Fig. 16-9. Zero-bound circuit no. 1 uses diodes in the same polarity as was used in Fig. 16-9, while zero-bound circuit no. 2 uses reverse polarity diodes (both shown in the insets to Fig. 16-13). In the case of the first circuit the $V+$ DC power supply is used as V_{ref}, while in the second the $V-$ DC supply is used as V_{ref}. In both cases the magnitude is the same, but the polarities are reversed. The difference in the threshold levels in this case is set by using different values of reference resistor: $3R$ in the first zero-bound circuit and $5R$ in the second zero-bound circuit. The result is the waveform of Fig. 16-12C.

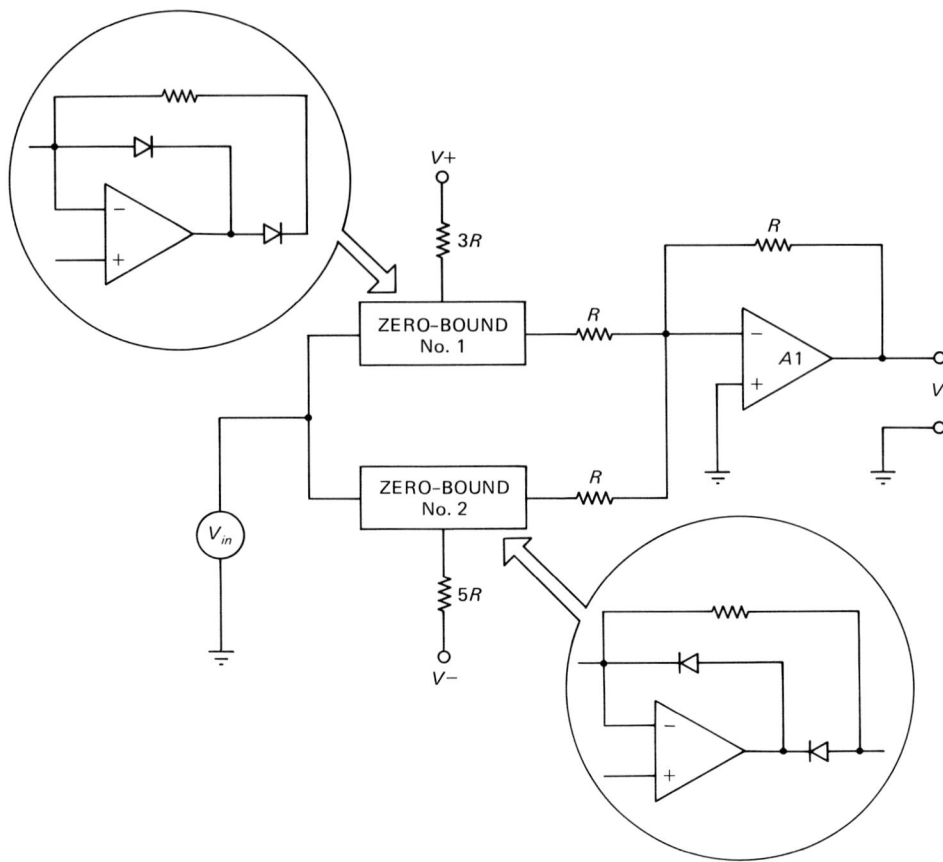

Fig. 16-13   Dead-band circuit made by summing to zero-bound circuits.

PEAK FOLLOWER CIRCUITS

The peak follower is a circuit that will output the highest value input voltage that was applied to it, regardless of what the input voltage does after that point. Figure 16-14 shows the action of a typical peak follower. Input voltage V_{in} varies over a wide range. The output voltage (shown by the heavy line profile), however, always remains at the highest value reached previously, and only increases if a new peak is encountered.

A typical peak follower is shown in Fig. 16-15. This circuit is basically a noninverting halfwave precise rectifier in which a unity-gain, noninverting buffer amplifier is inserted into the feedback loop between diode D2 and the feedback resistor. Also added to the circuit are a capacitor to hold the charge ($C1$) and a reset switch that is used to discharge the capacitor. The value of the capacitor is selected to be small enough to allow it to charge rapidly when an input signal is applied, but large enough to not saturate too quickly.

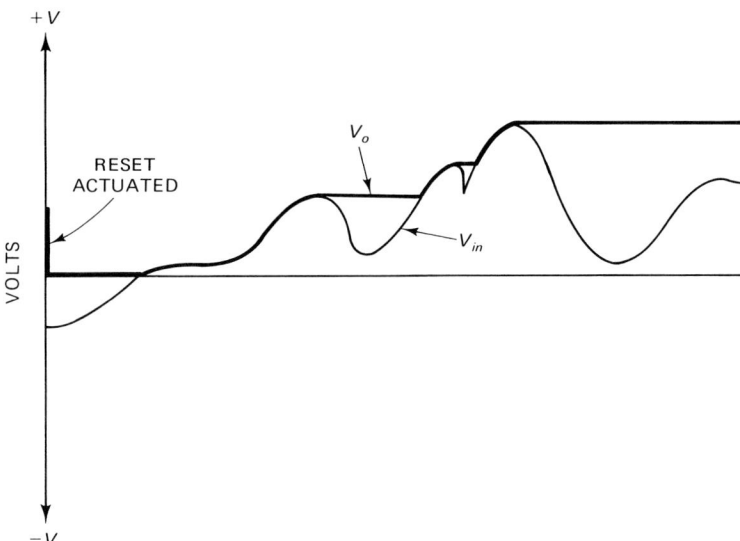

Fig. 16-14 Action of a peak follower.

When the input voltage is positive ($V_{in} > 0$), the output of $A1$ is also positive. This forces diode D1 to be reverse biased and D2 to be forward biased. At initial turn-on, or immediately after the reset switch is closed and then reopened, the capacitor is discharged. In this case $V_c = 0$. If a positive input voltage is applied to the input, this potential begins to charge $C1$, causing V_c to increase to V_{in}. When the voltage across $C1$ is equal to the output voltage of the amplifier, then current flow into the capacitor ceases, and the value of V_c is at the maximum value reached by V_{in}. Because amplifier $A2$ is a noninverting, unity-gain follower, the output voltage is equal to the voltage across the capacitor ($V_o = V_c$).

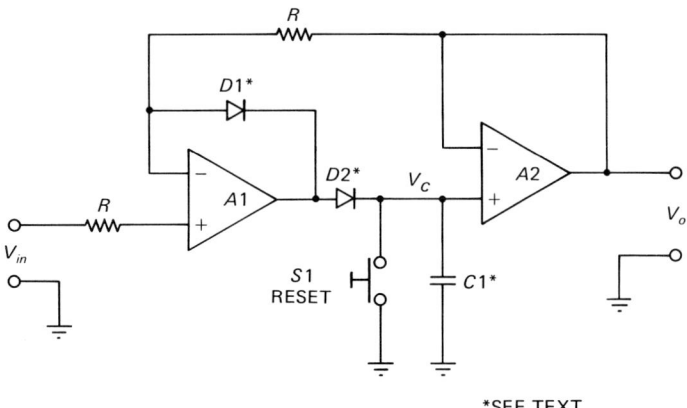

Fig. 16-15 Peak follower circuit.

As long as V_{in} is equal to the capacitor voltage, or less, the capacitor voltage remains unchanged. In other words, the capacitor voltage (hence also the output voltage) remains at the previous high value. But if the input voltage should rise to a point greater than the previous peak, then $V_{in} > V_c$, so a current will flow into $C1$ and cause its voltage to reach the new level. Again, the output voltage will track the previous high value of $V_{in(max)}$. Only after reset switch S1 is closed momentarily will the output return to zero.

For negative input voltages, diode D2 is reverse biased, so no current will pass either to or from the capacitor. This current ignores negative input potentials.

There are special precautions to take with respect to certain of the components in this circuit. Capacitors $C1$, for example, must be a very low leakage type. If there is a leakage current, which implies a shunt resistance across the capacitance, then the charge on the capacitor will bleed off with time. For the same reason, the input impedance of amplifier $A2$ must have an extremely high input impedance. For this reason, special premium operational amplifiers are selected. BiMOS, BiFET, and other very low input bias current models are preferred for this application. Also, the diode selected for D2 must have an extremely high reverse resistance. In other words, the leakage current that passes through D2 must be kept as low as possible. The reason for these precautions is to prevent the charge on $C1$ from bleeding off prematurely. The outward result of this circuit action is apparent "droop" of the output voltage V_o.

Sample-and-Hold Circuit

The peak follower circuit of Fig. 16-15 can be modified to form a sample-and-hold (S/H) circuit. By adding a series switch (S2) at the

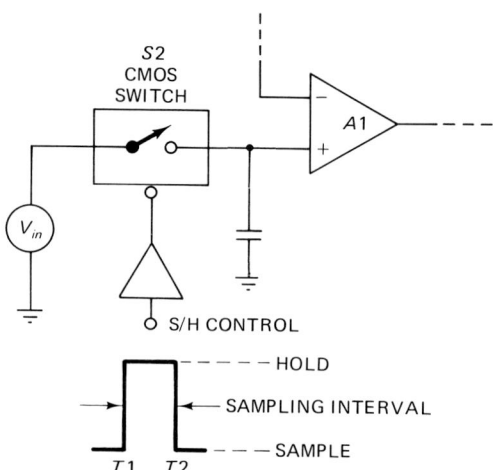

Fig. 16-16   Sample-and-hold circuit.

input (Fig. 16-16) the peak follower will admit signal only at a discrete time determined by S/H control signal. The switch is a CMOS electronic switch, and is used to allow a logic signal (as might be provided by a signal) to produce the S/H action. A similar switch can be used for S1 (Fig. 16-15). The correct action will first drive S1 closed

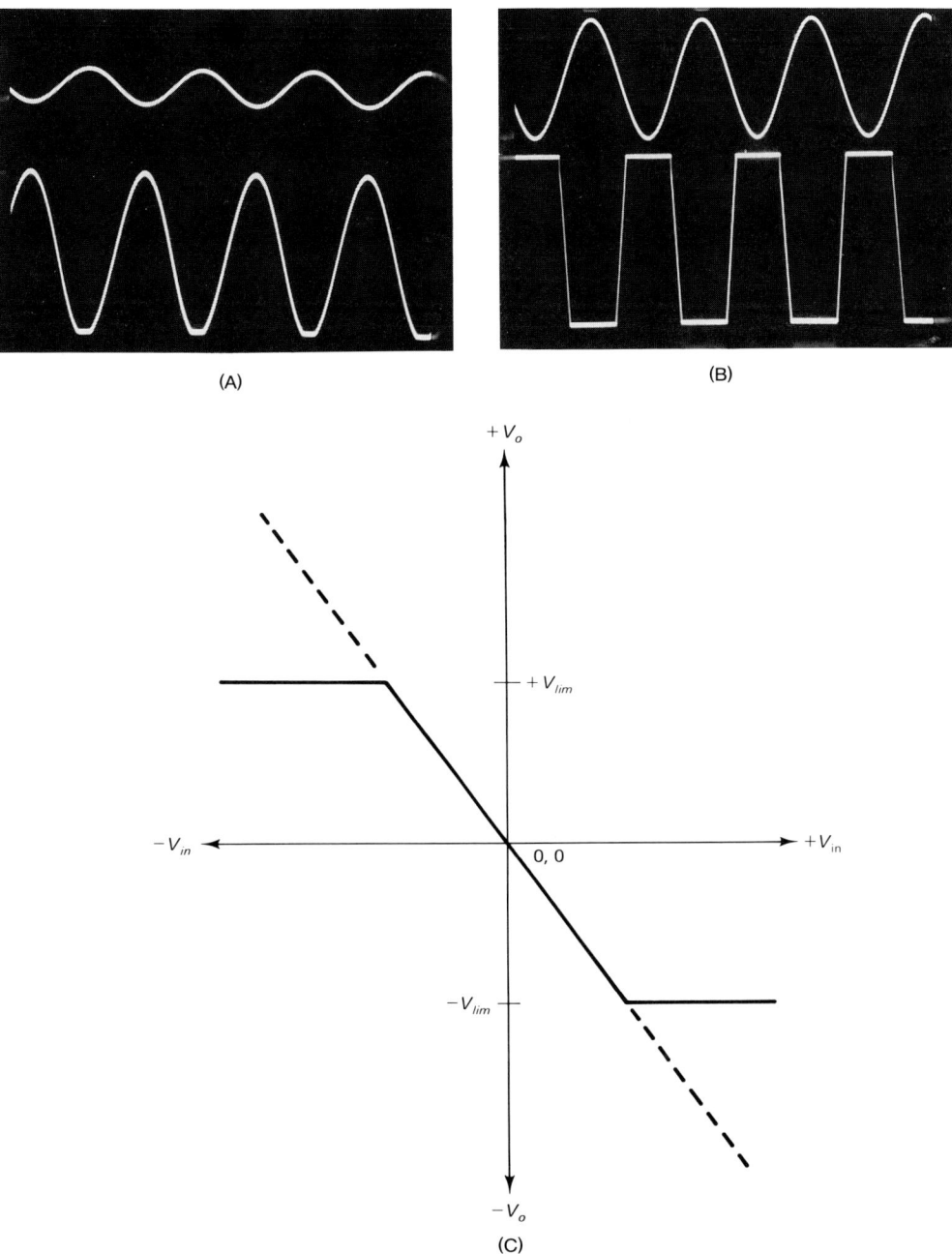

Fig. 16-17 Clipper circuit action: (A) input signal below limiting threshold, (B) input signal exceeds threshold, and (C) transfer characteristic of clipper.

to discharge $C1$, and then open $S1$ and close $S2$ to charge $C1$, with the maximum value of V_{in} reached during the $T2 - T1$ sampling interval.

CLIPPER CIRCUITS

Clipper (or clamp) circuits are the opposite of the dead-band circuit. In these circuits the output voltage will swing at will around zero provided that the input signal does not exceed a certain predetermined threshold. The positive peak, the negative peak, or both will be limited to a certain clamped value. Figure 16-17 shows a typical example. These waveforms were taken in an inverting follower circuit with a gain of (-100 kΩ/22 kΩ) or about 4.6. In Fig. 16-17A the input signal (upper trace) is close to, but below, the critical value. The output voltage is, therefore, unclipped on the positive peak and only moderately clipped on the negative peak. In Fig. 16-17B, however, the input signal is considerably increased, but the amplifier has no further output voltage to offer. As a result, the peaks of the amplifier output are clamped to a value determined by the DC power supply potential applied to the amplifier.

Figure 16-17C shows a transfer characteristic for an inverting clipper. Output voltage V_o is allowed to swing only between the lower and upper limits ($-V_{lim}$ and $+V_{lim}$, respectively). The dotted lines represent the output voltage that would exist in the absence of limiting.

It is not generally satisfactory to limit the DC power supply potentials to achieve clipping. The usual procedure is to use the full power supply potential, but to limit amplifier output voltage by certain circuit methods. Figure 16-18 shows one popular, but largely unsatisfactory, method for limiting the output swing. The feedback resistor is shunted by a pair of back-to-back zener diodes. On positive output

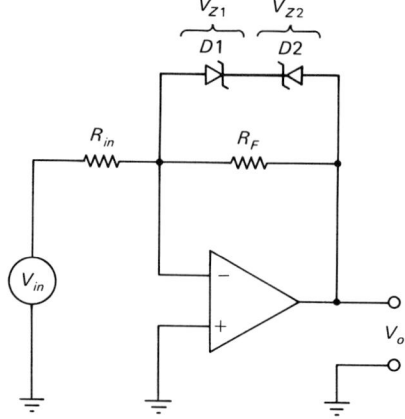

Fig. 16-18   Zener-diode attempt at making a clipper circuit.

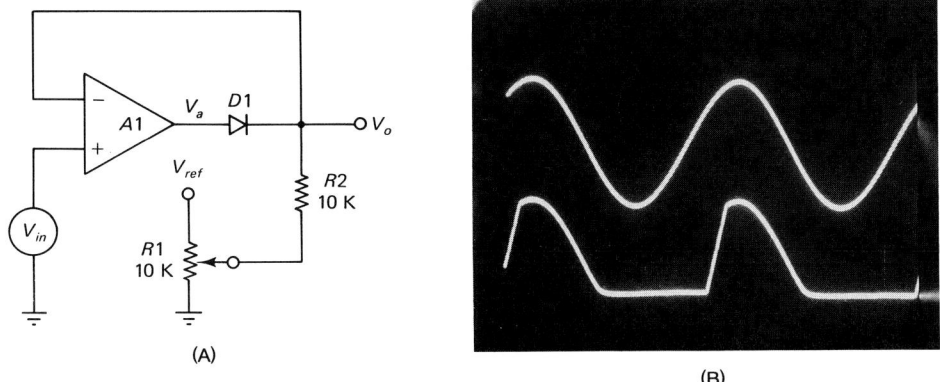

Fig. 16-19 (A) Simple positive-value clipper and (B) output and input signals.

voltages, D2 is forward biased and D1 is reverse biased. As long as $+V_o$ is less than V_{Z1} plus the forward drop across D2 (about 0.6 V), it follows the dictates of the usual transfer equation ($V_o = (V_{in}R_f/R_{in})$). At values greater than $V_{Z1} + 0.6$ V, however, the output is clamped. Similarly with negative output potentials. On negative swings of the output voltage the clamp occurs at $-V_{Z2} - 0.6$ V.

The circuit of Fig. 16-18 is not well regarded because it requires a relatively high signal level to sustain the zener diodes in the avalanche condition. A somewhat different approach is shown in Fig. 16-19. In this case the amplifier drives a diode (D1). In the absence of V_{ref}, the diode will be forward biased for positive values of V_{in}. But when V_{ref} is applied, it will reverse bias D1 and prevent an output voltage until V_a

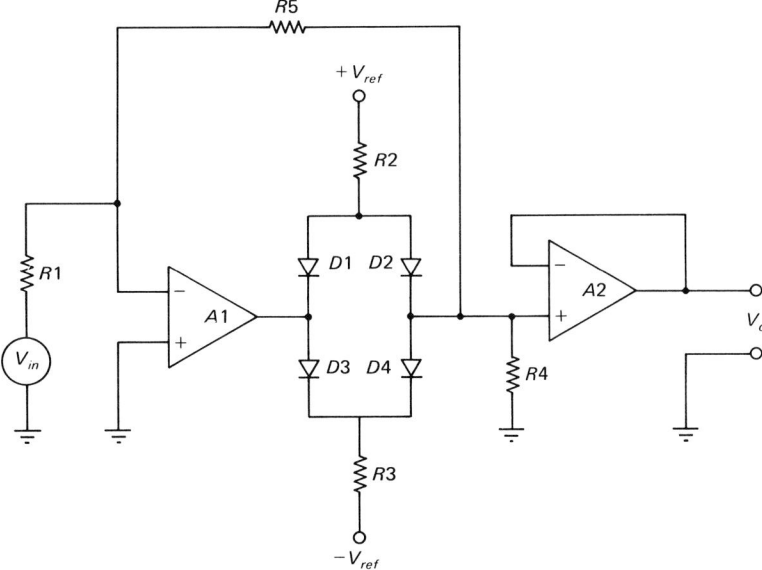

Fig. 16-20 Preferred clipper circuit.

overcomes the reference potential. The results of this circuit are also not very satisfactory (see Fig. 16-19B).

A more satisfactory approach is shown in Fig. 16-20. A diode bridge (D1–D4) is inserted into the feedback loop of amplifier $A1$. As a result, some of the nonlinearities that afflict the other circuits are "servoed-out" of this circuit by the $1/BA_v$ factor (in which B is the transfer equation of the feedback network). Two limiting conditions for this circuit are

1. To set $+V_{\text{lim}}$:

$$+V_{\text{lim}} = \frac{+V_{\text{ref}} R5 R4}{R5 R4 + R5 R2 + R4 R2} \qquad (16\text{-}15)$$

2. To set $-V_{\text{lim}}$:

$$-V_{\text{lim}} = \frac{-V_{\text{ref}} R5 R4}{R5 R4 + R5 R3 + R4 R3} \qquad (16\text{-}16)$$

17

Analog Signal Processing Circuits

Many electronic instruments require either the amplification or processing (or both) of analog electrical signals. In an era of massive computerization of such instruments, there is still a strong need for the analog subsystem. It might be necessary, for example, to boost an analog signal to the point where it can be input to the A/D converter connected to a computer. In addition to this simple scaling function, it is also often desirable to do some of the signal processing in the analog subsystem. Although this statement may seem almost heretical to computer oriented people, it is nonetheless often a reasonable trade-off. There may be a situation in which computer hardware or time-line constraints makes it less costly to use a simple analog circuit. It is often asserted that the computer solution is better than the analog circuit solution. While this claim is true much of the time, it is not universally true. As device manufacturer catalogs attest, analog signal processing is far from dead; it is alive, healthy, and larger than ever.

In this chapter we are going to take a look at standard laboratory amplifiers that are used for certain signal acquisition chores in electronic instrumentation. In addition, we will examine certain linear IC circuits used for analog signal processing.

The term *laboratory amplifiers* describes a wide range of instruments of many and varying capabilities. Although some are quite complex, many of them can be designed using simple linear IC devices. Some of these instruments are categorized according to several schemes. For example, one can divide them according to input coupling method: DC versus AC. In the case of AC amplifiers, there is often a frequency response characteristic that will take some of the burden of filtering in the system. We can also categorize the amplifiers according to gain:

- Low gain   1 to 100
- Medium gain   100 to 1000
- High gain   > 1000

These ranges are commonly accepted, but because they are popularly established rather than established through formal industry standards they may vary somewhat from one manufacturer to another. The three categories nonetheless serve as a reasonable context for our discussion.

Some amplifiers carry names that represent certain special applications. For example, the biopotentials amplifier is used to acquire natural electrical signals from living things and, because of certain practical problems, tend to have very high input impedances and certain other attributes that establish the class.

Laboratory amplifiers can be either free-standing models, part of a plug-in mainframe data logging or instrumentation system, or built into another instrument. In the following sections we will discuss some of the special forms of laboratory amplifier that may be useful in certain specific cases.

CHOPPER AMPLIFIERS

One of the unfortunate characteristics of simple DC amplifiers is that they may be noisy and possess a certain inherent thermal drift of both gain and DC offset baseline (especially the latter). In low and medium gain applications these problems are less important than in high-gain amplifiers, especially in the lower regions of those gain ranges. As gain increases, however, these problems loom much larger. For example, a drift of 50 μV/°C in an $\times 100$ medium-gain amplifier produces an output voltage change of

$$(50 \ \mu V/°C) \times 100 = 5 \ mV/°C$$

A drift of 5 mV/°C is certainly tolerable in most low-gain circuits, but in an $\times 20{,}000$ high-gain amplifier the output voltage would escalate to

$$(50 \ \mu V/°C) \times 20{,}000 = 1 \ V/°C$$

This level of drift will obscure real signals from most sources in a short period of time.

Similarly, noise can be a problem in high-gain applications, where it had been negligible in most low to medium-gain applications. Operational amplifier noise is usually specified in terms of nanovolts of noise per square root hertz [i.e., $NOISE_{(rms)} = nV/(Hz)^{1/2}$]. A typical low-cost operational amplifier has a noise specification of 100 $nV/(Hz)^{1/2}$, so at a bandwidth of 10 kHz the noise amplitude will be 0.00001 V. In an $\times 100$ amplifier without low-pass filtering, the output amplitude will be only 1 mV; but in an $\times 100{,}000$ amplifier it will be 1 V.

A circuit called a chopper amplifier can solve both problems because it makes use of a relatively narrow-band AC-coupled amplifier in which the advantages of feedback can be optimized. The drift problem is reduced significantly by two properties of AC amplifiers. One property is the inability to pass low-frequency (i.e., near-DC) changes such as those caused by drift. To the amplifier, drift looks like a valid low-frequency (sub-Hertzian) signal, so it is attenuated by the frequency response characteristic. The other property is the ability to regulate the amplifier through the use of negative feedback.

Many low-level analog signals are very low frequency—in the DC to 30-Hz range (for example, human electrocardiogram signals have subharmonic frequency components down to 0.05 Hz), so they will not pass through a narrow-band AC amplifier. The solution to this problem is to chop the signal at a higher frequency so that it passes through a narrow-band ᴬAC amplifier, and then to demodulate the amplifier output signal to recover the original wave shape, but at a higher amplitude.

Figure 17-1 shows a block diagram of the basic chopper amplifier circuit. The traditional chopper mechanism is a vibrator-driven SPDT switch (S1) connected so that it alternately grounds first the input and then the output of the AC amplifier. Modern choppers are made with field-effect transistor, optoisolators, PIN diodes, or other devices.

An example of a chopped waveform is shown in Fig. 17-2. A low-pass filter following the amplifier will filter out any residual chopper "hash" and any miscellaneous noise signals that may be present. Most old-fashioned mechanical choppers used a chop rate of either 60 Hz or 400 Hz, although 100-Hz, 200-Hz, and 500-Hz choppers are also found. The main criterion for the chop rate is that it be at least

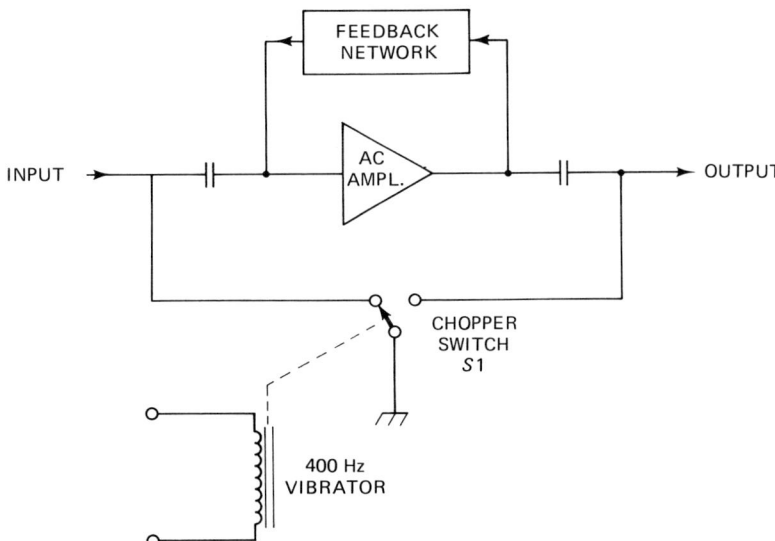

Fig. 17-1 Single-ended chopper amplifier.

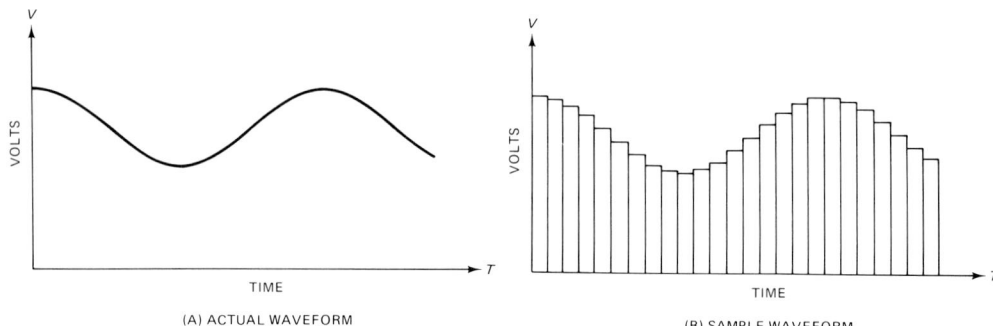

Fig. 17-2 (A) Actual time-domain waveform and (B) sampled version of the same waveform.

twice the highest component frequency that is present in the input waveform (Nyquist's sampling criterion).

A differential chopper amplifier is shown in Fig. 17-3. In this circuit an input transformer with a center-tapped primary is used. One input terminal is connected to the transformer primary center-tap, while the other input terminal is switched back and forth between the respective ends of the transformer primary winding. A synchronous demodulator following the AC amplifier detects the signal and restores the original, but now amplified, wave shape. Again, a low-pass filter smoothes out the signal.

The modern chopper amplifier does not use mechanical vibrator switches as the chopper. A pair of CMOS or JFET electronic switches driven out of phase with each other will perform exactly the same function. Other electronic switches used in commercial chopper amplifiers include PIN diodes, varactors, and optoisolators. Fig. 17-4 shows a modern electronically chopped amplifier that can be obtained in either IC or hybrid form.

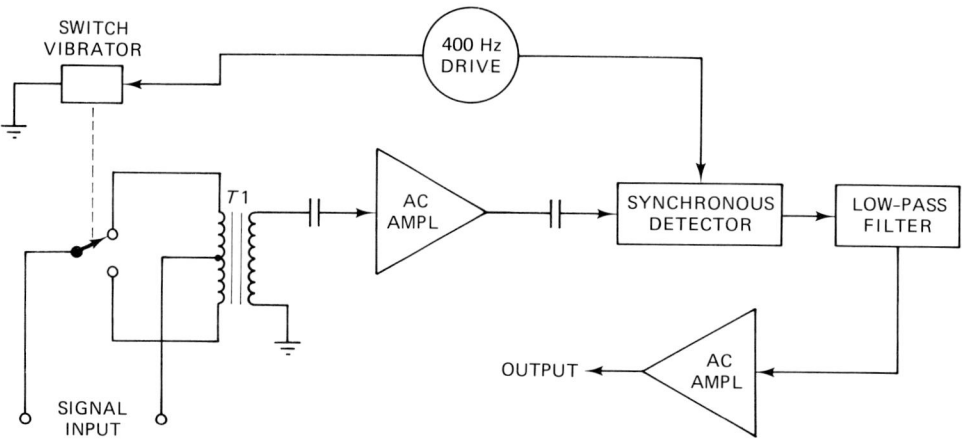

Fig. 17-3 Differential chopper amplifier.

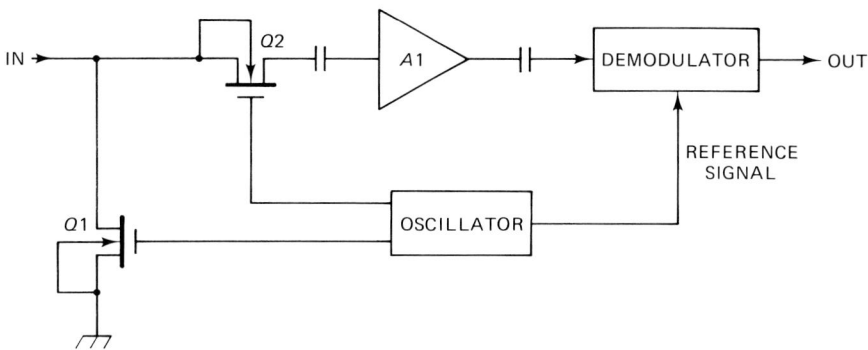

Fig. 17-4   Solid-state chopper amplifier.

Chopper amplifiers limit the noise both because of the low-pass filtering required and because the AC amplifier frequency response can be set to a narrow passband around the chopper frequency.

At one time the chopper amplifier was the only practical way to obtain low drift in high-gain situations. Modern IC and hybrid amplifiers, however, have such improved drift properties that no chopper is needed (especially in the lower end of the high-gain range). The Burr–Brown OPA-103 is a monolithic IC operational amplifier in a TO-99 eight-pin metal can package. This low-cost device exhibits a drift characteristic of 2 $\mu V/°C$. In addition, amplifiers are available (especially in hybrid form) that are actually electronic-chopper stabilized; but to the outside world the device simply looks like a low-drift amplifier. The Burr–Brown 3271/25 device is an example that exhibits a drift characteristic of 0.1 $\mu V/°C$.

CARRIER AMPLIFIERS

A carrier amplifier is any type of signal processing amplifier in which the signal carrying the desired information is used to modulate another (higher-frequency) signal—a carrier signal. The chopper amplifier is considered by many to fit this definition, but it is usually regarded as a unique type in its own right. The two principal carrier amplifiers are the DC-excited and AC-excited varieties.

Figure 17-5 shows a DC-excited carrier amplifier. A Wheatstone bridge transducer provides the input signal and is excited by a DC potential V. The output of the transducer is a low-level DC voltage that varies with the value of the stimulating parameter. The transducer signal is usually of very low amplitude, and it may be noisy. An amplifier increases the signal amplitude, and a low-pass filter removes much of the noise. In some models the first stage is actually a composite of these two functions, being essentially a frequency-selective filter with gain.

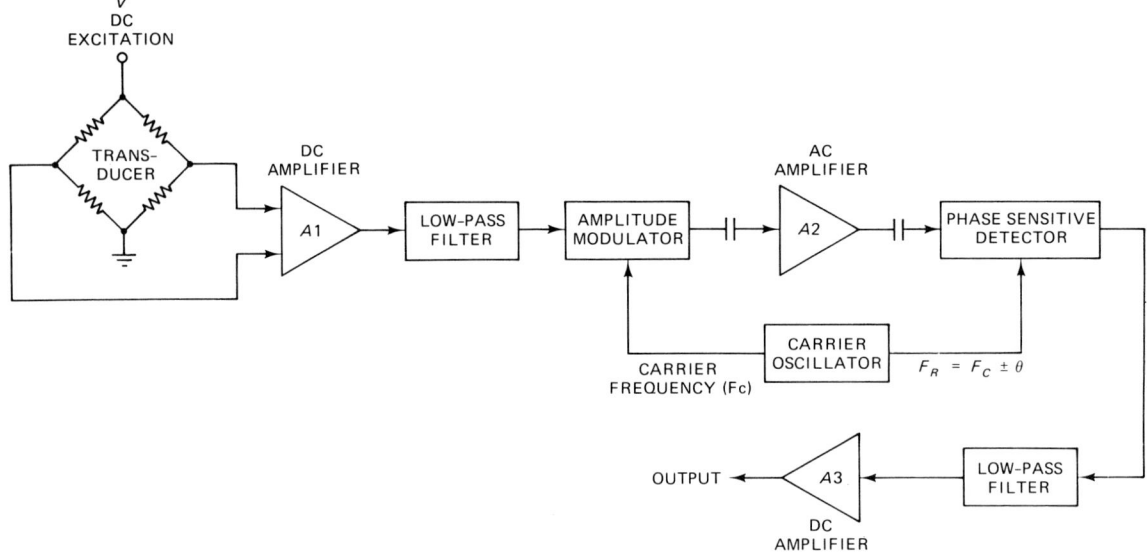

Fig. 17-5   DC-excited carrier amplifier.

The signal at the output of the amplifier-filter section is used to amplitude-modulate a carrier signal. Typical carrier frequencies range from 400 Hz to 25 kHz, with 400 Hz, 1 kHz, and 2.4 kHz being very common. The signal frequency response of a carrier amplifier is a function of the carrier frequency, and it is usually (at maximum) one-fourth of that carrier frequency. A carrier frequency of 400 Hz, then, is capable of signal frequency response of 100 Hz, while the 25-kHz carrier will support a frequency response of 6.25 kHz. Further amplification of the signal is provided by an AC amplifier.

The key to the performance of any carrier amplifier is the phase sensitive detector (PSD) that demodulates the amplified AC signal. Envelope detectors, while very simple and low cost, suffer from an inability to discriminate between the real signals and certain spurious signals. Phase-sensitive detectors overcome these problems.

The advantages of the PSD include the fact that it rejects signals not of the carrier frequency and certain signals that are of the carrier frequency. The PSD, for example, will reject even harmonics of the carrier frequency and those components that are out of phase with the reference signal. The PSD will, however, respond to odd harmonics of the carrier frequency. Some carrier amplifiers seem to neglect this problem altogether. But in most cases, manufacturers will design the AC amplifier section to be a bandpass amplifier with a response limited to $F_C \pm (F_C/4)$. This response will eliminate any third or higher order odd harmonics of the carrier frequency before they reach the PSD. It is then only necessary to assure that the reference signal has acceptable purity regarding total harmonic distortion and phase noise.

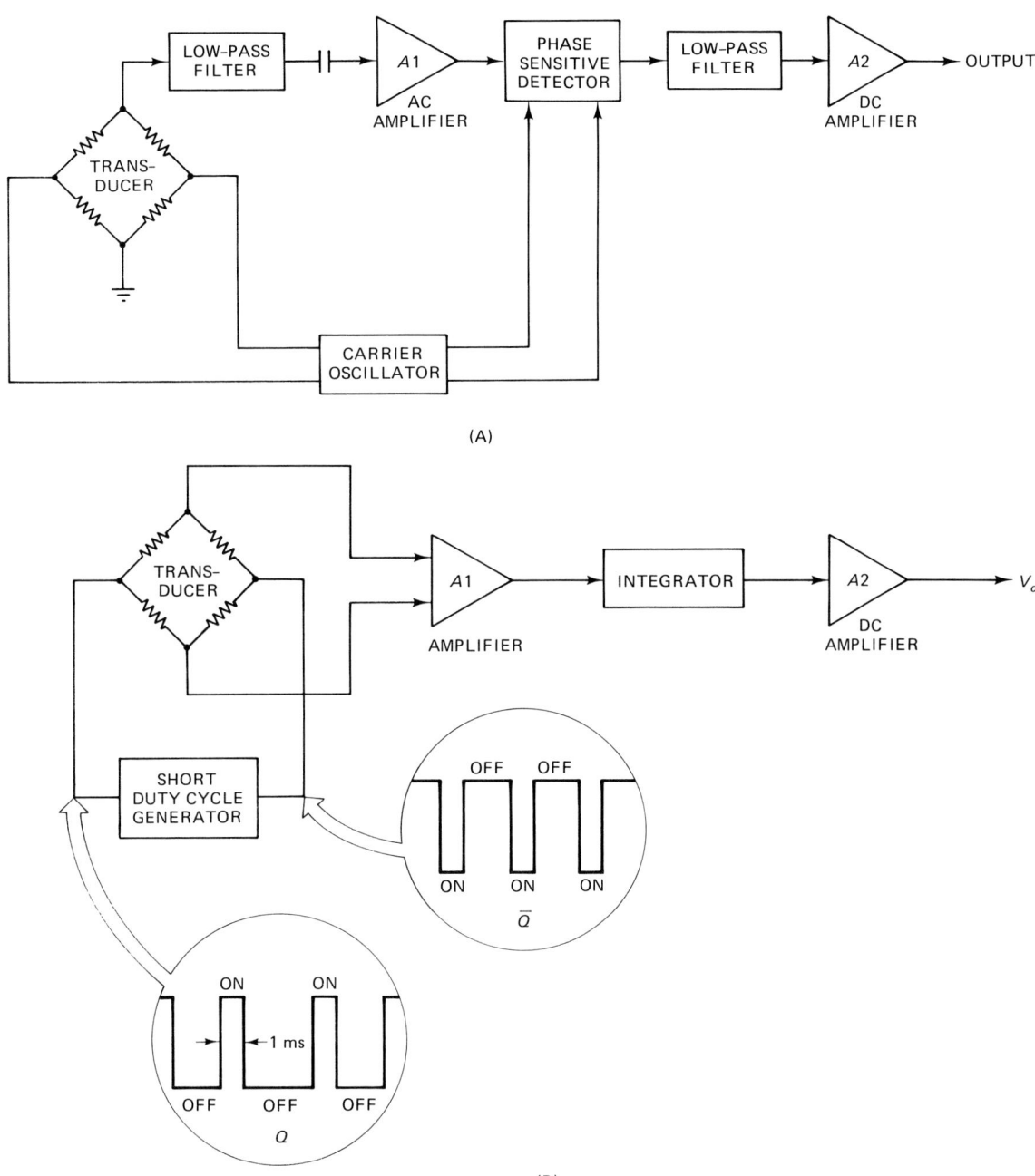

Fig. 17-6 (A) AC-excited carrier amplifier and (B) pulse-excited carrier amplifier.

An alternate, but very common, form of carrier amplifier is the AC-excited circuit shown in Fig. 17-6A. In this circuit the sensor is AC-excited by the carrier signal, eliminating the need for the amplitude modulator. The small AC signal from the transducer is amplified and filtered before being applied to the PSD circuit. Again, some designs use a bandpass AC amplifier to eliminate odd harmonic response. This circuit allows adjustment of transducer offset errors in the PSD circuit instead of in the transducer, by varying the phase of the reference signal.

Figure 17-6B is the block diagram to a transducer amplifier that uses pulsed excitation for the Wheatstone bridge. A short duty-cycle pulse generator produces either monopolar or (in the case shown) bipolar pulses that are used to provide the excitation potential. The advantage of the pulsed method is that the short duty cycle limits the amount of power dissipated in the transducer resistance elements, and therefore reduces the effects of transducer self-heating on the thermal drift. Several methods are used to demodulate the pulse waveform at the output of amplifier $A1$. A Miller integrator, which finds the time average of the signal, will create a DC potential that is proportional to the amplified transducer output signal. Alternative schemes use CMOS or other electronic switches to demodulate the signal.

Another advantage of the pulsed scheme is that an amplifier drift-cancellation circuit can be implemented. Switching is provided that shorts together the input of $A1$ during the off time of the pulse and connects a capacitor to the output of $A1$. When the capacitor is charged, it can be connected as an offset null potential to amplifier $A2$ during the on time of the pulse. One medical equipment manufacturer used this scheme as a baseline stablization method in a medical patient-monitor oscilloscope.

LOCK-IN AMPLIFIERS

The amplifiers discussed so far in this chapter produce relatively large amounts of noise and will respond to any noise present in the input signal. They suffer from the usual shot noise, thermal noise, H-field noise, E-field noise, ground loop noise, and so forth, that affects all amplifiers. The noise at the output is directly proportional to the square root of the circuit bandwidth. The lock-in amplifier is a special case of the carrier amplifier in which the bandwidth is very narrow. Some lock-in amplifiers use the carrier amplifier circuit of Fig. 17-6, but with an input amplifier having a very high Q bandpass characteristic. The carrier frequency will be between 1 and 200 kHz. The lock-in principle works because the information signal is made to contain the carrier frequency in a way that is easy to demodulate and interpret. The AC amplifier accepts only a narrow band of frequencies centered about the carrier frequency. The narrowness of the amplifier bandwidth, which makes possible the improved signal-to-noise ratio, also limits the lock-in amplifier to very low frequency input signals. Even then, it is sometimes necessary to integrate (i.e., time-average) the signal to obtain the needed data.

Lock-in amplifiers are capable of reducing the noise and retrieving signals that are otherwise buried below the noise level. Improvements

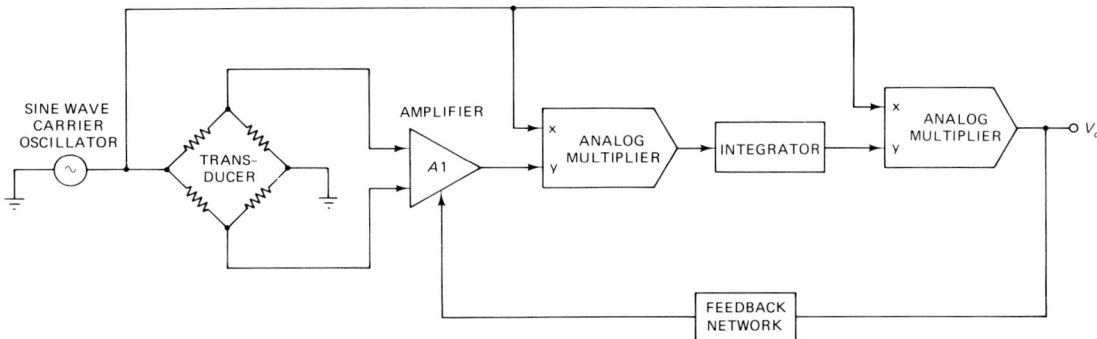

Fig. 17-7 Autocorrelation or "lock-in" amplifier.

of up to 85 dB are relatively easily obtained, and up to 100-dB reduction is possible if cost is less of a factor (please note the ambiguity inherent in the word *relatively*).

There are actually several different forms of lock-in amplifier available. The type discussed here is the simplest type. It is merely a very narrow-band version of the AC-excited carrier amplifier. The lock-in amplifier of Fig. 17-7, however, uses a slightly different technique: It is called an autocorrelation amplifier. The carrier is modulated by the input signal and then integrated. The output of the integrator is demodulated in a product detector circuit. The circuit of Fig. 17-7 produces very low output voltages for input signals that are not in phase with the reference signal, but it produces relatively high output voltages at the proper input frequency and phase.

ELECTRONIC INTEGRATORS AND DIFFERENTIATORS

Integration and differentiation are very important mathematical processes to electronic instrumentation and signal processing. These processes are inverses of each other, so a function that is first integrated and then differentiated, returns to the original function. A similar relationship occurs when a function is differentiated and is then integrated. Such is the normal nature of mathematically inverse processes.

These processes are seen elsewhere in electronics, but sometimes under different names. The differentiator is sometimes called a rate-of-change circuit, or, if the time constant is correct, a high-pass filter. Similarly, the integrator might be called a time-averager circuit or low-pass filter.

Let us consider an example of integration in electronic instruments. In Fig. 17-8 a voltage represents a pressure transducer output—in this particular case, the output of a human blood-pressure

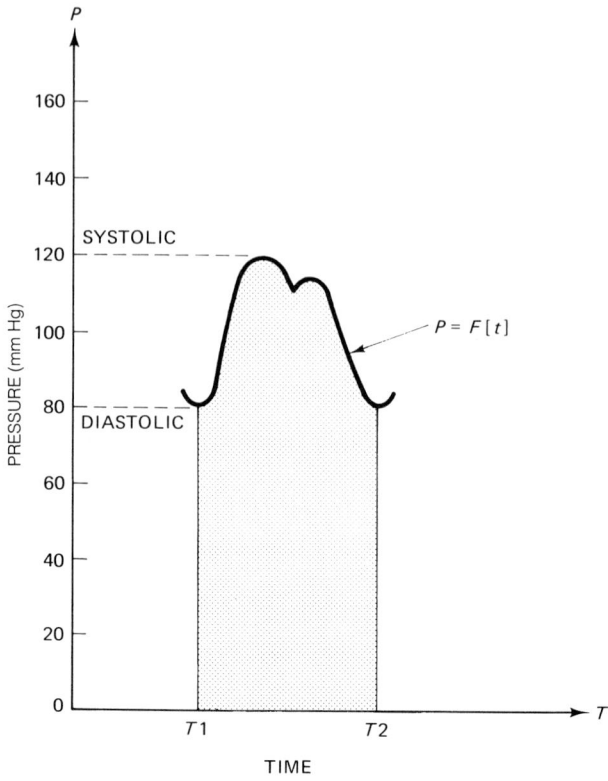

Fig. 17-8 In physiology the mean arterial blood pressure is found by integrating the blood-pressure waveform.

transducer. Such sensors are common in hospital intensive care units. Notice that the pressure voltage varies with time from a low (diastolic) to a high (systolic) between $T1$ and $T2$ (which interval represents one complete cardiac cycle). If you want to know the mean arterial blood pressure (MAP), you would want to find the area under the pressure-versus-time curve over one cardiac cycle.

An electronic integrator circuit serves to compute the time-average of the analog voltage waveform that represents the time-varying arterial blood pressure. In an electronic blood pressure monitoring instrument, a voltage serves to represent the pressure. If, for example, a scaling factor of 10 mV/mm Hg is used (as is commonly the case in medical devices), a pressure of 100 mm Hg is represented by a potential of 1000 mV, or 1.000 V. This voltage will vary over the range 800 mV to 1200 mV for the case shown in Fig. 17-8 (pressure varies from 80 to 120 mm Hg).

Electronic integrators and differentiators affect signals in different ways. Figure 17-9 shows the example of a square wave (Fig. 17-9A) applied to the inputs of an integrator and differentiator. The integrator output is shown at Fig. 17-9B, while the differentiator output is shown at Fig. 17-9C.

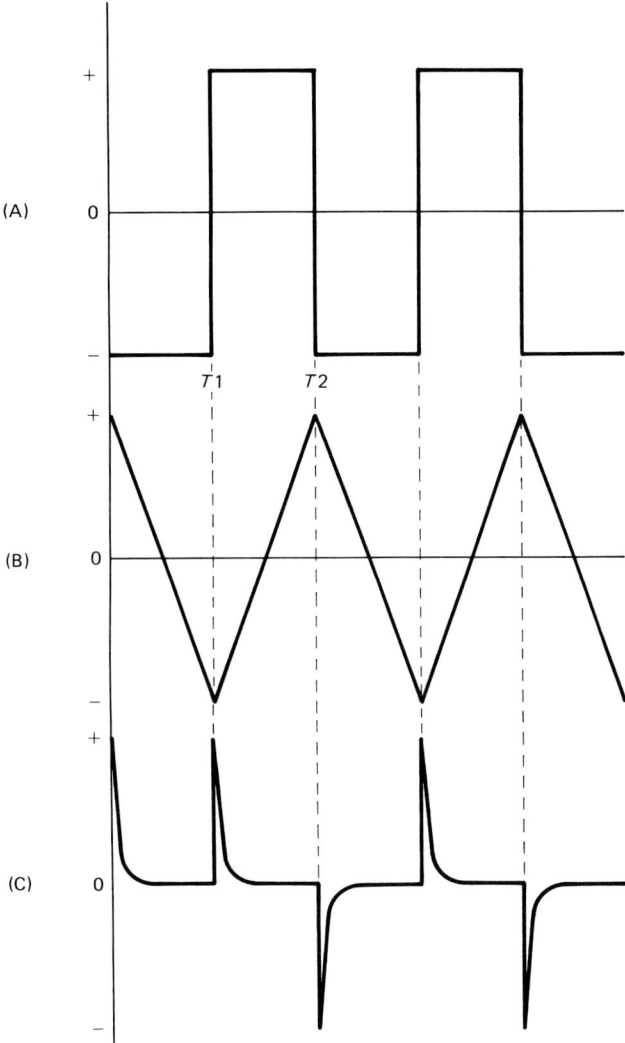

Fig. 17-9 Integrators and differentiators affect signals in different ways: (A) square wave input signal as seen at (B) the output of an integrator and (C) the output of the differentiator.

First consider the operation of the integrator circuit. The integrator output waveform in Fig. 17-9B shows a constant positive-going slope between $T1$ and $T2$. The steepness of the slope is dependent on the amplitude of the input square wave, but the line is linear. You can see from curve B in Fig. 17-9 that the square wave into the integrator produces a triangle waveform.

Now consider the operation of the differentiator circuit (see the output waveform Fig. 17-9C). At time $T1$ the square wave makes a positive-going transition to maximum amplitude. At this instant it has a very high rate of change, so the output of the differentiator is very high (see the waveform in Fig. 17-9C at $T1$). But then the amplitude of the input signal reaches maximum and remains constant until $T2$, when it drops back to its previous value. Thus, the differentiator will produce a sharp positive-going spike at $T1$ and a sharp negative-going spike at $T2$. In an ideal circuit there is no transition between these states, but in real circuits there is an exponential transition that is proportional to the RC time constant of the circuit and the rise time of the waveform. Differentiator output spikes are frequently used in circuits such as timers and zero-crossing detectors.

If a sine wave is applied to the inputs of either integrators or differentiators, the result is a sine-wave output that is shifted in phase 90°. The principal difference between the two forms of circuit is in the direction of the phase shift. Such circuits are frequently used to provide quadrature or sine–cosine outputs from a sine-wave oscillator.

RC Integrator Circuits

The simplest form of integrator and differentiator are simple resistor and capacitor circuits, such as shown in Fig. 17-10. The integrator is shown in Fig. 17-10A, and the differentiator is in Fig. 17-10B. The integrator consists of a resistor element in series with the signal line and a capacitor across the signal line. The differentiator is just the opposite: The capacitor is in series with the signal line and the resistor is in parallel with the line. These circuits are also known as lowpass and highpass RC filters, respectively. The lowpass case (integrator) has a -6 dB/octave falling characteristic frequency response, while the highpass case (differentiator) has a $+6$ dB/octave rising frequency response.

The operation of the integrator and differentiator is dependent on the time constant of the RC network (i.e., $R \times C$). The integrator time

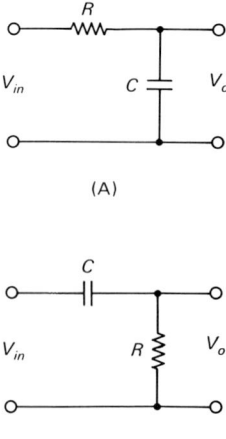

(A)

(B)

Fig. 17-10   Passive networks for integration and differentiation: (A) RC integrator and (B) RC differentiator.

constant is set long (i.e., $> 10 \times$) compared with the period of the signal being integrated, while in the differentiator the RC time constant is short (i.e., $< 1/10 \times$) compared with the period of the signal. Several integrators can be connected in cascade to increase the time-averaging effect or increase the slope of the frequency-response fall-off.

Active Differentiator and Integrator Circuits

The operational amplifier makes it relatively easy to build high-quality active integrator and differentiator circuits. Previously, one had to construct a stable, drift-free, high-gain transistor amplifier for this purpose. Figure 17-11 shows the basic circuit of the operational amplifier differentiator. Again the RC elements are used, but in a slightly different manner. The capacitor is in series with the op-amp's inverting input, and the resistor is the op-amp feedback resistor.

Analysis of the circuit to derive the transfer function follows a procedure similar to that followed for inverting and noninverting followers earlier in this book (Chapter 12).

From Kirchhoff's current law (KCL):

$$I2 = -I1 \tag{17-1}$$

From basic passive circuit theory, including Ohm's law:

$$I1 = C\,\Delta V_{in}/\Delta t \tag{17-2}$$

$$I2 = V_o/R \tag{17-3}$$

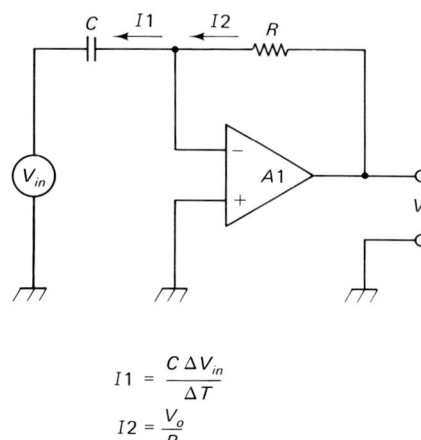

$$I1 = \frac{C\,\Delta V_{in}}{\Delta T}$$
$$I2 = \frac{V_o}{R}$$

Fig. 17-11 Active differentiator circuit.

Substituting (17-2) and (17-3) into (17-1):

$$V_o/R = -C\,\Delta V_{in}/\Delta t \qquad (17\text{-}4)$$

or, with the terms rearranged:

$$V_o = -RC(\Delta V_{in}/\Delta t) \qquad (17\text{-}5)$$

where V_o and V_{in} are in the same units (volts, millivolts, etc.), R in ohms, C in farads, and t in seconds.

Equation (17-12) is a mathematical way of saying that output voltage V_o is equal to the product of the RC time constant and the derivative of input voltage V_{in} with respect to time (dV_{in}/dt). Since the circuit is essentially a special case of the familiar inverting follower circuit, the output is inverted; hence the negative sign.

Figure 17-12 shows the classical operational amplifier version of the Miller integrator circuit. Again, an operational amplifier is the active element, while a resistor is in series with the inverting input and a capacitor is in the feedback loop. Notice that the placement of the capacitor and resistor elements are exactly opposite in both the RC and operational amplifier versions of integrator and differentiator circuits. In other words, the RC elements reverse roles between Figs. 17-11 and 17-12. That fact will tell the astute student quite a bit regarding the nature of integration and differentiation.

The output of the integrator is dependent on the input signal amplitude and the RC time constant. The transfer function for the Miller integrator is derived in a manner similar to that of the differentiator.

From KCL:

$$I2 = -I1 \qquad (17\text{-}6)$$

From Ohm's law:

$$I1 = V_{in}/R \qquad (17\text{-}7)$$

and

$$I2 = C(dV_o/dt) \qquad (17\text{-}8)$$

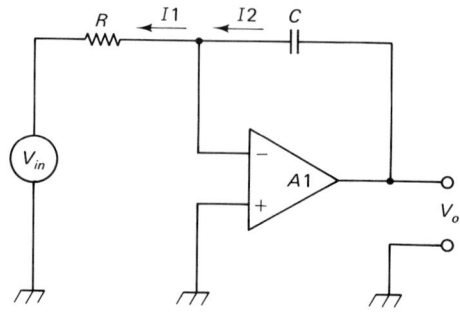

Fig. 17-12   Active integrator, also called the Miller integrator.

Substituting Eqs. (17-7) and (17-8) into Eq. (17-6):

$$C \, dV_o/dt = -V_{in}/R \qquad (17\text{-}9)$$

Integrating both sides:

$$\int \frac{C \, dV_o}{dt} = \int \frac{-V_{in}}{R} \, dt \qquad (17\text{-}10)$$

$$CV_o = \int \frac{-V_{in}}{R} \, dt \qquad (17\text{-}11)$$

Collecting and rearranging terms:

$$CV_o = \frac{-1}{R} \int V_{in} \, dt \qquad (17\text{-}12)$$

$$V_o = \frac{-1}{RC} \int V_{in} \, dt \qquad (17\text{-}13)$$

And accounting for initial conditions:

$$V_o = \frac{-1}{RC} \int V_{in} \, dt + K \qquad (17\text{-}14)$$

where: V_o and V_{in} are in the same units (volts, millivolts, etc.), R in ohms, C in farads, and t in seconds.

This expression is a way of saying that the output voltage is equal to the time-average of the input signal plus some constant K, which is the voltage that may have been stored in the capacitor from some previous operation (often zero in electronic applications).

Practical Circuits

The circuits shown in Figs. 17-11 and 17-12 are classic textbook circuits. Unfortunately, they do not work very well in some practical cases. The problem is that these circuits are too simplistic because they depend on the properties of ideal operational amplifiers. Unfortunately, real op-amps fall far short of the ideal in several important ways that affect these circuits. In real circuits, differentiators may ring, or oscillate, and integrators may saturate from their tendency to integrate bias currents and other inherent DC offsets very shortly after turn-on.

There is another problem with this kind of circuit, and it magnifies the problem of saturation. Namely, the integrator circuit of Fig. 17-12 has a very high gain with certain values of R and C. The voltage gain of this circuit is given by the term $-1/RC$, which, depending on the values selected for R and C, can be quite high.

In other words, with a gain of -10.000, a $+1$ V applied to the input will want to produce a $-10,000$-V output. Unfortunately, the operational amplifier output is limited to the range of approximately -10 to -20 V, depending on the device and the applied V DC power supply voltage. For this case, the operational amplifier will saturate very rapidly! To keep the output voltage from saturating, it is necessary to prevent the input signal from rising too high. If the maximum output voltage allowable is 10 V, the maximum input signal is 10 V/10,000 or 1 millivolt! Obviously, it is necessary to keep the RC time constant within certain bounds.

How to Solve the Problem

Fortunately, there are some design tactics that allow keeping the integration aspects of the circuit while removing the problems. A practical integrator is shown in Fig. 17-13. The heart of this circuit is an RCA BiMOS operational amplifier, type CA-3140, or an equivalent BiFET device. The reason why this works so well is that it has a low input bias current (being MOSFET input).

Capacitor $C1$ and resistor $R1$ in Fig. 17-13 form the integration elements and are used in the transfer equation to calculate performance. Resistor $R2$ is used to discharge $C1$ to prevent DC offsets on the input signal and the op-amp itself from saturating the circuit. The RESET switch is used to set the capacitor voltage back to zero (to prevent a K factor offset) before the circuit is used. In some measurement applications the circuit is initialized by closing S1 momentarily.

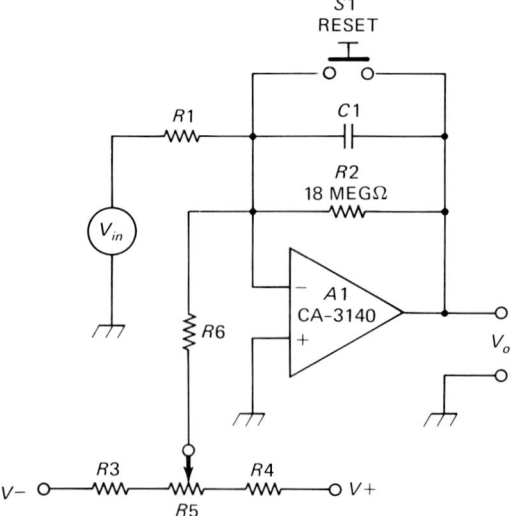

Fig. 17-13   Practical Miller integrator circuit.

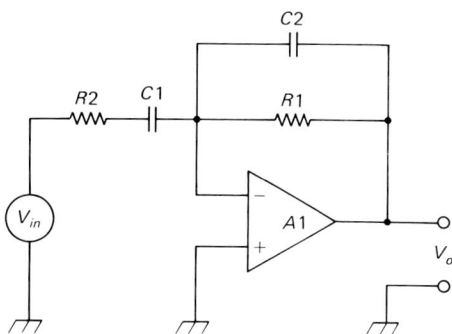

Fig. 17-14 Practical active differentiator circuit.

In actual circuits, S1 may be a mechanical switch, an electromechanical relay, a solid-state relay, or a CMOS electronic switch.

If there is still a minor drift problem in the circuit, potentiometer $R5$ can be added to the circuit to cancel it. This component adds a slight counter current to the inverting input through resistor $R6$. To adjust this circuit, set $R5$ initially to mid-range. The potentiometer is adjusted by shorting the V_{in} input to ground (setting $V_{in} = 0$) and then measuring the output voltage. Press S1 to discharge $C1$, and note the output voltage (it should go to zero). If V_o does not go to zero, then turn $R5$ in the direction that counters the change of V_o. This change can be observed after each time RESET switch S1 is pressed. Keep pressing S1 and then making small changes in $R5$, until the setting is found at which the output voltage stays very nearly zero, and constant, after S1 is pressed (there may be some very long-term drift).

Figure 17-14 shows the practical version of the differentiator circuit. The differentiation elements are $R1$ and $C1$, and the previous equation for the output voltage is used. Capacitor $C2$ has a small value (1 to 100 pF), and it is used to alter the frequency response of the circuit to prevent oscillation or ringing on fast rise-time input signals. Similarly, a snubber resistor ($R2$) in the input also limits this problem. The operational amplifier can be almost any type with a fast enough slew rate, and the CA-3140 is often recommended. The values of $R2$ and $C2$ are often determined by rule of thumb, but their justification is taken from the Bode plot of the circuit.

LOGARITHMIC AND ANTILOG CIRCUITS

Logarithmic amplifiers are often used in instrumentation circuits, especially where data compression is required. The overall transfer equation for an operational amplifier circuit is determined by the transfer equation of the feedback network. As might be guessed from

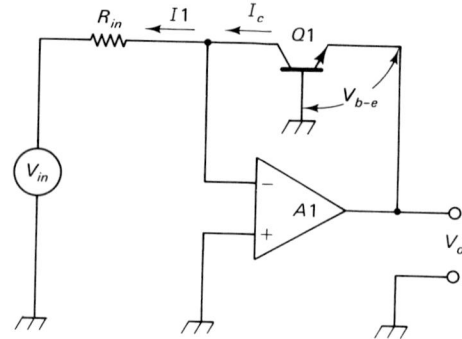

Fig. 17-15   Logarithmic amplifier circuit.

this fact, a logarithmic transfer equation can be created in an operational amplifier circuit by placing a nonlinear element in the negative feedback loop. An ordinary PN junction transistor meets this requirement. A logarithmic amplifier is one that has a transfer equation of the form of either

$$V_o = k \ln(V_{in}) \qquad (17\text{-}15)$$

or

$$V_o = k \log(V_{in}) \qquad (17\text{-}16)$$

Figure 17-15 shows the basic circuit for an inverting logarithmic amplifier. As with any inverting amplificr, we can assume that the summing junction potential is zero because the noninverting input is grounded. In basic transistor theory the base-emitter voltage of the transistor is given by

$$V_{b\text{-}e} = \frac{kT}{q} \ln\left(\frac{I_c}{I_s}\right) \qquad (17\text{-}17)$$

where $V_{b\text{-}e}$ is the base-emitter potential in volts, k Boltzmann's constant (1.38×10^{-23} J/K), T the temperature in Kelvin (K), and q the electronic charge [1.6×10^{-19} C (coulombs)]; ln indicates the natural (or "base-e") logarithms, I_c is the collector current of the transistor in amperes (A), and I_s the reverse saturation current of the transistor (approximately 10^{-13}A at 300 K)

Because the configuration of Fig. 17-15 makes $V_{b\text{-}e} = V_o$:

$$V_o = \frac{kT}{q} \ln\left(\frac{I_c}{I_s}\right) \qquad (17\text{-}18)$$

or, for those who prefer base-10 logarithms:

$$\frac{\log X}{\ln X} = 0.4343 \qquad (17\text{-}19)$$

and

$$\frac{\ln X}{\log X} = 2.3 \qquad (17\text{-}20)$$

so

$$V_{o} = \frac{2.3kT}{q} \log\left(\frac{I_{c}}{I_{s}}\right) \qquad (17\text{-}21)$$

When the constants kT/q are accounted for,

$$V_{o} = 60 \text{ mV} \log\left(\frac{I_{c}}{I_{s}}\right) \qquad (17\text{-}22)$$

Equation (17-22) demonstrates that the output voltage V_{o} is a logarithmic function. From KCL it is known that

$$I1 = -I_{c} \qquad (17\text{-}23)$$

And from Ohm's law:

$$I1 = V_{in}/R_{in} \qquad (17\text{-}24)$$

Substituting Eqs. (17-23) and (17-24) into Eq. (17-22) yields

$$V_{o} = 60 \text{ mV} \log\left(\frac{I1}{I_{s}}\right) \qquad (17\text{-}25)$$

or

$$V_{o} = 60 \text{ mV} \log\left(\frac{V_{in}/R_{in}}{I_{s}}\right) \qquad (17\text{-}26)$$

Thus, the output voltage V_{o} is proportional to the logarithm of the input voltage V_{in}. The simple circuit of Fig. 17-15 is the one usually published in textbooks, but in a practical sense it only works some of the time due to the realities of non-ideal operational amplifiers. But there is one common problem: oscillation. Unfortunately, the amplifier oscillates in the basic configuration.

A modified version of the circuit is shown in Fig. 17-16. In this circuit a compensation network ($R3/R4/C1$) is added to prevent the

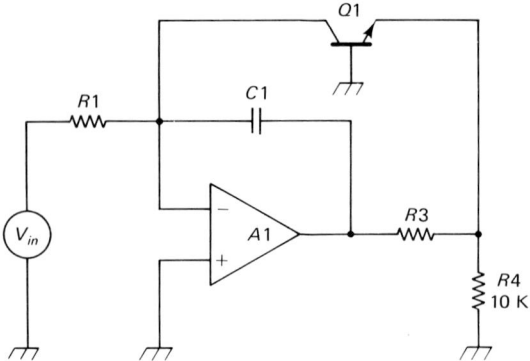

Fig. 17-16 Compensated logarithmic amplifier corrects defects of Fig. 17-15.

oscillation. The values of the network components (except $R4$) are found from empirical data based on the following approximations:

$$R3 = \frac{V_{o(max)} - 0.7}{\left(V_{in(max)}/R1\right) + \left(V_{o(max)}/R4\right)} \qquad (17\text{-}27)$$

and

$$C1 = 1/2\pi F R3 \qquad (17\text{-}28)$$

Another problem of the logarithmic amplifier is temperature sensitivity. Recall that the operant equation for the logarithmic amplifier is

$$V_o = \frac{kT}{q} \ln\left(\frac{(V_{in}/R_{in})}{I_s}\right) \qquad (17\text{-}29)$$

The T term in the equation is temperature in Kelvin. Temperature is a variable rather than a constant, so we can expect the output voltage to be a function of both the applied input signal voltage and the temperature of the b-e junction in the transistor. This temperature, in turn, is a function of ambient temperature. To prevent pollution of the output signal data it is necessary to temperature-compensate the logarithmic amplifier circuit. Figure 17-17 shows two approaches to the temperature compensation job. The value of $R1$ in Fig. 17-17A is approximately 15.7 times the temperature of the thermistor (R_t) at room temperature. The circuit of Fig. 17-17B is a little more complex, but it also offers greater dynamic range than the previous circuit.

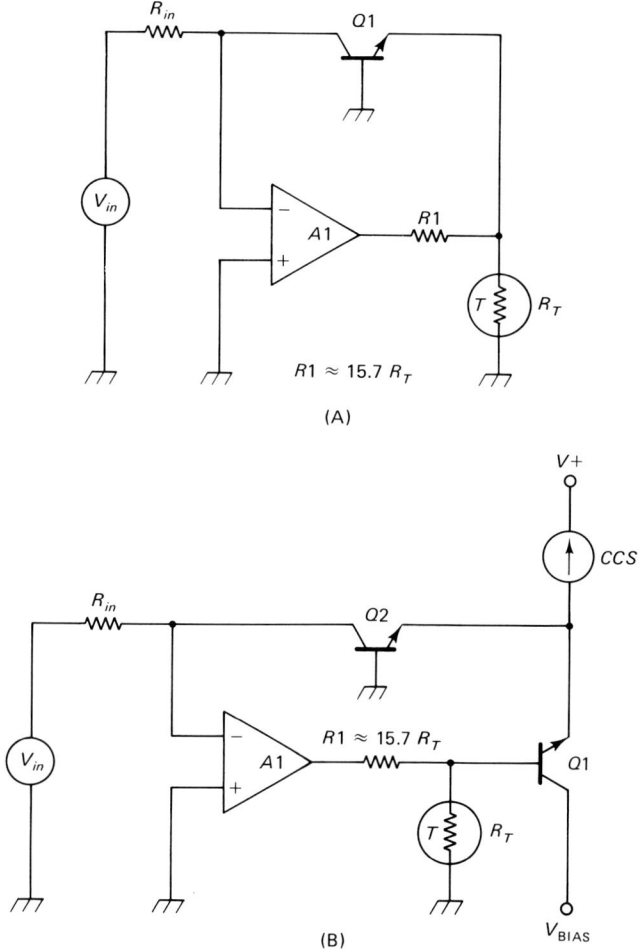

Fig. 17-17 Temperature-compensated logarithmic amplifier (A) common circuit and (B) improved circuit.

Antilog Amplifiers

The antilog amplifier performs the inverse function of the logarithmic amplifier. The output voltage from the antilog amplifier is

$$V_o = k \text{ antilog}(V_{in}) \qquad (17\text{-}30)$$

The simplified circuit for a conventional antilog amplifier is shown in Fig. 17-18. Note that, again, a PN junction from a bipolar transistor is used because of its logarithmic transfer function. But note here that the respective positions of the transistor and resistor are reversed from the logarithmic amplifier. From basic operational ampli-

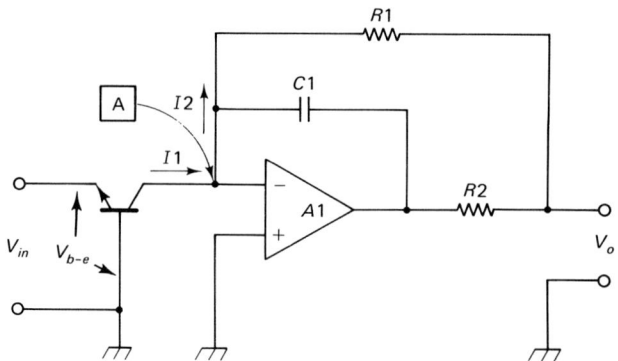

Fig. 17-18 Antilogarithmic amplifier circuit.

fier theory and Kirchhoff's current law (KCL):

$$|I1| = |I2| \tag{17-31}$$

Further, by Ohm's law and the fact that the summing junction (point A) is at virtual ground (zero) potential.

$$I2 = V_o R1 \tag{17-32}$$

In Fig. 17-18 voltage V_{in} is applied across the b-e junction of Q1, so $V_{b-e} = V_{in}$. Current $I1$ is the collector current of Q1, so $I_c = I1$. Recall Eq. (17-17).

$$V_{b-e} = \frac{kT}{q} \ln\left(\frac{I_c}{I_s}\right) \tag{17-33}$$

which can be rewritten to the form

$$V_{in} = \frac{kT}{q} \ln\left(\frac{I1}{I_s}\right) \tag{17-34}$$

Because $I1 = I2$,

$$V_{b-e} = \frac{kT}{q} \ln\left(\frac{(V_o/R1)}{I_s}\right) \tag{17-35}$$

Collecting terms:

$$\frac{qV_{in}}{kT} = \ln\left(\frac{(V_o/R1)}{I_s}\right) \tag{17-36}$$

Taking the antilog of Eq. (17-36):

$$\exp(qV_{in}/kT) = \frac{V_o R1}{I_s} \qquad (17\text{-}37)$$

$$\frac{I_s \exp(qV_{in}/kT)}{R1} = V_o \qquad (17\text{-}38)$$

Equation (17-38) is the transfer equation for the antilog amplifier.

Special Function Circuits

Circuit elements can be combined into a single circuit to form a special-function circuit. The combined circuit can be fabricated either in the monolithic IC form or as a hybrid. Although the range of possible special-function circuits is nearly limitless, a few examples serve to illustrate the concept.

Multi-Function Converter

Figure 17-19 shows the block diagram for a Burr–Brown 4301 or 4302 Multi-Function Converter device. This circuit can perform analog multiplication or division, plus serve such functions as squaring, square rooting, taking other roots, exponentiation, sine, cosine, arctan, and root sum squares (RSS).

The 4301/4302 devices consist of a logarithmic amplifier, a log-ratio amplifier, a summer, and an antilog amplifier (see Fig. 17-19).

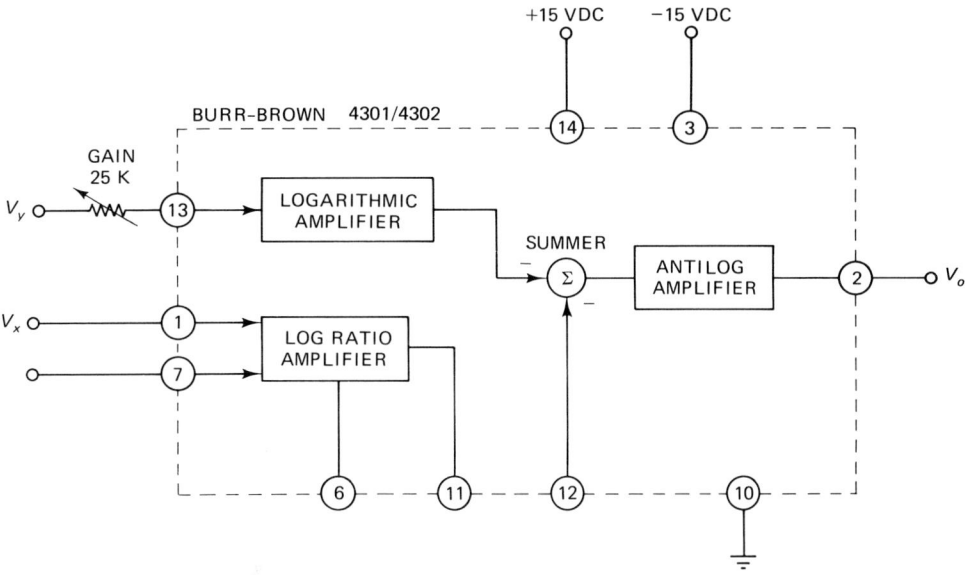

Fig. 17-19 Block diagram for the Burr–Brown 4301/4302 devices.

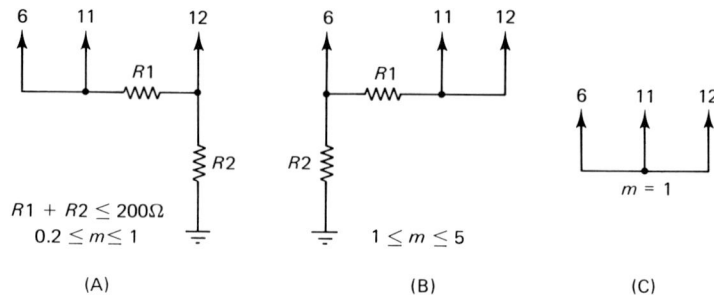

Fig. 17-20 Resistor networks for programming the 4301/4302 devices.

The transfer function for this circuit is

$$V_o = V_y \left[\frac{V_z}{V_x} \right]^m \tag{17-39}$$

The output signal can swing to $+10$-V output, and it will supply 5 mA of output current. The three input signal voltages (V_x, V_y, and V_z) must $+10$ V.

The function performed by these devices depends on the nature of the resistor network connected between pins 6, 11, and 12. Figure 17-20 shows the three types of resistor network used to program the 4301/4302. The exponent m [see Eq. (17-39)] determines the function performed. The value of m is set by the external resistors; it can be any value between 0.2 and 5. For root functions $m < 1$. For example, the square root function is $m = 0.5$. The resistor network that yields this function is shown in Fig. 17-20A. The values of the resistors for this type of function are set by ($m < 1$).

$$m = R2/(R1 + R2) \tag{17-40}$$

Values greater than one ($1 < m < 5$) allow the 4301/4302 devices to be used as squarers and exponentiators. For these functions use the resistor network in Fig. 17-20B. The values of the resistors are found from

$$m = (R1 + R2)/R2 \tag{17-41}$$

Using the 4301/4302 as an analog multiplier or divider requires no resistor networks at all. In these circuits $m = 1$, and all three terminals are strapped together (Fig. 17-20C). In this application, two inputs will represent signals, while the third represents a constant and is set to a fixed voltage. For example, to build an analog multiplier, V_y and V_z are used as signal inputs, while V_x is connected through a resistor to a fixed reference potential. The transfer function in that case is

$$V_o = kV_yV_z \tag{17-42}$$

where k is proportional to V_x.

Similarly, to make an analog divider, V_y becomes the constant (k) and the input signals are V_x and V_z. The transfer equation is

$$V_o = k(V_z/V_x) \qquad (17\text{-}43)$$

Devices such as the Burr–Brown 4301/4302 are usable for a wide variety of analog signal processing tasks.

Programmable Gain Amplifiers

Amplifier gains must sometimes be changed for different applications. In other situations the gain must be changed in an effort to calibrate a system. A programmable gain amplifier allows the gain to be set (or programmed) from an external source. In some simple cases the programming is done by setting a bias current. A single resistor is connected between the programming terminal and a reference DC power supply. This form of amplifier does not allow the type of control that is amenable to either computer control or control by other digitally oriented circuits.

There are two methods for digitally programming an amplifier. One is to use a multiplying digital-to-analog converter (MDAC). All DACs provide an output that is proportional to a binary word and a DC reference input. In a multiplying DAC, or MDAC, the DC reference is supplied from an external source. It is possible on some MDACs to use either a fixed DC source, a slowly varying nearly DC signal, an AC signal with a DC offset, or a symmetrical AC signal (in which case the offset is provided by the designer) as the reference. If the reference is the signal to be amplified, then the output will be a function of the reference signal input and the binary word applied to the digital inputs.

Another approach to designing the programmable amplifier is shown in Fig. 17-21. The amplifier circuit is the three-device instrumentation amplifier circuit. Gain is set by

$$A_v = \left(\frac{2R2}{R1} + 1 \right)\left(\frac{R6}{R4} \right) \qquad (17\text{-}44)$$

assuming $R2 = R3$, $R4 = R5$, and $R6 = R7$.

By making all resistors fixed, the programmable amplifier gain can be controlled solely by varying $R1$. In Fig. 17-21 this function is performed by shunting additional resistors across a relatively high value of fixed resistor that is always in the circuit. The switches (S1 through S3) are CMOS electronic switches with active LOW digital control terminals. When the control terminals (A, B, and C) are HIGH, the switch is open, and when the control terminal is LOW, the switch is closed. Gain is set by selecting any (or several) switches to be closed to alter the resistance of $R1$ as seen by the amplifier. The circuit of Fig. 17-21 is actually quite crude compared with programmable

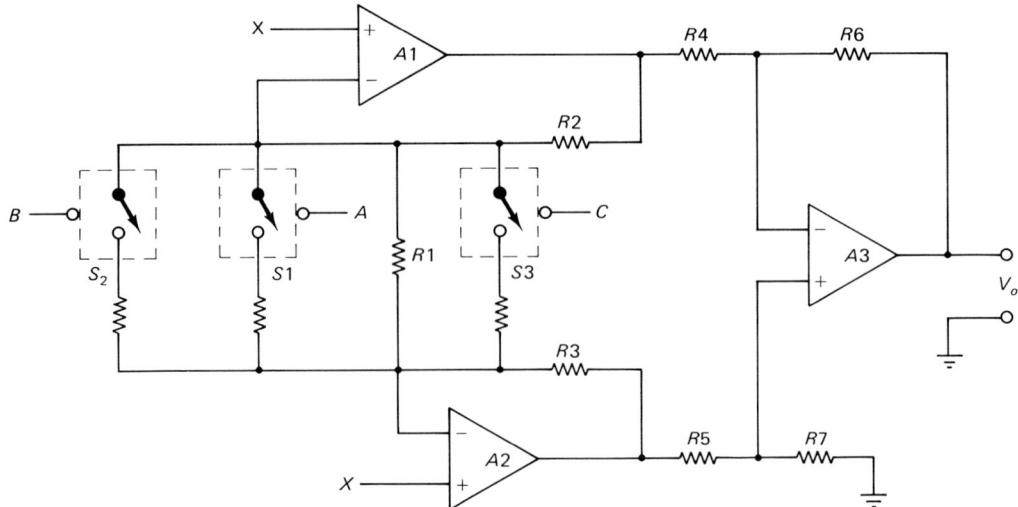

Fig. 17-21   Programmable gain instrumentation amplifier.

amplifiers offered by device manufacturers. In some of those products more than one resistance parameter can be varied, resulting in gain settings of small increments from unity to 1024.

Thermocouple Amplifiers

A thermocouple is a junction of two dissimilar metals. These devices are used to measure temperature over a very wide range (> 2700°C), and they can be very precise when properly built and calibrated. The thermocouple works because of the Seebeck effect. All metals have a certain work function that defines how much energy is needed to loosen electrons from the atoms. If two metals with different work functions are joined together in a fused junction, then the difference between work functions, a voltage, is generated across the two ends of the wires. That voltage is proportional to the temperature of the junction and the work function differential. Thermocouples are given letter designations (e.g., K and T) to denote both a certain characteristic and a pair of specific metals used to form the junction.

The inverse function, called the Peltier effect, is responsible for the so-called solid-state refrigerators; applying a voltage across the loose ends of the thermocouple causes one wire to lose heat and the other to absorb heat.

It is common practice in thermocouple temperature measurements to use two thermocouples. One is placed in a 0°C ice bath and becomes a reference pole. Iced water is exactly 0°C when there is equilibrium between ice and liquid water in the bath. The other thermocouple is the measurement pole; it is connected in series with the ice-point thermocouple. A differential amplifier measures the difference potential between the two thermocouples, and the resulting output voltage (within the linearity of the system) is proportional to

TEMPERATURE MEASUREMENT COMPONENTS

Temperature Transducer Signal Conditioners

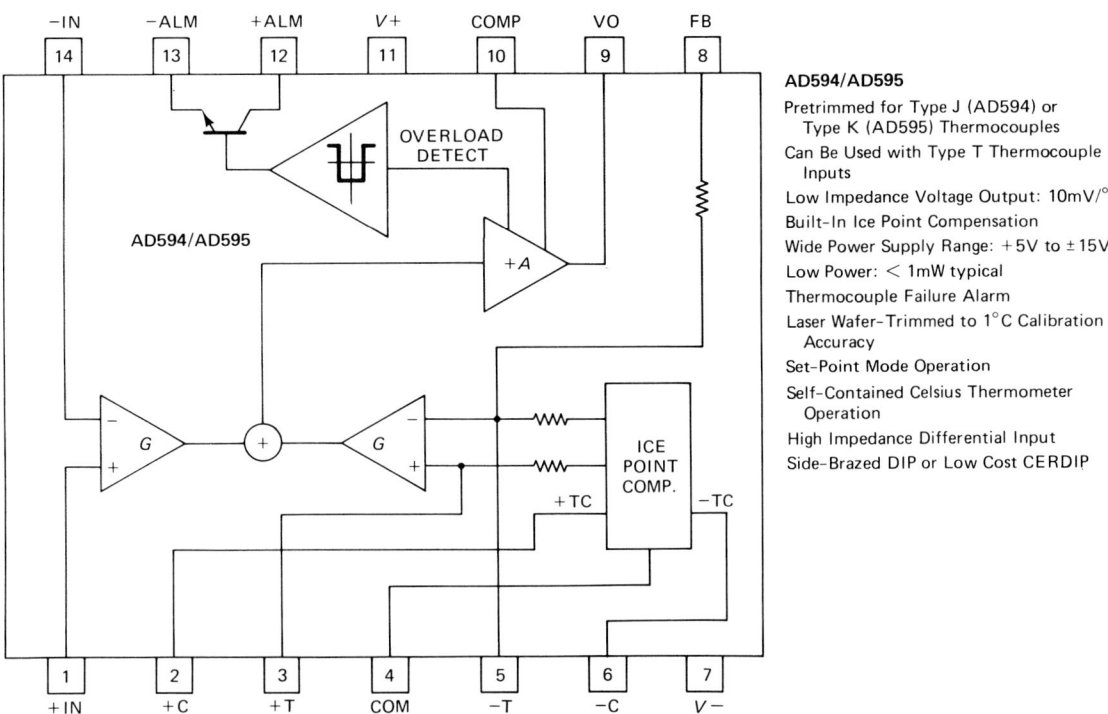

AD594/AD595

Pretrimmed for Type J (AD594) or
Type K (AD595) Thermocouples

Can Be Used with Type T Thermocouple
Inputs

Low Impedance Voltage Output: 10mV/°C

Built-In Ice Point Compensation

Wide Power Supply Range: +5V to ±15V

Low Power: < 1mW typical

Thermocouple Failure Alarm

Laser Wafer-Trimmed to 1°C Calibration
Accuracy

Set-Point Mode Operation

Self-Contained Celsius Thermometer
Operation

High Impedance Differential Input

Side-Brazed DIP or Low Cost CERDIP

Fig. 17-22 Temperature transducer signal-conditioner IC (courtesy of Analog Devices, Inc., Norwood, MA).

the temperature difference between 0°C and the measured temperature. This arrangement is called *ice-point compensation*.

Analog Devices Inc. makes a special-function IC device that is used as a transducer signal conditioner. A DC differential input amplifier converts the thermocouple signal into a single-ended output signal, V_o. The AD-594 and AD-595 devices (Fig. 17-22) contain all of the circuitry necessary to properly operate the thermocouple. These devices also contain electronic circuitry that performs the ice-point compensation function mentioned above. Both Type-T and Type-K thermocouples can be accommodated. The output of the IC will produce a scaling function of 10 mV/°C.

18

Signal Sources, Waveform Generators, and Wave-Shaping Circuits

Waveform generators are used to produce the large variety of electronic waveforms that are needed in many different circuits and applications. Some circuits are sine-wave oscillators, even though the word *oscillator* is also correctly applied to circuits that produce other waveforms. The astable multivibrator (AMV) may produce square waves, triangular waves, or other nonsinusoidal waveforms. Similarly, a digital clock is a special case of the astable multivibrator that is used in digital logic and computer circuits.

In general, the term oscillator may be used to denote all three cases, including sine-wave oscillators, astable multivibrators and digital clocks. The term oscillator can be defined as a circuit that will produce a periodic waveform (i.e., one that repeats itself). The output waveform can be a sine wave, square wave, triangle wave, sawtooth wave, pulses, or any of several other wave shapes. The important thing is that the waveform is periodic.

A class of waveform generator that is not an oscillator is the monostable multivibrator, or one-shot, circuit. This circuit produces only a single output pulse when triggered, so it is not periodic.

There are two basic forms of oscillator circuit: relaxation oscillators and feedback oscillators. The latter use an active device such as an amplifier, and then provide feedback in a manner that produces regeneration instead of degeneration. These circuits account for a large number of the oscillators used in practical electronic circuits.

Relaxation oscillators use any of several available breakdown devices (e.g., neon lamps) or negative resistance devices (e.g., tunnel diodes). Negative resistance devices operate according to Ohm's law under certain conditions, and opposite Ohm's law under other conditions. Breakdown relaxation oscillators use devices that pass little or no current at voltages below some threshold, and pass a large current at voltages above the threshold. Examples of these devices are neon glow-lamps and unijunction transistors (UJT).

There is also a subclass of oscillators that are based on IC devices such as voltage comparators, operational amplifiers, integrators, and so forth. It is these circuits that are discussed in this chapter. Because these circuits are based on the charge and discharge properties of resistor–capacitor networks, it is prudent to review the operation of simple *RC* networks.

REVIEW OF *RC* NETWORKS

Many of the waveform generators discussed in this chapter depend on the characteristics of the *RC* network for their operation. As a result, this section is provided as a brief review of *RC* network DC theory. Consider Fig. 18-1A. Assuming that the initial condition is as shown, switch S1 is in position A and is thus open-circuited. There is initially no charge stored in capacitor *C* (i.e., $V_c = 0$). If switch S1 is moved to position B, however, voltage *V* is applied to the *RC* network. The capacitor begins to charge with current from the battery, and V_c begins to rise towards *V* (see curve V_{cb} in Fig. 18-1B). The instantaneous capacitor voltage is found from

$$V_c = V(1 - e^{-T/RC}) \tag{18-1}$$

where V_c is the capacitor voltage, *V* the applied voltage from the source, *T* the elapsed time (in seconds) after charging begins, *R* is the resistance in ohms, and *C* is the capacitance in farads.

The product *RC* is called the time constant of the network, and is sometimes abbreviated (T). If *R* is in ohms, and *C* is in farads, then the product *RC* is specified in seconds. The capacitor voltage rises to approximately 63.2 percent of the final value after $1RC$, 86 percent after $2RC$, and > 99 percent after $5RC$. A capacitor in an *RC* network is considered fully charged by definition after five time-constants.

If switch S1 in Fig. 18-1A is next set to position C, the capacitor will begin to discharge through the resistor. In the discharge condition:

$$V_c = Ve^{-T/RC} \tag{18-2}$$

Voltage V_c drops to 36.8 percent of the full charge level after one time-constant ($1RC$) and to very nearly zero after $5RC$. Next consider Fig. 18-1C. This graph represents a situation commonly encountered in waveform generator circuits. In this graph the capacitor is required to charge from some initial condition (V_{C1}), which may or may not be 0 volts, to a final condition (V_{C2}), which may or may not be the fully charged $5RC$ point, in a specified time interval *T*. The question asked

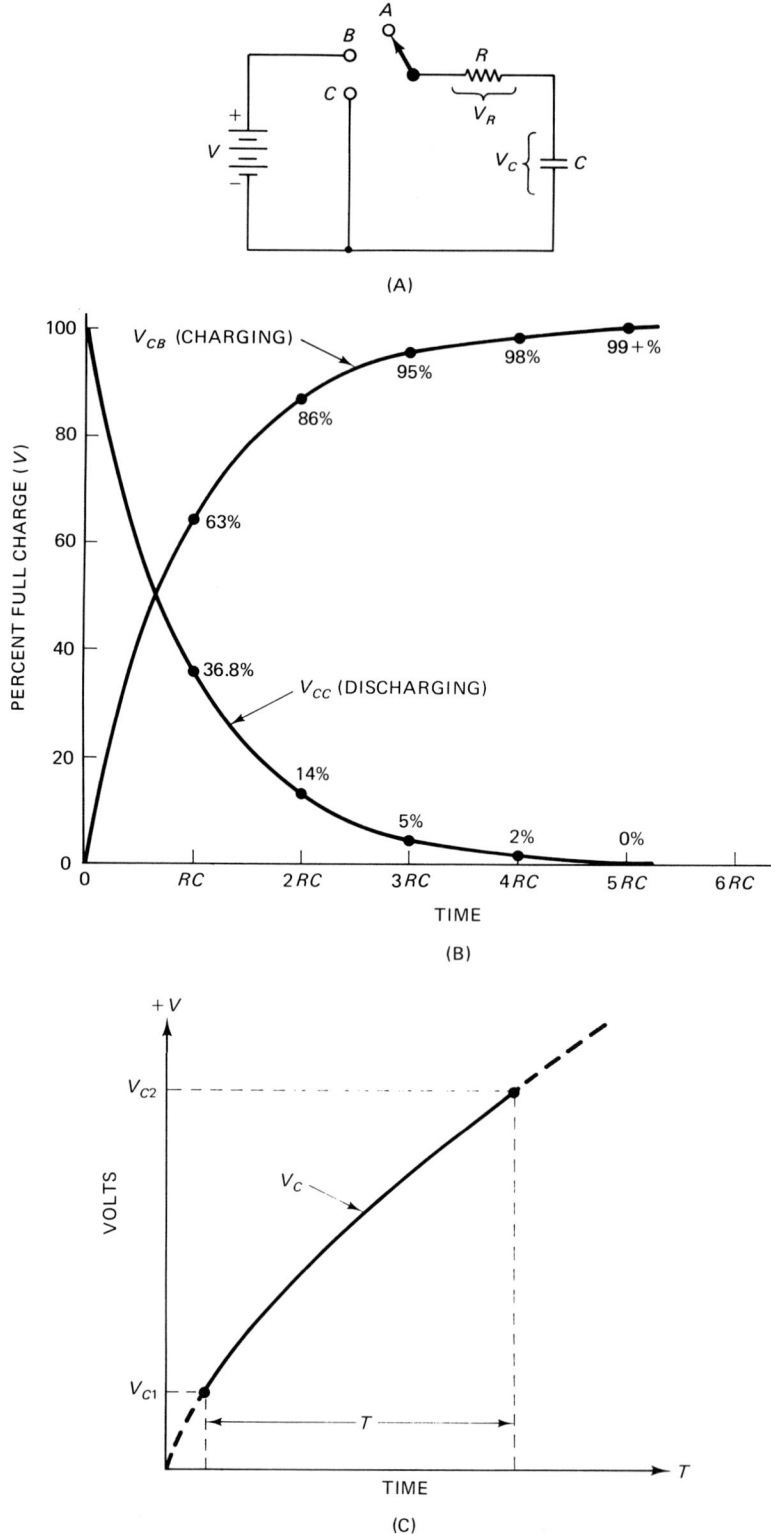

Fig. 18-1 (A) Simple RC charge/discharge circuit, (B) timing waveforms, and (C) charging from a given starting point to a specified ending point.

is "What RC time constant will force V_{C1} to rise to V_{C2} in time T?" Assuming that $V_{C1} < V_{C2} < V$,

$$V - V_{C2} = (V - V_{C1})\left[e^{-T/RC}\right] \tag{18-3}$$

$$\frac{V - V_{C2}}{V - V_{C1}} = e^{-T/RC} \tag{18-4}$$

$$\ln\left(\frac{V - V_{C2}}{V - V_{C1}}\right) = \frac{-T}{RC} \tag{18-5}$$

or, rearranging terms,

$$RC = \frac{-T}{\ln\left(\dfrac{V - V_{C2}}{V - V_{C1}}\right)} \tag{18-6}$$

Example

An RC network is connected to a $+12$ V DC source. What RC product will permit voltage V_c to rise from $+1$ V DC to $+4$ V DC in 200 ms? Note: $V = -12$ V DC, $V_{C2} = +4$ V DC, and $V_{C1} = +1$ V DC.

Solution

$$RC = \frac{-T}{\ln\left(\dfrac{V - V_{C2}}{V - V_{C1}}\right)}$$

$$RC = \frac{-\left(200 \text{ ms} \times \dfrac{1 \text{ s}}{1000 \text{ ms}}\right)}{\ln\left(\dfrac{12 - 4}{12 - 1}\right)}$$

$$RC = \frac{-0.200 \text{ s}}{\ln\left(\dfrac{8}{11}\right)}$$

$$RC = \frac{-0.200 \text{ s}}{\ln(0.727)}$$

$$RC = (-0.200 \text{ s})/(-0.319) = 0.627$$

Equation (18-3) can be used to derive the timing or frequency setting equations of many different RC-based waveform generator

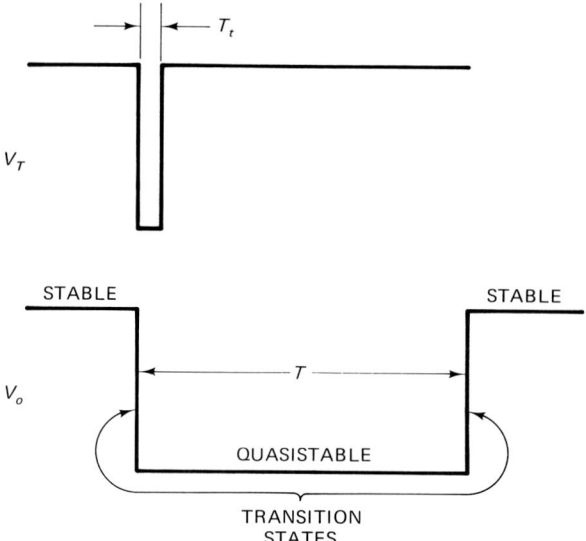

Fig. 18-2 Monostable multivibrator timing waveforms.

circuits. The key voltage levels will, most often, be device trip points or critical values set by the design of the circuit.

MONOSTABLE MULTIVIBRATOR CIRCUITS

The monostable multivibrator (MMV) has two permissible output states (HIGH and LOW), but only one of them is stable. The MMV produces one output pulse in response to an input trigger signal (Fig. 18-2). The output pulse (V_o) has a duration T in which the output is in the quasi-stable state. The MMV is also known under several alternate names: one-shot, pulse generator, and pulse stretcher. The latter name derives from the fact that the output duration T is longer than the trigger pulse ($T > T_c$).

Monostable multivibrators find a wide variety of applications in electronic circuits. Besides the pulse stretcher mentioned above, the MMV also serves to lock out unwanted pulses. Figure 18-3 shows that the output responds to only the first trigger pulse. The next two pulses occur during the active time T, so they are ignored. Such an MMV is said to be nonretriggerable. A common application of this feature is in switch contact debouncing. All mechanical switch contacts bounce a few times on closure, creating a short run of exponentially decaying pulses. If an MMV is triggered by the first pulse from the switch, and if the MMV remains quasi-active long enough for the bouncing pulses to die out, then the MMV output signal becomes the debounced switch closure. The main requirement is that the MMV duration be longer than the switch contact bounce pulse train (5 ms is generally considered adequate for most switch types).

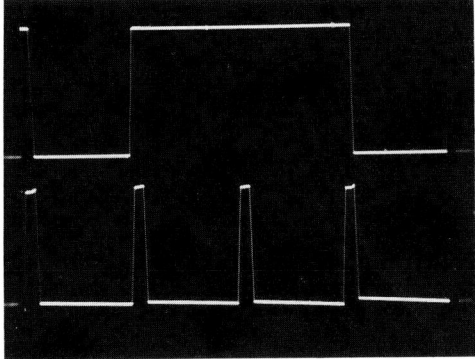

Fig. 18-3 Nonretriggerable monostable multivibrators do not respond to further input triggers until after the circuit "times-out."

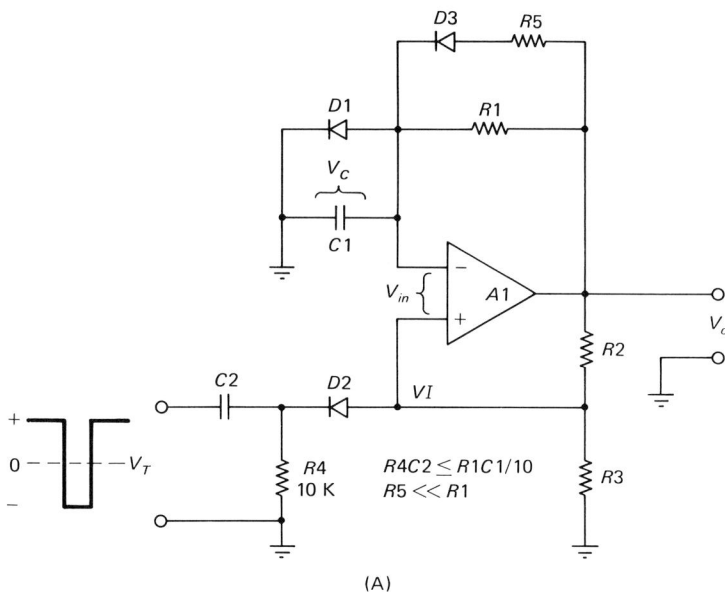

(A)

Fig. 18-4 (A) Circuit for an op-amp monostable multivibrator and (B) timing waveforms. (*Figure continues.*)

The range of possible MMV applications is too broad for detailed discussion here, so only a general set of categories can be presented. These include pulse generation, pulse stretching, contact debouncing, pulse signal clean-up, switching, and synchronization of circuit functions (especially digital).

Figure 18-4A shows the circuit for a nonretriggerable monostable multivibrator based on the operational amplifier. This circuit is based on the voltage comparator circuit. When there is no feedback, the effective voltage gain of an op-amp is its open-loop gain (A_{vol}). When both $-$IN and $+$IN are at the same potential, the differential input voltage (V_{id}) is zero, so the output is also zero. But if $V_{(-\text{IN})}$ does not equal $V_{(+\text{IN})}$, the high gain of the amplifier forces the output to either

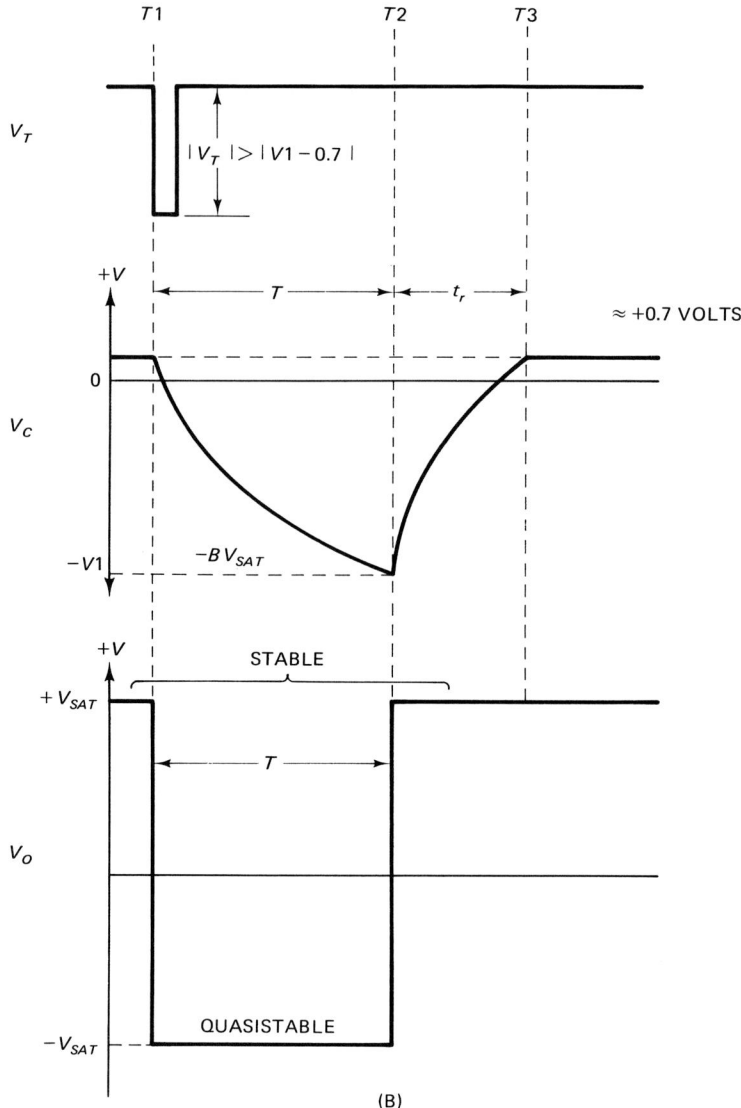

Fig. 18-4 (*continued*)

its positive or its negative saturation values. If $V_{(-IN)} > V_{(+IN)}$, the op-amp sees a positive differential input signal, so the output saturates at $-V_{sat}$. However, if $V_{(-IN)} < V_{(+IN)}$, the amplifier sees a negative differential input signal and the output saturates to $+V_{sat}$. The operation of the MMV depends on the relationship of $V_{(-IN)}$ and $V_{(+IN)}$.

Four states of the monostable multivibrator must be considered: stable state, transition state, quasi-stable state, and refractory state.

Stable State

The output voltage V_o is initially at $+V_{sat}$. Capacitor $C1$ will attempt to charge in the positive going direction because $+V_{sat}$ is applied to the

$R1C1$ network. But, because of diode D1 shunted across $C1$, the voltage across $C1$ is clamped to $+V_{D1}$. For a silicon diode such as the 1N914 or 1N4148, $+V_{D1}$ is about $+0.7$ V DC. Thus, the inverting input $(-IN)$ is held to $+0.7$ V DC during the stable state. The noninverting input $(+IN)$ is biased to a level $V1$, which is

$$V1 = R3(+V_{sat})/(R2 + R3) \qquad (18\text{-}7)$$

or, in the special case of $R2 = R3$,

$$V1 = +V_{sat}/2 \qquad (18\text{-}8)$$

The factor $R3/(R2 + R3)$ is often designated by the Greek letter beta (B), so

$$B = R3/(R2 + R3) \qquad (18\text{-}9)$$

Therefore,

$$V1 = B(+V_{sat}) \qquad (18\text{-}10)$$

The amplifier $(A1)$ sees a differential input voltage (V_{id}) of $(V1 - V_{D1})$, or $(V1 - 0.7)$ volts. Using the previous notation,

$$V_{id} = \frac{R3(+V_{sat})}{R2 + R3} - 0.7 \qquad (18\text{-}11)$$

As long as $V1 > V_{D1}$, the amplifier effectively sees a negative DC differential voltage at the inverting input, so (with its high open-loop gain, A_{vol}) will remain saturated at $+V_{sat}$. For purposes of this discussion the amplifier is a type 741 operated at DC power supply potentials of ± 12 V DC, so V_{sat} typically will be ± 10 V.

Transition State

The input trigger signal (V_t) is applied to the MMV of Fig. 18-4A through RC network $R4C2$. The general design rule for this network is that its time-constant should be not more than one-tenth the time-constant of the timing network.

$$R4C2 < (R1C1)/10 \qquad (18\text{-}12)$$

At time $T1$ (see Fig. 18-4B) trigger signal V_t makes an abrupt HIGH-to-LOW transition to a peak value that is less than $(V1 - 0.7)$ volts. Under this condition the polarity of V_{id} is now reversed and the inverting input now sees a positive voltage: $(V1 + V_t - 0.7)$ is less than V_{D1}. The output voltage V_o now snaps rapidly to $-V_{sat}$. The fall time of the output signal is dependent on the slew rate and the open-loop gain of the operational amplifier $A1$.

Quasi-Stable State

The output signal from the MMV is the quasi-stable state between $T1$ and $T2$ in Fig. 18-4B. It is called quasi-stable because it does not change over $T = T2 - T1$, but when T expires, the MMV "times out" and V_o reverts to the stable state $(+V_{sat})$.

During the quasi-stable time, D1 is reverse biased and capacitor $C1$ discharges from $+0.7$ V DC to zero and then recharges toward $-V_{sat}$. When $-V_o$ reaches $-V1$, however, the value of V_{id} crosses zero, and that change forces V_o to snap once again to $+V_{sat}$.

Appealing to Eq. (18-6) makes it possible to derive the timing equation for the MMV. The timing capacitor must charge from an initial value (V_{C1}) to a final value (V_{C2}) in time T. The question is "What value of $R1C1$ will cause the required transitions?" Consider the case $R2 = R3$ $(V1 = 0.5V_{sat})$:

$$R1C1 = \frac{-T}{\ln\left(\dfrac{V_{sat} - V_{C2}}{V_{sat} - V_{C1}}\right)} \tag{18-13}$$

$$R1C1 = \frac{-T}{\ln\left(\dfrac{V_{sat} - [(0.5)(V_{sat} + 0.7)]}{V_{sat} - 0.7}\right)} \tag{18-14}$$

and for the case where $V_{sat} = 10$ V DC:

$$R1C1 = \frac{-T}{\ln\left(\dfrac{10 \text{ V DC} - [(0.5)(10 + 0.7)]}{10 \text{ V DC} - 0.7 \text{ V}}\right)} \tag{18-15}$$

$$R1C1 = \frac{-T}{\ln\left(\dfrac{10 \text{ V DC} - 5.35 \text{ V}}{10 \text{ V DC} - 0.7 \text{ V}}\right)} \tag{18-16}$$

$$R1C1 = \frac{-T}{\ln\left(\dfrac{4.65}{9.3}\right)} \tag{18-17}$$

$$R1C1 = \frac{-T}{\ln(0.5)} \tag{18-18}$$

$$R1C1 = \frac{-T}{-0.69} \tag{18-19}$$

Thus,

$$T = 0.69R1C1 \tag{18-20}$$

Equation (18-20) represents the special case in which $B = \frac{1}{2}$ (i.e., $R2 = R3$). Although $R2 = R3$ may be the usual case for this class of circuit, $R2$ and $R3$ might not be equal in other cases. A more generalized expression is

$$RC = \frac{T}{\ln\left(\dfrac{1 + 0.7\ V/V_{\text{sat}}}{1 - B}\right)} \tag{18-21}$$

in which

$$B = R3/(R2 + R3) \tag{18-22}$$

When the quasi-stable state times out, the circuit status returns to the stable state (where it remains dormant until triggered again).

Refractory Period

At time $t2$ the output signal voltage V_o switches from $-V_{\text{sat}}$ to $+V_{\text{sat}}$. Although the output has timed out, the MMV is not yet ready to accept another trigger pulse. The refractory state between $t2$ and $t3$ is characterized by the output being in the stable state, but the input is unable to accept a new trigger input stimulus. The refractory period must await the discharge of $C1$ under the influence of the output voltage to satisfy $V1 < (V1 - 0.7)$ volts.

Retriggerable Monostable Multivibrators

The circuit of Fig. 18-4A is a nonretriggerable MMV. Once it is triggered the circuit will not respond to further trigger inputs until after both the quasi-stable and refractory states have been completed. This characteristic is used to advantage in some applications. But in other cases it might be desired to retrigger the MMV. A retriggerable monostable multivibrator (RMMV) one-shot is a circuit that will respond to further trigger signals.

Figure 18-5 shows the retriggerable MMV response. An initial trigger signal (V_t) is received at time $t1$. The output snaps LOW and, under normal circumstances, it would remain in this quasi-stable state until time $t3$ when the duration T expires. But at time $t2$ a second trigger pulse is received. The circuit is now retriggered for another duration T, so it will not time out until $t4$. The total time that the

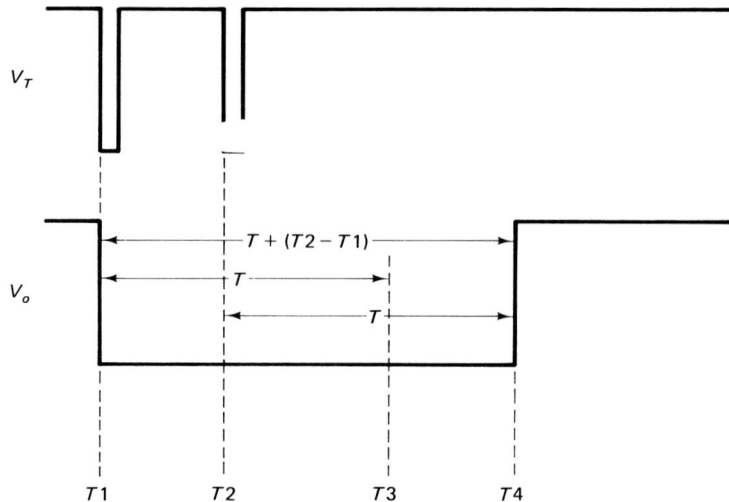

Fig. 18-5 Timing diagram for a retriggerable monostable shows that the circuit will respond to new trigger pulses that occur prior to original time-out ($t3$).

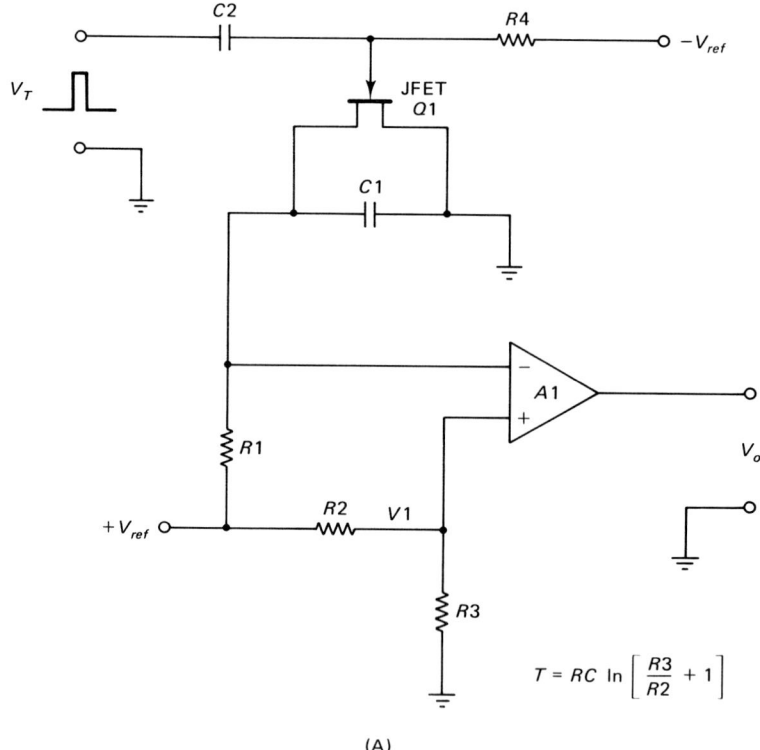

$$T = RC \ln \left[\frac{R3}{R2} + 1 \right]$$

(A)

Fig. 18-6 (A) Retriggerable monostable circuit uses a JFET electronic switch to discharge timing capacitor, (B) timing waveform for a single trigger pulse, and (C) timing waveform for a multiple trigger pulse. (*Figure continues on the following two pages.*)

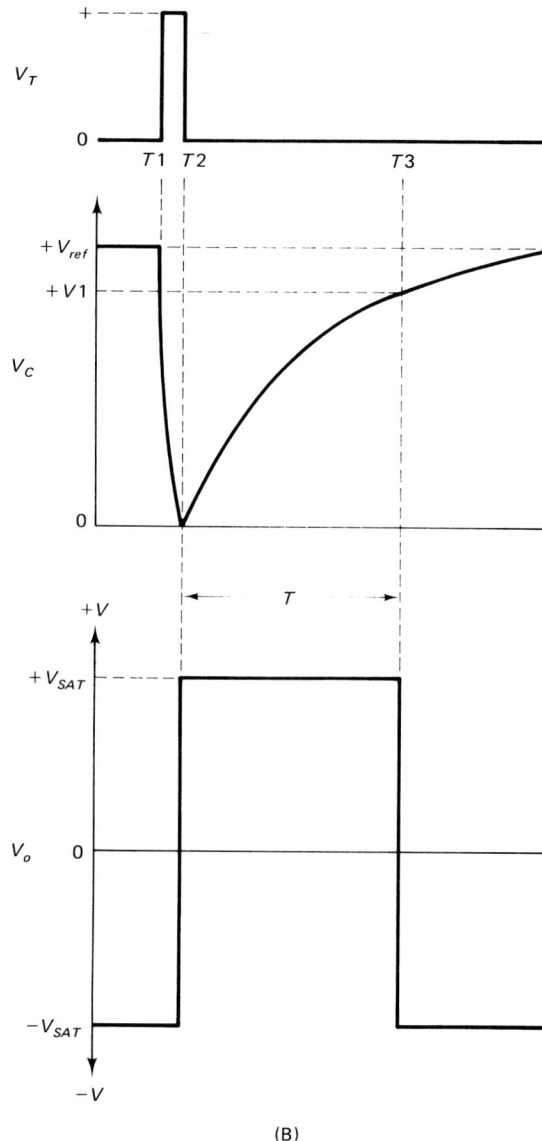

Fig. 18-6 (*continued*)

RMMV is in the quasi-stable state is $[T + (t2 - t1)]$. In other words, the RMMV output is active for the entire duration T plus that portion of the previous active time that expired when the next trigger pulse was received.

Figure 18-6A shows the circuit for a simple RMMV based on an operational amplifier. The two inputs are biased from a reference voltage source, $+V_{ref}$. The potential applied to $+IN$ is a fraction of $+V_{ref}$. That is, $[(R3)(+V_{ref})/(R2 + R3)]$. The potential applied to $-IN$ is a function of $+V_{ref}$ and time-constant $R1C1$. If the circuit is not triggered at turn-on, the capacitor ($C1$) charges up to $+V_{ref}$, so $-IN$ is more positive than $+IN$. This situation forces V_o to $-V_{sat}$, which is

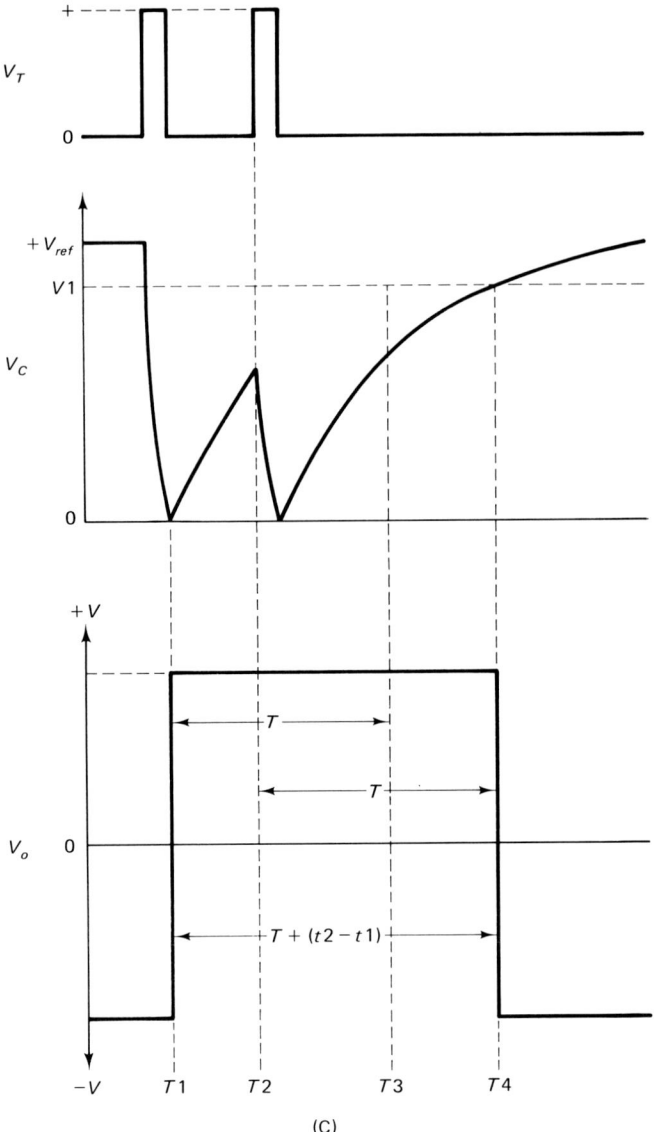

Fig. 18-6 (*continued*)

the stable state. When a positive-going trigger pulse (V_t) is received (see Fig. 18-6B), it biases the junction field effect transistor (JFET) Q1 hard on. The JFET drain-source channel resistance drops very low, causing C1 to discharge rapidly between $t1$ and $t2$. With V_c close to 0 V DC, $+$IN is more positive than $-$IN, so the output snaps abruptly to $+V_{sat}$ at time $t1$. During the interval $t2$ to $t3$, capacitor C1 begins charging towards $+V_{ref}$, and V_o remains at $+V_{sat}$. Once V_c reaches $+V1$, however, the output of A1 snaps back to $-V_{sat}$. The duration T is found from

$$T = R1C1 \ln\left(\frac{R3}{R2} + 1\right) \qquad (18\text{-}23)$$

The operation discussed here and depicted in Fig. 18-6B is for normal, nonretriggered operation. Figure 18-6C shows the retriggered case. The RMMV receives a second trigger pulse at time $t2$, which forces the JFET (Q1) to turn on again and to rapidly discharge $C1$. The charging process then starts over again and continues until the circuit times out, unless a further trigger pulse is received.

A common use for the RMMV is in alarm or sensing circuits. The RMMV is triggered by some external event, and it will continually retrigger as long as the external event keeps occurring. But if no event is sensed prior to time-out, the RMMV returns to the stable state and the following circuitry is triggered to alarm status. An example is a medical respirator alarm. A sensor in the respirator line senses variations in either pressure or air temperature caused by breathing. Each time a breath is sensed it retriggers the RMMV. But if the patient ceases breathing, the RMMV will time out and cause an alarm to nearby medical personnel.

ASTABLE (FREE RUNNING) CIRCUITS

The circuits discussed in the previous section are periodic; that is, an output pulse occurs only once in response to a stimulus or trigger. Such circuits are said to be monostable because they possess only one stable state. An astable multivibrator (AMV) is free-running. The output of the AMV is a pulse or wave train that is periodic. In a periodic signal the wave repeats itself indefinitely until the circuit is either turned off or otherwise inhibited.

Astable multivibrators are oscillators. Waveforms available from the AMV include square waves, triangle waves, and sawtooth waves. Sine waves are also available from oscillator circuits, but those circuits operate differently from the others and are handled separately.

NONSINUSOIDAL WAVEFORM GENERATORS

The nonsinusoidal AMV circuit produces square, triangular, or sawtooth waves. When it is combined with a monostable multivibrator (MMV), a pulse generator results. Because the square-wave generator is the most basic form, the discussion of AMV circuits begins with square waves.

Square-Wave Generators

Figure 18-7 shows the classical square wave. Each time interval of the wave is quasi-stable, so one may conclude that the square-wave generator has no stable states (hence is astable). The waveform snaps back

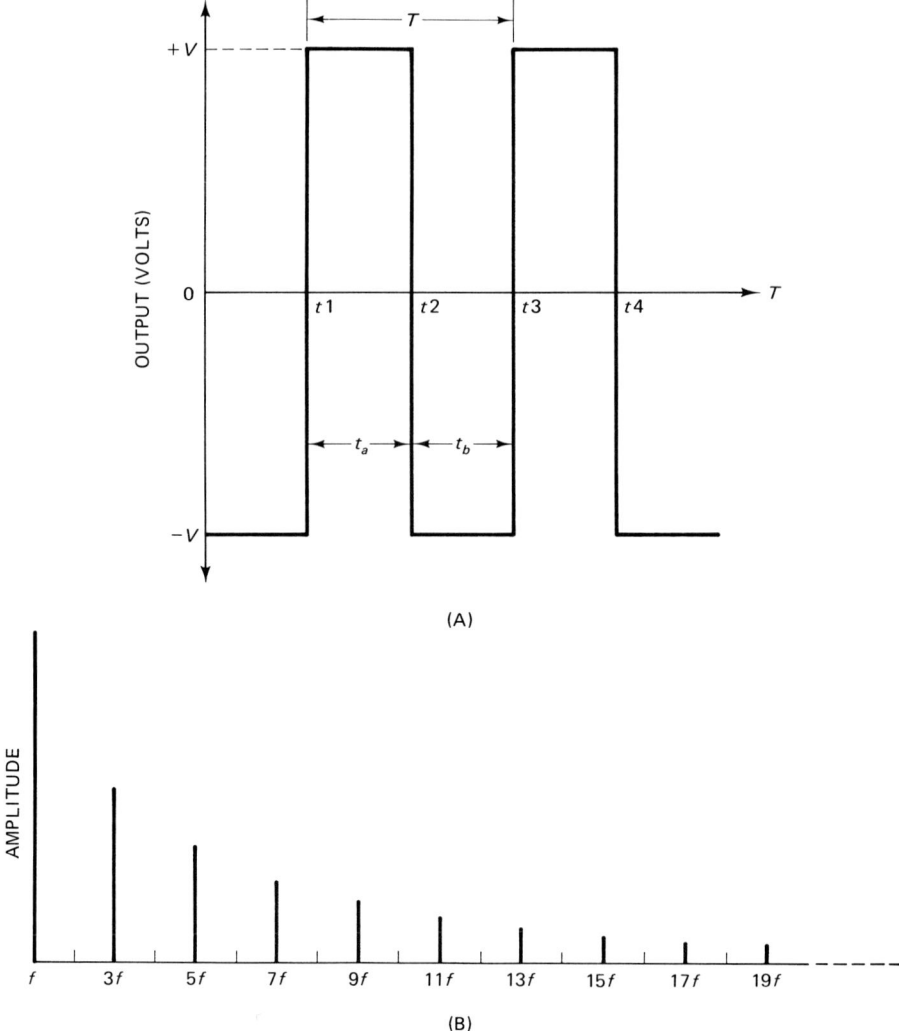

Fig. 18-7 (A) Time domain square-wave signal and (B) frequency domain square-wave spectrum.

and forth between $-V$ and $+V$, dwelling on each level of a duration of time (t_a or t_b). The period T is

$$T = t_a + t_b \qquad (18\text{-}24)$$

where T is the period of the square wave ($t1$ to $t3$), t_a the interval $t1$ to $t2$, and t_b the interval $t2$ to $t3$.

The frequency of oscillation (F) is the reciprocal of T:

$$F = 1/T \qquad (18\text{-}25)$$

The ideal square wave is both baseline and time-line symmetrical. That means that $|+V| = |-V|$ and $t_a = t_b$. Under time-line symmetry $t_a = t_b = t$, so $T = 2t$ and $f = 1/2t$.

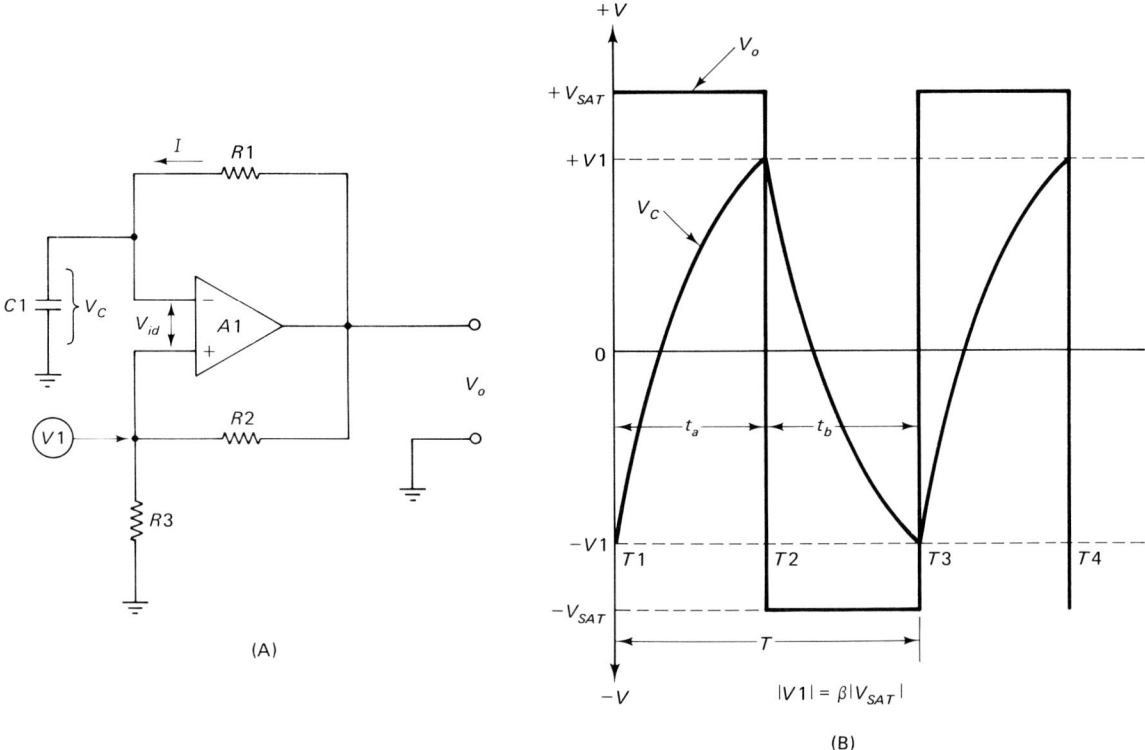

Fig. 18-8 (A) Op-amp square-wave generator circuit and (B) timing waveform.

In the ideal, symmetrical square wave, the Fourier spectrum (Fig. 18-7B) consists of the fundamental frequency (f) plus the odd order harmonics ($3f, 5f, 7f, \ldots$). Furthermore, the harmonics are in-phase with the fundamental. Theoretically, an infinite number of odd-number harmonics are present in the ideal square wave. However, in practical square waves the ideal is considered satisfied with harmonics to about $999f$. That ideal is almost never reached, however, due to the normal bandwidth limitations of the circuit. An indicator of harmonic content is the rise time of the square waves: The faster the rise time, the higher the number of harmonics.

The circuit for an operational amplifier square-wave generator is shown in Fig. 18-8A. The basic circuit is similar to the simple voltage comparator and the MMV. Like the MMV, the AMV operation depends on the relationship between $V_{(-\text{IN})}$ and $V_{(+\text{IN})}$. In the circuit of Fig. 18-8A the voltage applied to the noninverting input ($V_{+\text{IN}}$) is determined by a resistor voltage divider, $R2$ and $R3$. This voltage is called $V1$ in Fig. 18-8A and is

$$V1 = V_o R3 / (R2 + R3) \tag{18-26}$$

or, when V_o is saturated,

$$V1 = V_{\text{sat}} R3 / (R2 + R3) \tag{18-27}$$

Once again, the factor $R3/(R2 + R3)$ is often designated B.

$$B = R3/(R2 + R3) \tag{18-28}$$

Because Eq. (18-28) is always a fraction, $V1 < V_{sat}$ and $V1$ is of the same polarity as V_{sat}.

The voltage applied to the inverting input (V_{-IN}) is the voltage across capacitor $C1$, which is V_{C1}. This voltage is created when $C1$ charges under the influence of current I, which in turn is a function of V_o and the time constant of $R1C1$. Timing operation of the circuit is shown in Fig. 18-8B.

At turn-on, $V_{C1} = 0$ volts and $V_o = +V_{sat}$, so $V1 = +V1 = B(+V_{sat})$. Because $V_{C1} < V1$, the op-amp sees a negative differential input voltage so the output remains at $+V_{sat}$. During this time, however, V_{C1} is charging toward $+V_{sat}$ at a rate of

$$V_{C1} = V_{sat}(1 - e^{-t2/R1C1}) \tag{18-29}$$

When V_{C1} reaches $+V1$, however, the op-amp sees $V_{C1} = V1$, so $V_{id} = 0$. The output now snaps from $+V_{sat}$ to $-V_{sat}$ (time $t2$ in Fig. 18-8B). The capacitor now begins to discharge from $+V1$ toward zero, and then recharges toward $-V_{sat}$. When it reaches $-V1$, the inputs are once again zero, so the output again snaps to $+V_{sat}$. The output continuously snaps back and forth between $-V_{sat}$ and $+V_{sat}$, thereby producing a square-wave output signal.

Again appealing to Eq. (18-6) to find the time constant required to charge from an initial voltage V_{C1} to an end voltage V_{C2} in time t is defined by

$$RC = \frac{-T}{\ln\left(\dfrac{V - V_{C2}}{V - V_{C1}}\right)} \tag{18-30}$$

In Fig. 18-8A the RC time constant is $R1C1$. From Fig. 18-8B it is apparent that, for interval t_a, $V_{C1} = -BV_{sat}$, $V_{C2} = -BV_{sat}$, and $V = V_{sat}$. To calculate the period T:

$$2R1C1 = \frac{-T}{\ln\left(\dfrac{V_{sat} - BV_{sat}}{V_{sat} - (-BV_{sat})}\right)} \tag{18-31}$$

or, rearranging Eq. (18-31),

$$-T = 2R1C1 \ln\left(\frac{V_{sat} - BV_{sat}}{V_{sat} - (-B_{sat})} \right) \qquad (18\text{-}32)$$

$$-T = 2R1C1 \ln\left(\frac{1 - B}{1 + B} \right) \qquad (18\text{-}33)$$

$$T = 2R1C1 \ln\left(\frac{1 + B}{1 - B} \right) \qquad (18\text{-}34)$$

Because $B = R3/(R2 + R3)$,

$$T = 2R1C1 \ln\left(\frac{1 + [R3/(R2 + R3)]}{1 - [R3/(R2 + R3)]} \right) \qquad (18\text{-}35)$$

which reduces to

$$T = 2R1C1 \ln\left(\frac{2R2}{R3} \right) \qquad (18\text{-}36)$$

Equation (18-36) defines the frequency of oscillation for any combination of $R1$, $R2$, $R3$, and $C1$. In the special case $R2 = R3$,

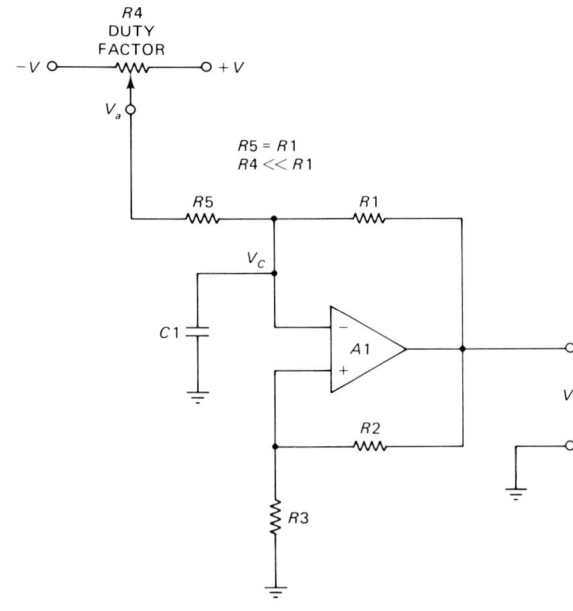

Fig. 18-9 Potentiometer $R4$ allows a variable duty factor to be created.

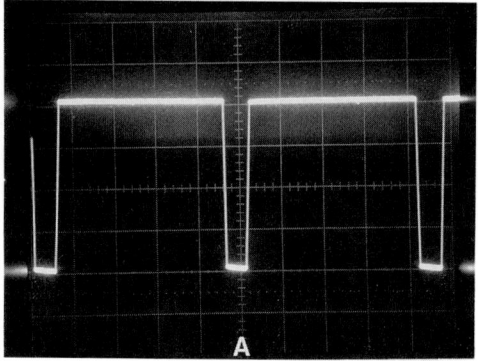
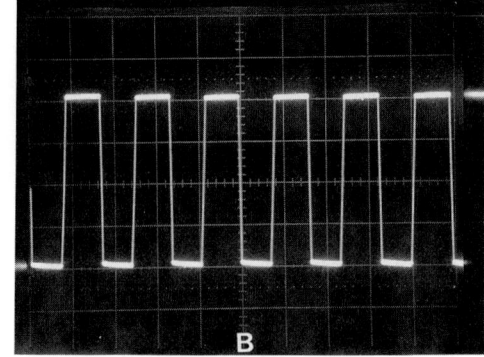
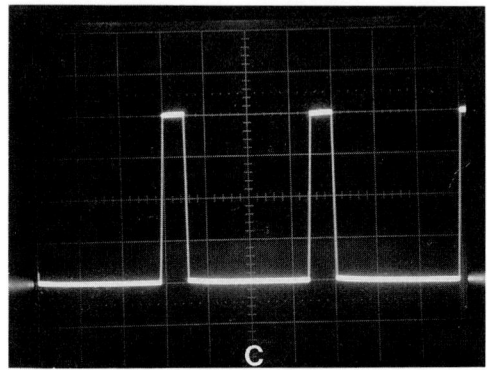

Fig. 18-10 Three different positions of $R4$ produce (A) long duty factor (> 50 percent), (B) 50-percent duty factor, and (C) short duty factor (< 50 percent).

B = 0.5, so

$$T = 2R1C1 \ln\left(\frac{1 + 0.5}{1 - 0.5}\right) \tag{18-37}$$

$$T = 2R1C1 \ln\left(\frac{1.5}{0.5}\right) \tag{18-38}$$

$$T = 2R1C1 \ln(1.09) = 2.2R1C1 \tag{18-39}$$

The circuit of Fig. 18-8A produces time-line symmetrical square waves (i.e., $t_a = t_b$). If time-line asymmetrical square waves are required, then a circuit such as either Fig. 18-9 or 18-11A is required. The circuit in Fig. 18-9 uses a potentiometer ($R4$) and a fixed resistor ($R5$) to establish a variable duty cycle asymmetry. The circuit is similar to Fig. 18-8A, but with an offset circuit ($R4/R5$) added. The assumptions are $R5 = R1$ and $R4 \ll R1$. If V_a is the potentiometer output voltage, $C1$ charges at a rate of $(R1/2)C1$ toward a potential of $(V_a + V_{sat})$. After output transition, however, the capacitor discharges at the same $(R1/2)C1$ rate toward $(V_a - V_{sat})$. The two interval times are therefore different: t_a and t_b are no longer equal.

Figure 18-10 shows three extremes of V_a: $V_a = +V$ (Fig. 18-10A), $V_a = 0$ (Fig. 18-10B), and $V_a = -V$ (Fig. 18-10C). These traces represent very long, equal, and very short duty cycles, respectively.

The circuit of Fig. 18-11A also produces asymmetrical square waves, but the duty cycle is fixed instead of variable. Once again the basic circuit is like Fig. 18-8A, but with added components. In Fig. 18-11A the RC timing network is altered such that the resistors are different on each swing of the output signal. During t_a, $V_a = +V_{sat}$; so diode D1 is forward biased and D2 is reverse biased. For this interval,

$$t_a = (R1A)(C1)\ln\left(1 + \frac{2R2}{R3}\right) \qquad (18\text{-}40)$$

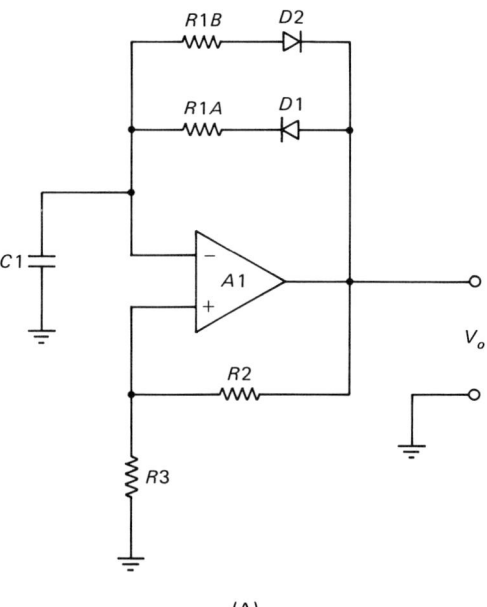

(A)

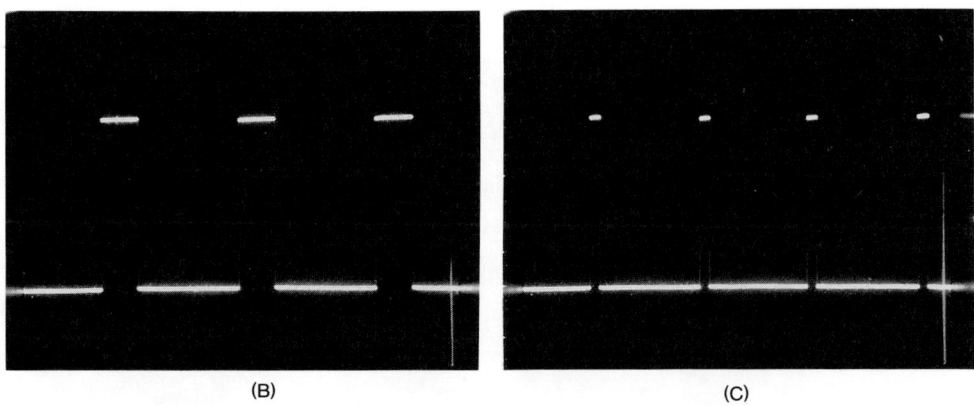

(B)   (C)

Fig. 18-11 (A) Duty factor can also be altered by using different resistors for charge and discharge cycles, as selected by switching diodes D1 and D2; (B) output waveform for $R1B = 3 \times R1A$; and (C) output waveform for $R1B = 10 \times R1A$.

During the alternate half-cycle (t_b), the output voltage V_o is at $-V_{sat}$; so D1 is reverse biased and D2 is forward biased. During this interval R1B is the timing resistor, while R1A is effectively out of the circuit. The timing equation is

$$t_b = (R1B)(C1)\ln\left(1 + \frac{2R2}{R3}\right) \qquad (18\text{-}41)$$

The total period, T is $t_a + t_b$, so

$$T = (R1A)(C1)\ln\left(1 + \frac{2R2}{R3}\right)$$

$$+ (R1B)(C1)\ln\left(1 + \frac{2R2}{R3}\right) \qquad (18\text{-}42)$$

Collecting terms,

$$T = (R1A + R1B)(C1)\ln\left(1 + \frac{2R2}{R3}\right) \qquad (18\text{-}43)$$

Equation (18-43) defines the oscillation frequency of the circuit in Fig. 18-11A. Figures 18-11B and 18-11C show the effects of two values of $R1A/R1B$ ratio. In Fig. 18-11B the ratio $R1A/R1B = 3:1$, while in Fig. 18-11C the ratio $R1A/R1B = 10:1$.

The effect of this circuit on capacitor charging can be seen in Fig. 18-12. A relatively low $R1A/R1B$ ratio is seen in Fig. 18-12A. Notice in the lower trace that the capacitor charge time is long compared with the discharge time. The effect is seen more clearly for the case of a high $R1A/R1B$ ratio (Fig. 18-12B).

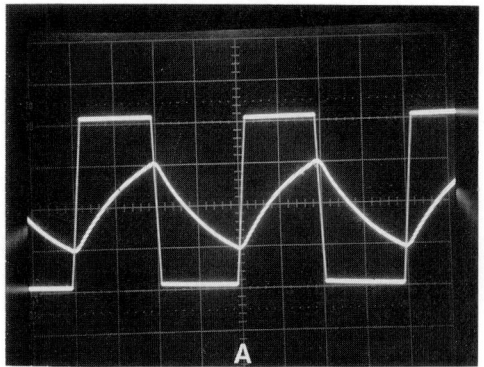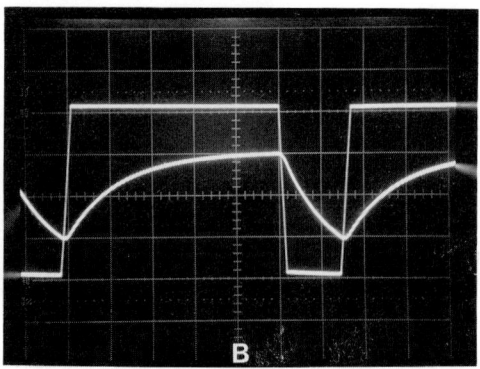

Fig. 18-12 Timing waveforms for the two ratios of $R1A$ and $R1B$: (A) $R1A$ and $R1B$ approximately equal and (B) $R1B \gg R1A$.

Output Voltage Limiting

The standard op-amp MMV or AMV circuit sometimes produces a relatively sloppy square output wave. By adding a pair of back-to-back zener diodes (Fig. 18-13A) across the output, however, the signal can be cleaned up (although at the expense of amplitude). For each polarity the output signal sees one forward-biased and one reverse-biased zener diode. On the positive swing, the output voltage is clamped at $(V_{Z1} + 0.7)$ volts. The 0.7-V factor represents the normal junction potential across the forward-biased diode (D2). On negative swings of the output signal, the situation reverses. The output signal is clamped to $[-(V_{Z2} + 0.7)]$ volts.

Figure 18-13B shows the unclamped output signal of a square-wave generator. The signal swings ± 10 V. Figure 18-13C shows the output of the same circuit when a pair of 5.6-V zener diodes are

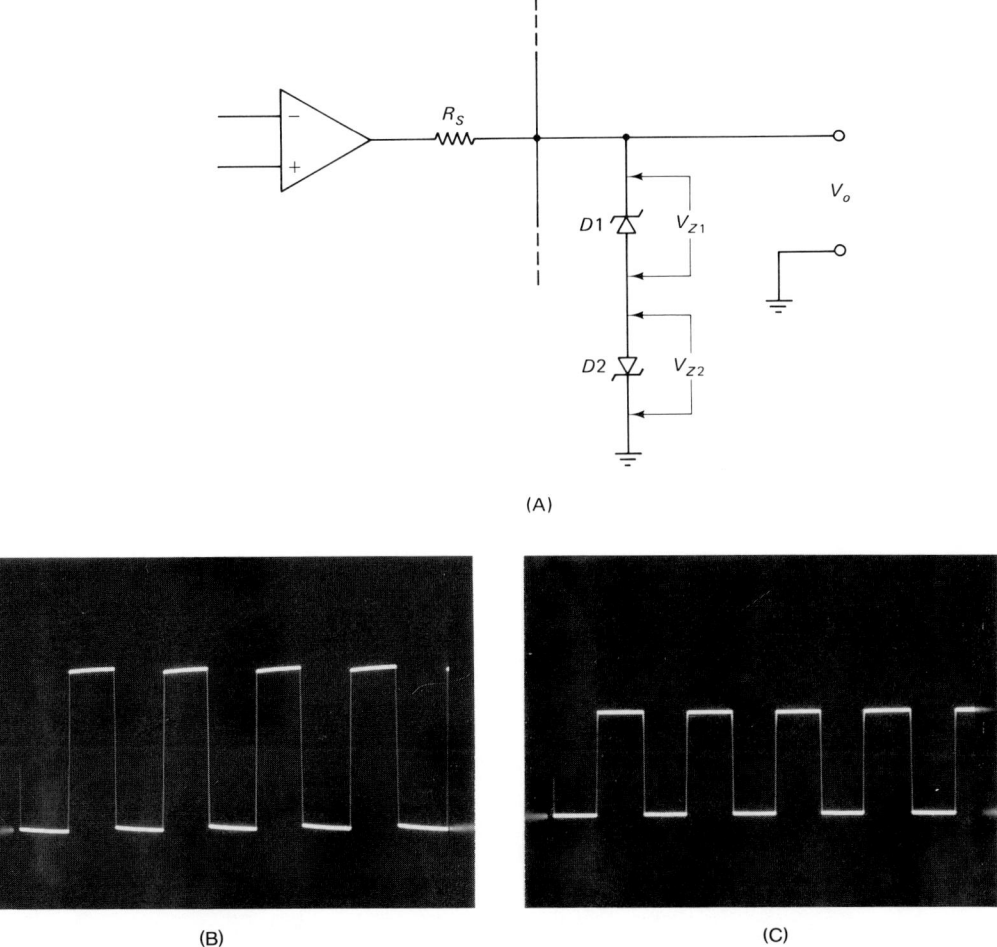

(A)

(B)   (C)

Fig. 18-13 (A) Zener diode output limiting, (B) unlimited output waveform, and (C) limited output waveform.

connected across the output. The output is reduced to $\pm(5.6 + 0.7)$ volts, but the corners are sharper.

Square Waves from Sine Waves

Figure 18-14 shows a method for converting sine waves to square waves. The circuit is shown in Fig. 18-14A and the waveforms are shown in Fig. 18-14B. The circuit is an operational amplifier connected as a comparator. Because the op-amp has no negative feedback path, the gain is very high (i.e., A_{vol}). In op-amps, gains of 20,000 to 2,000,000 (250,000 typical) are found. Thus, a voltage difference across the input terminals of only a few millivolts will saturate the output. From this behavior, the operation of the circuit and the waveform in Fig. 18-14B can be understood.

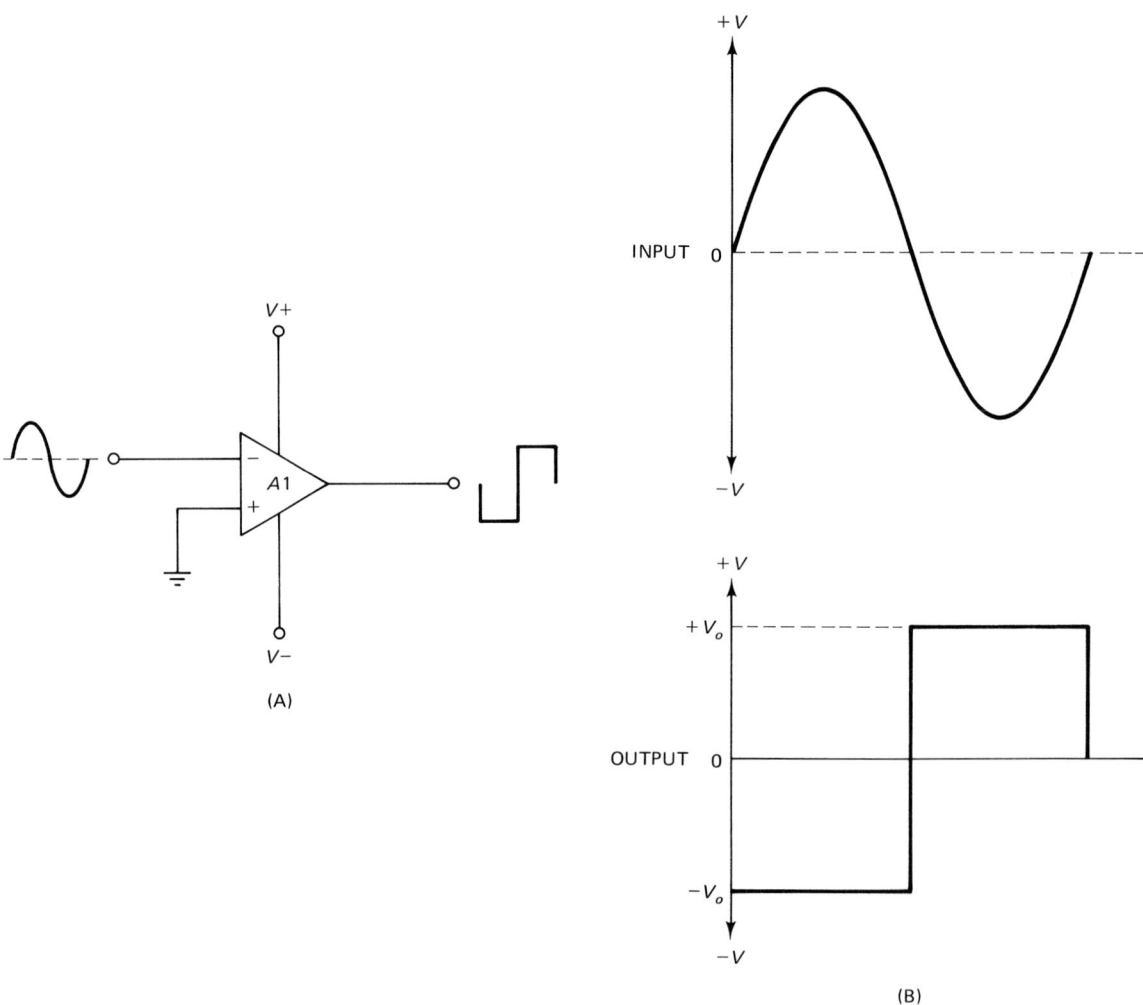

Fig. 18-14 (A) Voltage comparator used to produce a square wave and (B) timing waveforms.

The input waveform is a sine wave. Because the noninverting input is grounded (Fig. 18-14A), the output of the op-amp is zero only when the input signal voltage is also zero. When the sine wave is positive, the output signal will be at $-V_o$; when the sine wave is negative, the output signal will be at $+V_o$. The output signal will be a square wave at the sine-wave frequency, with a peak-to-peak amplitude of $[(+V_o) - (-V_o)]$.

TRIANGLE AND SAWTOOTH WAVEFORM GENERATORS

Triangle and sawtooth waveforms (Fig. 18-15) are examples of periodic ramp functions. The sawtooth (Fig. 18-15A) is a single-ramp waveform. The voltage begins to rise linearly at time $t1$. At time $t2$ the waveform abruptly drops back to zero, where it again starts to ramp up

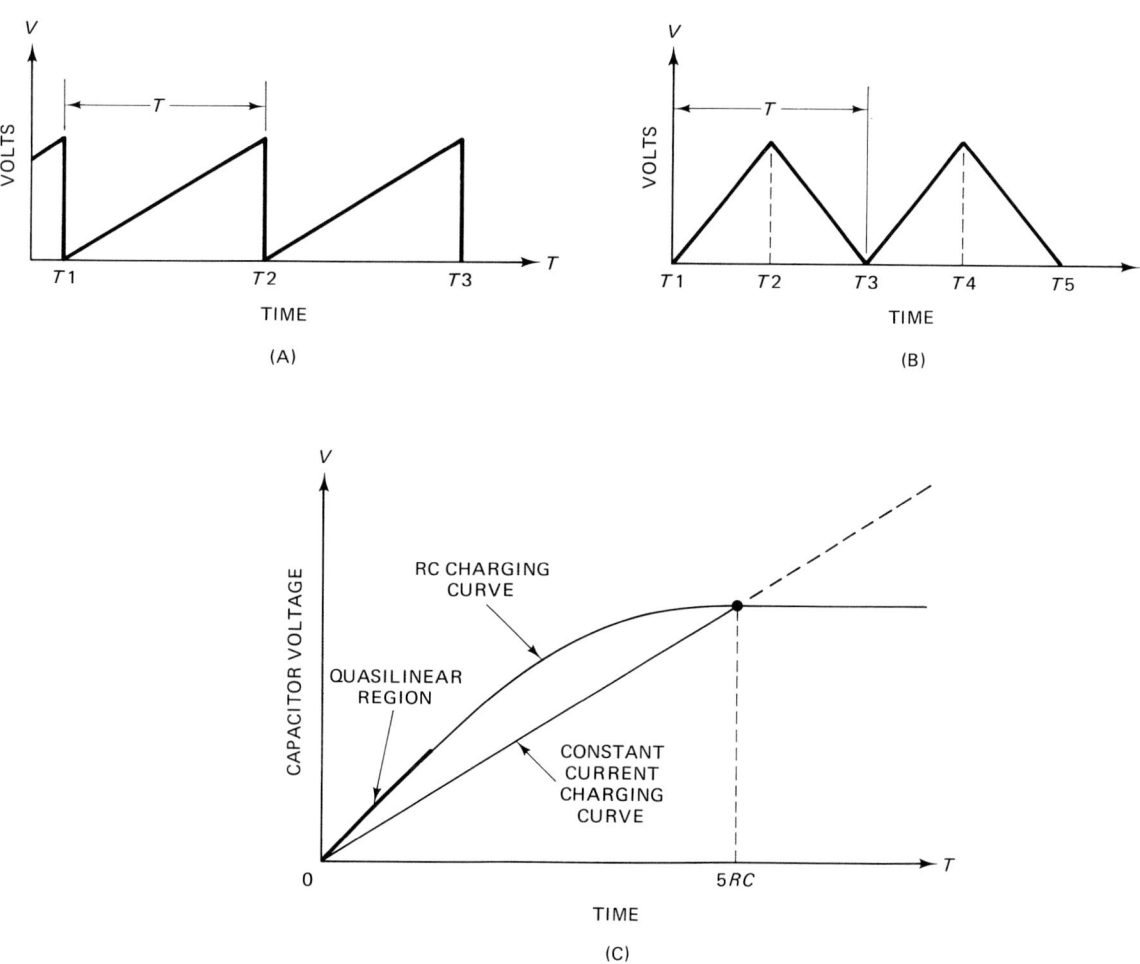

Fig. 18-15 (A) Sawtooth waveform, (B) triangle waveform, and (C) using an RC charging waveform to produce sawtooth waveform.

linearly. The sawtooth is usually periodic (although single-sweep variants are sometimes used), and the period is designated T (see Fig. 18-15A). The frequency is therefore $1/T$.

The triangle waveform (Fig. 18-15B) is a double ramp. The waveform begins to ramp up linearly at time $t1$. It reverses direction at time $t2$ and then ramps downward linearly until time $t3$. At time $t3$ the waveform again reverses direction and begins ramping upwards. The period of the triangle waveform (T) is $T1-T3$.

Ramp generators are derived from capacitor charging circuits. The familiar RC charging curve was discussed earlier in this chapter; it is reproduced in simplified form in Fig. 18-15C. The RC charging waveform has an exponential shape, so it is not well suited to generating a linear ramp function. There are two approaches to forcing the capacitor charging waveform to be more linear. The first is to limit the charging time to the short quasi-linear segment shown in Fig. 18-15C. The ramp thus obtained is not very linear, is limited in amplitude to a small fraction of $V1$, and has a relatively steep slope that may or not be useful for any given application. A superior method is to charge the capacitor through a constant current source (CCS). Using the CCS to charge the capacitor results in the linear ramp shown in Fig. 18-15C.

Triangle and sawtooth waveform oscillators create the constant current form of ramp generator by using a Miller integrator circuit to charge the capacitor (Fig. 18-16A). When a Miller integrator is driven

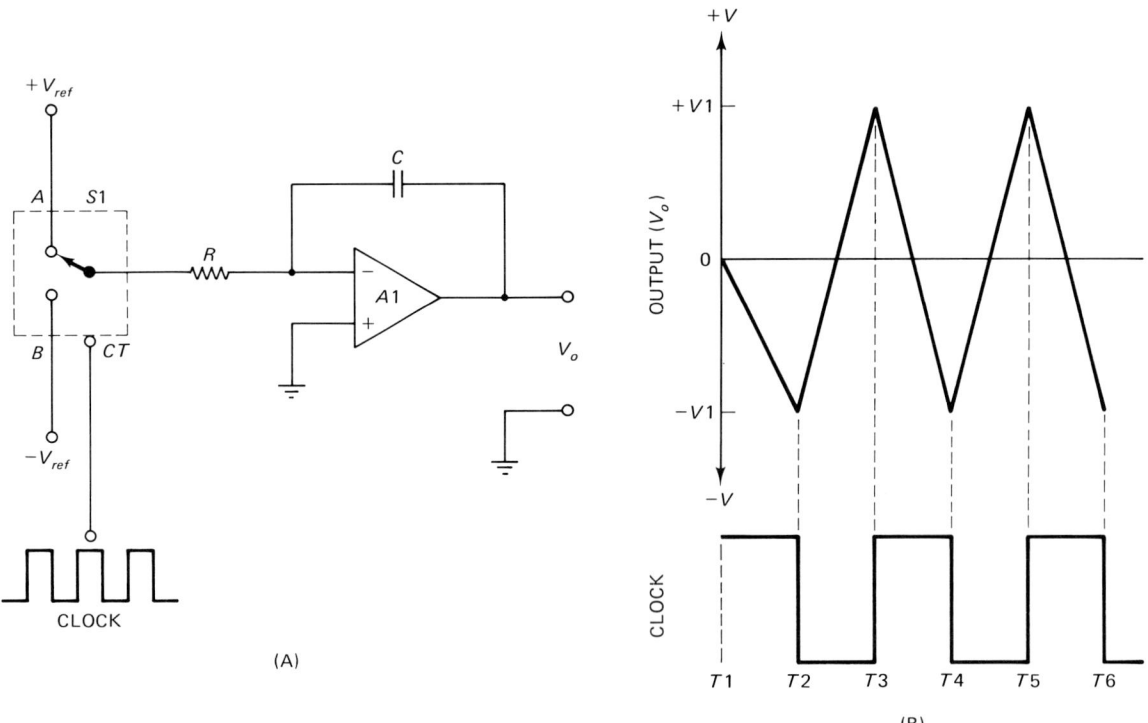

Fig. 18-16 (A) Simple triangle waveform circuit and (B) timing waveforms.

by a stable reference voltage, the output is a linearly rising ramp. The ramp voltage (V_o) is

$$V_o = V_{ref}/T \qquad (18\text{-}44)$$

or, because $T = RC$,

$$V_o = \frac{V_{ref}}{RC} \qquad (18\text{-}45)$$

If $V_{ref} = +10$ V DC, and the RC time constant is $T = RC = 0.001$ seconds, the ramp slope is

$$V_o = \frac{10 \text{ volts}}{0.001 \text{ second}} \qquad (18\text{-}46)$$

$$V_o = 0.10 \text{ volts/second} \qquad (18\text{-}47)$$

Triangle Generators

Figure 18-16A shows a simplified circuit model of a triangle waveform generator. This circuit consists of a Miller integrator as the ramp generator and an SPDT switch (S1) that can select either positive ($+V_{ref}$) or negative ($-V_{ref}$) reference voltage sources. For purposes of this discussion switch S1 is an electronic switch that is toggled back and forth between positions A and B by a square wave applied to the control terminal (CT). Assume an initial condition at time $t2$ (at which point $V_o = -V1$) and the input of the integrator is connected to $-V_{ref}$. At time $t2$ the square-wave switch driver changes to the opposite state, so S1 toggles to connect $+V_{ref}$ to the integrator input. The ramp output will rise linearly at a rate of $-V_{ref}/RC$ until the switch again toggles at time $t3$. At this point, the ramp is under the influence of $-V_{ref}$, so it drops linearly from $+V1$ to $-V1$. The switch continuously toggles back and forth between $-V_{ref}$ and $+V_{ref}$, so the output (V_o) continuously ramps back and forth between $-V1$ and $+V1$.

The circuit of Fig. 18-16A is not practical, but serves as an analogy for the actual circuit. Figure 18-17A shows the circuit for a triangle waveform generator in which a Miller integrator forms the ramp generator and a voltage comparator serves as the switch. The comparator uses the positive feedback configuration, so it operates as a noninverting Schmitt trigger. Such a circuit snaps HIGH ($V_B = +V_{sat}$) when the input signal crosses a certain threshold voltage in the positive-going direction. It will snap LOW again ($V_B = -V_{sat}$) when the input signal crosses a second threshold in a negative-going direction. The two thresholds are not always the same potential.

Because zener diodes D1 and D2 are in the circuit, the maximum allowable value of $+V_B$ is $(V_{ZD1} + 0.7)$ volts, while the limit for $-V_B$ is $-(V_{ZD2} + 0.7)$ volts. If $V_{ZD1} = V_{ZD2}$, then $|+V_B| = |-V_B|$. These potentials represent $\pm V_{ref}$ discussed in the analogy presented above, so they are the potentials that affect the ramp generator input. Consider an initial state in which V_B is at the negative limit $-V_B$. The output V_o

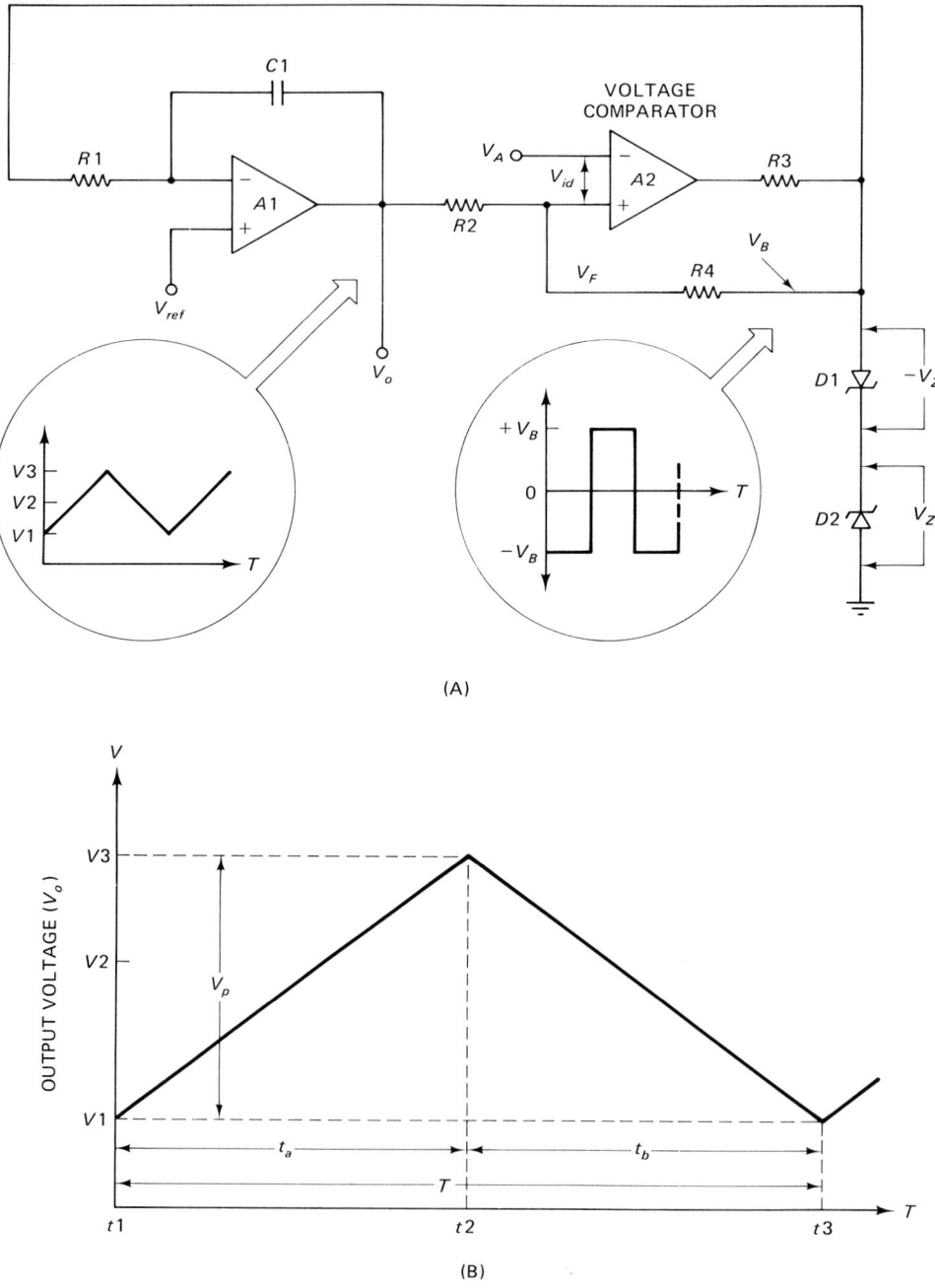

(A)

(B)

Fig. 18-17 (A) Automatic switching improves triangle generator circuit; (B) timing waveform.

will begin to ramp upwards from a minimum voltage of

$$V1 = \frac{V_A(R2 + R4)}{R4} - \frac{V_B R2}{R4} \tag{18-48}$$

The output will continue to ramp upwards toward a maximum value of

$$V3 = \frac{V_A(R2 + R4)}{R4} + \frac{V_B R2}{R4} \tag{18-49}$$

causing a peak swing voltage of

$$V_p = V3 - V1 \tag{18-50}$$

$$V_p = \left(\frac{V_A(R2 + R4)}{R4} + \frac{V_B R2}{R4} \right)$$

$$- \left(\frac{V_A(R2 + R4)}{R4} - \frac{V_B R2}{R4} \right) \tag{18-51}$$

$$V_P = \frac{V_B R2}{R4} + \frac{V_B R2}{R4} = \frac{2V_B R2}{R4} \tag{18-52}$$

Switching of the comparator occurs when the differential input voltage V_{id} is zero. The inverting input ($-$IN) voltage is V_A, which is a fixed reference potential. The noninverting input ($+$IN) is at a voltage (V_F) that is the superposition of two voltages, V_o and V_B.

$$V_F = \frac{V_o R4}{R2 + R4} + \frac{\pm V_B R2}{R2 + R3} \tag{18-53}$$

If $+V_B = -V_B$, the positive and negative thresholds are equal. The duration of each ramp (t_a and t_b) can be found from

$$t_{a,b} = \frac{V_p}{\left[\dfrac{V_B}{R1C1} \right]} \tag{18-54}$$

The value of V_B is selected from $-V_B$ or $+V_B$ as needed. In Eq. (18-52) it was found that $V_p = 2V_B R2/R4$, so

$$t_{a,b} = \frac{\left[\dfrac{2V_B R2}{R4}\right]}{\left[\dfrac{V_B}{R1C1}\right]} \tag{18-55}$$

$$t_{a,b} = \left[\frac{R1C1}{V_B}\right]\left[\frac{2V_B R2}{R4}\right] \tag{18-56}$$

$$t_{a,b} = R1C1\left[\frac{2V_B R2}{R4}\right] \tag{18-57}$$

or, in the less general (but more common) case of $t_a = t_b$:

$$T = 2R1C1\left[\frac{2R2}{R4}\right] \tag{18-58}$$

The frequency of the triangle wave is the reciprocal of the period $(1/T)$, so

$$F = \frac{1}{T} \tag{18-59}$$

$$F = \frac{1}{\left[\dfrac{4R1C1R2}{R4}\right]} \tag{18-60}$$

$$F = \frac{R4}{4R1C1R2} \tag{18-61}$$

Sawtooth Generators

The sawtooth wave (Fig. 18-15A) is a single-slope ramp function. The wave ramps linearly upwards (or downwards) and then abruptly snaps back to the initial baseline condition. Figure 18-18A shows a simple model of a sawtooth generator circuit. A constant current source charges a capacitor in a manner that generates the linear ramp function (Fig. 18-18B). When the ramp voltage (V_c) reaches the maximum point (V_p) switch S1 is closed, forcing V_c back to zero by discharging the capacitor. If switch S1 remains closed, the sawtooth is terminated. If S1 reopens, however, a second sawtooth is created as the capacitor recharges.

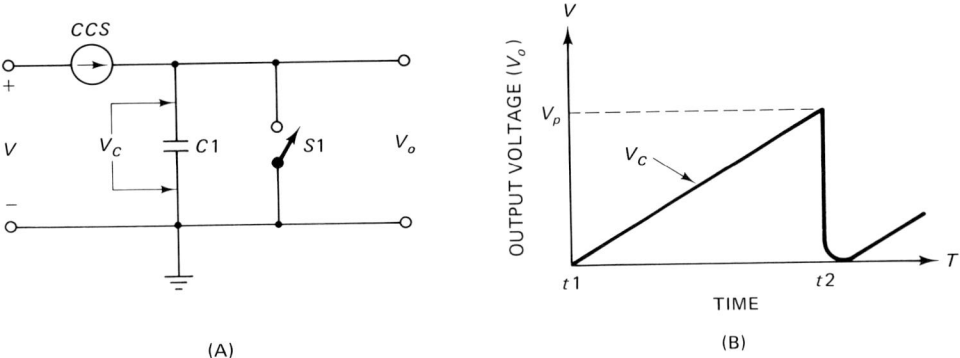

Fig. 18-18 (A) Simple RC sawtooth generator circuit and (B) waveform.

Figure 18-19A shows the circuit for a periodic sawtooth oscillator. It is similar to Fig. 18-18A except that a junction field effect transistor (JFET) Q1 is used as the discharge switch. When Q1 is turned off, the output voltage ramps upwards (see Fig. 18-19B). When the gate is pulsed hard on, the drain-source channel resistance drops from a very high value to a very low value, forcing $C1$ to discharge rapidly. In the absence of a gate pulse, however, the channel resistance remains very high. At time $t1$ the gate is turned off, so V_c begins ramping upwards. At $t2$ the JFET gate is pulsed, so $C1$ rapidly discharges back to zero. When the pulse $(t2–t3)$ ends, however, Q1 turns off again and the ramp starts over. The same circuit can be used for single-sweep operation by replacing the pulse train applied to the gate of Q1 with the output of a monostable multivibrator.

The circuit of Fig. 18-20A shows a sawtooth generator that uses a Miller integrator ($A1$) as a ramp generator; it replaces the discharge switch with an electronic switch that is driven by a voltage comparator and one-shot circuit. The timing diagram for this circuit is shown in

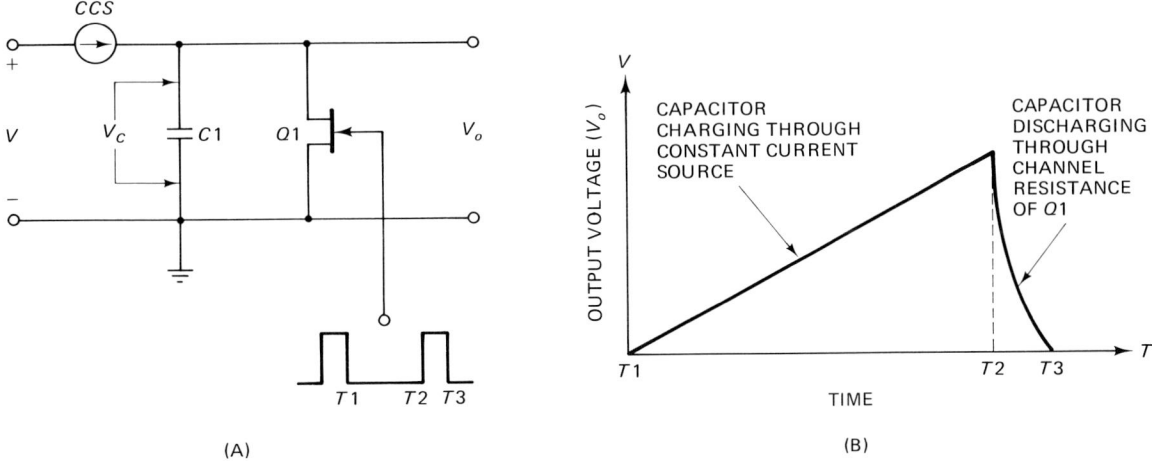

Fig. 18-19 (A) JFET-switched RC sawtooth generator and (B) output waveform.

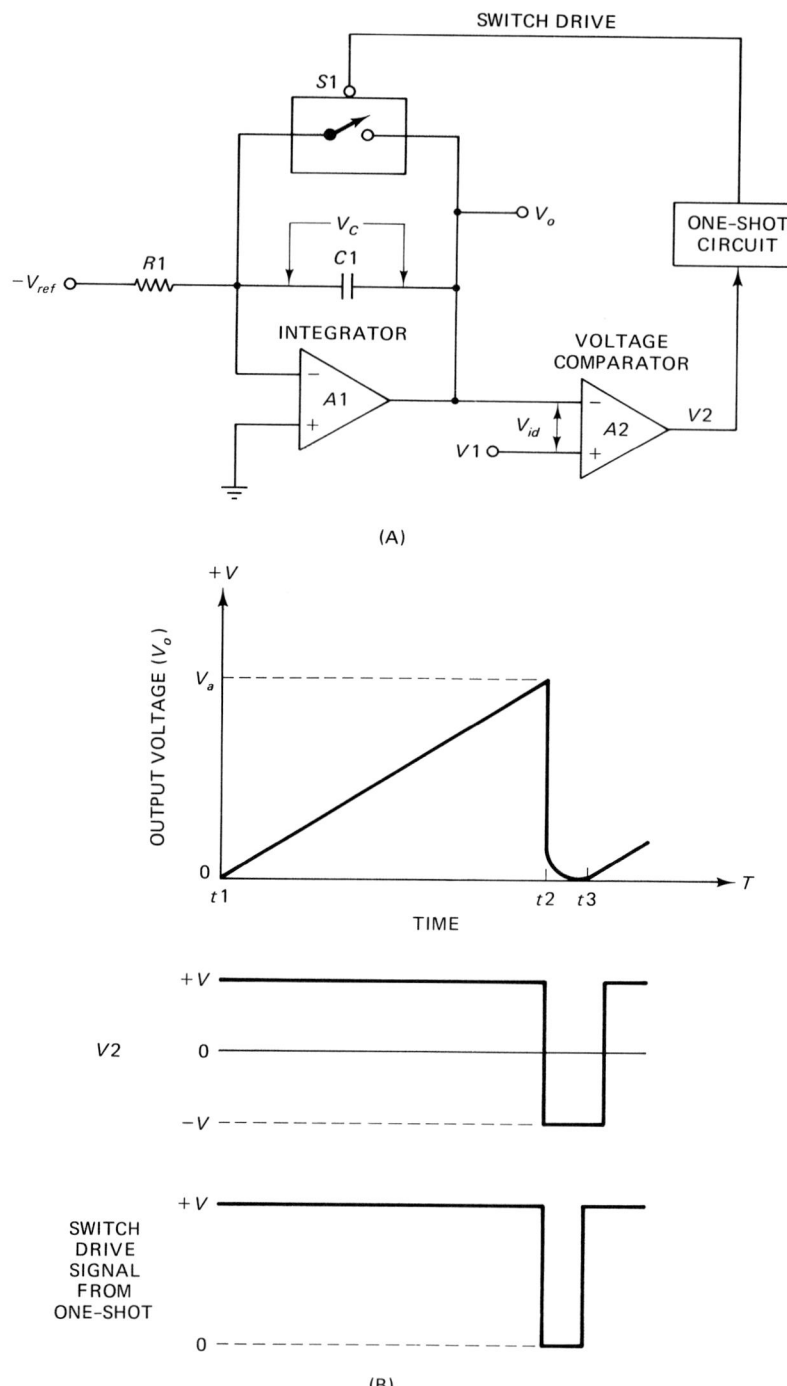

Fig. 18-20 (A) Block diagram for an op-amp sawtooth generator circuit and (B) timing waveforms.

Fig. 18-20B. Under the initial conditions, at time $t1$, the output voltage (V_o) ramps upwards at a rate of $[-(-V_{ref})/R1C1]$. The voltage comparator ($A2$) is biased with the noninverting input (+IN) set to $V1$ and the inverting input at V_o. The comparator differential input voltage $V_{id} = (V1 - V_o)$. As long as $V1 > V_o$ the comparator sees a negative input, so it produces a HIGH output of $+V_{sat}$. At the point where $V1 = V_o$ the differential input voltage is zero, so the output of $A2$ (voltage $V2$) drops LOW (i.e., $-V_{sat}$). The negative-going edge of

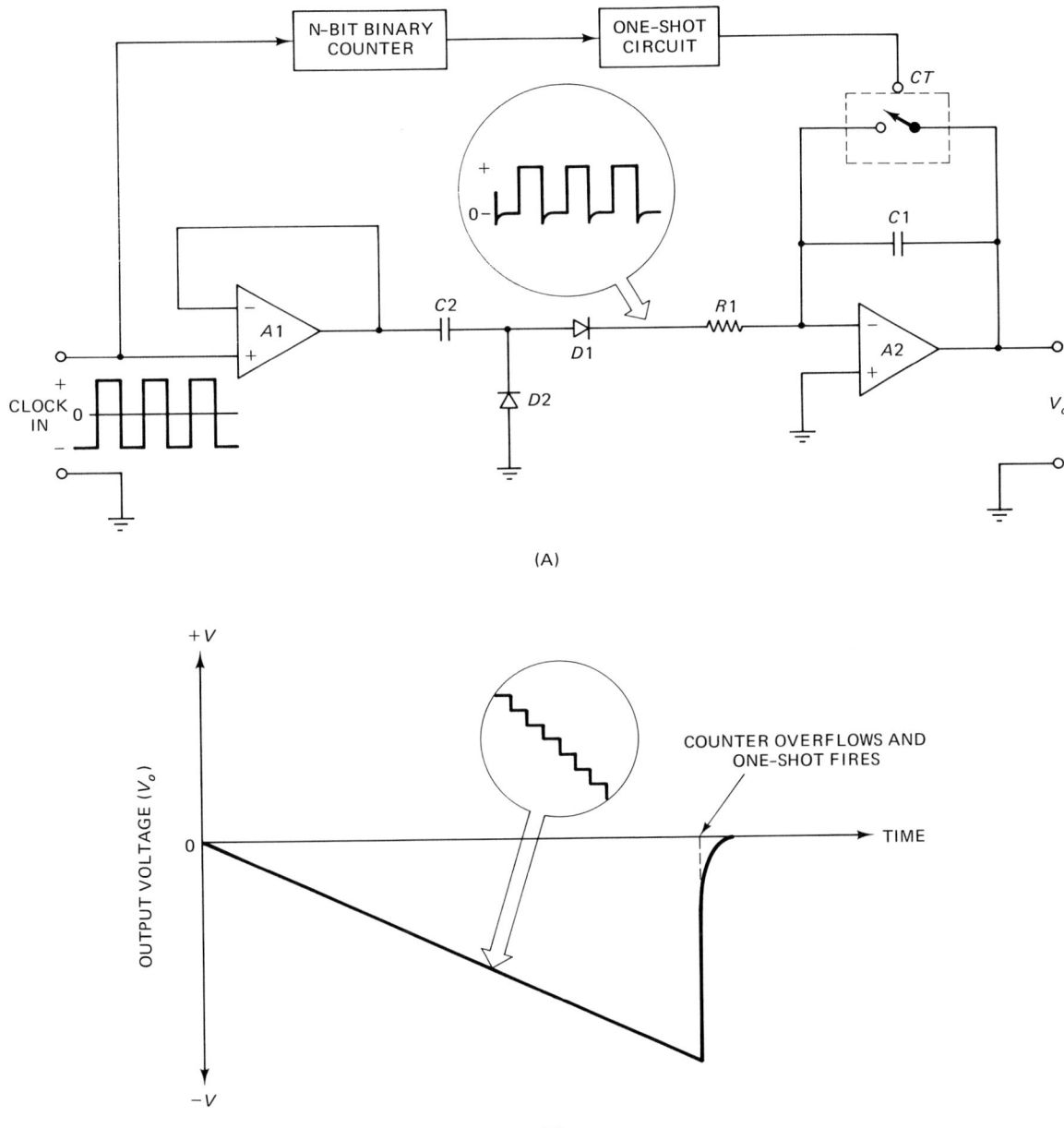

(A)

(B)

Fig. 18-21 (A) Alternate circuit for a sawtooth generator and (B) output waveform.

$V2$ at time $t2$ triggers the one-shot circuit. The output of the one-shot briefly closes electronic switch S1, causing the capacitor to discharge. The one-shot pulse ends at time $t3$, so S1 reopens and allows V_o to again ramp upwards.

A variant of the sawtooth generator circuit is the staircase generator of Fig. 18-21A. The input amplifier ($A1$) provides buffering. A square-wave clock signal applied to the input of $A1$ is passed through capacitor $C2$ to a diode clipping network (D1, D2). The clipping circuit removes the negative excursions of the square wave (see inset to Fig. 18-21A). The remaining positive polarity pulses are applied to the input of the inverting Miller integrator ramp-generator circuit. Each pulse adds a slight step increase to the capacitor charge voltage, so (unless there is significant droop between pulses) the output will ramp up to a negative potential in the staircase fashion shown in the inset to Fig. 18-21B.

The reset circuitry in this circuit is a little different. Although the comparator method of Fig. 18-20A would also work, this circuit takes advantage of the input square wave to provide the period timing of the sawtooth. The square waves are applied to the input of an N-bit binary digital counter circuit. When 2^N pulses have passed, the counter overflows on $2^N + 1$ and triggers a one-shot circuit. As in the previous case, the one-shot output pulse momentarily closes the electronic reset switch shunted across capacitor $C1$.

FEEDBACK OSCILLATORS

A feedback oscillator (Fig. 18-22) consists of an amplifier with an open-loop gain of A_{vol} and a feedback network with a gain or transfer function B. It is called a "feedback oscillator" because the output

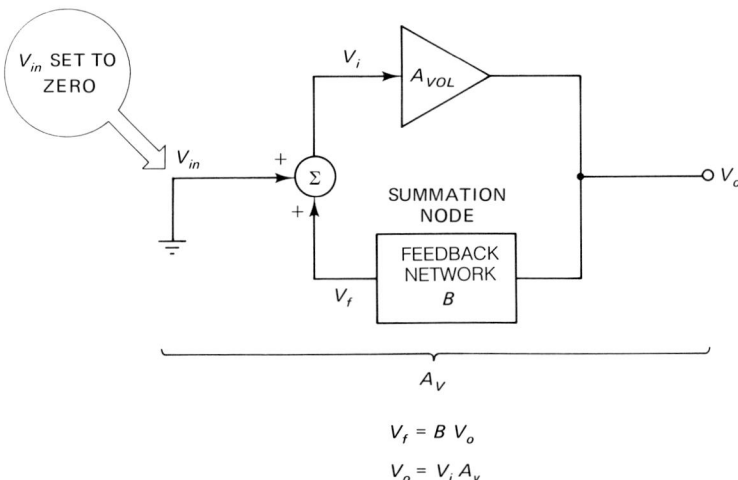

Fig. 18-22 Block diagram model for a feedback oscillator circuit.

signal of the amplifier is fed back to the amplifier's own input by way of the feedback network. Figure 18-22 is a block diagram model of the feedback oscillator. It is no coincidence that it bears more than a superficial resemblance to a feedback amplifier. Indeed, as anyone who has misdesigned or misconstructed an amplifier knows all too well, a feedback oscillator is an amplifier in which special conditions prevail. These conditions are called Barkhausen's criteria for oscillation:

1. Feedback voltage V_F must be in-phase (360°) with the input voltage, and
2. The loop gain BA_{vol} must be unity (1).

The first of these criteria means that the total phase shift from the input of the amplifier to the output of the amplifier, around the loop, and back to the input must be 360° (2π radians) or an integer (N) multiple of 360° (i.e., $N2\pi$ radians).

The amplifier can be any of many different devices. In some circuits it will be a common-emitter bipolar transistor (NPN or PNP devices). In others it will be a junction field-effect transistor (JFET) or metal oxide semiconductor field-effect transistor (MOSFET). In older equipment it was a vacuum tube. In modern circuits the active device will probably be either an integrated circuit operational amplifier or some other form of linear IC amplifier.

The amplifier is most frequently an inverting type, so the output is out of phase with the input by 180°. As a result, to obtain the required 360° phase shift, an additional phase shift of 180° must be provided in the feedback network, at the frequency of oscillation only. If the network is designed to produce this phase shift at only one frequency, then the oscillator will produce a sine-wave output on that frequency.

Before considering a specific sine-wave oscillator circuits, let us examine Fig. 18-22 more closely. Several things can be determined about the circuit.

$$V_i = V_{in} + V_F \tag{18-62}$$

so

$$V_{in} = V_{in} - V_F \tag{18-63}$$

and also,

$$V_F = BV_o \tag{18-64}$$

$$V_o = V_i A_{vol} \tag{18-65}$$

The transfer function (or gain) A_v is

$$A_v = \frac{V_o}{V_{in}} \tag{18-66}$$

Substituting Eqs. (18-63) and (18-65) into Eq. (18-66),

$$A_v = \frac{V_i A_{vol}}{V_i - V_F} \tag{18-67}$$

From Eq. (18-64), $V_F = BV_o$, so

$$A_v = \frac{V_i A_{vol}}{V_i - BV_o} \tag{18-68}$$

But Eq. (18-65) shows $V_o = V_i A_{vol}$, so Eq. (18-68) can be written:

$$A_v = \frac{V_i A_{vol}}{V_i - BV_i A_{vol}} \tag{18-69}$$

and, dividing both numerator and denominator by V_i,

$$A_v = \frac{A_{vol}}{1 - BA_{vol}} \tag{18-70}$$

Equation (18-70) serves for both feedback amplifiers and oscillators. But in the special case of an oscillator $V_{in} = 0$, so $V_o \to \infty$. Implied, therefore, is that the denominator of Eq. (18-70) must also be zero.

$$1 - B_{vol} = 0 \tag{18-71}$$

Therefore, for the case of the feedback oscillator,

$$BA_{vol} = 1 \tag{18-72}$$

BA_{vol} is the loop gain of the amplifier and feedback network, so Eq. (18-72) meets Barkhausen's second criterion.

SINE-WAVE OSCILLATORS

Sine-wave oscillators produce an output signal that is sinusoidal. Such a signal is ideally very pure; if indeed it is perfect, then its Fourier spectrum will contain only the fundamental frequency and no harmonics. It is the harmonics in a nonsinusoidal waveform that give it a characteristic shape. The active element in the circuits described in this circuit is the operational amplifier. However, any linear amplifier will work in place of the operational amplifier. The one circuit that shows the principles most clearly is the *RC* phase shift oscillator; so it is with that circuit that the discussion starts.

Stability in oscillator circuits can refer to several different phenomena. First is frequency stability, which refers to the ability of the oscillator to remain on the design frequency over time. Several different factors affect frequency stability, but the most important are variations in temperature and power supply voltage. Another form of stability is amplitude stability. Because sine-wave oscillators do not operate in the saturated mode, it is possible for minor variations in circuit gain to affect the amplitude of the output signal. Again the factors most often cited for this problem include variations in temperature and DC power supply. The latter is overcome by using regulated DC power supplies for the oscillator. The former is overcome by either temperature-compensated design or by maintaining a constant operating temperature. Some variable sine-wave oscillators will exhibit amplitude variation of the output signal when the operating frequency is changed. In these circuits either a self-compensation element is used or an *automatic level control* amplifier stage is used.

Still another form of stability regards the purity of the output signal. If the circuit exhibits spurious oscillations, these will be superimposed on the output signal. As with any circuit containing an op-amp, or any other high-gain linear amplifier, it is necessary to properly decouple the DC power supply lines. It may also be necessary to frequency-compensate the circuit.

RC Phase-Shift Oscillator Circuits

The *RC* phase-shift oscillator is based on a three-stage cascade resistor–capacitor network such as shown in Fig. 18-23A. An *RC* network will exhibit a phase shift ϕ (Fig. 18-23B) that is a function of resistance (R) and capacitive reactance (X_c). Because X_c is inversely proportional to frequency ($1/2\pi fC$), the phase angle is therefore a function of frequency. The goal in designing the *RC* phase-shift oscillator is to create a phase shift of 180° between the input and output of the network at the desired frequency of oscillation. It is conventional practice to make the three stages of the network identical, so that each provides a 60° phase shift. Although it is common practice, it is also not strictly necessary, provided that the total phase shift is 180°. One reason for using identical stages, however, is that it is possible for the nonidentical designs to have more than one frequency for which the total phase shift is 180°. This phenomenon can lead to undesirable multimodal oscillation.

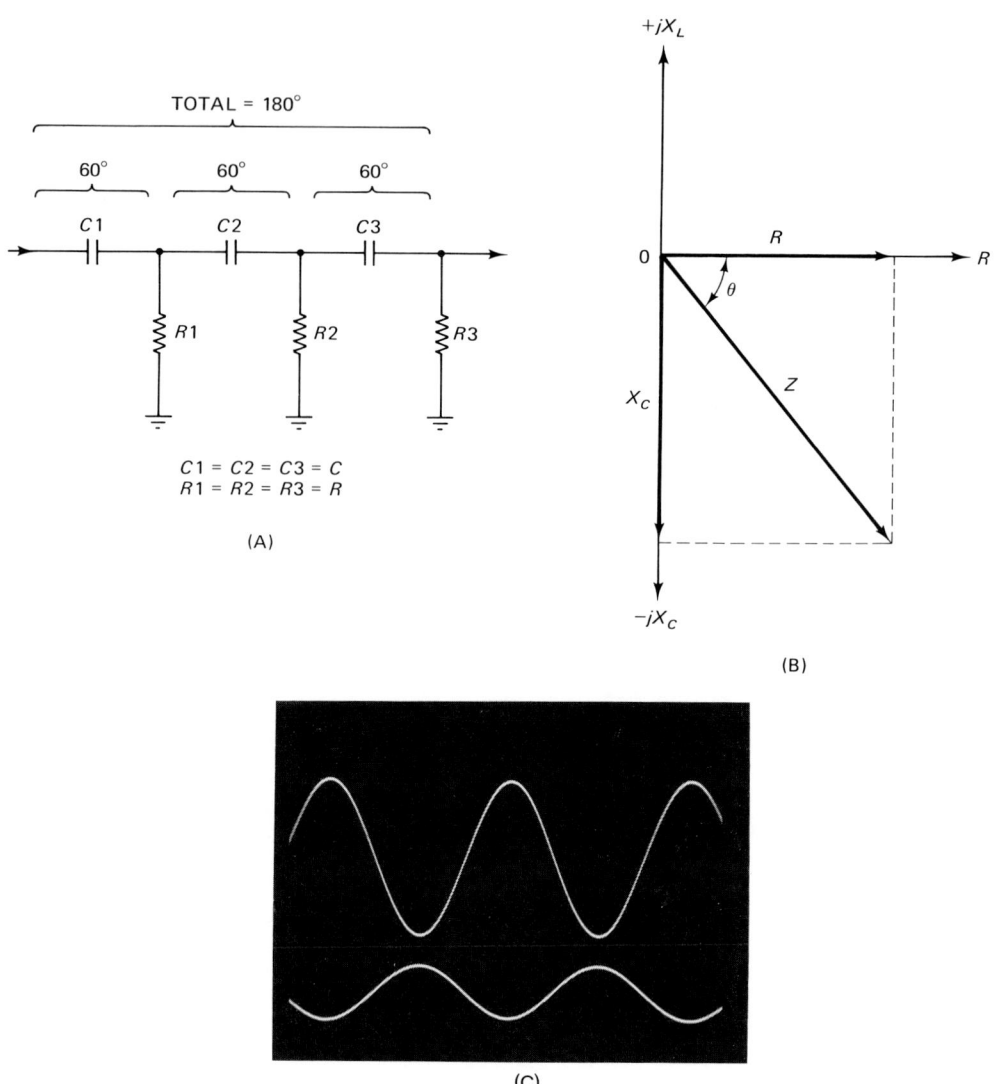

(A)

(B)

(C)

Fig. 18-23 (A) Three-stage *RC* phase-shift network produces a total phase change of 180° at only one frequency, (B) vector representation of circuit relationships, and (C) the sinusoidal response. Top trace is input, bottom is output. Note 180° phase difference; also be aware that amplitude reduction is actually 1/29 because oscilloscope scale factors were different.

Figure 18-23C shows the input and output waveforms of an *RC* network in which three stages of 10 kΩ/0.01 μF were used with an input frequency of 650 Hz. Note the 180° phase shift in the lower trace. Also, be aware that the vertical input scale factors for these two traces are different. The peak-to-peak amplitude of the upper trace is 7.8 V while that of the lower trace is 0.268 V, or 1/29 of the input amplitude. This attenuation factor is important because it establishes the minimum gain requirement in the amplifier.

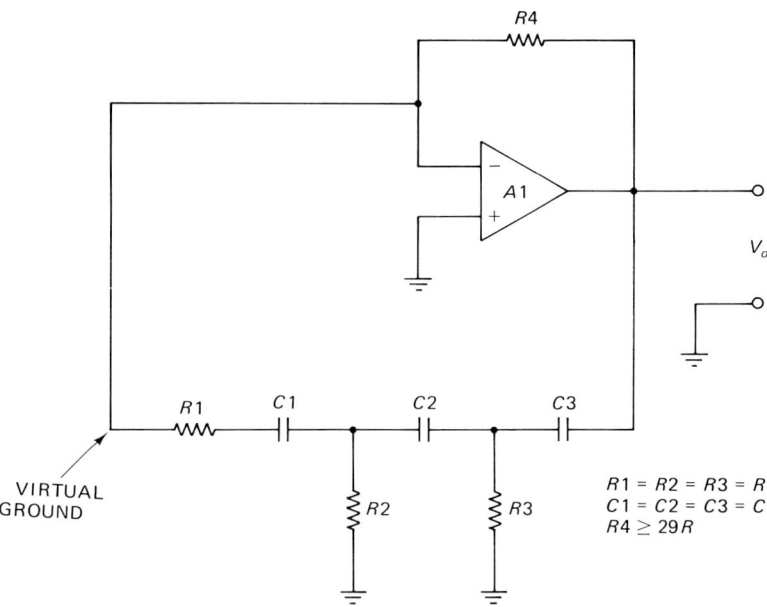

Fig. 18-24 *RC* phase-shift oscillator based on an operational amplifier.

Figure 18-24 shows the circuit for an operational amplifier *RC* phase-shift oscillator. The cascade phase-shift network *R1 R2 R3/ C1 C2 C3* provides 180° of phase shift at a specific frequency, while the amplifier provides another 180° (because it is an inverting follower). The total phase shift is therefore 360° at the frequency for which the *RC* network provides a 180° phase shift. The frequency of oscillation (f) for this circuit is given by

$$f = \frac{1}{2\pi RC(6)^{1/2}} \tag{18-73}$$

where f is in hertz (Hz), R in ohms (Ω), and C in farads (F).

It is common practice to combine the constants in Eq. (18-73) to arrive at a simplified expression:

$$f = 1/15.39RC \tag{18-74}$$

Because the required frequency of oscillation is usually determined from the application, it is necessary to select an *RC* time constant to force the oscillator to operate as needed. Also, because capacitors come in fewer standard values, it is common practice to select an arbitrary trial value of capacitance and then select the resistance that will cause the oscillator to produce the correct frequency. Also, to make the calculations simpler, it is prudent to express the equation in such a way that permits specifying the capacitance (C)

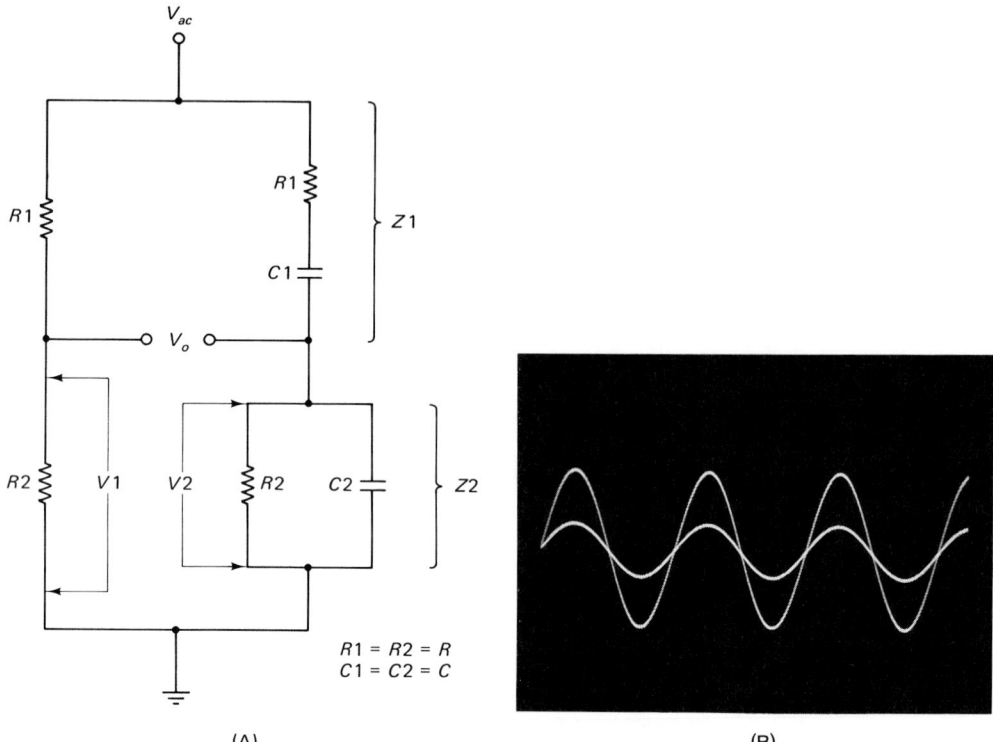

Fig. 18-25   (A) Wien bridge circuit and (B) V AC superimposed on $V2$.

in microfarads. As a result Eq. (18-74) is sometimes rewritten as

$$R = 1,000,000/15.39 C_{\mu F} f \qquad (18\text{-}75)$$

The attenuation through the feedback network must be compensated by the amplifier if loop gain is to be unity or greater. At the frequency of oscillation, the attenuation is $1/29$. The loop gain must be unity, so the gain of amplifier $A1$ must be at least 29 to satisfy $AB = 1$. For the inverting follower (as shown), $R1 = R$ and $A_v = R4/R1$. Therefore, it can be concluded that $R4 \geq 29R$ to meet Barkhausen's criterion for loop gain.

Wien Bridge Oscillator Circuits

The Wien bridge circuit is shown in Fig. 18-25A. Like several other well-known bridge circuits, the Wien bridge consists of four impedance arms. Two of the arms ($R1$, $R2$) form a resistive voltage divider that

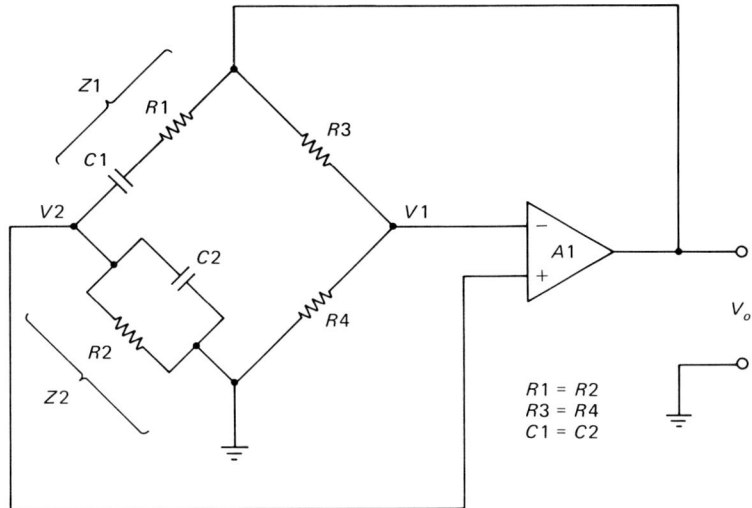

Fig. 18-26   Wien bridge oscillator circuit.

produces a voltage $V1$ of

$$V1 = \frac{V_{ac}R2}{R1 + R2} \tag{18-76}$$

The remaining two arms ($Z1$, $Z2$) are complex RC networks that each consist of one capacitor and one resistor each. Impedance $Z1$ is a series RC network and $Z2$ is a parallel RC network. The voltage and phase shift produced by the $Z1/Z2$ voltage divider are functions of the RC values and the applied frequency. Figure 18-25B shows V_{ac} superimposed on $V2$. Note that $V2 = V_{ac}/3$ and that $V2$ and V_{ac} are in-phase with each other.

Figure 18-26 shows the circuit for a Wien bridge oscillator. The resistive voltage divider supplies $V1$ to the inverting input ($-$IN), while $V2$ is applied to the noninverting input ($+$IN). In Fig. 18-26 the bridge signal source is the output of the amplifier ($A1$). The AC signal is applied to $+$IN, so the gain it sees is found from

$$A_v = R3/R4 + 1 \tag{18-77}$$

The AC feedback applied to $+$IN is

$$B = Z2/(Z1 + Z2) \tag{18-78}$$

At resonance, $B = 1/3$; so (as shown in Fig. 18-25B),

$$V2 = V_0/3 \tag{18-79}$$

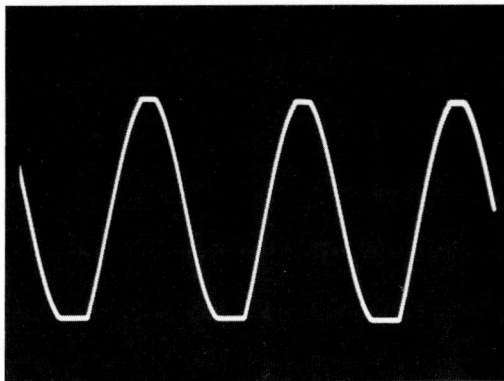

Fig. 18-27   Output sine wave is distorted in the basic Wien bridge circuit.

Because $A_v = V_o/V2$ by definition, satisfying Barkhausen's loop gain criterion $(-A_vB = 1)$ requires that $A_v = V_o/V2 = 3$. Using this result in Eq. (18-77):

$$A_v = R3/R4 + 1 \qquad (18\text{-}80)$$

or,

$$R3 = 2R4 \qquad (18\text{-}81)$$

If $R1 = R2 = R$ and $C1 = C2 = C$, the resonant frequency of the Wien bridge is

$$f = 1/2\pi RC \qquad (18\text{-}82)$$

For the standard Wien bridge oscillator, in which $R1 = R2 = R$ and $C1 = C2 = C$, and $R3 = 2R4$, a sine-wave output will result on frequency f.

Amplitude Stability

The oscillations in the Wien bridge oscillator circuit want to build up without limit when the gain of the amplifier is high. Figure 18-27 shows the result of the gain being only slightly above that required for stable oscillation. Note that some clipping is beginning to appear on the sine-wave peaks. At even higher gains the clipping becomes more severe, and it will eventually look like a square wave. Figure 18-28 shows several methods for stabilizing the waveform amplitude. Figure 18-28A shows the use of small signal diodes such as the 1N914 and 1N4148 devices. At low signal amplitudes the diodes are not sufficiently biased, so the gain of the circuit is

$$A_v = (R1 + R3)/R2 + 1 \qquad (18\text{-}83)$$

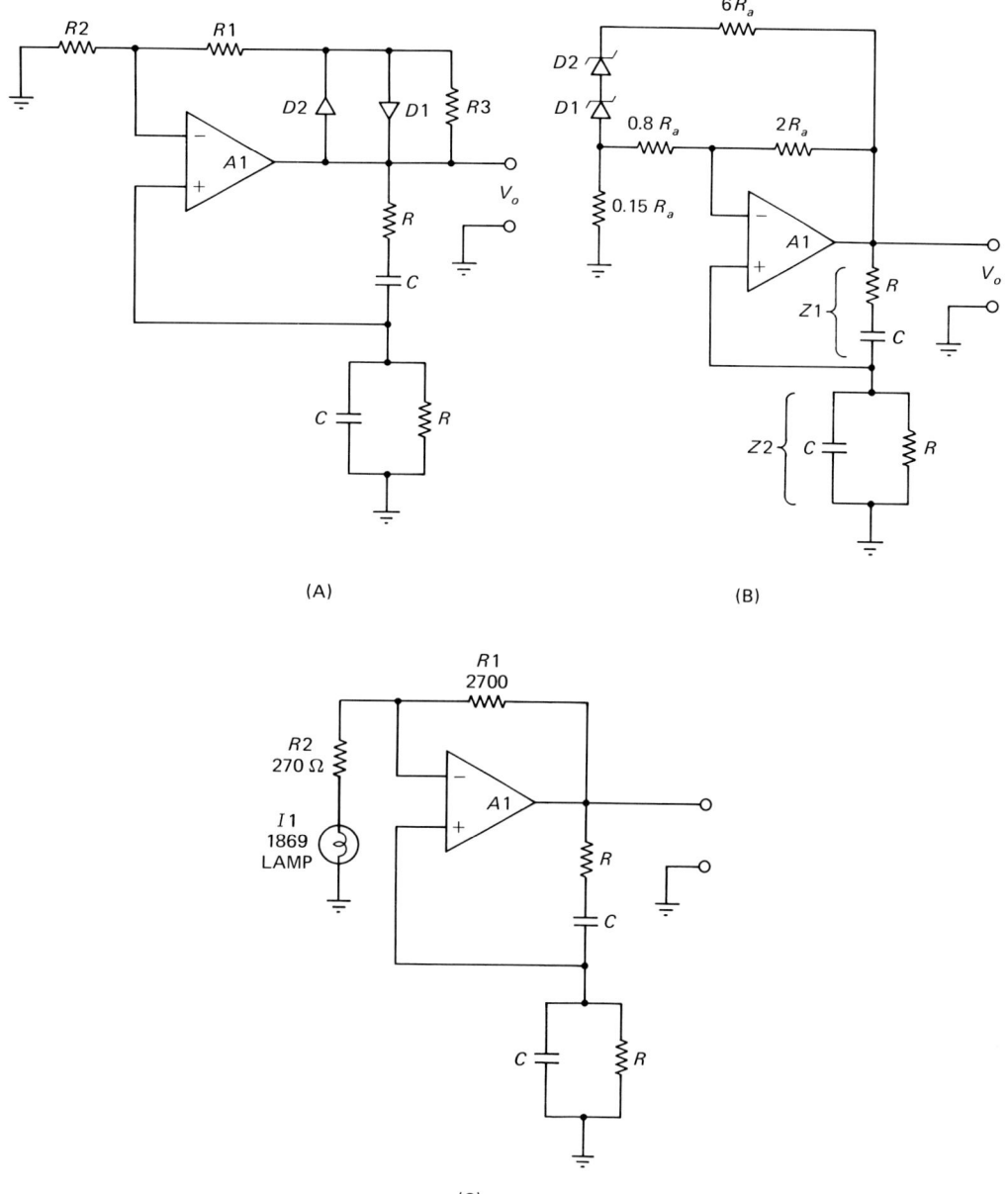

Fig. 18-28 (A) Ordinary signal diodes used to stabilize output amplitude, (B) zener diode output stabilization, and (C) use of an incandescent lamp to stabilize output amplitude.

As the output signal voltage increases, however, the diodes become forward biased. D1 is forward biased on negative peaks of the signal, while D2 is forward biased on positive peaks. Because D1 and D2 are shunted across $R3$, the total resistance $R3'$ is less than $R3$. By inspection of Eq. (18-83) one can determine that reducing $R3$ to $R3'$ reduces the gain of the circuit. The circuit is thus self-limiting.

Another variant of the gain-stabilized Wien bridge oscillator is shown in Fig. 18-28B. In this circuit a pair of back-to-back zener diodes provide the gain limitation function. With the resistor ratios shown, the overall gain is limited to slightly more than unity; so the circuit will oscillate. The output peak voltage of this circuit is set by the zener voltages of D1 and D2 (which should be equal for low-distortion operation).

One final version of the gain-stabilized oscillator is shown in Fig. 18-28C. In this circuit a small incandescent lamp is connected in series with resistor $R2$. When the amplitude of the output signal tries to increase above a certain level, the lamp will draw more current, causing the gain to reduce. The lamp-stabilized circuit is probably the most popular form where stable outputs are required. A thermistor is sometimes substituted for the lamp.

Quadrature and Biphasic Oscillators

Signals that are in quadrature are of the same frequency but are phase shifted 90° with respect to each other. An example of quadrature signals are sine and cosine waves (Fig. 18-29A). Applications for the quadrature oscillator include demodulation of phase-sensitive detector signals in data acquisition systems. The sine wave has an instantaneous voltage $v = V \sin(\omega_o t)$, while the cosine wave is defined by $v = V \cos(\omega_o t)$. Note that the distinction between sine and cosine waves is meaningless unless either both are present or some other timing method is used to establish when zero degrees is supposed to occur. Thus, when sine and cosine waves are called for it is in the context of both being present, and a phase shift of 90° is present between them.

The circuit for the quadrature oscillator is shown in Fig. 18-29B. It consists of two operational amplifiers, $A1$ and $A2$. Both amplifiers are connected as Miller integrators, although $A1$ is a noninverting type while $A2$ is an inverting integrator. The output of $A1(V_{o1})$ is assumed to be the sine-wave output. To make this circuit operate, a total of 360° of phase shift is required between the output of $A1$, around the loop, and back to the input of $A1$. Of the required 360° phase shift, 180° are provided by the inversion inherent in the design of $A2$ (it is in the inverting configuration). Another 90° obtains from the fact that $A2$ is an integrator, which inherently causes a 90° phase shift. An additional 90° phase shift is provided by RC network $R3C3$. If $R1 = R2 = R3 = R$, and $C1 = C2 = C3 = C$, then the frequency of oscillation is given by

$$f = 1/2\pi RC \qquad (18\text{-}84)$$

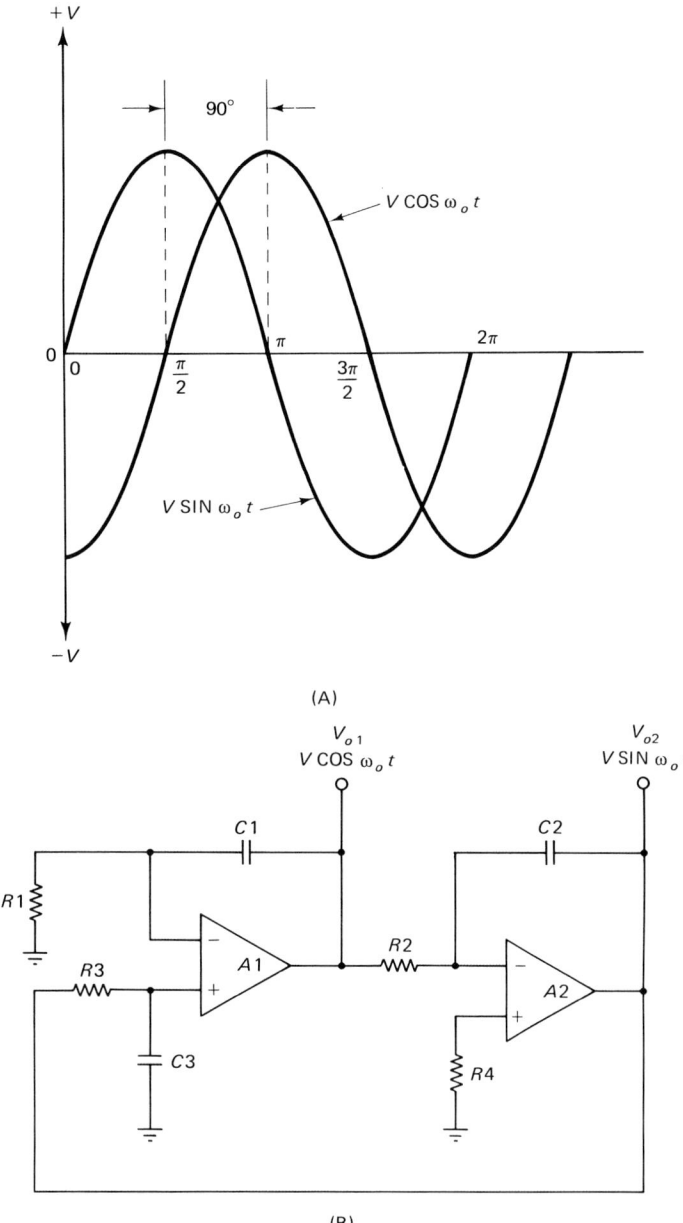

Fig. 18-29 (A) Quadrature sine-wave signals are 90° out of phase with each other; (B) op-amp quadrature oscillator.

The cosine output (V_{o2}) is taken from the output of amplifier $A2$. The relative amplitudes are approximately equal, but the phase is shifted 90° between the two stages.

A biphasic oscillator is a sine-wave oscillator that outputs two identical sine-wave signals that are 180° out of phase with each other.

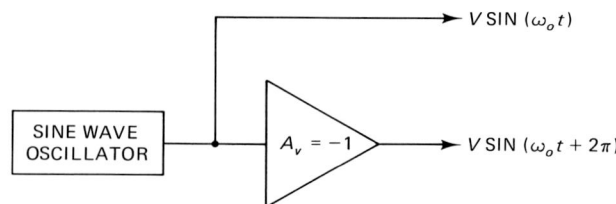

Fig. 18-30 A single inverter will give complementary sine-wave outputs (180° phase shift).

The basic circuit is simple; it is shown in block diagram form in Fig. 18-30. The biphasic oscillator consists of a sine-wave oscillator followed by an inverting amplifier that has a gain of one. The output of the sine-wave oscillator is $V \sin(\omega_o t)$, while the output of the inverter is $V \sin(\omega_o t + 2\pi)$. Biphasic oscillators are sometimes used in transducer excitation applications in carrier amplifiers.

19

Analog Multiplier and Divider Circuits

Analog multiplier and divider circuits are available in both monolithic integrated circuit and hybrid circuit forms. Analog multipliers produce an output voltage V_o that is the product of two input voltages, V_x and V_y. The general form of the multiplier transfer function is

$$V_o = KV_xV_y \qquad (19\text{-}1)$$

where V_o is the output potential in volts, V_x the potential (in volts) applied to the X-input, V_y the potential (in volts) applied to the Y-input, and K a constant (usually $1/10$).

If the proportionality constant K is $1/10$, then Eq. (19-1) becomes

$$V_o = V_xV_y/10 \qquad (19\text{-}2)$$

There are several different basic designs for analog multiplier circuits. The logarithmic amplifier was discussed in Chapter 17, and its use as a multiplier will be reviewed briefly here. When the outputs of two logarithmic amplifiers are first summed together and then applied to an antilog amplifier, the output of the antilog amplifier is proportional (via scale factor K) to the product of the two input voltages.

Transconductance amplifiers (Chapter 13) can also be used to make an analog multiplier. In Chapter 13 an example was presented of a multiplier based on the operational transconductance amplifier (OTA) IC device. There is also a type called the transconductance cell analog multiplier. Other varieties of multiplier circuit will also be examined in this chapter.

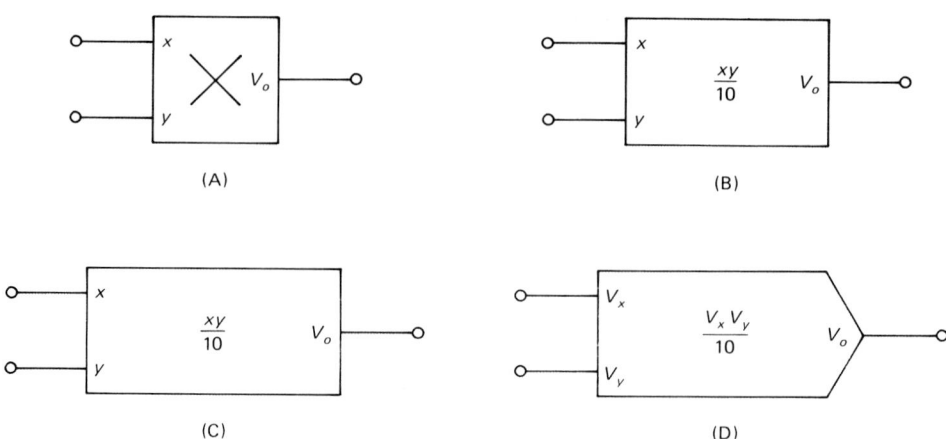

Fig. 19-1   Schematic symbols for the analog multiplier.

ANALOG MULTIPLIER CIRCUIT SYMBOLS

Figure 19-1 shows typical symbols used to represent analog multiplier and divider circuits in schematic diagrams. Although there are standards for circuit symbols, the multiplier is one type of device in which corporate, IEEE, and military standards all seem to be used simultaneously. The symbols shown in Fig. 19-1 are several of those commonly found, and the reader should be aware that other symbols (as well as variations of these) may well be used in actual practice.

ONE-, TWO-, AND FOUR-QUADRANT OPERATION

Analog multipliers and dividers are classified according to the number of quadrants in which they will operate. These are the four quadrants of the standard X-Y cartesian coordinate system. Figure 19-2A illustrates one-quadrant operation. In this type of system both input voltages must be positive ($V_x \geq 0$, $V_y \geq 0$). The only possible output voltage polarity is positive. At least one commercial hybrid multiplier operates in one quadrant, but with both input voltages negative. Again, the only permissible output voltage is positive. That type of operation is a rarity, however. The least complex multipliers based on logarithmic amplifiers are normally one-quadrant devices.

A second form of multiplier is the two-quadrant form (Fig. 19-2B). These circuits operate in a manner that allows the output voltage to be either positive or negative, but there are constraints on the allowable input voltage polarities. One input voltage will be limited to positive values only, while the other can be either positive or negative.

Four-quadrant operation (Fig. 19-3) is the most flexible because it allows operation with any combination of input signal polarity. The

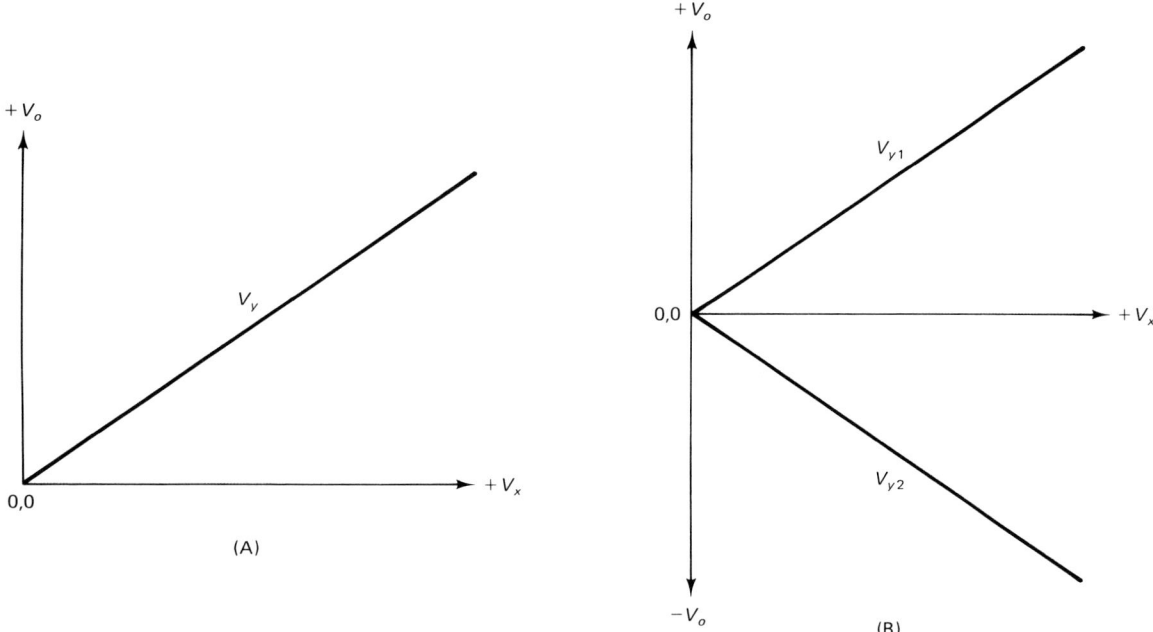

Fig. 19-2 (A) One-quadrant multiplier operation and (B) two-quadrant operation.

output signal can be either positive or negative, as can either (or both) input signal voltages. Figure 19-4 shows the relationship between input and output polarities, relating these to the quadrant of operation. These limits are summarized as follows:

QI: $V_x \geq 0,$ $V_y \geq 0,$ $0 < V_o < +V$

QII: $V_x \leq 0,$ $V_y \geq 0,$ $-V < V_o < 0$

QIII: $V_x \leq 0,$ $V_y \leq 0,$ $0 < V_o < +V$

QIV: $V_x \geq 0,$ $V_y \leq 0,$ $-V < V_o < 0$

In most cases, the voltage ranges of the two inputs are symmetrical with each other. For example, a typical range is $V_{x(\text{max})} = V_{y(\text{max})} = \pm 10$ V.

Another category of device is the multiplier/divider shown in Fig. 19-5. These devices have a transfer function of the form

$$V_o = V_x V_y / V_z \qquad (19\text{-}3)$$

In the multiplier mode, the X and Y inputs are used, and the usual scaling factor K is set in this case by applying a voltage to the Z input. In the division mode, either the X and Z or Y and Z inputs are used for the signal voltages, while the scaling factor is set by applying a fixed voltage to the remaining input.

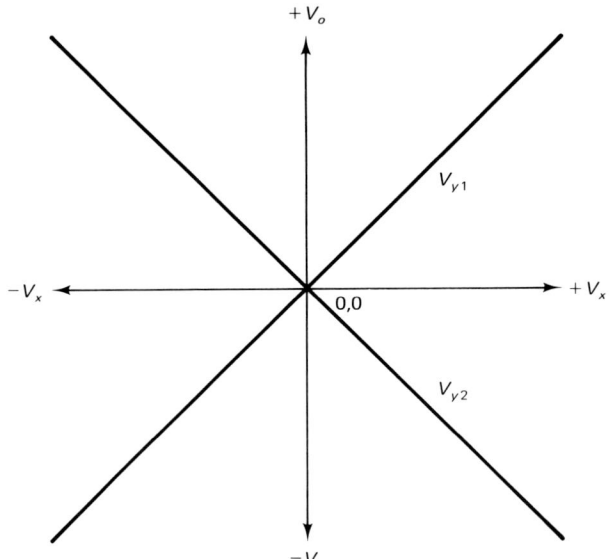

Fig. 19-3 Full four-quadrant operation.

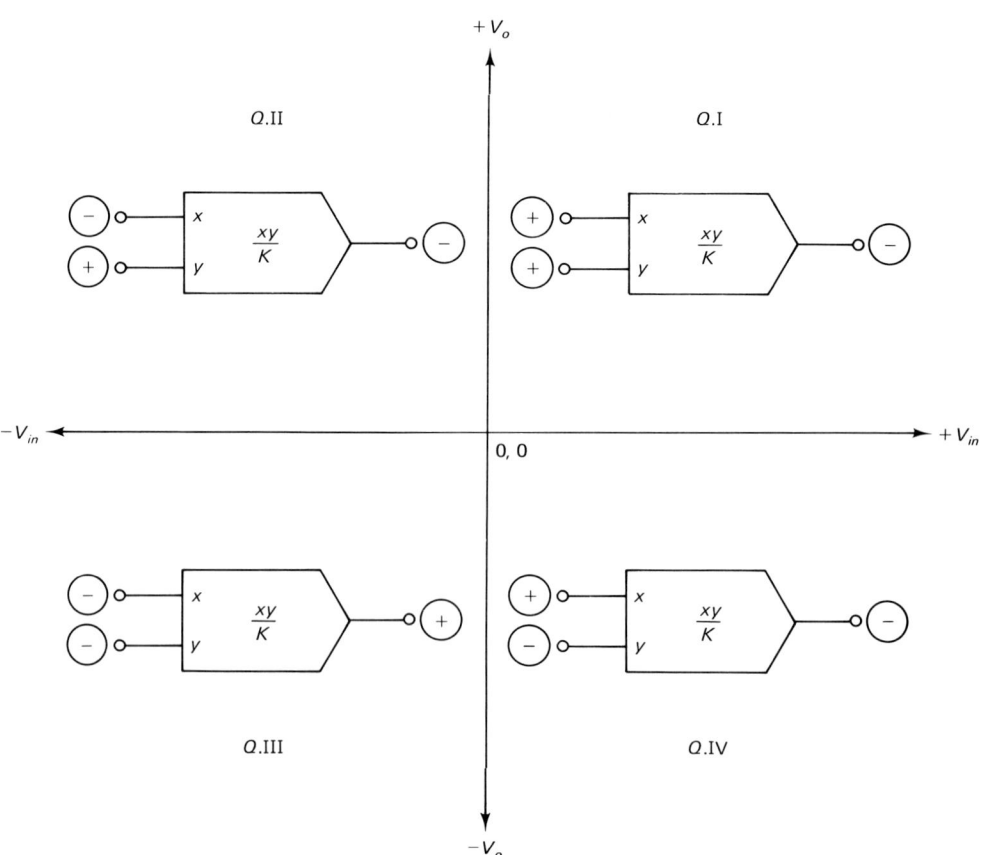

Fig. 19-4 Signal polarities permissible in four quadrants of operation.

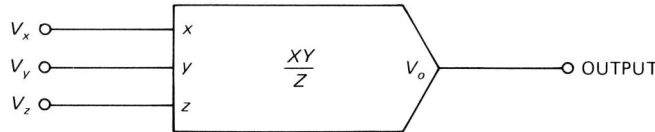

Fig. 19-5 XY/Z analog multiplier/divider.

DIFFERENTIAL INPUT MULTIPLIERS AND DIVIDERS

Most of the multiplier and divider circuits presented in this chapter are single-ended devices. That is, the input signals are measured between the input terminal and ground, and there is but one input line each for V_x and V_y. A differential input multiplier and/or divider circuit is shown in Fig. 19-6. In this circuit the differential input voltages V_{dx} and V_{dy} are defined as

$$V_{dx} = X1 - X2 \qquad (19\text{-}4)$$

$$V_{dy} = Y1 - Y2 \qquad (19\text{-}5)$$

The multiplier transfer function is

$$V_o = K V_{dx} V_{dy} \qquad (19\text{-}6)$$

or

$$V_o = K(X1 - X2)(Y1 - Y2) \qquad (19\text{-}7)$$

In most cases, $K = 1/10$, so these expressions become

$$V_o = V_{dx} V_{dy}/10 \qquad (19\text{-}8)$$

$$V_o = [(X1 - X2)(Y1 - Y2)]/10 \qquad (19\text{-}9)$$

The differential input multiplier is particularly useful in at least two different situations: First, where a Wheatstone bridge or other

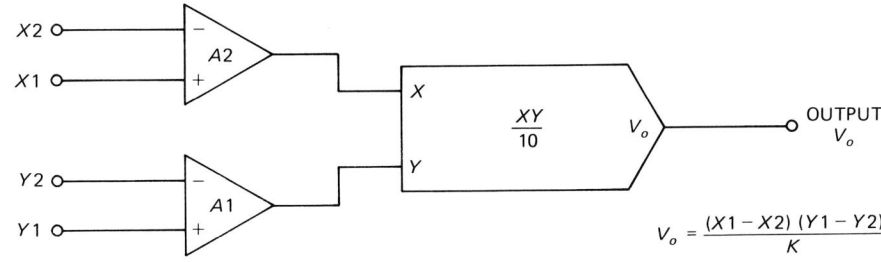

Fig. 19-6 Differential input analog multiplier.

balanced output signal source supplies one or both multiplier input signals; and second, in cases where the signal processing requirements are such that an input to the multiplier is the difference between source signals. For example, if a fluid pressure transducer measures a pressure in which a variation rides on top of a certain large offset minima pressure (e.g., the human blood pressure, in which the varying pressure wave rides atop the diastolic static pressure), then the input signal will be a slowly varying DC wave on top of a large DC static offset potential. If the DC offset can be subtracted prior to being applied to the multiplier, the overall dynamic range of the system is improved.

Either input of Fig. 19-6 can be made single-ended by the simple expedient of grounding the unwanted input.

TYPES OF MULTIPLIER CIRCUIT

In this section we will examine several popular approaches to multiplier design. Some of them are currently used in state-of-the-art analog multipliers, while others are now nearly obsolete. The reason for including the nearly obsolete circuits is to demonstrate the process of multiplication and the range of possibilities. Besides, it is generally an error to disdain older methods because the same fundamentals are often resurrected when newer methods can make their implementation better. For example, the incandescent lamp multiplier circuit below is now obsolete, but the JFET manifestation of the same concept is currently used. Yet it is nonetheless profitable to examine the lamp version because it is inherently easier to understand.

Matched Resistor and Lamp Multiplier Divider

A simple and effective multiplier and divider circuit can be constructed from a pair of operational amplifiers and a special gain setting block (see Fig. 19-7A) consisting of a pair of matched photoresistors ($R1$ and $R2$), both of which are illuminated by the same incandescent lamp ($IL1$). Because $R1$ and $R2$ are matched for any given light level, it is true that $R1 = R2 = R$. Because $R1$ is in the feedback loop of amplifier $A2$, nonlinearities in the voltage-versus-brightness ratio of $IL1$ are effectively servoed-out.

In previous chapters we have used an analysis method based on Ohm's law and Kirchhoff's current law (KCL). That practice will be continued here. In Fig. 19-7A, the input bias current of each amplifier (I_{b1} and I_{b2}) are each zero. In addition, the noninverting inputs of both amplifiers are grounded, so the two summing junctions (A and B) are at zero volts (ground) potential. By Ohm's law, and considering that

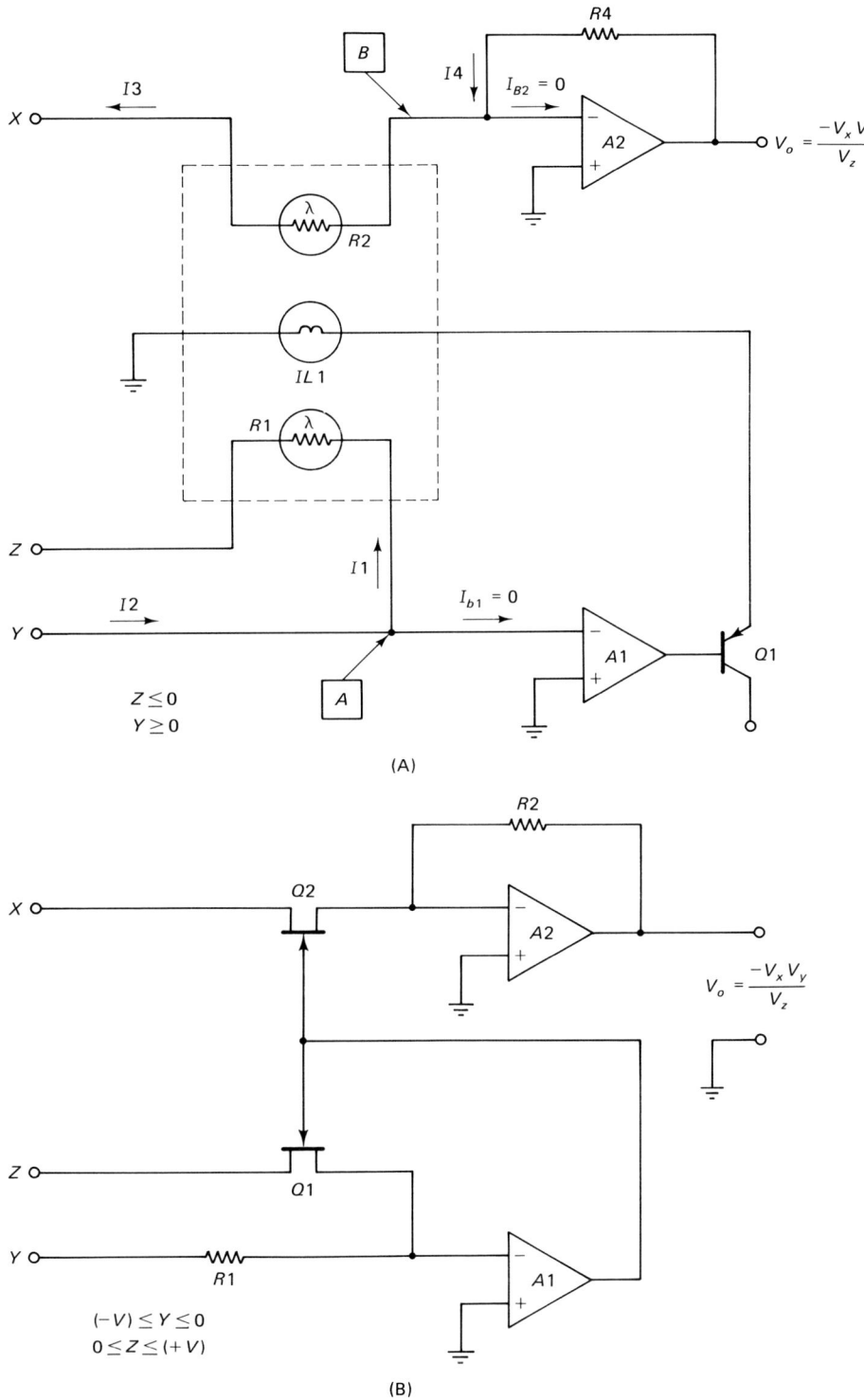

Fig. 19-7   (A) Matched resistors multiplier and (B) JFET version.

$R1 = R2 = R,$

$$I1 = V_z/R \tag{19-10}$$

and

$$I2 = V_y/R3 \tag{19-11}$$

By KCL:

$$I1 = -I2 \tag{19-12}$$

Substituting Eqs. (19-10) and (19-11) into Eq. (19-12):

$$V_y/R3 = -V_z/R \tag{19-13}$$

and rearranging terms:

$$R = -V_z R3/V_y \tag{19-14}$$

Current $I3$ is, by Ohm's law,

$$I3 = V_x/R \tag{19-15}$$

So, by substituting Eq. (19-14) into Eq. (19-15):

$$I3 = \frac{V_x}{\left(-V_z R3/V_y\right)} \tag{19-16}$$

and then rearranging terms:

$$I3 = -V_x V_y/V_z R3 \tag{19-17}$$

Because $I_{b1} = 0$, and because point B is at ground potential, by KCL it can be stated:

$$I3 = -I4 \tag{19-18}$$

From Ohm's law:

$$I4 = V_o/R4 \tag{19-19}$$

By substituting Eqs. (19-17) and (19-19) into Eq. (19-18):

$$-V_o/R4 = -V_x V_y/V_z R3 \tag{19-20}$$

or, by rearranging terms:

$$V_o = V_x V_y R4/V_z R3 \tag{19-21}$$

Accounting for the fact that V_z is restricted to negative values:

$$V_o = -V_x V_y R4 / V_z R3 \qquad (19\text{-}22)$$

Placing Eq. (19-22) into the standard form:

$$V_o = -K V_x V_y / V_z \qquad (19\text{-}23)$$

in which $K = R4/R3$.

The circuit of Fig. 19-7A is somewhat impractical today, but it was once implemented in discrete form in analog circuitry. Figure 19-7B shows a modern version in which the lamp and resistors have been replaced with junction field-effect transistors (JFET). The circuit of Fig. 19-7B works on the basis of the channel resistance of the JFETs. At gate potentials below the pinch-off voltage, the JFET drain-source resistance is a function of the gate voltage. Thus, Q1 and Q2 operate as voltage-controlled resistors.

Logarithmic Amplifier-Based Multipliers

Many popular forms of analog multiplier–divider circuit are based on the properties of the logarithmic and antilog amplifiers. A logarithmic amplifier has a transfer function of the form either

$$V_o = k \ln(V_{in}) \qquad (19\text{-}24)$$

or

$$V_o = k \log(V_{in}) \qquad (19\text{-}25)$$

One of the properties of logarithms is that their use converts multiplication operations into addition, and division operations are converted into subtraction. Therefore,

$$XY = \log X + \log Y \qquad (19\text{-}26)$$

and

$$X/Y = \log X - \log Y \qquad (19\text{-}27)$$

Equations (19-26) and (19-27) provide us the basis for designing an analog multiplier–divider circuit; Figure 19-8 shows the block diagram of such a circuit. The X-input (V_x) is applied to the input of LOGAMP-A to produce signal $A = \log(V_x)$. Similarly, the Y-input (V_y) is applied to the input of LOGAMP-B to produce a signal $B = \log(V_y)$. These signals are each applied to a summer amplifier (in the multiplication case) or a difference amplifier (in the division case). The output

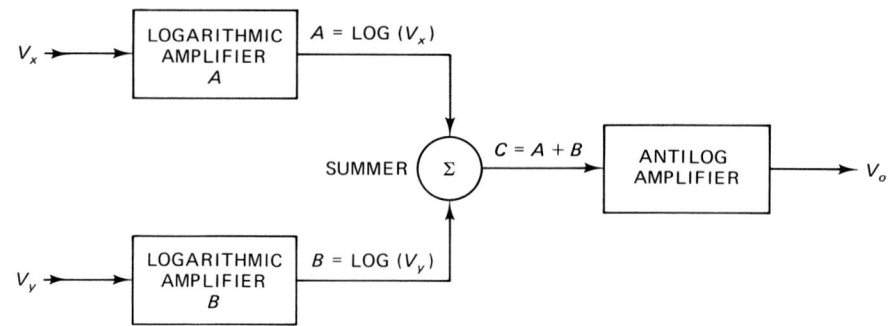

Fig. 19-8   Logarithmic amplifier type of analog multiplier.

of the summer circuit is $C = A + B$, or $C = \log(V_x) + \log(V_y)$. By passing this signal through an antilog amplifier it is possible to construct the product of X and Y:

$$V_o = \log^{-1}(C) \tag{19-28}$$

$$V_o = \log^{-1}\left[\log(V_x) - + \log(V_y)\right] \tag{19-29}$$

$$V_o = V_x V_y \tag{19-30}$$

The logarithmic and antilog amplifiers were discussed earlier. That discussion revealed that an uncompensated LOG/ANTILOG amplifier exhibits a strong temperature dependence. To prevent error, temperature compensation must therefore be incorporated into the multiplier circuit as well.

Quarter Square Multipliers

Figure 19-9 shows the block diagram of a quarter square multiplier. Consider the following expression:

$$V_o = \frac{(X + Y)^2 - (X - Y)^2}{4} \tag{19-31}$$

Expanding this polynomial results in

$$V_o = \frac{(X^2 - X^2) + (Y^2 - Y^2) + 2XY + 2XY}{4} \tag{19-32}$$

$$V_o = \frac{2XY + 2XY}{4} \tag{19-33}$$

$$V_o = XY \tag{19-34}$$

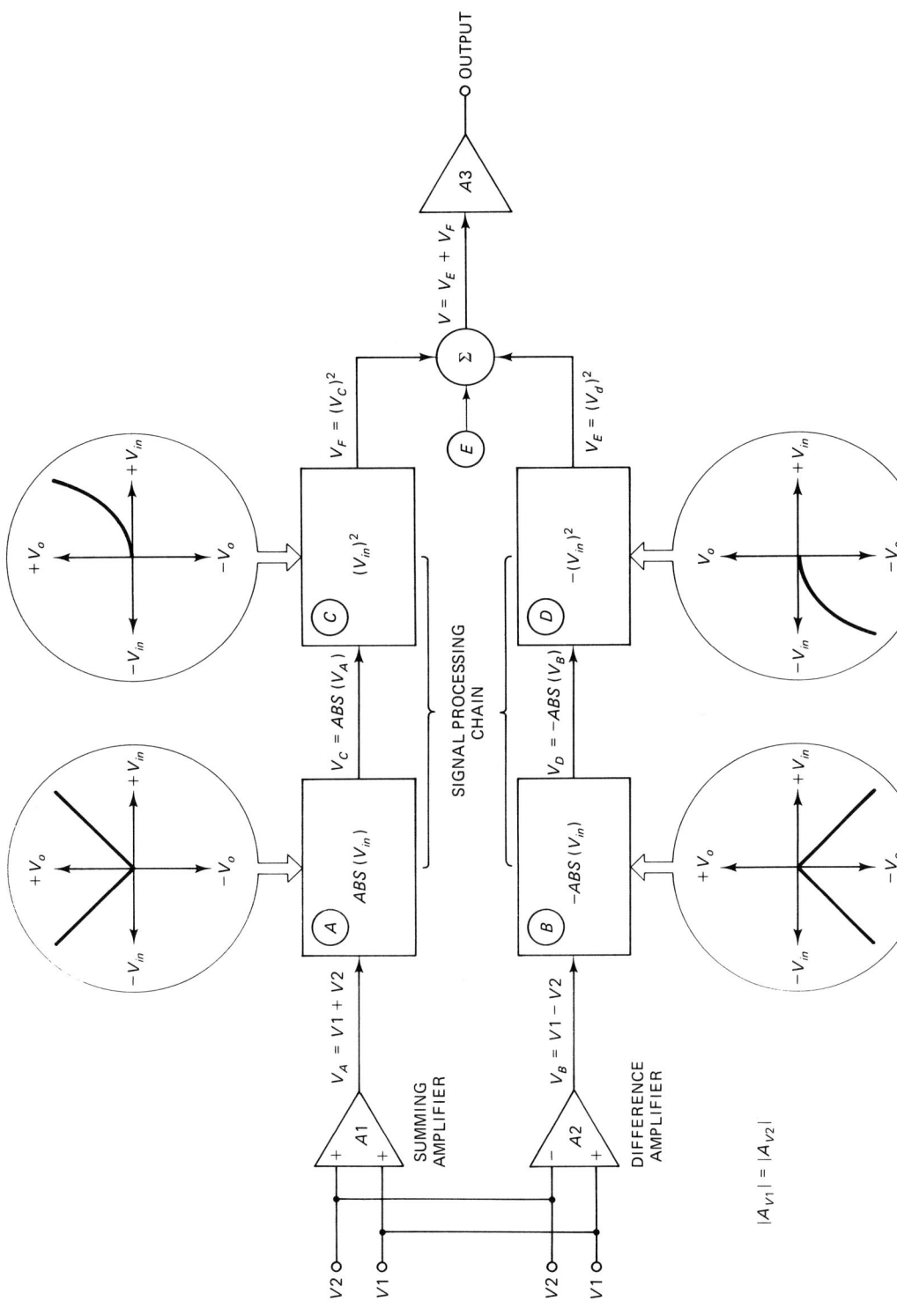

Fig. 19-9 Quarter square analog multiplier.

Figure 19-9 shows the block diagram of a circuit that will implement Eq. (19-31). There are two chains of circuits in this multiplier. Each chain contains a two-port input amplifier: $A1$ is a summation amplifier and produces an output voltage $V_a = (V1 + V2)$; $A2$ is a difference amplifier and produces an output $V_B = (V1 - V2)$.

Each signal processing chain also contains an absolute value amplifier and a squaring circuit. The absolute value circuit is also called a full wave precise rectifier. The only difference between these two chains is that A is noninverting while B is inverting. Thus,

$$V_C = \text{ABS}(V_a) \tag{19-35}$$

$$V_C = \text{ABS}(V1 + V2) \tag{19-36}$$

and

$$V_D = -\text{ABS}(V_B) \tag{19-37}$$

$$V_D = -\text{ABS}(V1 - V2) \tag{19-38}$$

The squarers (C and D) produce an output signal that is proportional to the square of the input voltage $V_o = (V_{in})^2$. In analog multipliers the squarers are diode breakpoint generators. Ordinary PN junction diodes have a square-law region in their operating characteristic. Using as few as 10 PN junction diodes, each biased to slightly different points, produces a breakpoint generator with a linearity approaching ± 0.1 percent of full scale. The outputs of the squarers are

$$V_E = (V_D)^2 \tag{19-39}$$

$$V_F = (V_C)^2 \tag{19-40}$$

Combining these voltages in a summer produces

$$V_o = (V_C)^2 + (V_D)^2 \tag{19-41}$$

Relating Eq. (19-41) to (19-31) is (as they say in college books) "a good exercise for the reader" (if you're interested).

Transconductance Multipliers

The most commonly used form of IC analog multiplier is the transconductance type. The word *transconductance* indicates that an output current (I_o) is controlled by an input voltage (V_{in}):

$$G_m = dI_o / dV_{in} \tag{19-42}$$

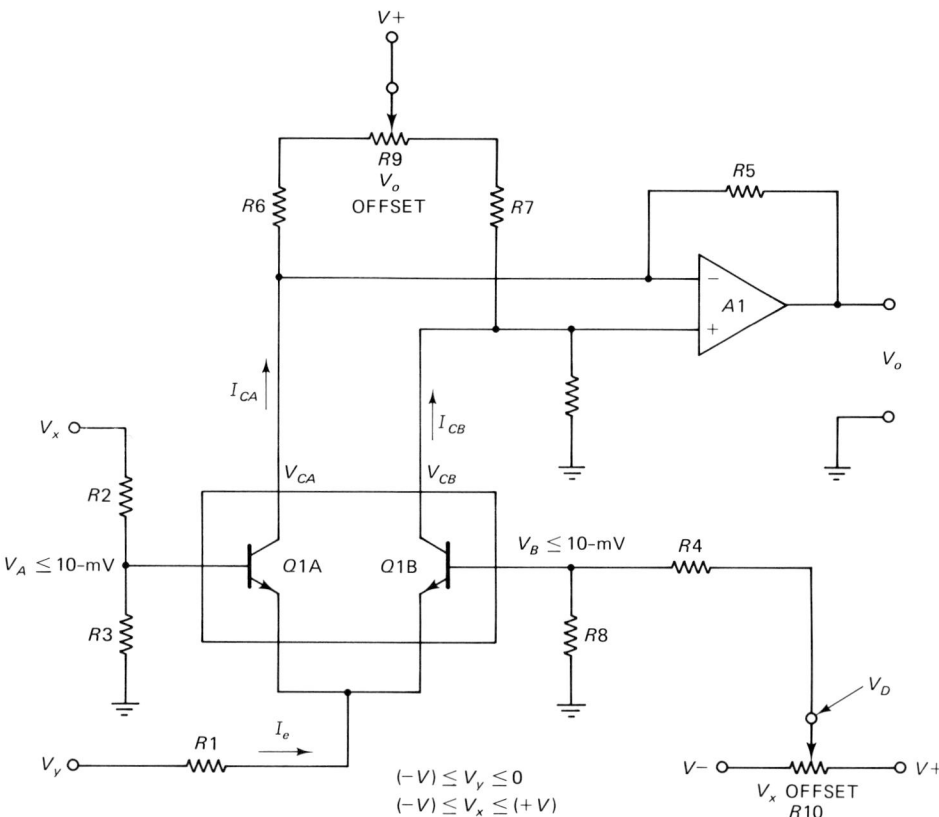

Fig. 19-10 Simple transconductance multiplier.

In older texts the unit of conductance was the descriptive term *mho* (ohm spelled backwards), but today the siemens is used for conductance. (Note: It is only a name change, for 1 siemens = 1 mho.)

Figure 19-10 shows the circuit for a simple op-amp based transconductance multiplier. The basis for the circuit is a dual NPN transistor such as the MAT-01, MAT-02, or LM-114 devices. It is important that the two transistors be part of the same substrate to maintain thermal tracking between the two devices. For very low signal levels (10 mV or so), the following relationship is true:

$$I_{ca} = kI_e V_x \qquad (19\text{-}43)$$

where I_{ca} is the collector current of Q1A, I_e the emitter current, V_x the voltage applied to the base-emitter junction of Q1A, $k = q/2KT$, q is the electronic charge 1.6×10^{-19} coulombs (C), K the Boltzman constant (1.38×10^{-23} J/K), and T the temperature in Kelvins (K).

Output voltage V_o is

$$V_o = I_{ca} R5 \qquad (19\text{-}44)$$

The emitter current is supplied from input voltage V_y, so

$$I_e = V_y/R1 \tag{19-45}$$

By substituting Eqs. (19-43) and (19-45) into (19-44):

$$V_o = kV_xV_y(R5/R1) \tag{19-46}$$

or, because $R5/R1$ is also a constant,

$$V_o = k_{tot}V_xV_y \tag{19-47}$$

The transconductance cell forms the basis for most easily available IC multiplier–divider circuits. Figure 19-11 shows a typical IC

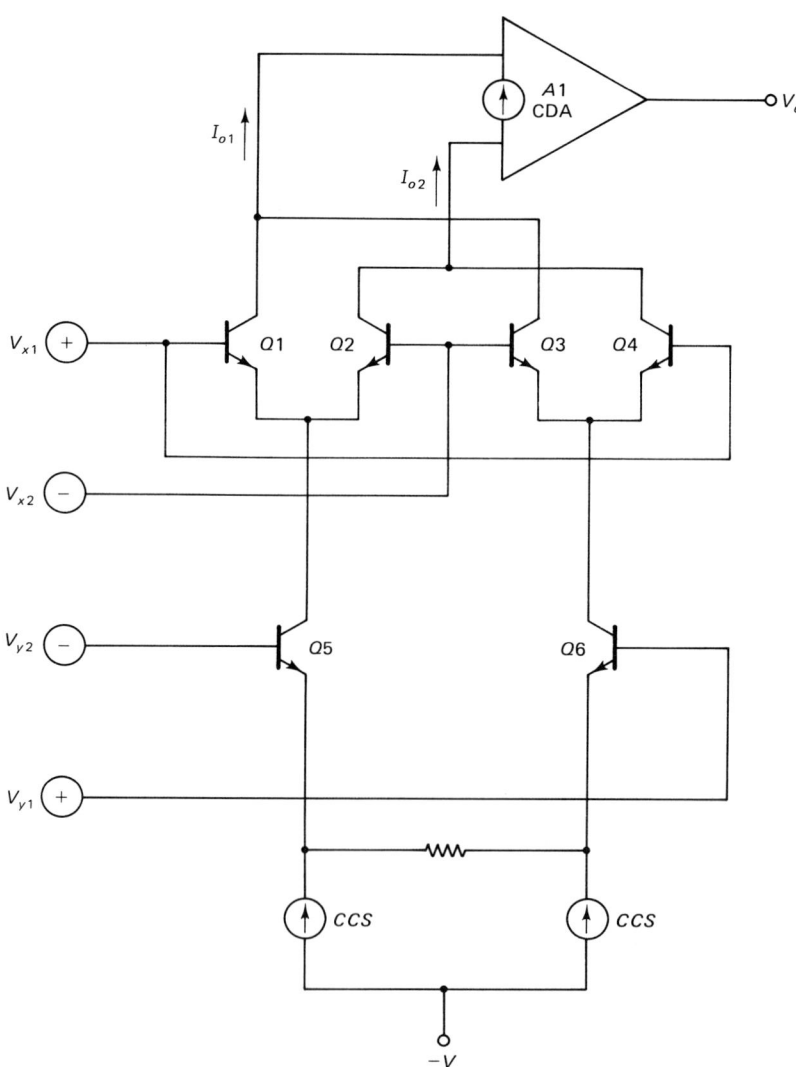

Fig. 19-11   Basic transconductance cell used in IC multipliers.

multiplier circuit based on the transconductance cell. The output is a current, so it must be transformed into an output voltage in a current difference amplifier.

THE AD-533: A PRACTICAL ANALOG MULTIPLIER / DIVIDER

The Analog Devices Inc. AD-533 is a low-cost analog multiplier–divider circuit in integrated circuit form. Although these chips tend to be more expensive than the discrete circuits discussed above, they are also often more cost effective because they tend to work better and require less tweaking to make them operate properly.

The pinouts for the 14-pin DIP version of the AD-533 are shown in Fig. 19-12. The AD-533 contains a transconductance multiplier for the X and Y inputs, and a summing junction at an operational amplifier for a Z input. The AD-533L version of this chip is capable of a full-scale linearity error of only 0.5 percent. In addition, a low-temperature coefficient of 0.01%/°C is provided. The AD-533 device is capable of operating to a small-signal bandwidth of 1 MHz, a full-power bandwidth of 750 kHz, and a slew rate of 45 V/μS. The AD-533 will multiply in four quadrants with a transfer function of

$$V_o = V_x V_y / 10 \text{ V} \tag{19-48}$$

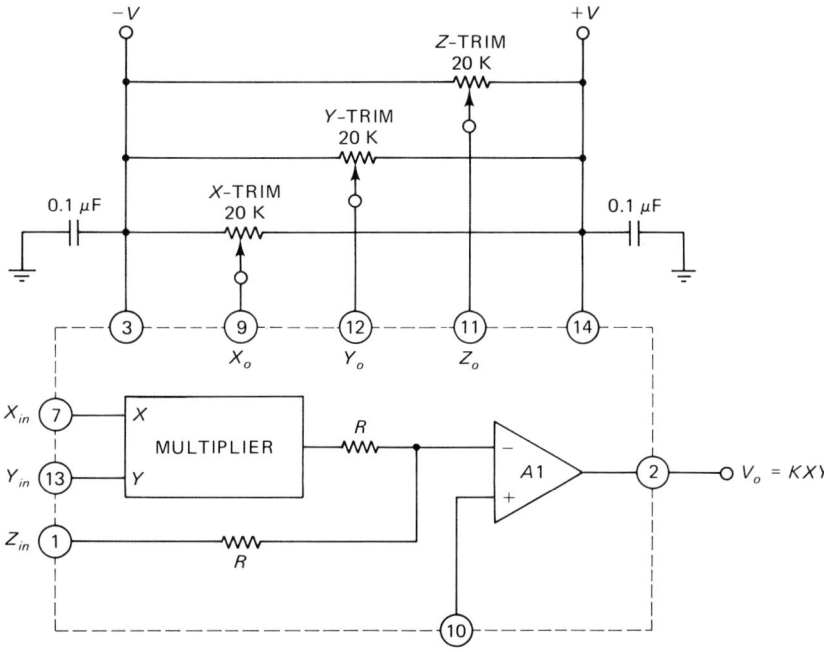

Fig. 19-12 Internal circuitry of the Analog Devices Inc. AD-533 multiplier.

For division, the AD-533 will operate in two quadrants with a transfer function of

$$V_o = 10V_z/V_x \text{ V} \qquad (19\text{-}49)$$

Multiplier Operation of the AD-533

Figure 19-13 shows the connection of the AD-533 for straight multiplier operation. The V_x and V_y inputs are used for the operands, while the V_z input is fed back to the output. A scaling control circuit ($R1$ and $R2$) is used to set the output scale factor. The trimming circuit shown in Fig. 19-12 is also used, but is not shown here for the sake of simplicity. The range of V_x, V_y, and V_o is ± 10 V.

The trim procedure for the AD-533 multiplier is simple; it involves only three external multiturn potentiometers. The procedure is also follows:

1. Set $V_x = V_y = 0$ V, and then adjust Z-TRIM for $V_o = 0$ V.
2. Set $V_x = 0$ V and then apply a signal of 50 Hz at 20 V peak-to-peak; adjust X-TRIM for minimum signal output.
3. Set $V_y = 0$ and then apply a signal of 50 Hz at 20 V peak-to-peak; adjust Y-TRIM for minimum signal output.
4. Readjust Z-TRIM for $V_o = 0$ V DC output.
5. Set $V_x = +10$ V, and set V_y to 50 Hz at 20 V peak-to-peak; adjust $R1$ for $V_o = V_y$.

Rearranging the connection of the AD-533 will render it a divider, squarer, squarerooter, and other function circuits.

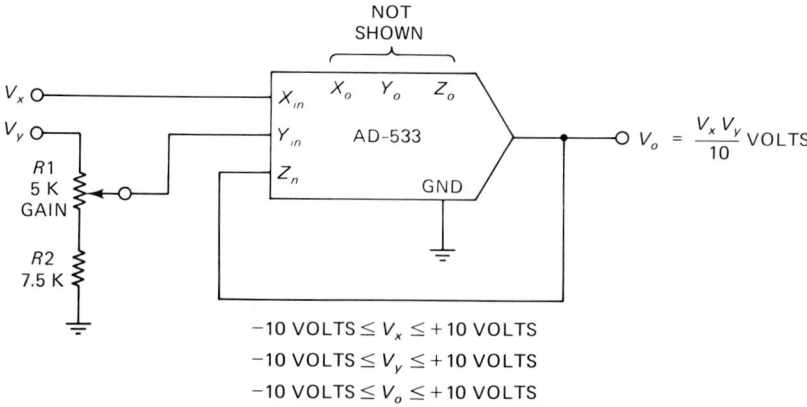

Fig. 19-13 AD-533 connected as a multiplier.

Divider Operation of the AD-533

The AD-533 can also be used as an analog divider circuit with the transfer function:

$$V_o = \frac{10V_z}{V_x} \text{ V} \tag{19-50}$$

The circuit for the analog divider configuration is shown in Fig. 19-14. The X- and Z-inputs are used, with feedback going to the Y-input. As was true in the previous case, the trimming circuits are not shown for sake of simplicity, but are used in real circuits nonetheless. The trim procedure for the analog divider is as follows:

1. Set all trimmer potentiometers at mid-scale.
2. Set $V_z = 0$, adjust Z-TRIM so that V_o remains constant as V_x is varied from -10 V DC through -1 V DC.
3. With $V_z = 0$, $V_x = -10$ V DC, set Y-TRIM for $V_o = 0$ V DC.
4. Set $V_z = V_x$ and then adjust X-TRIM for the minimum variation of V_o as V_x is adjusted from $+10$ to -1 V DC.
5. Repeat steps 2 and 3 for minimum interaction.
6. Set $V_z = V_x$, and then trim the gain control ($R1$) for V_o to have the closest approach to ± 10 V DC output as V_x is varied over the range -10 to -3 V DC.

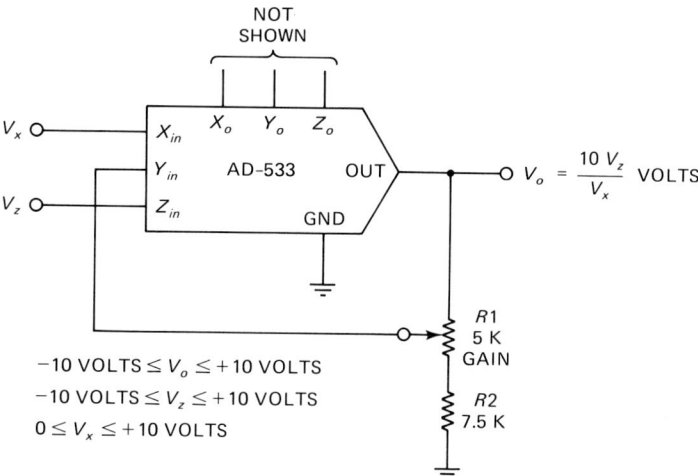

Fig. 19-14 AD-533 connected as a divider.

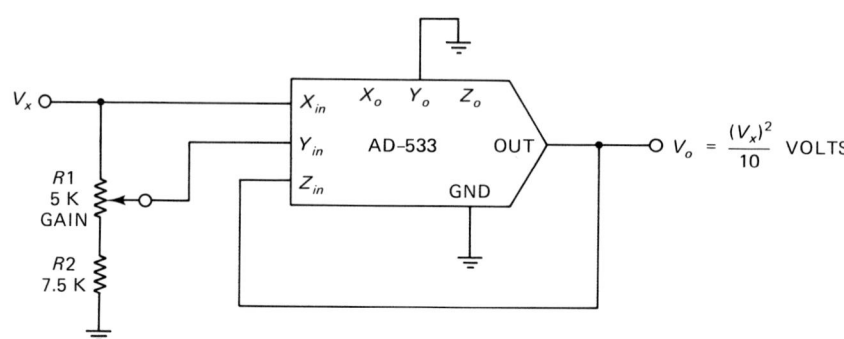

Fig. 19-15 AD-533 connected as a squarer.

Operation of the AD-533 as a Squarer

A squarer is a circuit that outputs a voltage that is the square of the input voltage. Prior to the invention of good analog multiplier circuits, the standard means for squaring a voltage in analog circuits was in a diode breakpoint generator. In those circuits, a series of diodes operating in the square law region of their characteristic are used to fit an output voltage to the square of its input voltage. Although the method worked, it suffered from a severe temperature sensitivity and also from the fact that only a limited number of breakpoints were possible. In the analog multiplier version of a squarer circuit, however, it is possible to simply apply the same voltage to both V_x and V_y inputs: If $V_x = V_y = V$, then $V_x V_y = V^2$.

Figure 19-15 shows a circuit for a squarer based on the AD-533 analog multiplier–divider. In this circuit the Y_o input, which is normally connected to the Y-TRIM potentiometer, is grounded. The X-TRIM and Z-TRIM potentiometers are as shown earlier, but they are deleted from Fig. 19-15 for simplicity. When properly trimmed, the output of the circuit in Fig. 19-15 is

$$V_o = (V_x)^2/10 \qquad (19\text{-}51)$$

The analog squarer circuit will operate in the two quadrants where both of the input voltages are either positive or negative. The output voltage is, of course, only found in the one positive quadrant (QI).

Operation of the AD-533 as a Frequency Doubler

There is a special case of the operation of the squarer that allows frequency doubling. When a sine-wave signal is applied to the input of a squarer, the output frequency will be doubled. The AD-533 makes a nearly ideal frequency doubler because it does not require a tuned input or output network. The tuning circuits used on other analog frequency multipliers makes them single-frequency only. The frequency doubler based on the AD-533 is able to operate on any frequency within its range. Consider the trigonometric identity:

$$\sin(A)\sin(B) = \tfrac{1}{2}[\cos(A - B) - \cos(A + B)] \qquad (19\text{-}52)$$

If A and B each represent the same signal, then we can set $B = A$, and the identity takes the form

$$\sin(A)\sin(A) = \tfrac{1}{2}[\cos(A - A) - \cos(A + A)] \qquad (19\text{-}53)$$

or, collecting terms, and redefining A as $2\pi ft$:

$$\sin(2\pi ft)^2 = \frac{1}{2} - \frac{\cos(4\pi ft)}{2} \qquad (19\text{-}54)$$

Rewriting Eq. (19-54) in the form that accounts for the standard IC multiplier that has a transfer function of

$$V_o = (V_x)^2/10 \qquad (19\text{-}55)$$

and because $V_x = V\,2\pi ft$:

$$V_o = (V\,2\pi ft)^2/20 \qquad (19\text{-}56)$$

Using the identity of Eq. (19-54):

$$V_o = \frac{V^2}{20}[1 - \cos(4\pi ft)] \qquad (19\text{-}57)$$

Operation of the AD-533 as a Phase Detector and Meter

A multiplier driving a low-pass filter (Fig. 19-16) will operate as a phase meter. The output of this circuit is a DC level that is propor-

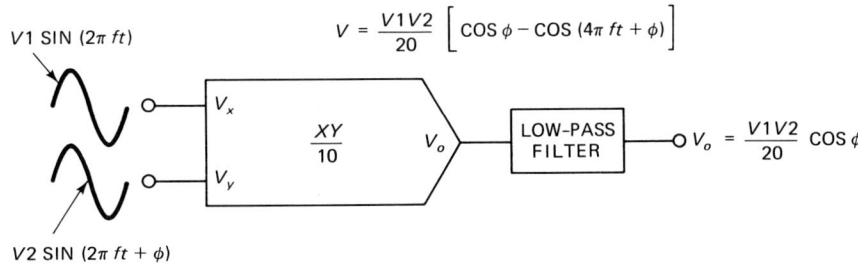

Fig. 19-16 AD-533 as a phase detector.

tional to the difference in phase angle (ϕ) between two different signals of the same frequency. Given a multiplier with a transfer function $V_o = V_x V_y/10$ and low-pass filter that passes only signals with frequencies below the input frequency, only the DC level remains in the output; that DC level is proportional to phase difference. The input signals to the circuit of Fig. 19-16 are

$$V_x = V1 \sin(2\pi ft) \qquad (19\text{-}58)$$

and

$$V_y = V2 \sin(2\pi ft + \phi) \qquad (19\text{-}59)$$

Appealing to the identity of Eq. (19-52):

$$\sin(V_x)\sin(V_y) = \tfrac{1}{2}\left[\cos(V_x - V_y) - \cos(V_x + V_y)\right] \qquad (19\text{-}60)$$

and for a multiplier with the transfer function $V_x V_y/10$, and substituting Eqs. (19-58) and (19-59):

$$V_o = \frac{V1V2}{20}\left[\cos(\phi) - \cos(4\pi ft + \phi)\right] \qquad (19\text{-}61)$$

The expression of Eq. (19-61) represents the output of the analog multiplier. When the low-pass filter removes the AC term, the equation becomes

$$V_o = \frac{V1V2}{20}\cos(\phi) \qquad (19\text{-}62)$$

Solving Eq. (19-62) for the phase angle (ϕ) results in

$$\phi = \cos^{-1}\left(\frac{20V_o}{V1V2}\right) \qquad (19\text{-}63)$$

The phase detector operation described here assumes sinusoidal input signals. The circuit will also work with square waves or pulses; however, in that case the output of the multiplier will be a series of pulses of width proportional to the phase difference. When the pulses are passed through the low-pass filter, the effect is to time-average them and thereby produce a DC output that is zero when the signals are in quadrature, maximum positive at 0 degrees, and maximum negative at 180° phase difference.

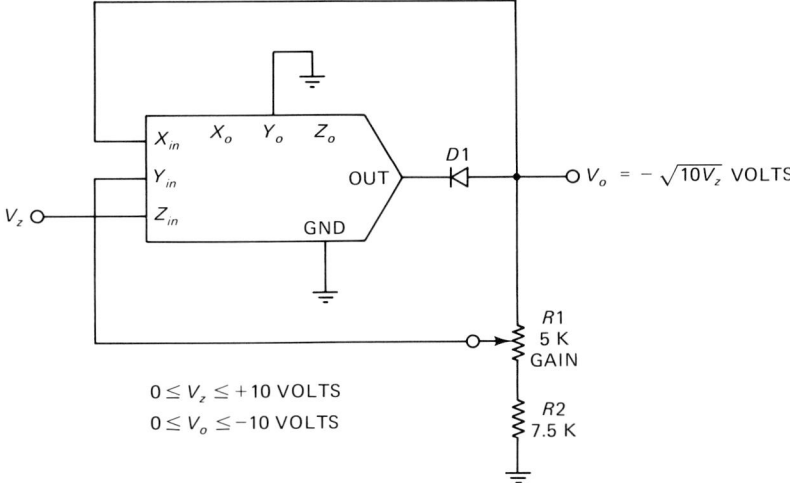

Fig. 19-17 AD-533 as a squarerooter.

Operation of the AD-533 as a Squarerooter

A squarerooter is a circuit that will produce an output voltage that is the square root of the input voltage. Figure 19-17 shows the connection of the AD-533 as a squarerooter. In this circuit, the Y_o input is grounded, as was true with the squarer, but the X-TRIM and Z-TRIM external potentiometers are used (but not shown). The input signal is applied to the Z-input, while the Y- and X-inputs are fed back via a diode (D1).

The circuit of Fig. 19-17 will accommodate input signals of 0 to $+10$ V, but outputs a negative voltage in the range 0 to -10 V. Therefore, the transfer function for this circuit is

$$V_o = -(10V_z)^{1/2} \qquad (19\text{-}64)$$

If a positive output voltage is needed, the output of Fig. 19-17 can be inverted in an operational amplifier circuit.

THE MULTIPLIER AS AN AUTOMATIC GAIN CONTROL

An analog voltage multiplier can be used as an automatic gain control (AGC) if the right external circuitry is added. An AGC is a circuit that is designed to maintain a constant output signal level despite changes in the amplitude of the input signal. Examples of the use of AGC circuits include radio receivers and signal generators. The receiver uses the AGC (also sometimes called automatic volume control or AVC) to maintain a level output signal despite the fact that various radio

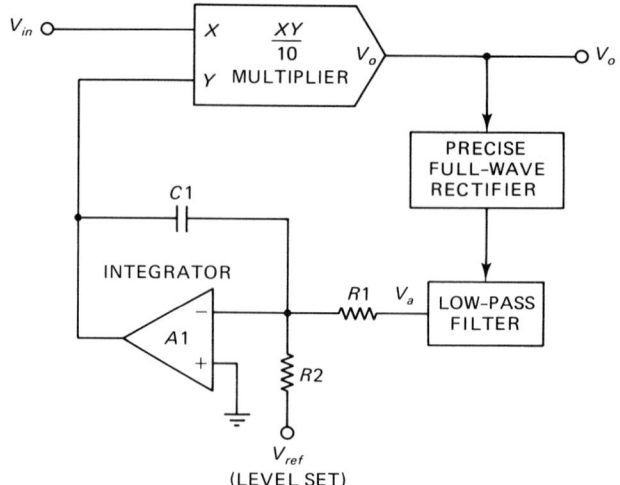

Fig. 19-18 AD-533 as an automatic gain control.

stations being received vary considerably in strength. A signal generator might use the AGC circuit to maintain a level output despite the
fact that variable tuned oscillators tend to output different signal levels
at different frequencies.

An AGC circuit based on an analog multiplier is shown in Fig.
19-18. In this circuit the multiplier acts essentially as a gain-controlled
amplifier. The output of the multiplier is $V_o = V_x V_y / 10$. The input
signal is applied to the X-input and a DC feedback signal is applied to
the Y-input. The feedback signal is formed by sampling the output
signal V_o, rectifying it in a precise full-wave rectifier, and then filtering
the rectifier output to form a DC signal (V_a) that is proportional to the
output signal amplitude. This DC control signal is applied to the input
of a Miller integrator, where it is compared to a reference level (V_{ref})
that determines the set-point of the circuit. The integrator output is
applied to the Y-input of the multiplier. The Y-input of the multiplier
acts as a gain control signal. If an external level-set is needed, the
reference voltage V_{ref} can be made variable.

VOLTAGE-CONTROLLED EXPONENTIATOR CIRCUIT

In Chapter 17, as part of the discussion of logarithmic amplifiers, the
Burr–Brown 4301/4302 multifunction modules were discussed. These
hybrid circuits have a transfer function of the form

$$V_o = \left(\frac{V_y V_z}{V_x} \right)^m \qquad (19\text{-}65)$$

In the transfer equation of the multifunction module in Eq.
(19-65), the exponent m is normally set by an external resistor voltage

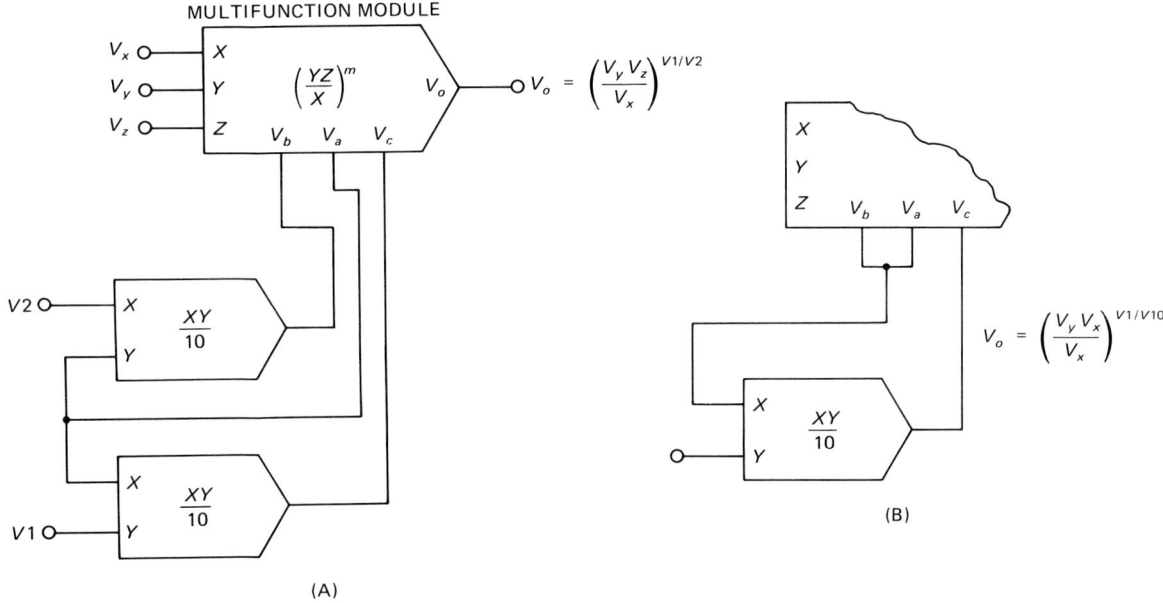

Fig. 19-19 AD-533 as a voltage-controlled exponentiator: (A) differential circuit and (B) single-ended circuit.

divider connected to the V_a, V_b, and V_c inputs (see Fig. 19-19A). If $m = 1$, then the circuit operates as an ordinary multiplier or divider, depending on which inputs are used. For example, when V_x is set to a constant reference of 10 V, the multifunction module ($m = 1$) will operate as an $XY/10$ multiplier. Similarly, when V_z is set to a constant 10 V, the device functions as a divider of the form $10Y/X$. If $m < 1$, the module operates as a rooter (e.g., a squarerooter when $m = 1/2$). If $m > 1$, the module operates as an exponentiator, including a squarer. If m is set by a dynamic external voltage source, however, the multifunction module operates as a voltage-controlled exponentiator.

In Fig. 19-19A the V_b and V_c inputs are driven by the outputs of a pair of $XY/10$ multipliers, which are in turn driven by control voltage $V1$ and $V2$. The V_a input is connected to the alternate inputs of the multipliers. The transfer function of the circuit in Fig. 19-19A is

$$V_o = \left[\frac{V_y V_z}{V_x} \right]^{V1/V2} \tag{19-66}$$

A single control voltage version of the exponentiator circuit is shown in Fig. 19-19B (it is a partial version of Fig. 19-19A). Signals V_a and V_b are connected together and applied to the X-input of an external multiplier. The Y-input of the multiplier becomes the control voltage, forming a transfer function of

$$V_{o} = \left(\frac{V_{y}V_{z}}{V_{x}} \right)^{V1/10} \tag{19-67}$$

Similarly, if the output of the multiplier is connected to the V_{b} input, and V_{a} and V_{c} are connected to the X-input of the multiplier, the transfer function becomes

$$V_{o} = \left(\frac{V_{y}V_{z}}{V_{x}} \right)^{10/V1} \tag{19-68}$$

20

Analog Active Filter Circuits

An electronic frequency selective filter is a circuit that favors some frequencies and discriminates against certain other frequencies (or bands of frequencies). In other words, a filter circuit will pass some favored frequencies and sharply attenuate or reject altogether other frequencies. Frequencies that pass through the filter with little attenuation are said to be in the passband, and frequencies that are heavily attenuated are said to be in the stopband.

Filter circuits can be classified several ways: passive or active, analog or digital or software; by frequency range (e.g., audio, RF, or microwave); or by passband characteristic. Passive filters are made of combinations of passive components such as resistors (R), capacitors (C), and inductors (L). In general, passive filters are lossy and not very flexible. An active filter, on the other hand, is one based on an active device such as a transistor or an operational amplifier along with passive components that determine frequency. In most cases, the passive components are resistors and capacitors (although a few with inductor-based circuits are known).

Active filters use linear circuit techniques such as those found throughout this book. Digital filters use digital IC devices; they are often based on switching techniques. Software filters implement solutions to frequency selective equations using computer programming techniques.

Filters can also be classified by frequency range. Audio filters operate from the sub-audio to ultrasonic ranges (near-DC to about 20 kHz). RF filters operate at frequencies above 20 kHz, up to about 900 MHz. Microwave filters operate at frequencies > 900 MHz. These range designations are not absolute, but they do serve to indicate approximate points at which a change of design techniques usually takes place. For example, filters can be made frequency selective using inductors. But in the audio range the inductance values are typically very large, so the inductors are bulky, costly, and lossy. In addition, inductors also produce stray magnetic fields that can interfere with other nearby circuits. On the other hand, inductors are the elements of choice in the RF region. But once frequencies approach several hundred megahertz, the inductance values required become too low for practical use and other techniques are required. In the microwave and high UHF region transmission-line and cavity techniques are used.

Finally, filters may be classified by their frequency response characteristic. This method of categorizing filters takes note of the filter's passband. In this chapter we will examine low-pass filters, high-pass filters, bandpass filters, and stopband filters. We will also examine a related circuit called the all-pass phase shifter.

Filter Characteristics

Figure 20-1A shows in general terms the characteristics of theoretically ideal filters. These curves will be discussed in greater detail later. A low-pass filter has a passband from DC to a specified cutoff frequency ($F1$). All frequencies above the cutoff frequency are attenuated, so they are in the stopband.

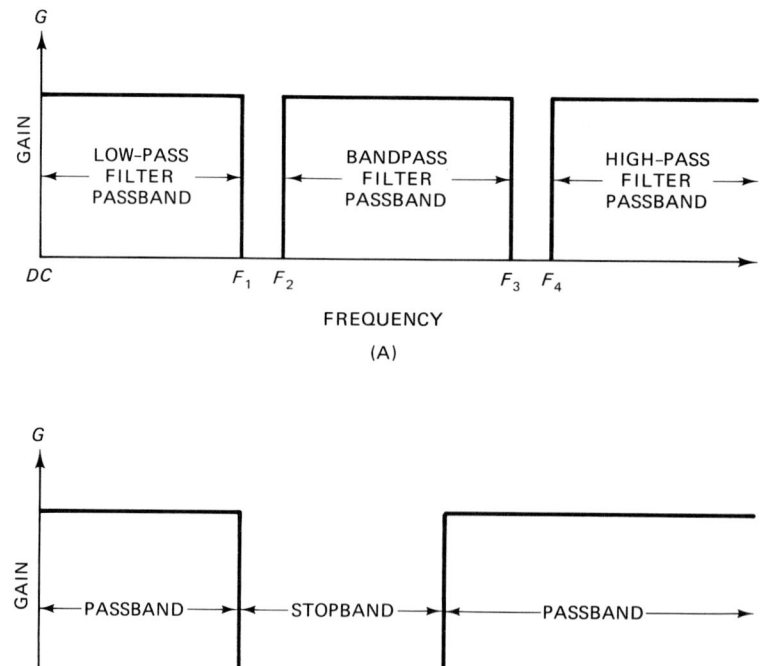

Fig. 20-1 (A) Idealized frequency response for low-pass, bandpass, and high-pass filters; and (B) idealized response for the stopband filter.

A bandpass filter has a passband between a lower limit ($F2$) and an upper limit ($F3$). All frequencies lower than $F2$ or greater than $F3$ are in the stopband. A high-pass filter has a stopband from DC to a certain lower limit ($F4$). All frequencies $> F4$ are in the passband.

A stopband filter response is shown in Fig. 20-1B. This filter severely attenuates frequencies between lower and upper limits ($F5$ to $F6$), but passes all others. When the stopband is very narrow, the stopband filter is called a notch filter. Such filters are often used to remove a single, unwanted frequency. An example of such an application is removal of unwanted 60-Hz interference caused by local power lines.

Ideal versus Practical Filter Response Curves

The response curves shown in Fig. 20-1 are unrealistic ideal generalizations. It might seem that a step-function cutoff is desirable, but in actual practice such a response is neither attainable nor desirable. The reason it is undesirable is that the ideal response actually causes a problem because the filter may ring when a fast rise-time signal is applied. This phenomenon is especially likely to occur on narrow bandpass filters. Although sinusoidal signals do not usually pose a problem, noise impulses or transient step functions can easily cause unwanted ringing.

There are several common filter responses: Butterworth, Chebyshev, Cauer (also sometimes called elliptic), and Bessel. The basic Butterworth response is shown in Fig. 20-2A. The noteworthy properties of the Butterworth filter are that both the passband and stopband are relatively flat, as is the transition region slope between them.

It is standard practice in filter design to specify the passband between points where the response falls off -3 dB. Therefore, for the low-pass filter shown in Fig. 20-2A the cutoff frequency is at the point where gain ($A1$) falls off to 0.707 times the low-frequency gain ($A2$) (i.e., the -3 dB point).

At frequencies $f > f_c$, the gain falls off linearly at a rate that depends on the order of the filter. The slope (S) of the fall-off is measured in either decibels per octave (a $2:1$ frequency change) or decibels per decade (a $10:1$ frequency change). Note that these two systems can be scaled: -6 dB/octave is the same slope as -20 dB/decade. The slopes shown in Fig. 20-2A cover three Butterworth cases. A first-order filter offers a roll-off of -20 dB/decade, a second order filter offers a roll-off of -40 dB/decade, and a third-order filter rolls at -60 dB/decade. These correspond to 6, 12, and 18 dB/octave, respectively.

On first glance, it might appear that only third-order filters would be used because their transition from passband to stopband is more rapid. But higher order response is obtained at the cost of more complexity, greater component-value error-sensitivity, and more difficult design procedures. Some higher order filter designs are also more

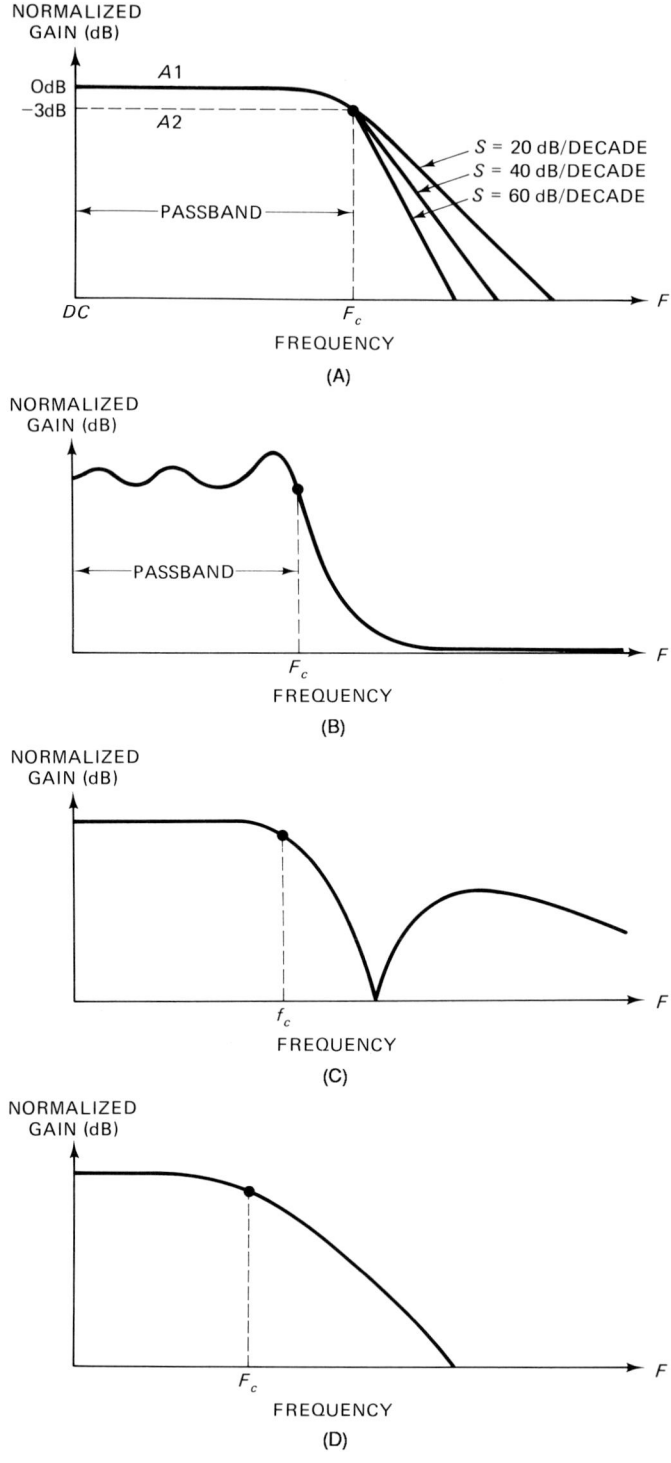

Fig. 20-2 (A) Low-pass filter Butterworth response for three different orders; (B) Chebyshev response, (C) Cauer response, and (D) Bessel response.

likely to oscillate than lower order equivalents. The selection of filter "order" is a trade-off between system requirements and complexity. If a large, undesired signal is expected at a frequency up to, say, the third harmonic of f_c, then a higher order response might be indicated. Conversely, weaker stopband signals in that region may permit the use of a lower-order filter.

The steepness and shape of the roll-off curve is a function of the filter damping factor. As a class, Butterworth filters tend to be heavily damped, which explains the gradual roll-off in the response curve. The Chebyshev filter response (Fig. 20-2B) is lightly damped, so has a variation (or ripple) within the passband. The Chebyshev filter offers a generally faster roll-off than the Butterworth filter, but at the cost of less flatness within the passband.

The Cauer or elliptic filter response curve (Fig. 20-2C) offers the fastest roll-off for frequencies close to the cutoff frequency, as well as relatively good flatness within the passband. Notches of -40 to -60 dB can be achieved close to f_c, but only at a cost of less attenuation further into the stopband. A typical Cauer filter has a deep notch close to f_c, rises to a peak at some high frequency removed from f_c, and then gradually falls off at even higher frequencies at a rate of -20 to -40 dB/decade.

The response of the Bessel filter is shown in Fig. 20-2D. Although it appears similar to the Butterworth response, it is not maximally flat within the passband, and falls off somewhat less rapidly. The benefit of the Bessel filter is a flat phase response across the passband. In some applications the phase response is the overriding consideration, so the Bessel response is preferred.

Filter Phase Response

Most frequency selective filter circuits exhibit a phase change over the passband. The responses for two different filters, Butterworth and Bessel, are shown in Fig. 20-3. Note that the maximally flat Butterworth exhibits a decidedly nonlinear phase response in both passband and stopband. The frequency-dependent phase shift of a low-pass filter is $-45°$ at f_c, and increases by a factor of $-45°$ for each additional increase of -20 dB/decade in the roll-off slope. Thus, a first-order low-pass filter phase shift is $-45°$, the second-order phase shift (for -40 dB/decade roll-off) is $-90°$, and for a third-order filter (-60 dB/decade) it is $-135°$. The high-pass response is of the same magnitude for each order of filter, but the sign is opposite. Thus, a first-order high-pass filter shows a $+45°$ phase shift, a second-order filter shows $+90°$, and a third-order high-pass filter shows $+135°$.

The Bessel filter also shows a phase shift over the passband, but it is nearly linear. A useful feature of this characteristic is that it allows a uniform time delay all across the passband. As a result, the Bessel filter

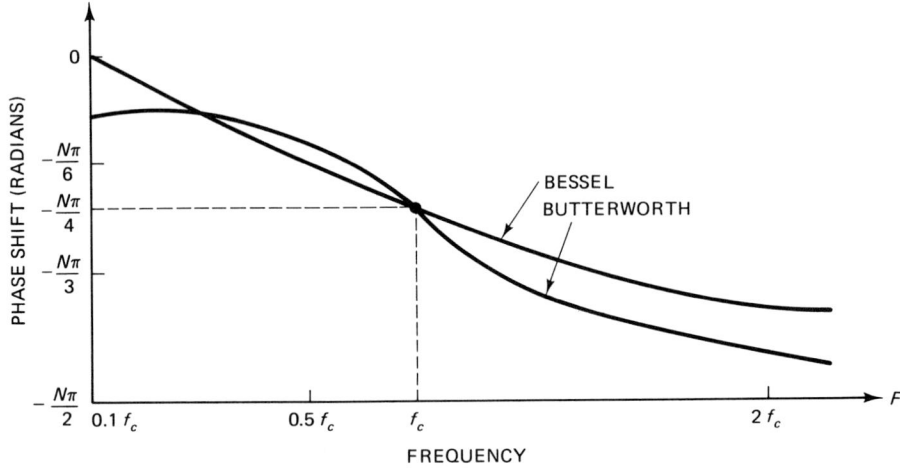

Fig. 20-3 Phase response for Butterworth and Bessel filters.

offers the ability to pass transient pulse waveforms with minimum distortion. For the Bessel filter, the phase shift maxima occurs at

$$\Delta\phi_{max} = -n\pi/2 \qquad (20\text{-}1)$$

where $\Delta\phi_{max}$ is the maximum phase shift and n is the order of the filter (i.e., number of poles).

In a properly designed Bessel filter, the cutoff frequency f_c occurs at a point where the phase shift is half the maximum phase shift, or

$$\Delta\phi_{fc} = -n\pi/4 \qquad (20\text{-}2)$$

The Bessel filter is said to work best at the frequency where $f = f_c/2$.

LOW-PASS FILTERS

In this section and the sections to follow, practical design of active filters will be discussed. The model for the filter is shown in Fig. 20-4. This filter is called the voltage-controlled voltage source (VCVS) filter of Sallen–Key filter. The basic configuration is a noninverting follower operational amplifier ($A1$). The op-amp selected should have a high-gain bandwidth product, relative to the cutoff frequency, to permit the filter to operate properly. The gain of the circuit is given by

$$A_v = \frac{R_f}{R_{in}} + 1 \qquad (20\text{-}3)$$

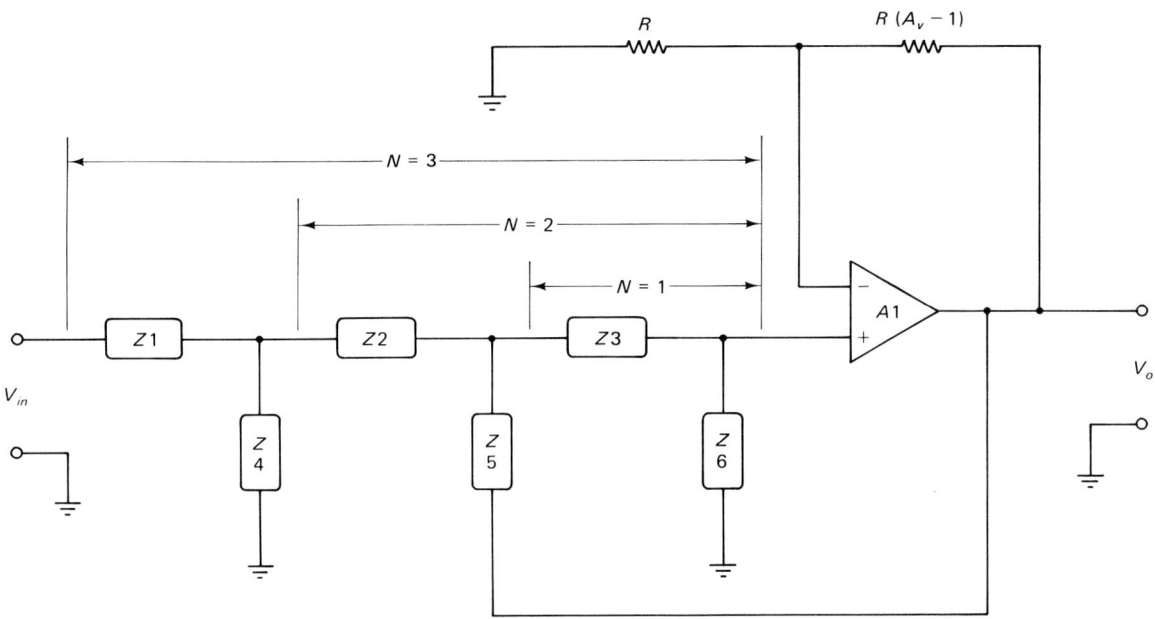

Fig. 20-4 Circuit model for the VCVS filter.

So, if $R_{in} = R$, we may deduce:

$$R_f = R(A_v - 1) \qquad (20\text{-}4)$$

In some other circuits, the gain may be unity. In those circuits the resistor voltage-divider feedback network is replaced with a single connection between output and the inverting input.

The input circuitry of the generic VCVS filter consists of a network of impedances labeled $Z1$ through $Z6$. Each of these blocks will be either a resistance (R) or a complex capacitive reactance $(-jX_c)$. Which element becomes which type of component is determined by whether the filter is a low-pass or a high-pass type.

The order of the filter, denoted by n, refers to the number of poles in the design or, in practical terms, the number of RC sections. A first-order filter $(n = 1)$ consists of $Z3$ and $Z6$, a second-order filter $(n = 2)$ consists of $Z2$, $Z3$, $Z5$, and $Z6$, a third-order filter $(n = 3)$ consists of all six impedance $(Z1 - Z6)$. Higher order filters $(n > 3)$ can also be built, but they are not discussed here.

In a low-pass filter $Z1$ through $Z3$ are resistances, while $Z4$ through $Z6$ are capacitances. The component roles are reversed in high-pass filters. Now that we have laid a foundation, let us review the properties of the low-pass filter and then learn to design first, second, and third-order filters.

By way of review, Fig. 20-5 shows the low-pass Butterworth filter response curve. This type of filter is maximally flat within the pass-band, and passes all frequencies below a certain critical frequency (F_c).

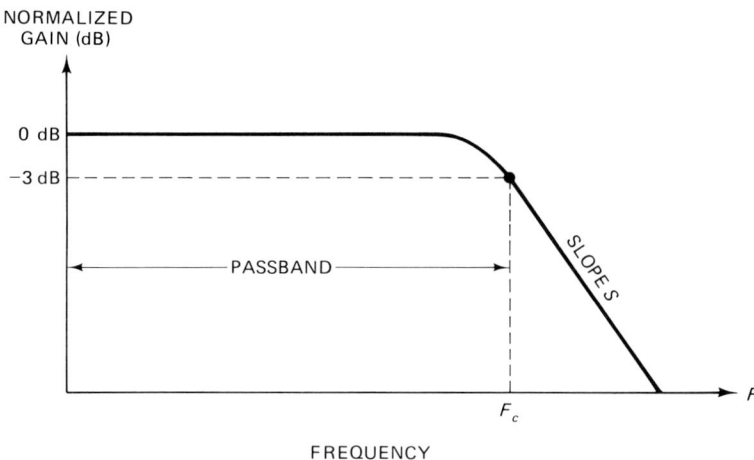

Fig. 20-5   Butterworth VCVS low-pass filter response.

The breakpoint between the passband and the stopband is the point at which the gain of the circuit has dropped off -3 dB from its lower frequency value. Above the critical frequency the gain falls off at a certain slope. The steepness of the slope is usually given in terms of decibels (dB) of gain per octave of frequency (an octave is a $2:1$ change in frequency); alternatively, dB/decade (a decade is a $10:1$ change in frequency) is sometimes used.

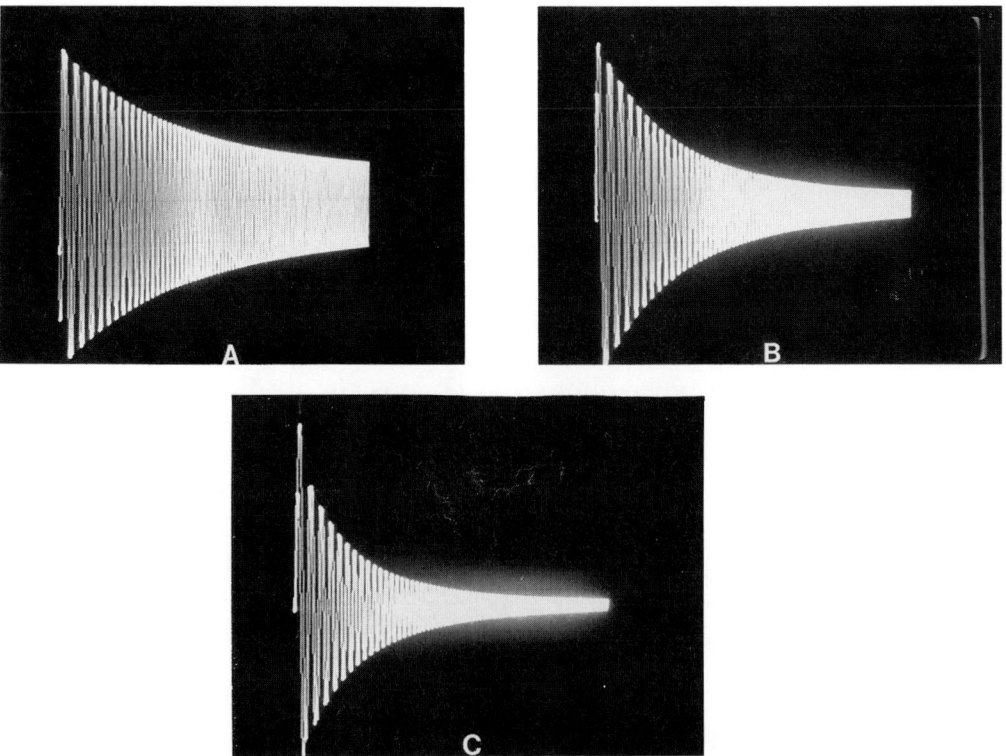

Fig. 20-6   Frequency response for three different orders of low-pass filters: (A) first-order, (B) second-order, and (C) third-order.

Frequency sweeps of a set of low-pass filters are shown in Fig. 20-6. These oscilloscope traces were created using a sweep generator that varies a sine-wave signal from 1 to 10 kHz. The -3 dB frequency of the filters was 3 kHz. The Y-input of the oscilloscope displays the output amplitude of the filter, while the X-input was externally swept using the same sawtooth that was used to sweep the signal generator. Thus, the X-Y oscilloscope trace is a plot of the frequency response of the filter under test. Figure 20-6A shows the response of a first-order filter in the area from just below f_c up to 10 kHz. The response of the second-order filter is shown in Fig. 20-6B and that of the third-order filter in Fig. 20-6C. All three traces were taken under the same conditions, so one can see how the attenuation of frequencies greater than f_c is greater in the higher-order filters.

First-Order Low-Pass Filters (-20 dB/Decade)

The first-order low-pass filter is shown in Fig. 20-7A, and its response curve is shown in Fig. 20-7B. The filter consists of a single-section RC low-pass filter driving the noninverting input of an operational amplifier. The gain of the op-amp is $[(R2/R3) + 1]$. The high input impedance of $A1$ prevents loading of the RC network. The general form of the transfer equation for the amplitude versus frequency response for the first-order filter is

$$A_{dB} = 20 \log(A_v) - 20 \log\left[\left(1 + (\alpha_o)^2\right)\right]^{1/2} \qquad (20\text{-}5)$$

where A_{dB} is the gain of the circuit in decibels, A_v the voltage gain within the passband, log denotes the base-10 logarithms, and α_o is the ratio of the input frequency to the cutoff frequency ($f_o = f/f_c$).

The voltage at the output of the RC network (V_a) is found from the voltage divider equation:

$$V_a = \frac{-jX_c V_{in}}{R - jX_c} \qquad (20\text{-}6)$$

where $-jX_c = 1/j2\pi fC$ and j is the imaginary operator ($\sqrt{-1}$).

Substituting the value for $-jX_c$:

$$V_a = \frac{\dfrac{V_{in}}{j2\pi fC}}{R + \dfrac{1}{j2\pi fC}} \qquad (20\text{-}7)$$

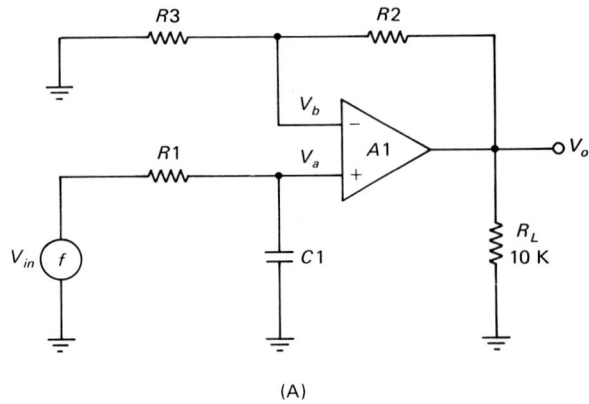

(A)

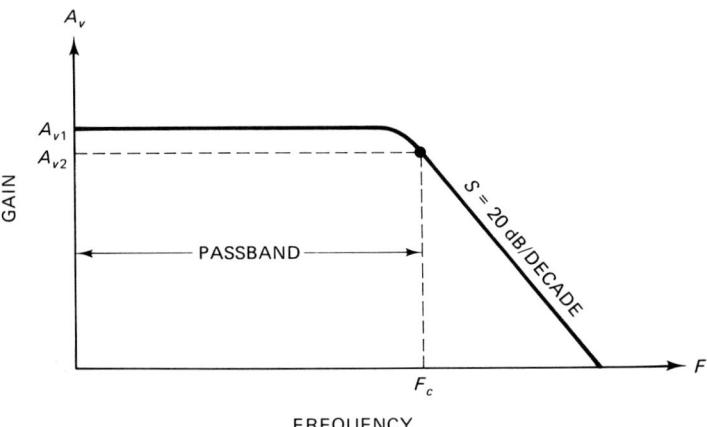

$$A_{v1} = \frac{R2}{R3} + 1$$

$$A_{v2} = A_{v1} - 3\,dB = 0.707\,A_{v1}$$

(B)

Fig. 20-7   (A) First-order low-pass VCVS filter and (B) response curve for the first-order filter.

which simplifies to

$$V_a = \frac{V_{in}}{1 + 2\pi fCR} \tag{20-8}$$

If the transfer function of the noninverting follower is

$$V_o = V_{in}\left(\frac{R2}{R3} + 1\right) \tag{20-9}$$

and since $V_{in} = V_a$ [Eq. (20-7)],

$$V_o = \left(\frac{V_{in}}{1 + 2\pi fCR}\right)\left(\frac{R2}{R3} + 1\right) \tag{20-10}$$

Equation (20-9) can be put into a more generic transfer equation of the form

$$\frac{V_o}{V_{in}} = \frac{A_v}{1 + j(f/f_c)} \tag{20-11}$$

where v_o is the output signal voltage, V_{in} the input signal voltage, A_v the passband gain $[(R2/R3) + 1]$ [see Eq. (20-8)], f is the signal frequency, and f_c is the -3 dB frequency $(1/2\pi RC)$.

The filter parameters are required to define the operation of any particular circuit. The gain magnitude and phase shift are found from the following equations:

Gain magnitude

$$\left|\frac{V_o}{V_{in}}\right| = \frac{A_v}{\left[1 + (f/f_c)^2\right]^{1/2}} \tag{20-12}$$

and phase shift angle (in radians),

$$\phi = -\tan^{-1}(f/f_c) \tag{20-13}$$

Because the filter characterization depends in part on the ratio f/f_c, the equations take different forms at different values of f and f_c. These can be reduced to

At low frequencies well within the passband $(f < f_c)$

$$\left|\frac{V_o}{V_{in}}\right| = A_v = \frac{R2}{R3} + 1 \tag{20-14}$$

At the -3 dB cutoff frequency $(f = f_c)$

$$\left|\frac{V_o}{V_{in}}\right| = 0.707A_v \tag{20-15}$$

At a high frequency well above the -3 dB frequency $(f > f_c)$

$$\left|\frac{V_o}{V_{in}}\right| < A_v \tag{20-16}$$

Table 20-1 shows the characteristics of first-order filters at several different ratios of f/f_c.

Table 20-1
Filter Characteristics

Applied	Gain magnitude		Phase shift
10	2.00	6.02	-0.57
20	1.999	6.018	-1.15
50	1.998	6.009	-2.86
80	1.994	5.993	-4.57
100	1.990	5.977	-5.71
200	1.961	5.850	-11.31
500	1.789	5.052	-26.57
800	1.561	3.872	-38.66
1000	1.414	3.010	-45
2000	0.894	-0.969	-63.43
5000	0.392	-8.129	-78.69
8000	0.248	-12.109	-82.87
10000	0.199	-14.023	-84.29
20000	0.0998	-20.011	-87.14
50000	0.0399	-27.960	-88.85
80000	0.0249	-32.042	-89.28
100000	0.0199	-33.980	-89.43

Design Procedure for a First-Order Low-Pass Filter

There are two basic ways to design a low-pass filter: ground-up and frequency scaling. In this section the ground-up method is discussed.

Procedure

1. Select the -3 dB cutoff frequency (f_c) from consideration of the circuit requirements and applications.
2. Select a standard value capacitance ($C \leq 1 \mu F$).
3. Calculate the required resistance from

$$R1 = \frac{1}{2\pi f_c C}$$

4. Select the passband gain for $f < f_c$.
5. Select a value for resistor R.
6. Calculate R_f from

$$R_f = R(A_v - 1)$$

Design by Normalizing Model

The filter design can be simplified by using a normalized model. First, design a filter for a standardized frequency (e.g., 1 Hz, 10 Hz, 100 Hz, 1000 Hz, or 10,000 Hz) and list the component values. The values

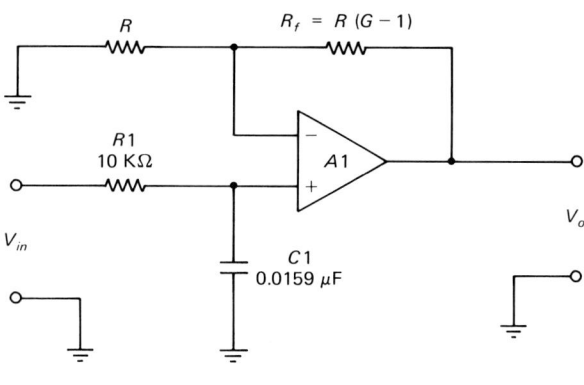

F_c NORMALIZED TO 1 KHz

Fig. 20-8 1-kHz normalized VCVS low-pass filter.

for any other frequency can then be computed by a simple ratio and proportion.

An example of a first-order low-pass Butterworth filter is shown in Fig. 20-8. The component values shown in Fig. 20-8 are normalized for 1 kHz. The actual required component values ($R1'$ and $C1'$) are found by dividing the normalized values shown by the desired cutoff frequency in kilohertz.

$$C1' = \frac{(C1)(1\ \text{kHz})}{F} \tag{20-17}$$

or

$$R1' = \frac{(R1)(1\ \text{kHz})}{F} \tag{20-18}$$

Leave one of the values alone, and calculate the other. In general, it is easier to obtain precision resistors in unusual values, so it is common practice to select a standard capacitance and calculate the new resistance.

A generalized form of the equations is

$$A' = A(f_c/f) \tag{20-19}$$

where A' is the new component value for either $C1$ or $R1$, A the original component value for either $C1$ or $R1$, f_c the filter -3 dB cutoff frequency, and f the new design frequency.

Second-Order Low-Pass Filters (-40 dB/Decade)

The circuit for a second-order low-pass filter is shown in Fig. 20-9A, while the response curve is shown in Fig. 20-9B. Note that this circuit

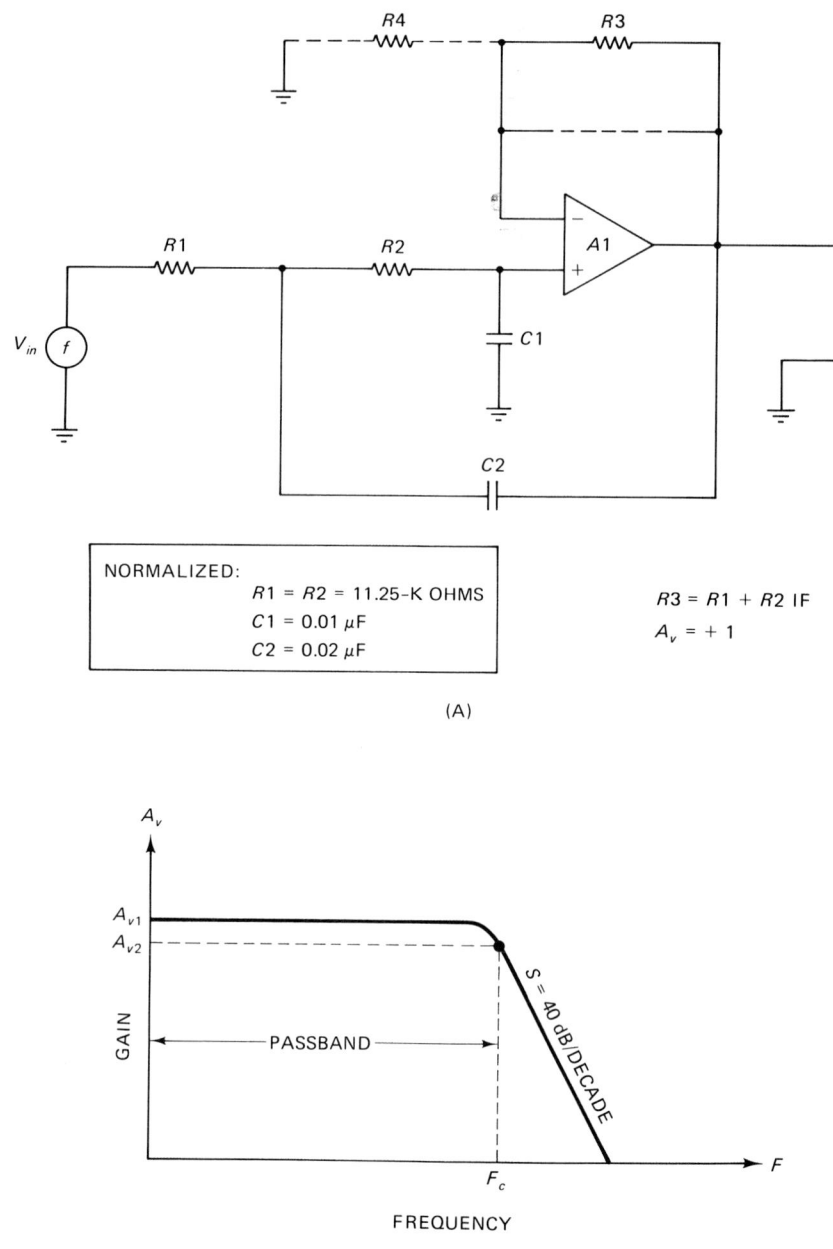

Fig. 20-9 (A) Basic second-order VCVS filter and (B) response curve.

is similar to the first-order filter, but with an additional RC network in the frequency-selective portion of the circuit.

The particular version of this circuit shown in Fig. 20-9A is connected in the unity-gain configuration. The purpose of $R3$ is to help counteract the DC offset at the output of the operational amplifier that is created by input bias currents charging the capacitors in the frequency-selective network. The value of $R3$ in the unity-gain situation is $2R$, where R is the value of the resistors in the frequency selective network. In cases where DC offset is not a problem, resistor $R3$ can be replaced with a short circuit between the op-amp output and the inverting input. If passband gain is required, resistors $R3$ and $R4$ are used.

The second-order VCVS filter is by far the most commonly used type. It is -40 dB/decade roll-off, coupled with a high degree of stability, results in a generally good trade-off between performance and complexity.

The general form of the second-order filter transfer equation is similar to the expression for the first-order filter:

$$A_{dB} = 20 \log(A_v)$$

$$- 20 \log\left[(\alpha_o)^4 + (a^2 - 2)(\alpha_o)^2 + 1\right]^{1/2} \quad (20\text{-}20)$$

where a is the damping factor of the circuit and the other terms are as defined earlier for the first-order case.

The damping factor term (a) is determined by the form of filter circuit. For the Butterworth design, which is used in most of the examples of filters in this chapter, the value of $a = [2]^{1/2}$, or 1.414.

The passband gain for this circuit is the normal gain for any noninverting follower/amplifier. If the output is strapped directly to the inverting input, or if $R3$ (but not $R4$) is used in the feedback network, then the gain is unity ($A_v = +1$). For gains greater than unity ($A_v > 1$), the following is true:

$$A_v = (R3/R4) + 1 \quad (20\text{-}21)$$

The cutoff frequency (f_c) is the frequency at which the voltage gain drops -3 dB from the passband gain. This gain is found from

$$A_v = \frac{1}{2\pi(R1\,R2\,C1\,C2)^{1/2}} \quad (20\text{-}22)$$

The gain magnitude $\text{ABS}(V_o/V_{in})$ is found in a manner similar to the first-order case.

$$\left|\frac{V_o}{V_{in}}\right| = \frac{A_v}{\left[1 + (f//f_c)^4\right]^{1/2}} \quad (20\text{-}23)$$

There is no requirement in VCVS filters that like components (R or C) in the frequency selective network be made equal, but such a step simplifies the design procedure. It does, however, make the design other than Butterworth. If $R1 = R2 = R$, and $C1 = C2 = C$, then:

$$f_c = \frac{1}{2\pi RC} \qquad (20\text{-}24)$$

A constraint on this simplification is that the response is guaranteed only if $A_v \leq 1.586$.

Design Procedure

1. Select the -3 dB cutoff frequency (f_c) from consideration of the circuit requirements and applications.
2. Select a standard value capacitance (30 pF $\leq$ C $\leq$ 1 μF).
3. Calculate the required resistance from

$$R1 = \frac{1}{2\pi f_c C} \qquad (20\text{-}25)$$

4. Select the passband gain for $f < f_c$.
5. Select a value for resistor $R4$.
6. Calculate $R3$ from

$$R3 = R4(A_v - 1) \qquad (20\text{-}26)$$

The normalized 1 kHz trial values for doing scaling design of the second-order low-pass filter are shown in Fig. 20-9 as an inset. The design here is based on a more complex arrangement whereby $C2 = 2C1$. Some authorities maintain that this is the superior design. The same scaling rule is applied to the second-order filter as was used in the first-order filter.

Third-Order Low-Pass Filters (-60 dB/Octave)

A third-order filter has a frequency roll-off slope of -60 dB/decade, or -18 dB/octave. There are two main forms of third-order filter. One type is similar to the first and second-order filters, except for an extra low-pass RC filter in the frequency selective network.

The other method is to cascade first and second-order filters. Figure 20-10A shows the circuit for the former type, along with the response curve shown in Fig. 20-10B. This circuit is the normalized version. In past examples, filters were normalized using frequency f_c as the determining factor, so in this example we will use the radian form in which frequency ω_c is the cutoff frequency in radians per second and is equal to $2\pi f_c$. The filter circuit shown in Fig. 20-10A is normalized to one radian per second.

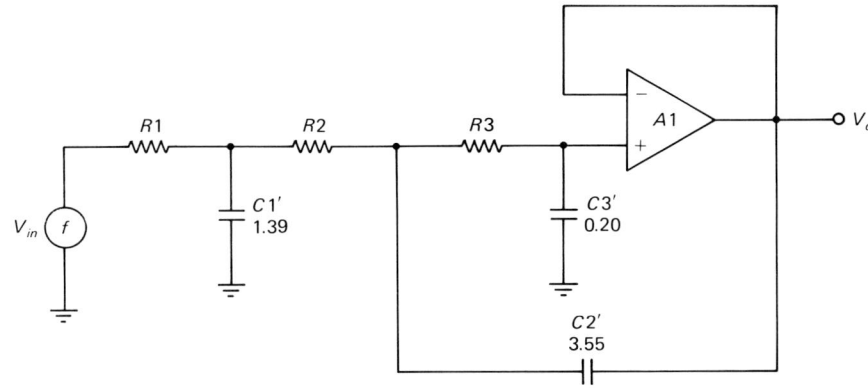

NORMALIZED TO 1 RADIAN/SECOND (1 Hz)
R1 = R2 = R3 = R

(A)

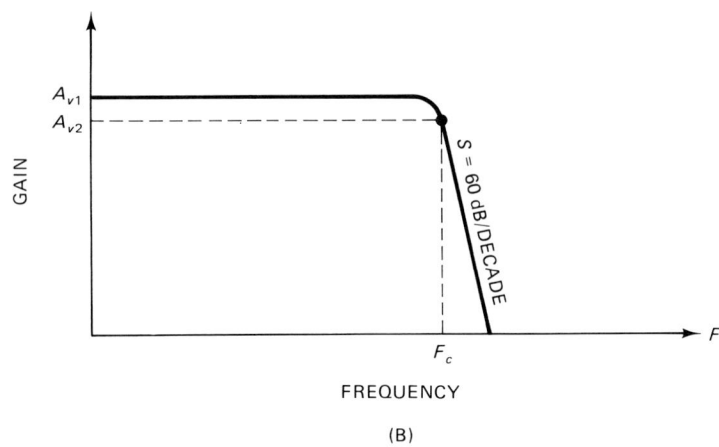

(B)

Fig. 20-10 (A) Third-order VCVS low-pass filter circuit and (B) response curve.

The values assigned to the capacitor values in Fig. 20-10A are parametric values; they are used in a two-step process for finding the actual final values $C1$, $C2$, and $C3$. The parameters are $C1' = 1.39$, $C2' = 3.55$, and $C3 = 0.20$ (for Butterworth filters).

A second method for designing a third-order low-pass filter is to cascade first-order and second-order filters of the same cutoff frequency (Fig. 20-11). The roll-off of the first-order filter is -20 dB/decade, and for the second-order filter it is -40 dB/decade. The combined roll-off is $-(20 + 40) = -60$ dB/decade. The components in this circuit have the following relationships: $R1 = R2 = R3 = R$.

$$R = \frac{1{,}000{,}000}{2\pi f_\text{c} C3_{\mu\text{F}}} \qquad (20\text{-}27)$$

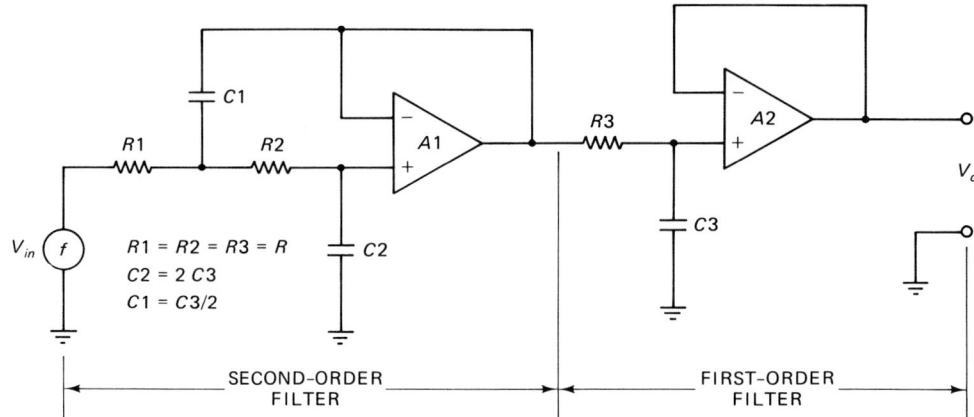

Fig. 20-11   Cascaded LPF and HPF filters form a wide bandpass filter.

$C3$ is selected for convenient value.

$$C1 = C3/2 \tag{20-28}$$

$$C2 = 2C3 \tag{20-29}$$

The design procedure calls for selecting a reasonable trial value for $C3$ and then calculating $C1$ and $C2$. If these values are not standards, set another trial value of $C3$ and try again. Repeat the procedure until a good set of values is obtained. Finally, calculate R.

HIGH-PASS FILTERS

The high-pass filter is the inverse of the low-pass filter, so one can reasonably expect its frequency response characteristic to mirror that of the low-pass filter response. Figure 20-12 shows the basic high-pass filter response with roll-off slopes of -20, -40, and -60 dB/decade. The passband of the high-pass filter are all frequencies above the cutoff frequency f_c. As in the low-pass case, f_c is the frequency at which passband gain drops -3 dB; that is, $A_{vc} = 0.707A_v$.

The cutoff frequency phase shift in a high-pass filter has the same magnitude as the low-pass case, but the sign is opposite. At f_c, the high-pass filter exhibits a phase shift of $+45°$ per 20 dB/decade of roll-off. Put another way, the phase shift is $n \times 45°$, where n is the order of the filter.

The high-pass versions of the VCVS filters are of the same form as the low-pass filter. That form was laid out in Fig. 20-4 earlier. In the case of the high-pass filter, however, impedances $Z1$ through $Z3$ are capacitances, while $Z4$ through $Z6$ are resistances. In the high-pass filter the roles of the resistors and capacitors are reversed.

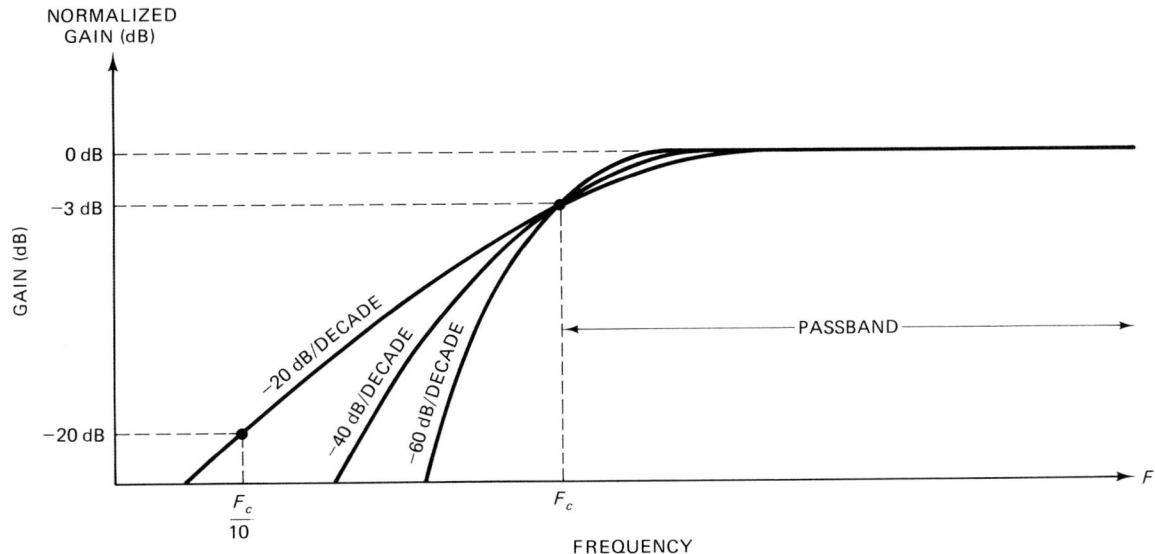

Fig. 20-12 High-pass filter Butterworth responses.

First-Order High-Pass VCVS Filters
(−20 dB/Decade)

The circuit for a first-order high-pass filter is shown in Fig. 20-13. This circuit is identical to the first-order low-pass filter in which the roles of $C1$ and $R1$ are interchanged. The filter shown here is normalized for 1 kHz. Passband gain of this circuit is

$$A_v = (R_f/R_{in}) + 1 \qquad (20\text{-}30)$$

The voltage at the noninverting input of the operational amplifier (V_a) is developed across resistor $R1$ and is given by

$$V_a = \frac{j2\pi f R1 C1 V_{in}}{1 + j2\pi f R1 C1} \qquad (20\text{-}31)$$

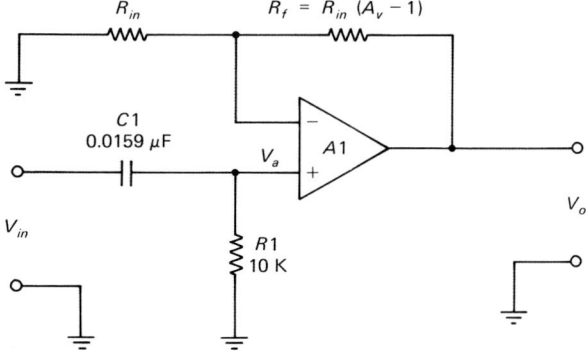

Fig. 20-13 First-order high-pass VCVS filter.

The transfer equation for the circuit is

$$V_o = A_v V_a \tag{20-32}$$

$$V_o = \left(\frac{R_f}{R_{in}} + 1 \right)\left(\frac{j2\pi fR1C1V_{in}}{1 + j2\pi fR1C1} \right) \tag{20-33}$$

And, in the traditional form, the equation becomes

$$\frac{V_o}{V_{in}} = \frac{A_v j(f/f_c)}{1 + j(f/f_c)} \tag{20-34}$$

As in the previous cases, the cutoff frequency f_c is found from

$$f_c = \frac{1}{2\pi R1C1} \tag{20-35}$$

The gain magnitude of this circuit is the absolute value of the traditional form of the transfer equation.

$$\left| \frac{V_o}{V_{in}} \right| = \frac{A_v(f/f_c)}{\left[1 + (f/f_c)^2 \right]^{1/2}} \tag{20-36}$$

The VCVS high-pass filter shown in Fig. 20-12 is normalized to 1 kHz. The same scaling technique is used for this circuit as was used for the low-pass filters discussed earlier.

Second-Order High-Pass Filter (−40 dB/Decade)

The second-order high-pass filter offers a roll-off slope of −40 dB/ decade. This VCVS filter circuit (Fig. 20-14) is, like its low-pass counterpart, probably the most commonly used form of high-pass

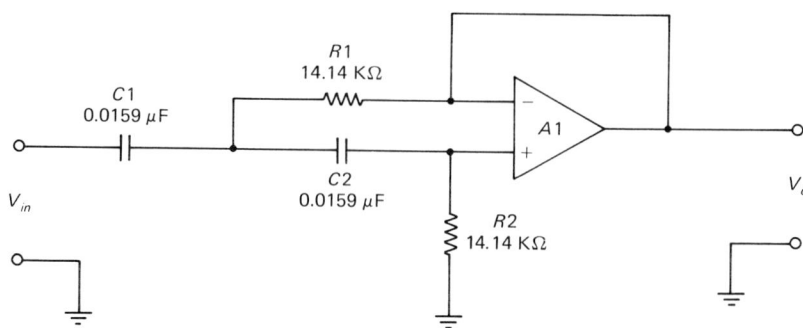

Fig. 20-14 Second-order high-pass VCVS filter.

filter. The circuit is similar to the low-pass design except for a reversal of the roles of capacitors and resistors. The cutoff frequency is the frequency at which gain falls off -3 dB; it is found from

$$f_c = \frac{1}{2\pi[R1R2C1C2]^{1/2}} \tag{20-37}$$

or, in the case where $R1 = R2 = R$ and $C1 = C2 = C$,

$$f_c = \frac{1}{2\pi RC} \tag{20-38}$$

The gain magnitude of the circuit is found from

$$\left|\frac{V_o}{V_{in}}\right| = \frac{A_v}{\left[1 + (f_c/f)^4\right]^{1/2}} \tag{20-39}$$

Multiple Feedback Path Filters

A different form of active filter, the multiple feedback path (MFP) circuit, is shown in Fig. 20-15. The low-pass version is shown in Fig. 20-15A and the high-pass in Fig. 20-15B. The values are normalized for 1 kHz, and we find the actual values in the manner previously described. Change either the capacitor or the resistor values, but not both.

BANDPASS FILTERS

The bandpass filter is a circuit that has a passband between an upper limit and a lower limit. Frequencies above and below these limits are in the stopband. There are two basic forms of bandpass filter: wide bandpass and narrow bandpass. These two types are sufficiently different that they offer different responses. The wide bandpass filter may have a passband that is wide enough to be called a bandpass amplifier rather than a filter. The wide bandpass filter response is shown in Fig. 20-16A, while the narrow response is shown in Fig. 20-16B. The passband is defined as the frequency difference between the upper -3 dB point ($F2$), and the lower -3 dB point ($F1$). The bandwidth BW is

$$\text{BW} = F2 - F1 \tag{20-40}$$

The center frequency f_c of the bandpass filter is usually symmetrically placed between $F1$ and $F2$, or $(F2 - F1)/2$. If the filter is a

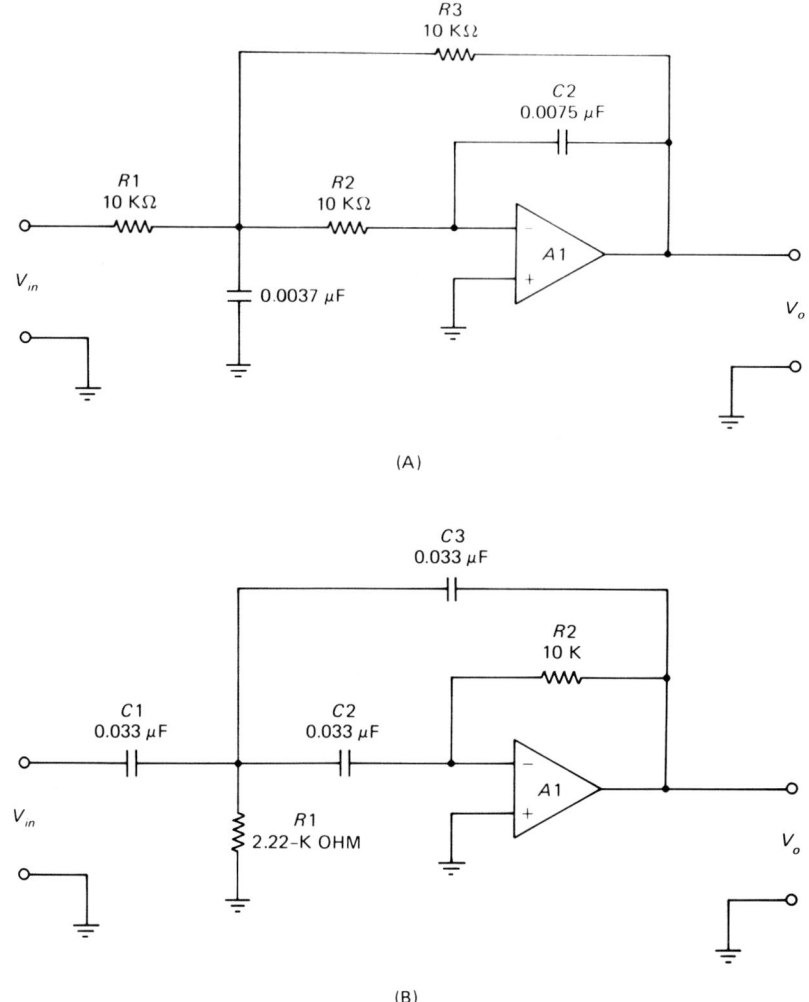

Fig. 20-15 Multiple feedback-path bandpass filters: (A) low-pass and (B) high-pass.

very wide band type, however, the center frequency is

$$f_c = [F1F2]^{1/2} \tag{20-41}$$

Bandpass filters are sometimes characterized by the *figure of merit*, or Q. The Q is a factor that describes the sharpness of the filter; it is found from

$$Q = \frac{f_c}{BW} \tag{20-42}$$

$$Q = \frac{f_c}{F2 - F1} \tag{20-43}$$

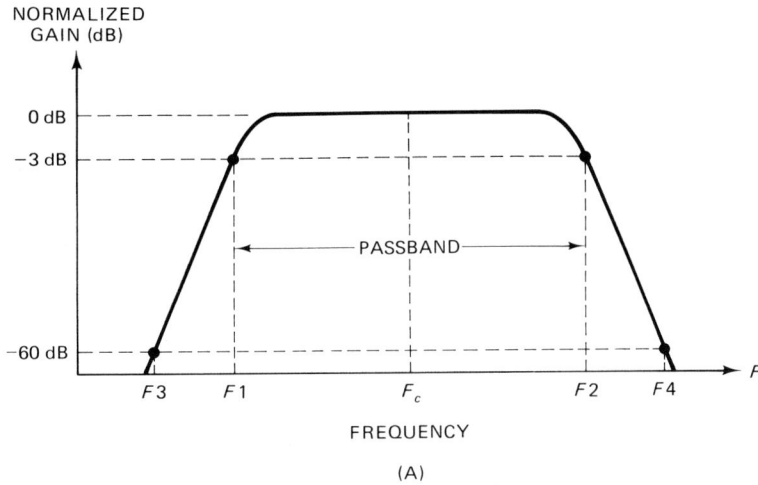

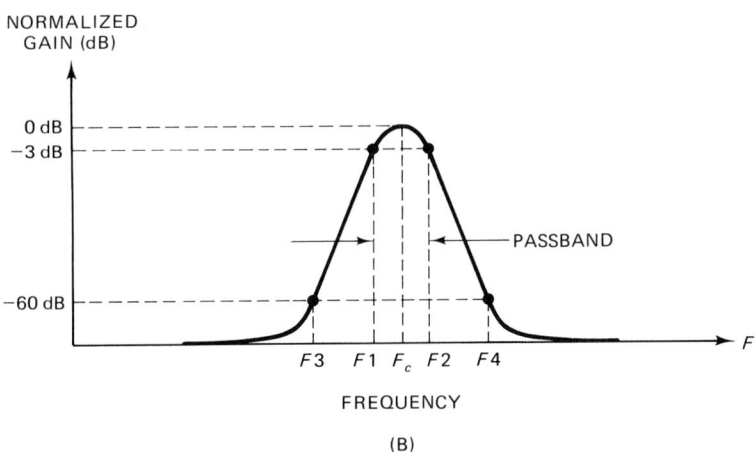

Fig. 20-16 Bandpass filter responses, (A) wide and (B) narrow.

The Q of the filter tells us something of the passband characteristic. Wide-band filters generally have a $Q < 10$, while narrow-band filters have a $Q > 10$.

Shape Factor

The shape factor (SF) of the filter characterizes the slope of the roll-off curve, so it is obviously related to the order of the filter. The shape factor is defined as the ratio of the -60 dB bandwidth to the -3 dB bandwidth:

$$SF = \frac{BW_{-60\,dB}}{BW_{-3\,dB}} \qquad (20\text{-}44)$$

First-Order VCVS Bandpass Filters

A wide-band first-order bandpass filter response is obtained by cascading first-order high-pass and low-pass filter circuits, as shown in Fig. 20-17A. This arrangement overlays, or superimposes, the frequency response characteristics of both filter stages. Figure 20-17B shows the situation when cascade high and low-pass filters are used. The low-pass filter response (solid line) is from DC to the -3 dB point at $F2$. The high-pass filter response is from the highest possible frequency within the range of the circuit down to the -3 dB point at $F1$. The passband is the intersection of the two sets, high and low-pass characteristics, which falls between $F1$ and $F2$.

The gain of the overall bandpass filter within the passband is the product of the two individual gains: $A_{vt} = A_{vl}A_{vh}$. The gain magnitude term of this form of filter is found by

$$\left| \frac{V_o}{V_{in}} \right| = \frac{A_{vt}(f/F1)}{\left[(1 + (f/F1)^2(1 + (f/F2)^2)\right]^{1/2}} \qquad (20\text{-}45)$$

where V_o is the output signal voltage, V_{in} the input signal voltage, f the applied frequency, $F1$ the lower -3 dB point frequency, $F2$ the

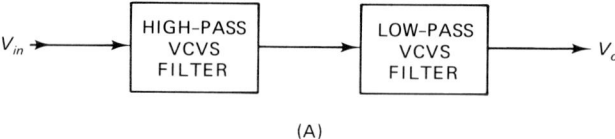

(A)

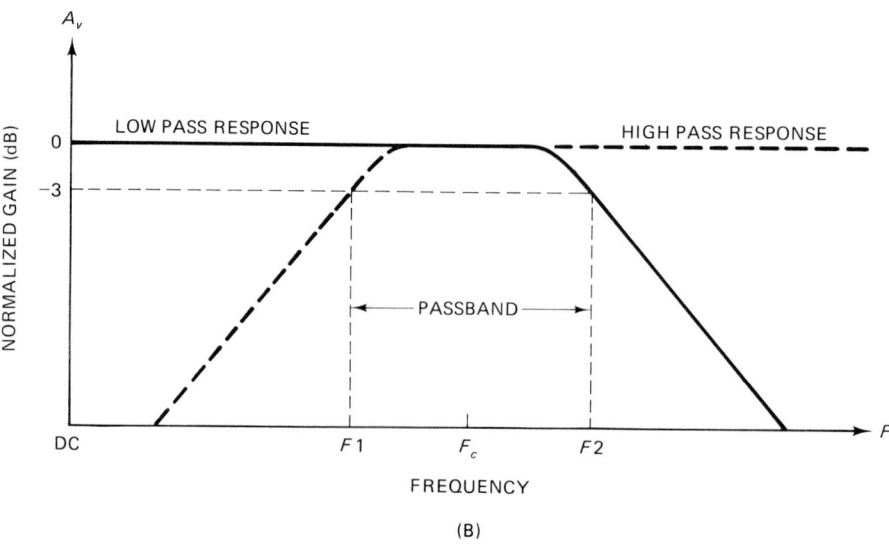

(B)

Fig. 20-17 Composite of LPF and HPF to form bandpass filter response.

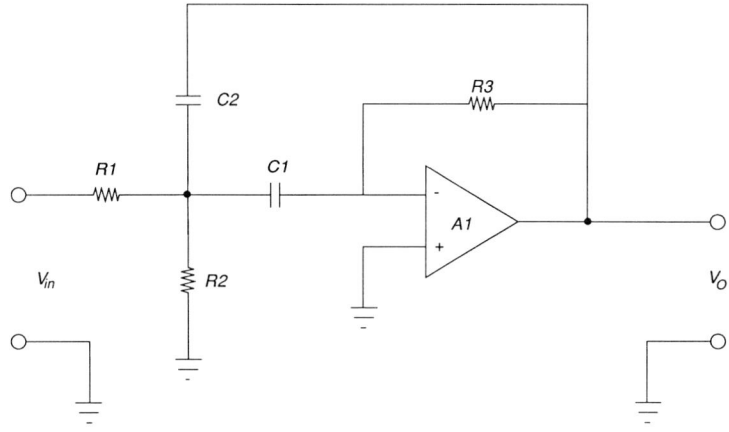

Fig. 20-18   MFP bandpass filter circuit.

upper -3 dB point frequency, and A_{vt} the total cascade gain of the filter.

Cascading low and high-pass filter sections can be used to make wide-band filters, but because of component tolerance and other problems it becomes less useful as Q increases above about 10 or so. For narrow-band filters a multiple feedback path (MFP) filter circuit such as Fig. 20-18 can be used. This filter offers first-order performance and relatively narrow bandpass. The circuit will work for values of $10 \leq Q \leq 20$ and for gains up to about 15. The center frequency of the MFP bandpass filter is

$$f_c = \frac{1}{2\pi} \sqrt{\frac{1}{R3C1C2}\left(\frac{1}{R1} + \frac{1}{R2}\right)} \qquad (20\text{-}46)$$

To calculate the resistor values, it is necessary to first set the passband gain (A_v) and the Q. It is the general practice to select values for $C1$ and $C2$ and then calculate the required resistances for the specified values of f_c, A_v, and Q. The resistor values are

$$R1 = \frac{1}{2\pi A_v C2 f_c} \qquad (20\text{-}47)$$

$$R2 = \frac{1}{2\pi f_c(2Q^2 - A_v)C2} \qquad (20\text{-}48)$$

$$R3 = \frac{2Q}{2\pi f_c C2} \qquad (20\text{-}49)$$

and the gain

$$A_v = \frac{R3}{R1\left[1 + \dfrac{C2}{C1}\right]} \qquad (20\text{-}50)$$

These equations can be simplified if the two capacitors are made equal ($C1 = C2 = C$) and if it is assumed that $Q > (A_{\mathrm{v}}/2)^{1/2}$.

$$R1 = \frac{Q}{2\pi f_{\mathrm{c}} A_{\mathrm{v}} C} \tag{20-51}$$

$$R2 = \frac{Q}{2\pi f_{\mathrm{c}} C(2Q^2 - A_{\mathrm{v}})} \tag{20-52}$$

$$R3 = \frac{2Q}{2\pi f_{\mathrm{c}} C} \tag{20-53}$$

$$A_{\mathrm{v}} = \frac{R3}{2R1} \tag{20-54}$$

The MFP bandpass filter can be tuned using only one of the resistors. If $R2$ is varied, the center frequency will shift; but the bandwidth, Q, and gain will remain constant. To scale the circuit to a new center frequency using only $R2$ as the change element, select a new value of $R2$ according to

$$R2' = R2\left[\frac{f_{\mathrm{c}}}{f_{\mathrm{c}'}}\right]^2 \tag{20-55}$$

BAND REJECT (NOTCH) FILTERS

A band reject or notch filter is used to pass all frequencies except a single frequency or small band of frequencies. An application for this circuit is to remove 60-Hz interference from electronic instruments. The medical electrocardiograph machine, for example, often suffers 60-Hz interference because the input leads are unshielded at the tips. These machines often include a switch-selectable 60-Hz notch filter to remove the 60-Hz artifact that could result.

Figure 20-19A shows a typical active notch filter, and Fig. 20-19B shows the frequency response for the circuit. Note that the gain is constant throughout the frequency spectrum except in the immediate vicinity of f_{c}. The depth of the notch is infinite in theory, but in practical circuits precision-matched components will offer -60 dB of suppression, while ordinary bench-run components (not precision) can offer -40 to -50 dB of suppression. The resonant frequency of this notch filter is found from

$$f_{\mathrm{c}} = 1/2\pi RC \tag{20-56}$$

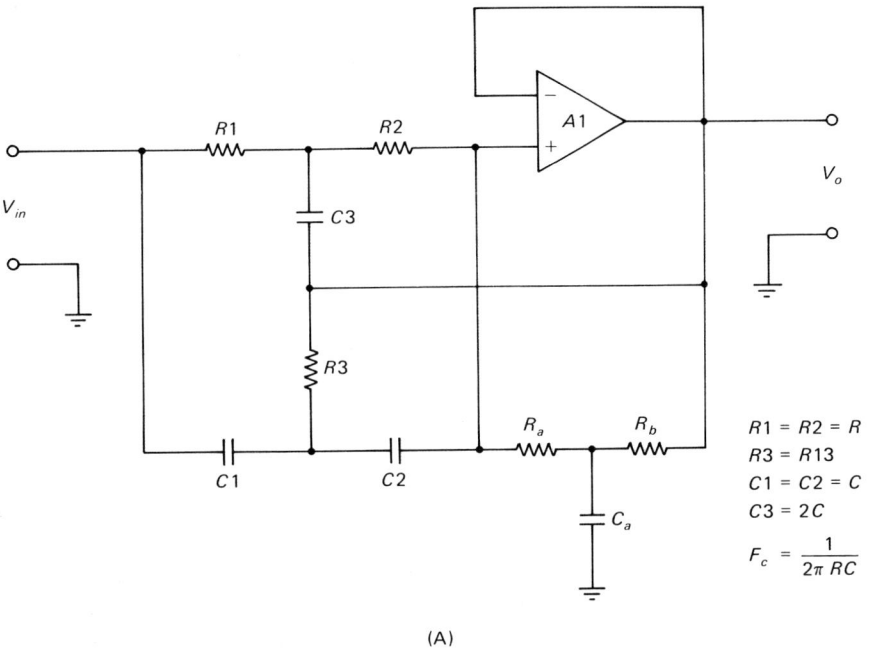

$R1 = R2 = R$
$R3 = R13$
$C1 = C2 = C$
$C3 = 2C$

$F_c = \dfrac{1}{2\pi RC}$

(A)

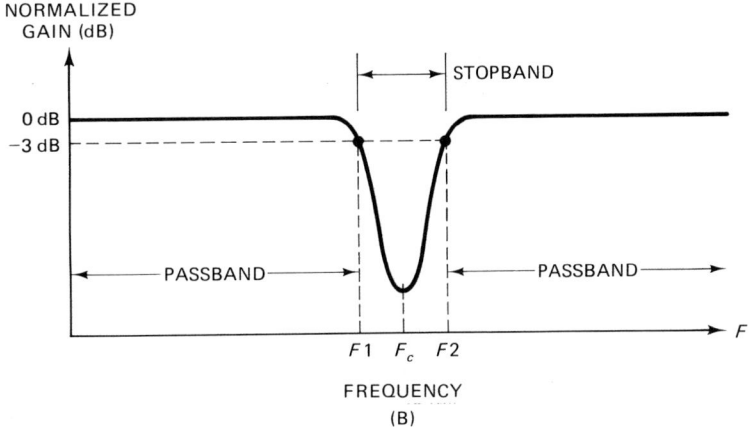

(B)

Fig. 20-19   (A) Active notch filter circuit and (B) frequency response.

The gain of the circuit is unity, but the Q can be set according to the following equations:

$$Q = R_a/2R \tag{20-57}$$

or

$$Q = C/C_a \tag{20-58}$$

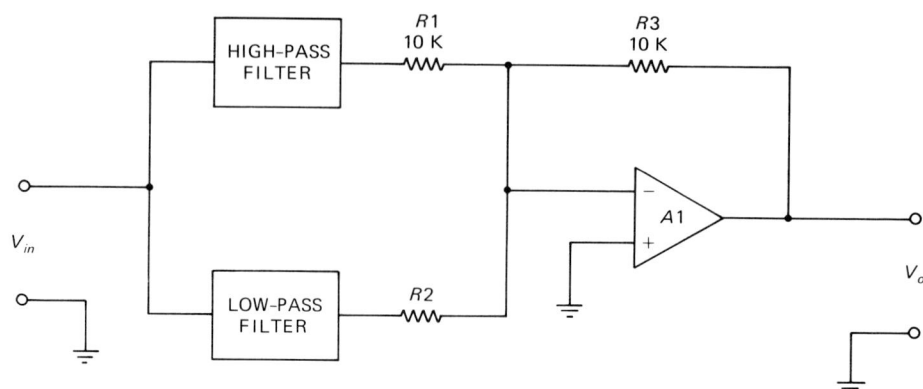

Fig. 20-20   Bandstop filter circuit.

A bandstop filter is an example of a notch filter with a wide stopband. Just as the wide-band pass filter can be made by cascading high and low-pass filter, the wide-band notch (or stopband) filter can be made by paralleling high and low-pass filter sections. Figure 20-20 shows a band stop filter in which the outputs of a high-pass filter and a low-pass filter are summed together in a two-input unity-gain inverting follower amplifier circuit. The frequency response curves of the two filter sections are superimposed to eliminate the undesired band. Make the -3 dB point of the high-pass filter equal to the upper end of the stopband, and make the -3 dB point of the low-pass filter equal to the lower end of the stopband.

ALL-PASS PHASE-SHIFT FILTERS

The all-pass phase-shifter (APPS) is a special category of filter in which all frequencies (within the ability of the op-amp) are passed, but they are shifted in phase a specified amount. Figure 20-21A shows the circuit for an APPS that will exhibit phase shift between input and output of $-180°$ to $0°$. If the roles of $R1$ and $C1$ are reversed, then the phase shift will be $-360°$ to $-180°$. The gain response of this circuit is

$$A_v = \frac{1 - 2\pi f R1 C1}{1 + 2\pi f R1 C1} \tag{20-59}$$

and the amount of phase shift is

$$\Delta\phi = -2\tan^{-1}(2\pi f R1 C1) \tag{20-60}$$

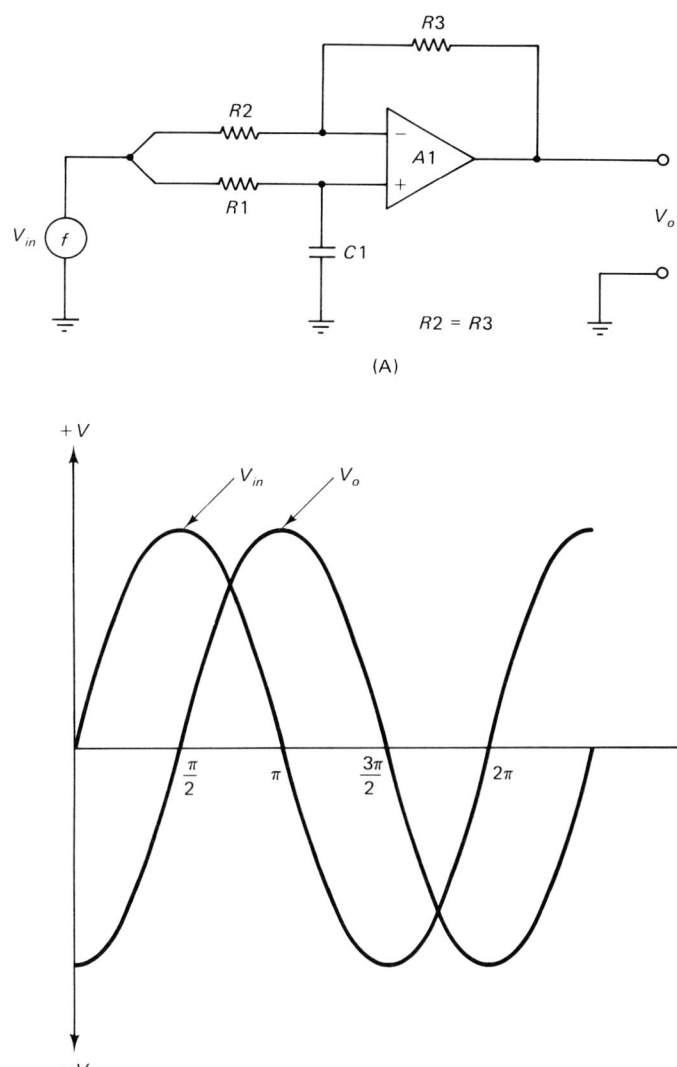

Fig. 20-21 (A) All-pass phase shifter filter and (B) input and output waveforms superimposed to show phase shift.

STATE-VARIABLE ANALOG FILTERS

The state-variable filter is a variation on the multiple feedback path design using three operational amplifiers. While the circuit is more complex than the other circuits presented in this chapter, it is also more versatile. The state-variable filter is capable of simultaneously providing bandpass, low-pass, and high-pass responses. Figure 20-22 shows the block diagram of the state-variable filter. The constituent parts include a summing amplifier, two integrators, and a damping

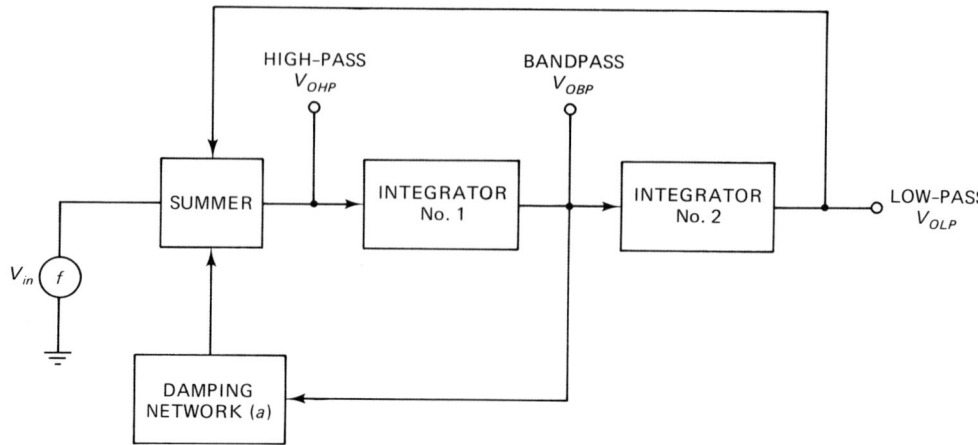

Fig. 20-22 Block diagram of the state-variable filter.

network. The integrators are inverting Miller integrators (Chapter 11). The summing amplifier is also a simple op-amp summer, and the damping network is a simple resistor voltage divider.

The damping factor (a) is the same for all three outputs; it is defined as the reciprocol of Q.

$$a = \frac{1}{Q} \qquad (20\text{-}61)$$

The state-variable filter shown in Fig. 20-23 is normalized for 1 kHz, $C = 0.0159 \ \mu\text{F}$, and $R = 10 \ \text{k}\Omega$. Designing for other frequen-

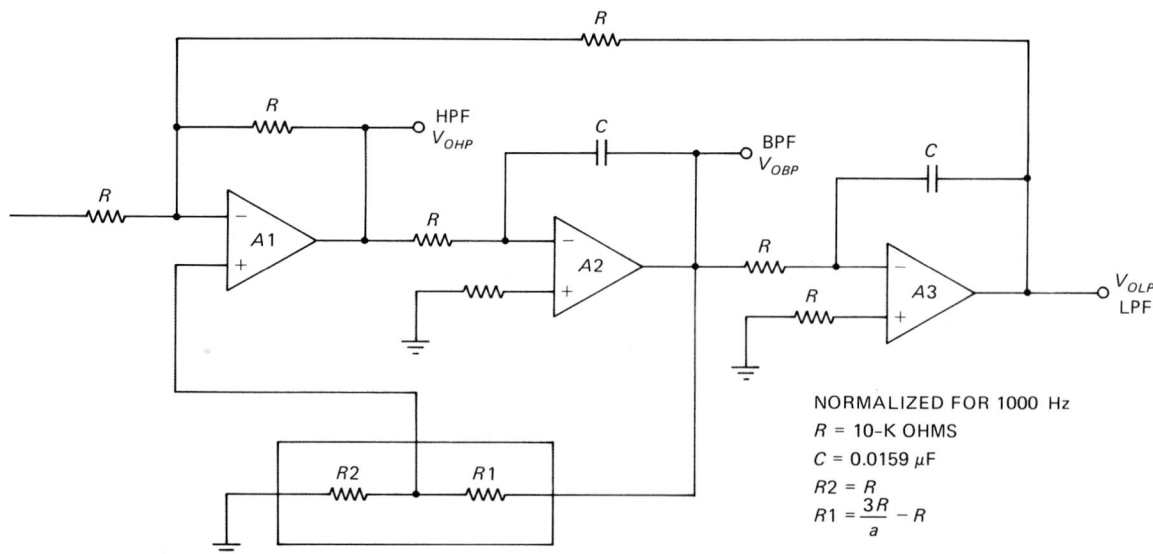

Fig. 20-23 Circuit for a practical state-variable filter circuit.

cies is a matter of scaling these values to the new values required for the new frequency. As in the previous examples, change either R or C, but not both. The value of components for the damping network are found from

$$R1 = \frac{3R}{a} - R \tag{20-62}$$

and

$$R2 = R \tag{20-63}$$

VOLTAGE-TUNABLE FILTERS

A voltage-tunable filter is one in which the -3 dB response point is a function of an input control voltage. The analog multiplier and divider can be used to make a simple voltage-tunable filter in either low-pass (Fig. 20-24A), high-pass (Fig. 20-24B), or bandpass (Fig. 20-24C) versions.

The low-pass filter is shown in Fig. 20-24A. This circuit consists of a Miller integrator with an analog multiplier in a second negative-feedback path through $R2$. The X-input of the multiplier is the output of the integrator, while the Y-input is the control voltage ($V1$) that tunes the filter. The output voltage of this circuit is

$$V_o = \frac{-V_{in}}{1 + (2\pi f R1 C1 / V1)} \tag{20-64}$$

The high-pass filter circuit is shown in Fig. 20-24B. This circuit consists of an analog divider driving an RC differentiator ($R1C1$). The output of the differentiator is summed with the input signal. For $R2 = R3 = R$, $C1 = C$, and $R1 < R$, the following transfer equation obtains.

$$V_o = -V_{in}\left(1 + \frac{2\pi f RC}{V1}\right) \tag{20-65}$$

The bandpass circuit of Fig. 20-24C is based on the analog multiplier and a pair of integrator circuits. This circuit produces a gain of -1 and a Q of

$$Q = \sqrt{-10V1} \tag{20-66}$$

The center frequency is

$$f_c = \frac{1}{2\pi RC}\sqrt{\frac{-V1}{10}} \tag{20-67}$$

assuming a control voltage $V1 \leq 0$.

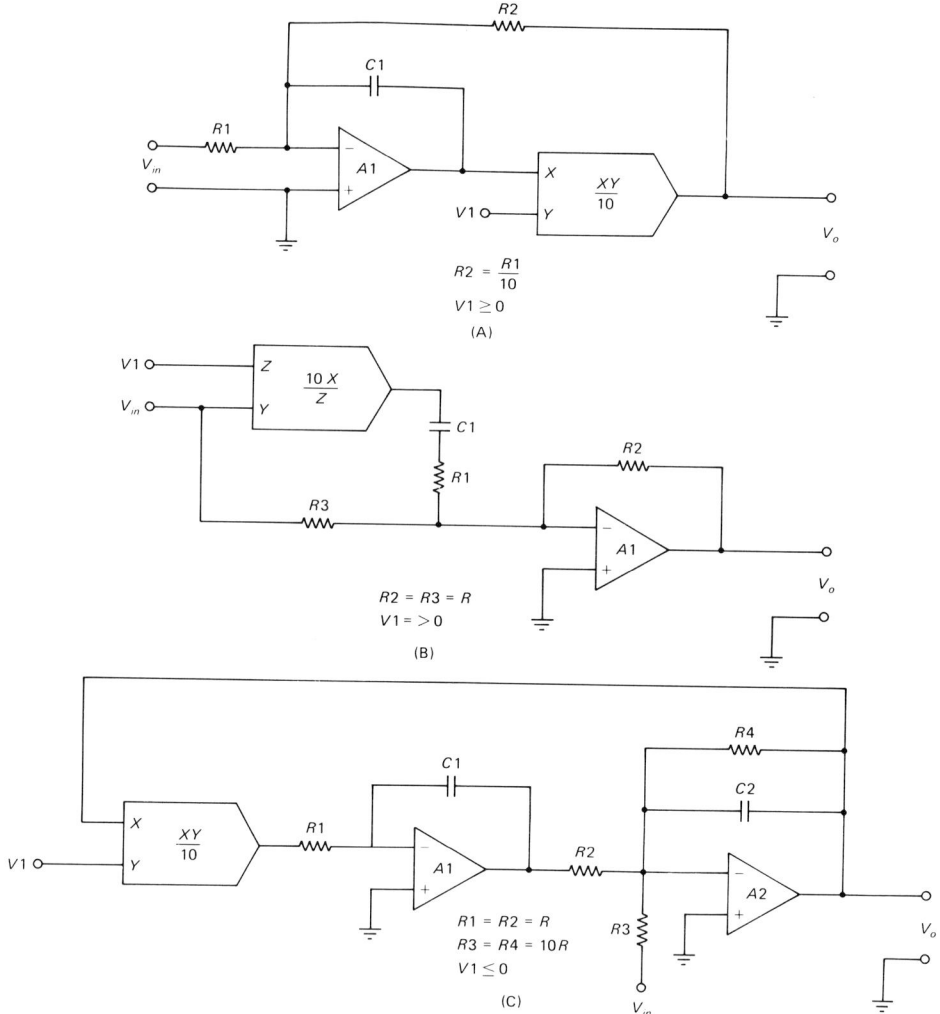

Fig. 20-24 Voltage-controlled filters: (A) low-pass, (B) high-pass, and (C) bandpass.

The bandwidth is

$$BW = 1/20\pi RC \qquad (20\text{-}68)$$

All three of these multiplier or divider based circuits are dependent on the properties of the device used. The multiplier gain, error, linearity, and response time determine the properties of the filter and the tuning rate.

21

IC and Digital Timer Circuits

Timers are monostable or astable circuits that are used in either timing or synchronization applications. While there are numerous discrete or op-amp timer circuits, there is also a class of integrated circuit dedicated timers (of which, the 555 device is the most popular). In this chapter we will explore the basic operation of IC timer devices and circuits.

INTRODUCTION TO THE 555
FAMILY OF IC TIMERS

The integrated circuit (IC) timer represents a class of chips that are extraordinarily well behaved and easy to apply. These timers are based on the properties of the series RC timing network and the voltage comparator. In many ways these devices are similar to RC circuits discussed in Chapter 18, but operational amplifiers are not explicitly used. Indeed, a combination of voltage comparator circuits and digital circuits are used inside these chips. Although several devices are on the market, the most common and best known is the type 555 device. The 555 is now made by a number of different semiconductor manufacturers, but it was originated by Signetics, Inc. in 1970. Today the 555 remains one of the most widespread IC devices on the market, rivaling even some general purpose operational amplifiers in numbers.

The original Signetics products included the SE-555, which operated at a temperature range of -55 to $+125°C$, and the NE-555, which operated over the range 0 to $+70°C$. Several different designations are now commonly used for the 555 made by other makers, including simply 555 and LM-555, or some variant of these. A dual 555-class timer is also marketed under the number 556. There is also a low-power CMOS version of the 555 marketed as the LMC-555.

The 555 is a multipurpose chip that will operate at DC power supply potentials from $+5$ to $+18$ V DC. The temperature stability of these devices is on the order of 50 PPM/°C (i.e., 0.005%/°C). The output of the 555 can either sink or source up to 200 mA of current. It is compatible with TTL devices (when the 555 is operated from $+5$ V DC power supply), CMOS devices, operational amplifiers, other linear IC devices, transistors, and most classes of solid-state devices. The 555 will also operate with most passive electronic components. Several factors contribute to the popularity of the 555 device. Besides the versatile nature of the device, it is well-behaved in the sense that operation is straightforward and circuit designs are generally simple. Like the general purpose operational amplifier, the 555 usually works in a predictable manner, according to the standard published equations.

The 555 operates in two different modes: monostable (one-shot) and astable (free-running). Figure 21-1A shows the astable mode output from pin no. 3 of the 555. The waveform is a series of square waves that can be varied in duty cycle over the range 50 to 99.9 percent and in frequency from less than 0.1 Hz to more than 100 kHz. Monostable operation (Fig. 21-1B) requires a trigger pulse applied to pin no. 2 of the 555. The trigger must drop from a level $> 2(V+)/3$ down to $< (V+)/3$. Output pulse durations from microseconds up to hours are possible. The principal constraint on longer operation is the leakage resistance of the capacitor used in the external timing circuit.

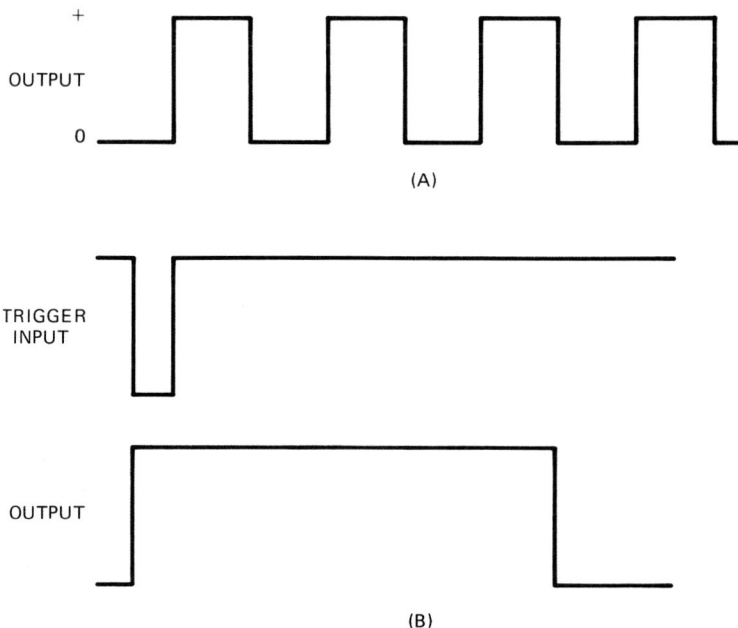

Fig. 21-1 555 IC timer output signals: (A) astable configuration and (B) monostable configuration.

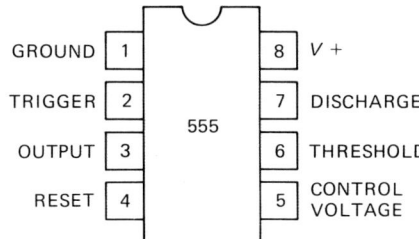

Fig. 21-2   555 IC timer in eight-pin miniDIP package.

PIN-OUTS AND INTERNAL CIRCUITS OF THE 555 IC TIMER

The package for the 555 device is shown in Fig. 21-2. Most 555s are sold in the eight-pin miniDIP package as shown, although some are found in the eight-pin metal-can IC package. The latter are mostly the military specification temperature range SE-555 series. The pin-outs are the same on both miniDIP and metal-can versions. The internal circuitry is shown in block form in Fig. 21-3. The following stages are found: two voltage comparators (COMP1 and COMP2), a reset-set (RS) control flip-flop (which can be reset from outside the chip through pin no. 4), an inverting output amplifier (A1), and a discharge transis-

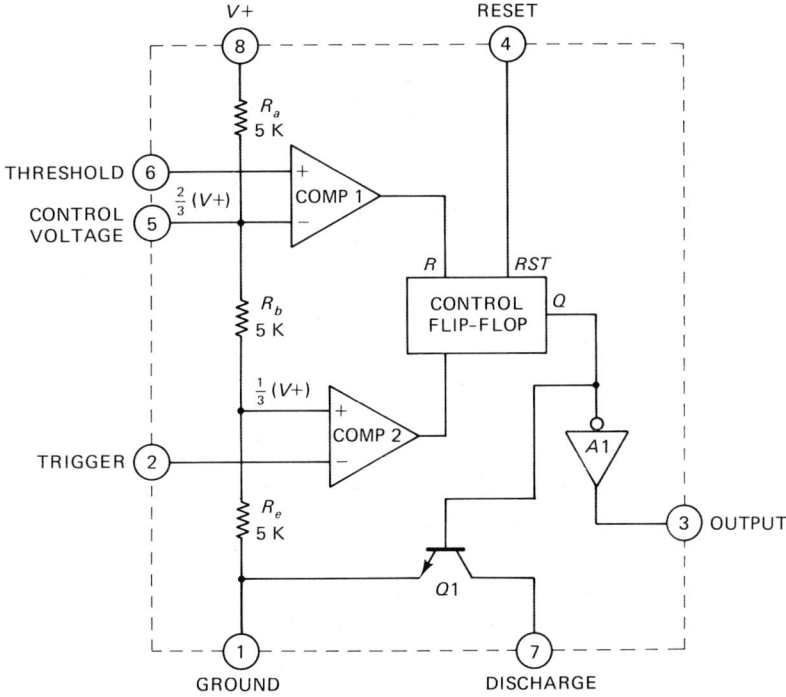

Fig. 21-3   Internal circuitry of the 555 timer.

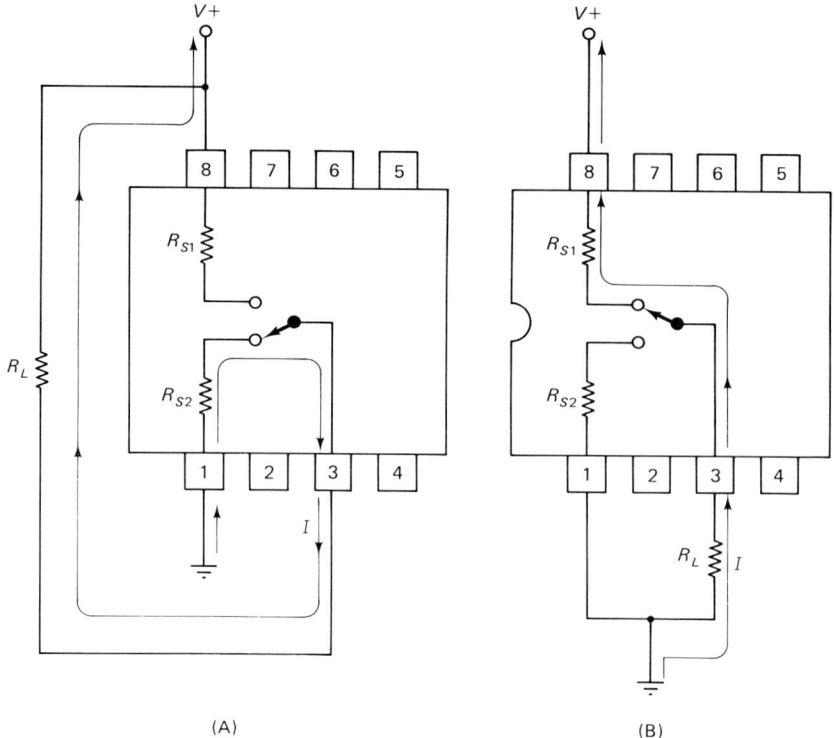

Fig. 21-4 Equivalent circuits for 555 output states: (A) V_o = LOW, and (B) V_o = HIGH.

tor (Q1). The bias levels of the two comparators are determined by a resistor voltage divider (R_a, R_b, and R_c) between $V+$ and ground. The inverting input of COMP1 is set to $2(V+)/3$, and the noninverting input of COMP2 is set to $(V+)/3$. Figures 21-3 and 21-4 show the pin-outs of the 555. In the following descriptions the term HIGH implies a level $> 2(V+)/3$ and LOW implies a grounded condition ($V = 0$), unless otherwise specified in the discussion. These pins serve the following functions:

Ground (Pin No. 1) This pin serves as the common reference point for all signals and voltages in the 555 circuit both internal and external to the chip.

Trigger (Pin No. 2) The trigger pin is normally held at a potential $> 2(V+)/3$. In this state the 555 output (pin no. 3) is LOW. If the trigger pin is brought LOW to a potential $< (V+)/3$, the output (pin no. 3) abruptly switches to the HIGH state. The output remains HIGH as long as pin no. 2 is LOW, but the output does not necessarily revert back to LOW immediately after pin no. 2 is brought HIGH again (see operation of the Threshold input below).

Output (Pin No. 3) The output pin of the 555 is capable of either sinking or sourcing current up to 200 mA. This operation is in contrast to other IC devices in which the outputs of various devices will either sink or source current, but not both. Whether the 555 output operates as a sink or a source depends on the configuration of the external load. Figure 21-4 shows both types of operation.

In Fig. 21-4A the external load R_L is connected between the 555 output and $V+$. Current only flows in the load when pin no. 3 is LOW. In that condition the external load is grounded through pin no. 1 and a small internal source resistance, R_{S1}. In this configuration the 555 output is a current sink.

The operation depicted in Fig. 21-4B is for the case where the load is connected between pin no. 3 of the 555 and ground. When the output is LOW the load current is zero. When the output is HIGH, however, the load is connected to $V+$ through a small internal resistance R_{S2} and pin no. 8. In this configuration the output serves as a current source.

Reset (Pin No. 4)   The reset pin is connected to a preset input of the 555 internal control flip-flop. When a LOW is applied to pin no. 4 the output of the 555 (pin no. 3) switches immediately to a LOW state. In normal operation it is common practice to connect pin no. 4 to $V+$ to prevent false resets from noise impulses.

Control Voltage (Pin No. 5)   This pin normally rests at a potential of $2(V+)/3$ due to an internal resistive voltage divider (see R_a through R_c in Fig. 21-3). Applying an external voltage to this pin, or connecting a resistor to ground, will change the duty cycle of the output signal. If not used, pin no. 5 should be decoupled to ground through a 0.01 to 0.1-μF capacitor.

Threshold (Pin No. 6)   This pin is connected to the noninverting input ($+$IN) of comparator COMP1 and is used to monitor the voltage across the capacitor in the external RC timing network. If pin no. 6 is at a potential of $< 2(V+)/3$, the output of the control flip-flop is LOW and the output (pin no. 3) is HIGH. Alternatively, when the voltage on pin no. 6 is $\geq 2(V+)/3$, then the output of COMP1 is HIGH and chip output (pin no. 3) is LOW.

Discharge (Pin No. 7)   The discharge pin is connected to the collector of NPN transistor Q1, and the emitter of Q1 is connected to the ground pin (no. 1). The base of Q1 is connected to the NOT-Q output of the control flip-flop. When the 555 output is HIGH, the NOT-Q output of the control flip-flop is LOW; so Q1 is turned off. The c-e resistance of Q1 is very high under this condition, so does not appreciably affect the external circuitry. When the control flip-flop NOT-Q output is HIGH, however, the 555 output is LOW and Q1 is biased hard on. The c-e path is in saturation, so the c-e resistance is very low. Pin no. 7 is effectively grounded under this condition.

V+ Power Supply (Pin No. 8)   The DC power supply is connected between ground (pin no. 1) and pin no. 8, with pin no. 8 being positive. In good practice a 0.1 to 10 μF decoupling capacitor will normally be used between pin no. 8 and ground.

MONOSTABLE OPERATION
OF THE 555 IC TIMER

A monostable multivibrator (MMV), also called the one-shot circuit, produces a single output pulse of fixed duration when triggered by an input pulse. Operational amplifier versions of the monostable circuit were discussed in Chapter 18. The output of the one-shot will snap HIGH following the trigger pulse and will remain HIGH for a fixed, predetermined duration. When this time expires the one-shot is timed-out, so it snaps LOW again. The output of the one-shot will remain LOW indefinitely unless another trigger pulse is applied to the circuit. The 555 can be operated as a monostable multivibrator by suitable connection of the external circuit.

Figure 21-5 shows the operation of the 555 as a monostable multivibrator. To make the operation of the circuit easier to under-

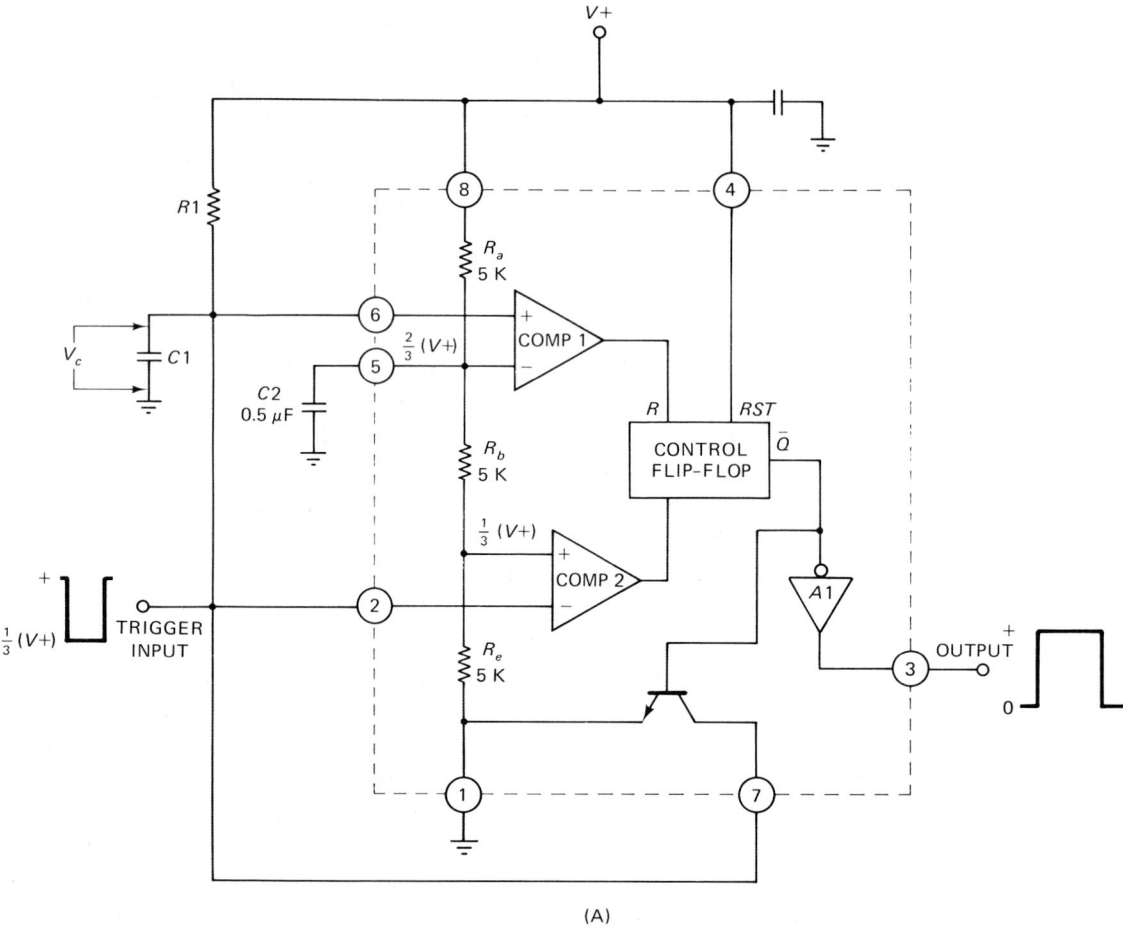

(A)

Fig. 21-5 (A) Monostable circuit for the 555 IC timer in block form, (B) timing diagram, and (C) circuit diagram as it appears in schematics. (*Figure continues.*)

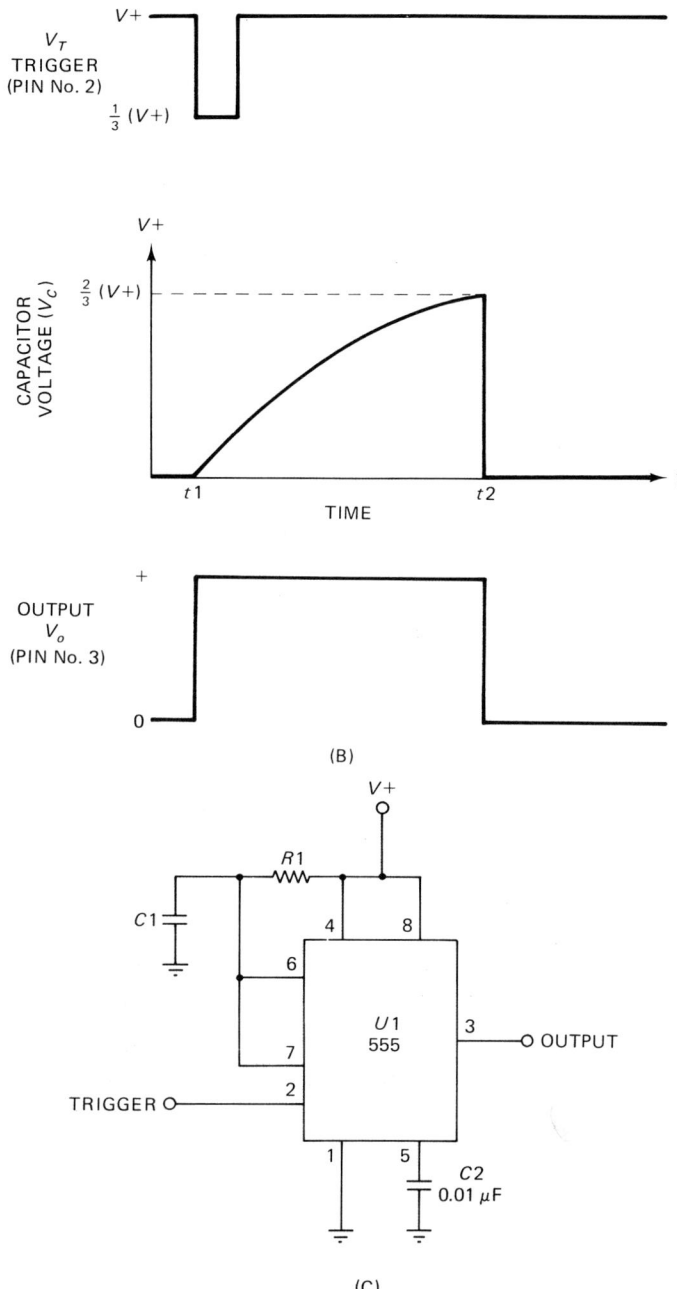

Fig. 21-5 *(continued)*

stand, Fig. 21-5A shows the internal circuitry as well as the external circuitry; Figure 21-5B shows the timing diagram for this circuit; and Fig. 21-5C shows the same circuit in the more conventional schematic diagram format.

The two internal comparators are biased to certain potential levels by a series voltage divider consisting of resistors R_a, R_b, and R_c. The inverting input of voltage comparator COMP1 is biased to $2(V+)/3$, while the noninverting input of COMP2 is biased to $(V+)/3$. It is these levels that govern the operation of the 555 device in whichever mode is selected. An external timing network ($R1C1$) is connected between $V+$ and the noninverting input of COMP1 via pin no. 6. Also connected to pin no. 6 is 555 pin no. 7, which has the effect of connecting the transistor across capacitor $C1$. If the transistor is turned on, then the capacitor looks into a very low resistance short circuit through the c-e path of the transistor.

When power is initially applied to the 555, the voltage at the inverting input of COMP1 will go immediately to $2(V+)/3$, and the noninverting input of COMP2 will go to $(V+)/3$. The control flip-flop is in the reset condition, so the NOT-Q output is HIGH. Because this flip-flop is connected to output pin no. 3 through an inverting amplifier ($A1$), the output is LOW at this point. Also, because NOT-Q is HIGH, transistor Q1 is biased into saturation, creating a short circuit to ground across external timing capacitor $C1$. The capacitor remains discharged in this condition ($V_c = 0$).

If a trigger pulse is applied to pin no. 2 of the 555, and if that pulse drops to a voltage that is $< (V+)/3$, as shown in Fig. 21-5B, then comparator COMP2 sees a situation where the inverting input is less positive than the noninverting input, so the output of COMP2 snaps HIGH. This action sets the control flip-flop, forcing the NOT-Q output LOW and, therefore, the 555 output HIGH. The LOW at the output of the control flip-flop also means that transistor Q1 is now unbiased, so the short across the external capacitor is removed. The voltage across $C1$ begins to rise (see Figs. 21-5B and 21-5D). The voltage will continue to rise until it reaches $2(V+)/3$, at which time comparator COMP1 will snap HIGH, causing the flip-flop to reset. When the flip-flop resets, its NOT-Q output drops LOW again, terminating the output pulse and returning the capacitor voltage to zero. The 555 will remain in this state until another trigger pulse is received.

The timing equation for the 555 can be derived in exactly the same manner as the equations used with the operational amplifier MMV circuits. The basic equation was discussed in Chapter 18; it relates the time required for a capacitor voltage to rise from a starting point (V_{C1}) to an end point (V_{C2}) with a given RC time constant.

$$T = -RC \ln\left(\frac{V - V_{C2}}{V - V_{C1}}\right) \qquad (21\text{-}1)$$

In the 555 timer the voltage source is $V+$, the starting voltage is zero, and the trip-point voltage for comparator COMP1 is $2(V+)/3$.

Equation (21-1) can therefore be rewritten as

$$T = -R1C1 \ln\left(\frac{V - V_{C2}}{V - V_{C1}}\right) \qquad (21\text{-}2)$$

$$T = -R1C1 \ln\left(\frac{(V+) - 2(V+)/3}{V+}\right) \qquad (21\text{-}3)$$

$$T = -R1C1 \ln(1 - 0.667) \qquad (21\text{-}4)$$

$$T = -R1C1 \ln(0.333) \qquad (21\text{-}5)$$

$$T = 1.1\,R1C1 \qquad (21\text{-}6)$$

Input Triggering Methods
for the 555 MMV Circuit

The 555 MMV circuit triggers by bringing pin no. 2 from a positive voltage down to a level $< (V+)/3$. Triggering can be accomplished by applying a pulse from an external signal source or through other means. Figure 21-6 shows the circuit for a simple pushbutton switch trigger circuit. A pull-up resistor ($R2$) is connected between pin no. 2 and $V+$. If normally open (N.O.) pushbutton switch S1 is open, then the trigger input is held at a potential very close to $V+$. But when S1 is closed, pin no. 2 is brought LOW to ground potential. Because pin no. 2 is now at a potential less than $(V+)/3$, the 555 MMV will trigger. This circuit can be used for contact debouncing.

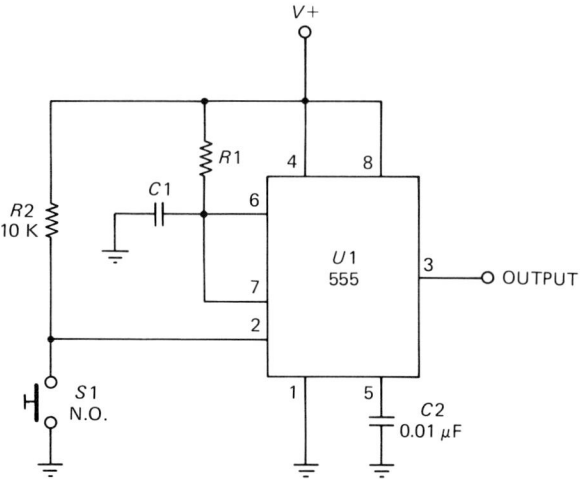

Fig. 21-6 Pushbutton triggering of the 555 IC timer.

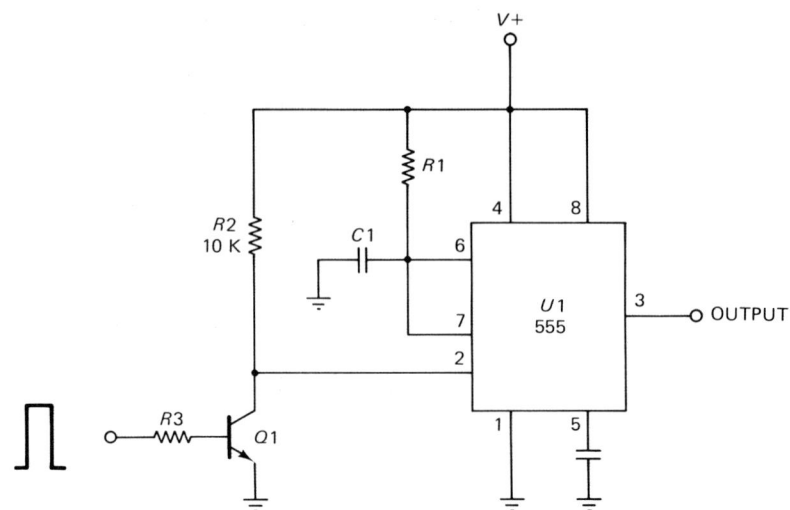

Fig. 21-7   Triggering the 555 IC timer with a positive pulse.

A circuit for inverting the trigger pulse applied to the 555 is shown in Fig. 21-7. In this circuit an NPN bipolar transistor is used in the common emitter mode to inverting the pulse. Again, a pull-up resistor is used to keep pin no. 2 at $V+$ when the transistor is turned off. But when the positive polarity trigger pulse is received at the base of transistor Q1, the transistor saturates; this forces the collector (and pin no. 2 of the 555) to near ground potential.

Figure 21-8 shows two AC-coupled versions of the trigger circuit. In these circuits a pull-up resistor keeps pin no. 2 normally at $V+$. But when a pulse is applied to the input end of capacitor $C3$, a differentiated version of the pulse is created at the trigger input of the 555. Diode D1 clips the positive-going spike to 0.6 or 0.7 V, passing only the negative-going pulse to the 555. If the negative-going spike can counteract the positive bias provided by $R2$ sufficiently to force the voltage lower than $(V+)/3$, then the 555 will trigger. A pushbutton switch version of this same circuit is shown in Fig. 21-8B.

A touchplate trigger circuit is shown in Fig. 21-9A. The pull-up resistor $R2$ has a very high value (22 MΩ shown here). The touchplate consists of a pair of closely spaced electrodes. As long as there is no external resistance between the two halves of the touchplate, the trigger input of the 555 remains at $V+$. But when a resistance is connected across the touchplate, the voltage $(V1)$ drops to a very low value. If the average finger resistance is about 20 kΩ, the voltage drops to

$$V1 = \frac{(V+)(20K)}{R2 + 20K} \tag{21-7}$$

which, when $R2 = 22$ MΩ, is $0.0009(V+)$—which is certainly less than $(V+)/3$.

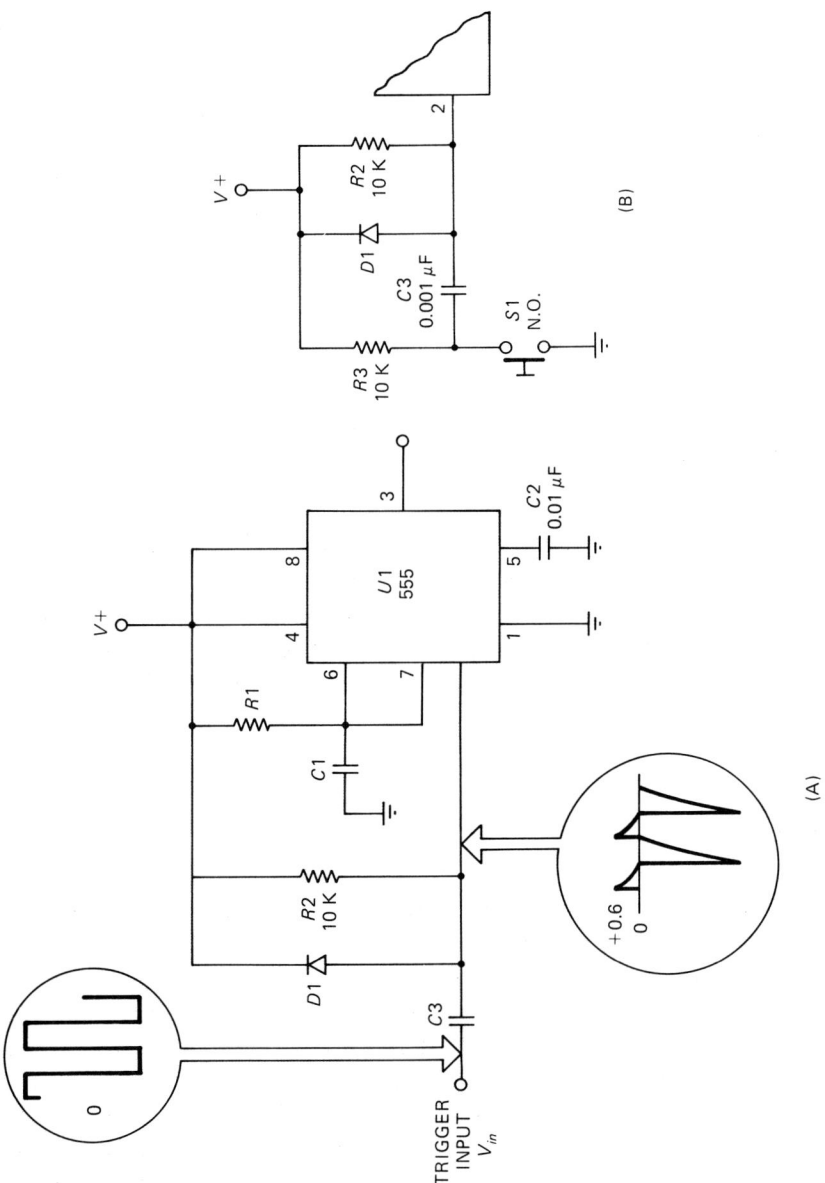

Fig. 21-8 (A) Capacitor-coupled triggering and (B) pushbutton version.

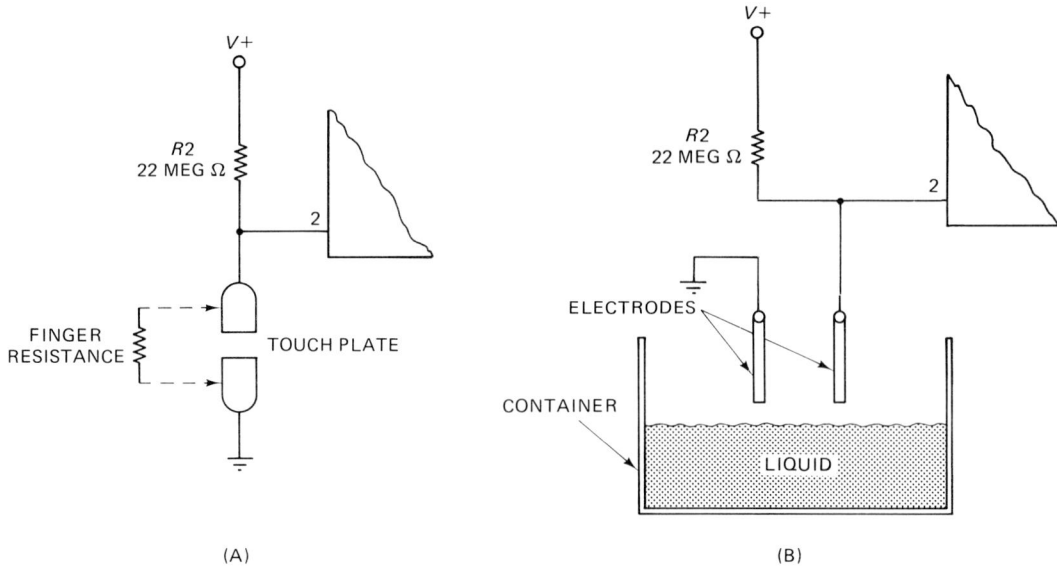

Fig. 21-9   (A) Finger resistance across touchplate allows touch-triggering of the 555 IC timer; (B) liquid level detector depends on liquid shorting electrodes together.

The same concept is used in the liquid level detector shown in Fig. 21-9B. Once again a 22 MΩ pull-up resistor is used to keep pin no. 2 at $V+$ under normal operation. When the liquid level rises sufficiently to short out the electrodes, however, the voltage on pin no. 2 ($V1$) drops to a very low level, forcing the 555 to trigger.

Retriggerable Operation of the 555 MMV Circuit

The 555 is a nonretriggerable monostable multivibrator. If additional trigger pulses are received prior to the time-out of the output pulse, then the additional pulses have no effect on the output. But the first pulse after time-out occurs will cause the output to again snap HIGH.

The circuit in Fig. 21-10 will permit retriggering of the 555 device. An external NPN transistor (Q2 in Fig. 21-10B) is connected with its c-e path across timing capacitor $C1$. In this sense it mimics the internal discharge transistor seen earlier. A second transistor, Q1, is connected to the trigger input of the 555 in a manner similar to Fig. 21-7 (discussed earlier). The bases of the transistors form the trigger input. When a positive pulse is applied to the combined trigger line, both transistors become saturated. Any charge in $C1$ is immediately discharged, and pin no. 2 of the 555 is triggered by the collector of Q1 being dropped to less than $(V+)/3$. As long as no further trigger pulses are received, this circuit behaves like any other 555 MMV circuit. But if a trigger pulse is received prior to the time-out defined by Eq. (21-6),

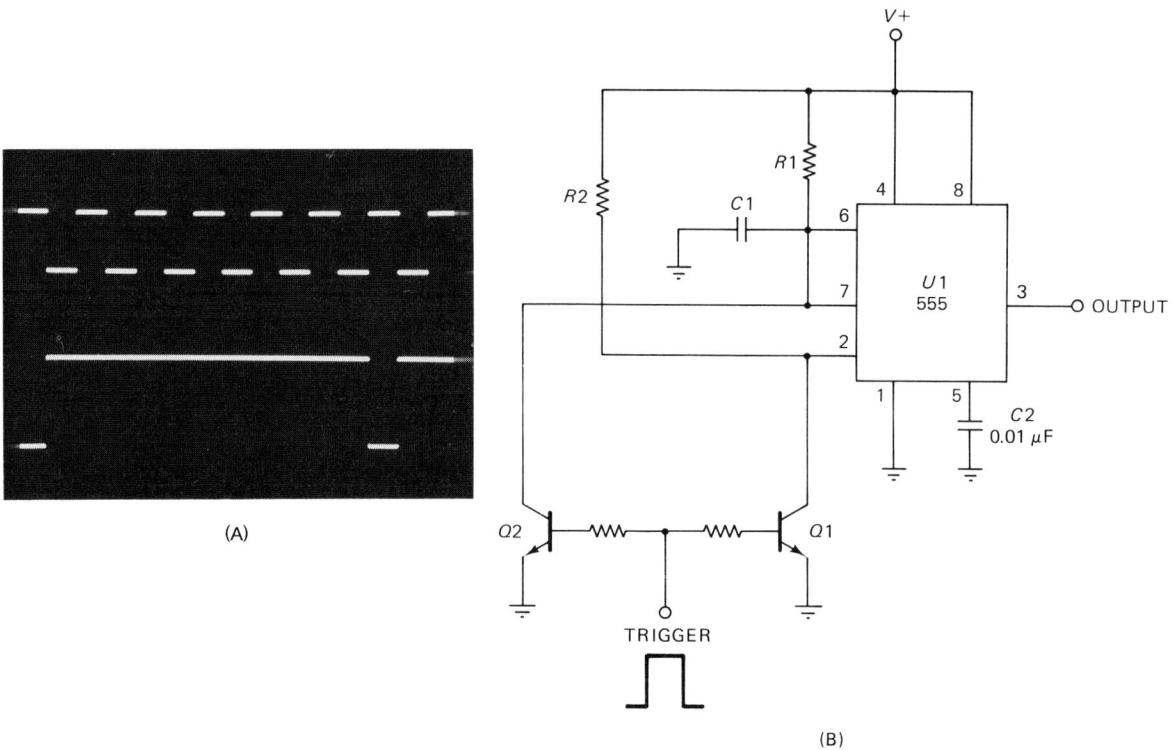

Fig. 21-10 (A) Nonretriggerable operation of the 555 results in this timing waveform (trigger pulses upper, output pulse lower) and (B) retriggerable 555 monostable circuit.

the transistors are forward biased once again. Q1 retriggers the 555, and Q2 dumps the charge built up in the capacitor. Thus, the 555 retriggers.

Applications for the 555 One-Shot Circuit

The MMV is a one-shot circuit that produces a single output pulse for every trigger input pulse, except for those that fall inside the output pulse and any associated refractory period. There are numerous potential applications for these circuits, of which a few are presented in this section.

Missing Pulse Detector

A missing pulse detector circuit remains dormant as long as a series of trigger pulses are received, but will produce an output pulse when an expected pulse is missing. These circuits are used in a variety of applications including alarms. For example, in a bottling plant soft-drink cans are packaged into six-packs. As each can passes a photocell a pulse is generated to the input of a missing-pulse detector.

If a pulse is not received, however, the machine knows that the count is one can short and issues an alarm or corrective action. Similarly, in a wildlife photography system. An infrared light-emitting diode (LED) is modulated or chopped with a pulse waveform. As long as the pulse is received at the sensor, the circuit is dormant. But if an animal passes through the IR beam even briefly a missing pulse detector will sense its presence and issue an output that fires a camera flashgun and electrical shutter control.

Figure 21-11A shows the circuit for a missing pulse detector based on the 555 IC timer, while Fig. 21-11B shows the timing waveforms.

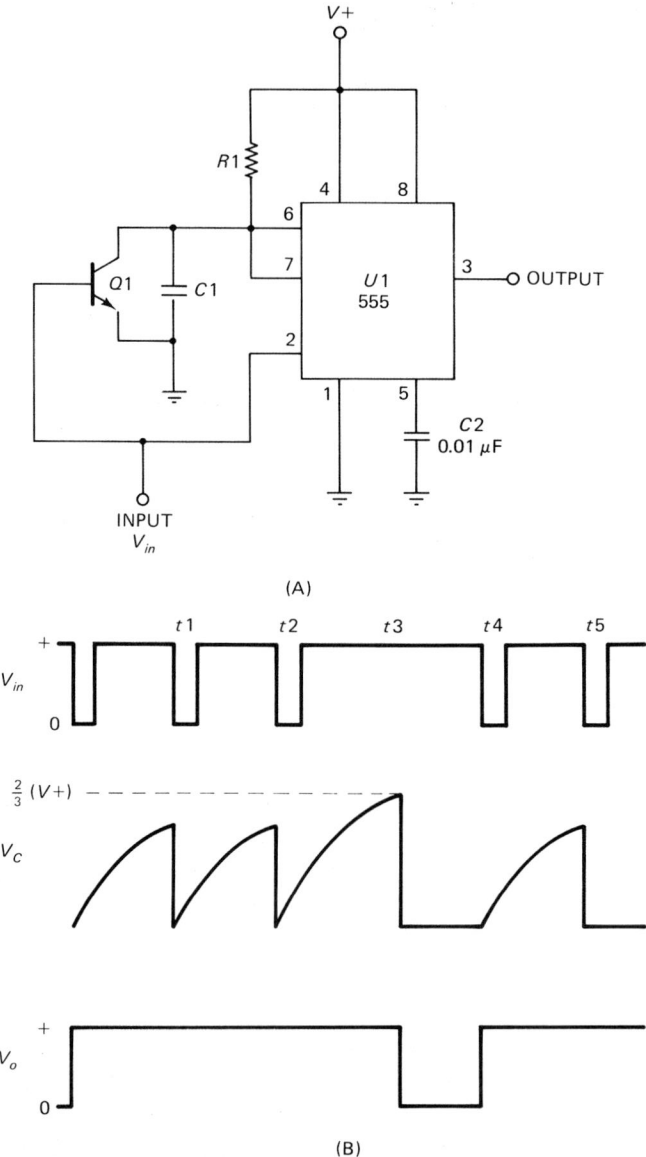

Fig. 21-11 (A) Missing pulse detector circuit and (B) timing waveforms.

This circuit is the standard 555 MMV, except that a discharge transistor is shunted across capacitor $C1$. When a pulse is applied to the input it will trigger the 555 and turn on Q1, causing the capacitor to discharge. After the first input pulse the output of the 555 snaps HIGH and remains HIGH until a missing pulse is detected.

Circuit action can be seen in Fig. 21-11B. At times $t1$ and $t2$, input pulses are received. As long as $(t2 - t1)$ is less than the time required for $C1$ to charge up to $2(V+)/3$, the 555 will never time-out. But if a pulse is missing, as at $t3$, the capacitor voltage continues to rise to the critical $2(V+)/3$ threshold value. When V_c reaches this point the 555 will time-out, forcing its output LOW. The output remains LOW until a subsequent input pulse is received ($t4$), at which time Q1 turns on again and forces the capacitor to discharge. The cycle can then continue as before.

Pulse Position Circuit

A pulse positioner is a circuit that will allow adjustment of the timing of a pulse to coincide with some external event. For example, in some instrumentation circuits a short pulse must be positioned to a certain point on a sine wave (e.g., the peak). The pulse positioner could be triggered from the zero-crossing of the sine wave and then adjusted to place the output pulse where it is needed.

Figure 21-12A shows the concept of pulse positioning using two one-shot circuits, labeled OS1 and OS2. The circuit is shown in Fig. 21-12B. The repositioned pulse is not actually the original pulse, but rather it is a recreated pulse with similar characteristics. The input pulse is used to trigger OS1. The duration of this one-shot circuit is fixed to the delay required of the repositioned pulse. If the delay must be variable, then resistor R1 is made variable. When OS1 times-out, it will trigger OS2. The output pulse of OS2 is set to the parameters of the original input pulse. An inverter circuit is used to make the output of OS2 have the same polarity as the trigger pulse at the input of OS1. To an outside observer the pulse appears to have been repositioned, although in fact it was merely recreated at time T (the delay period in Fig. 21-12B).

Tachometry

The word tachometry is used to designate the measurement of a repetition rate. In the automotive tachometer, for example, the instrument counts the pulses produced by the ignition coil to measure the engine speed in RPM. In medical instruments it is often necessary to measure factors such as heart or respiration rate electronically using tachometry circuits. A heart rate meter (cardiotachometer) measures the heart rate in beats per minute (BPM), while the respiration meter (pneumotachometer) measures breathing rate in breaths per second.

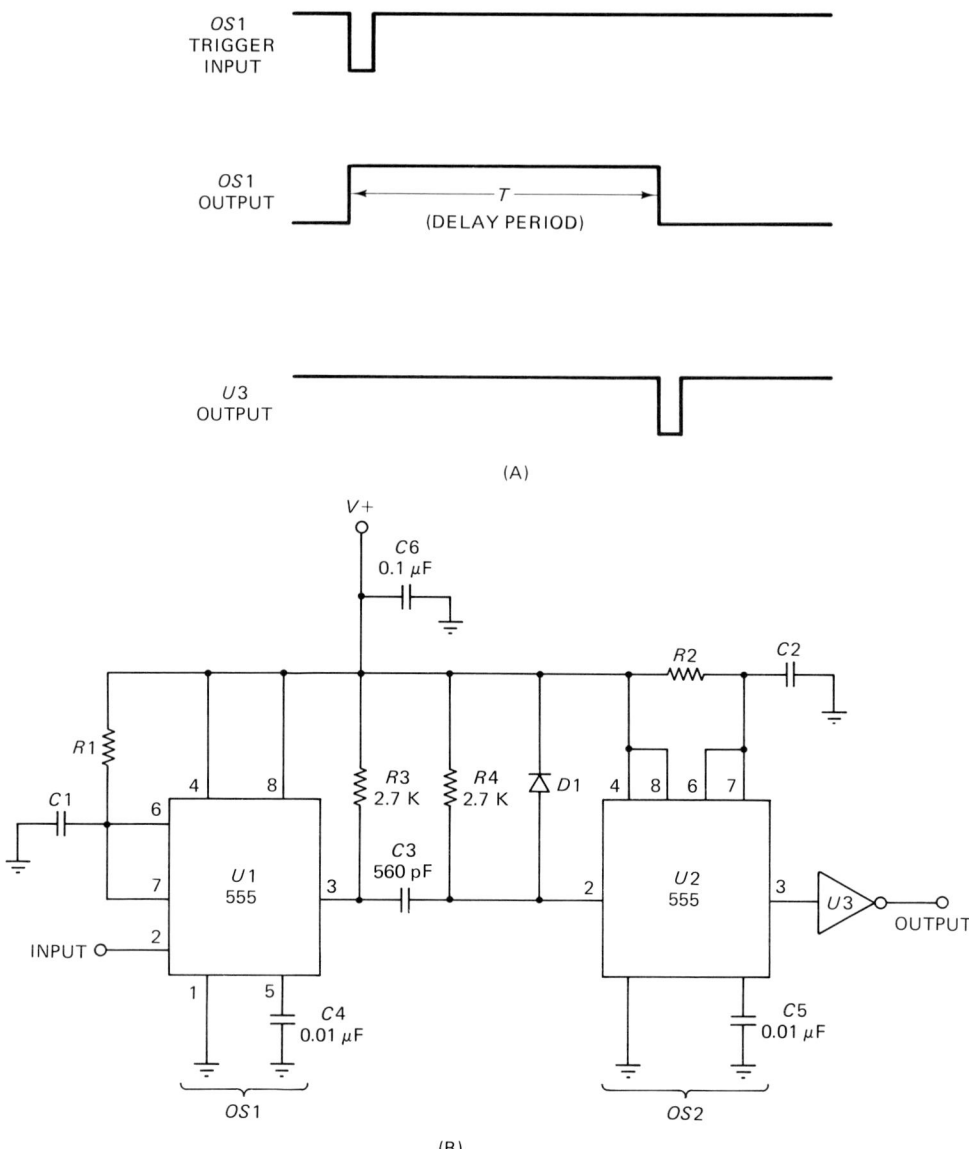

Fig. 21-12 (A) Using a pair of one-shot circuits will allow apparent pulse delay; (B) pulse delay circuit.

There is a certain commonality among nondigital tachometer circuits. It doesn't matter whether the rate is audio or sub-audio, or even above the audio rate—the basic circuit design is the same.

Figure 21-13 shows the basic tachometer circuit in block diagram form. Not all of the stages will be present in all circuits, but some of them are basic to the problem, so they are universally found. The AC amplifier and Schmitt trigger will be used only when needed. These stages are used for input signal conditioning, so they are used only where such conditioning is needed. The one-shot circuit and the Miller integrator are basic, however, so they are used for all such circuits.

Fig. 21-13 Basic tachometer circuit (block diagram).

The essential idea is to convert a frequency or repetition rate to an analog voltage. This is done by first converting the signal to pulse form. The AC input amplifier is used only if it is necessary to scale the input signal to a level where it will drive a Schmitt trigger or other squaring circuit. The purpose of the following stage is to produce a square-wave output signal at the same frequency as the input signal. The purpose of the stages in Fig. 21-13 is to produce a DC voltage output that is proportional to the input frequency or pulse repetition rate. The integrator is designed to produce an output voltage that is the time-average of the input signal. That is, the integrator output is proportional to the area under the input signal. The job of the tachometer designer is to create a situation in which the only variable is the frequency or repetition rate of the input signal. Variation obscures the results.

The output pulse of a one-shot circuit has a constant amplitude and constant duration. The area under the pulse is the product of the amplitude and duration, so from pulse to pulse the area does not change. If the one-shot is constantly retriggered by the input signal, the total area under the resultant pulse train is a function of only the number of pulses. Therefore, the time-average of the integrator output will be a DC voltage that is proportional to the input frequency.

Figure 21-14 shows a practical application of the tachometer principle. The circuit was used to demodulate the audio frequency modulated signal from an instrumentation telemetry set. A similar circuit (but not based on the 555) was once popular as a coilless FM detector in communications and broadcast receivers. These pulse counting detectors operate at 10.7 MHz (a commonly used FM IF frequency in receivers). The circuit shown in Fig. 21-14 was used to demodulate a human electrocardiograph (ECG) signal transmitted over telephone lines. The ECG, an analog voltage waveform, and was used to frequency-modulate an audio voltage controlled oscillator (VCO) at the transmit end. Normally, the ECG has too low a Fourier frequency content (0.05 to 100 Hz) to pass over the restricted passband of the telephone lines (300 to 3000 Hz). When used to frequency-modulate a 1500-Hz carrier, however, the signal passed easily over telephone circuits.

The circuit for the demodulator circuit is shown in Fig. 21-14A. The input waveshaping function is performed by an LM-311 voltage comparator. The job of the LM-311 is to square the 200-mV peak-to-peak sine-wave input signal so that it is capable of triggering the 555

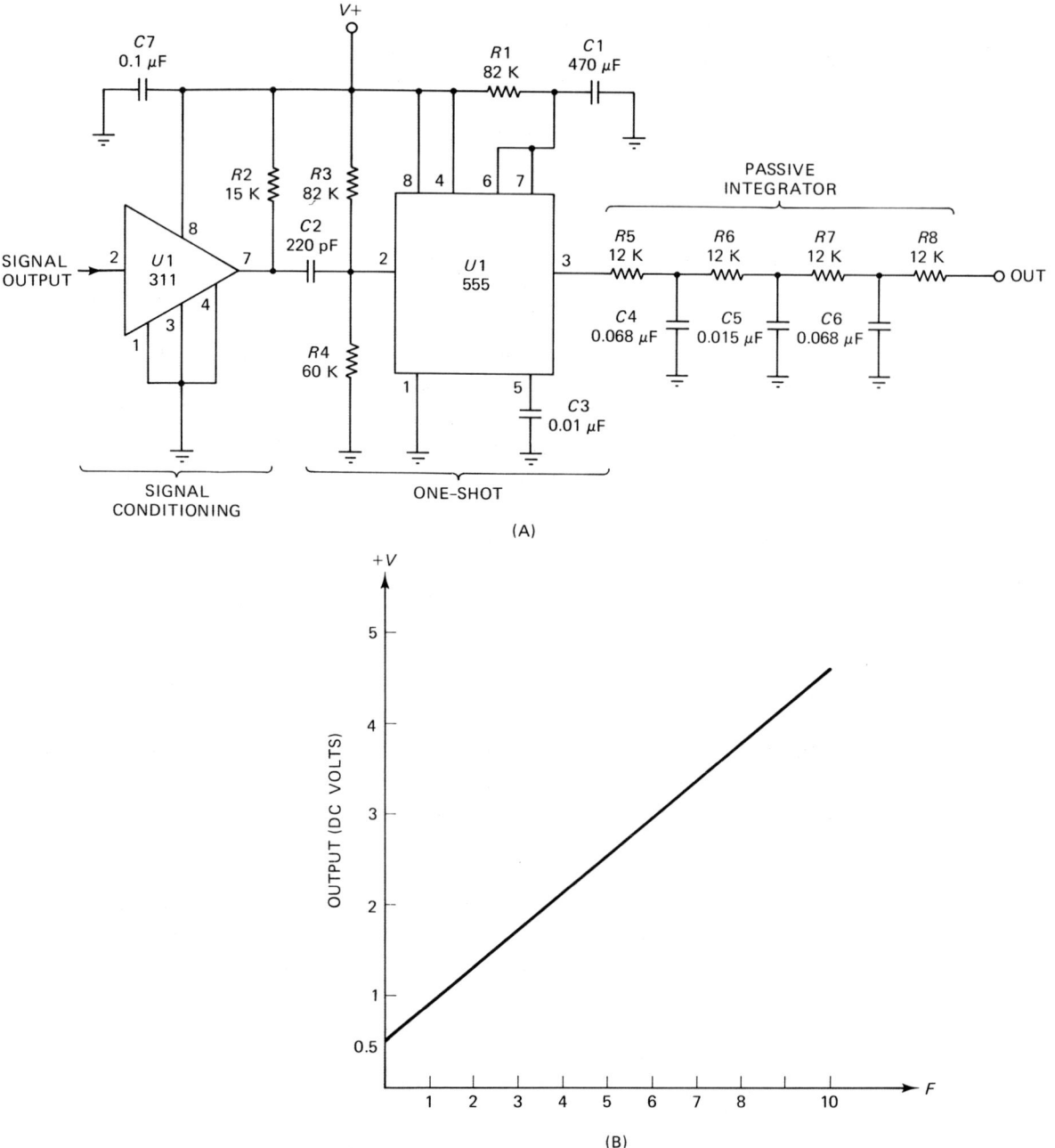

Fig. 21-14 Practical tachometer circuit: (A) circuit diagram and (B) output transfer function.

(U2). In this mode the LM-311 is operating basically as a zero-crossing detector circuit. The output of the 555 is a pulse train that has constant amplitude and duration. These pulses vary only in repetition rate, which is the same as the frequency of the input signal. The 555 output pulses are integrated in a passive *RC* integrator (R5-R7/C4-C6). The output of the integrator is a DC voltage that is a linear function of

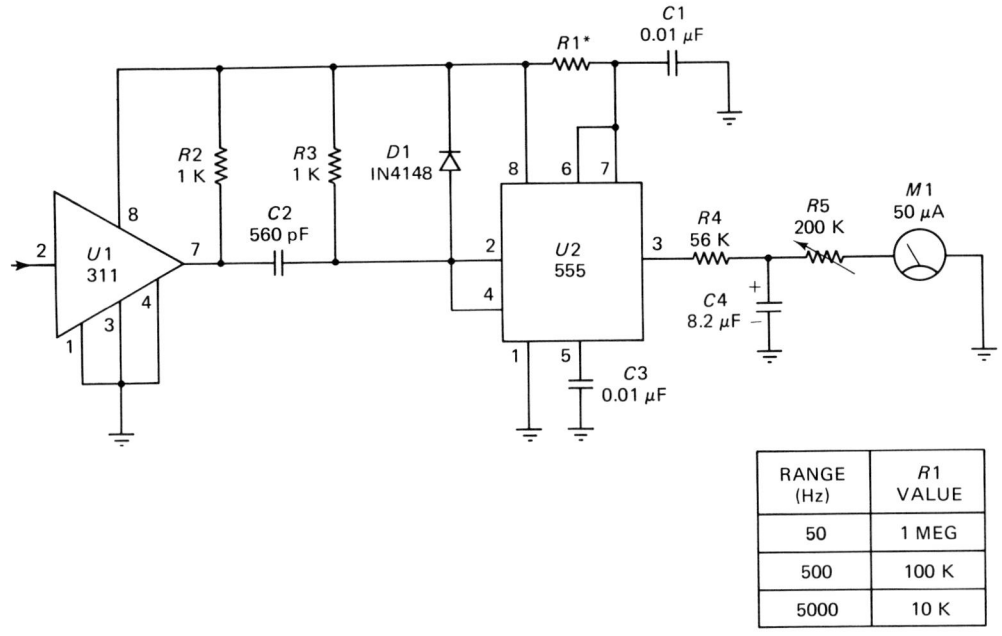

Fig. 21-15   Audio-frequency meter circuit.

RANGE (Hz)	R1 VALUE
50	1 MEG
500	100 K
5000	10 K

input frequency (see Fig. 21-14B). This DC voltage can be scaled, if necessary, to any desired level.

A related circuit is shown in Fig. 21-15. This 555-based tachometer is used to measure audio frequency over three ranges: DC to 50 Hz, DC to 500 Hz, and DC to 5000 Hz. The circuit uses the same form of input signal conditioning as the previous circuit, and it uses a 555 as the one-shot circuit. The integration function is taken up by the combination of RC network $R4/C4$ and the mechanical inertia of the meter ($M1$) movement.

ASTABLE OPERATION OF THE 555 IC TIMER

An astable multivibrator (AMV) is a free-running circuit that produces a square-wave output signal. The 555 can be connected to produce a variable duty cycle AMV circuit (Fig. 21-16). A version of the circuit showing the internal stages of the 555 is shown in Fig. 21-16A, while the circuit as it normally appears in schematic drawings is shown in Fig. 21-16B. The factor that makes this circuit an AMV is that the threshold and trigger pins (6 and 2) are connected together, forcing the circuit to be self-retriggering.

Under initial conditions at turn-on the voltage across timing capacitor $C1$ is zero, while the biases on COMP1 and COMP2 are (as usual) set to $2(V+)/3$ and $(V+)/3$, respectively, by the internal

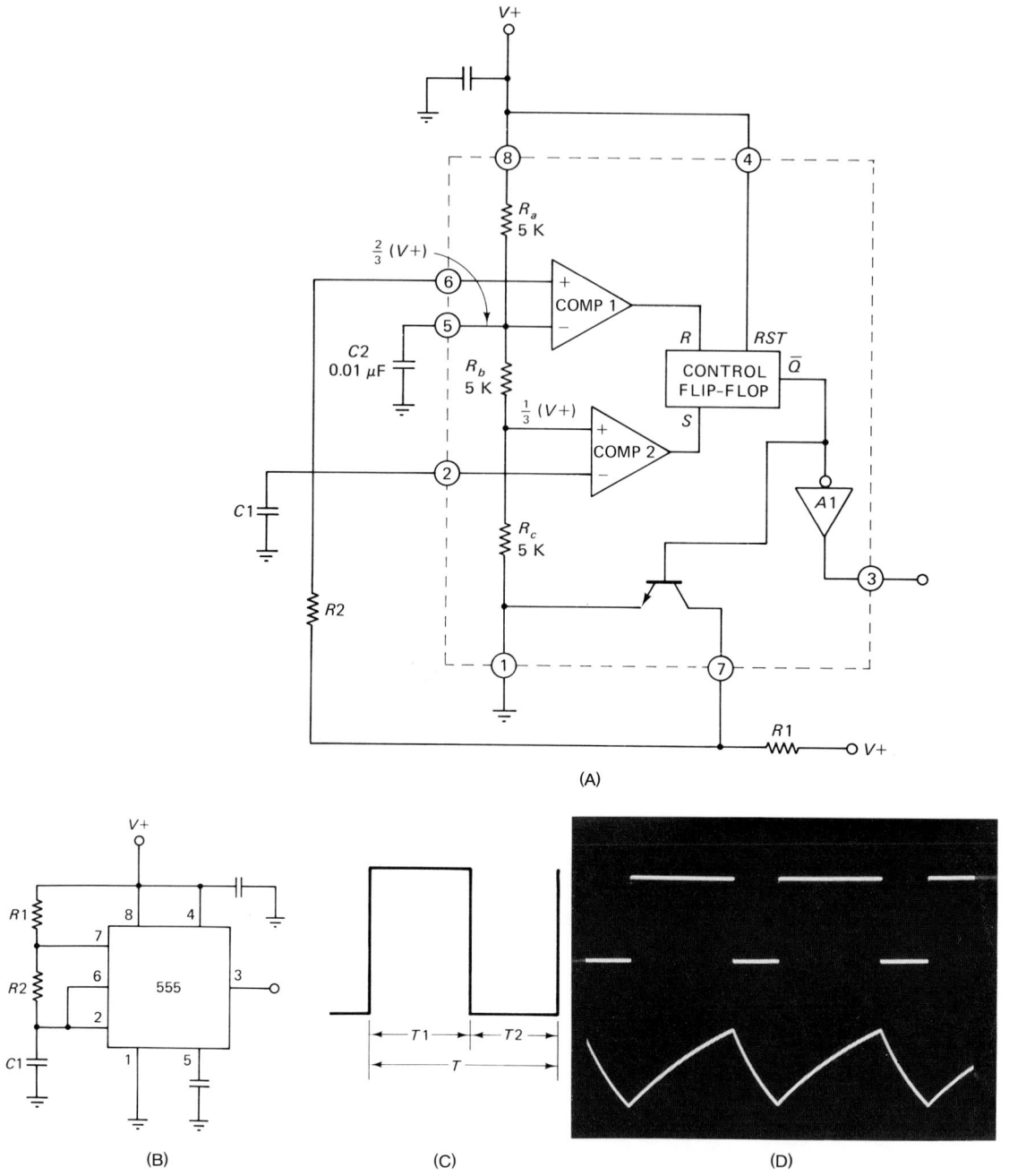

Fig. 21-16 Astable operation of the 555 IC timer: (A) block diagram form for the circuit; (B) schematic form of the circuit, (C) output waveform definitions, and (D) actual timing waveform for an operating 555 astable multivibrator.

resistor voltage divider (R_a, R_b, and R_c). The output of the 555 is HIGH under this condition, so $C1$ begins to charge through the combined resistance ($R1 + R2$). On discharge, however, transistor Q1 shorts the junction of $R1$ and $R2$ to ground, so the capacitor discharges only through $R2$. The result is the waveform shown in Fig. 21-16C. The time that the output is HIGH is $t1$, and the LOW time is $t2$. The period (T) of the output square wave is the sum of these two durations: $T = (t1 = t2)$. As with all similar RC-timed circuits, the equation that sets oscillating frequency is determined from Eq. (21-8).

$$T = -RC \ln\left(\frac{V - V_{C2}}{V - V_{C1}}\right) \qquad (21\text{-}8)$$

For the case where the output is HIGH ($t2$ in Fig. 21-16C), the resistance R is ($R1 + R2$), and the capacitance (C) is $C1$. Because of the internal biases of the voltage comparator stages of the 555, the capacitor will charge from $(V+)/3$ to $2(V+)/3$ and will then discharge back to $(V+)/3$ on each cycle. Thus, Eq. (21-8) can be rewritten.

$$t1 = -(R1 + R2)C1 \ln\left(\frac{(V+) - 2(V+)/3}{(V+) - (V+)/3}\right) \qquad (21\text{-}9)$$

or, once the algebra is done,

$$t1 = 0.695(R1 + R2)C1 \qquad (21\text{-}10)$$

By similar argument it can be shown that

$$t2 = 0.695R2C1 \qquad (21\text{-}11)$$

For the total period T,

$$T = t1 + t2 \qquad (21\text{-}12)$$

$$T = [0.695(R1 + R2)C1] + [0.695R2C1]$$

$$T = 0.695(R1 + 2R2)C1 \qquad (21\text{-}13)$$

Equation (21-13) defines the period of the output square wave. To find the frequency of oscillation, take the reciprocal of Eq. (21-13).

$$F = 1/T \qquad (21\text{-}14)$$

$$F = \frac{1.44}{(R1 + 2R2)C1} \qquad (21\text{-}15)$$

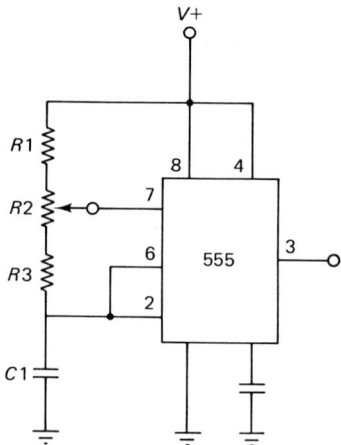

Fig. 21-17   Variable duty cycle astable circuit.

Duty Cycle of 555 Astable Multivibrator

Time segments $t1$ and $t2$ are not equal in most cases, so the charge and discharge times for capacitor $C1$ are also not equal (see Fig. 21-16D). The duty cycle of the output signal is the ratio of the HIGH period to the total period ($t1/T$). Expressed as a percentage,

$$\%DC = \frac{R1 + R2}{R1 + 2R2} \tag{21-16}$$

Various method are used for varying the duty cycle. First, a voltage can be applied to pin no. 5 (control voltage). Second, a resistance can be connected from pin no. 5 to ground. Both of these tactics have the effect of altering the internal bias voltages applied to the comparator.

Alternatively, one can also divide the external resistances $R1$ and $R2$ into three values. Figure 21-17 shows a variable duty factor 555 AMV that uses a potentiometer ($R2$) to vary the ratio of the charge and discharge resistances.

Synchronized Operation of the 555 Astable Multivibrator

A synchronized AMV operates with its operating frequency locked to an external frequency. The horizontal and vertical deflection oscillators in a television receiver operate in this manner because they are locked to the sync pulses transmitted by the TV broadcast station. A method for locking-in the oscillating frequency of the 555 is shown in Fig. 21-18. In this circuit a 7400 TTL NAND gate is used to sample both

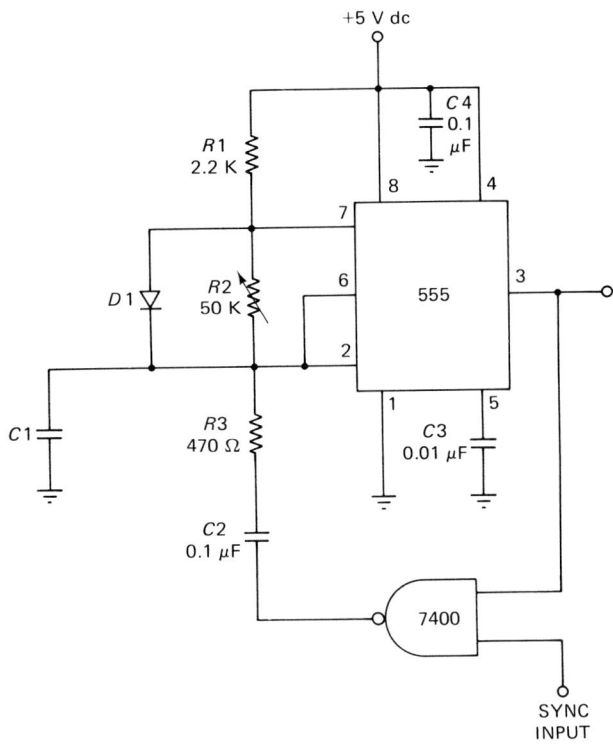

Fig. 21-18   Frequency-locking the 555 to an external oscillator.

the 555 AMV output signal and to the input sync signal. The properties of the NAND gate are these:

1. If either input is LOW, then the output is HIGH.
2. Both inputs must be HIGH for the output to be LOW.

Because the output of the NAND gate is applied to the timing circuit of the 555 (through wave-shaping network $R3/C2$), it will affect the relative timing of the circuit. This circuit is analogous to a mechanical pendulum oscillator that has an external nonresonant forcing frequency applied. The circuit will lock to the new frequency if it is reasonably close to the natural oscillating frequency or to an integer harmonic of the natural frequency. Students interested in modern chaos theory might want to investigate the behavior of this and similar circuits at sync frequencies away from the natural frequency or its harmonics and subharmonics.

A 555 Sawtooth Generator Circuit

A sawtooth waveform (Fig. 21-19A) rises linearly to some value, and then drops abruptly back to the initial conditions. The circuit for a

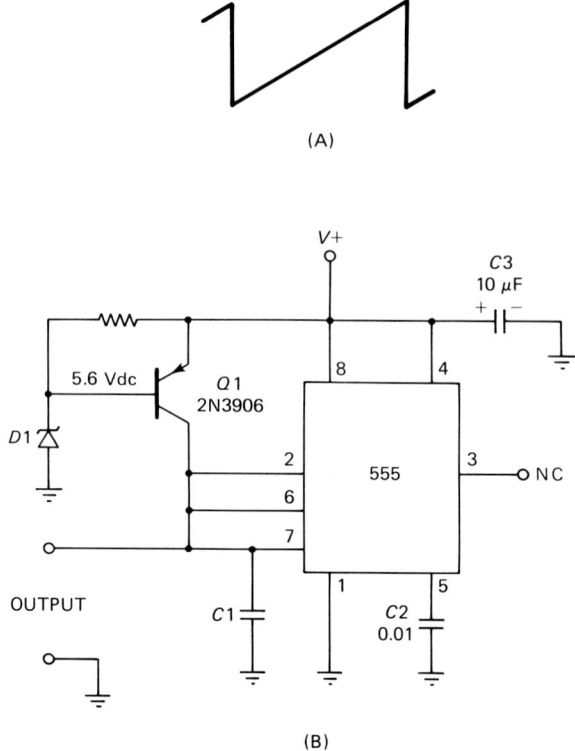

Fig. 21-19 (A) Sawtooth waveform and (B) 555 sawtooth generator circuit. (Note that the output is taken across timing capacitor $C1$.)

555-based sawtooth generator (Fig. 21-19B) is simple; it is based on the 555 timer IC. The basic circuit is the monostable multivibrator configuration of the 555, in which one of the timing resistors is replaced with a transistor operated as a current source (Q1). Almost any audio small signal PNP silicon replacement transistor can be used, although for this test the 2N3906 device was used here. The zener diode is a 5.6 V DC unit. Note that the output is taken from pins 6-7, rather than the regular chip output pin no. 3—which is not used.

The circuit as shown is a one-shot multivibrator. Triggering occurs in the 555 when pin no. 2 is brought to a potential less than 2/3 the supply potential. When a pulse is applied to pin no. 2 through differentiating network $R1C1$, the device will trigger because the negative-going slope meets the triggering criteria. To make an astable sawtooth multivibrator, drive the input of this circuit with either a square wave or pulse train that produces at least one pulse for each required sawtooth. Being a nonretriggerable monostable multivibrator, the circuit of Fig. 21-19B will ignore subsequent trigger pulses during the one-shot's refractory period.

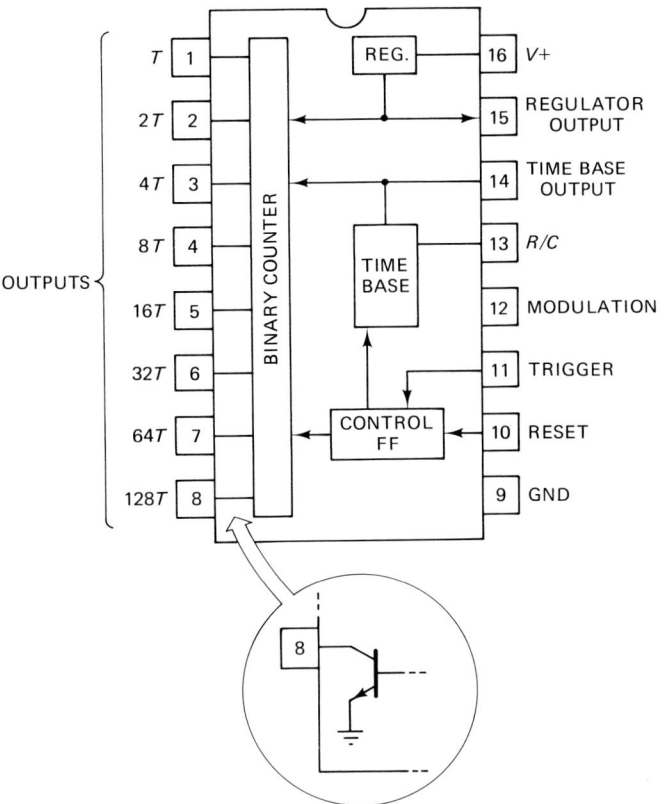

Fig. 21-20 Internal circuitry for the XR-2240 IC timer.

XR-2240 FAMILY OF IC TIMERS

The XR-2240 (also designated 8240) IC timer (Fig. 21-20) is based on the same circuit concept as the 555 device, that is, a window comparator that sets and resets a control flip-flop. The timebase circuit must receive its power from outside the chip, so in normal operation a 20-kΩ resistor is connected between the regulator output (pin no. 15) and timebase output (pin no. 14). This feature allows external timebase circuits to be used.

The timebase section of the XR-2240 is basically the 555-style timer modified to remove the constant 1.1 from the timing equation [see Eq. (21-6)]. The constant is in the 555 because the resistors in the internal voltage divider that biases the comparators are equal. But in the XR-2240 nonequal resistances are used, making the reference levels $0.27(V+)$ and $0.73(V+)$ instead of $0.33(V+)$ and $0.67(V+)$ as used in the 555. This change was made to simplify the timing equation to

$$T = RC \qquad (21\text{-}17)$$

The values of the timing components are $1 \text{ k}\Omega \le R \le 10 \text{ M}\Omega$, and $0.05 \text{ } \mu\text{F} \le C \le 1000 \text{ } \mu\text{F}$. The XR-2240 is packaged in a 16-pin DIP case; it will operate over the range $+4.5$ to $+18$ V DC. The XR-2240 differs from the 555 in that the trigger and reset pulses are positive-going rather than negative-going. The minimum amplitude of the trigger and reset pulses is approximately two PN-junction voltage drops, or about 1.4 V DC. As a practical matter, however, a minimum 3-V DC level is recommended to guard against trigger failures caused by negative noise impulses.

There are other differences between the 555 device and the XR-2240. The XR-2240 contains an eight-bit binary counter with open-collector NPN transistor outputs (see inset to Fig. 21-20). If these outputs are wired in the logical OR configuration, times from $1T$ to $255T$ can be programmed, where T is defined by Eq. (21-17). Each output is connected to $V+$ through a 10-kΩ pull-up resistor. The use of the open-collector configuration makes the output active-LOW. The combination of the large range of timer RC components and an eight-bit binary counter makes it possible to create very long duration timer circuits that are essentially as stable as the RC network. It is also possible to use the binary outputs independently of each other if each is supplied with its own 10-kΩ pull-up resistor. Examples of such applications include binary address sequences for digital circuits, and oscillators with base-2 related synchronized output frequencies. Other, less common, versions of the timer include the XR-2250 (or 8250), which offers binary coded decimal (BCD) outputs instead of binary, and the XR-2260, which uses BCD outputs but limits the most significant digit to 6.

Figure 21-21 shows the circuit for both the monostable and astable multivibrator configurations for the XR-2240. One interesting design feature of the XR-2240 is that the sole difference between the monostable and astable modes is a single-feedback resistor ($R3$) that automatically retriggers the XR-2240 at the end of each output cycle.

The timer is set into operation by applying a positive-going pulse to the trigger input (pin no. 11). This pulse is routed to the control logic, and performs several jobs simultaneously: resetting the binary counter to 00000000_2, driving all outputs HIGH and enabling the timebase circuit. As was true with the 555 timer, the XR-2240 works by charging capacitor $C1$ through resistor $R1$ from positive voltage source $V+$.

The open-collector configuration permits different multiples of the basic timer duration to be programmed by connecting the required outputs together in a wire OR configuration. For example, if a 57-second timer is needed, it is possible to use a 1-MΩ resistor and a 1-μF capacitor ($T = 1$ second) and connect together the $1T$, $8T$, $16T$, and $32T$ outputs: $(1 + 8 + 16 + 32)T = 57T$. In that circuit, pins 1, 4, 5, and 6 will be connected together and to $V+$ through a single 10-kΩ

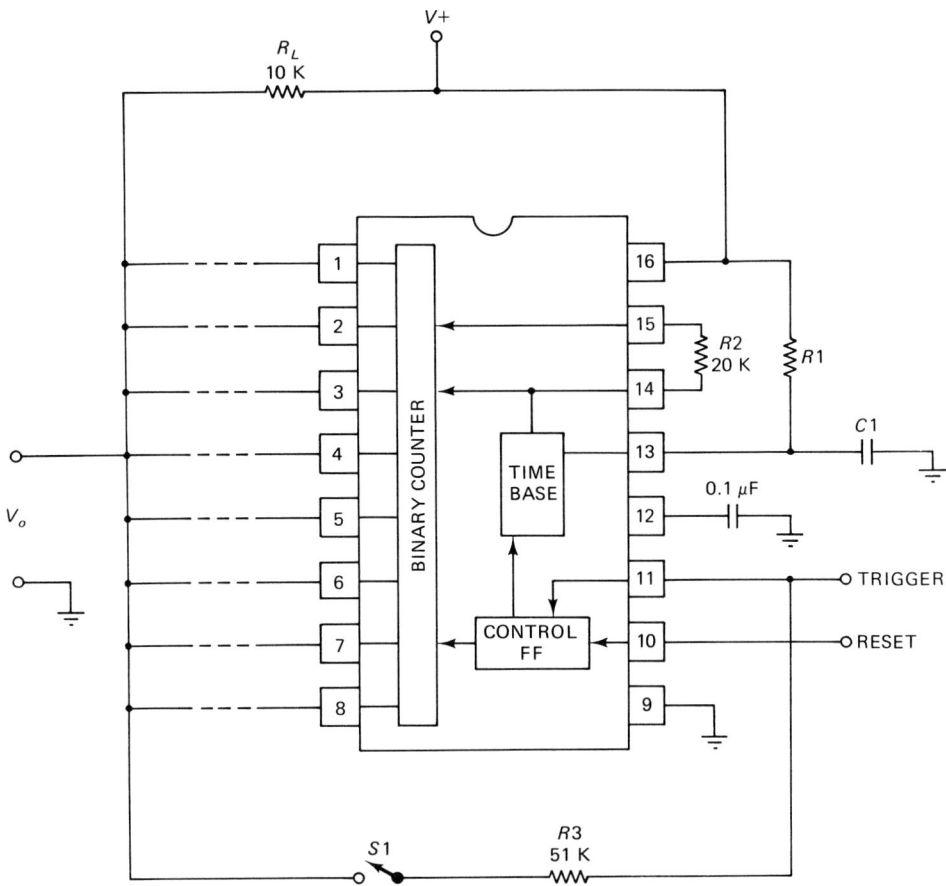

Fig. 21-21 Basic circuitry for the XR-2240 timer. Close S1 for astable operation.

resistor. The output will remain LOW for 57 seconds following trigger-ing, and then it will return to the HIGH state.

The XR-2240 is considered more flexible than the 555 because the duration (monostable mode) or period (astable mode) can be set either by the *RC* timing network or through selection of which outputs are wired together at the output of the circuit.

Synchronization to an external timebase, or modulation of the pulse width, is possible by manipulation of the modulation input (pin no. 12). In normal operation, the modulation input is bypassed to ground through a 0.01-μF capacitor so that noise signals will not disrupt operation of the device. A voltage applied to pin no. 12 will modulate the pulse width of the timebase output signal. This modulat-ing voltage should be between +2 and +5 volts for a change factor of 0.4 to 2.25.

Synchronization of the XR-2240 to an external time base is accomplished through a series *RC* network consisting of a 5.1-kΩ resistor and a 0.1-μF capacitor (see Fig. 21-22) connected to pin no.

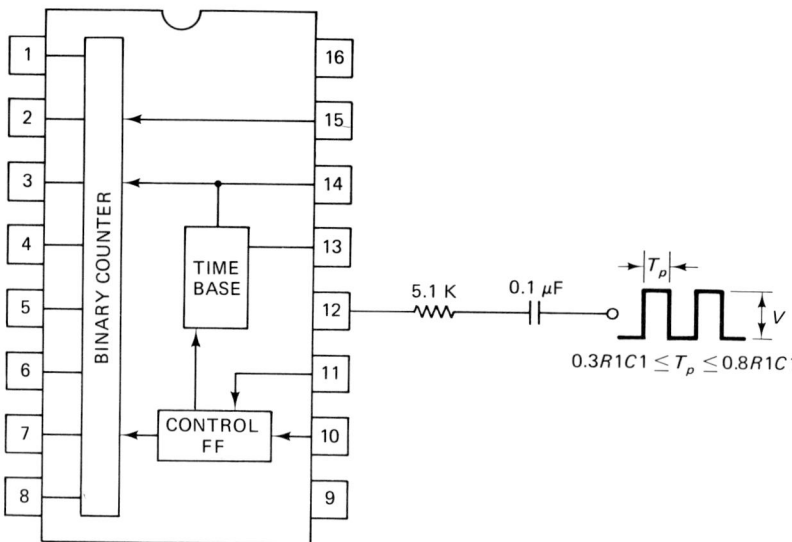

Fig. 21-22   Adding external synchronization to the XR-2240.

12. This network differentiates the input pulse. The synchronization pulse should have an amplitude of at least $+3$ V and a period of 0.3 to $0.8T$. Another method of synchronization is to use an external time-base connected directly to pin no. 14.

Very Long Duration Timers

The long-duration timer presents special problems to the electronic circuit designer. The drift and other errors of some components tend to accumulate, so the total error becomes large over longer time periods. In fact, whenever the errors are a function of time the long-duration timer will suffer markedly. Any long duration RC network, for example, will suffer from several time-related problems.

A viable alternative is to use an RC timer. But most high-value capacitors are electrolytic types, and therefore have a tolerance of -20 to $+100$ percent of the label capacitance. In addition, some capacitors show capacitance variation with both time and applied voltage. In addition, many large-value capacitors exhibit considerable shunt resistance that must be considered in timing networks. Because of these problems the 555 is limited to practical durations of about 100 seconds, if precision is a consideration.

The timebase section of the XR-2240 timer suffers from the same problems as does the 555, but the RC values needed for very long duration monopulse operations are much lower than the 555 because of the built-in binary counter. If the outputs of the XR-2240 are wired OR together, then times from 1 to $255RC$ are possible. For example, with an RC time constant of 10 seconds, a single XR-2240 can be

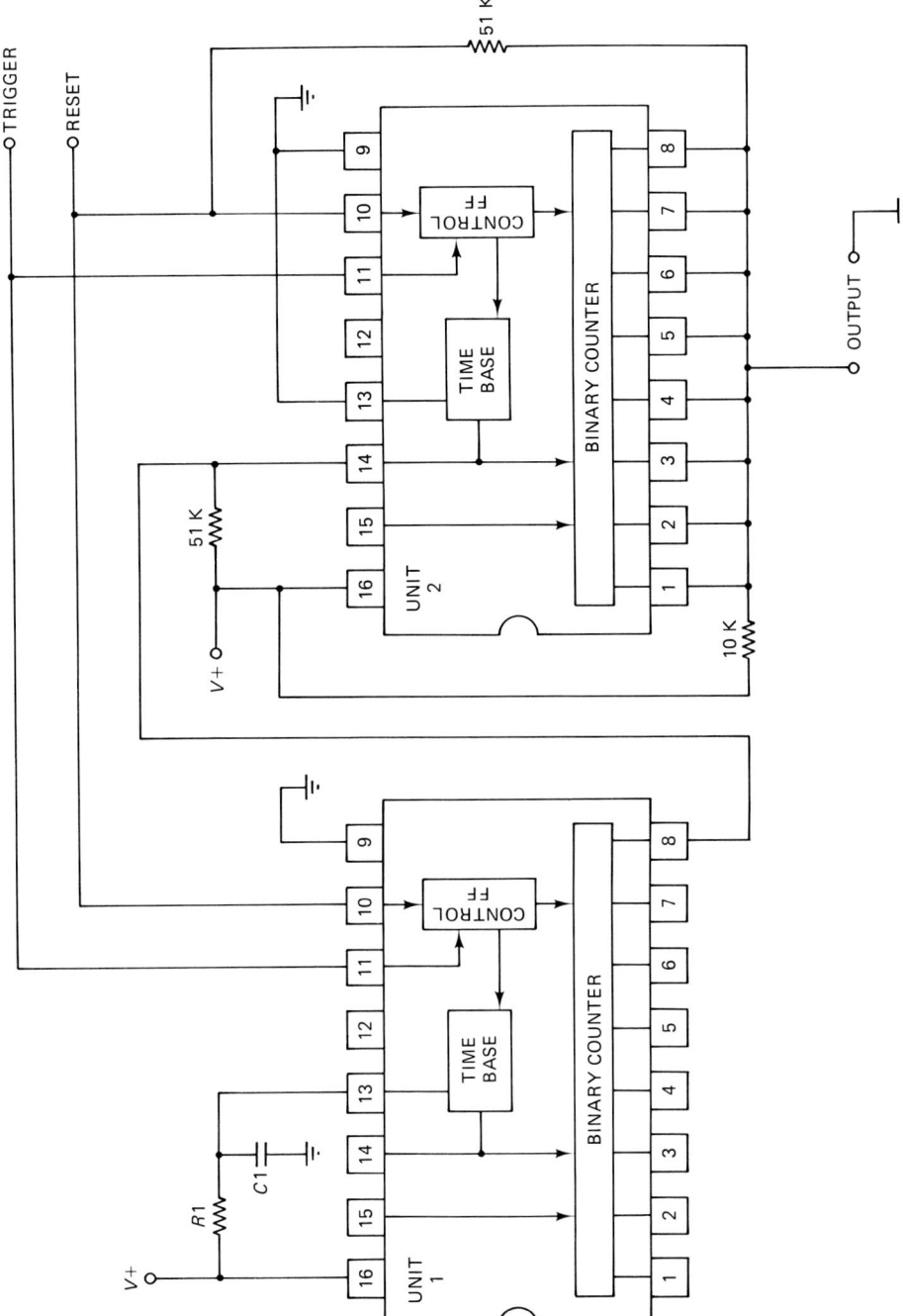

Fig. 21-23 A long-duration timer circuit based on the XR-2240.

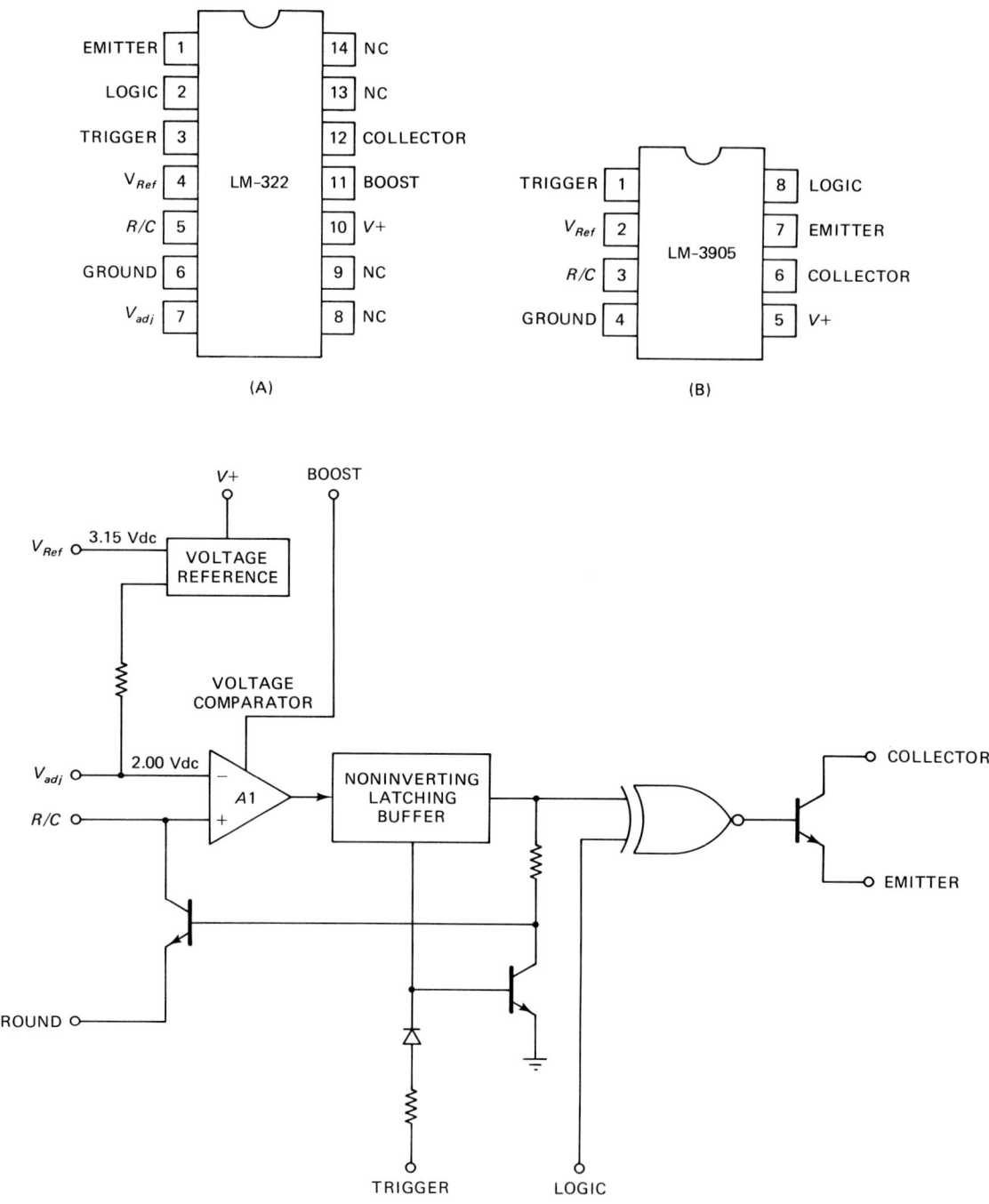

Fig. 21-24 (A) Internal circuitry of the LM-322 IC timer, (B) pin-outs for the LM-3905 timer, and (C) internal circuitry for the LM-3905 device.

used for up to 2550 seconds (42.5 minutes). Greater duration times can be accommodated by cascading two or more XR-2240 devices.

Figure 21-23 shows the use of two XR-2240 timers in cascade to produce a very long duration timer. Unit 1 is used as the timebase; it has a frequency set by the expression $1/RC$. Only the eight-bit of the binary counter ($128T$) is used. The output of Unit 1 becomes the timebase of Unit 2, which has all of its outputs wired OR together. The total period is $65,536T$. When the RC time constant is 1 second, the output duration is 65,536 seconds (18.2 hours!) even though the stability and accuracy attributes are those of a 1-second timer.

LM-122, LM-322, LM-2905, AND LM-3905 TIMERS

The LM-X22 and LM-X905 IC timers are precision devices that will operate with DC power supply voltages of $+4.5$ to $+40$ V DC to produce durations of microseconds to hours. The LM-122 and LM-322 device package is shown in Fig. 21-24A, and that of the LM-2905 and 3905 is shown in Fig. 21-24B. The LM-2905/3905 devices are identical to the LM-122/322 except that the BOOST and V_{adj} pins are not available. The LM-122 and LM-2905 operate over the temperature range -55 to $+125°C$, and the LM-322 and LM-3905 operate over 0 to $+70°C$. The internal circuitry of these timers is shown in Fig. 21-24C. The RC timing network is monitored by the noninverting input of a voltage comparator ($A1$); the inverting input is biased to $+2.00$ V DC. The output of the comparator is routed to a noninverting latching buffer, which in turn drives an exclusive-OR (XOR) gate. The alternate input of the XOR gate is connected to the LOGIC pin on

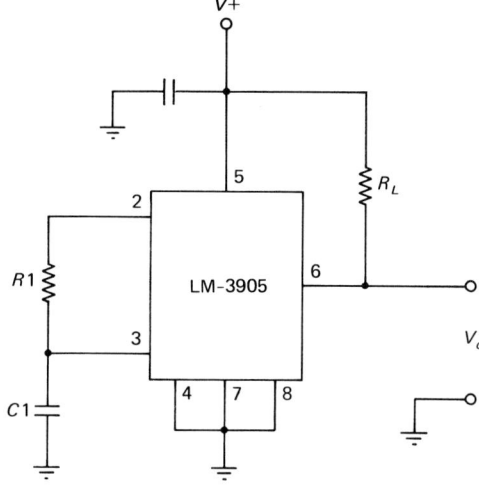

Fig. 21-25 Circuit using the LM-3905 timer.

the device. The output of these devices is a floating emitter and floating collector transistor. Both ground referenced and floating loads, at potentials up to 40 V DC, can be accommodated with this arrangement. The V_{adj} pin can be used in the LM-122 and LM-322 devices to vary the timing ratio up to 50 : 1 by using an external voltage. The circuit for the basic timer is shown in Fig. 21-25. The output duration is set by $T = R1C1$.

22

IC Data Converters and Their Application

One of the most powerful methods for using sensors and analog electronics is to form the front-end to input data into a digital computer.[1] The personal computer revolution made it possible to place small computers, often with extremely high computing capacity, into the cabinet of an electronic instrument. In still other cases, the personal computer *is* the instrument. For example, IBM-XT/AT compatible machines have become the mainstays of process control, data acquisition, and monitoring systems.

The analog signals output by most sensors and the signals provided by analog electronics are not compatible with the digital computer. As a result, data converters are needed. In this chapter, we will take a look at the basic methods of data conversion. If you are not familiar with the nature of, and problems with, sampled signals, then you are referred to Chapter 2 for a review of the subject.

In Chapter 23 we will take a look at some practical examples of commercial IC data converters, and in Chapter 24 we will discuss the methods of interfacing data converters between the analog world and the personal computer. First, however, we will become familiar with the basic methods of performing digital-to-analog and analog-to-digital conversions.

BASICS OF DATA CONVERSION

The data converter does one of two jobs, it either (1) converts a binary digital word to an equivalent analog current or voltage, or (2) converts

---

[1]Binary (base-2) numbers are used in discussing digital computers and digital circuits. The notation adopted in this text is to add a subscript 2 to any number in binary. Thus, "1000" denotes the decimal quantity "one-thousand" and "1000_2" denotes the binary number that is equivalent to decimal "eight." Where used, hexadecimal (base-16) numbers are denoted with an "H" as part of the number.

an analog current or voltage to an equivalent binary word. The former are digital-to-analog converters (DACs), and the latter are analog-to-digital converters (A/D or ADC). These devices form the interface between digital computers and the analog world. This chapter will examine the basic functioning of these data-conversion building blocks.

APPROACHES TO DATA CONVERSION

There are several approaches to converting data from electronic circuits and scientific instruments into the digital binary numbers required by a digital computer. This job requires an A/D converter. The opposite type of converter (DAC) does the opposite work. It converts the binary output from a computer into an analog voltage or current. In discussing data converters it is best to begin with a discussion of DACs, because many A/D circuits contain embedded DACs.

DIGITAL-TO-ANALOG CONVERTERS

There are several different approaches to DAC design, but all of them are varieties of a weighted current or voltage system that generates binary words by appropriate switch contacts. The most common example is the R-$2R$ ladder shown in Fig. 22-1. The active element, $A1$, is an operational amplifier in a unity-gain inverting follower configuration. In the circuit of Fig. 22-1 the digital inputs are shown as mechanical switches, but in a real data-converter circuit the switches would be replaced by electronic switching devices (e.g., transistors).

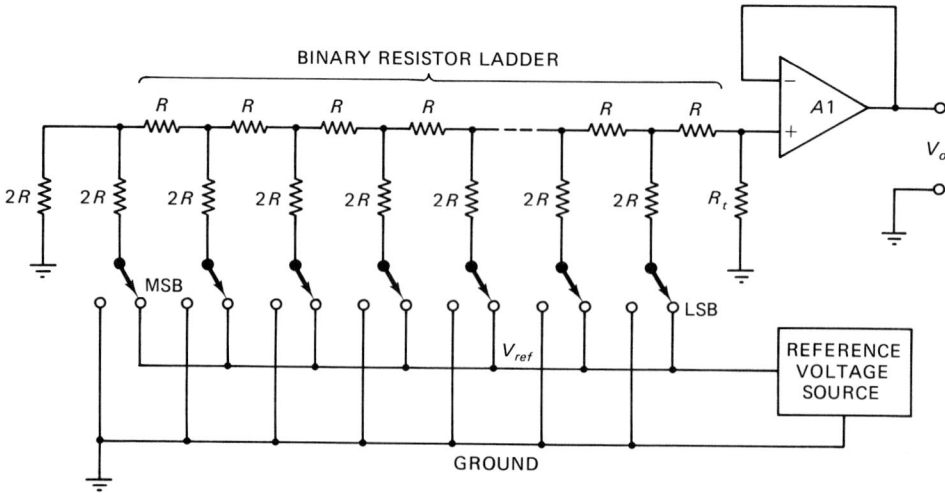

Fig. 22-1 R-$2R$ ladder network is the basis for many digital-to-analog converters.

The electronic switches are driven by either a binary counter or an N-bit parallel data line.

A precision reference voltage (V_{ref}) source is required for accurate data conversion. This voltage is most often $+2.56$ V, $+5.00$ V, or $+10.00$ V in practical circuits. Other voltages can also be used, however. The accuracy of the converter is dependent on the precision of the reference voltage source. There are other sources of error; but if the reference voltage accuracy is poor, there is no hope for any other factors to be effective in improving the performance of the circuit. Although almost any voltage regulator can be pressed into service as the reference, it is prudent to select a precision, low-drift model.

Returning to Fig. 22-1, consider the circuit action under circumstances where various binary bits are either HIGH or LOW. If all bits are LOW, then the output voltage will be zero. The value of the output voltage is given by the product IR, and when all bits are LOW this current is zero. In practical circuits, though, there might be some output voltage under these circumstances due to offsets in the operational amplifier, the R-$2R$ ladder, and the electronic switches. These offsets can be nulled to zero output voltage when all bits are intentionally set to zero (or ignored, if negligible).

The unterminated R-$2R$ ladder produces an output current. Some commercial IC DACs are current output models and have no output amplifier. If there is a terminating resistor (R_t) shunting the output terminals of the DAC, then the circuit produces an output voltage $I_o R_t$. The output impedance of such a circuit tends to be high, so some of these DACs use an output amplifier to produce a low-impedance voltage output. The transfer function of the R-$2R$ ladder type of DAC is

$$V_o = V_{ref} A / 2^N \qquad (22\text{-}1)$$

where V_o is the output potential, V_{ref} the reference potential, A the decimal value of the applied binary word, and N is the number of bits in the applied binary word.

If the most significant bit (MSB) is made 1 (i.e., HIGH), the output voltage will be approximately $V_{ref}/2$. Similarly, if the next most significant bit is turned on (set to HIGH) and all others are LOW, the output will be $V_{ref}/4$. The least significant bit (LSB) would contribute $V_{ref}/2^N$ to the total output voltage. For example, with an eight-bit DAC, the LSB changes the output $V_{ref}/2^8$, or $V_{ref}/256$. This change is called the "1-LSB" value.

Figure 22-2 graphs the output of a voltage DAC in response to the entire range of binary numbers applied to the digital inputs. The result is a staircase waveform that rises by the 1-LSB value for each 1-LSB change of the binary word. This step height represents the minimum discernible resolution of the circuit.

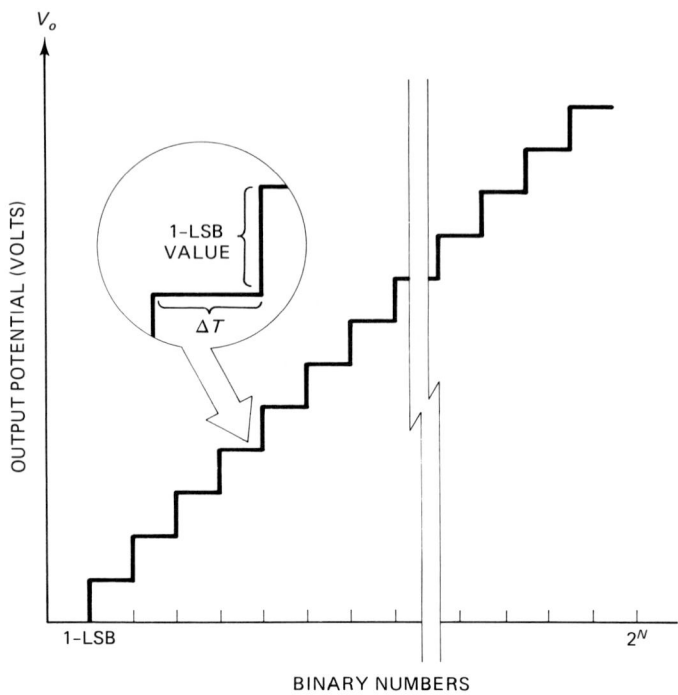

Fig. 22-2 Change of the least significant bit (LSB) causes the DAC output voltage to change by the 1-LSB equivalent voltage.

The reference source can be either internal or external to an integrated circuit DAC. If the reference voltage or current source is external, the DAC is said to be a multiplying DAC or MDAC. The multiplication takes place between the analog reference and the fraction defined as $A/2^N$ in Eq. (22-1). If the reference source is completely internal and not adjustable (except for fine trimming), the DAC is said to be a nonmultiplying DAC or simply DAC.

A practical, low-cost DAC example is presented in Chapter 23.

Coding Schemes

There are several different coding methods for defining the transfer function of the DAC. The most common of these are unipolar positive, unipolar negative, symmetrical bipolar, and asymmetrical bipolar.

The unipolar coding schemes provide an output voltage of one polarity only. These circuits usually produce 0 volts for the minimum and some positive or negative value for the maximum. Because one binary number state represents 0 V DC, there are $(2^N - 1)$ states to represent the analog voltages within the range. For example, in an eight-bit system there are 256 states, so if one state (00000000_2) represents 0 V DC, there are 255 possible states for the nonzero voltages. Thus, the maximum output voltage is always 1-LSB less than the reference voltage. For example, in a 10.00-V DC system, the maximum output voltage is 10.00 V/256 = 0.039 V = 39 mV less. Thus, the maximum output voltage will be approximately 9.96 V DC. The unipolar positive coding scheme is shown in the table here.

Unipolar output voltage	Binary input word
0.00 V DC	00000000_2
$V_{max}/2$	10000000_2
V_{max}	11111111_2

The negative version of this coding scheme (unipolar negative) is identical, except that the mid-scale voltage is $-V_{max}/2$ and the full-scale output voltage is $-V_{max}$. A variant on the theme inverts the definition so that the coding scheme is as shown on this table.

Unipolar output voltage	Binary input word
0.00 V DC	11111111_2
$V_{max}/2$	10000000_2
V_{max}	00000000_2

The bipolar coding scheme faces a difficulty that requires a trade-off in the design. There is an even number of output states in a binary system. For example, in the standard eight-bit system there are 256 different output states. If one state is selected to represent 0 V DC, there are 255 states left to represent the voltage range. As a result, there is an uneven number of states to represent positive and negative states on either side of 0 V DC. For example, 127 states might be assigned to represent negative voltages and 128 to represent positive voltages. In the asymmetrical bipolar coding, therefore, the pattern might look as shown in this table.

Bipolar output voltage	Binary input word
$-V_{max}$	00000000_2
0.00 V DC	10000000_2
$+V_{max}$	11111111_2

A decision must be made regarding which polarity will lose a small amount of dynamic range.

The other bipolar coding system is the symmetrical bipolar scheme. The decision in the symmetrical scheme is that each polarity will be represented by the same number of binary states on either side of 0 V DC. But this scheme does not permit a dedicated state for zero. The scheme is

Bipolar output voltage	Binary input word
$-V_{max}$	00000000_2
$-$ Zero (-1 LSB)	01111111_2
0.00 V DC	(disallowed)
$+$ Zero ($+1$ LSB)	10000000_2
$+V_{max}$	11111111_2

The state *plus zero* is more positive then the 1-LSB value than 0 V DC, while the *minus zero* state is more negative than 0 V DC by the same 1-LSB value.

ANALOG-TO-DIGITAL CONVERTERS

The analog-to-digital converter (A/D) is used to convert an analog voltage or current input to an output binary word that can be used by a computer. Of the many techniques that have been published for performing an A/D conversion, only a few are of interest to us; so we will consider only the voltage-to-frequency, single-slope integrator, dual-slope integrator, counter (or servo), successive approximation, and flash methods.

Integration A/D Methods

Most digital panel meters (DPM) and digital multimeters (DMM) use either the single or dual-slope integration methods for the A/D conversion process. An example of a single-slope integrator A/D converter is shown in Fig. 22-3A. The single-slope integrator is simple, but it is limited to those applications that can tolerate accuracy of one or two percent.

The single-slope integrator A/D converter of Fig. 22-3A consists of five basic sections: ramp generator, comparator, logic, clock, and output encoder. The ramp generator is an ordinary operational amplifier Miller integrator with its input connected to a stable, fixed, reference voltage source. This makes the input current I_{ref} essentially constant; so the voltage at point B will rise in a nearly linear manner, creating the voltage ramp.

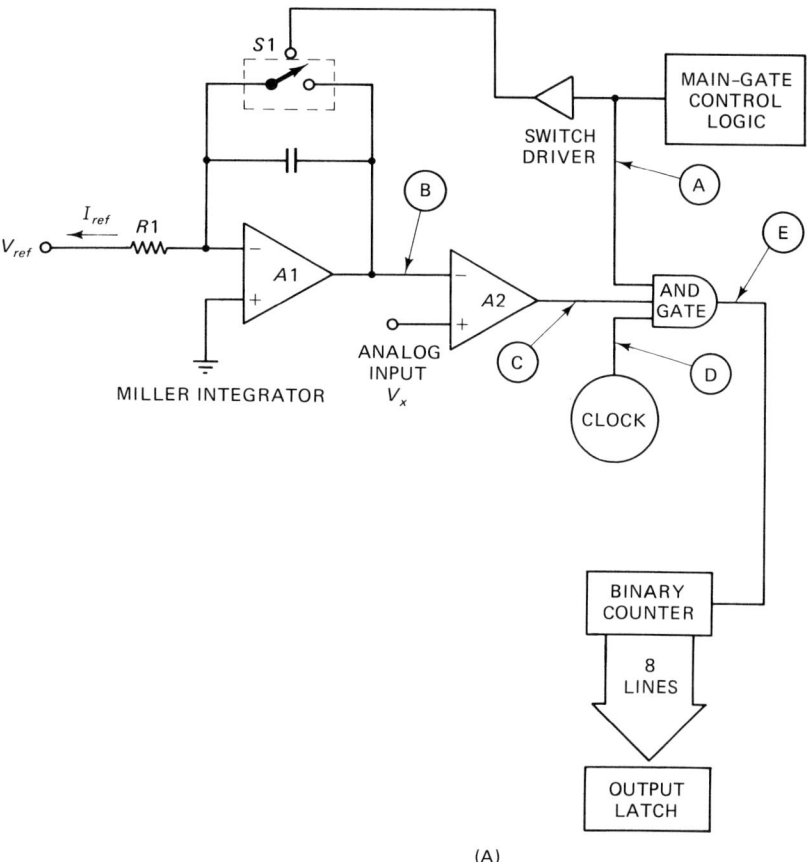

(A)

Fig. 22-3 (A) Single-slope integrator A/D converter circuit and (B) timing diagram. (*Figure continues.*)

The comparator is an operational amplifier that has no feedback loop. The circuit gain is the open-loop gain (A_{vol}) of the device selected —typically very high even in low-cost operational amplifiers. When the analog input voltage V_x is greater than the ramp voltage, the output of the comparator is saturated at a logic-HIGH level.

The logic section consists of a main AND gate, a main-gate generator, and a clock. The waveforms associated with these circuits are shown in Fig. 22-3B. When the output of the main-gate generator is LOW, switch S1 remains closed, so the ramp voltage is zero. The main-gate signal at point A is a low-frequency square wave with a frequency equal to the desired time-sampling rate. When point A is HIGH, S1 is open; so the ramp will begin to rise linearly. When the ramp voltage is equal to the unknown input voltage V_x, the differential voltage seen by the comparator is zero; so its output drops LOW.

The AND gate requires all three inputs to be HIGH before its output can be HIGH also; from times $T0$ to $T1$, the output of the AND gate will go HIGH every time the clock signal is also HIGH.

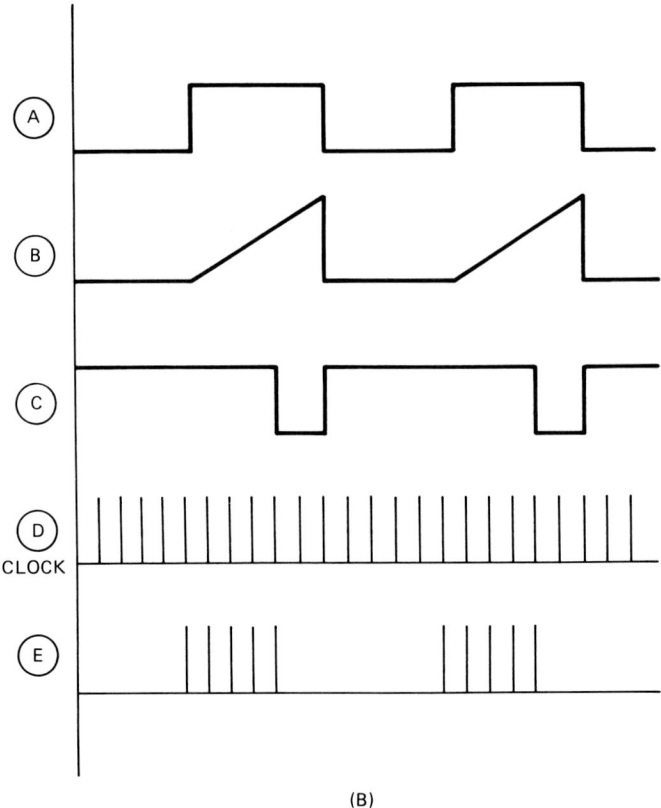

(B)

Fig. 22-3 (*continued*)

The encoder, in this case an eight-bit binary counter, will then see a pulse train with a length proportional to the amplitude of the analog input voltage. If the A/D converter is designed correctly, the maximum count of the encoder will be proportional to the maximum range (full-scale) value of V_x.

Several problems are found in single-slope integrator A/D converters.

- The ramp voltage may be nonlinear.
- The ramp voltage may have too steep or too shallow a slope.
- The clock pulse frequency could be wrong.
- The clock pulse frequency may be prone to changes in apparent value of V_x caused by noise.

Many of these problems are corrected by the dual-slope integrator of Fig. 22-4A. This circuit also consists of five basic sections: integrator, comparator, control logic section, binary counter, and a reference current or voltage source. An integrator is made with an operational amplifier connected with a capacitor in the negative feedback loop, as was the case in the single-slope version. The comparator in this circuit is also the same sort of circuit as was used in the previous example. In

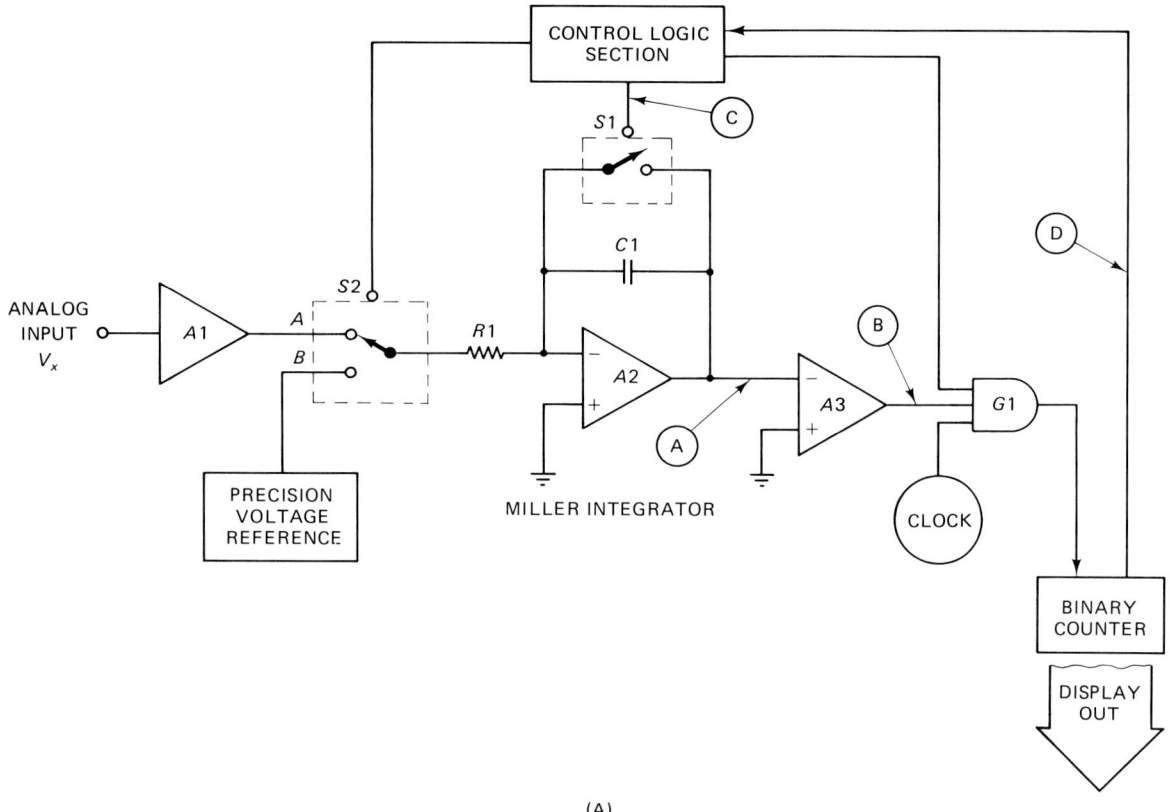

(A)

Fig. 22-4 (A) Dual-slope integrator A/D converter circuit and (B) timing diagram. (*Figure continues.*)

this case, though, the comparator is ground-referenced by connecting +IN to ground.

When a start command is received, the control circuit resets the counter to 00000000_2, resets the integrator to 0 V (by discharging $C1$ through switch S1), and sets electronic switch S2 to the analog input (position A). The analog voltage creates an input current to the integrator, which causes the integrator output to begin charging capacitor $C1$; the output voltage of the integrator will begin to rise. As soon as this voltage rises a few millivolts above ground potential (0 V DC), the comparator output snaps HIGH-positive. A HIGH comparator output causes the control circuit to enable the counter, which begins to count pulses.

The counter is allowed to overflow and this output bit sets switch S2 to the reference source (position B). The graph of Fig. 22-4B shows the integrator charging during the interval between start and the overflow of the binary counter. At time $T2$ the switch changes the integrator input from the analog signal to a precision reference source. Meanwhile, at time $T2$ the counter had overflowed, and again it has an output of 00000000_2 (maximum count plus one more count is the same as the initial condition). It will, however, continue to increment

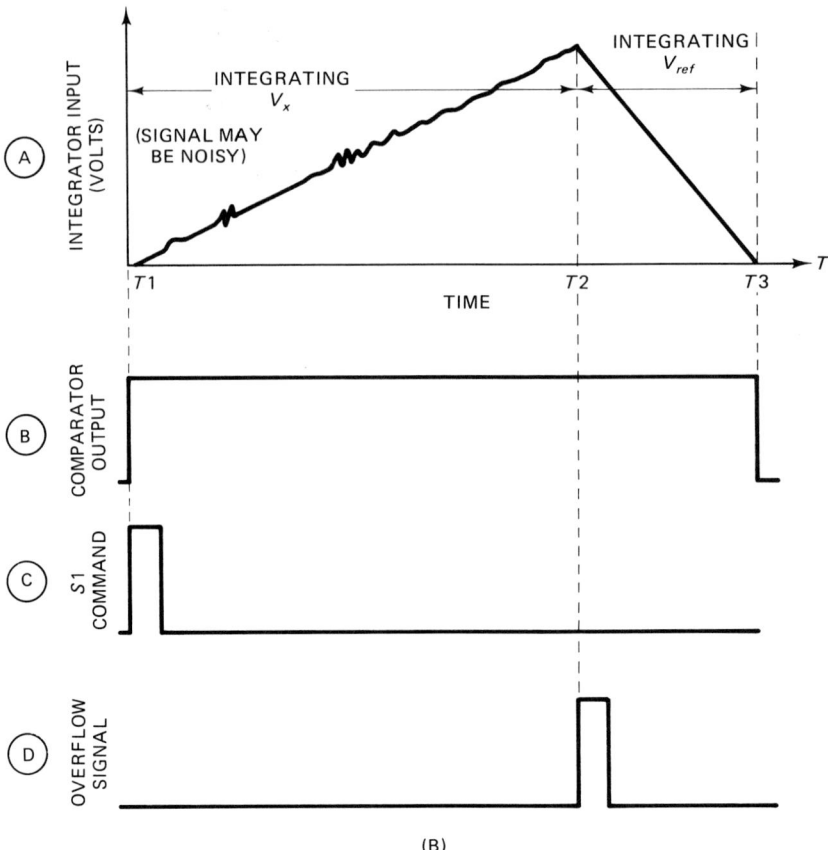

Fig. 22-4 (*continued*)

so long as we have a HIGH comparator output. The charge accumulated on capacitor $C1$ during the first time interval is proportional to the average value of the analog signal that existed between $T1$ and $T2$.

Capacitor $C1$ is discharged during the next time interval ($T2-T3$). When $C1$ is fully discharged the comparator will see a ground condition at its active imput, so it will change state and make its output LOW. Even though this causes the control logic to stop the binary counter, it does not reset the binary counter. The binary word at the counter output at the instant it is stopped is proportional to the average value of the analog waveform over the interval ($T1-T2$). An end of conversion (EOC) signal is generated to notify the computer that the output data is both stable and valid (therefore ready for use).

VOLTAGE-TO-FREQUENCY CONVERTERS

These circuits are not A/D converters in the strictest sense, but they are very good for representing analog data in a form that can be tape recorded on a low-cost machine. The V/F converter output can also be used for direct input to a computer if a binary counter is used to measure the output frequency. Two forms of V/F converter are common. One is a voltage-controlled oscillator (VCO); that is, a regular oscillator circuit in which the output frequency is a function of an input control voltage. If the VCO is connected to a binary or binary coded decimal (BCD) counter, the VCO becomes a V/F form of A/D converter.

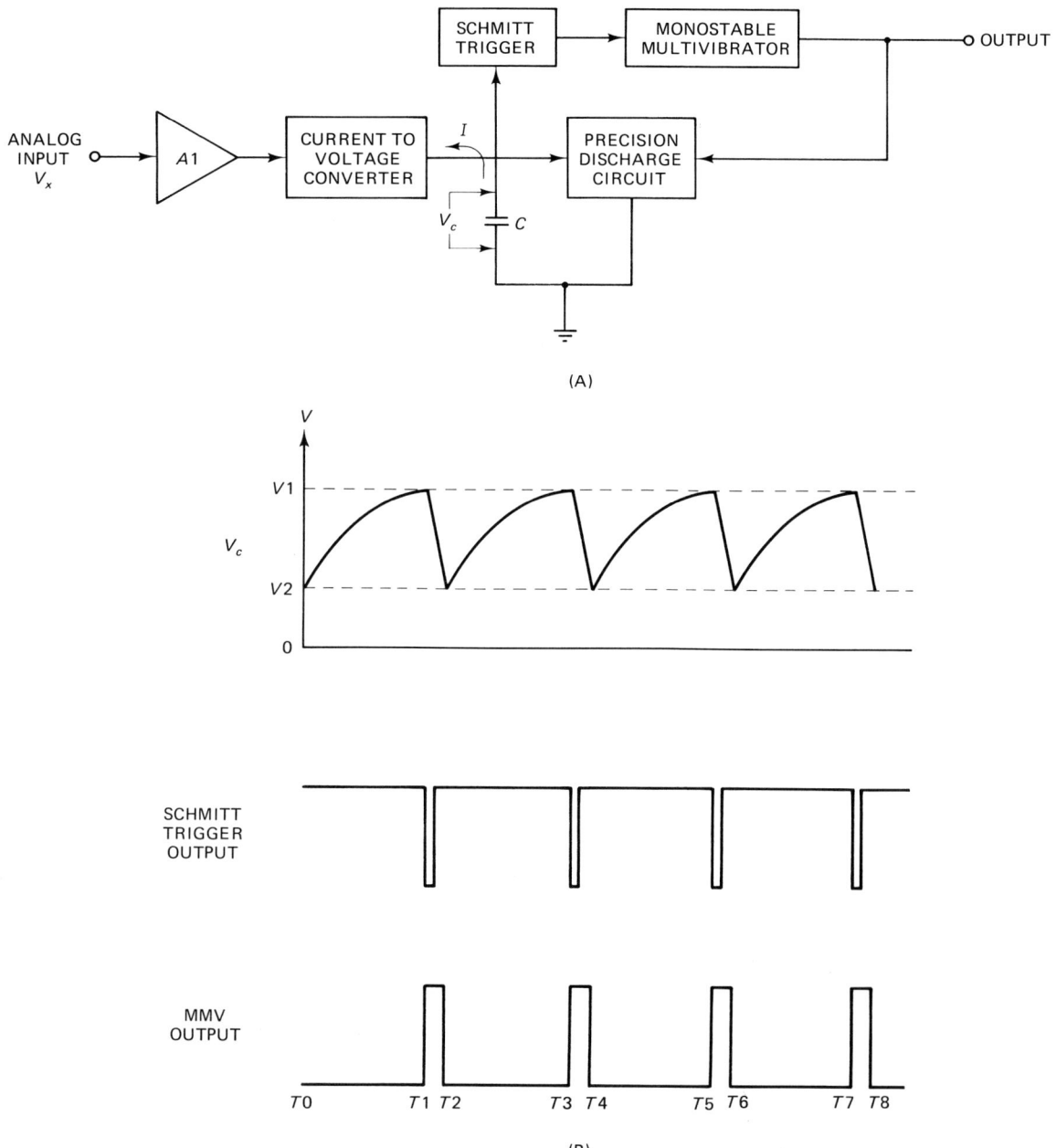

Fig. 22-5 (A) Voltage-to-frequency converter circuit and (B) timing diagram.

The type of V/F converter shown in Fig. 22-5 is superior to the VCO method. The circuit is shown in Fig. 22-5A, and the timing waveforms are shown in Fig. 22-5B. The operation of this circuit is dependent on the charging of a capacitor, although not an RC network as in the case of some other oscillator or timer circuits. The input voltage signal (V_x) is amplified (if necessary) by $A1$ and then converted to a proportional current level in a voltage-to-current converter stage. If the voltage applied to the input remains constant, so will the current output of the V-to-I converter (I).

The current from the V-to-I converter is used to charge the timing capacitor C. The voltage appearing across this capacitor (V_c) varies with time as the capacitor charges (see the V_c waveform in Fig. 22-5B). The precision discharge circuit is designed to discharge capacitor C to a certain level ($V2$) whenever the voltage across the capacitor reaches a predetermine value ($V1$). When the voltage across the capacitor reaches $V2$, a Schmitt trigger circuit is fired that turns on the precision discharge circuit. The precision discharge circuit, in its turn, will cause the capacitor to discharge rapidly but in a controlled manner to value $V1$. The output pulse snaps HIGH when the Schmitt trigger fires (i.e., the instant V_c reaches $V1$) and drops LOW again when the value of V_c has discharged to $V2$. The result is a train of output pulses whose repetition rate is exactly dependent on the capacitor charging current, which, in turn, is dependent on the applied voltage. Hence, the circuit is a voltage-to-frequency converter.

Like the VCO circuit, the output of the V/F converter can be applied to the input of a binary counter. The parallel binary outputs become the data lines to the computer. Alternatively, if the frequency is relatively low the computer can be programmed to measure the period between pulses. Also, certain interface devices such as the 6522 and Z80-CTC chips have built-in timers that can measure the period.

BINARY COUNTER (SERVO) A/D CONVERTERS

A counter type A/D converter (also called servo or ramp A/D converter) is shown in Fig. 22-6. It consists of a comparator, voltage output DAC, binary counter, and the necessary control logic. When the start command is received, the control logic resets the binary counter to 00000000_2, enables the clock, and begins counting. The counter outputs control the DAC inputs; so the DAC output voltage will begin to rise when the counter begins to increment. As long as analog input voltage V_{in} is less than V_{ref} (the DAC output), the comparator output is HIGH. When V_{in} and V_{ref} are equal, however, the comparator output goes LOW, which turns off the clock and stops

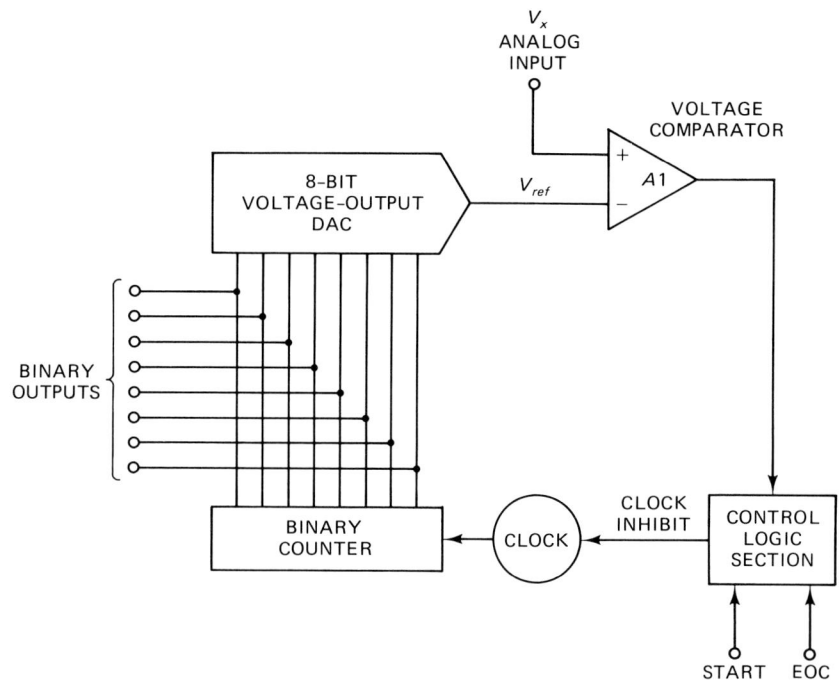

Fig. 22-6 Binary counter A/D converter.

the counter. The digital word appearing on the counter output at this time represents the value of V_{in}.

Both slope and counter A/D converters take too long for many applications—on the order of 2^N clock cycles (where N = number of bits). Conversion time becomes critical if a high-frequency component of the input waveform is to be faithfully reproduced. Nyquist's criteria require that the sampling rate (i.e., conversions per second) be at least twice the highest frequency to be recognized.

SUCCESSIVE APPROXIMATION A/D CONVERTERS

Successive approximation A/D conversion is best suited for many applications where speed is important. This type of A/D converter requires only $N + 1$ clock cycles to make the conversion; some designs allow truncation of the conversion process after fewer cycles if the final value is found prior to $N + 1$ cycles.

The successive approximation converter operates by making several successive trials at comparing the analog input voltage with a reference generated by a DAC. An example is shown in Fig. 22-7. This circuit consists of a comparator, control logic section, a digital shift register, output latches, and a voltage output DAC.

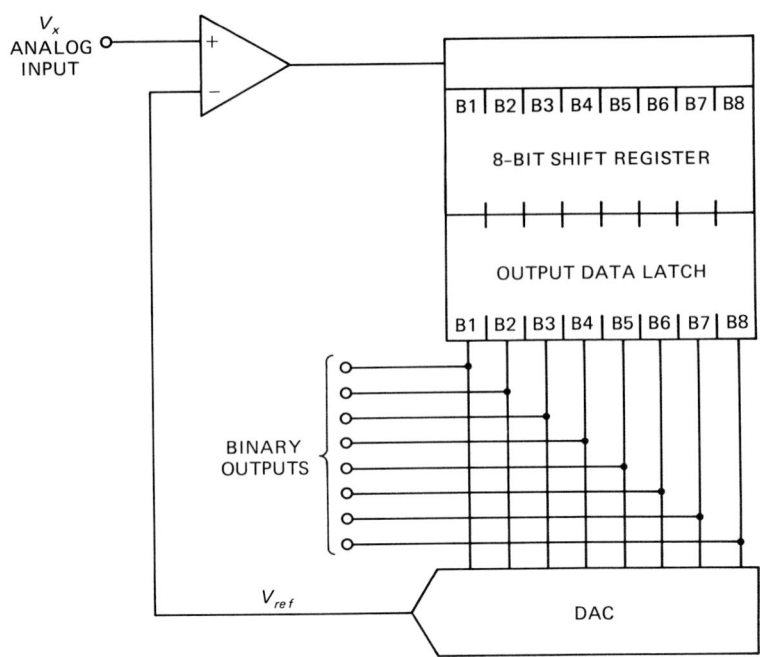

Fig. 22-7 Successive approximation A/D converter.

When a START command is received, a binary-1 (HIGH) is loaded into the MSB of the shift register, and this sets the output of the MSB latch HIGH. A HIGH is the MSB of a DAC will set the output voltage V_{ref} to half-scale. If the input voltage V_{in} is greater than V_{ref}, the comparator output stays HIGH and the HIGH in the shift register MSB position shifts one-bit to the right, and therefore occupies the next most significant bit (bit 2). Again the comparator compares V_{in} with V_{ref}.

If the reference voltage from the DAC is still less than the analog input voltage, the process will be repeated with successively less significant bits until either a voltage is found that is equal to V_{in} (in which case the comparator output drops LOW) or the shift register overflows.

If, on the other hand, the first trial with the MSB indicates that V_{in} is less than the half-scale value of V_{ref}, the circuit continues making trials below V_{ref}. The MSB latch is reset to LOW and the HIGH in the MSB shift register position shifts one-bit to the right to the next most significant bit (bit 2). Here the trial is repeated again. This process will continue as before until either the correct level is found or overflow occurs. At the end of the last trial (bit 8 in this case) the shift register overflows and the overflow bit becomes an end-of-conversion (EOC) flag to tell the rest of the world that the conversion is completed.

This type, and most other types of A/D converters, requires a starting pulse and signals completion with an EOC pulse. This requires the computer or other digital instrument to engage in bookkeeping to repeatedly send the start command and look for the EOC pulse. If the start input is tied to the EOC output, conversion is continuous and the computer need only look for the periodic raising of the EOC flag to know when a new conversion is ready. Such operation is said to be asynchronous.

PARALLEL (FLASH) A/D CONVERTERS

The parallel A/D converter (Fig. 22-8) is probably the fastest A/D circuit known. Indeed, the very fastest ordinary commercial products use this method. Some sources call the parallel A/D converter the flash circuit because of its inherent high speed.

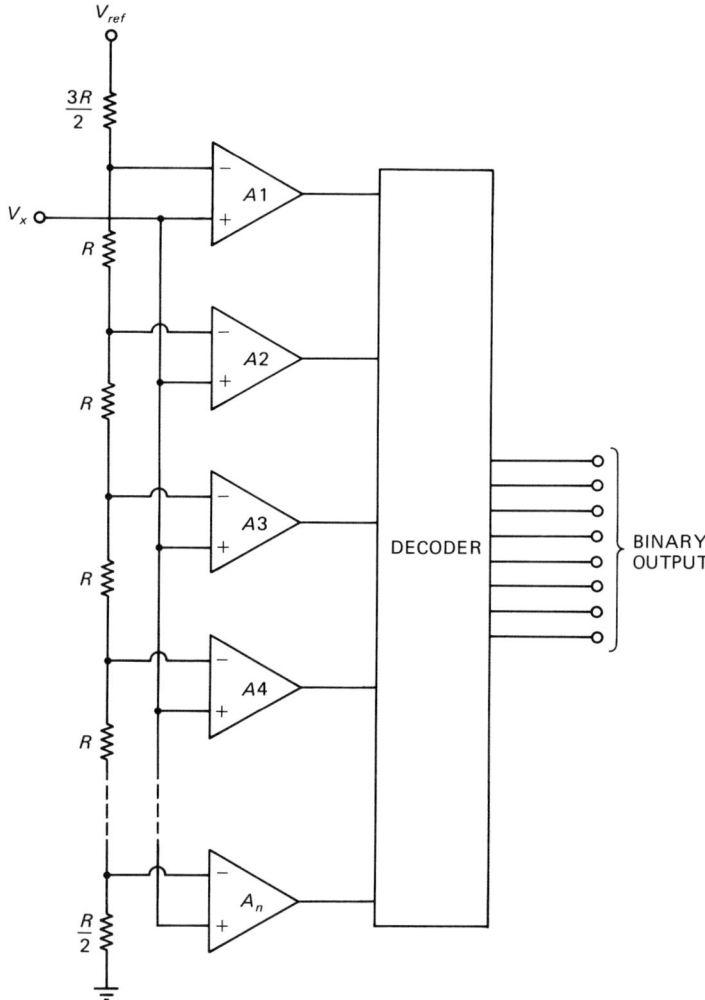

Fig. 22-8 Flash A/D converter.

The parallel A/D converter consists of a bank of $[2^N - 1]$ voltage comparators biased by reference potential V_{ref} through a resistor network that keeps the individual comparators 1-LSB apart. Since the input voltage is applied to all the comparators simultaneously, the speed of conversion is limited essentially by the slew rate of the slowest comparator in the bank, and also by the decoder circuit propagation time. The decoder converts the output code to binary code needed by the computer.

Multiple-Input A/D Converters

Many systems require conversion of more than one analog voltage at a time. These systems can be accommodated either by using a separate A/D converter for each input (which is a possibility now that costs are down), or by using a multiple-input A/D converter. In the sense used here, multiple-input actually means a single-input A/D converter in which the input is time multiplexed between a number of sources. In the simplest form, an electronic switch bank alternately selects the various input signals.

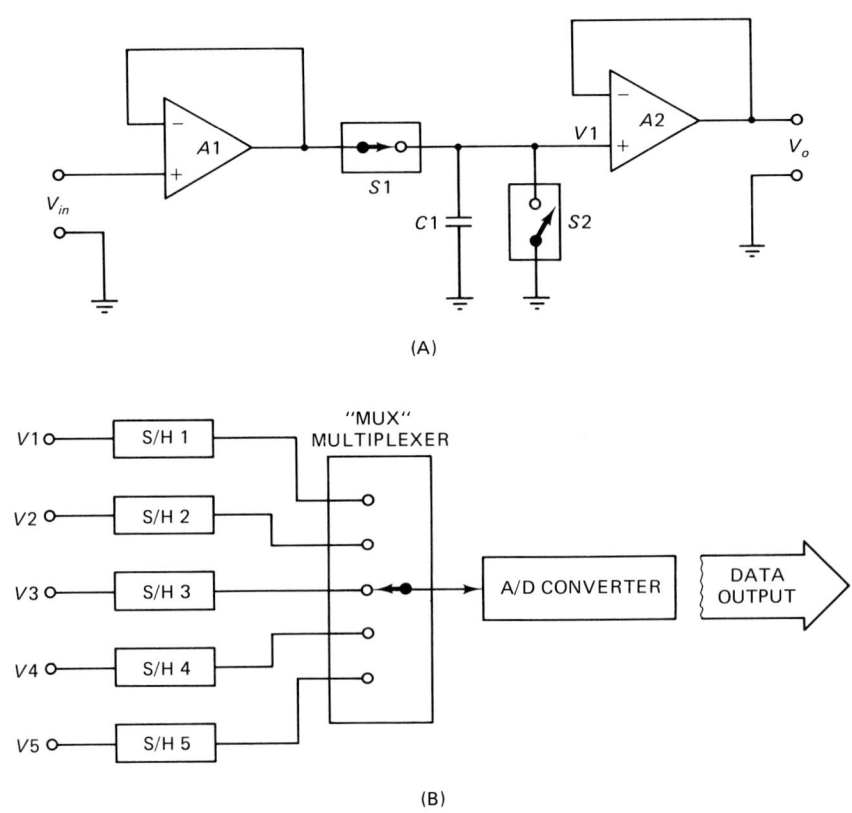

Fig. 22-9 (A) A bank of sample-and-hold (S/H) circuits, and (B) the arrangement of S/H circuits at the input of the A/D converter.

It is often the case, however, that one needs to know the value of two or more analog voltages at the same time. The simultaneity makes it necessary to use a bank of sample-and-hold (S/H) circuits (Fig. 22-9A). The S/H circuit uses a capacitor to temporarily store a charge that is created by the input signal V_{in}. In one version of the A/D converter, switch S2 initially closes and then opens to discharge $C1$. Following the reset of $C1$, switch S1 closes to recharge $C1$ with the new value.

Figure 22-9B shows the arrangement of several S/H circuits at the input of the A/D converter. The multiplexer (MUX) is an electronic CMOS switch that steps through the various inputs in sequence.

One problem seen with the multiple input A/D converter is that the values stored in the S/H circuits tend to droop somewhat, distorting the data. Guarding against this problem requires such tactics as (1) using only very low leakage resistance capacitors for $C1$, (2) using a very high input impedance op-amp for $A2$ (BiMOS or BiFET), and (3) using an A/D converter with a fast conversion time.

CONCLUSION

In this chapter, we have discussed the fundamentals of data conversion. Now that a foundation is established, let us go on to examine practical data converters.

23

Practical Data Converter Circuits

The previous chapter discussed the various methods for implementing data converters. In this chapter, the topic is extended by showing practical examples of data converter. The IC devices selected were chosen for both universality and for ease of interfacing to the personal computer (see Chapter 24). The first data converter that we will examine is an R-$2R$ ladder digital-to-analog converter (DAC), called the DAC-08.

A PRACTICAL IC DAC EXAMPLE

A number of different manufacturers offer low-cost IC DACs that contain almost all of the circuitry needed for the process, except possibly the reference source (although some devices do contain the reference source also) and some operational amplifiers for either level-shifting or current-to-voltage conversion.

In this chapter the DAC-08 device is used as a practical circuit example. This eight-bit DAC is now something of an industry standard; it is available from several sources (under DAC-08 or similar type numbers). The DAC-08 is a later generation version of the Motorola MC-1408 device. This DAC is sometimes designated LM-DAC-0800. An easily available, and closely related, device is the DAC-0806.

Figure 23-1A shows the basic circuit configuration for the DAC-08. In subsequent circuits, the power supply terminals are deleted for simplicity's sake; they will always be the same as shown here. The internal circuitry of the DAC-08 is the R-$2R$ ladder shown in the previous section, but it has two outputs: I_o and NOT-I_o (Note: a bar over I_o in the illustration denotes NOT-I_o—that is, the complement of I_o). These current outputs are unipolar and complementary (Fig. 23-

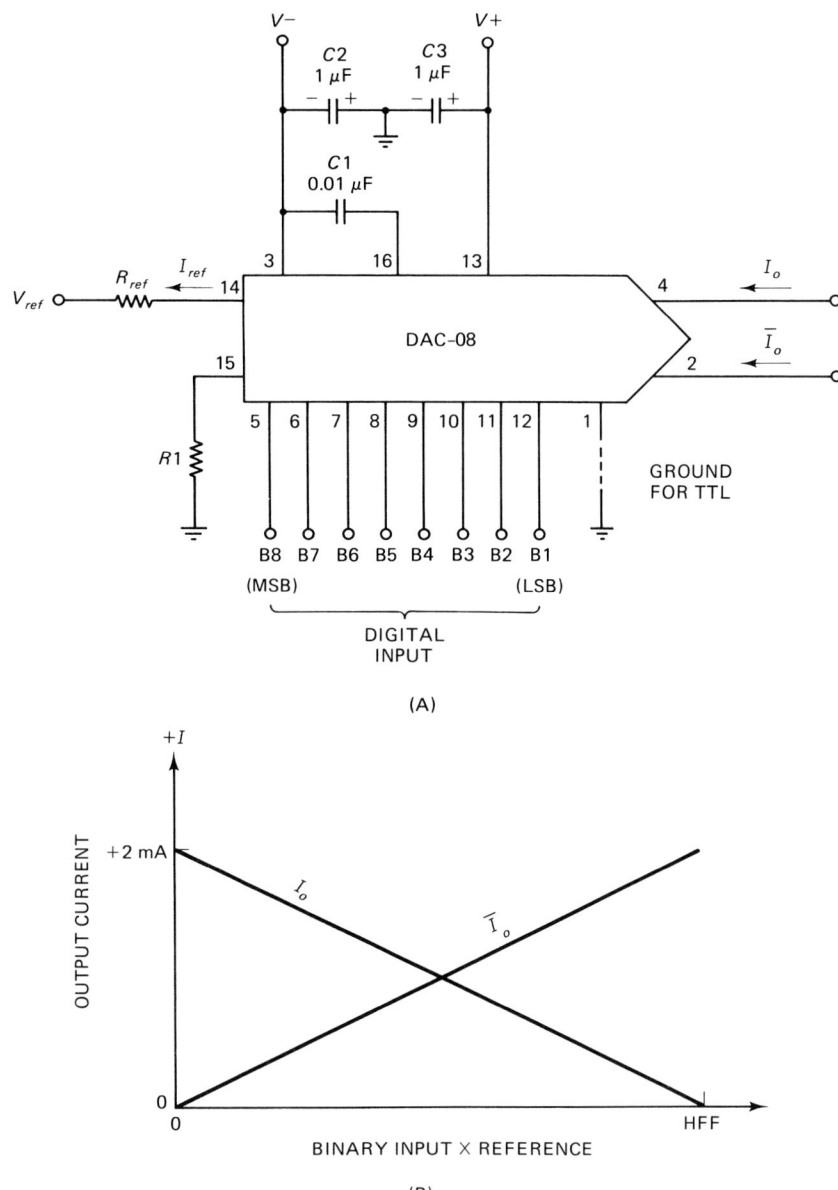

Fig. 23-1 (A) DAC-08 circuit and (B) output current functions for the DAC-08.

1B); if the full-scale output current is I_{max}, then $I_{max} = [I_o + \text{NOT-}I_o]$. The specified value of I_{max} on the DAC-08 family of DACs is 2 mA.

Two types of input signal are required to make this DAC work: an analog reference and an eight-bit digital signal. The analog signal is the reference current I_{ref} applied through pin no. 14. This current may be generated by combining a precision reference voltage source with a precision, low-temperature-coefficient resistor to convert V_{ref} to I_{ref}. Alternatively, a constant-current source may be used to provide I_{ref}.

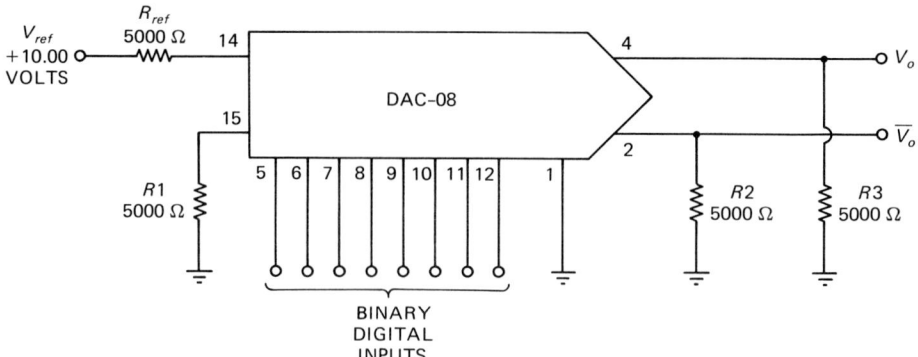

Fig. 23-2 Simple voltage-output circuit for the DAC-08.

For TTL compatibility of the binary inputs, make V_{ref} = 10.000 V and R_{ref} = 5000 Ω.

The other type of input is the eight-bit digital word, which is applied to the IC at pins 5 through 12, as shown. The logic levels that operate these inputs can be preset by the voltage applied to pin no. 1 (for TTL operation, pin no. 1 is grounded). In the TTL-compatible configuration shown, LOW is 0 to 0.8 V and HIGH is +2.4 to +5 V.

Figure 23-2 shows the connection of the DAC-08 (less power supply and reference input) required to provide the simplest form of unipolar voltage output operation over the range of approximately 0 to 10 V. When the input word is 00000000_2, then the DAC output is 0 V plus or minus the DC offset error. A half-scale voltage (-5 V) is given when the input word is 10000000_2. This situation occurs when the MSB is HIGH and all other digital inputs are LOW. The full-scale output will exist only when the input word is 11111111_2 (all HIGH). The output under full-scale conditions will be -9.96 V rather than 10 V, as might be expected. (Note: 9.96 V is 1-LSB less than 10 V.)

The circuit in Fig. 23-2 works by using resistors $R2$ and $R3$ as current-to-voltage converters. When currents I_o and NOT-I_o pass through these resistors, a voltage drop of IR, or $5.00 \times I_o$ (mA), is created. A problem with this circuit is that it has a high source impedance (5 kΩ with the values shown for $R2/R3$).

Figure 23-3 shows a simple method for converting I_o to an output voltage (V_o) with a low output impedance (less than 100 Ω) by using an inverting follower operational amplifier. The output voltage is simply the product of the output current and the negative feedback resistor:

$$V_o = RI_o \tag{23-1}$$

As in the case previously described, a 5000-Ω resistor and a 10.00-V DC reference voltage will produce a 9.96-V output voltage when the DAC-08 is set up for TTL inputs and 2.0 mA $I_{o(max)}$.

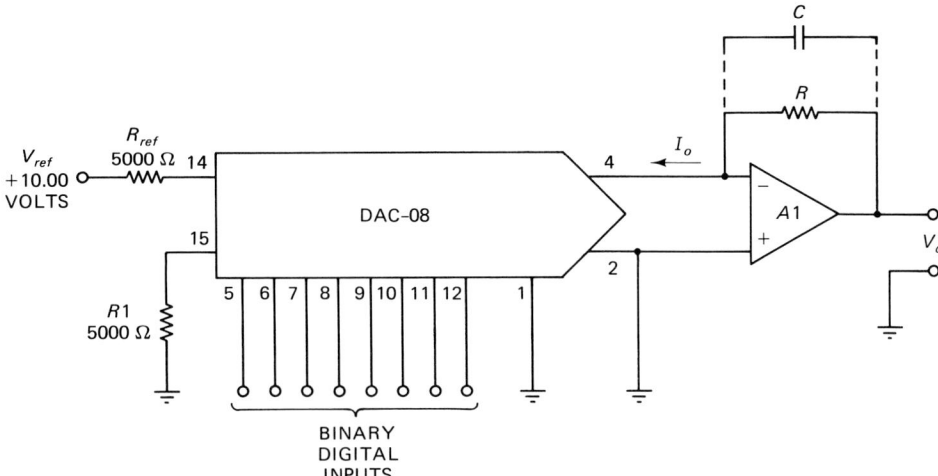

Fig. 23-3   Low-impedance version of the voltage-output circuit (inverting type).

The frequency response of the DAC circuit can be tailored to meet certain requirements. The normal output waveform of the DAC is a staircase when the digital input increments up from 00000000_2 to 11111111_2 in a monotonic manner. To make the staircase into an actual ramp function, a low-pass filter is needed at the output to remove the "stepness" of the normal waveform. A capacitor shunted across the feedback resistor (R) will offer limited filtering on the order of -6 dB/octave above a cutoff frequency of

$$F = \frac{1,000,000}{6.28\,RC_{\mu F}} \tag{23-2}$$

where F is the -3 dB frequency in hertz (Hz), R in ohms (Ω), and $C_{\mu F}$ in microfarads (μF).

In most practical circuits the required value of F is known from the application. It is the highest-frequency Fourier component in the input waveform. It is necessary to calculate the value of capacitor needed to achieve that cutoff frequency, so we swap the F and C terms in Eq. (23-2).

$$C_{\mu F} = 1,000,000/6.28\,RF \tag{23-3}$$

A related method shown in Fig. 23-4 produces an output voltage of the opposite polarity from that of Fig. 23-3. The circuit of Fig. 23-4 is connected to a noninverting unity-gain follower at the output. The output voltage is the product of I_0 and $R2$. If a higher output voltage is needed, the circuit variant shown in the inset to Fig. 23-4 can be used.

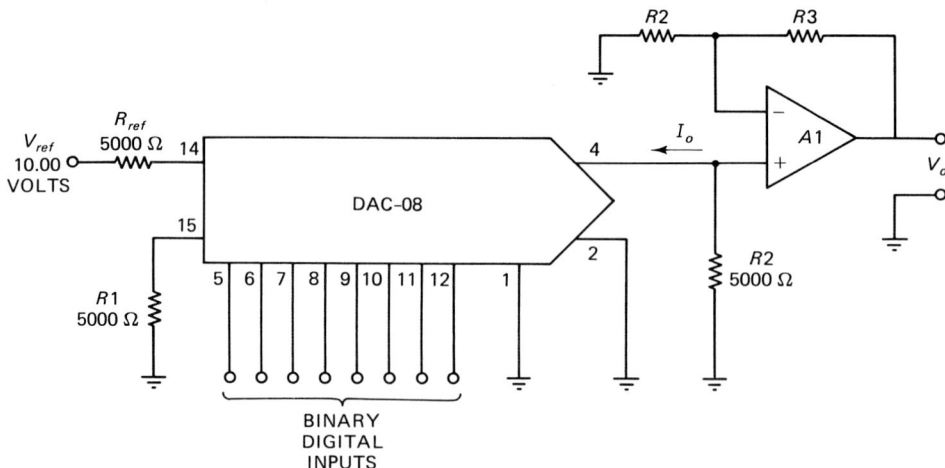

Fig. 23-4 Noninverting low-impedance voltage-output circuit.

In this case, the output amplifier has gain, so the output voltage will be

$$V_o = I_o R2 \left(\frac{R4}{R3} + 1 \right) \tag{23-4}$$

One of the ways to achieve bipolar binary operation is shown in Fig. 23-5. In this circuit the output amplifier is a DC differential amplifier, and both current outputs of the DAC-08 are used. Note that the maximum and minimum voltages are positive and negative. The zero selected can be either (+)zero (+ 1-LSB voltage) or (−)zero (− 1-LSB voltage). It cannot be exactly zero because an even number of output codes are equally spaced around zero. In other words, the absolute value of $FS(-)$ is equal to the absolute value of $FS(+)$. There

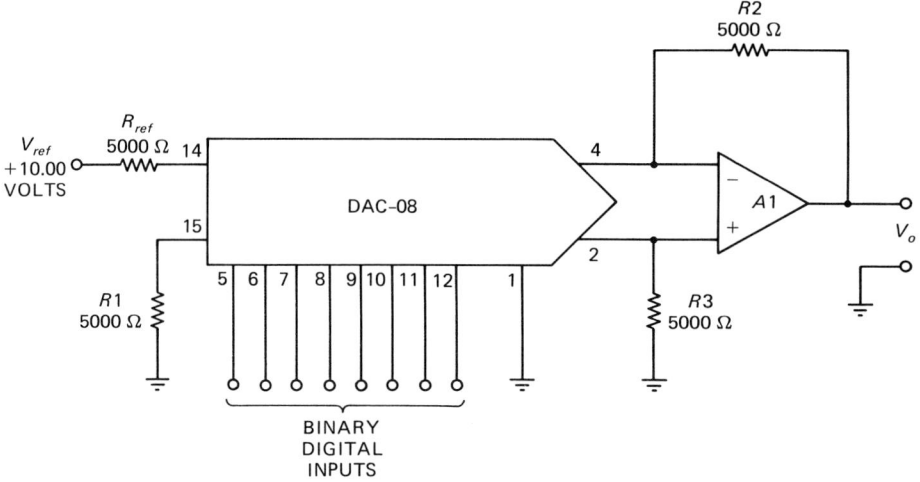

Fig. 23-5 Simple DAC-08 circuit using TTL-compatible digital inputs.

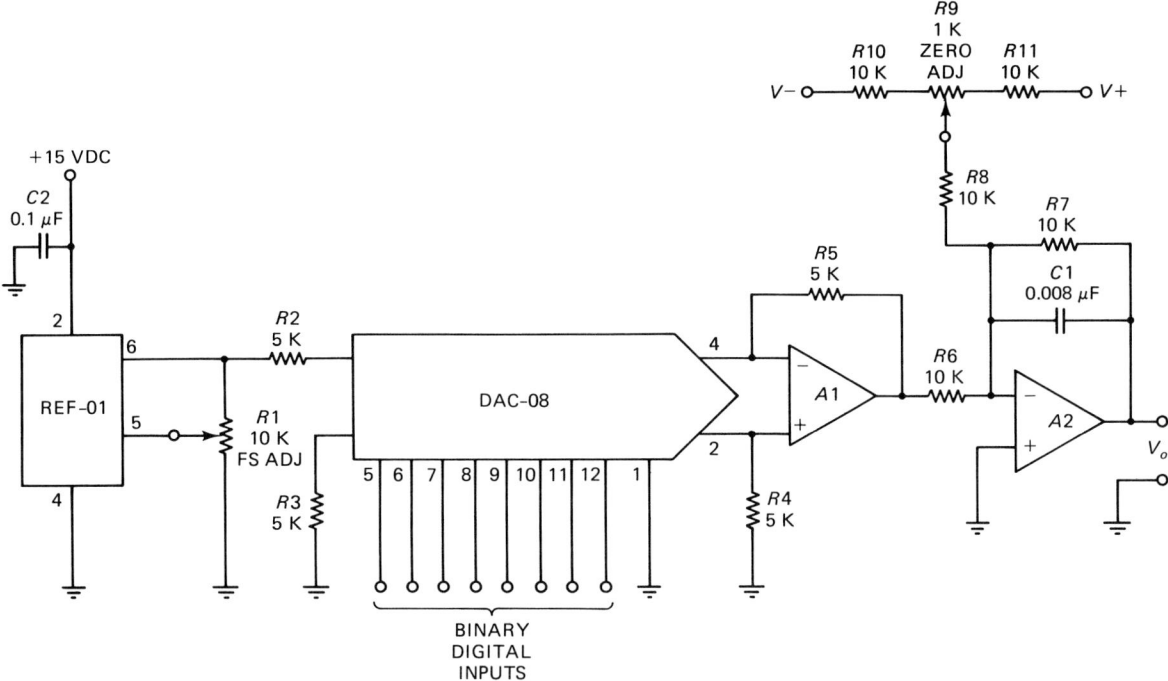

Fig. 23-6 Practical DAC-08 circuit for voltage-output applications.

are also circuits that make zero = zero, but at the expense of uneven ranges for $FS(-)$ and $FS(+)$.

A practical DAC circuit is shown in Fig. 23-6. This circuit combines the circuit fragments shown earlier to make a complete circuit that can be used. The power connections are not shown. The heart of this circuit is a DAC-08 connected in the bipolar binary circuit discussed above.

The reference potential in Fig. 23-6 is a REF-01 10.000-V IC reference voltage source. Potentiometer $R1$ adjusts the value of the actual voltage, and it also serves as a full-scale adjustment for the output voltage V_o.

The output amplifier can be a 741-class operational amplifier, or any other form; the need is not critical. Potentiometer $R9$ acts as a zero adjustment for V_o. The capacitor across $R7$ limits the frequency response to 200 Hz with the value shown. This frequency limit can be changed with the equation given earlier.

Adjustment of the DAC Circuit

1. Set the binary inputs all LOW (00000000_2).
2. Adjust $R9$ for $V_o = 0.00$ V.
3. Set all binary inputs HIGH (11111111_2).
4. Adjust potentiometer $R1$ for $V_o = 9.96$ V.

The DAC-08, its antecedents (e.g., MC-1408), and clones are perhaps the most widely used eight-bit DAC chips on the market. They have the ability to produce usable results over a wide range of applications.

A PRACTICAL IC A / D EXAMPLE

The ADC0808 is produced by National Semiconductor; it is a micro-processor-compatible eight-channel, eight-bit, successive-approximation analog-to-digital converter (see Fig. 23-7 for the internal functions). Internal trimming eliminates the need for external range and zero adjustments, provided that 1/2-LSB error (1-LSB on some versions) is acceptable. With a single +5 V DC power supply (the type available on all computers), the ADC0808 can handle a 0 to 5-V analog input range. The ADC0808 comes in a 28-pin DIP package (Fig. 23-8). Conversion time is 100 μs, with a 15-mW power consumption on a 500-kHz clock frequency.

The internal circuit is shown in functional block diagram form in Fig. 23-7. The central A/D converter is a successive approximation, similar to those discussed in Chapter 22. The inputs to the A/D section include the single-channel analog data from the input multiplexer, a differential voltage reference [$V_{ref(-)}$ and $V_{ref(+)}$], the start command, and the control signals from the internal control logic

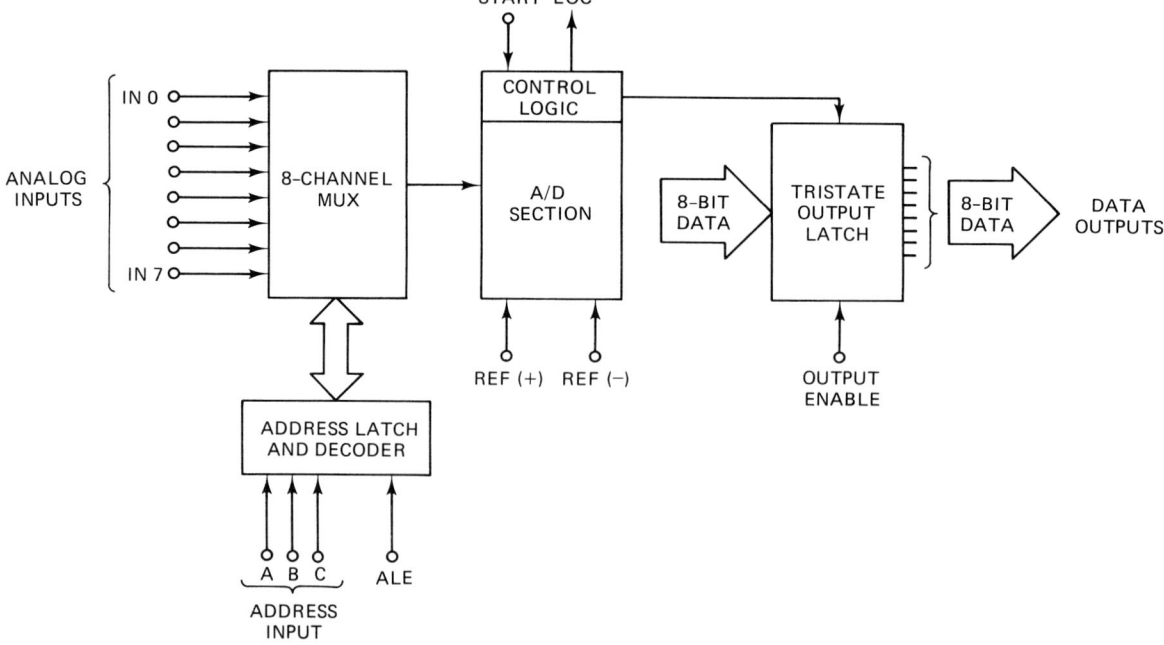

Fig. 23-7   Block diagram for the ADC0808 A/D converter.

section. The output includes the eight output data lines and the end-of-conversion (EOC) signal.

The outputs of the A/D are coupled to a tri-state output data latch. A tri-state device is one in which the output floats at a high impedance until commanded to either HIGH or LOW state in conjunction with an output-enable (OE) command. When the OE is active, the tri-state latch transfers the input data to the output lines. The use of tri-state logic allows the ADC0808 to be connected across a microprocessor or microcomputer data bus without the need for intervening interface circuitry.

The eight analog input channels are connected to an eight-channel multiplexer (8-CHAN MUX) that selects from among them in accordance with the commands received from the address latch and decoder section. For a three-bit address (ADD A, ADD B, ADD C), the eight channels are selected according to the protocol:

| Address bits | | | Analog input |
C	B	A	selected
0	0	0	IN0
0	0	1	IN1
0	1	0	IN2
0	1	1	IN3
1	0	0	IN4
1	0	1	IN5
1	1	0	IN6
1	1	1	IN7

The address on the input data bus must be latched into the decoder by making the address latch enable (ALE) input active.

Figure 23-9 shows a typical connection arrangement for the ADC0808. The versatility of this device makes it possible to interface with a wide variety of computers and microprocessor chips. To facilitate the discussion, it is assumed that the machine selected has NOT-READ and NOT-WRITE signals and an interrupt (INT) capability. These signals are nearly universal, and are also available on various plug-in protyping cards for Apple II and IBM-PC series machines.

The ADC0808 only uses three address input lines, while the typical computer has either 16 or 20 bits on the address bus. It is typically the case that the ADC0808 bits (A, B, C) are used as the low-order three bits of a larger address, while the computer supplies a decode signal (shown as NOT-ADR) to indicate when the correct higher-order address is present on the line. This function is usually performed on Apple II and IBM-PC series machines by the logic on the plug-in card.

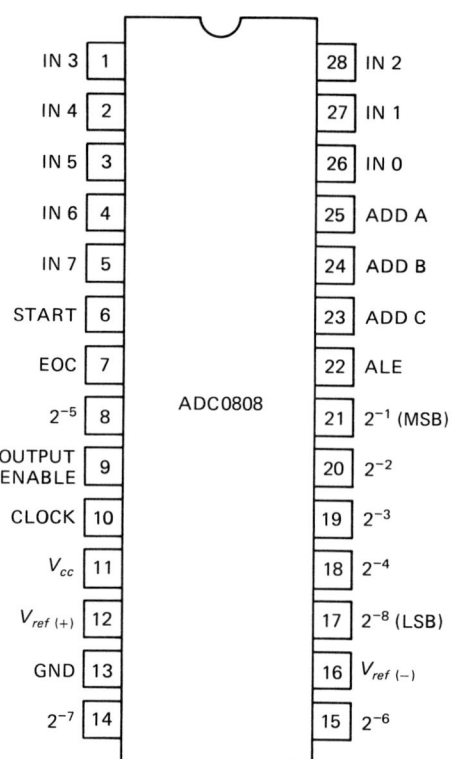

Fig. 23-8 ADC0808 pinouts.

Two 5-V DC power supplies are used in this configuration. The +5 V DC (DC power) line is the power supply of the computer. This is the power for the ADC0808 circuits. The other 5-V supply ($V_{ref(+)}$ 5.00 V DC) is a reference voltage. This voltage must be separate from the other supply or errors will be created. It must also be a precision source (see Chapter 25). The +5-V DC power supply might be anything between +4.9 and +5.2-V and still be in tolerance, so it could create a tremendous error in the output data—not to mention that this power supply is noisy. The reference 5.00-V supply should be 5.00 ± 0.001 V.

A pair of NOR gates (G1 and G2) are used to decode the status of the NOT-WRITE and NOT-READ lines. The start, OE, and ALE lines are active-HIGH; the control lines from the computer are active-LOW. The two-input NOR gates used follow these rules:

1. If either input is HIGH, then the output is LOW.
2. If both inputs are LOW, then the output is HIGH.

Therefore, both inputs of either G1 or G2 must be LOW for the associated output to be HIGH—which is the state required to activate the ADC0808 inputs.

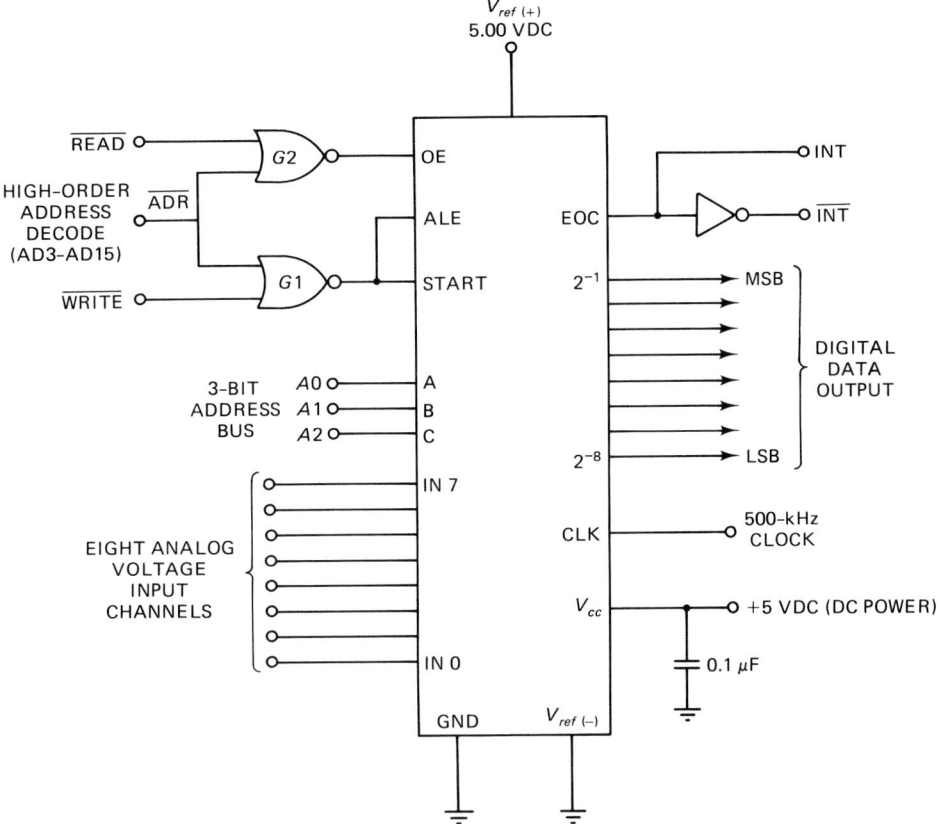

Fig. 23-9 Practical ADC0808 circuit.

When the computer wants to start a conversion it will place the address of the A/D converter on the address bus. The low-order three bits are applied to the ADC0808 address inputs. A separate address decoder outputs a NOT-ADR signal when the correct address is present. At the same time, a NOT-WRITE signal is output. These two active-LOW signals will force both inputs of G1 LOW, so the output of G1 snaps HIGH. The HIGH on the output of G1 is the correct active signal to turn on the start function and to input the low-order bits (which select the analog input) to the address decoder (ALE is active).

When the A/D converter is finished making a conversion, it signals the computer that the data is ready by making the interrupt line (INT) high. Because many computers use active-LOW INT signals (i.e., NOT-INT), an inverter is also used to supply this form of signal. The computer responds to the interrupt by creating a NOT-READ signal, which along with the NOT-ADR signal places a HIGH on the output of G2, forcing output enable (OE) HIGH. The data is then transferred to the digital output lines.

24

Data Converter Interfacing

Before the data converter can be used in most practical instrumentation systems, it must be interfaced to a computer. In this chapter we will take a look at the basics of interfacing. Understand that this discussion is generic in nature, and one must consult a specific computer's technical data references to adapt these methods. For people using IBM-PC or Apple II computers, interface cards are available. Both ready-made versions that include the A/D converter and all control circuit (Fig. 24-1) and blank prototyping versions are available.

INTERFACING DACs

The digital-to-analog converter (DAC) is a device that will convert a binary word applied to its inputs to a proportional voltage or current at the output. Both voltage- and current-output DACs exist. There are two basic methods for interfacing DACs: I/O based and memory mapped.

The I/O-based method is shown in Fig. 24-2. Since several aspects of DAC interfacing are common to both methods, we will show them only in this figure. One facet is current-to-voltage conversion; another is low-pass filtering. The I-to-V conversion is used to produce a voltage output (which is what is needed by oscilloscopes and most strip-chart paper recorders) from the output of the current type of DAC. In Fig. 24-2, the DAC output current (I_o) is converted to a voltage by passing it through a fixed-precision resistor R. The output voltage will be, according to Ohm's law,

$$V_o = I_o R$$

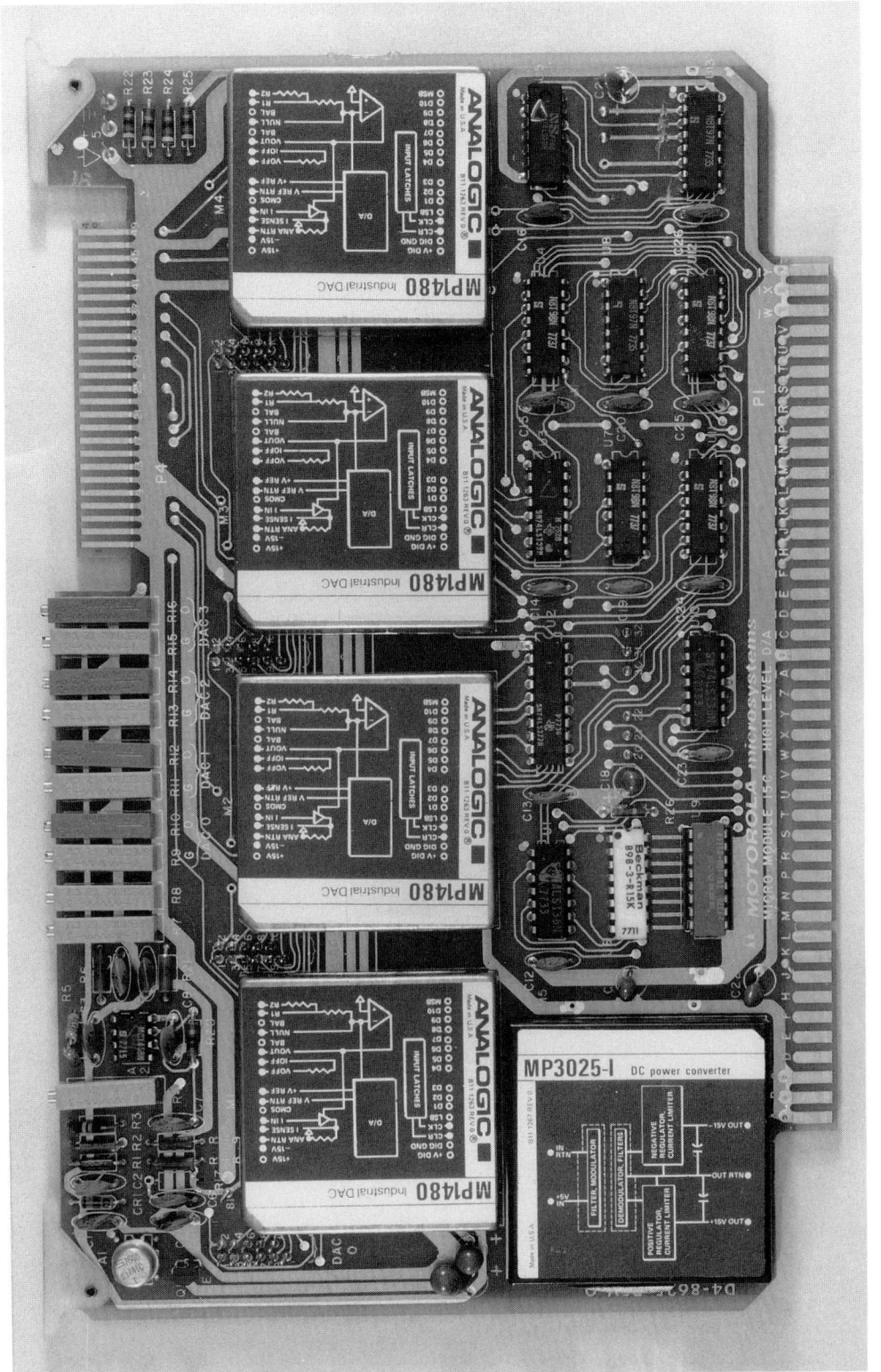

Fig. 24-1 Commercial A/D interface card.

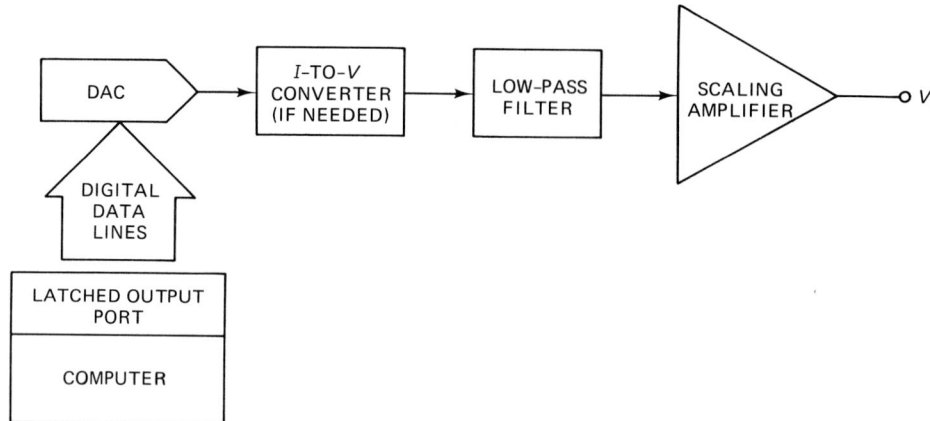

Fig. 24-2 Generic DAC interface circuit.

The low-pass filter is used to smooth the output waveform. The DAC can produce only certain discrete output levels, so instead of a ramp function it will produce an equivalent staircase function. The low-pass filter will smooth the staircase to make it look more like a ramp.

The interfacing method shown in Fig. 24-2 is used when there is an output port available that is latched (as are most); that is, the output port will contain the last valid data even after the computer has gone on to other chores. In that case, it is merely necessary to connect the output port bits to the DAC input bits on a one-for-one basis.

In a memory-mapped system, the DAC (or other peripheral; the method is not limited to data converters) is treated as a memory location and is assigned an address. Figure 24-3 shows the basic memory-mapped system. The OUT N signal is an output device-select pulse. For the case of a microcomputer that uses such a system, the elements that go into forming the OUT N signal are those that form a memory write operation. When the computer executes a write to the memory location defined in the OUT N signal, data on the data bus are transferred to the output of the eight-bit data latch. The outputs of the latch are used to drive the inputs of the DAC and are updated whenever the computer writes a new value to the memory location defined by the OUT N operation. Several IC devices on the market will do this will job.

The circuits shown thus far in this section works when used with eight-bit DACs connected to eight-bit data bus computers. But how do we connect a DAC that uses more than eight bits? The eight-bit computer can easily handle greater than eight-bit input words because it can use double-precision programming techniques. We will, therefore, occasionally see the need for interfacing a larger than eight-bit DAC to an eight-bit microcomputer.

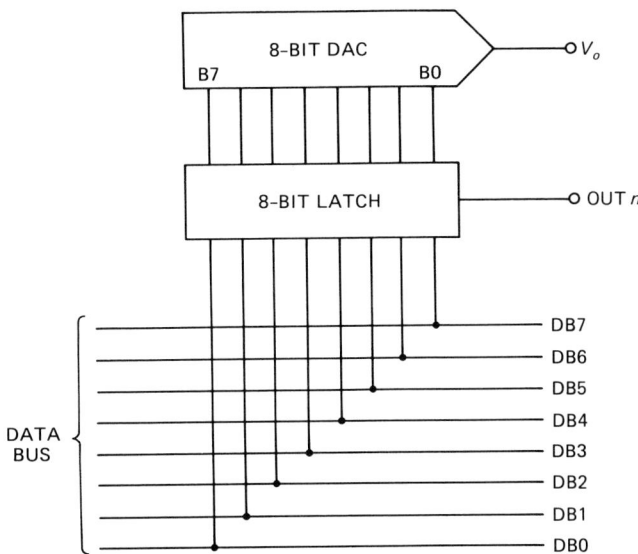

Fig. 24-3 Memory-mapped eight-bit DAC interface.

Figure 24-4A shows the method for connecting the large-length DAC to an eight-bit output port or data bus (which, of course, depends on whether memory-mapping is used); it is called the double-buffered method. For any word greater than eight-bits in length, we can output the entire word using more than one output operation. For example, up to 16 bits can be handled by two successive output operations. If we wanted to output, say, a 12-bit word, we could output the lower-order eight bits on the first operation and the high-order four bits on the second operation. This is the basis for operation of Fig. 24-4A.

Let us assume that the circuit is in the memory-mapped mode, as shown in Fig. 24-4A. The OUT1, OUT2, and OUT3 signals are device-select pulses from the computer or plug-in card logic. The lower order eight bits of the 12-bit data word are output on the eight-bit data bus, and an OUT1 signal is generated by the CPU. This signal will cause IC1, and eight-bit data latch, to input and hold the signal. Thus the lower order eight bits of the required 12 will be stored at the output of IC1 after the OUT1 signal disappears.

On the next operation, the high-order four bits of the 12-bit data word will appear on the lower four bits of the data bus while simultaneously an OUT2 signal is generated. The effect of the OUT2 signal is to cause IC3 to input and hold the lower order four bits; only four of the eight bits are used on this operation. Hence, after the OUT2 signal disappears, we will have the lower order eight bits stored in IC1 and the higher order four bits of the 12-bit data word stored in IC3.

The DAC is now ready to receive the entire 12 bits. If an attempt had been made to apply any of the data to the DAC prior to this time, the DAC would temporarily see an incorrect data word for part of the operation. Now that the entire 12 bits are available at the outputs of

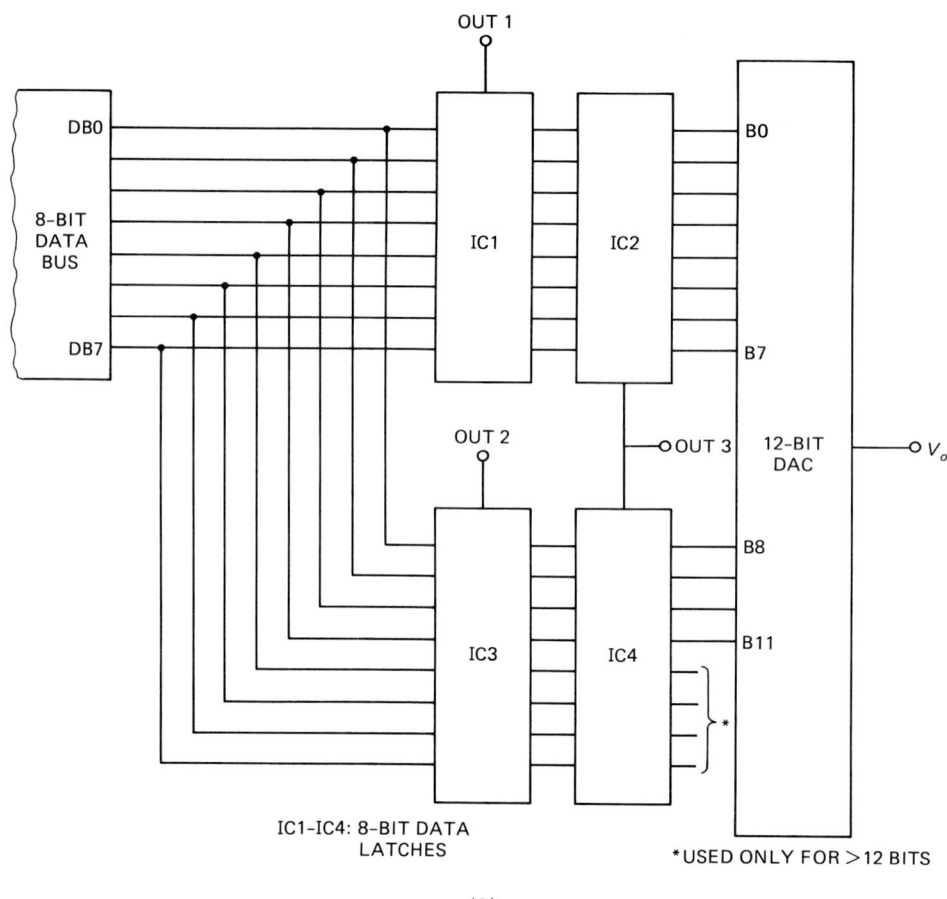

Fig. 24-4 (A) Memory-mapped > 8-bit DAC interface to eight-bit data bus and (B) I/O oriented version. (*Figure continues.*)

IC1 and IC3, we can crank the data into the 12-bit DAC-driver register consisting of IC2 and IC4. An OUT3 pulse will turn on both IC2 and IC4 and thereby transfer the data that are on the outputs of IC1 and IC3 to the DAC inputs. The DAC will now have an entire 12-bit data word on its inputs.

The specific circuit shown in Fig. 24-4A will accommodate up to 16 bits because each latch is an eight-bit latch. Sixteen-bit applications can be accommodated either by using a 16-bit latch or a pair of eight-bit latches, as discussed here.

Figure 24-4B shows a variation of the basic circuit that allows interfacing with a pair of eight-bit output ports. If you are using a commercially available computer or intend to use one of the commercial "no frills" single-board computers that are frequently sold as controllers, it may be more cost-effective to use extra I/O ports for this application rather than design a memory-mapped add-on. Two output ports are needed, here designated as ports 1 and 2 (any designation could be used). The output lines from port 1 are connected to the input

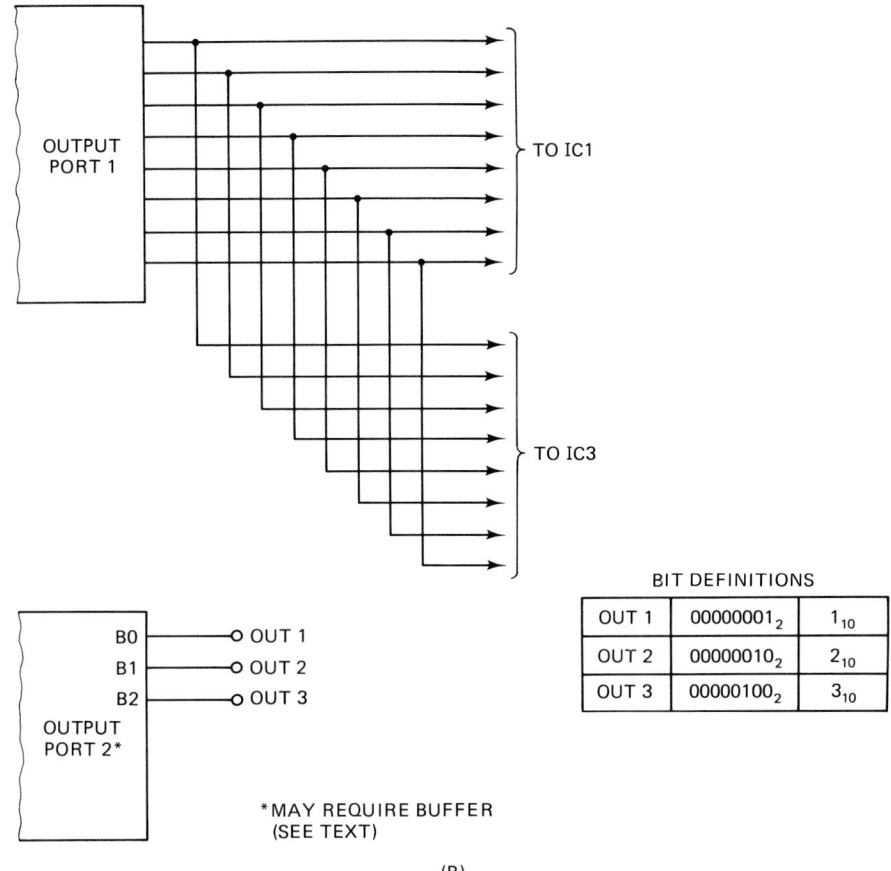

BIT DEFINITIONS

OUT 1	00000001_2	1_{10}
OUT 2	00000010_2	2_{10}
OUT 3	00000100_2	3_{10}

*MAY REQUIRE BUFFER
(SEE TEXT)

(B)

Fig. 24-4 (*continued*)

lines of the latches designated IC1 and IC3 in Fig. 24-4A. The OUT1, OUT2, and OUT3 signals are taken from three bits of a second output port; we must write an appropriate data word to that port that will cause the correct bit to go HIGH for a short period and then drop LOW again. For example, the line for the OUT1 signal is connected to bit B0 HIGH and all others LOW (i.e., 00000001_2). A typical program sequence will follow these steps:

1. Write 01 hex to port 2.
2. Jump to a timer subroutine that will provide a short delay (1 ms is usually a good selection, and the 74100 devices will react to much faster pulses).
3. Clear OUT2 by writing 00 hex to port 2.

In the initialization section of the main program, it will be necessary to ensure that port 2 is reset (i.e., 00000000_2) or there may be unwanted HIGH conditions on some of the lines at certain times following power up. The binary words needed to create the three select signals are given in the inset table in Figure 24-4B.

Because of the limitation of output port drive current, it may be that bit B2 of port 2, which drives the OUT3 signal, will require a high-current buffer stage. This is because there are four TTL inputs connected to this line (i.e., pins 12 and 23 on IC2 and IC4), and many microcomputer output devices are limited to 3.6 mA, which will support only two TTL inputs. This extra buffer will not be used if the output port device used in some particular application will support four TTL lines.

TYPICAL A/D CONVERTER SIGNALS

There are two basic methods of providing A/D converter control systems. Most A/D converters have a start line, which will cause the converter to initiate the conversion process when it is made active. It is in the signal that tells the outside world when the data are valid that the various converters differ. Figure 24-5 shows both systems. In Fig. 24-5A we see the timing diagram for the system that uses an end-of-conversion (EOC) pulse. The data output lines (B0–B7) may contain invalid data after the initiation of the start pulse, and these data cannot be used. When the conversion process is completed, however, the data on B0 to B7 are valid, so an EOC pulse is issued. The period between DOC pulses (t_1 to t_0) is the conversion time T_c.

The second method uses a status signal as shown in Fig. 24-5B. The status line may also be called the busy signal. It will be HIGH when the data are valid and drop LOW when the conversion is being made and the data are invalid.

Figure 24-6 shows a method (see Chapter 23, Fig. 23-9 for an alternate method) for converting an EOC pulse system to a status line system. This conversion may be required in some cases when an A/D converter is being interfaced to an existing computer that uses software that wants to see a status signal.

The basis of this circuit is an R-S (reset-set) flip-flop that is made from a pair of cross-connected NAND gates. When the start pulse is received, the status line will drop LOW. The start pulse need only be momentarily present, so the circuit is ideal for catching that temporary pulse. Similarly, when the EOC pulse is received, the status line goes HIGH again.

Another problem is that the A/D converter will be dormant between conversions. The device must be tickled by the start pulse, which is generated by the computer, before it can begin it work. But suppose we want the A/D converter to make continuous conversions, that is, to operate asynchronously. In that case (see Fig. 24-7A) we can connect the EOC output to the START input. When the EOC pulse

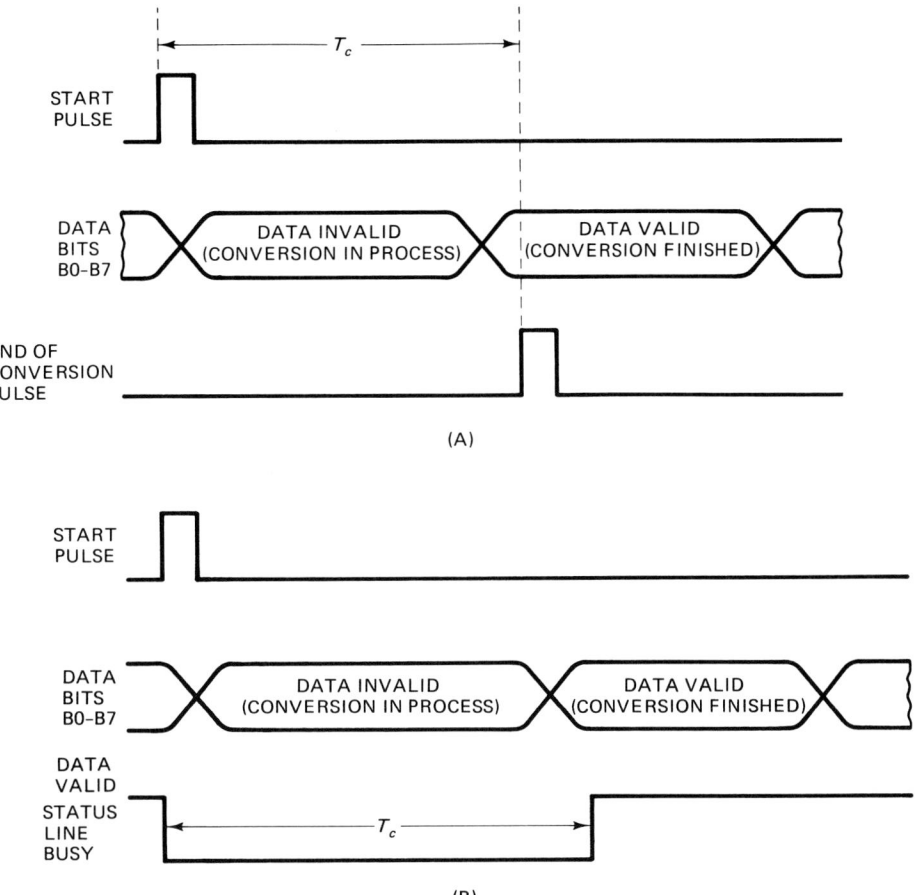

Fig. 24-5 (A) A/D timing signals using start and EOC pulses and (B) timing signals when status line is used.

occurs, it automatically tells the A/D converter to begin again the conversion process.

A problem that is sometimes experienced with asynchronous conversion, however, is that the data are valid for only one clock pulse. If that period is too fast, some data may be lost because the computer cannot input them fast enough. A solution to this problem (Fig. 24-7B)

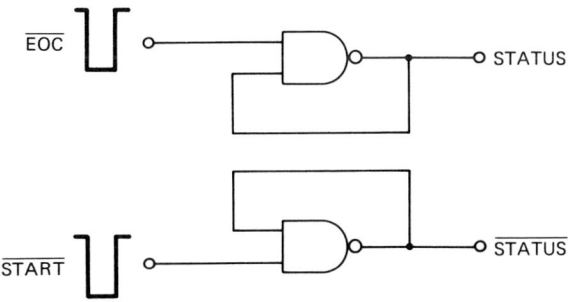

Fig. 24-6 Generating STATUS signals from EOC and START signals.

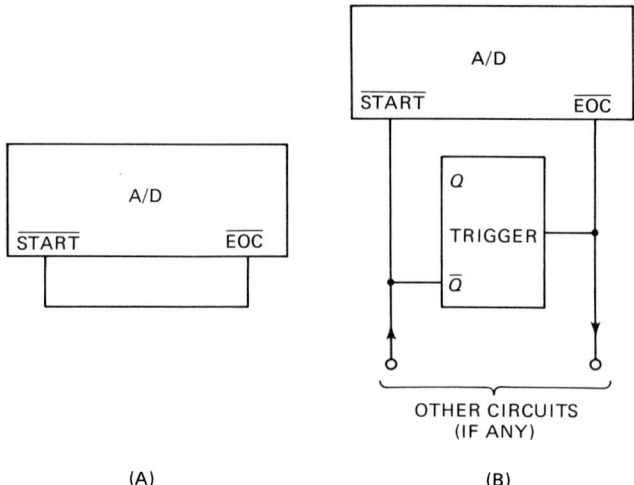

Fig. 24-7   (A) Simple asynchronous A/D circuit and (B) with brief EOC pulse delay.

is to connect a monostable multivibrator (one-shot) stage between the EOC output and START input. The one-shot will insert a delay equal to its period between the two events.

INTERFACING A/Ds

Perhaps the simplest method for interfacing an A/D converter to a microcomputer is shown in Fig. 24-8; Fig. 24-8A shows the method used when the operation is synchronized under program control, and Fig. 24-8B shows the method used when the A/D converter operates asynchronously. In both cases, the eight data lines from the A/D converter are connected directly to the eight lines of an eight-bit parallel input port.

Figure 24-8A shows the circuit for the case when the A/D converter is under direct program control (i.e., the program issues the start pulse that begins the conversion process). The start line of the A/D converter is connected to one bit (B0 selected here) of output port 1. Any bit or port could be selected, and the unused bit remaining can be used for other applications.

The EOC (end of conversion) pulse is applied to one bit of a second input port (other than the data input port). A typical program sequence would be as follows:

1. Write 01 hex to port 2. This step causes B0 of port 2 to be HIGH.
2. Reset port 2 by writing 00 hex to port 2.
3. Loop until bit B0 of input port 1 is made HIGH.
4. Input data on port 1.

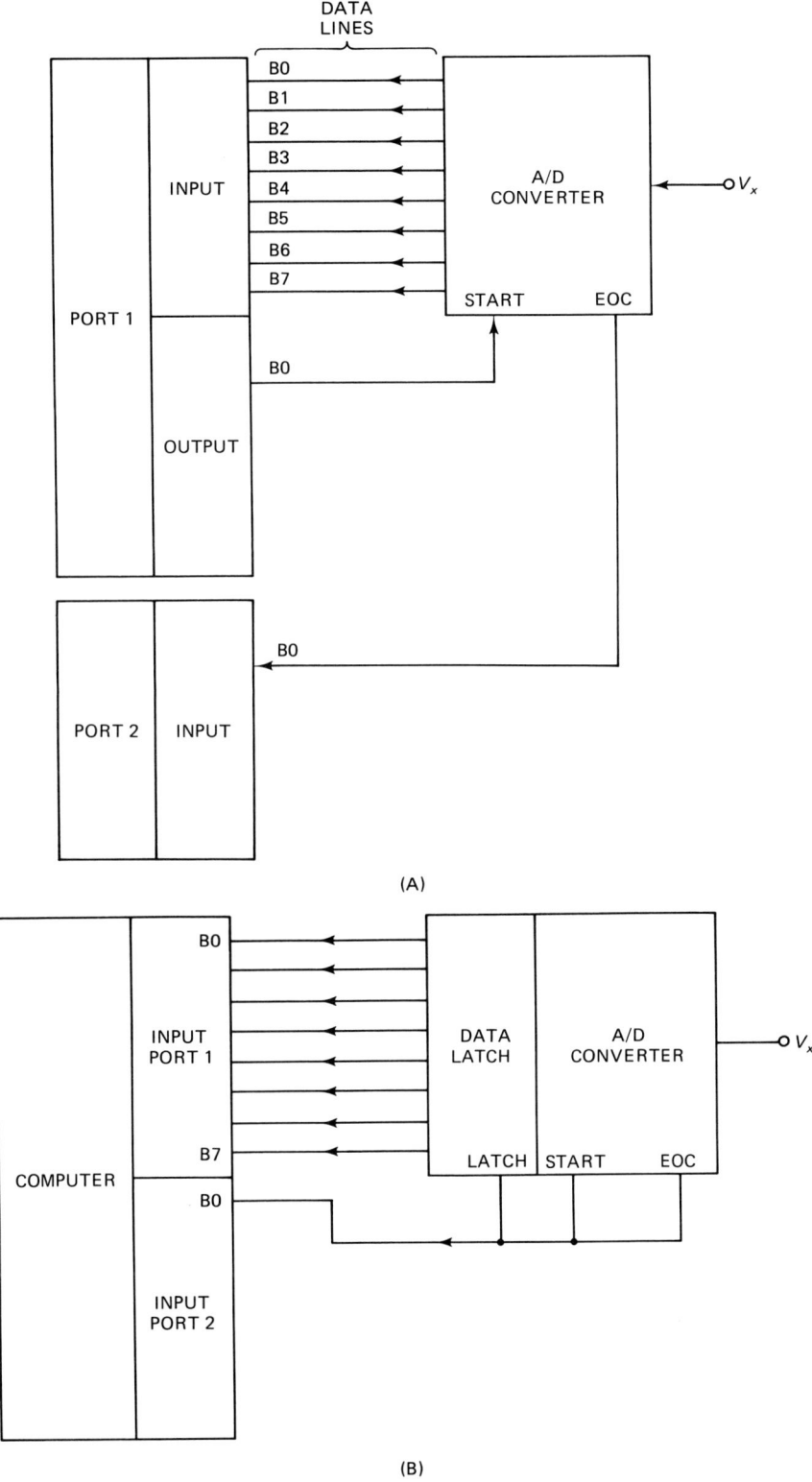

Fig. 24-8 (A) A/D interfaced to I/O ports, (B) data-latched I/O interfacing, and (C) externally latched A/D interfacing. (*Figure continues.*)

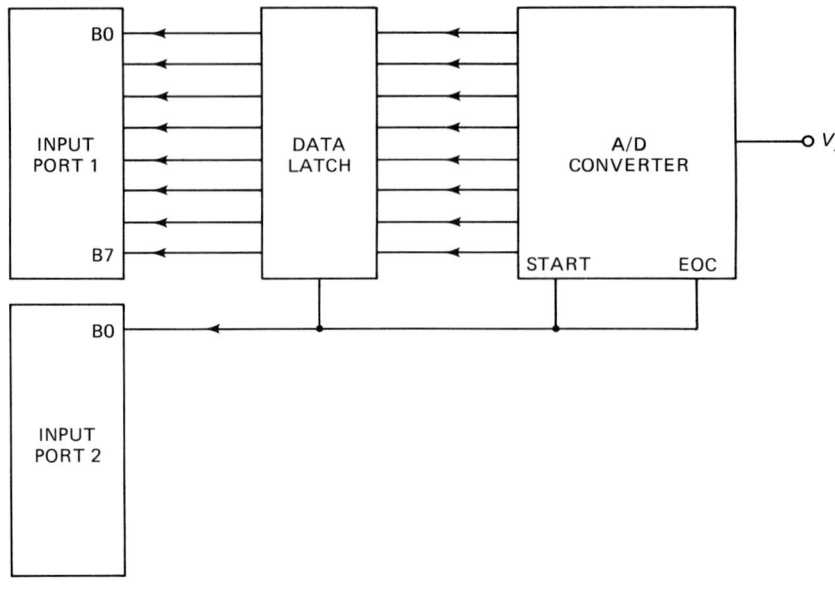

(C)

Fig. 24-8 (*continued*)

The method shown in Fig. 24-8A is wasteful of one output port (i.e., the port used for the start pulse), and it requires the program to be continuously dedicated to that task. The method of Fig. 24-8B is asynchronous and will free up the computer somewhat, provided that the A/D converter uses a latched output stage and the conversion time is sufficiently long. The asynchronism is gained by the simple expedient of connecting the EOC and start lines together. The EOC pulse becomes the start pulse for the next conversion cycle. The computer will loop until it sees the EOC pulse on bit B0 of input port 2. Again, the assumption is made that the A/D converter has a latched output stage.

We can add a latched output stage to an A/D converter that lacks such capability by using the circuit of Fig. 24-8C. The data latch is a dual-quad latch (or some similar chip). The two halves of the latch are each activated by a separate strobe terminal (pins 12 and 23), which are here wired together to accommodate the eight-bit word length. When the EOC/start pulse is generated, indicating that the data on the output of the A/D converter are valid, the data will be transferred from the inputs of the latch to the respective outputs.

Figure 24-9 shows a method that assumes the availability of an eight-bit output port and an eight-bit input port. In addition, the start (NON-ST) and end of conversion (NOT-EOC) signals are active-LOW. The A/D converter is connected to require an initial start pulse from the computer, through bit B0 of the output port, and then run asynchronously thereafter. This method of connection ensures that the

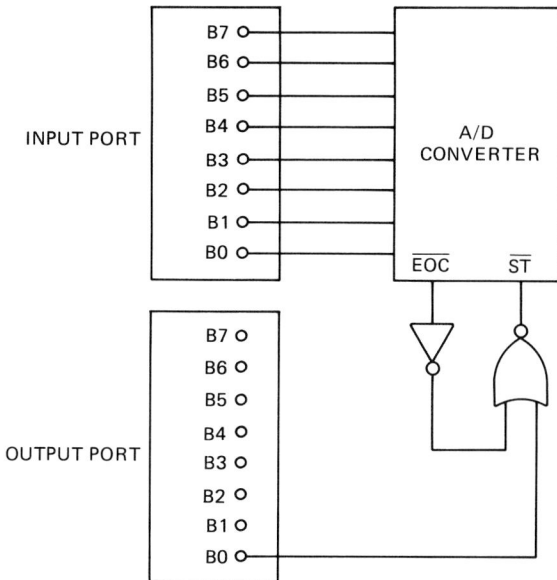

Fig. 24-9 Simple interface to kick -off a synchronous A/D converter at turn-on.

A/D receives the initial pulse and starts conversion. When the computer is initializing itself, it outputs a brief HIGH to the NOR gate. A HIGH at one input of a NOR gate forces the output LOW, so the start (NOT-ST) criterion is satisfied. When the A/D issues its EOC pulse, it is inverted and applied to the other input of the NOR gate. From then on, the operation is asynchronous.

Serial A/D Interfacing

The methods shown thus far in this chapter assume either access to the computer data bus or the availability of a parallel input/output (I/O) port. For some applications, however, these are either not available or would be costly to add. If the computer has an RS-232C serial asynchronous communications port (different brands call it by variations of this name), one can use a system such as Fig. 24-10. The eight-bit output data from the A/D converter is applied to the transmitter input side of a universal asynchronous receiver transmitter (UART) chip. These devices convert parallel data to serial format in the transmitter side and serial data to parallel format on the receiver side.

The serial data line from the typical UART will be TTL compatible, which means that binary 1 is a positive voltage of $+2.4$ V or more, while binary 0 is zero volts. But the RS-232C system wants to see

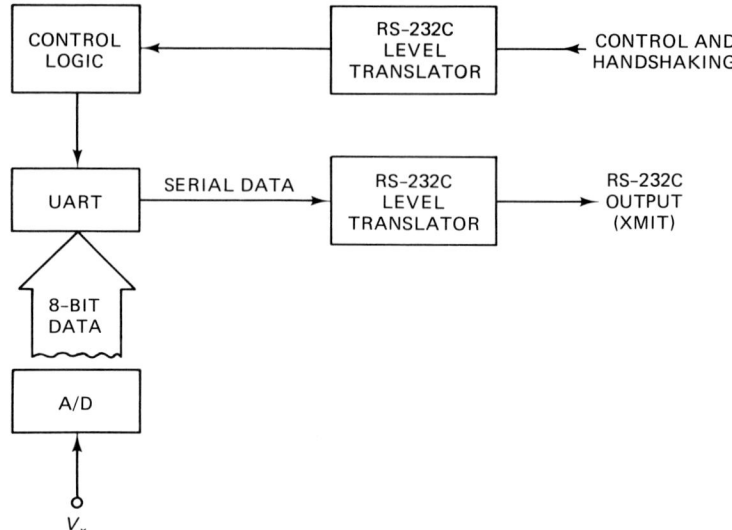

Fig. 24-10   RS-232C serial interfacing of parallel format A/D converters.

$+12/-12$ volts for the logic levels. As a result, we need to use an RS-232C level translator chip (e.g., 1488 and 1489) between the RS-232C connector and the circuit.

If more than purely asynchronous operation is required, the control and handshaking signals from the RS-232C line must be decoded in the control logic section and used to control the UART.

25

Design and Construction of DC Power Supplies

The DC power supply is important to the success of any electronic circuit or equipment. The power supply converts the alternating current available from the power mains to the direct current needed to operate electronic circuits. A chapter on DC power supplies was deemed useful by some advisors because these supplies are both critical to the proper functioning of circuits and are usually overlooked as somehow incidental to the design process. Some of the material in this chapter is included by way of review.

The typical DC power supply consists of several different components: transformer, rectifier, ripple filter, and (in some designs) voltage regulator. The transformer scales the AC voltage from the power lines up or down as needed for the particular application. The job of the rectifier is to convert the bidirectional AC into unidirectional pulsating DC, while the ripple filter smoothes the pulsating DC into nearly pure DC. The voltage regulator is used to stabilize the voltage in the face of changing load currents and AC input voltage.

Also part of some DC power supplies are functions such as overvoltage protection and current limiting. These circuits, as well as the main components of the DC power supply are discussed in detail in this chapter. You will learn the fundamentals of DC power supply design, especially the regulated, low-voltage DC power supplies that are typically used with circuits containing linear integrated circuit elements.

RECTIFIERS

Rect-i-fy: "To make right; remove impurities." The purpose of a rectifier in a DC power supply circuit is to remove the impurities of the

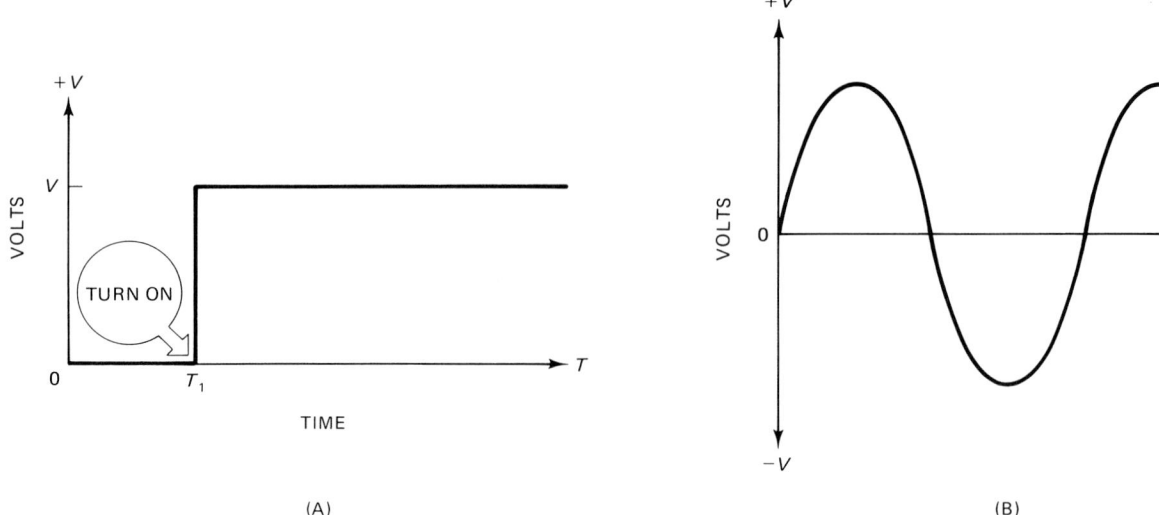

Fig. 25-1 (A) Unidirectional direct current (DC) and (B) bidirectional alternating current (AC).

AC line current and make it right for DC-craving electronic circuits that require pure, or nearly pure, direct current.

Before discussing the details of solid-state rectifiers, let us review the two basic forms of electrical current in the context of rectification. DC and AC (see Fig. 25-1). Direct current (DC) is graphed in Fig. 25-1A. The key feature of this form of electrical current is that it is unidirectional—current flows through the circuit in only one direction. It will be zero until turned on (time $T1$), and it will then rise to a certain level and remain there. Electrons flow from the negative terminal to the positive terminal of the power supply, and that polarity never reverses direction.

Alternating current, on the other hand, is bidirectional (see Fig. 25-1B). On one half-cycle the current flows in one direction. Then the power supply polarity reverses, so the current flows in the opposite direction. The electrons still flow from negative to positive, but since the positive and negative poles have switched places, the physical direction of current flow has reversed. In the normal AC power mains the voltage and current waveforms vary as a sine wave. By convention, flow in the positive direction is graphed above the zero-volts (or zero amperes) line, and flow in the negative direction is graphed below the zero line. Raw AC is incompatible with nearly all electronic circuits, so must be changed to DC by a rectifier and a ripple filter. The main requirement for a rectifier is that it convert bidirectional AC into a unidirectional form of current. Although industry once used rotary mechanical switches, synchronous vibrators, and vacuum tubes to accomplish the rectification job, all modern circuits rely on solid-state PN junction rectifiers.

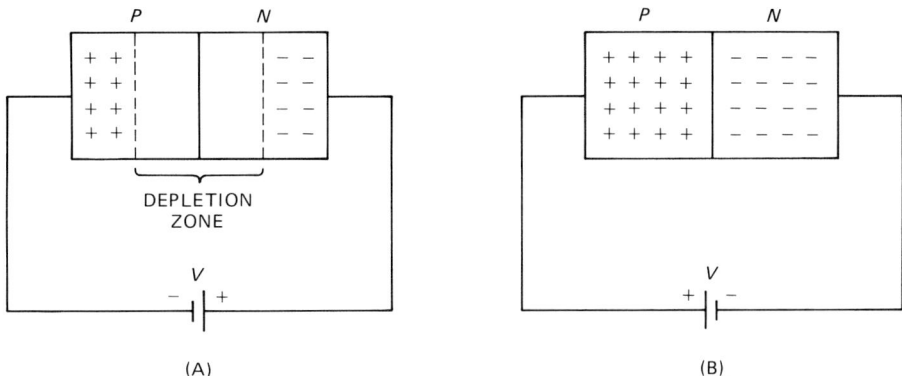

Fig. 25-2 PN junction diodes (A) reverse biased and (B) forward biased.

PN JUNCTION DIODE RECTIFIERS

The modern solid-state rectifier really isn't so new after all. Various versions of the rectifier date back to the dawn of the radio—indeed the electrical—age. All common rectifier diodes in use today, however, are silicon PN junction diodes (shown schematically in Fig. 25-2). The PN junction diode rectifier (Fig. 25-2) consists of a silicon semiconductor material that is doped with impurities to form N-type material at one end and P-type material at the other end. The charge carriers (which form the electrical current) in the N-type material are negatively charged electrons, and the charge carriers in the P-type material are positively charged "holes."

The reverse bias situation is shown in Fig. 25-2A. In this case the negative terminal of the voltage source (V) is connected to the P-type material and the positive terminal is connected to the N-type material. Positive-charge carriers are thus attracted away from the PN junction toward the negative-voltage terminal, and negative-charge carriers are drawn away toward the positive terminal. That leaves a charge-free depletion zone in the region of the junction that contains no carriers. Under this condition, there is little or no current flow across the junction. Theoretically, the junction current is zero, although in real diodes there is always a tiny leakage current across the junction.

The forward-biased case is shown in Fig. 25-2B. Here the polarity of voltage source V is reversed from Fig. 25-2A. The positive terminal is applied to the P-type material, and the negative terminal is applied to the N-type material. Because like charges repel, the charge carriers in both P- and N-type material are driven away from the power supply terminals toward the junction. The depletion zone disappears, allowing positive and negative charges to get close to the boundary between regions. As these opposite charges attract each other across the junction, a current flows in the circuit.

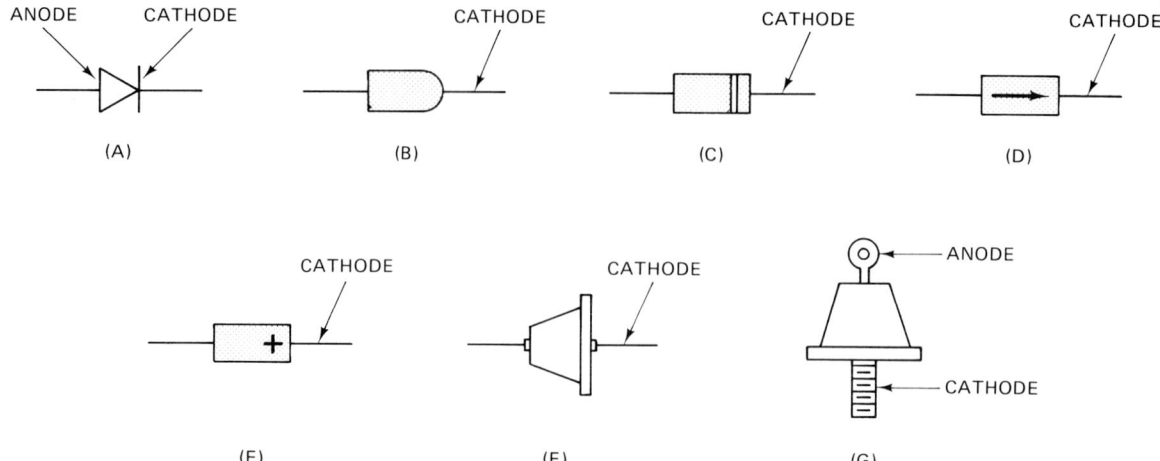

Fig. 25-3 Rectifier diodes: (A) circuit symbol, (B–E) various epoxy case rectifiers, (F) top-hat rectifier, and (G) stud-mounted rectifier.

From the above description it is apparent that a PN junction diode is able to convert bidirectional AC into unidirectional current because it allows current to flow in only one direction. Thus, it is a rectifier. However, the rectifier output current is not pure DC (as from batteries), but rather it is pulsating DC.

Figure 25-3 shows the standard circuit symbol for the solid-state rectifier diode (Fig. 25-3A), along with the common shapes of some actual diodes. The input side where AC is applied is the anode, and the DC output is the cathode. The diodes shown in Figs. 25-3B through 25-3G are positioned so that the respective anodes and cathodes are aligned with those of the circuit symbol in Fig. 25-3A. Rectifiers 25-3B through 25-3E are epoxy package devices, which are the type seen most often. The cathode end will be marked either with a rounded end (25-3B), a line (25-3C), a diode arrow (25-3D), or a plus sign (25-3E).

The diode shown in Fig. 25-3F is the old-fashioned (now obsolete) top-hat type. Unless otherwise specified, the top-hat type can safely pass a current of 500 mA; those in Figs. 25-3B through 25-3E generally pass 1 A (or more, for larger-sized but similar packages). The stud-mounted type shown in Fig. 25-3G is a high-current model. These diodes are rated at currents from 6 A and up (50- and 100-A models are easily obtained). These diodes are mounted using a threaded screw at one end, which also forms one electrical connection. The other electrical connection is the solder terminal at the other end. Unless otherwise specified, the solder terminal is the anode and the stud-mount is the cathode terminal. Exceptions to the polarity rule are sometimes seen. The reverse-polarity diodes will have either an arrow symbol pointing in the opposite direction (the arrow always points to the cathode) or an R suffix on the type number (e.g., 1NxxxxR instead of 1Nxxxx).

Rectifier Specifications

The proper use of solid-state rectifiers requires consideration of several key specifications: forward current, leakage current, surge current, junction temperature, and peak inverse voltage (PIV)—also called peak reverse voltage (PRV). The forward current is the maximum constant current that the diode can pass without damage. For the 1N400x series of rectifiers the forward specified current is one ampere (1 A).

The leakage current is the maximum current that will flow through the reverse biased PN junction. In an ideal PN junction diode, the leakage current is zero; in high-quality practical diodes it is typically very low compared to the forward current.

The surge current is typically very much larger than the forward current, and it is sometimes erroneously taken to be the operating current of the diode (a mistake that is made but once). Surge current is defined as the maximum short-duration current that will not damage the diode. Short-duration typically means one AC cycle (1/60 second in a 60-Hz system). Don't use the surge current as if it were the forward current.

The specified junction temperature is the maximum allowable operating temperature of the PN junction. The actual junction temperature depends on the forward current and how well the package (and environment) rids the diode of internal heat. Although typical maximum junction temperatures range up to $+125°$ Celsius for silicon devices, good design requires as low a temperature as possible. One reliability guide requires that the junction temperature be held to a maximum of $+110°$ Celsius.

The peak inverse voltage (PIV) is the maximum allowable reverse-bias voltage that will not damage the diode. This rating is usually the limiting rating in certain power supply designs.

RECTIFIER CIRCUITS

Figure 25-4 shows a solid-state rectifier diode (D1) in a simple halfwave rectifier circuit. In Fig. 25-4A the diode is forward biased: The positive terminal of the voltage source is connected to the anode of the rectifier. Current (I) flows through the load resistance (R). In Fig. 25-4B the opposite situation is found: The negative terminal of the voltage source is applied to the anode, so the diode is reverse-biased and no current flows.

The circuit of Fig. 25-4 is called a halfwave rectifier for reasons that immediately become apparent from Fig. 25-4C. In this figure the output current through the load (R) is graphed as a function of time when an AC sine wave is applied. From time $T1$ to $T2$ the diode is forward biased, so current flows in the load (also from $T3$ to $T4$). But

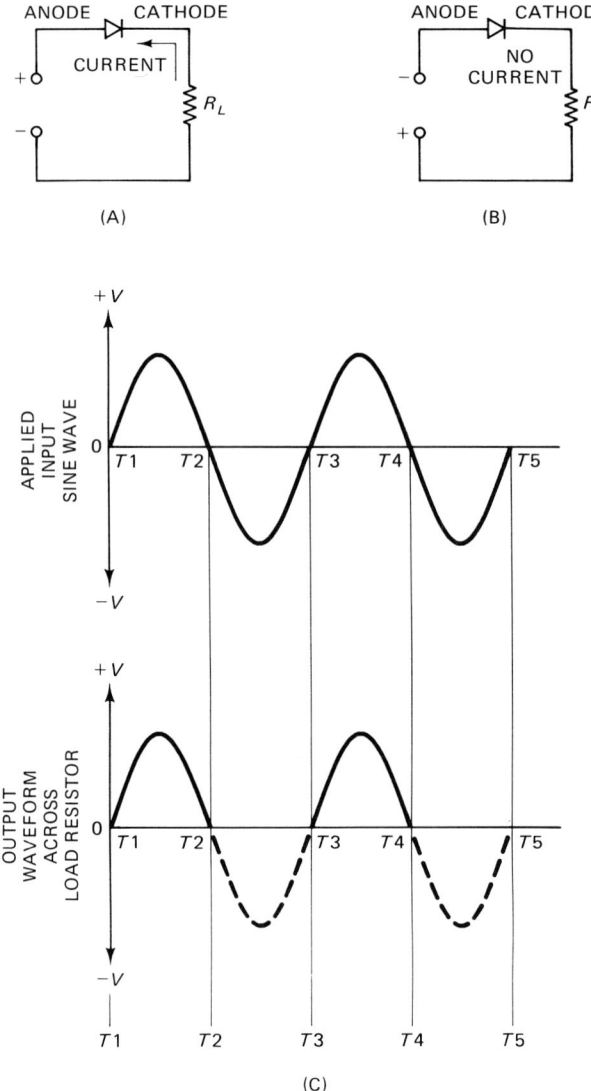

Fig. 25-4 Half-wave rectification: (A) diode is forward biased; (B) diode is reverse biased; and (C) associated waveforms.

during the period $T2$ to $T3$ the diode is reverse biased, so no current flows. Because the entire sine wave takes up the period $T1$ to $T3$, only half of the input sine wave is used. The output waveform shown in Fig. 25-4C is called a halfwave rectified pulsating DC wave.

The halfwave rectifier is low cost, but it wastes energy due to its use of only one-half of the input AC waveform. Efficiency is increased by making use of the entire waveform in a fullwave rectifier.

Figure 25-5A shows the standard fullwave rectifier. This circuit uses a transformer that has a center-tapped secondary winding. Because the center tap (CT) is used as the zero volts reference (in most circuits it is grounded), the polarities at the ends of the secondary are

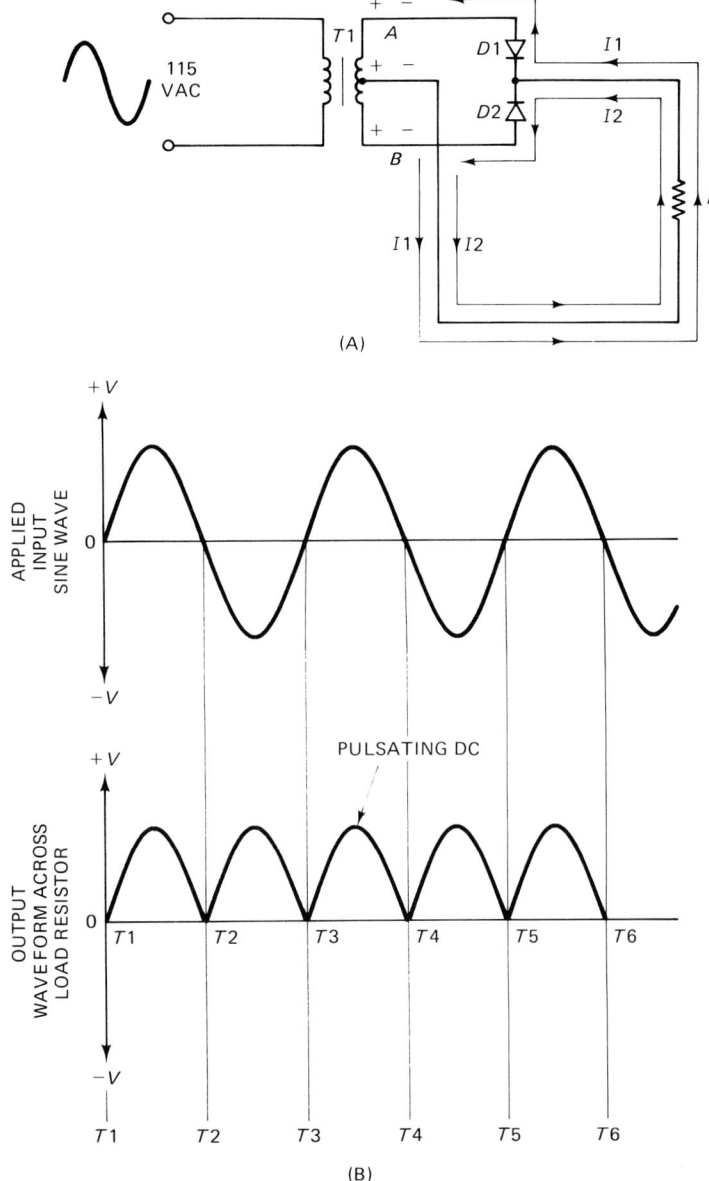

Fig. 25-5 (A) Full-wave rectifier circuit and (B) associated input and output waveforms.

always opposite each other (i.e., 180°). On one half-cycle, point A is positive with respect to the CT while point B is negative. On the next half-cycle, point A is negative and point B is positive with respect to the CT. This situation makes D1 forward biased on one half-cycle while D2 is reverse biased. Alternatively, on the next half-cycle D1 is reverse biased and D2 is forward biased.

Follow the circuit of Fig. 25-5A through one complete AC cycle (times $T1$ through $T3$ in Fig. 25-5B). On the first half-cycle ($T1$–$T2$),

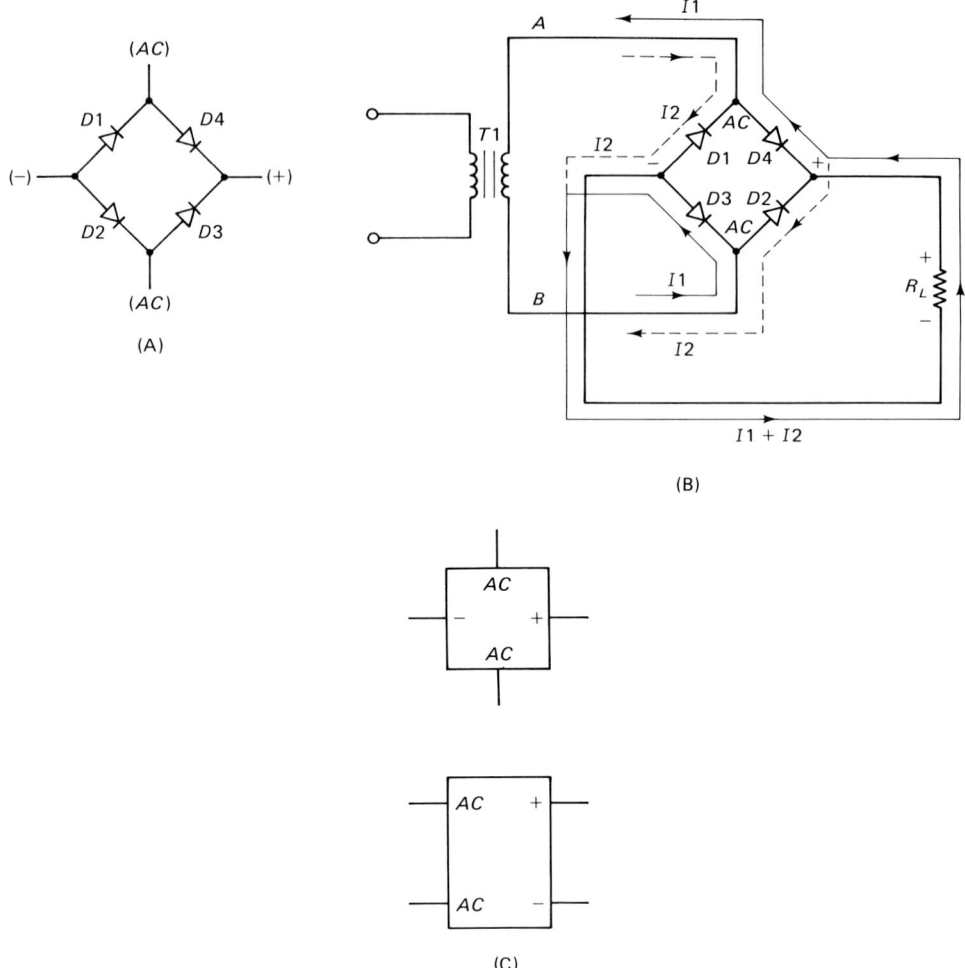

Fig. 25-6 (A) Bridge rectifier, (B) rectifier circuit, and (C) various common circuit symbols.

point A is positive, so D1 is forward biased and conducts current; D2 is reverse biased. Current $I1$ flows from the CT, through load R, through diode D1, and back to the transformer at point A. On the alternate half-cycle, current $I2$ flows from the CT, through load R, through diode D2, and back to the transformer at point B. Now notice what happened: $I1$ and $I2$ are equal currents, generated on alternate half-cycles, and they *flow in load R in the same direction*. Thus, a unidirectional current is flowing through load R on both halves of the AC sine wave. The waveform resulting from this action is shown in Fig. 25-5B; it is called fullwave rectified pulsating DC.

The center tap on the secondary winding of the transformer can be eliminated by using the fullwave bridge rectifier circuit of Fig. 25-6A. This circuit requires twice as many rectifier diodes as the other form of fullwave rectifier, but it allows the use of a simpler transformer (no center tap). The operation, however, is similar (see Fig. 25-6B). On

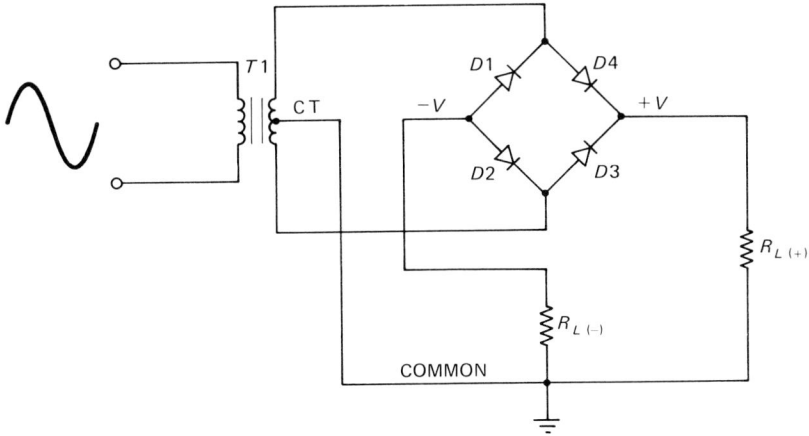

Fig. 25-7 Dual polarity output power supply.

one half-cycle, point A is positive and point B is negative. Current I1 flows from the transformer at point B, through D4, load R, diode D1, and back to the transformer at point A. On the alternate half-cycle, point A is negative and point B is positive. In this case, current $I2$ flows from point A, through diode D3, load R (in the same direction as $I1$), diode D2, and back to the transformer at point B.

A bridge rectifier can be built using four discrete diodes (D1–D4). In most modern equipment, however, a bridge stack is used. These components are bridge rectifiers built into a single package with four leads coming out. Figure 25-6C shows various alternate schematic diagram circuit symbols for bridge rectifier stacks. Sometimes, a sine-wave symbol is used for the AC power connections instead of the AC letters shown.

Figure 25-7 shows a half-bridge fullwave rectifier. This circuit is used more today because of the dual polarity power supplies used in a lot of equipment. Operational amplifiers and some CMOS devices typically require ± 12-V or ± 15-V bipolar DC power supplies. A fullwave bridge rectifier stack coupled to a center-tapped transformer will create a pair of fullwave rectified DC power supplies. The CT is the common (or ground); the bridge positive terminal supplies the positive output voltage and the negative terminal supplies the negative output voltage. Both outputs from the half-bridge rectifier are fullwave-rectified pulsating DC.

Selecting Rectifier Diodes

The two parameters most often used to specify practical power supply diodes are the forward current and peak inverse voltage. The forward current rating of the diode must be at least equal to the maximum current load that the power supply must deliver. In practical circuits,

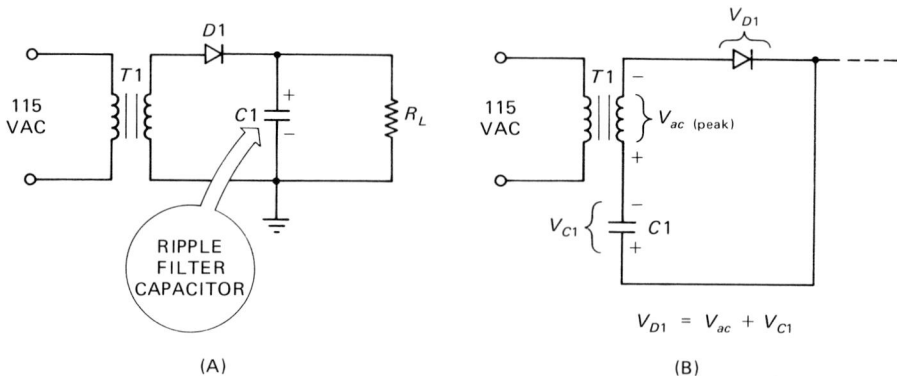

Fig. 25-8 (A) Single-section "brute-force" filter and (B) circuit redrawn.

however, there is also a real need for a safety margin to account for tolerances in the diodes and variations of the real load (as opposed to the calculated load). It is also true that making the rating of the diode somewhat larger than the load current will greatly improve reliability. A good rule of thumb is to select a diode with a forward current rating of 1.5 to 2 times the calculated (or design goal) load current—or more if available. Selecting a diode with a very much larger forward current rating (e.g., 200 A for a 100-mA circuit) is both wasteful and likely to make the diode not work exactly like a rectifier diode. The general rule is to make the rating as high as feasible. The 1.5-to-2 times rule, however, should result in a reasonable margin of safety.

The peak inverse voltage (PIV) rating can be a little more complicated. In unfiltered, purely resistive, circuits the PIV rating need only be greater than the maximum peak applied AC voltage (1.414 times RMS). But if a 20-percent safety margin is desired (a good idea), make it 1.7 times RMS voltage. Most rectifiers in practical DC power supplies are used in ripple-filtered circuits (e.g., Fig. 25-8A), and that makes the PIV problem different. Figure 25-8B shows the simple halfwave rectifier capacitor-filtered circuit redrawn to better illustrate the circuit action. Keep in mind that capacitor $C1$ is charged to the peak voltage with the polarity shown. This peak voltage is 1.414 times the RMS voltage. The peak voltage across the transformer secondary voltage (V) is in series with the capacitor voltage. When voltage V is positive, the transformer voltage and capacitor voltage cancel out, making the diode reverse voltage nearly zero. But when the transformer voltage (V) is negative, the two negative voltages (V and V_c) add algebraically to twice the peak voltage.

$$V_{D1} = V_{ac\,(peak)} + V_{C1} \qquad (25\text{-}1)$$

but

$$V_{C1} = V_{ac\,(peak)} \qquad (25\text{-}2)$$

so

$$V_{D1} = V_{ac\,(peak)} + V_{ac\,(peak)} \qquad (25\text{-}3)$$

$$V_{D1} = 2V_{ac\,(peak)} \qquad (25\text{-}4)$$

Because $V_{ac\,(peak)} = 1.414V_{ac\,(rms)}$,

$$V_{D1} = 2(1.414V_{ac\,(rms)}) \qquad (25\text{-}5)$$

$$V_{D1} = 2.828V_{ac\,(rms)} \qquad (25\text{-}6)$$

The reverse voltage across the diode is approximately 2.83 times the RMS voltage. Therefore, the absolute minimum value of PIV rating for the diode in a filtered DC supply is 2.83 times the applied RMS. If a 20% safety margin is preferred, the diode PIV rating should be 3.4 times the applied RMS voltage (or more).

Using Rectifier Diodes

In most cases, especially low-voltage power supplies, diodes can be used as shown in the foregoing circuits. In Fig. 25-9A, however, the proper way to use the solid-state diode rectifier is shown. The resistor

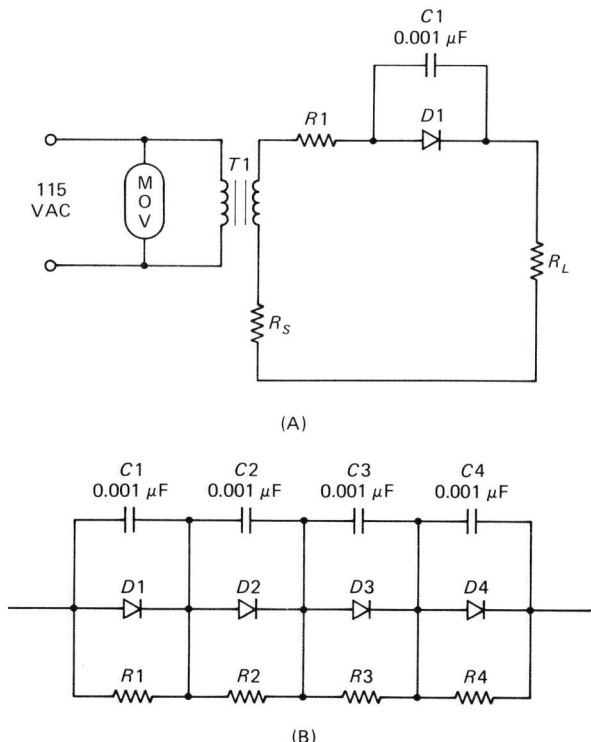

Fig. 25-9 (A) Circuit protections for rectifier diode and (B) increasing voltage rating of the diodes.

in series with diode D1 (i.e., $R1$) is used to limit the forward current. Many circuits, especially those with capacitor-input filter circuits, exhibit a large surge current at initial turn-on. This current can sometimes destroy the diode, so $R1$ is used to limit the possible damage. The resistance value of $R1$ is typically 5 to 20 Ω. In most cases, however, $R1$ can be eliminated by using a diode with a current rating significantly larger than the load current (for example, the two-times rule). Also, the transformer secondary winding resistance (R_s) serves the current-limiting function of $R1$ in many circuits. Capacitor $C1$ in Fig. 25-9A is used to bypass high-voltage transient spikes around the diode. These spikes could possibly blow the diode PN junction by severely exceeding the PIV and forward voltage ratings. High-voltage line spikes are a frequent source of damage to rectifier diodes. Placing a small-value capacitor in parallel with the diode will eliminate that problem. The DC working voltage (WVDC) of the capacitor should be equal to or greater than the PIV rating of the diode. By use of 1000-V PIV diodes (even in low-voltage circuits), much of the damage caused by transients is avoided.

The capacitors can usually be eliminated if a metal oxide varistor (MOV), or some other high-voltage spike suppressor device, is shunted across the AC supply voltage in the primary circuit of the power transformer.

Figure 25-9B shows the proper method for connecting several diodes in series to increase the PIV rating. Assuming that the PIV ratings of the diodes are equal, the overall rating is four times the rating of one diode. If 1000-V PIV diodes are used in this circuit, the total PIV rating of the assembly is 4000 V.

The capacitors used in Fig. 25-9B are for exactly the same purpose as in Fig. 25-9A. The resistors, however, are needed for a different purpose: They equalize the forward voltage drop across each diode. A 470-kΩ 2-W resistor is typically used for 1000-V PIV diodes. The 2-W rating is required not because of the power dissipation of the resistors, but rather for their voltage rating (yes, resistors do have voltage ratings).

Figure 25-10 shows the proper method for mounting an axial lead rectifier on a prototyping "perfboard" or printed circuit board. This method is used anytime except where excessive vibration is expected, which includes most sedentary circuits or equipment. The space be-

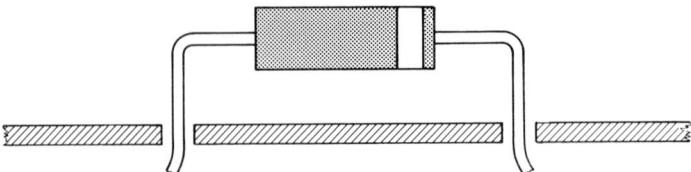

Fig. 25-10 Proper mounting of the axial lead epoxy rectifier.

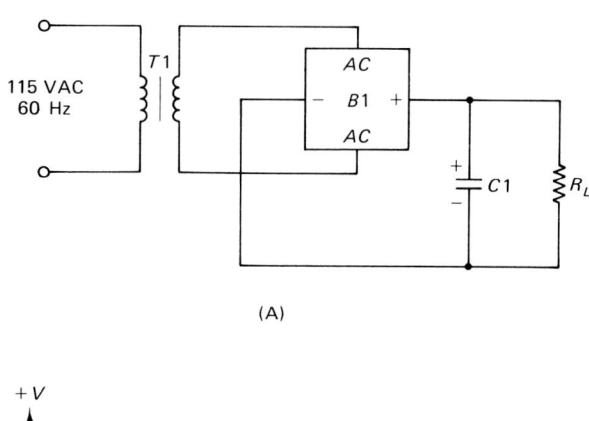

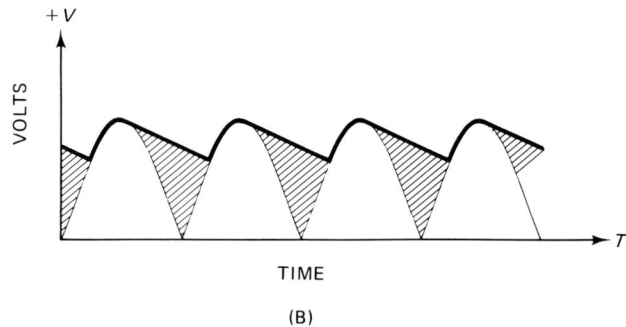

Fig. 25-11 (A) Simple DC power supply filter and (B) circuit action.

neath the diode body allows air to circulate (keeping the diode cooler) and prevents diode heat from damaging the board.

RIPPLE-FILTER CIRCUITS

The pulsating DC output from either fullwave or halfwave rectifiers is almost as useless for some electronic circuits as the AC input waveform. A ripple-filter circuit is used to smooth out the pulsating DC into a cleaner DC. Figure 25-11A shows the simplest form of filter circuit: a single capacitor ($C1$) connected in parallel with the load. Circuit action is shown in Fig. 25-11B. The job of the capacitor is to store electrical charge on voltage peaks and then dump that charge into the load when the voltage drops between peaks. The shaded area of Fig. 25-11B shows the filling in caused by the capacitor charge. The output voltage is the sum of both rectifier and capacitor contributions; it is represented by the heavy line in Fig. 25-11B.

The value of the filter capacitor is determined by the amount of ripple factor that can be accepted. The ripple factor (r) is defined as the ratio of the ripple voltage amplitude to the average voltage of the rectified waveform. Values tend to be in the range 0.01 to 1.0 for common electronic circuits. The rule of thumb for ripple factor for

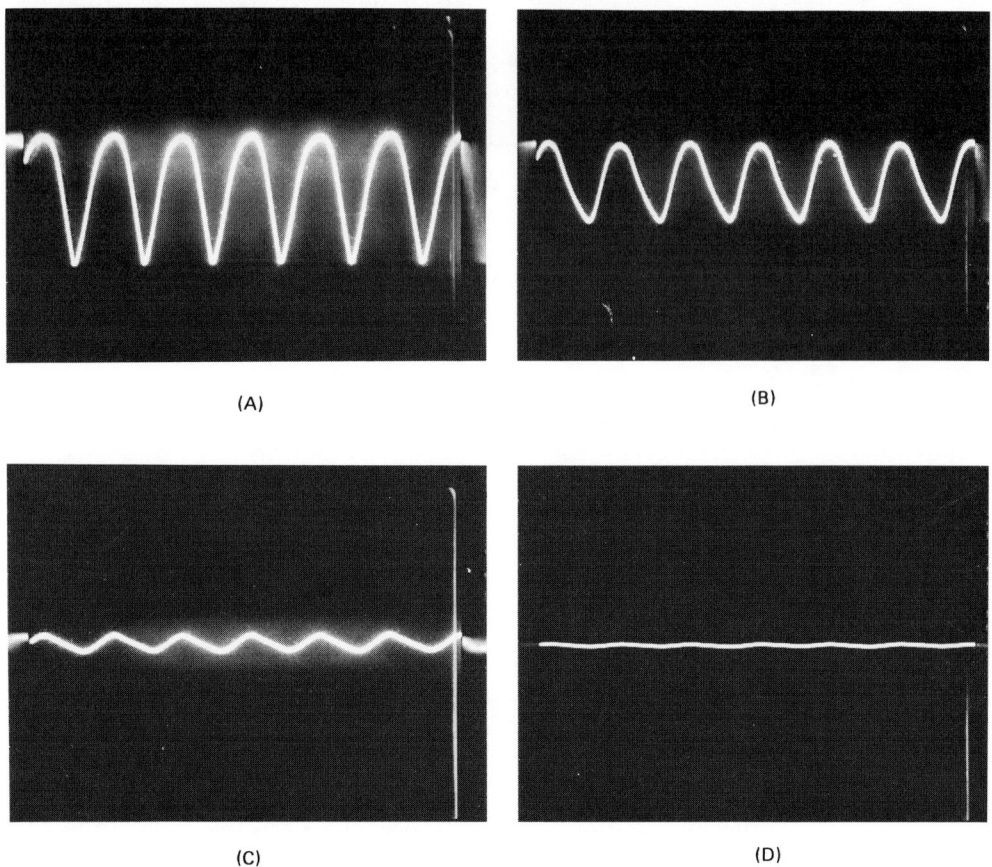

Fig. 25-12 (A) Full-wave rectified pulsating DC, (B) ripple for 150 μF, (C) ripple for 1000 μF, and (D) ripple for 6800 μF.

60-Hz circuits is
  Half-wave rectified circuits

$$r = \frac{1,000,000}{208 C1 R_{\mathrm{L}}} \tag{25-7}$$

Full-wave rectified circuits

$$r = \frac{1,000,000}{416 C1 R_{\mathrm{L}}} \tag{25-8}$$

where: $C1$ is the capacitor value in microfarads and R_{L} is the load resistance in ohms. The load resistance is the ratio of the output voltage to the output current: $V_{\mathrm{o}}/I_{\mathrm{o}}$. Consider a full-wave example.

So what do these figures mean in practical terms? A few oscilloscope waveform photos (Fig. 25-12) illustrate the effect of adding capacitance to the filter circuit. In all cases the AC applied to the rectifier was nominally 12 V AC, and the load resistance was 25 Ω.

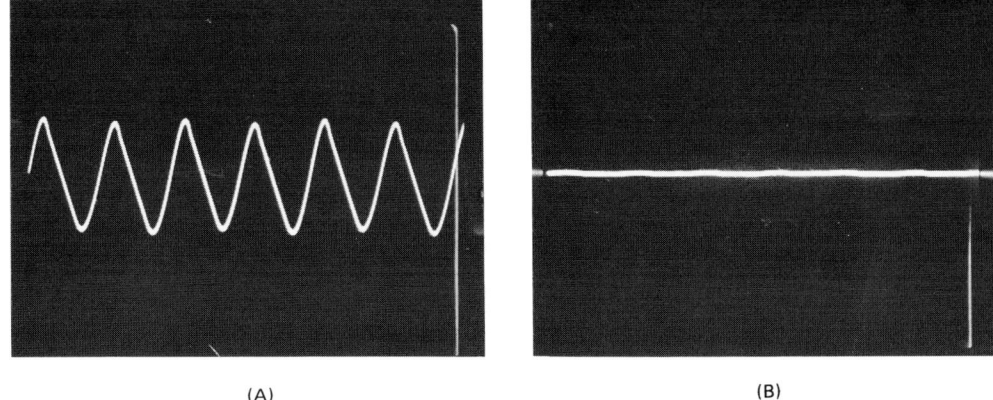

(A)   (B)

Fig. 25-13   (A) Ripple at input of the regulator and (B) ripple at the output.

These photos were taken with the 'scope AC-coupled to permit expansion of the ripple in the presence of the large DC offset. The waveforms shown in Fig. 25-12 ride on top of the DC output applied to the load, but the DC offset is suppressed by the 'scope input circuit. The 'scope sensitivity was 5 V/cm.

The unfiltered, full-wave rectified pulsating DC (Fig. 25-12A) was measured with a DC voltmeter and found to be 9.25 V DC. This value is not the peak voltage (which was close to 12 V DC), but rather an average value caused by the fact that the DC meter naturally tends to integrate (time-average) the reading. When a 150-μF capacitor is connected across the 25-Ω load the ripple reduces to 0.64 (Fig. 25-12B), and the output voltage rises to 10.86 V DC. Note that the filtering action is just beginning to take place. Connecting a 1000-μF capacitor across the load further drops the ripple factor to 0.096 (Fig. 25-12C); the voltage rises to 11.8 volts. Finally, a 6800-μF capacitor is connected across the load. The ripple factor drops to 0.014 (Fig. 25-12D). This filtered DC is nearly pure, but it still contains a small ripple factor.

Next, the sensitivity of the 'scope was increased from 5 V/cm to 0.1 V/cm (a 50 × increase) and the same 6800-μF capacitor was used in parallel with the 25-Ω load. The ripple shows up a lot better under this condition (Fig. 25-13A). Recall from the foregoing that this waveform represents a ripple factor $r = 0.014$.

A voltage regulator tends to smooth out the ripple considerably. Power supply makers sometimes claim in advertising that their products have the equivalent of one farad of output capacitance. What they mean is that the ripple is reduced by a voltage regulator the same as a one-farad (1,000,000 μF) capacitor across the load. Want proof? Examine Fig. 25-13B. With the same 0.1 V/cm 'scope sensitivity and the same power supply, this waveform shows a remarkable drop of ripple compared with Fig. 25-13A. The lesson here is to use a voltage regulator in circuits where it is necessary to reduce the ripple to a very low value.

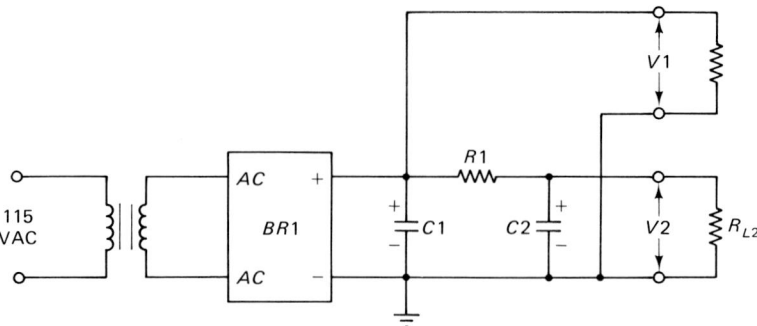

Fig. 25-14   Two-section *RC* ripple-filter circuit.

Pi-Network Filter Circuits

Another form of filter circuit is the *RC* Pi-network shown in Fig. 25-14. Output voltage *V*1 represents a circuit similar to those described, but output voltage *V*2 has a lower voltage and substantially lower ripple factor. For full-wave circuits the ripple factor is

$$r = \frac{2.5 \times 10^{-6}}{C1\,C2\,R1\,R_{\mathrm{L}}} \tag{25-9}$$

Capacitor Working Voltage

The other significant rating for filter capacitors, besides capacitance, is the DC working voltage (WVDC). This rating specifies the maximum voltage that the capacitor can sustain safely on a continuous basis (not a transient peak voltage). Because AC line voltages can vary ± 15 percent, and the WVDC rating tolerance may have a ± 20-percent tolerance, it is prudent to use a capacitor with a WVDC rating that is at least half again higher than the normal output voltage expected. That is, the minimum WVDC rating of the capacitor should be 1.5 times the maximum output voltage of the rectifier. If a filter capacitor has a WVDC rating close to the power supply output voltage, then a short life and a spectacular end may be the fate of the power supply.

VOLTAGE REGULATION

The output voltage of ordinary rectifier/filter DC power supplies is not stable—it may vary considerably over time. There are two main sources of variation in the output of this type of power supply. First, there is always a certain fluctuation of the AC input voltage. Ordinary

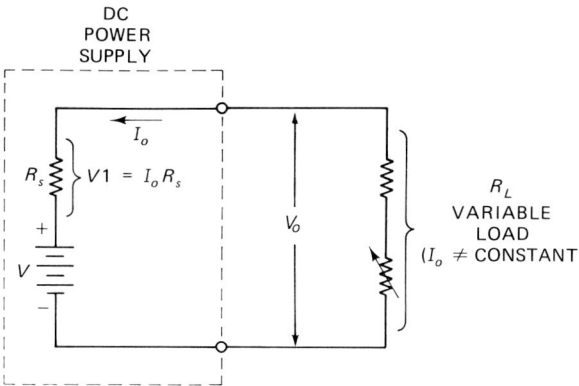

Fig. 25-15 Equivalent circuit of a DC power supply showing internal resistance.

commercial power lines vary from 105 to 120 V AC (rms) normally, and it may droop to less than 100 V during power brownouts. The second source of variation is created by load variation (see Fig. 25-15). The problem arises because real DC power supplies are not ideal. The ideal textbook power supply has zero ohms internal resistance, while real power supplies have a certain amount of internal resistance (represented by R_s in Fig. 25-15). When current is drawn from the power supply there is a voltage drop ($V1$) across the internal resistance, and this voltage is subtracted from the available voltage (V). In an ideal power supply, output voltage V_o is the same as V, but in real supplies V_o is equal to ($V - V1$). Because $V1$ varies with changes in the load current I_o, the output voltage will also vary with changes in current demand.

The goodness or badness of a power supply can be defined in terms of its percentage of regulation. This specification is a measure of how badly the voltage changes under changes of load current; it is found from

$$\%\text{REG} = (V - V_o)(100\%)/V \qquad (25\text{-}10)$$

where V is the open-terminal (no output current) output voltage, V_o the output voltage under full-load current, and %REG the percentage of regulation.

Many electronic circuits do not work properly under varying supply-voltage conditions. Oscillators and some waveform generators, for example, tend to change frequency if the DC power supply voltage changes. Obviously, some means must be provided to stabilize the DC voltage. The zener diode is perhaps the simplest such voltage regulator device.

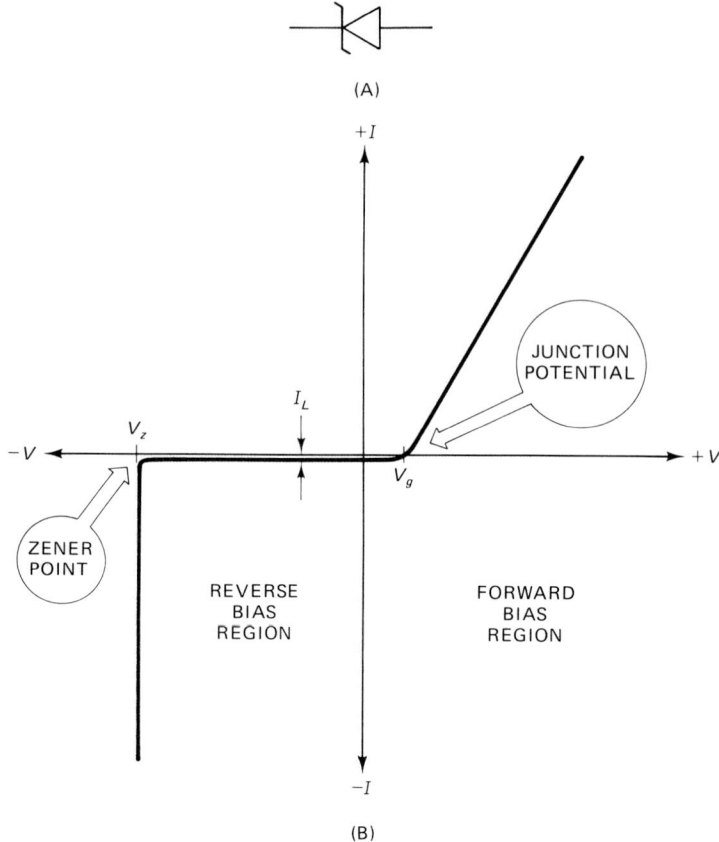

Fig. 25-16 (A) Zener diode symbol and (B) zener diode *I*-versus-*V* curve.

Zener Diodes

The zener ("zen-ner") *diode* is a special case of the PN junction diode; Figure 25-16 shows both the circuit symbol (Fig. 25-16A) and *I*-versus-*V* curve (Fig. 25-16B) for a zener diode. In the forward-bias region operation of the zener diode is the same as for other PN junction diodes. For this case, the anode is positive with respect to the cathode; so a forward-bias current ($+I$) flows. For voltages greater than V_g, which is approximately 0.6 to 0.7 V, the current flow increases approximately linearly with increasing voltage. At potentials less than V_g, the current increases from a small reverse leakage current (I_L) at $V = 0$, to a small forward current at $+V_g$. The diode is, like all other PN junction diodes, nonlinear in this low-voltage region.

The zener diode also acts like any other PN junction diode in the reverse bias region between $V = 0$ and the zener potential $V = -V_z$. In this region, only the small reverse leakage current flows.

At an applied potential of $-V_z$ or greater, the zener diode breaks down and allows a large reverse current to flow. Note in Figure 25-16B

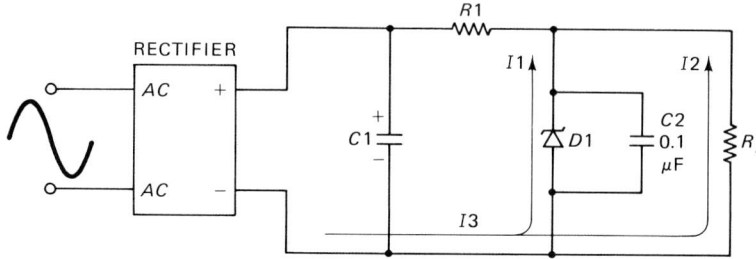

Fig. 25-17 Typical zener voltage-regulator circuit.

that further increase in $-V$ does not cause an increased voltage drop across the diode. Thus, the zener diode regulates the voltage to its zener potential by clamping action.

Zener Diode Voltage-Regulator Circuits

A zener diode operates as a parallel or shunt regulator because it is connected in parallel with the load. It regulates by clamping the output voltage across the load to the zener potential. Figure 25-17 shows a typical zener diode regulator circuit that takes advantage of these attributes.

In Fig. 25-17, resistor R_L represents the load placed across the power supply, that is, the circuits that draw current from the supply. The value of the load resistance is V_z/I_o. Resistor $R1$ is used as a series current limiter to protect the diode. Refer back to Fig. 25-16B to see why it is needed. Note that $-I$ increases sharply when $-V$ reaches the zener potential. If $R1$ is not used, this current will destroy the diode. One of the tasks in designing a zener diode voltage-regulator circuit is selecting the resistance value and power rating of resistor $R1$.

Capacitor $C2$ is optional; it is used to suppress the hash noise generated by the zener diode. The zener process is an avalanche phenomenon, so it is inherently noisy. In fact, certain RF and audio noise generators use a zener diode to create the noise signal. The other capacitor in the circuit, $C1$, is the regular filter capacitor used in any rectifier/filter DC power supply. Its purpose is to smooth the pulsating DC into nearly pure DC. It doesn't really serve a function in the zener regulator circuit, except that the power supply should be filtered prior to the regulator circuit.

The main current drawn from the rectifier ($I3$) is broken into two branches: $I1$ flows through the zener diode and $I2$ flows through the load; $I1$ usually is approximately 10 percent of $I2$. According to Kirchhoff's law, the relationship between currents is $I3 = I1 + I2$.

Table 25-1
Design Equations for Circuit Conditions in Zener Diodes

Condition 1: Variable V_{in}, constant I_o

$$R1 = \frac{V_{in(min)} - V_z}{1.1(I2)}$$

Condition II: Constant V_{in}, variable I_o

$$R1 = \frac{V_{in(max)} - V_z}{1.1(I2_{(max)})}$$

Condition III: V_{in} and I_{in} both variable

$$R1 = \frac{V_{in(max)} - V_z}{1.1(I2_{(max)})}$$

Power dissipations for all three conditions

Diode dissipation

$$P_{D1} = \frac{\left[V_{in(max)} - V_z\right]^2}{R1} - [(I2)V_z]$$

Resistor dissipation

$$P_{R1} = P_{D1} + [(I2)V_z]$$

Designing Zener Diode Voltage-Regulator Circuits

When designing zener diode regulators it is necessary to know certain circuit conditions and then to use them to specify (a) the resistance of $R1$, (b) the power rating of $R1$, and (c) the power rating of zener diode $D1$. There are three circuit conditions, which are designated I, II, and III. The properties of these three conditions are

Condition I: Variable supply voltage with constant load current.
Condition II: Constant supply voltage with variable load current.
Condition III: Variable supply voltage with variable load current.

Table 25-1 shows the design equations for all three conditions. Note that the power dissipation expressions for $R1$ and $D1$ are the same for all three conditions. Of course, V_{in} and $V_{in(max)}$ are the same for the constant supply voltage case.

INCREASING VOLTAGE-REGULATOR OUTPUT CURRENT

The output current that can be supplied by a zener diode voltage regulator is somewhat limited. In cases where a larger output current is needed, it is possible to amplify the effect of the zener diode by using it to control the base of a series-pass transistor (Q1 in Fig. 25-18). The output voltage produced by this circuit is approximately 0.6 to 0.7 V

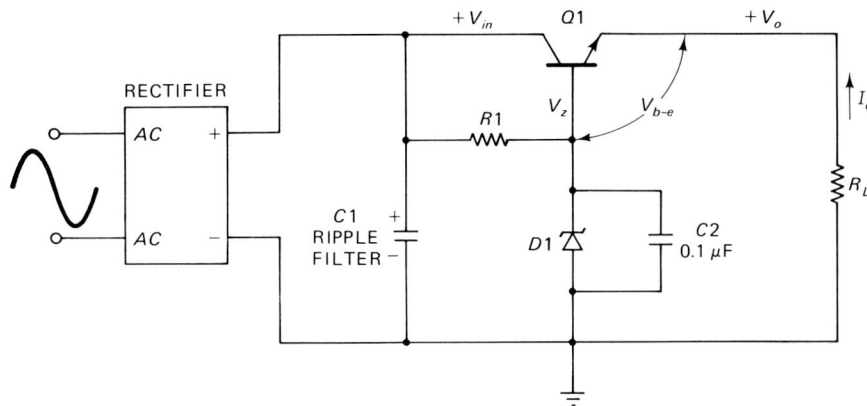

Fig. 25-18   Series-pass regulator circuit.

less than the zener potential. This reduction is accounted for by the base-emitter potential of the transistor, V_{b-e}. The output current rating is limited by the collector current rating of the transistor, with due regard for the collector dissipation. The collector will dissipate a power of $(V_{in} - V_o) \times I_o$. If there is a large difference between V_{in} and V_o, it is possible to exceed the maximum collector dissipation rating of Q1 even if less than the maximum collector current flows in the circuit. The output current must be limited to a value less than that required to exceed the transistor collector dissipation under the voltage difference conditions established in the circuit.

Another series-pass voltage regulator circuit is the feedback regulator shown in Fig. 25-19. In this circuit, a sample of the output voltage and a reference potential are applied to the differential inputs of

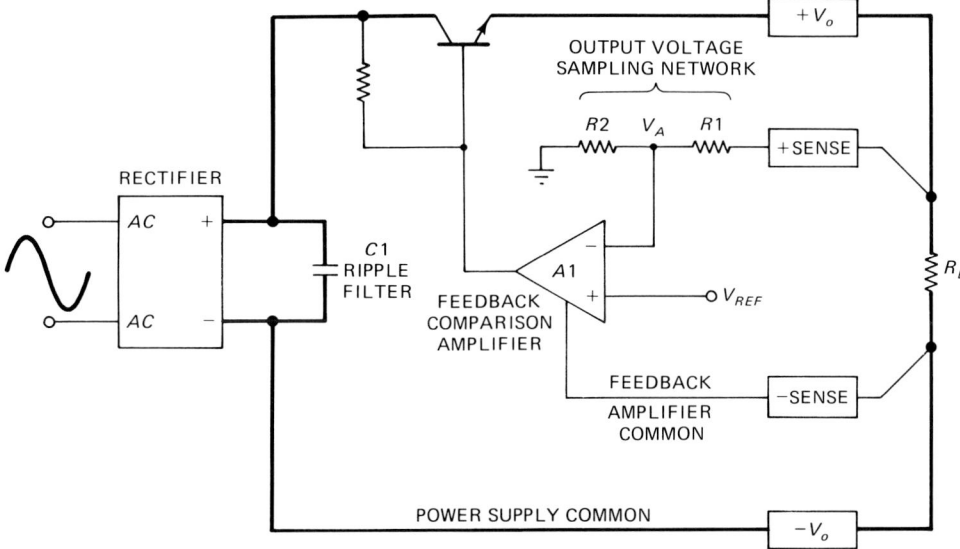

Fig. 25-19   Feedback amplifier voltage-regulator circuit (with sense lines).

a feedback amplifier ($A1$). When the difference between V_A and V_{ref} is nonzero, the amplifier drives the base of transistor Q1 harder, thereby increasing the output voltage. The actual voltage will be stable at a point determined by V_{ref}.

The circuit in Fig. 25-19 shows a feature that is very useful in high-current DC power supplies, especially where the power supply must be operated more than a few inches from the load. The voltage divider $R1/R2$ takes the sample of output voltage V_o that drives $A1$. The lines from the positive output and negative output to the voltage divider are separate from the main current-carrying lines. This arrangement makes it possible to place these sense lines at the points in the actual circuit where the value of V_o must be maintained at a precise value. For example, in a microcomputer that uses high-current TTL devices, it matters little that $+5$ V DC is maintained at the output of the DC power supply; it matters a lot, however, that the voltage at the microcomputer printed circuit board is $+5$ V DC. If the ($+$)SENSE line is connected to the $+5$ V DC bus of the computer and the ($-$)SENSE line is connected to the ground bus, the feedback power supply will keep the voltage at the rated value at the PCB and not at the power supply. This method servos out $I \times R$ drop in the power supply lines.

THREE-TERMINAL IC VOLTAGE REGULATORS

Voltage regulators for low current levels (up to 5 A) are reasonably simple to build now that simple three-terminal IC regulators are available. The circuit used with positive three-terminal regulators is shown in Fig. 25-20, and typical package styles are shown in Fig. 25-21. Capacitor $C1$ is the normal ripple-filter capacitor; it should have a value of 1000 μF per ampere of load current (some authorities insist on 2000 μF/A). Capacitor $C4$ is used to improve the transient response to sudden increases in current demand (something that

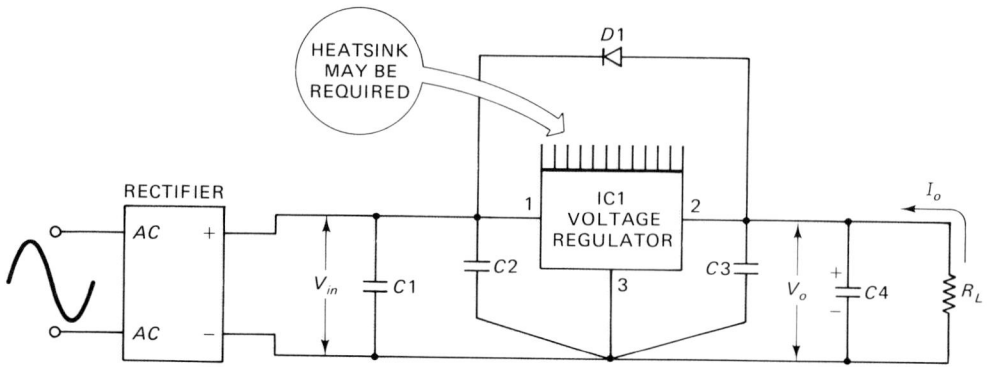

Fig. 25-20 Three-terminal IC voltage-regulator circuit.

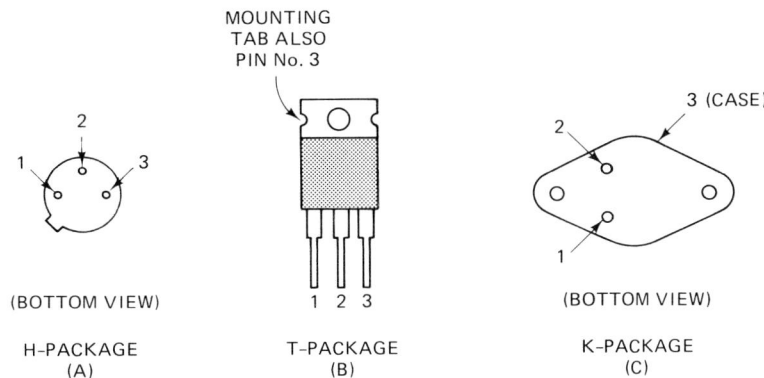

Fig. 25-21 Package styles for three-terminal IC regulators: (A) H-package, (B) T-package, and (C) K-package.

happens in digital circuits). Capacitor $C4$ should have a value of approximately $100\text{-}\mu\text{F}/\text{A}$ load current. Capacitors $C2$ and $C3$ are used to improve the immunity of the voltage regulator to transient noise impulses. These capacitors are usually 0.1 to 1 μF; they are to be mounted as close as possible to the body of the voltage regulator $IC1$.

Diode D1 is not shown in a lot of circuits, but is highly recommended for applications where $C4$ is used. If the diode is not present, then charge stored in $C4$ would be dumped back into the regulator when the circuit is turned off. That current has been implicated in poor regulator reliability. The mechanism of failure is that the normally reverse-biased PN junction formed by the IC regulator substrate and the circuitry become forward biased under these conditions. This situation allows a destructive current to flow. The diode should be a 1-A type at power supply currents up to 2 A and larger for larger current levels. For most low-voltage, 1-A or less, supplies a 1N400x is sufficient.

Several three-terminal IC voltage regulator packages are shown in Fig. 25-21. The "H" package (Fig. 25-21A) is used at currents up to 100 mA, the TO-220 "T" package (Fig. 25-21B) at currents up to 750 mA, and the TO-3 "K" package (Fig. 25-21C) at currents to 1 A.

There are two general families of IC regulator. One is designated 78xx, in which the xx is replaced with the fixed output voltage rating. Thus, a 7805 is a 5-V regulator, and a 7812 is a 12-V regulator. The LM-340y-xx series is also used. The y is the package style (H, K, or T) and the xx is the voltage. Thus, an LM-340K-05 is a 1-A, 5-V regulator in a "similar-to-TO3" type-K package; an LM-340T-12 is a 12-V, 750-mA regulator in a plastic TO-220 power transistor package.

Negative polarity versions of these regulators are available under the 79xx and LM-320y-xx designations. Figure 25-22 shows the typical circuit symbol. Note that the pin-outs on the voltage regulator device are different from those of the positive regulator.

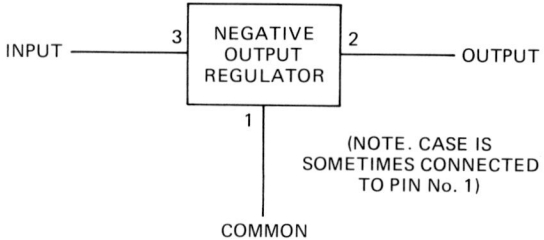

Fig. 25-22   Negative polarity three-terminal IC regulator.

The minimum input voltage to the three-terminal IC voltage regulator is usually 2.5 V higher than the rated output voltage. Thus, for a +5-V regulator, the minimum allowable input voltage is 7.5 V DC. The power dissipation is proportional to the voltage difference between this input potential and the rated output potential. For a 1-A regulator, therefore, the dissipation will be 2.5 W if the minimum voltage is used and considerably higher if a higher voltage is used. It is recommended that an input voltage close to the minimum allowable voltage be used. For +5-V supplies used in digital projects, a standard 6.3-V AC transformer is sufficient. When fullwave rectified and filtered with 1000 μF/A or more, the output voltage will be approximately +8 V DC. The student is encouraged to work out the arithmetic to prove this statement.

Figure 25-23 shows a dual-polarity DC power supply such as might be used in operational amplifier circuits, some microcomputers, and many other applications. The voltage-regulator portion of the circuit is a combination of positive and negative output versions of Fig. 25-20. The transformer/rectifier section bears some explanation, however. The rectifier is a 1-A bridge stack, but it is not used as a regular bridge. The center-tap on the secondary of transformer T1 establishes a zero-reference, so the bridge actually consists of a pair of conventional fullwave bridges connected to the same AC source. Thus, the

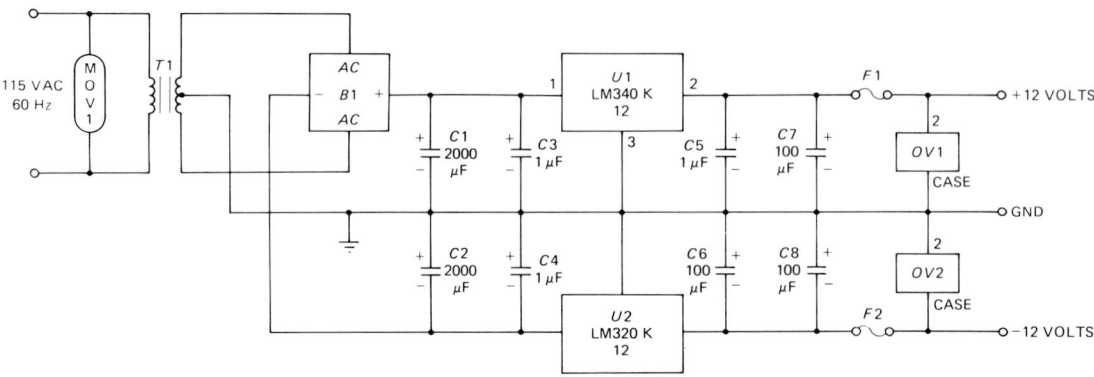

Fig. 25-23   Dual-polarity regulated power supply.

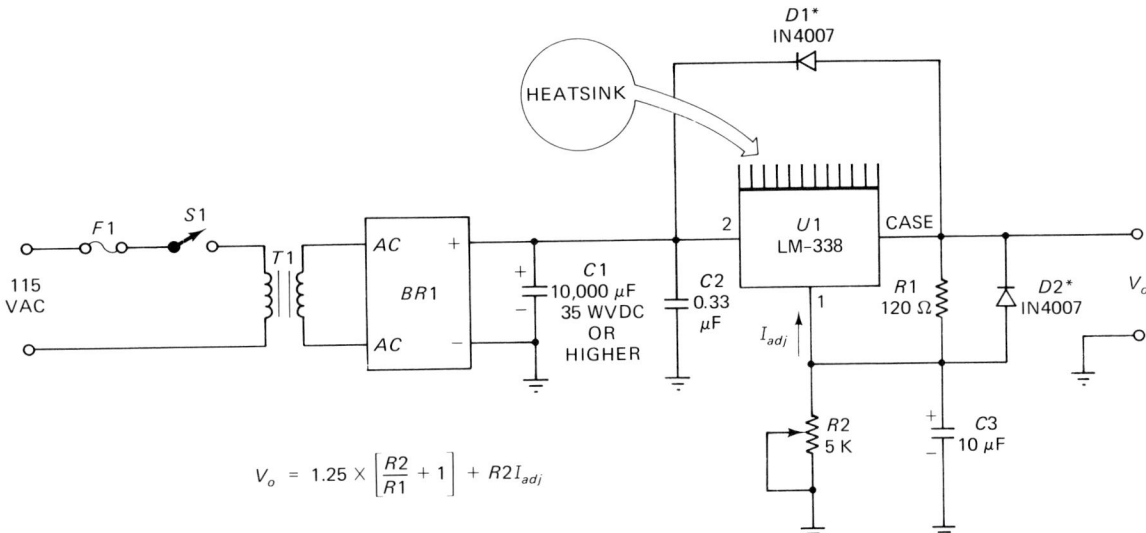

Fig. 25-24 High-current, variable-voltage power supply.

$(-)$ terminal of the bridge stack drives the negative voltage regulator, and the $(+)$ terminal drives the positive voltage regulator. This rectifier is sometimes called a half-bridge rectifier.

Adjustable IC Voltage Regulators

The IC voltage regulators discussed so far are fixed-output voltage types. They offer an output voltage that is predetermined and unchangeable without extraordinary effort. A variable-output voltage regulator, on the other hand, can be programmed to any voltage desired within its range. These devices can be used for either variable DC power supplies or to supply custom output voltages other than those allowed by the standard fixed voltages. Two similar models are considered here as examples.

The LM-317 and LM-338 are variable DC voltage regulators that are capable of delivering up to 1.5 and 5 A, respectively, at voltages up to +32 V DC. Figure 25-24 shows a typical circuit for these regulators. The input voltage must be 3 V higher than the maximum output voltage. The output voltage is set by the ratio of two resistors, $R1$ and $R2$, according to the equation:

$$V_{\mathrm{o}} = (1.25 \text{ V})\left|\frac{R2}{R1} + 1\right| \qquad (25\text{-}11)$$

An example from the National Semiconductor, Inc., *Linear Databook* (now in three volumes) shows 120 Ω for $R1$ and a 5-kΩ potentiometer for $R2$. This combination produces a variable output

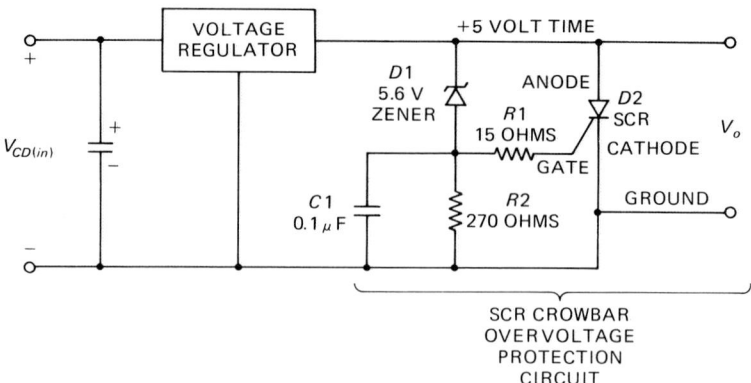

Fig. 25-25   SCR crowbar overvoltage protection circuit.

voltage of 1.2 V DC to 25 V DC, when V_{in} is > 28 V DC. Diode D1 can be any of the series 1N4002 through 1N4007 for LM-317 supplies and any 3-A type for LM-338 supplies.

OVERVOLTAGE PROTECTION

If the series-pass transistor shorts collector to emitter, or if the base control circuit fails, the input voltage will be applied to the output terminal of the voltage regulator. Because this voltage is often considerably higher than the regulated output voltage, serious damage can result to the electronic circuitry powered by the regulator. A protective circuit called an SCR crowbar provides overvoltage protection to such circuits.

The SCR crowbar (Fig. 25-25) is shunted across the output of the power supply (V is the power supply output voltage). A $+5$-V DC supply is used as an example. Diode D1 is a zener diode that has a zener potential that is a little higher than the rated output voltage. Zener voltage from 5.6 to 6.8 V DC can be used to protect a $+5$-V DC power supply. Power supplies with other output voltages than $+5$ V DC can be protected by scaling the zener diode voltage proportionally.

Diode D2 is a silicon controlled rectifier (SCR). This type of diode is open-circuited (i.e., has a high resistance in both directions) until a current is caused to flow in the gate terminal. When this occurs, the SCR breaks over and operates as any ordinary PN junction diode. The gate terminal of diode D2 is controlled by the network around diode D1. When the power supply voltage V exceeds the zener potential of $D1$, a current is conducted through $D1$ creating a voltage drop across $R2$. This voltage drop becomes the source for the gate current that flows in $R1$ and the gate of $D2$. At this instant, the SCR becomes conductive and shorts out the power supply line. Either a fuse or fuse resistor can be connected in series with the DC line; it will open-circuit when $D2$ conducts. The fuse should be in series with the positive input line of the voltage regulator. In a few cases, there is no fuse. In those circuits, the SCR ($D2$) must have a tremendous current rating because it must carry the short-circuit current of the power supply. It will clamp the output to near ground level until the circuit is turned off.

Lambda Electronics, Inc. offers overvoltage protection integrated circuits. Current levels available are 2, 6, 12, 20, and 35 amperes, with available voltage levels being 5, 6, 12, 15, 18, 20, 24, 28, and 30 volts. Several package styles are used, which also indicate the ampere level. A TO-66 power transistor case is used for 2 A and the TO-3 power transistor case is used for 6 A. Higher current levels are packaged in epoxy cases. The Lambda overvoltage protection modules are designated with a type number of the form L-y-OV-xx, in which y indicates the ampere level and xx indicates the voltage. Thus, an L-6-OV-5 device is a 5-V, 6-A model.

CURRENT LIMITING

Another catastrophe that can befall a DC power supply is an output short circuit. For unprotected power supplies, such an event will result in destruction of the circuit. It is possible to place a circuit in the power supply that will provide a current knee above which the supply shuts down. Figure 25-26 shows a representative circuit. Transistor Q1 in Fig. 25-26 is the series-pass transistor in a regulated power supply, and Q2 is the sense transistor that determines when the

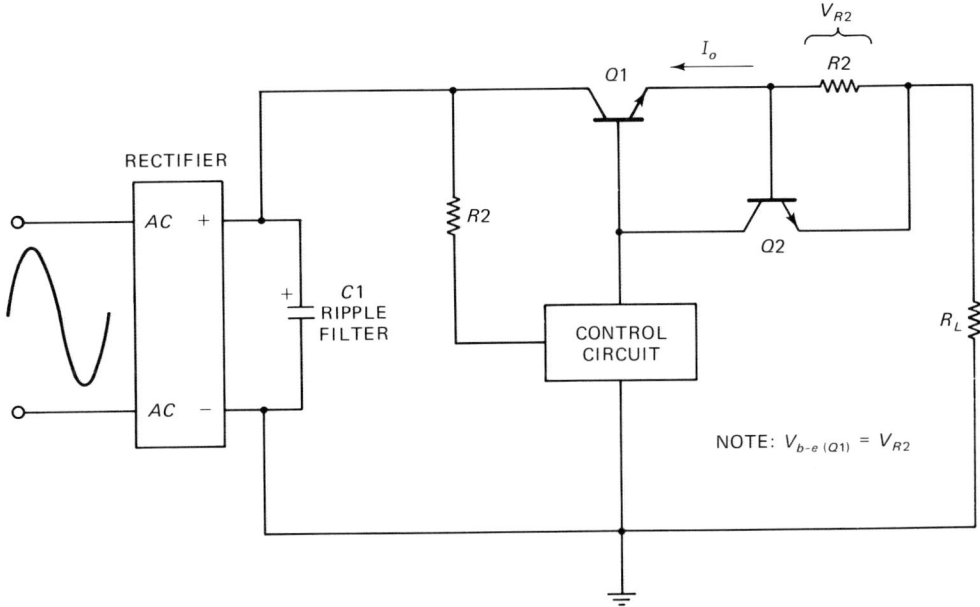

Fig. 25-26 Output current-limiting circuit.

current flow is too high. Some IC voltage regulators can also be used with this circuit (and some three-terminal types have the circuit built-in) if they have a sense terminal or some other provision. Resistor $R2$ is used to sense the level of current flow. The voltage drop across this resistor provides forward bias to Q2; it is proportional to the current flow:

$$V_{R2} = I_o R2 \qquad (25\text{-}12)$$

For silicon transistors, the forward-bias voltage required to saturate the transistor is approximately 0.6 V. When V_{R2} exceeds this critical voltage, transistor Q2 is heavily forward biased so its $V_{c\text{-}e}$ drops to a very low value, essentially shorting the base-emitter terminals of Q1. This actions turns off the power transistor. The value of $R2$ is, therefore,

$$R2 = 0.6 \text{ V DC}/I_{o(max)} \qquad (25\text{-}13)$$

Consider a practical example. Suppose a computer power supply delivers 10 A maximum. The value of $R2$ would be

$$R2 = 0.6 \text{ V DC}/I$$

$$= 0.6 \text{ V DC}/10 \text{ A}$$

$$= 0.06 \ \Omega$$

A value of 0.06 Ω (60 mΩ) seems somewhat difficult to achieve, but such a resistor can be made from fine wire. Alternatively, several wirewound power resistors or fuse resistors can be connected in parallel to form the low value required. For example, a 0.33-Ω resistor is often used as a fusistor in auto radios or as the emitter resistor in audio power amplifiers. Five of these resistors in parallel produce very nearly 60 mΩ. Various values of fusistors are available from 0.09 to 1.5 Ω; these can be paralleled in assorted combinations to produce the required resistance.

HIGH VOLTAGE TRANSIENT PROTECTION

Experts warn that 20 to 500 μs transient pulses of 1500 V or more strike residential and small business power lines several times per day. In industrial facilities that number may be considerably greater because of the heavy electrical machinery that is often in use. Until digital electronics devices, including computers, were widespread, however, this fact was interesting but somewhat trivial. But high voltage transient pulses can seriously disrupt digital circuits. The circuit may simply fail to operate correctly, or it can be damaged by the transient

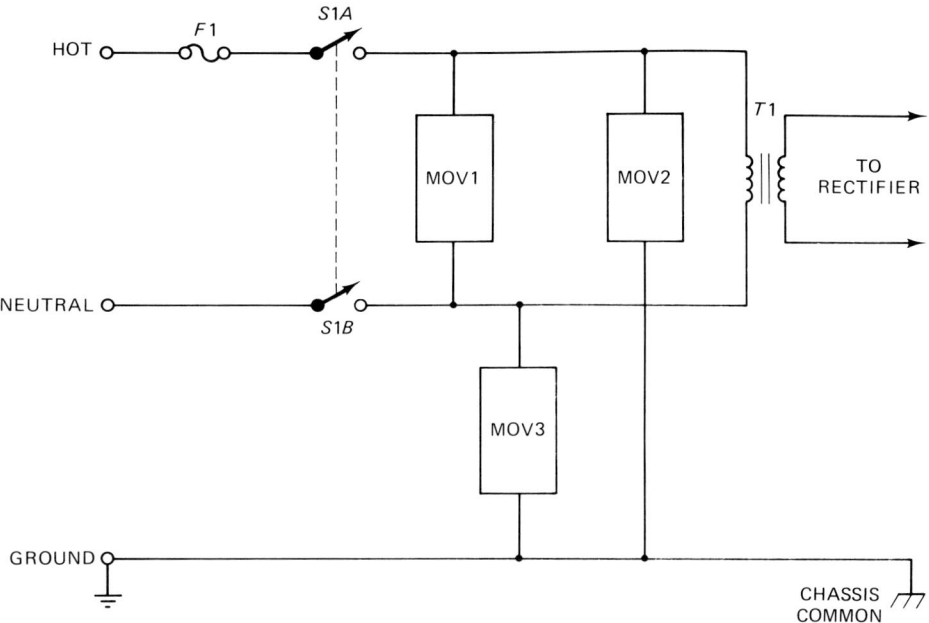

Fig. 25-27 Transient suppression using MOV devices across the AC line.

pulse. If a computer seems to occasionally bomb out while executing a program that ran properly only a short while ago, then suspect these transient pulses as the root cause.

Figure 25-27 shows metal oxide varistor (MOV) devices shunted across the power lines. Normally, only MOV1 will be needed, but MOV2 and MOV3 are recommended in serious cases. These devices, made by General Electric (similar devices are made by others) can be modeled as a pair of back-to-back zener diodes with a V_z rating of about 180°. The purpose of the MOV devices is to clip transient pulses over 180°.

Some applications require an LC low-pass filter on the AC power lines. In some very severe transient cases where the MOV is not

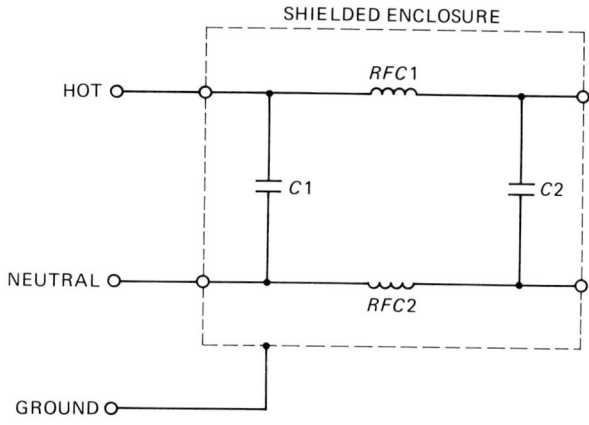

Fig. 25-28 EMI filter for AC input to power supply.

sufficient, or in cases in which a strong RF field is present (as in a radio transmitter), or in cases where the digital device creates RFI, the LC filter of Fig. 25-28 might be indicated. This filter should be mounted as close as possible to the point where the AC power cord enters the equipment.

Several manufacturers offer RFI filters that are shielded and especially suited for this service. Some models are molded inside of a chassis-mounted AC receptacle.

PRECISION VOLTAGE AND CURRENT SOURCES

How do you know the new digital multimeter that you just bought for the laboratory is accurate? Or, alternatively, has it maintained its calibration over the years? The same question can also be asked of oscilloscopes, laboratory power supplies, amplifiers, and other DC measuring devices. Do you have a microcomputer with an A/D converter in it? If the voltage reference source for an A/D or D/A converter is not as good as the bit-length of the converter, it is not useful. One quick way to get four-bit performance out of an eight-bit data converter is to use a shabby voltage reference supply! This section discusses how to design and build simple but effective reference supplies.

PRECISION ZENER DIODES

The simplest device that can be used for voltage regulation, hence also for some low-precision reference applications, is the ordinary zener diode. There are some problems with the basic zener diode; these problems become especially acute when it is used as a reference source (which implies accuracy). First, the voltage is only nominal. In other words, a "6.8-V" zener produces a voltage close to 6.8 V DC, but rarely is it exactly 6.8 V DC! Another problem is that the voltage drifts somewhat with temperature—hardly a characteristic desired in a reference supply.

Figure 25-29A shows a crude attempt at stabilizing the temperature drift problem. In this circuit, a number of zener diodes are connected in a series-parallel arrangement. Each series string produces a voltage drop ($V1$ and $V2$), so the differential output voltage is ($V1 - V2$). The idea here is that all diodes, assuming they are identical and in the same thermal environment, will drift approximately the same amount; so the differential effects of drift are zero.

A superior idea is shown in Fig. 25-29B. The device shown here is the National Semiconductor LM-199 (or LM-399) device. It consists of

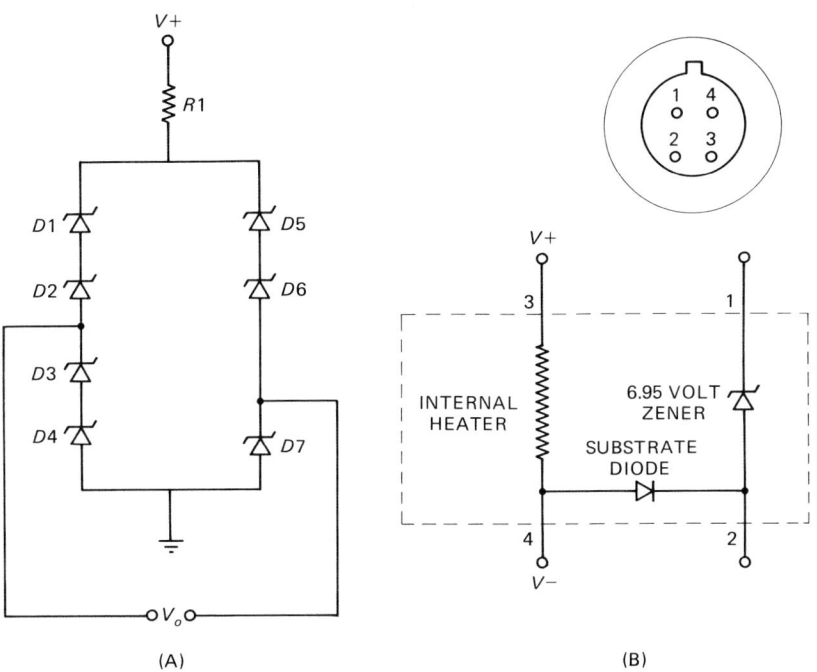

Fig. 25-29   (A) A crude attempt at stabilizing the temperature drift problem and (B) a superior attempt.

a 6.95-V zener diode embedded in an electrical heater device. One source told me that the heater was little more than a Class-A amplifier with the input shorted and that the zener is built on the same substrate, so it shares the thermal environment. The heater acts to keep the diode at a constant temperature somewhat above ambient room temperature. With the temperature constant, the diode voltage drop will not drift. The LM-199/LM-399 devices offer some startling voltage drift specifications. There are also other (similar) devices on the market.

Op-Amp Voltage Reference Source

Even the LM-199 device produces only a nominal output voltage. Although that voltage remains constant, it may be a little different from the rated 6.95 V. The circuit in Fig. 25-30 will adjust the voltage to any desired value and will make it precise. In addition, the operational amplifier serves to buffer the reference supply against changes in the load conditions.

The basic circuit of Fig. 25-30 is the noninverting follower with gain op-amp configuration. The LM-199 device is used to supply the

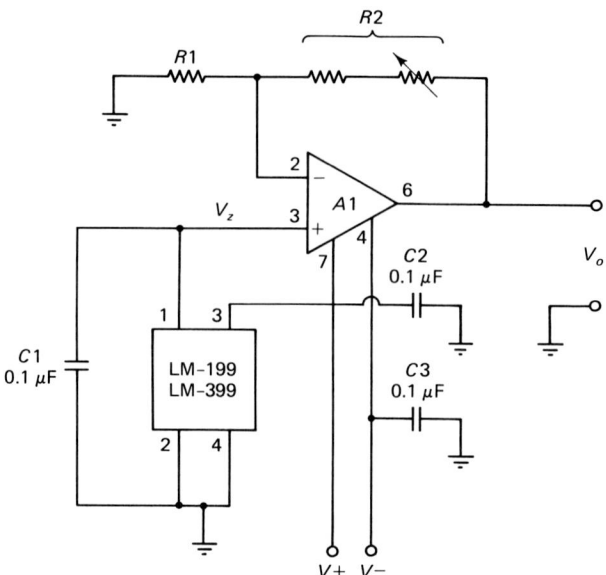

Fig. 25-30   Circuit for LM-199 device.

input voltage on pin no. 3, so the output voltage will be

$$V_{\text{o}} = V_z \left(\frac{R2}{R1} + 1 \right) \tag{25-14}$$

Selection of appropriate values of $R2$ and $R1$ will produce the desired output voltage. If we make $R1 = 1000$ Ω, a 10.00-V power supply can be made if $R2$ is 438.8 Ω. In most cases, $R2$ will be a combination of a fixed resistor (low temperature coefficient!) and a

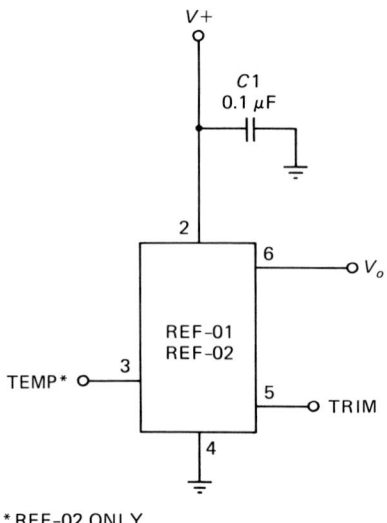

* REF-02 ONLY

Fig. 25-31   REF-01 and REF-02 devices.

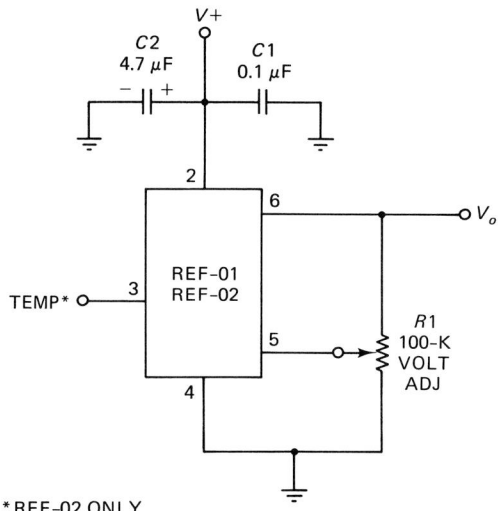

Fig. 25-32 Operating circuit for the REF-01 and REF-02 devices.

multiturn trimmer potentiometer. The trimpot is adjusted for the desired output voltage.

IC REFERENCE SOURCES

A very popular device for use in reference supplies is the integrated circuit reference supply. Although there are many different types on the market, we will use the REF-01 and REF-02 devices as our example (see Fig. 25-31). The REF-02 device is a +5.00-V reference source, and the REF-01 is a 10.000-V unit. Both REF-01 and REF-02 are packaged in an eight-pin metal IC can, and both use the pin-out definitions shown in Fig. 25-31. The supply voltage is applied across pins 2 and 4, and the output voltage is taken across pins 6 and 4 (pin 4 is common/ground). Pin no. 5 is used as a trim/adjust input.

The REF-02 uses pin no. 3 in a unique manner: It is an electronic thermometer transducer. The voltage appearing at pin no. 3 will have a value of 2.1 mV/K ambient temperature. It can be used to form an electronic thermometer.

Figure 25-32 shows the usual operating circuit for the REF-01 and REF-02 devices. The trimmer circuit consists of a linear taper potentiometer that selects a sample of the output voltage and inputs it to the trim circuit. This potentiometer should be a multiturn type to closely set the output voltage.

Index

618